Nanotubes and Nanosheets

Focusing on the research and development of the renowned two-dimensional nanomaterial, white graphene, also known as boron nitride (BN) nanosheets, this book covers a comprehensive range of topics. These include the structures, production methods, properties, and exciting applications in energy storage, thermal management, and environmental solutions of nanotubes and nanosheets.

Since the publication of the first edition, BN nanosheets have seen remarkable progress in research and application, leading to breakthroughs in discovering new properties, such as high modulus, thermal conductivity, and catalytic capabilities. Written by internationally acclaimed researchers with extensive expertise in BN nanomaterials, each chapter explores the unique properties of atomic-thin nanosheet structures and their surface chemistry. Later chapters highlight the diverse applications of BN nanosheets in areas including batteries, gas and oil adsorption and separation, and heat management. Readers will gain insights into the fundamental science behind this remarkable two-dimensional material, along with real-world applications in advanced batteries, heat dissipation, gas storage, and separation technologies.

This well-structured arrangement aims to cater to a wide readership, attracting both university researchers in chemistry, materials science, and energy storage and engineers from various industries. Additionally, this book serves as an invaluable reference for university students, offering an opportunity to explore new chemistry, novel material structures, and the immense potential of two-dimensional materials in addressing critical industrial challenges.

Ying Ian Chen is a Deakin Extinguished Professor and the Director of the Australian Research Council's Research Hub in New Safe and Reliable Energy Storage at Deakin University. He has been working on BN nanomaterials and applications for 30 years. His research achievement is recognized as a Clarivate Highly Cited Researcher for three years.

Srikanth Mateti is a Research Fellow at Deakin University, and he specializes in nanomaterial production for gas storage and carbon capture. His breakthroughs include isotropic thermal conductivity in BN nanosheets, patents licensed to White Graphene Pty. Ltd., and multiple accolades, such as Deakin's 2023 Best Research Award and international recognition.

Qiran Cai is a Research Fellow at the Institute for Frontier Materials, Deakin University, Australia, where he obtained his Ph.D. in 2017. He has been awarded the Australian Research Council Discovery Early Career Researcher Award (DECRA). His research centers on investigating fundamental properties and applications of BN nanosheets.

Nanotubes and Nanosheets
The Science and Applications of White Graphene

Second Edition

Edited by
Ying (Ian) Chen, Srikanth Mateti, and Qiran Cai

CRC Press
Taylor & Francis Group
Boca Raton London New York

CRC Press is an imprint of the
Taylor & Francis Group, an **informa** business

Front cover image: © American Chemical Society and Elsevier, used with permission.

Second edition published 2026
by CRC Press
2385 NW Executive Center Drive, Suite 320, Boca Raton FL 33431

and by CRC Press
4 Park Square, Milton Park, Abingdon, Oxon, OX14 4RN

CRC Press is an imprint of Taylor & Francis Group, LLC

First edition published by CRC Press 2021

ISBN: 978-1-041-08977-3 (hbk)
ISBN: 978-1-041-08968-1 (pbk)
ISBN: 978-1-003-64781-2 (ebk)

DOI: 10.1201/9781003647812

Typeset in Kepler
by codeMantra

Contents

SECTION III Keeping It Cool: Advanced Thermal Management with BN Nanosheets

SECTION IV The Energy Applications: Boron Nitride Nanosheets in Storage and Harvesting

Contributors

Arvind Agarwal
Mechanical and Materials Engineering
Florida International University
Miami, Florida

Qiran Cai
Institute for Frontier Materials
Deakin University
Waurn Ponds, Victoria, Australia

Ying Ian Chen
Institute for Frontier Materials
Deakin University
Waurn Ponds, Victoria, Australia

Alireza Dibaji
School of Life and Environmental Sciences
Deakin University
Waurn Ponds, Victoria, Australia

Siyuan Ding
Tsinghua Shenzhen International Graduate School
Tsinghua University
Shenzhen, China

Aijun Du
Centre for Materials Science
Queensland University of Technology
Brisbane, Queensland, Australia

Alexey Falin
Institute for Frontier Materials
Deakin University
Waurn Ponds, Victoria, Australia

Jiangfeng Fan
Shenzhen Institutes of Advanced Technology
Chinese Academy of Sciences
Beijing, China

Ye Fan
Institute for Frontier Materials
Deakin University
Waurn Ponds, Victoria, Australia

Dulini Fernando
School of Life and Environmental Sciences
Deakin University
Waurn Ponds, Victoria, Australia

Wei Gan
Institute of Physical Science and Information Technology and Information Materials and Intelligent Sensing Laboratory of Anhui Province
Anhui University
Hefei, China

Rui Han
School of Civil and Environmental Engineering
University of Technology Sydney
Sydney, NSW, Australia

Qi Han
School of Science
RMIT University
Melbourne, Victoria, Australia

Shaoming Huang
Nanomaterials and Chemistry Key Laboratory
Wenzhou University
Wenzhou, China

Zhenguo Huang
School of Civil and Environmental Engineering
University of Technology Sydney
Sydney, NSW, Australia

Hongbo Jiang
Institute for Frontier Materials
Deakin University
Waurn Ponds, Victoria, Australia

Xiangfen Jiang
State Key Laboratory of Mechanics and Control of Mechanical Structures, Key Laboratory for Intelligent Nano Materials and Devices of the Ministry of Education, College of Material Science and Engineering
Nanjing University of Aeronautics and Astronautics
Nanjing, China

Donggun Kim
Institute for Frontier Materials
Deakin University
Waurn Ponds, Victoria, Australia

Weiwei Lei
Institute for Frontier Materials
Deakin University
Waurn Ponds, Victoria, Australia
and
School of Science
RMIT University
Melbourne, Victoria, Australia

Lu Hua Li
Institute for Frontier Materials
Deakin University
Waurn Ponds, Victoria, Australia

Feng Liu
School of Materials Science and Engineering
Shenyang University of Technology
Shenyang, China

Dan Liu
Institute for Frontier Materials
Deakin University
Waurn Ponds, Victoria, Australia
and
School of Science
RMIT University
Melbourne, Victoria, Australia

Yuchen Liu
Institute for Frontier Materials
Deakin University
Waurn Ponds, Victoria, Australia

Zhen Liu
School of Life and Environmental Sciences
Deakin University
Waurn Ponds, Victoria, Australia

Srikanth Mateti
Institute for Frontier Materials
Deakin University
Waurn Ponds, Victoria, Australia

Motilal Mathesh
School of Life and Environmental Sciences
Deakin University
Waurn Ponds, Victoria, Australia

Yuhang Meng
National Laboratory of Solid State Microstructures (NLSSM), Collaborative Innovation Center of Advanced Microstructures, Jiangsu Key Laboratory of Artificial Functional Materials, College of Engineering and Applied Sciences
Nanjing University
Nanjing, China

Sohail M.A.K. Mohammed
Mechanical and Materials Engineering
Florida International University
Miami, Florida

Pranjal Nautiyal
Mechanical and Aerospace Engineering
Oklahoma State University
Stillwater, Oklahoma

Kazue Orikasa
Mechanical and Materials Engineering
Florida International University
Miami, Florida

Selvakannan Periasamy
School of Science
RMIT University
Melbourne, Victoria, Australia

Si (Alex) Qin
Institute for Frontier Materials
Deakin University
Waurn Ponds, Victoria, Australia

Ling Qiu
Tsinghua Shenzhen International Graduate School
Tsinghua University
Shenzhen, China

Md Mokhlesur Rahman
Institute for Frontier Materials
Deakin University
Waurn Ponds, Victoria, Australia

Dandan Sun
National Laboratory of Solid State Microstructures (NLSSM), Collaborative Innovation Center of Advanced Microstructures, Jiangsu Key Laboratory of Artificial Functional Materials, College of Engineering and Applied Sciences
Nanjing University
Nanjing, China

Rong Sun
Shenzhen Institutes of Advanced Technology
Chinese Academy of Sciences
Beijing, China

Khagesh Tanwar
Institute of Molecular Science
University of Valencia
Paterna, Spain

Tao Tao
Institute for Frontier Materials
Deakin University
Waurn Ponds, Victoria, Australia

Jiangting Wang
Institute for Frontier Materials
Deakin University
Waurn Ponds, Victoria, Australia

Xuebin Wang
National Laboratory of Solid State Microstructures (NLSSM), Collaborative Innovation Center of Advanced Microstructures, Jiangsu Key Laboratory of Artificial Functional Materials, College of Engineering and Applied Sciences
Nanjing University
Nanjing, China

Dehong Yang
National Laboratory of Solid State Microstructures (NLSSM), Collaborative Innovation Center of Advanced Microstructures, Jiangsu Key Laboratory of Artificial Functional Materials, College of Engineering and Applied Sciences
Nanjing University
Nanjing, China

Guoliang Yang
Institute for Frontier Materials
Deakin University
Waurn Ponds, Victoria, Australia

Keqin Yang
Nanomaterials and Chemistry Key Laboratory
Wenzhou University
Wenzhou, China

Lifeng Wang
Institute for Frontier Materials
Deakin University
Waurn Ponds, Victoria, Australia

Wenrong Yang
School of Life and Environmental Sciences
Deakin University
Waurn Ponds, Victoria, Australia

Baozhi Yu
Institute for Frontier Materials
Deakin University
Waurn Ponds, Victoria, Australia

Xiaoliang Zeng
Shenzhen Institutes of Advanced Technology
Chinese Academy of Sciences
Beijing, China

Chunmei Zhang
Centre for Materials Science
Queensland University of Technology
Brisbane, Queensland, Australia

Preface

This book is among the first to focus exclusively on the research and development of boron nitride (BN) nanosheets, also known as white graphene, a remarkable two-dimensional nanomaterial. It provides a comprehensive exploration of their structure, synthesis methods, properties, and diverse applications in energy storage, thermal management, and environmental solutions. Each chapter is authored by internationally recognized experts with extensive experience in BN nanomaterials.

Structurally similar to graphene, BN nanosheets possess unique properties, including exceptional thermal and chemical stability, electrical insulation with high thermal conductivity, and distinctive catalytic surface chemistry. BN nanosheets have paved the way for numerous groundbreaking applications and offer innovative solutions to critical industrial and environmental challenges—providing advantages that graphene alone cannot achieve. Significant efforts have been made to explore the mass production of these BN materials. Notably, two companies—BNNT Tech Ltd. and White Graphene—have been established.

Following the success of a previously published volume on BN nanotubes, this book introduces BN nanosheets to a broad audience. It serves as an essential resource for researchers in chemistry, materials science, and energy storage and engineers across various industries. This book provides both fundamental scientific insights and practical knowledge on real-world applications, including battery safety, next-generation lithium–sulfur batteries, thermal management of electronic devices, and advancements in gas storage and energy-efficient separation technologies.

Additionally, this book serves as a valuable reference for university students, offering an opportunity to explore novel chemistries, unique material structures, and the immense potential of two-dimensional materials in solving major industrial challenges.

The editors express their gratitude to all contributors for their efforts in shaping this book. Special thanks go to our families, colleagues, and students for their support, as well as to the staff at CRC Press, Taylor & Francis Group, for their assistance in bringing this work to publication.

Section I

New Structure Creation

Synthesis of Boron Nitride at the Nanoscale

Chapter 1

Synthesis of Boron Nitride Nanomaterials

Rui Han and Zhenguo Huang

1.1 INTRODUCTION

Research on hexagonal boron nitride (*h*-BN) has surged since the successful isolation of single-layer graphene using the Scotch tape method in 2004 [1], given that *h*-BN is isostructural to graphene. The synthesis of *h*-BN materials parallels that of their carbon counterparts, with practical methods typically categorized into top-down and bottom-up strategies. Each synthesis approach offers distinct advantages, making them suitable for different applications. For instance, mechanical exfoliation produces highly crystalline *h*-BN nanosheets, though the lateral sizes depend on the bulk *h*-BN starting materials. However, achieving deep exfoliation with uniform thickness remains challenging due to the stronger interlayer forces in *h*-BN compared to graphite. Thus, mechanically exfoliated *h*-BN nanosheets are ideal as composite fillers but not for elaborate electronics. In contrast, chemical vapor deposition (CVD) offers precise control over thickness and lateral size, making it suitable for producing films used as dielectric layers in field-effect transistors.

1.2 SYNTHESIS OF BORON NITRIDE NANOSHEETS

The synthesis of *h*-BN nanosheets generally falls into two broad categories: top-down and bottom-up approaches. Top-down approaches typically involve the mechanical exfoliation of bulk *h*-BN or sonication-assisted exfoliation to produce nanosheets. In contrast, bottom-up synthesis involves processes such as CVD, physical vapor deposition (PVD), and high-temperature pyrolysis, each of which allows for the controlled assembly of *h*-BN at the atomic level.

1.2.1 Top-Down Synthesis

Top-down synthesis of *h*-BN involves breaking down the van der Waals forces that hold the layers of bulk *h*-BN together, enabling the extraction of thin nanosheets. This approach relies on physically or chemically overcoming the interlayer interactions without disrupting the inherent two-dimensional structure of the material. Two common techniques used in this approach are mechanical exfoliation and sonication-assisted exfoliation.

DOI: 10.1201/9781003647812-2

1.2.1.1 Mechanical Exfoliation

Mechanical exfoliation for obtaining *h*-BN nanosheets is inspired by the Scotch tape mechanical cleavage technique used for isolating atomic layers of graphene. In this process, *h*-BN layers are peeled from raw materials using adhesive tapes [2], which overcomes the van der Waals forces that hold the layers together. Repeating this process can yield thin or even monolayer *h*-BN nanosheets. However, due to the stronger interlayer forces in *h*-BN, exfoliation efficiency is lower than in graphene. The quality of the exfoliated nanosheets depends largely on the starting material. These nanosheets, with high crystallinity and low defect density, are ideal for characterizing physical and optical properties. However, the repetitive nature of this method makes it labor-intensive and difficult to scale up. Motorized equipment, such as ball mills or vortex fluidic devices, can enhance exfoliation efficiency.

In ball milling, the type of mill, operational parameters, and process control agents can influence the quality of the exfoliated nanosheets. High-energy ball mills generate large mechanical forces, but these forces are often irregular. As the *h*-BN raw materials are exfoliated, the in-plane structure is also damaged, resulting in significantly reduced crystal size and amorphization. Ghosh et al. [3] observed that the initial stage of milling induces lattice strains and expedites the propagation of deformation stacking faults and twin faults, which are present in the raw materials. During the later stage, twin faults have rapidly soared compared to deformation stacking faults. Both effects contribute to the decrease in crystal size and amorphization of the nanosheets. Planetary ball mills, which primarily generate shear forces, can achieve more effective exfoliation without compromising crystallinity when combined with process control agents. One challenge in ball milling is cold welding, which causes exfoliated sheets to restack. Li et al. [4] found that the cold-welding effect could be mitigated by adding benzyl benzoate to the milling process. This is owing to benzyl benzoate's similar surface tension to *h*-BN. Moreover, benzyl benzoate's higher viscosity reduces the grinding media's impact on *h*-BN, which is helpful for minimizing contamination.

Vortex fluidic devices, which are based on the rapid rotation of tubular containers at an angle, can also produce shear forces to cleave layered materials such as graphene and *h*-BN. The force originates from the synergy between centrifugal and gravitational forces. Chen et al. reported successful exfoliation of graphene and *h*-BN using a vortex fluidic device. *h*-BN starting material is dispersed in N-methyl-pyrrolidone in a glass tube inclined at 45°. Monolayers (<4%) of *h*-BN nanosheets were obtained at 8000 rpm. As illustrated in Figure 1.1 [5], when spinning at high rotational speed, the liquid flows upward along the internal wall while downward at the top

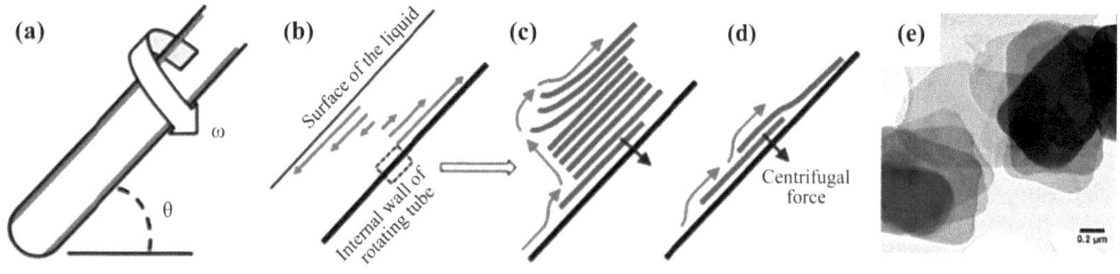

Figure 1.1 Schematic illustration of exfoliation of *h*-BN by the vortex fluidic device: (a) schematic illustration of the vortex fluidic device setup. (b) Schematics of the liquid flow. (c) and (d) Exfoliation process highlighting the shearing process and the centrifugal force. (e) TEM image of the partially stacked *h*-BN sheets [5]. (Reproduced from Ref. [5] with permission from the Royal Society of Chemistry.)

surface. The centrifugal force presses the *h*-BN against the walls of the tube, and shear forces are parallel to the walls of the tube. Transmission electron microscopy (TEM) studies (Figure 1.1e) reveal that partially stacked BN sheets originating from the same bulk material manifest the effectiveness of this technique in exfoliating layered materials.

1.2.1.2 Sonication-Assisted Exfoliation

Sonication-assisted exfoliation is a common technique in nanosheet synthesis. This method subjects a liquid suspension of starting material to high-frequency sound waves or ultrasound. The sound waves create rapid and intense agitation of the liquid, leading to the formation, growth, and implosive collapse of bubbles called cavitation bubbles [6]. The collapse of these cavitation bubbles generates localized high temperatures up to hundreds of degrees and high pressures (hundreds of bars) during very short lifetimes of a few microseconds. This creates microenvironments around the cavitation bubbles with high dynamics and impacts that can melt most metals [6]. The intense agitation caused by cavitation results in the mechanical fragmentation of precursor materials into smaller particles or sheets. In the case of layered materials such as graphene or transition metal dichalcogenides, sonication treatment can exfoliate stacked layers into thin nanosheets. The mechanical forces generated by cavitation can overcome the van der Waals forces holding the layers together, leading to the separation of nanoscale sheets. Depending on the liquid media and surfactants used during the treatment, sonication can also grant *h*-BN nanosheets with functional groups, giving rise to interesting chemical properties.

It is believed that a liquid medium with surface energy matching the van der Waals force holding layered materials together is advantageous for the sonication-assisted exfoliation. N,N-dimethylformamide is a polar solvent with a surface energy similar to *h*-BN's interlayer van der Waals force. Moreover, the strong interaction between polar N,N-dimethylformamide and BN's surface could also contribute to the exfoliation [7]. Similarly, other polar solvents, such as N-methyl-2-pyrrolidone, dimethylacetamide, 1,2-dichlorobenzene, and ethylene glycol, also enhance the cleavage of *h*-BN during sonication treatment [8]. In contrast, sonication treatment in nonpolar solvents, such as cyclohexane and toluene, is less effective in exfoliating *h*-BN. The dispersion and exfoliation of nanosheets are effective in liquid media whose surface energies are similar to those of the nanosheets [9]. For *h*-BN, the solvents with a surface energy of $65\,mJ\,m^{-2}$, such as 2-propanol and N-methyl-pyrrolidone, would be optimal for breaking down the binding force between layers. Mixing different solvents to achieve an overall surface energy close to $65\,mJ\,m^{-2}$ is also conducive to producing thin nanosheets. For example, the mixture of water and ethanol proves more efficacious than either solvent by itself. Exfoliating *h*-BN using water does not seem to be a good strategy at first glance, as *h*-BN is hydrophobic. However, experiments showed that sonication treatment in water exfoliated *h*-BN sheets quite effectively, with efficiency close to that in N,N-dimethylformamide [10]. This is because *h*-BN underwent hydrolysis with the assistance of sonication. Hydroxyl group (-OH) and ammonia gas were detected after the treatment, which validates the existence of the hydrolysis reaction. The hydrolysis reaction acts like scissors cutting through the *h*-BN planes, leading to smaller and thinner nanosheets, the schematics of which are shown in Figure 1.2.

1.2.2 Bottom-Up Synthesis

Unlike the top-down approach, the bottom-up method refers to the synthesis of *h*-BN nanosheets via self-assembly of atomic or molecular building blocks. For the top-down method, the mother materials largely determine the chemical and physical structures of nanosheets. However, for the bottom-up

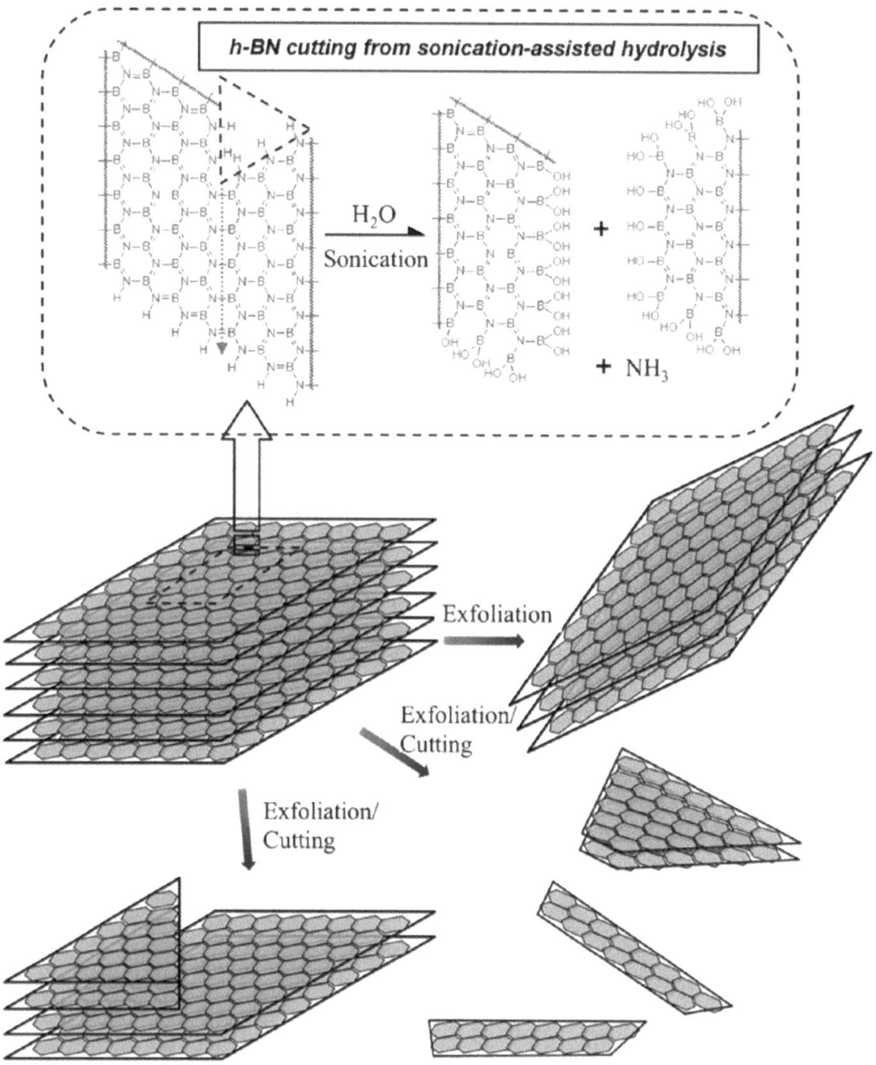

Figure 1.2 Schematics of exfoliation of *h*-BN in water [10]. (Adapted with permission from Ref. [10]. Copyright 2011, American Chemical Society.)

method, the structure can be designed or modulated at the atomic level by carefully choosing building blocks. Common bottom-up methods include CVD, PVD, and high-temperature pyrolysis.

1.2.2.1 CVD

CVD is a prominent bottom-up approach for synthesizing *h*-BN nanosheets via chemical reaction of gaseous precursors on a heated substrate. This technique offers precise control over growth parameters, such as temperature, pressure, and gas flow rate, enabling the production of nanosheets with desired thickness, uniformity, and crystallinity. The *h*-BN nanosheets synthesized through CVD are known for their exceptional thermal conductivity, electrical insulation, and mechanical strength, making them highly desirable for applications in electronics, nanocomposites, and advanced coatings. The first CVD synthesis of *h*-BN dates to the 1960s. Rand and Roberts [11] first reported the growth of vitreous BN films (rather than nanosheets) with thicknesses up to 6 μm on various substrates, including Si, Ta, Mo, Ge, and fused silica at temperatures of 600°C–1000°C under N_2 or H_2. Ammonia is the nitrogen source, with diborane being the boron source. The overall reaction is shown below (Equation 1.1). During growth, the substrates are heated under the ammonia flow, and then diborane is introduced.

$$B_2H_6 + 2NH_3 \rightarrow 2BN + 6\,H_2 \tag{1.1}$$

In 1995, Nagashima et al. synthesized *mo*nolayer *h*-BN that was epitaxially grown on Ni (111), Pd (111), and Pt (111) under ultrahigh vacuum (UHV) with a base pressure of 1×10^{-8} Pa [12]. Instead of using two separate molecules as sources of boron and nitrogen, they used borazine ($B_3N_3H_6$) with a 1:1 stoichiometric ratio between B and N. Since then, there has been a decade of silence in the CVD synthesis of *h*-BN nanosheets, which experienced resurgence after the famous Scotch tape isolation of monolayer graphene. Now, CVD has become the most effective method for growing *h*-BN nanosheets with large lateral sizes and desired thickness. As of now, high-quality *h*-BN nanosheets can be successfully obtained under milder conditions, such as low-pressure (LP) and atmospheric-pressure (AP) conditions. *h*-BN thin films used as dielectric layers in nanoelectronics are usually produced by CVD. This section will introduce the mechanism of growth, common precursors, and substrates for the CVD synthesis of *h*-BN nanosheets.

1.2.2.1.1 Precursors

The precursor for CVD growth of *h*-BN nanosheets should satisfy the following requirements: (1) contain B and N elements; (2) can be readily vaporized; and (3) undergo polymerization to form extended *h*-BN sheets at high temperatures. Borazine, a cyclic 6-membered compound with a chemical formula $B_2H_6N_3$, is widely used for CVD growth of *h*-BN. At elevated temperatures and catalyzed by substrates, borazine undergoes intermolecular dehydrogenation and fuses together to form planar *h*-BN nanosheets. However, borazine degrades in air due to reaction with moisture, forming solid polyborazylene [13], and it is normally stored under cold conditions in an inert atmosphere. β-trichloroborazine, which is borazine's derivative, can also be used as a precursor, but it again reacts with moisture, releasing corrosive HCl gas. Boron and nitrogen can also come from two different compounds, such as B_2H_6 and NH_3, providing B and N, respectively.

Ammonia borane is a solid with relatively good stability under ambient conditions. When heated to 110°C, it decomposes into borazine, hydrogen, and polyaminoborane. Polyaminoborane with a high molecular weight can be easily blocked from entering the CVD furnace using a glass fiber filter [14]. This way, ammonia borane serves as an indirect precursor to provide borazine when heated. Common precursors are listed in Table 1.1.

1.2.2.1.2 Substrate

Substrates play key roles in the CVD synthesis, since they have a catalytic effect at high temperatures to fuse the molecular precursors together, and their surfaces affect the morphology of the film. Similar to graphene, Cu and Ni are the most commonly used substrates for the CVD growth of *h*-BN nanosheets. Especially, *h*-BN has low lattice mismatches with Cu (111) and Ni (111), being 2.4% and –0.4%, respectively [20]. Using single-crystalline substrates is advantageous for growing large, high-quality nanosheets. Wang et al. reported the epitaxial growth of 100 cm^2 single-crystal monolayer *h*-BN nanosheets on a single-crystal Cu (110) substrate using low pressure CVD with

TABLE 1.1 COMMON PRECURSORS FOR CVD GROWTH OF H-BN NANOSHEETS, AND CORRESPONDING EXPERIMENT CONDITIONS AND FILM PROPERTIES

Precursor	Growth temperature (°C)	Pressure	Substrate	Thickness & lateral sizes	References
Ammonia borane	700–1100	AP, LP	Cu, Ni, Fe, Ag	Single/multilayers; Up to a few cm	[15]
Borazine	700–1000	UHV, AP, LP	Cu, Ni, Fe, Ag	Single/multilayers; 0.3 mm	[12]
β-trichloroborazine	723	UHV	Ni, Pt	Single layer; N/A	[16]
Diborane and ammonia	600–1025	LP	Ni, Sapphire	Single/multilayers; N/A	[17]
BCl_3, NH_3, N_2, H_2	1000	AP	Si	Single/multilayers; Up to 500 nm	[18]
Trimethyl ammonia borane	1050	AP	Cu	Single layer; Up to 100 μm	[19]

ammonia borane as the precursor [21]. According to SEM images, the *h*-BN domains grown on the Cu (110) were all triangular shaped and aligned in the same direction across the substrate. SEM in situ observation suggests that unidirectional alignment of the *h*-BN domains was due to step-edge-mediated nucleation. The triangular domains merge into a single crystal when coalesced, as the single-crystal lattice is the most stable structure. Even though monolayer is the 'neatest' form of 2D materials and suitable for physical property characterization, multilayer counterparts are advantageous for practical applications such as the dielectric for field-effect transistors, as multilayer nanosheets are more mechanically robust and can also mitigate the negative effects from substrates (impurities, surface roughness, *etc.*). Ma et al. reported wafer-scale single-crystal tri-layer growth of *h*-BN nanosheets on a single-crystal Ni (111) substrate at a growth temperature of 1220°C with a borazine precursor [22]. Similarly, unidirectionally aligned *h*-BN domains were found during the initial stage of growth, and then domains merged into a continuous film after 60 minutes of growth. The *h*-BN domains were tri-layer, and the thickness remained the same when they coalesced to fully cover the substrate. It is suggested that a $Ni_{23}B_6$ solid solution was formed at high temperatures due to the high solubility of B in Ni, and the tri-layer *h*-BN formed on top of the $Ni_{23}B_6$ layer via a surface-mediated process.

Using an alloyed substrate can reap the benefits of both worlds. It is reported that high-quality multilayer *h*-BN nanosheets with a centimeter scale and 5 nm thickness were successfully grown on a polycrystalline Fe–Ni alloy foil [23]. The combination of Fe and Ni can prevent the local segregation of *h*-BN, which typically occurs in pure Fe, and also inhibit the structural transformation of Fe at high temperatures, leading to the uniform growth of multilayer *h*-BN nanosheets [24].

The crystal orientation of the substrate significantly impacts the growth rate of *h*-BN nanosheets. For instance, *h*-BN grows faster on Cu (111) than on Cu (110) [25]. Not only do the substrate type and crystal orientation affect nanosheet growth, but the crystallinity and phase of the substrates also play a crucial role in CVD growth. Point defects and grain boundaries serve as preferential sites for nucleation. Since the initial growth of *h*-BN nanosheets on a substrate occurs via heterogeneous nucleation, a higher concentration of point defects, impurities, and grain boundaries provides more nucleation sites, resulting in smaller single-crystal domains of *h*-BN. Annealing substrates at elevated temperatures can smooth their surfaces and reduce surface defects, ultimately enhancing the quality of the *h*-BN film. Heating the substrates to their melting point can further eliminate volatile impurities. In addition, molten substrates at high temperatures create a dynamic environment that suppresses nucleation, thereby improving the crystallinity of the *h*-BN nanosheets [14,26]. Lee et al. [27] reported wafer-scale single-crystal growth on a molten gold (Au) on a tungsten support layer at 1100°C. The aim of using this substrate is to create a highly flat Au surface with high surface tension for strong adhesion of the borazine precursor. Unlike the single-crystal *h*-BN growth on solid substrates of Cu (110) and Ni (111) [21,22], *h*-BN grains are of circular shape. As time progressed, more grains were deposited on the liquid substrate without a change in size. Interestingly, the grains rotated and merged because of Coulomb interaction, leading to a seamless stitching without grain boundaries.

1.2.2.1.3 Growth Mechanism

For borane-based precursors such as ammonia borane, heating leads to the formation of borazine, which adsorbs onto the substrate surface. The catalytic effect of the substrate promotes dehydrogenation and subsequent polymerization of borazine at elevated temperatures.

Early research indicates that *h*-BN exhibits surface-mediated growth behavior on Cu substrates, as evidenced by the observation that *h*-BN single crystals can grow across Cu grain boundaries with minimal distortion [28,29]. This is in contrast to epitaxial growth, where strong interaction between the substrate and the film is observed. However, later findings revealed that *h*-BN growth on Cu involves more than surface-mediated growth alone. Kidambi et al. [30] demonstrated that at growth temperatures, borazine molecules dissociate into B and N atoms, with B atoms dissolving into Cu, as evidenced by the lattice dilation of Cu and the precipitation of B during postgrowth cooling. In contrast, N has significantly lower solubility in Cu, limiting its dissolution. The dissolved B forms a reserve within the Cu substrate, which later diffuses out and contributes to CVD growth. B is supplied from both the substrate surface and the internal B reserve, while N is only supplied from the substrate surface, as depicted in Figure 1.3 [30]. Theoretical studies have also been conducted to explain *h*-BN growth at the molecular level. For further details, readers may refer to Islam's review [31].

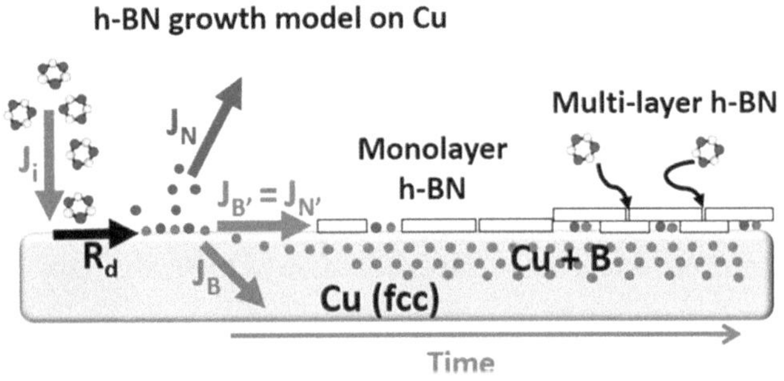

Figure 1.3 Schematic illustration of *h*-BN nanosheets grown on Cu substrates. J_B represents the diffusion flux of B into the Cu; J_N represents the diffusion flux of N out of Cu; $J_{B'}$ and $J_{N'}$ represent the diffusion flux to form *h*-BN nanosheets on Cu [30]. (Adapted with permission from Ref. [30]. Copyright 2014, American Chemical Society.)

1.2.2.2 PVD

PVD is a versatile and widely used technique for creating thin films on various substrates. In this process, material is physically transferred from a condensed phase (solid or liquid) to a vapor phase and then condensed back into a thin film. PVD techniques are typically carried out under UHV conditions and can deposit metals, ceramics, and composites with high precision and uniformity. Vapor is commonly generated using two methods: thermal evaporation and sputtering. In thermal evaporation, a focused high-energy electron beam heats the target material to produce vapor. In sputtering, accelerated gas ions, such as Ar^+, bombard the source material to generate vapor. Compared to CVD, where precursors are more complex, the source material for growing *h*-BN films in PVD is *h*-BN itself, which is relatively benign and inexpensive.

Ion beam sputtering deposition can grow large, single-crystalline monolayer *h*-BN on nickel substrates. The size of these domains is highly dependent on the surface roughness of the Ni substrate. Research shows that reducing the root-mean-square surface roughness of Ni from 36 to 7.5 nm increases the *h*-BN domain size by fivefold. With extended growth times, *h*-BN domains coalesce, eventually forming a continuous film [32]. Additionally, irradiating a pyrolytic *h*-BN target with a CO_2-pulsed laser deposition system has been shown to successfully deposit *h*-BN nanosheets onto substrates [33].

1.2.2.3 High-Temperature Pyrolysis

High-temperature pyrolysis is a highly effective bottom-up approach for synthesizing *h*-BN nanosheets, involving the thermal decomposition of molecular precursors in an inert atmosphere. In this process, boron- and nitrogen-containing compounds are heated to high temperatures, triggering chemical reactions that result in the formation of *h*-BN nanosheets. This method enables cost-effective, large-scale production of nanosheets with highly porous structures. The resulting product is typically in powder form, ready for postprocessing without requiring additional transfers, making it suitable for various applications. Due to the dynamic nature of high-temperature environments, the synthesized products often exhibit a turbostratic structure, characterized by small crystal domains and poor crystallinity. Despite this, high-temperature pyrolysis serves as a valuable complement to CVD and sonication-assisted synthesis techniques.

The high-temperature pyrolysis method for preparing *h*-BN nanomaterials typically involves boron-containing molecules such as boron oxide (B_2O_3) and boric acid (H_3BO_3), and common nitrogen-containing compounds such as melamine ($C_3H_6N_6$) and urea ($CO(NH_2)_2$). In some cases, ammonia (NH_3) can also be used as a nitrogen source. These widely available and inexpensive chemicals make this method ideal for large-scale industrial applications.

During the heating process, gases such as CO_2, CO, and NH_3 are released, contributing to the high specific surface area of the final product. As shown in Figure 1.4, *h*-BN nanosheets produced from boric acid and urea through pyrolysis typically consist of 2–3 atomic layers, exhibit a high defect density, and possess a specific surface area of up to 1900 $m^2 g^{-1}$ [34].

1.3 SYNTHESIS OF BORON NITRIDE NANOTUBES

Unlike *h*-BN with its sheet-like structure that can exist stably in bulk crystalline form, larger-dimension *h*-BN tubes are uncommon. Breaking down bulk BN to directly yield *h*-BN nanotubes (BNNTs) is less energy favorable than forming *h*-BN nanosheets. As a result, top-down approaches to synthesize BNNTs by breaking down larger *h*-BN structures are rare. Instead, BNNTs are

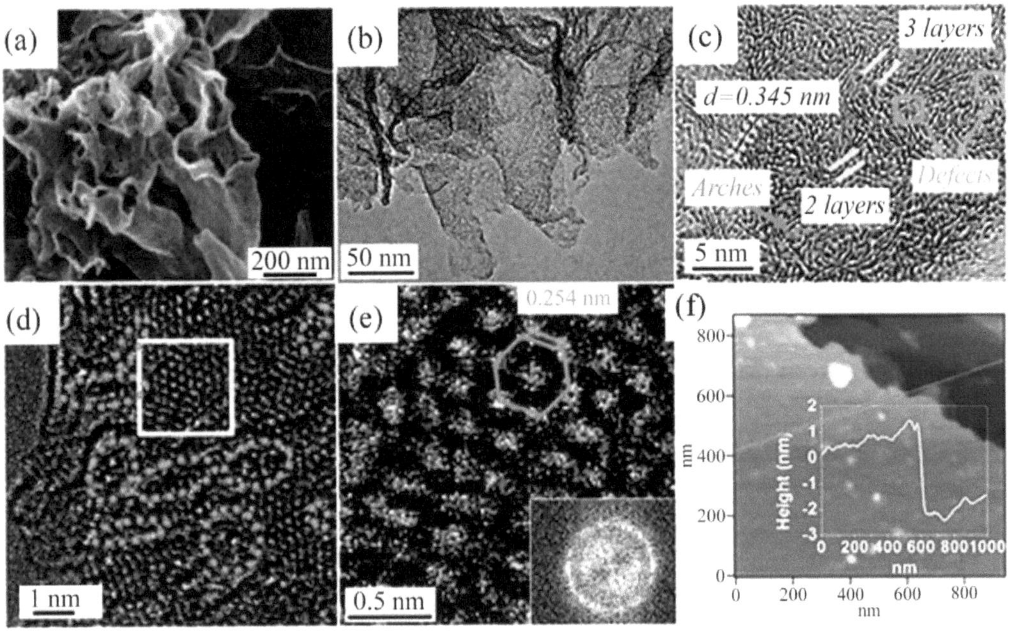

Figure 1.4 Morphologies and structure of *h*-BN nanosheets obtained by high-temperature pyrolysis [34]. (Reproduced from Ref. [34] with permission from the Royal Society of Chemistry.)

typically produced through bottom-up synthesis, involving reactions between boron- and nitrogen-containing atomic or molecular precursors. This section will focus on bottom-up methods for synthesizing BNNTs, such as ball milling with annealing, arc discharge, laser ablation, and CVD. These techniques allow for precise control over the dimensions and properties of BNNTs by adjusting atomic-level processing conditions, making them valuable for producing high-quality BNNTs tailored for use in electronics, nanocomposites, and other advanced materials.

1.3.1 Ball Milling and Annealing

The ball milling method is a versatile technique for synthesizing BNNTs. It involves the mechanical grinding of boron and nitrogen precursors in a high-energy ball mill, where repeated impacts and friction facilitate the formation of BNNT precursors. Postmilling annealing promotes the growth of BNNTs. This method is advantageous due to its simplicity, cost-effectiveness, and scalability, making it appealing for both research and industrial applications. Unlike traditional high-temperature methods, ball milling and annealing allow BNNT synthesis under milder conditions, creating opportunities for material innovation.

BNNTs can be synthesized by ball-milling elemental boron under an ammonia atmosphere (300 kPa) at room temperature for 150 hours, followed by annealing under N_2 gas at 1000 °C [35]. The high-energy ball milling process induces a nitriding reaction between boron and ammonia, forming a metastable material consisting of disordered BN and nanocrystalline boron. This nitriding reaction is likely catalyzed by iron particles, a common contaminant in ball milling. After milling, annealing promotes BNNT growth. Ball milling creates microstrain and structural defects, such as grain boundaries, which are critical for boron–nitrogen bonding at elevated

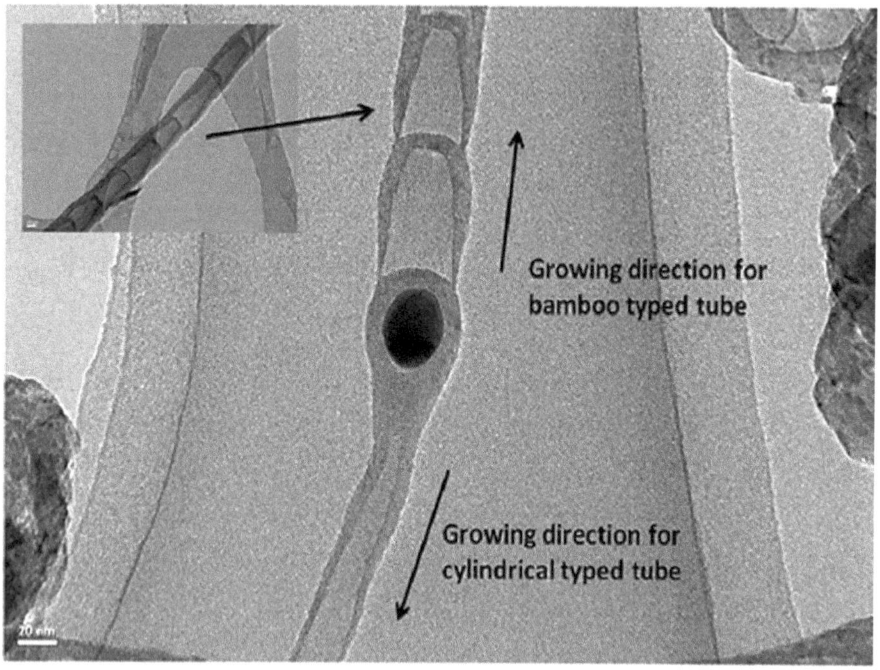

Figure 1.5 TEM image of BNNTs synthesized by the ball mill plus annealing process showing both bamboo and cylindrical shaped tubes [36]. (Reprinted from Ref. [36] with permission from Elsevier Science Ltd, 2002.)

temperatures. However, it has been reported that the nitriding reaction is not essential for BNNT growth; instead, the amorphization of boron on the surface of iron particles during milling may be the source of BNNT growth during annealing [36]. The resulting nanotubes can grow into cylindrical or bamboo-like shapes (Figure 1.5) and are typically multiwalled, with 30–100 nanowalls.

Although ball milling does not allow precise control over particle size and morphology, single-walled carbon nanotubes can be used as a sacrificial template to guide the growth of single-walled BNNTs. Wang et al. demonstrated the synthesis of single-walled BNNTs using a combination of high-energy ball milling, arc discharge, and template methods [37]. In their experiment, single-walled carbon nanotubes and boron powder were milled with Al_2O_3 balls in a modified planetary ball mill with an arc discharge device to convert N_2 into reactive N^- species. After 20 hours of milling at 500°C, the powders were washed and oxidized at 800°C to remove the CNT template, yielding single-walled BNNTs. This innovative method offers a promising approach for overcoming the unpredictability of morphology in typical ball mill synthesis and has the potential for scalable production of single-walled BNNTs.

1.3.2 Arc Discharge

The arc discharge method involves generating a high-temperature plasma arc between two electrodes, typically made of boron or a boron-containing compound and a nitrogen source, in an inert gas atmosphere such as argon or nitrogen. The intense heat from the arc vaporizes the electrode material, and as the vapor condenses, *h*-BNNTs are formed. This technique is favored for its simplicity and ability to produce high-quality nanotubes with uniform dimensions and fewer defects. Originally developed for carbon nanotube synthesis, the arc discharge method has been adapted for *h*-BNNTs using similar principles. It is particularly known for producing nanotubes with high crystallinity and desirable electronic and thermal properties, making it a valuable method for advanced nanotechnology applications.

BNNTs were first synthesized using the arc discharge method in 1995, employing a BN-packed tungsten rod and a cooled copper electrode under helium gas [38]. The BN-packed tungsten rod (a BN-filled tungsten electrode with a diameter of 3.17 mm) was used because bulk BN's insulating nature makes the arc discharge process challenging. BNNTs were deposited onto the cathode during the process. The melting of tungsten during growth indicated that temperatures exceeded 3700 K. The resulting multiwalled BNNTs had outer diameters of 6–8 nm and inner diameters of 1–3 nm, with an interlayer distance consistent with bulk *h*-BN. Tungsten particles found at the end of each nanotube suggested a growth mechanism: the outer layers of the BN coating may detach from the inner layer and expand outward to form the tube, with growth stopping when a metal particle collides with the tube's open end and attaches to its dangling bonds. This mechanism is particularly effective when the metal forms stable nitrides and borides, as tungsten does. Electron energy-loss spectroscopy analysis confirmed that the BNNTs are sp^2-bonded *h*-BN with a stoichiometric ratio of BN.

Later, highly conductive borides such as YB_6 and HfB_2 were employed for BNNT growth in a nitrogen (N_2) atmosphere. In the case of YB_6, the process begins with the generation of ionized and radical gases containing yttrium (Y), boron (B), and nitrogen (N) through arc melting [39]. Specifically, Y and B ions may form a semi-liquid Y+B compound, and as the system cools, B becomes supersaturated and separated from the compound's surface. The nitrogen in the atmosphere then reacts with the separated boron to form BN nanotubes. The resulting compound, YB_{6-x}, is more thermodynamically stable than YB_6. Figure 1.6. illustrates this growth mechanism of BN nanotubes using YB_6. However, scaling up the arc discharge process remains a challenge, as the reaction zone is confined to the small space at the arc core [40].

1.3.3 Laser Ablation

The laser ablation method for synthesizing BNNTs is a prominent technique known for producing high-quality nanotubes with minimal impurities. It involves using a high-power laser to ablate a boron-containing target (such as boron or boron nitride) in the presence of a nitrogen source, typically nitrogen gas or ammonia. The intense laser pulses vaporize the target material, generating a high-temperature plasma plume where boron and nitrogen atoms react to form BNNTs. This process occurs in an inert or nitrogen-rich atmosphere, preventing oxidation and facilitating the growth of nanotubes. Laser ablation allows for precise control over key parameters, including laser energy and pulse duration, enabling fine-tuning of nanotube properties. The technique is valued for producing BNNTs with high crystallinity and uniformity, making it ideal for advanced nanotechnology and materials science applications.

In 1996, Golberg et al. reported the successful synthesis of BNNTs using a laser ablation method under nitrogen pressures of 5–15 GPa [41]. The CO_2 laser, with power up to 240 W, targeted single-crystal cubic BN in a diamond anvil cell. Local temperatures exceeding 5000 K were reached. TEM and electron energy-loss spectroscopy were used to characterize the multiwalled nanotubes, which displayed both circular and polygonal cross-sections with 3–8 shells and outer diameters of 3–15 nm. The BNNTs were found to grow either in melted cubic boron nitride or hexagonal/amorphous boron nitride (*h*-BN), which recrystallized from the fluid phase on the specimen's surface

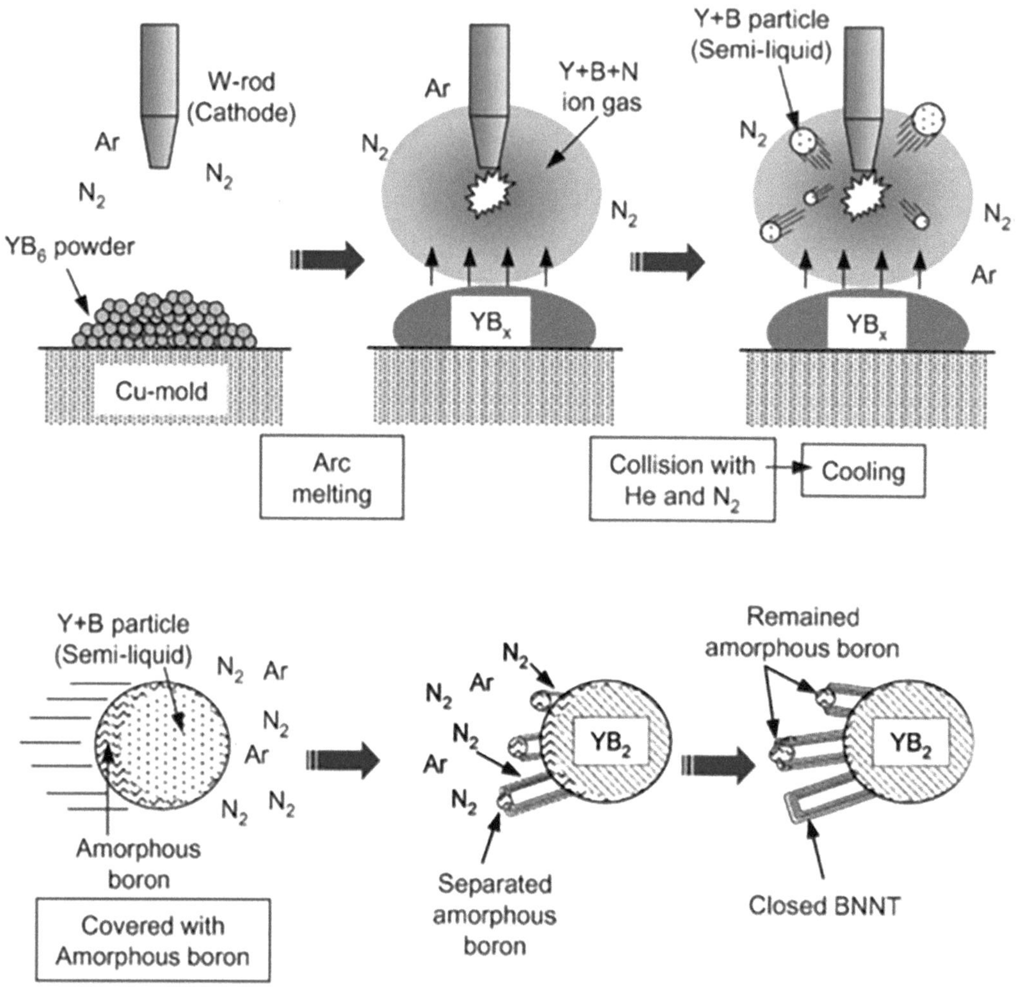

Figure 1.6 Schematic illustration of the growth mechanism of the BNNTs using YB_6 [39]. (Reprinted from Ref. [39] with permission from Elsevier Science Ltd, 2002.)

depending on reaction temperature and nitrogen pressure. The absence of metal inclusions in BNNTs synthesized via laser ablation represents a significant improvement over traditional methods, such as arc discharge, which often results in metallic particles at the nanotube ends.

Even though BNNTs can be synthesized without a catalyst, the reaction conditions are quite harsh, typically requiring high pressures. A subsequent study by Yu et al. in 1998 explored BNNT synthesis using excimer laser ablation at 1200°C [42]. Unlike Golberg's earlier work [41], nano-sized nickel (Ni) and cobalt (Co) powders were mixed with BN powders and hot pressed to form the target. This method predominantly produced single-walled BNNTs with diameters ranging from 1.5 to 8 nm. The choice of carrier gases (argon, nitrogen, helium) greatly influenced the nanotube morphology and yield. Argon and helium promoted single-walled structures, while nitrogen produced double-walled tubes. The authors suggested that the size and charge of carrier gas atoms may affect the mean-free path or lifetime of activated boron and nitrogen species ejected from the target by the laser. The absence of metallic particles and the clean surfaces of the nanotubes highlighted the efficiency of this method in producing high-purity BNNTs.

In 1999, Zhou et al. further investigated the role of catalysts in BNNT synthesis using laser ablation [43]. Their study showed that catalysts like nanosized Ni and Co powders influenced the thickness, length, surface smoothness, and tip morphology of BNNTs. Without catalysts, BNNTs typically had thicker walls, irregular tips, and rougher surfaces. In contrast, catalytically grown nanotubes were predominantly single-walled, with smoother surfaces and more regular tip structures. Catalysts were proposed to stabilize the dangling bonds at the tube ends, promoting axial growth and preventing the formation of amorphous clusters.

Laser ablation for BNNT synthesis offers the advantage of producing high-purity nanotubes with minimal impurities due to the absence of metal catalysts and precise control over synthesis parameters. This method allows fine-tuning of BNNT properties, such as flat caps and polygonal terminations. However, the technique is costly and complex, requiring high-power lasers and controlled environments, and often yields lower quantities compared to other methods like CVD or arc discharge. While it provides superior purity and structural control, its limited scalability makes it less suitable for large-scale production.

1.3.4 CVD

Similar to *h*-BN nanosheets, CVD is a widely used method for synthesizing BNNTs, offering precise control over structural properties and high purity. In the CVD process, gaseous precursors containing boron and nitrogen decompose at high temperatures within a reaction chamber, where they react and deposit BNNTs onto a substrate. This method allows for fine-tuning of reaction parameters like temperature, pressure, and gas flow, which are crucial for optimizing BNNT growth. The ability of CVD to control diameters, lengths, and wall structures makes it a key technique for developing advanced nanomaterials used in electronics, composite materials, and thermal management systems.

Early CVD growths of BNNTs or BN filaments used B_2H_6 (diborane) and NH_3 (or N_2) as precursors over a boride catalyst at temperatures around 1100°C [44]. The chemical reactions governing BNNT formation are:

$$B_2H_6 + 2NH_3 \rightarrow 2BN + 6H_2 \tag{1.2}$$

or

$$B_2H_6 + N_2 \rightarrow 2BN + 3H_2 \tag{1.3}$$

Similar to carbon nanotube growth, Gleize et al. suggested that BNNT growth follows a dissolution–precipitation mechanism [44]. ZrB_2 was chosen as a catalyst because it dissolves boron and nitrogen simultaneously. The resulting BNNTs had lengths of 1–50 μm and widths of 0.1–0.3 μm. It was observed that BNNTs grew from ZrB_2 grains, and many of them contained Zr-based particles inside their hollow tubes (Figure 1.7). Interestingly, the boron supply initially came from the ZrB_2 catalyst rather than the boron precursor (B_2H_6).

In later developments, borazine ($B_3N_3H_6$) emerged as a single precursor for synthesizing BNNTs via CVD. Lourie et al. synthesized BNNTs by heating a mixture of $(NH_4)_2SO_4$ and $NaBH_4$ to generate borazine, which was fed into a tube furnace reactor by a carrier gas. TEM and electron energy-loss spectroscopy confirmed the successful growth of BNNTs with lengths up to 5 μm, albeit with irregular caps.

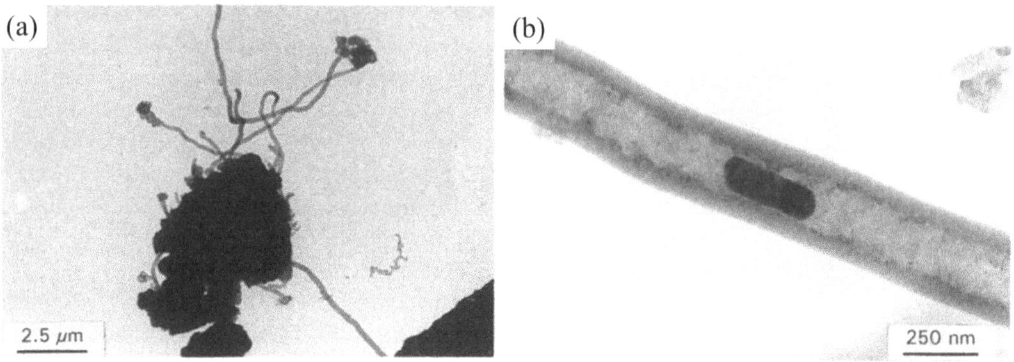

Figure 1.7 (a) BNNTs grown from ZrB_2 grain and (b) a hollow BNNT with a Zr-based core [44]. (Reproduced from Ref. [44] from Spring Nature with permissions, 1994.)

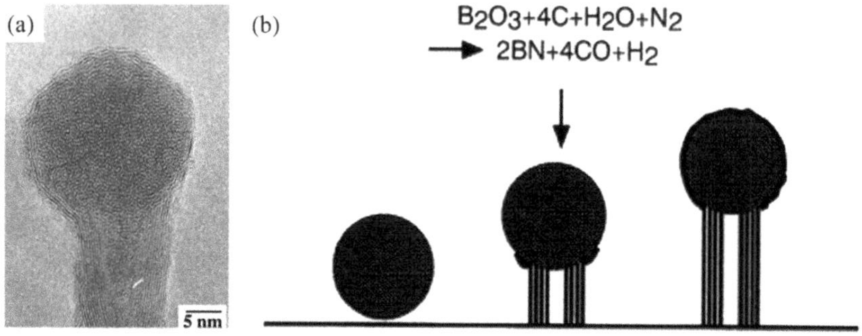

Figure 1.8 Noncatalytic growth of BNNTs using a B–N–O containing precursor: (a) BNNT with a bulbous tip; (b) possible growth mechanism [45]. (Reproduced from Ref. [45] with permission from Elsevier Science Ltd, 2001.)

Interestingly, BNNTs can also be grown via CVD without using a catalyst. Ma et al. demonstrated the noncatalytic growth of BNNTs using a B–N–O-containing precursor derived from boric acid and melamine [45]. The precursors were rapidly heated to 1700°C under a nitrogen (N_2) carrier gas. Deposits of white powders were collected on the crucible and graphite susceptor, where the local temperature was around 1200°C. Characterization revealed that the resulting multiwalled BNNTs had an sp^2 hybridized *h*-BN structure. Each BNNT featured a bulbous tip, which was a boron nitride cage encapsulating an amorphous B–N–O cluster.

The growth process involves B–N–O clusters carried by the carrier gas from the high-temperature region (1700°C) to the lower-temperature regions (around 1200°C), where they condense on the surfaces of the crucible and graphite susceptor. Concurrently, a vapor-phase reaction occurs:

$$B_2O_3 + 4C + H_2O + N_2 \rightarrow 2BN + 4CO + H_2 \tag{1.4}$$

BN diffuses out from the amorphous clusters and forms nanotubes (Figure 1.8). This method offers a solution for the bulk production of BNNTs without the need for a catalyst, making it a promising technique for large-scale synthesis [45].

1.4 CONCLUSIONS

In conclusion, this chapter has highlighted several synthesis methods for boron nitride nanostructures, including mechanical exfoliation, CVD, ball milling, arc discharge, laser ablation, and more. Each technique offers unique advantages in terms of controlling the size, morphology, and functionality of BNNS and BNNTs. These advancements in synthesis methodologies not only enhance the scalability and quality of BNNS and BNNT production but also expand their potential applications across various sectors such as electronics, photonics, and biomedical engineering. As research in this field continues to progress, the role of boron nitride nanomaterials in driving technological innovation is set to grow, solidifying their significance in the future of nanotechnology.

REFERENCES

1. K. S. Novoselov et al., Electric field effect in atomically thin carbon films. *Science* **306**, 666–669 (2004).
2. D. Pacilé, J. C. Meyer, Ç. Ö. Girit, A. Zettl, The two-dimensional phase of boron nitride: Few-atomic-layer sheets and suspended membranes. *Applied Physics Letters* **92**, 133107 (2008).
3. J. Ghosh, S. Mazumdar, M. Das, S. Ghatak, A. K. Basu, Microstructural characterization of amorphous and nanocrystalline boron nitride prepared by high-energy ball milling. *Materials Research Bulletin* **43**, 1023–1031 (2008).
4. L. H. Li et al., Large-scale mechanical peeling of boron nitride nanosheets by low-energy ball milling. *Journal of Materials Chemistry* **21**, 11862–11866 (2011).
5. X. Chen, J. F. Dobson, C. L. Raston, Vortex fluidic exfoliation of graphite and boron nitride. *Chemical Communications* **48**, 3703–3705 (2012).
6. K. S. Suslick, Sonochemistry. *Science* **247**, 1439–1445 (1990).
7. C. Zhi, Y. Bando, C. Tang, H. Kuwahara, D. Golberg, Large-scale fabrication of boron nitride nanosheets and their utilization in polymeric composites with improved thermal and mechanical properties. *Advanced Materials* **21**, 2889–2893 (2009).
8. K. Zhang, Y. Feng, F. Wang, Z. Yang, J. Wang, Two dimensional hexagonal boron nitride (2D-hBN): Synthesis, properties and applications. *Journal of Materials Chemistry C* **5**, 11992–12022 (2017).
9. J. N. Coleman et al., Two-dimensional nanosheets produced by liquid exfoliation of layered materials. *Science* **331**, 568–571 (2011).
10. Y. Lin et al., Aqueous dispersions of few-layered and monolayered hexagonal boron nitride nanosheets from sonication-assisted hydrolysis: Critical role of water. *The Journal of Physical Chemistry C* **115**, 2679–2685 (2011).
11. M. J. Rand, J. F. Roberts, Preparation and properties of thin film boron nitride. *Journal of the Electrochemical Society* **115**, 423 (1968).
12. A. Nagashima, N. Tejima, Y. Gamou, T. Kawai, C. Oshima, Electronic structure of monolayer hexagonal boron nitride physisorbed on metal surfaces. *Physical Review Letters* **75**, 3918–3921 (1995).
13. J. Li, S. Bernard, V. Salles, C. Gervais, P. Miele, Preparation of polyborazylene-derived bulk boron nitride with tunable properties by warm-pressing and pressureless pyrolysis. *Chemistry of Materials* **22**, 2010–2019 (2010).
14. M. H. Khan et al., Synthesis of large and few atomic layers of hexagonal boron nitride on melted copper. *Scientific Reports* **5**, 7743 (2015).
15. M. S. Bresnehan et al., Integration of hexagonal boron nitride with quasi-freestanding epitaxial graphene: Toward wafer-scale, high-performance devices. *ACS Nano* **6**, 5234–5241 (2012).
16. W. Auwärter, H. U. Suter, H. Sachdev, T. Greber, Synthesis of one monolayer of hexagonal boron nitride on Ni (111) from B-trichloroborazine $(ClBNH)_3$. *Chemistry of Materials* **16**, 343–345 (2004).
17. A. Ismach et al., Toward the controlled synthesis of hexagonal boron nitride films. *ACS Nano* **6**, 6378–6385 (2012).
18. C. Zhang, X. Hao, Y. Wu, M. Du, Synthesis of vertically aligned boron nitride nanosheets using CVD method. *Materials Research Bulletin* **47**, 2277–2281 (2012).

19. R. Y. Tay et al., Trimethylamine borane: A new single-source precursor for monolayer *h*-BN single crystals and h-BCN thin films. *Chemistry of Materials* **28**, 2180–2190 (2016).
20. S. Joshi et al., Boron nitride on Cu (111): An electronically corrugated monolayer. *Nano Letters* **12**, 5821–5828 (2012).
21. L. Wang et al., Epitaxial growth of a 100-square-centimetre single-crystal hexagonal boron nitride monolayer on copper. *Nature* **570**, 91–95 (2019).
22. K. Y. Ma et al., Epitaxial single-crystal hexagonal boron nitride multilayers on Ni (111). *Nature* **606**, 88–93 (2022).
23. S. Fukamachi et al., Large-area synthesis and transfer of multilayer hexagonal boron nitride for enhanced graphene device arrays. *Nature Electronics* **6**, 126–136 (2023).
24. Y. Uchida et al., Controlled growth of large-area uniform multilayer hexagonal boron nitride as an effective 2D substrate. *ACS Nano* **12**, 6236–6244 (2018).
25. J. K. Hite, Z. R. Robinson, C. R. Eddy, Jr., B. N. Feigelson, Electron backscatter diffraction study of hexagonal boron nitride growth on Cu single-crystal substrates. *ACS Applied Materials & Interfaces* **7**, 15200–15205 (2015).
26. M. H. Khan et al., Carbon- and crack-free growth of hexagonal boron nitride nanosheets and their uncommon stacking order. *Nanoscale* **8**, 15926–15933 (2016).
27. J. S. Lee et al., Wafer-scale single-crystal hexagonal boron nitride film via self-collimated grain formation. *Science* **362**, 817–821 (2018).
28. K. K. Kim et al., Synthesis and characterization of hexagonal boron nitride film as a dielectric layer for graphene devices. *ACS Nano* **6**, 8583–8590 (2012).
29. Q. Yu et al., Control and characterization of individual grains and grain boundaries in graphene grown by chemical vapour deposition. *Nature Materials* **10**, 443–449 (2011).
30. P. R. Kidambi et al., In situ observations during chemical vapor deposition of hexagonal boron nitride on polycrystalline copper. *Chemistry of Materials* **26**, 6380–6392 (2014).
31. M. S. Islam et al., Growth mechanisms of monolayer hexagonal boron nitride (*h*-BN) on metal surfaces: Theoretical perspectives. *Nanoscale Advances* **5**, 4041–4064 (2023).
32. H. Wang et al., Synthesis of large-sized single-crystal hexagonal boron nitride domains on nickel foils by ion beam sputtering deposition. *Advanced Materials* **27**, 8109–8115 (2015).
33. M. Sajjad, W. M. Jadwisienczak, P. Feng, Nanoscale structure study of boron nitride nanosheets and development of a deep-UV photo-detector. *Nanoscale* **6**, 4577–4582 (2014).
34. P. Wu et al., A template-free solvent-mediated synthesis of high surface area boron nitride nanosheets for aerobic oxidative desulfurization. *Chemical Communications* **52**, 144–147 (2016).
35. Y. Chen, J. Fitz Gerald, J. S. Williams, S. Bulcock, Synthesis of boron nitride nanotubes at low temperatures using reactive ball milling. *Chemical Physics Letters* **299**, 260–264 (1999).
36. J. Kim et al., Synthesis and growth of boron nitride nanotubes by a ball milling–annealing process. *Acta Materialia* **59**, 2807–2813 (2011).
37. S. Wang et al., Multitudinous industrialized fabrication of BNNT with high-energy ball milling and arc-discharge-aided template methods. *Journal of Nanoparticle Research* **19**, 352 (2017).
38. N. G. Chopra et al., Boron nitride nanotubes. *Science* **269**, 966–967 (1995).
39. I. Narita, T. Oku, Synthesis of boron nitride nanotubes by using YB_6 powder. *Solid State Communications* **122**, 465–468 (2002).
40. J. H. Kim, T. V. Pham, J. H. Hwang, C. S. Kim, M. J. Kim, Boron nitride nanotubes: Synthesis and applications. *Nano Convergence* **5**, 17 (2018).
41. D. Golberg et al., Nanotubes in boron nitride laser heated at high pressure. *Applied Physics Letters* **69**, 2045–2047 (1996).
42. D. P. Yu et al., Synthesis of boron nitride nanotubes by means of excimer laser ablation at high temperature. *Applied Physics Letters* **72**, 1966–1968 (1998).
43. G. W. Zhou, Z. Zhang, Z. G. Bai, D. P. Yu, Catalyst effects on formation of boron nitride nano-tubules synthesized by laser ablation. *Solid State Communications* **109**, 555–559 (1999).
44. P. Gleize, M. C. Schouler, P. Gadelle, M. Caillet, Growth of tubular boron nitride filaments. *Journal of Materials Science* **29**, 1575–1580 (1994).
45. R. Ma, Y. Bando, T. Sato, CVD synthesis of boron nitride nanotubes without metal catalysts. *Chemical Physics Letters* **337**, 61–64 (2001).

Chapter 2

Synthesis of Pristine, Doped, and Hybrid Boron Nitride Nanosheets Using Mechanochemistry

Srikanth Mateti, Ying Ian Chen, Yuchen Liu, and Qiran Cai

2.1 INTRODUCTION

Synthesizing BN nanosheets typically involves methods that aim to exfoliate bulk BN into thin nanosheets. BN is a compound composed of boron and nitrogen, and it has a structure similar to graphene, which makes it a promising material for various applications due to its unique properties, such as high thermal conductivity, high mechanical strength, and chemical stability [1]. Synthesizing BN nanosheets using ball milling involves the mechanical exfoliation of bulk BN into thin layers under the impact and shear forces generated by milling balls in a ball milling apparatus. This method is relatively simple and scalable, making it attractive for producing BN nanosheets. Various exfoliating agents (solid and liquid, such as methylenediformamide (MDF), N-Methyl-2-pyrrolidone (NMP), urea etc.) [2–4] have been explored to produce nanosheets; these can effectively reduce the structural damage of nanosheets, but post-treatment is required to wash out or a purification process to remove unreacted exfoliating agents. This post-treatment process introduces contamination, which affects the structure and properties of nanosheets. This chapter discusses and synthesizes pristine and doped, composite nanosheets in various sections, respectively.

2.2 SYNTHESIS OF NANOSHEETS USING MECHANOCHEMISTRY

Mechanical forces like friction, stress, and strain can cause wear and tear at both the macroscopic and molecular levels. On a molecular scale, these forces can initiate chemical reactions in solids by disrupting symmetry and destabilizing bonds through shear stress, a process referred to as mechanochemistry [5]. Unlike traditional methods such as photochemistry, thermochemistry, or electrochemistry—which depend on light, heat, or electrical energy—mechanochemistry harnesses mechanical energy to drive chemical reactions, often occurring in the solid state [6–9]. An AI-generated scheme is shown in Scheme 2.1: Mechanical energy from ball milling induces shear stress, causing bond disruption and the formation of new molecular structures.

DOI: 10.1201/9781003647812-3

Scheme 2.1 Mechanochemical Process: Visualizing Molecular Bond Disruption and Energy Transfer *via* Mechanical Forces.

Mechanochemistry presents notable advantages over solution-based synthesis techniques like sol-gel, coprecipitation, or sonochemical methods, especially for reactions involving insoluble solids or those that are sensitive to solvents. By removing the need for hazardous solvents, mechanochemistry provides a safer and more environmentally friendly alternative. The mechanical action in processes like high-energy ball milling reduces particle size, generates active surfaces, creates radicals, and promotes bond formation, making it particularly effective for facilitating solid-state reactions.

This approach has garnered interest for synthesizing nanomaterials due to its capability to break down crystalline structures into nanometer-sized seeds [10]. These seeds can then develop into nanostructures or convert inorganic materials into new compositions, often at room temperature [11]. During ball milling, particles experience flattening, fracturing, welding, and re-welding, which creates defects and refines crystallite sizes [12]. The process is influenced by factors such as the ball-to-powder ratio, milling speed, duration, and atmosphere. This precise control over reaction kinetics enables the tailored synthesis of advanced materials with enhanced properties, showcasing the transformative potential of mechanochemistry [13–15].

2.2.1 Pristine BN Nanosheets

In an experiment, a ball mill jar (Figure 2.1a) was loaded with several grams of bulk BN powder with several steel balls filled with 300 kPa pressure of a gas and sealed the jar. The gas did not adsorb onto the BN powder without milling action, which means no pressure change was observed. After milling for 15 hours, there was a pressure reduction of 60 kPa, and it was 80 kPa for 20 hours. The XRD patterns of BN milled in NH_3 for different milling times are shown in Figure 2.1b. These patterns (sharp (002) peak) confirm the retention of the crystal structure of BN until 70 hours. Further, SEM images show (Figure 2.1c) that the large micrometer-sized starting BN powder exfoliated into thin layers/nanosheets of BN after milling for 20 hours. Transmission electron microscopy (TEM) analysis also confirms that the BN nanosheets produced by milling in NH_3 have an excellent hexagonal structure (Figure 2.1d and e). The selected area electron diffraction patterns contain multiple sets of dots with a six-fold symmetry revealing a fair and crystalline in-plane structure [13].

Several different gases (such as C_2H_4, Ar, O_2, CH_4) have also been tested to determine whether they have similar effects on BN. The corresponding XRD patterns are shown in Figure 2.1f. BN milled in NH_3 and C_2H_4 has sharp (002) diffraction peaks; also, a c-direction peak (004) has been observed. Whereas other gases such as Ar, O_2, and CH_4 do not have similar effects as NH_3 and C_2H_4; this means that BN nanosheets are produced only when milled in an NH_3 and C_2H_4 atmosphere. Optimize the milling parameters such as milling time, rotation speed, and ball-to-powder ratio based on the desired thickness and quality of the BN nanosheets. Longer milling times and higher rotation speeds generally result in more extensive exfoliation but may also lead to increased defects and impurities. In the mechanism behind producing these BN nanosheets, gas molecules (NH_3) are decomposed to NH_x groups under the stress, which is created by balls during milling, and chemisorbed onto the defects and edges, saturating the dangling bonds, thus preventing cross-linking of nanosheets. Ball milling offers a straightforward and cost-effective

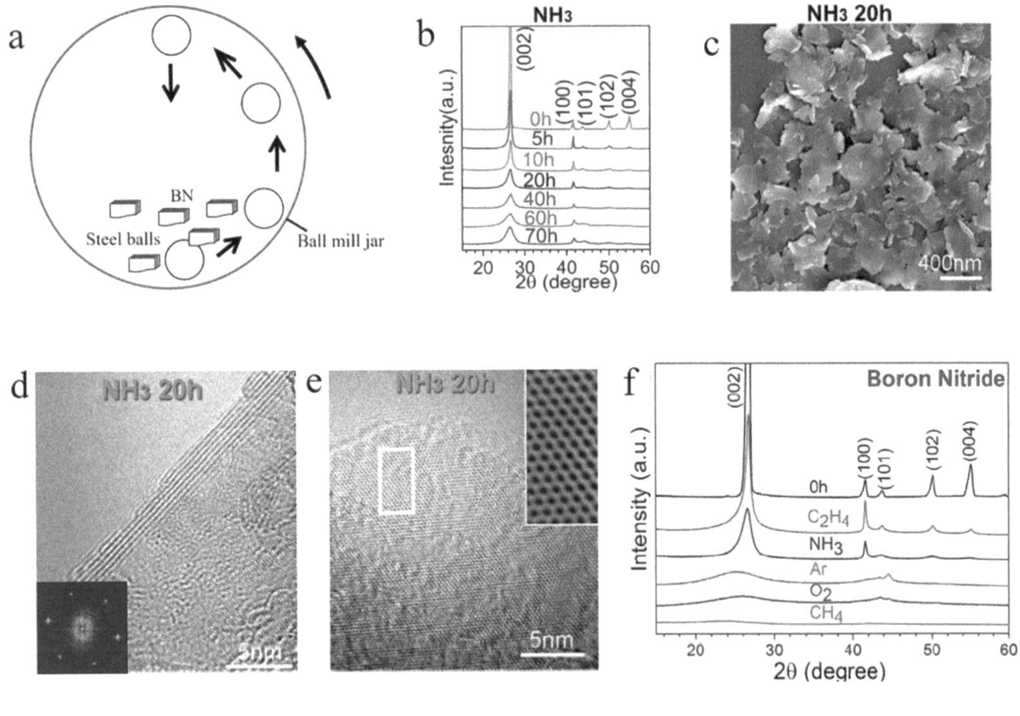

Figure 2.1 (a) Ball mill schematic. (b) XRD patterns of BN milled in NH_3 for different time intervals. (c) SEM images of BN milled in NH_3 for 20 hours. HR-TEM images of BN nanosheets: (d) edges and (e) in-plane structure. (f) XRD patterns of BN milled in different gases for 20 hours. (Reprinted with permission from Ref. [13], Copyright 2016, Springer Nature.)

approach to producing BN nanosheets with controllable thickness and properties. However, optimization of milling parameters and careful characterization are crucial to achieving the desired nanosheet quality and performance [13].

2.2.2 Carbon-Doped Nanosheets

Incorporating foreign atoms into materials can alter their electronic setup and adjust their chemical and physical characteristics. It's crucial to devise an effective doping method for various applications involving two-dimensional (2D) nanomaterials. This section presents an in situ approach capable of generating and doping carbon simultaneously into several 2D materials like BN using mechanochemistry. This innovative method yields substantial amounts of 2D materials with precisely controlled doping levels, thereby introducing novel properties [16].

Hexagonal BN particles underwent milling in ethylene (C_2H_4) for varying time spans to introduce carbon (C) into BN nanosheets. Analysis of XRD patterns (Figure 2.2a) reveals that the structure remains unchanged even after 30 hours of milling. The SEM image (Figure 2.2b) illustrates the layered structure of these thin sheets. X-ray photoelectron spectroscopy spectra (Figure 2.2c) obtained from BN milled for 30 hours in C_2H_4 demonstrate a carbon content of 32.4 at%, with identifiable C–N bonds at 285.5 eV and N=C–O at 288.8 eV. Energy-filtered TEM images combined with overlay mapping (Figure 2.2d, e) exhibit a uniform distribution of

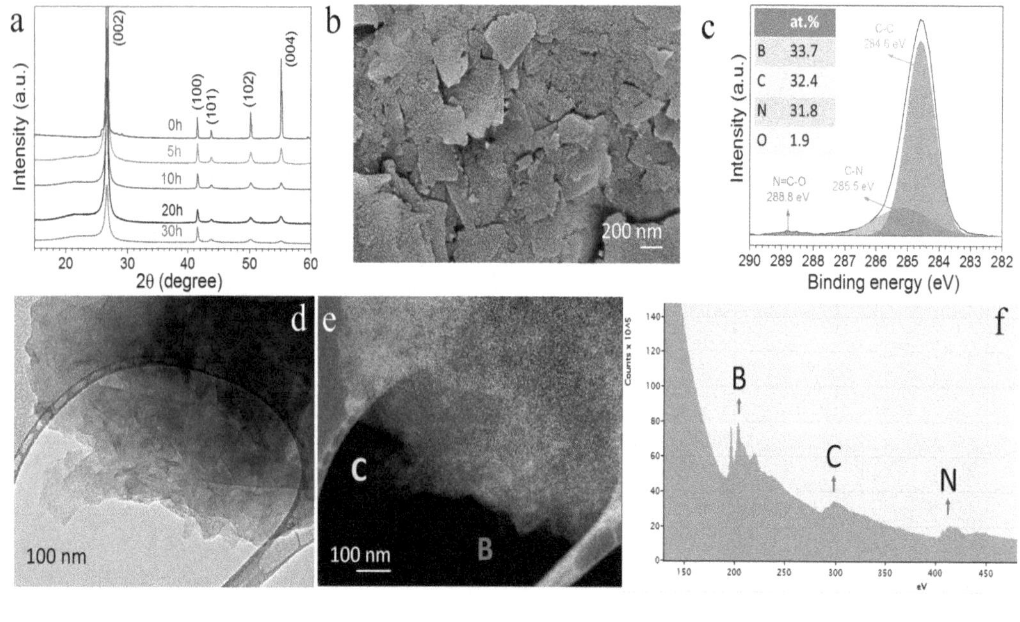

Figure 2.2 XRD patterns of BN subjected to milling in C_2H_4 for various durations (a), alongside the SEM image of the BN sample milled for 30 hours (b). Additionally, X-ray photoelectron spectroscopy spectra of C 1s for BN following 30 hours of milling are shown (c), along with an energy-filtered TEM image (d), elemental mapping (e), and EELS spectra (f) of BN post 30-hour milling. (Reprinted with permission from Ref. [16], Copyright 2019 Royal Society of Chemistry.)

boron and carbon, corroborated by electron energy loss spectroscopy (EELS) spectra depicted in Figure 2.2f. Clear indications of the B-K edge at 188 eV, C-K edge at 283.8 eV, and N-K edge at 401.6 eV are observed [17].

During high-energy impacts, C_2H_4 underwent dissociation and chemisorbed onto the defects and edges of nanosheets, thereby preventing cross-linking among the nanosheets. Throughout the milling procedure, notable pressure decreases of C_2H_4 gases were noted within the sealed milling chamber. This reduction in pressure could be attributed to gas absorption onto the freshly generated surfaces of the milled materials due to high-energy impacts, a phenomenon validated by the presence of nitrogen and carbon in the milled samples exhibiting escalating contents of the dopant with increasing levels. Other gases like CH_4 and C_3H_8 also stand as potential doping agents. Elevated milling intensity generates greater fresh surface area, active edges, and defects in the milled samples, thereby furnishing more sites for gas adsorption, decomposition, and chemical bond formation. Augmenting the dopant content can be achieved through intensified milling energy or employing a high-speed milling system featuring a high impact frequency. This novel in situ doping process ensures a uniform distribution of the dopant within the materials, presenting a straightforward and cost-effective approach for producing doped nanosheets in substantial quantities.

2.2.3 Hybrid Composite Nanosheets

Combining graphene and BN nanosheets into composite structures offers the potential to surpass the limitations of graphene alone. Notably, BN nanosheets exhibit superior thermal and chemical stability compared to graphene, along with distinct surface properties. While both materials are recognized as effective lubricants individually, their composite forms have not been thoroughly explored as lubricants. This necessitates experimental validation, which constitutes the primary focus of the current research endeavor [18].

Although many types of composite nanosheets can be grown on substrates using these and other methods, such as chemical vapor deposition, hydrothermal methods, and mechanical transfer, they usually face serious problems in scaling up [19,20]. Recently, controlled ball milling using ammonia or hydrocarbon gases as exfoliation agents has been developed for the large-scale synthesis of pure nanosheets such as graphene, BN, MoS_2, and WS_2 [13]. This method has been very successful in the preparation of nanosheets with minimal defects, keeping them flat and single-crystalline even after long milling. The NH_3 and C_2H_4 molecules act as lubricants in this process, minimizing milling-induced damage by interacting with the nanosheet surfaces. These gas molecules preferentially attach to edges and defects, engaging in chemical reactions that help repair the structural imperfections by saturating the dangling bonds.

In an experiment, commercial graphite and *h*-BN powders with diameters smaller than 40 μm were selected as initial materials. An optimized high-energy ball milling technique was employed to create composite nanosheets of graphite and *h*-BN within an ammonia environment. A steel vial containing several grams of a 1:1 mixture of graphite and *h*-BN was utilized. Steel balls measuring 25 mm in diameter were placed in the vial, maintaining a ball-to-powder weight ratio of 64:1, while the rotation speed was set at 140 rpm. The vial was then sealed and pressurized with 310 kPa of ammonia gas, a crucial atmosphere for effective nanosheet production [18].

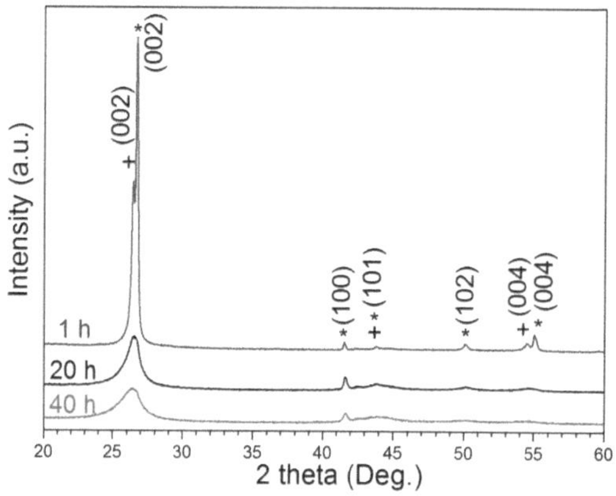

Figure 2.3 XRD patterns of BN and graphite with different milling times, +: graphite, *: BN. (Reprinted with permission from Ref. [18], Copyright 2018 Wiley-VCH GmbH, Weinheim.)

Figure 2.3 displays the XRD spectra of graphite and BN powder blends subjected to varying milling durations. Following ball milling, these spectra were obtained. After milling for an hour, the XRD patterns of the graphite and BN blend exhibit two prominent (002) characteristic peaks falling within the range of 26°–27°. By comparing these patterns with those of commercially available graphite and BN, it becomes evident that peaks at approximately 26.5° and 54.5° correspond to graphite, while those at around 26.7° and 56° are indicative of the *h*-BN phase. As the milling duration increases, the characteristic peaks associated with hexagonal structures ((002) and (004)) diminish in intensity and broaden due to the reduction in particle size. Peaks such as (100), (101), and (004) nearly disappear, implying a reduction in the thickness-to-size ratio of the initial particles and the exfoliation of nanosheets from bulk materials after 20 or 40 hours of ball milling. Moreover, the individual (002) and (004) peaks merge into a single peak, indicating the uniform integration of graphene and BN nanosheets into composite graphene/BN nanosheets. SEM images depicting the morphologies of composite nanosheets produced at various durations of ball milling are illustrated in Figure 2.4.

To mitigate the charging effect arising from the electrically insulating BN samples, a thin carbon coating was applied. Following one hour of ball milling, the particles appeared thick and blocky, with diameters exceeding 2 μm (Figure 2.4a). However, with prolonged milling, the layered structure underwent extensive exfoliation. After 20 hours of milling, nanosheets with substantial surface areas and uniform sizes became apparent (Figure 2.4b). Many composite nanosheets were loosely stacked with distinct edges and had diameters exceeding 200 nm. By extending the milling time to 40 hours, the diameter of individual graphene/BN composite nanosheets was reduced to approximately 100 nm (Figure 2.4c). These nanosheets were densely packed together, and individual ones that were initially loosely dispersed tended to aggregate into clusters. It is established that the particles experience shear and fragmentation forces during the ball milling process. When the shear force surpasses the weak interlayer Van der Waals bonds, the bulk materials are fragmented into nanosheets. Consequently, prolonged milling durations, such as 40 hours, lead to a reduction in the size of individual nanosheets and promote their tight agglomeration [21,22].

Figure 2.4 SEM images of the graphite and BN mixture samples milled for different times: (a) 1 hour; (b) 20 hours; (c) 40 hours; (d) the 20 hours sample without the oleic acid and carbon coating. (Reprinted with permission from Ref. [18], Copyright 2018 Wiley-VCH GmbH, Weinheim.)

To gain deeper insights into the microstructure and elemental distribution, graphene/BN composite nanosheets subjected to 20 hours of milling were examined using TEM and EELS. Figure 2.5a displays the edge of a composite nanosheet suspended on a holey carbon film, where the nanosheet edges are clearly visible. The random stacking of multiple nanosheets can be observed, with their transparency indicating the number of layers stacked. The thickness of these nanosheets was estimated from high-magnification TEM images, as illustrated in Figure 2.5b, c. Each graphene or BN nanosheet consists of approximately 10 monolayers, with a typical thickness of less than 10 nm. The EELS spectrum in Figure 2.5 highlights the elemental edges of B, C, and N, confirming the contributions of graphene and BN to the formation of the composite nanosheets.

Energy-filtered TEM imaging was carried out to observe the elemental distribution in a selected area of the sample suspended over a hole in a holey carbon film (Figure 2.6a). The energy-filtered image shown in Figure 2.6b colorizes the elemental distribution of boron and carbon in gray and black, respectively.

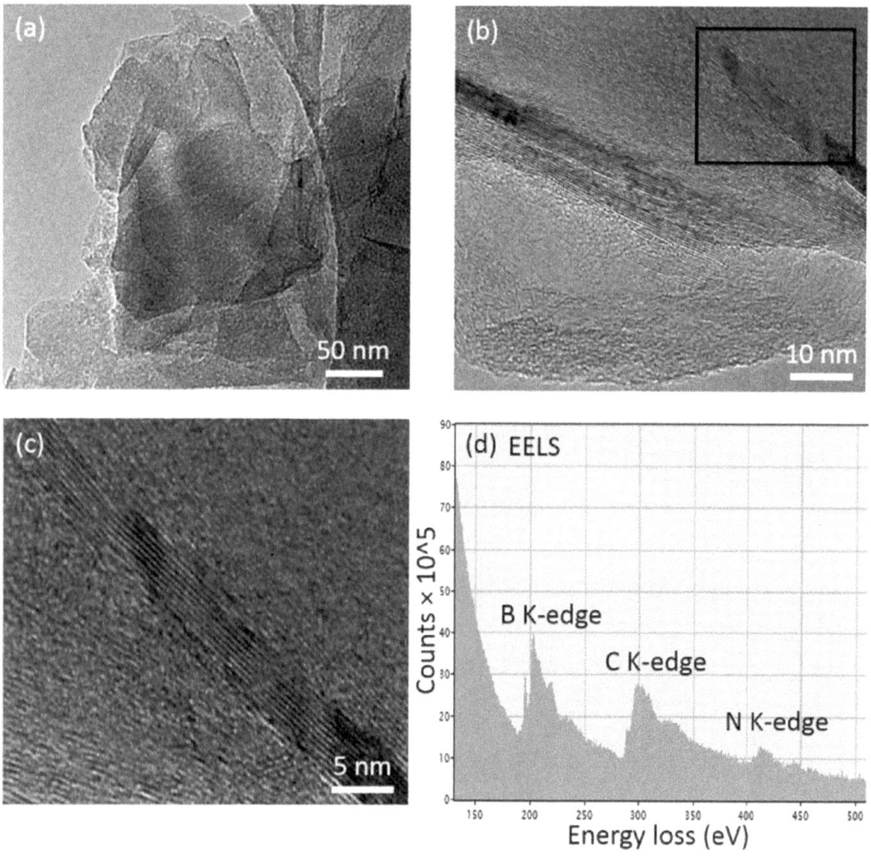

Figure 2.5 (a) TEM images of graphene/BN composite nanosheets produced by ball milling for 20 hours; (b) high-magnification TEM image; (c) enlarged TEM image of the selected area (indicated by the box); (d) EELS spectrum of graphene/BN composite nanosheets. (Reprinted with permission from Ref. [18], Copyright 2018 Wiley-VCH GmbH, Weinheim.)

Individual energy-filtered maps of boron and carbon are presented in Figure 2.6c and d, respectively. While the boron and carbon regions overlap, distinct graphene and BN nanosheets can also be discerned at the edges of the particles. This observation suggests that the composite nanosheets are formed by the stacking of multiple graphene and BN nanosheets together.

2.3 CONCLUSIONS

High-energy ball milling in the presence of NH_3 and C_2H_4 gases effectively protects BN nanosheets, whereas similar mechanical grinding in atmospheres such as Ar, CH_4, or O_2 leads to the formation of amorphous or highly disordered nanoparticles. This process also enables the in situ fabrication of nanosheets with controlled doping levels, including carbon doping of BN nanosheets *via* a mechanochemical method utilizing C_2H_4 gas. The robustness of 2D

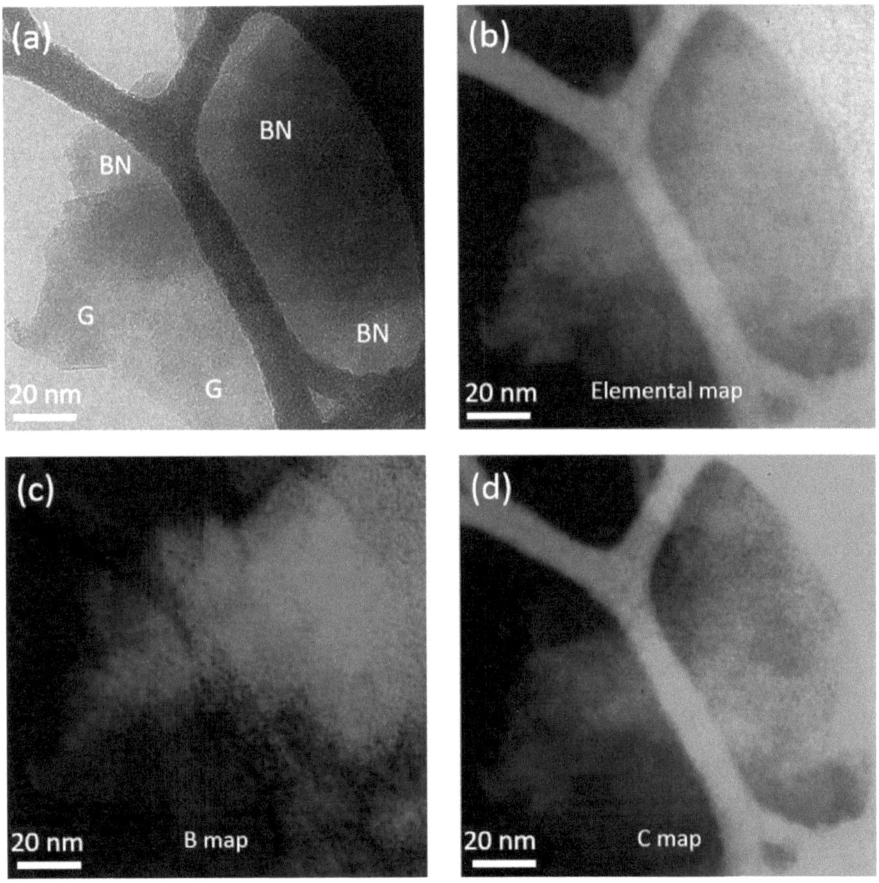

Figure 2.6 Energy-filtered TEM data: (a) an elastic TEM image; (b) an energy-filtered image of the same area with elemental contrast (boron–gray, carbon–black); (c) and (d) individual energy-filtered maps of boron and carbon. (Reprinted with permission from Ref. [18], Copyright 2018 Wiley-VCH GmbH, Weinheim.)

nanomaterials during intense milling in specific gases is attributed to the enhanced adsorption of ammonia and hydrocarbon gases. Mechanochemical reactions occur between reactive gases and the dangling bonds created during milling, alongside chemisorption of reactive species. These processes break bonds and prevent cross-linking by inhibiting the formation of bridging bonds between layers. Milling in reactive gas atmospheres has thus enabled the large-scale production of various individual, doped, and hybrid composite nanosheets.

REFERENCES

1. W. Lei, D. Portehault, D. Liu, S. Qin, Y. Chen, Porous boron nitride nanosheets for effective water cleaning. *Nature Communications* **4**, 1777 (2013).
2. K. F. Mak, C. Lee, J. Hone, J. Shan, T. F. Heinz, Atomically thin MoS_2: A new direct-gap semiconductor. *Physical Review Letters* **105**, 136805 (2010).

3. J. N. Coleman et al., Two-dimensional nanosheets produced by liquid exfoliation of layered materials. *Science* **331**, 568–571 (2011).
4. W. Lei et al., Boron nitride colloidal solutions, ultralight aerogels and freestanding membranes through one-step exfoliation and functionalization. *Nature Communications* **6**, 8849 (2015).
5. J. J. Gilman, Mechanochemistry. *Science* **274**, 65–65 (1996).
6. V. Šepelák, A. Düvel, M. Wilkening, K.-D. Becker, P. Heitjans, Mechanochemical reactions and syntheses of oxides. *Chemical Society Reviews* **42**, 7507–7520 (2013).
7. G. Heinicke et al., *Tribochemistry Akademie-Verlag.* (Wiley Online Library, 1984).
8. L. Takacs, The historical development of mechanochemistry. *Chemical Society Reviews* **42**, 7649–7659 (2013).
9. E. M. Gutman, *Mechanochemistry of Materials.* (Cambridge International Science Publishing, 1998).
10. R. Valiev, Nanostructuring of metals by severe plastic deformation for advanced properties. *Nature Materials* **3**, 511–516 (2004).
11. R. A. Buyanov, V. V. Molchanov, V. V. Boldyrev, Mechanochemical activation as a tool of increasing catalytic activity. *Catalysis Today* **144**, 212–218 (2009).
12. C. Suryanarayana, Mechanical alloying: A novel technique to synthesize advanced materials. *Research*, **2019**, 4219812 (2019).
13. T. Xing et al., Gas protection of two-dimensional nanomaterials from high-energy impacts. *Scientific Reports* **6**, 35532 (2016).
14. S. Mateti et al., Mechanochemistry: A force in disguise and conditional effects towards chemical reactions. *Chemical Communications* **57**, 1080–1092 (2021).
15. A. S. Rogachev, Mechanical activation of heterogeneous exothermic reactions in powder mixtures. *Russian Chemical Reviews* **88**, 875 (2019).
16. S. Mateti et al., In situ doping and synthesis of two-dimensional nanomaterials using mechano-chemistry. *Nanoscale Horizons* **4**, 642–646 (2019).
17. L. Hua Li et al., Photoluminescence of boron nitride nanosheets exfoliated by ball milling. *Applied Physics Letters* **100**, 261108 (2012).
18. Y. Liu et al., Synthesis of composite nanosheets of graphene and boron nitride and their lubrication application in oil. *Advanced Engineering Materials* **20**, 1700488 (2018).
19. R. M. Yunus, H. Endo, M. Tsuji, H. Ago, Vertical heterostructures of MoS_2 and graphene nanoribbons grown by two-step chemical vapor deposition for high-gain photodetectors. *Physical Chemistry Chemical Physics* **17**, 25210–25215 (2015).
20. C. Dean, A. Young, I. Meric, C. Lee, L. Wang, S. Sorgenfrei, K. Watanabe, T. Taniguchi, P. Kim, K. L. Shepard, J. Hone. Boron nitride substrates for high-quality graphene electronics. *Nature Nanotechnology* **5**, 722–726 (2010).
21. M. Yi, Z. Shen, A review on mechanical exfoliation for the scalable production of graphene. *Journal of Materials Chemistry A* **3**, 11700–11715 (2015).
22. L. H. Li et al., Large-scale mechanical peeling of boron nitride nanosheets by low-energy ball milling. *Journal of Materials Chemistry* **21**, 11862–11866 (2011).

Chapter 3

High-Concentration Dispersion of Boron Nitride Nanosheets and Their Applications

Qiran Cai, Hongbo Jiang, Srikanth Mateti, and Ying Ian Chen

3.1 INTRODUCTION

Since the successful isolation of graphene [1], two-dimensional (2D) materials have garnered significant attention due to their unique properties, which differ from their bulk counterparts, and their wide range of potential applications. Among these materials, BNNSs are notable for their honeycomb structure similar to graphene, but with distinct properties, such as a wide bandgap [2,3], excellent thermal stability and conductivity [4–6], high mechanical strength [7], and unique adsorption capabilities [8]. These qualities make BNNSs promising for use in various fields, including high-performance electronics, dielectric substrates, thermal management, sensors, and sorbents [9–13]. However, the low-yield production methods and poor solvent solubility of BNNSs limit their application in composites and devices [14]. Similar to graphene, BNNSs can be synthesized using two main approaches: bottom-up and top-down methods [15,16]. The bottom-up techniques, such as chemical vapor deposition and gas foaming, often result in high costs, low yield, and defects in the 2D materials. In contrast, top-down methods, like exfoliation from bulk hexagonal boron nitride (h-BN), are more suited for large-scale, high-quality production of BNNSs. Among various exfoliation methods, such as ultrasonication, hydrothermal techniques, and the Scotch tape method, controlled ball milling has emerged as one of the most effective approaches, balancing productivity, efficiency, quality, and purity [17,18].

However, the poor dispersibility of BNNSs in common solvents remains a major challenge for their use in advanced materials. To address this, surface functionalization or guest molecular wrapping is an effective solution [19,20]. By modifying the surface of BNNSs, it is possible to enhance their dispersibility. Functional groups and guest molecules can form hydrogen bonds with solvent molecules and/or generate electrostatic repulsion between BNNSs, reducing agglomeration and improving dispersion stability. For example, Lei et al. reported a dispersion of amino-functionalized BNNSs with an impressive 30 mg mL^{-1} concentration in water [21]. Chen et al. produced sucrose grafted BNNSs with concentrations of around 10 mg g^{-1} in water and 36 mg g^{-1} in ethanol–water [22]. Despite these advances, there is a strong need for an easy synthesis method to produce BNNSs with a large lateral size and simple functional groups. Therefore, this chapter will discuss the synthesis of amino- and/or hydroxyl-functionalized BNNSs, their dispersion mechanisms in common solvents, and their applications in acoustic energy harvesting, anti-corrosion coatings, and ion-selective adsorption [23,24].

DOI: 10.1201/9781003647812-4

3.2 FABRICATION OF BNNS DISPERSION

BNNS-NH_2 was synthesized using an ammonia-assisted ball milling method. The process employed a stainless-steel milling jar with stainless-steel balls, with commercial bulk *h*-BN as the starting material [23]. The ammonia pressure was maintained at 300 kPa, and the milling was conducted for 20 hours. No purification was necessary, as residual ammonia dissipated quickly during sample collection. In contrast, BNNS-OH was prepared using a sodium hydroxide (NaOH) aqueous solution (2.5 mol L^{-1})-assisted ball milling technique [23]. This method utilized a zirconium dioxide milling jar, zirconium milling balls, argon as a protective gas, and bulk *h*-BN as the precursor. The milling was performed with a FRITSCH P5 milling machine at 200 rpm for 20–40 hours. Afterward, the product was dialyzed until neutral and then dried at 60°C. During the ball milling process, shear forces generated by the milling balls induced preferential cleavage along the interplane bonds of *h*-BN, attributed to its weak interlayer van der Waals interactions. This process mechanically exfoliated BNNSs from bulk *h*-BN [17]. Simultaneously, defect sites at the edges and occasional basal plane defects caused by milling ball collisions facilitated the attachment of functional groups through the formation of new bonds, such as B–O–H or B–N–H.

Figure 3.1 illustrates a comparison between BNNS-NH_2 and BNNS-OH. Both samples exhibited a similar size distribution, averaging 600 nm, significantly reduced from the 25 µm size of the starting bulk *h*-BN (Figure 3.1a and, e). Atomic force microscopy measurements revealed average thicknesses of 2.8 nm for BNNS-NH_2 and 3.1 nm for BNNS-OH (Figure 3.1b and f). Transmission electron microscopy indicated high crystalline quality for both materials, with an interlayer distance of approximately 0.34 nm (Figure 3.1c,d, g–h). The X-ray diffraction patterns of BNNS-NH_2 and BNNS-OH showed reduced intensity and broader line widths after 20 hours of ball milling (Figure 3.2a), confirming the successful exfoliation of bulk *h*-BN. Attenuated total reflection Fourier-transform infrared spectroscopy detected a B-NH_2 vibration at 3420 cm^{-1} for

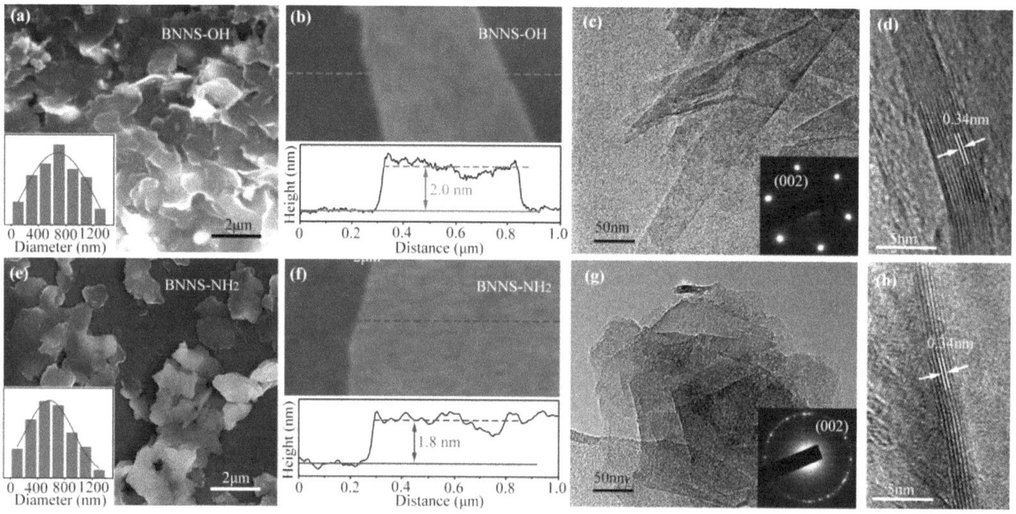

Figure 3.1 (a) SEM image with lateral size distribution, (b) atomic force microscopy image with corresponding height trace, (c) transmission electron microscopy image with the selected-area electron diffraction pattern, and (d) high-resolution transmission electron microscopy image of BNNS-OH. (e–h) Corresponding characterizations of BNNS-NH_2. (Reprinted with permission from [23]. Copyright 2021 American Chemical Society.)

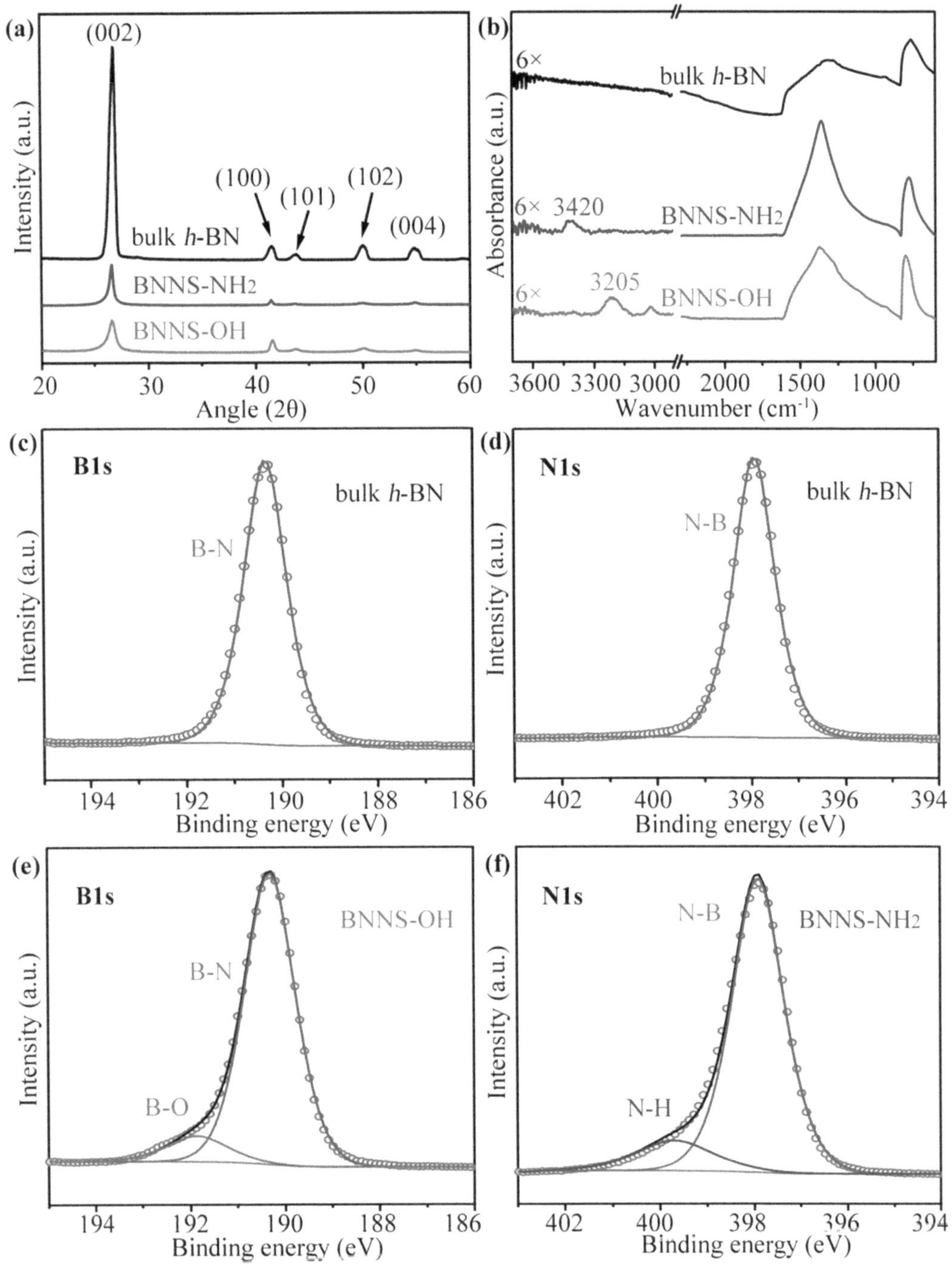

Figure 3.2 (a) X-ray diffraction patterns and (b) ATR–FTIR spectra of bulk *h*-BN, BNNS-OH, and BNNS-NH_2. (c) B 1s and (d) N 1s XPS spectra of bulk *h*-BN. (e) B 1s XPS spectrum of BNNS-OH. (f) N 1s XPS spectrum of BNNS-NH_2. (Reprinted with permission from [23]. Copyright 2021 American Chemical Society.)

$BNNS-NH_2$ and a B-OH vibration at 3205 cm^{-1} for BNNS-OH (Figure 3.2b) [25,26], verifying the success of functionalization. X-ray photoelectron spectroscopy (XPS) further corroborated this, identifying N–H bonds at 399.7 eV in the N 1s peak of $BNNS-NH_2$ and B–O bonds at 191.8 eV in the B 1s peak of BNNS-OH (Figure 3.2c–f) [21,22,27]. Functionalization with $-NH_2$ and -OH groups reduced the contact angle of BNNSs from ~95° to 48° and 45°, respectively, highlighting their hydrophilic nature. Thermogravimetric analysis indicated that both BNNS-OH and $BNNS-NH_2$ had a similar functional group density of approximately 1.2 mol% [23].

3.3 STABILITY MECHANISM OF BNNS DISPERSION

Zeta potential measurements were used to evaluate the stability of $BNNS-NH_2$ and BNNS-OH dispersions in water. Figure 3.3 shows the zeta potential of BNNS dispersions as a function of pH. At pH 7, BNNS-OH and $BNNS-NH_2$ exhibited similar zeta potentials of approximately −37 mV, indicating good stability for both dispersions [28,29]. In contrast, bulk *h*-BN platelets had a less negative zeta potential of around −20 mV [23]. The more negative zeta potential of the functionalized BNNSs can be attributed to two factors: (1) defective sites: ball milling introduced more defective sites, facilitating the adsorption of oxygen-containing groups (*e.g.*, -OH) [30,31]; (2) ionization of functional groups: the ionization of surface functional groups increased surface charge density. Due to the ionization of $-NH_2$ groups, $BNNS-NH_2$ exhibited a pH range of 8–9, higher than that of BNNS-OH, suggesting a higher surface charge density and improved stability. Additionally, an increased pH enhanced the ionization of functional groups, leading to a more negative zeta potential, while a lower pH reduced ionization and zeta potential. The effect of salt on the stability of BNNS dispersions was also studied using a 0.025 mol L^{-1} NaCl solution, as shown in Figure 3.3b. BNNS-OH experienced immediate coagulation upon the addition of NaCl, whereas $BNNS-NH_2$ showed slower precipitation under the same conditions. This behavior can be explained using the Derjaguin–Landau–Verwey–Overbeek theory. According to the theory, the total interaction potential ΔG(int) between BNNSs (parallel surface) is expressed by [29]:

$$\Delta G(\text{int}) = \Delta G(\text{repulsive}) - \Delta G(\text{attractive}) = (64c^0kT/\kappa)\exp(-\kappa H) - A_H/\left(12\pi H^2\right) \quad (3.1)$$

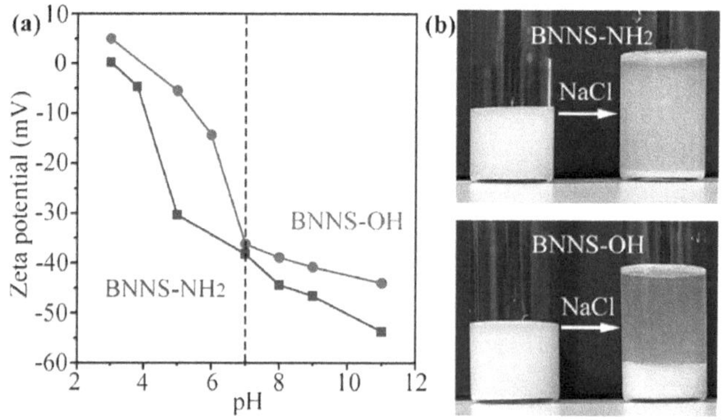

Figure 3.3 (a) pH-dependent zeta potential, and (b) salt effect of BNNS-OH and $BNNS-NH_2$ dispersed in water. (Reprinted with permission from [23]. Copyright 2021 American Chemical Society.)

where c^0 is the bulk electrolyte concentration, k is the Boltzmann constant, T is the temperature, $1/\kappa = (\varepsilon kT/e^2 \sum c_i z_i^2)^{1/2}$ is the thickness of the electrical double layer, A_H is the Hamaker constant, and H is the distance between two sheets. In addition to lower surface charge density, BNNS-OH exhibits stronger inter-sheet interactions through hydrogen bonding. This is due to the higher hydrogen bonding energy of O–H···O (−24 kJ mol^{-1}) compared to N–H···N (−7 kJ mol^{-1}) [32]. As a result, ΔG(attractive) for BNNS-OH is greater than that for BNNS-NH_2, leading to a lower coagulation energy barrier for BNNS-OH. With the addition of NaCl, the electrical double-layer thickness ($1/\kappa$) of both BNNS-OH and BNNS-NH_2 decreases rapidly, reducing the electrostatic repulsive force ΔG (repulsive) between sheets. Consequently, BNNS-OH readily forms aggregates through hydrogen bonding, as neighboring sheets attract each other and combine into larger structures. In contrast, BNNS-NH_2 is less prone to aggregation due to its lower hydrogen bonding energy, resulting in greater colloidal stability [23].

The dispersibility of BNNS-OH and BNNS-NH_2 in three commonly used solvents—deionized water, ethanol, and acetone—was evaluated. BNNS-OH or BNNS-NH_2 dispersions with various initial concentrations were prepared in these solvents, subjected to sonication, and allowed to settle. Over a seven-day period, precipitation typically occurred, and the supernatant reached a steady state. Figure 3.4a shows the stable concentration of BNNS-NH_2 in ethanol after seven days of settling, as a function of initial concentration. The stable concentration increased with the initial concentration before plateauing at approximately 46±2 mg mL^{-1}—the highest reported value—which remained consistent for more than three months [23]. The same procedure was applied to determine the maximum stable concentrations of BNNS-OH and BNNS-NH_2 in the other solvents, as summarized in Figure 3.4b. Among the solvents tested, BNNS-OH (the left bar for each solvent) exhibited the highest stable concentration of 38±2 mg mL^{-1} in acetone. This behavior was attributed to the strong interactions between BNNS-OH and acetone molecules [33], which reduced the Hamaker constant (A_H) and, consequently, the attractive potential

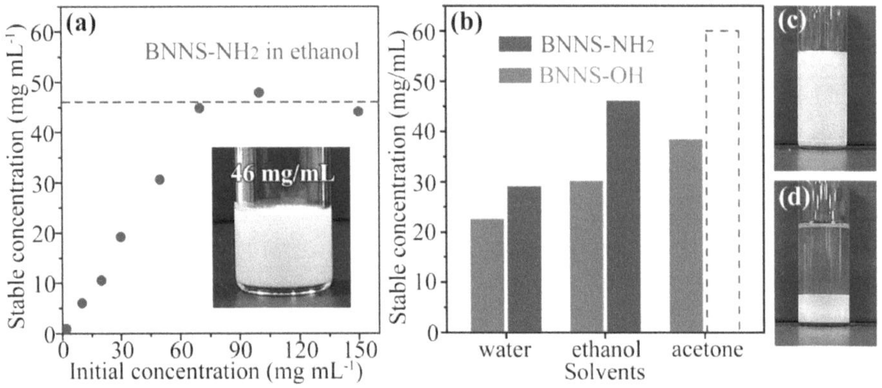

Figure 3.4 (a) Stable concentration of BNNS-NH_2 in ethanol as a function of initial concentration. The dashed line represents the average saturated concentration, and the inset shows a digital photo of the supernatant at a concentration of 46 mg mL^{-1}. (b) Maximum stable concentrations of BNNS-NH_2 and BNNS-OH in different solvents. The dashed bar indicates the expected stable concentration of BNNS-NH_2 in acetone, estimated based on the ratio of the stable concentrations of BNNS-NH_2 in ethanol and BNNS-OH (~1.5). Digital photographs of BNNS-NH_2 in acetone after seven days (c) and two months (d) at steady state. (Reprinted with permission from [23]. Copyright 2021 American Chemical Society.)

[ΔG(attractive)] between BNNS-OH sheets. Acetone's nonbonded electron pairs on oxygen atoms and its strongly polarized C=O bonds facilitated hydrogen bond formation with the -OH groups of BNNS-OH. In contrast to water and ethanol, which can form hydrogen bonds with themselves, acetone is an aprotic solvent [34]. As such, most acetone molecules are free to interact with BNNS-OH, reducing inter-sheet attraction and increasing the stable concentration.

Compared with BNNS-OH, BNNS-NH_2 dispersions (the right bar for each solvent in Figure 3.4b) showed higher stable concentrations. This result was attributed to the lower hydrogen bonding energy between BNNS-NH_2 sheets, which led to a higher coagulation energy barrier. BNNS-NH_2 was expected to achieve the highest stable concentration in acetone; however, the amino groups of BNNS-NH_2 reacted with acetone over time, producing a yellowish supernatant after a few days of settling (Figure 3.4c and d) [23]. In addition to the effects of functional groups and solvents, the size of BNNSs significantly influenced dispersion stability. According to Einstein's theory of Brownian motion for self-diffusivity [35,36]:

$$D = \frac{RT}{N_A} \frac{1}{6\pi\eta r_u} \tag{3.2}$$

where N_A is Avogadro's number, T is the temperature, η is the solvent viscosity, and r_u is the solute radius. Larger BNNSs diffuse more slowly and, consequently, aggregate more gradually. Thus, the combination of suitable functional groups (*e.g.*, -NH_2), an appropriate solvent, and an optimal BNNS size (~600 nm) contributed to the unprecedentedly high stable concentration of BNNS-NH_2 in ethanol, reaching 46 ± 2 mg mL^{-1}.

3.4 APPLICATIONS OF BNNS DISPERSION

As previously mentioned, the low-yield preparation methods and poor solvent solubility of BNNSs have greatly limited their applications in composites and devices. However, the significantly improved dispersibility and successful preparation of high-concentration BNNS dispersions, along with a deeper understanding of the dispersion mechanisms, have made such applications feasible. In this section, we will explore the potential applications of high-concentration BNNS dispersions in acoustic energy harvesting, anti-corrosion coatings, and ion-selective adsorption.

3.4.1 Acoustic Energy Harvesting

The term 'noise' refers to sound with a sound pressure level (SPL) of 80 decibels (dB) or higher, typically perceived as unwanted, disruptive, or loud. It can emanate from various sources such as machinery, traffic, or environmental factors. Besides its disruptive effects, noise carries a significant amount of energy, sometimes comparable to the power needed for microelectronics [37]. For instance, sound with an SPL of 100 dB, akin to the noise level of an approaching subway train, traveling through a 1 m^2 surface carries a power of 9.68 mW, exceeding the operating power of a miniaturized temperature sensor, typically 1.65 mW [38]. Consequently, there has been a growing interest in converting ambient noise energy into electricity due to its potential as a sustainable and renewable energy source. Acoustic energy harvesters have been developed based on three principles: electromagnetic, piezoelectric, and triboelectric effects [24]. However, while electromagnetic and piezoelectric devices, leveraging Faraday's law and piezoelectric properties, respectively, tend to generate relatively low voltage outputs [39,40], traditional triboelectric devices can produce much higher voltage outputs, reaching hundreds of volts [41]. Yet, their operation relies on the 'contact-separation' process between triboelectric components driven by mechanical energy, necessitating precise distance control between these components. This requirement poses challenges in fabrication, increases costs, and raises the risk of device damage under severe vibrations and shocks [40]. To address these challenges associated with 'precise distance control' in triboelectric devices, we propose a novel acoustoelectric device based on 'endogenous triboelectricity' [24] In addition, BNNSs have demonstrated their potential in optimizing the performance of triboelectric nanogenerators, benefiting from their piezoelectricity, charge affinity, and high dielectric constant [42–46]. However, studies focusing on their incorporation into PVDF nanofiber—a promising material in converting acoustic energy into electricity—based acoustoelectric device utilizing endogenous triboelectricity remain limited and unexplored. In this subsection, we will explore the application of BNNSs in acoustoelectric devices based on endogenous triboelectricity without 'precise distance control' [24].

Initially, a single-layer composite membrane—referred to as +PVDF/-BN—was created using a combination of electrospinning and electrospraying techniques. Electrospinning was employed to fabricate the PVDF nanofiber network, while electrospraying simultaneously produced uniformly distributed BNNS aggregates (BNNS-A) throughout the PVDF network, as depicted in Figure 3.5, utilizing previously mentioned BNNS dispersion as the starting material. The stable, high-concentration BNNS dispersion ensures the uniformity of the electrosprayed BNNS-A. In the +PVDF/-BN membrane, BNNSs-A, with an average size of 1.1 μm, are uniformly dispersed throughout the entire PVDF fiber network, which has an average diameter of 485 nm. For comparison, a pure PVDF nanofiber film without BNNS-A was prepared using the same electrospinning and electrospraying process, designated as +PVDF/-ethanol. Subsequently, acoustoelectric devices sized at $4 \times 3\,cm^2$ were constructed by positioning the composite film within two stainless-steel meshes. These meshes possessed a porosity of about 50%, allowing sound waves to interact with the composite film, and acted as electrodes (Figure 3.6a). Figure 3.6b illustrates a schematic diagram of the acoustoelectric device and the sound energy conversion system used to evaluate the energy conversion performance of the devices. When subjected to a sound with an SPL of 115 dB at a constant frequency of 230 Hz, the +PVDF/-BN acoustoelectric device generated periodic electric outputs, with a peak voltage output of 174.2 V and a peak current of 19.2 μA

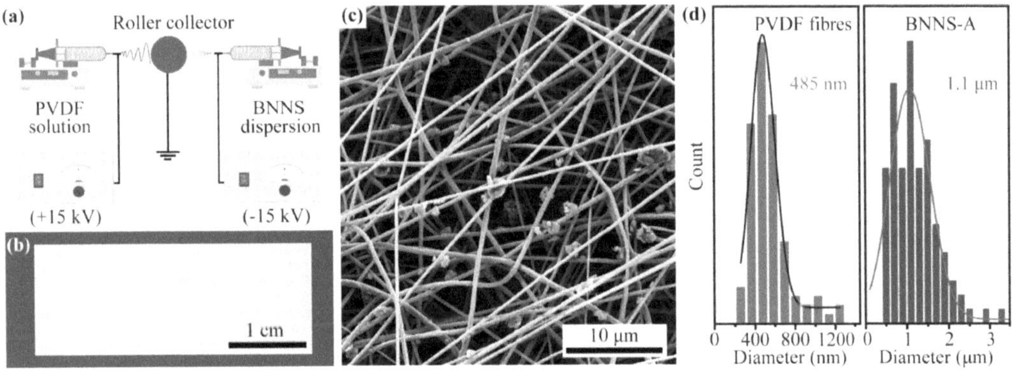

Figure 3.5 (a) Schematic diagram of the combined electrospinning and electrospraying techniques. (b) Digital photo and (c) SEM image of the as-fabricated +PVDF/-BN composite membrane. (d) Diameter distribution of PVDF fibers (left) and size distribution of BNNS-A (right) [24].

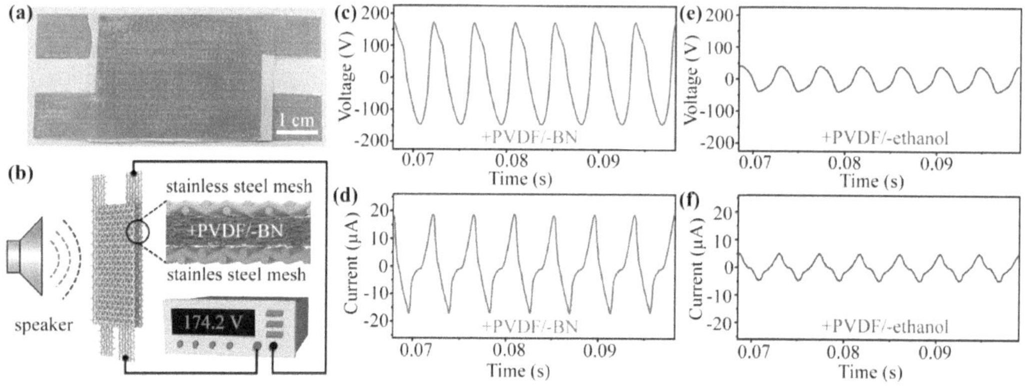

Figure 3.6 (a) Digital photo of the +PVDF/-BN device. (b) Schematic diagram of the +PVDF/-BN device and the sound energy conversion test system. (c) Voltage and (d) current outputs of the +PVDF/-BN acoustoelectric device under a 115 dB, 230 Hz sound input. (e) Voltage and (f) current outputs of the +PVDF/-ethanol device for comparison [24].

(Figure 3.6c and d). The peak voltage declined with reduced SPL and increased sound frequency. Nonetheless, under a 115 dB sound, it still produced a 4.6 V output at 1000 Hz and a 1.2 V output at 2000 Hz, indicating the suitability of the +PVDF/-BN composite membrane for converting noise into electricity for sound energy harvesting applications, given that prevalent noises in daily life typically fall below 2000 Hz [47]. In contrast, the +PVDF/-ethanol device exhibited much lower noise-to-electricity conversion efficiency, featuring a peak voltage output of 41.1 V (4.2 times lower) and a peak current output of 5.2 μA (3.7 times lower) under the same sound conditions (Figure 3.6e and f). Moreover, the maximum power output of the +PVDF/-BN device was approximately 30-fold higher than that of its +PVDF/-ethanol counterpart.

The notable enhancement in the electric outputs of +PVDF/-BN was attributed to the endogenous triboelectricity induced by the inclusion of BNNS-A. Within the +PVDF/-BN film, a BNNS-A adheres firmly to a PVDF fiber and also comes into contact with one or more neighboring PVDF fibers due to the 'single-layer' structure of the composite film, as shown in Figure 3.7. The composite's structure is highly porous, with a density of approximately 0.4 g cm^{-3}, providing ample space for both BNNS-A decorated and bare PVDF fibers to vibrate freely in response to sound. Importantly, theoretical calculations indicate that a PVDF fiber with a BNNS-A experiences reduced vibration amplitude in the direction of sound propagation compared to its counterpart without BNNS-A (Figure 3.7b). Consequently, the 'contact-separation' process between a BNNS-A and its neighboring PVDF fiber occurs as a result of their asynchronous vibration amplitudes, leading to triboelectrification due to the difference in electron affinity between BN and PVDF (Figure 3.7c). This structure, featuring inherent distance control between dielectric components regulated by the difference in vibration amplitude of PVDF fibers with and without BNNS-A, can be readily produced through electrospinning and electrospraying without requiring additional 'precise distance control', thereby simplifying the fabrication of acoustoelectric devices with high sound-to-electricity conversion efficiency. When operated under 115 dB, 230 Hz sound, the +PVDF/-BN device is capable of powering up to 80 LED lights, a digital calculator, or charging capacitors. This study offers a deep insight into the mechanism of endogenous triboelectricity in nanocomposites and proposes a promising method for creating high-performance nanofiber acoustoelectric devices [24].

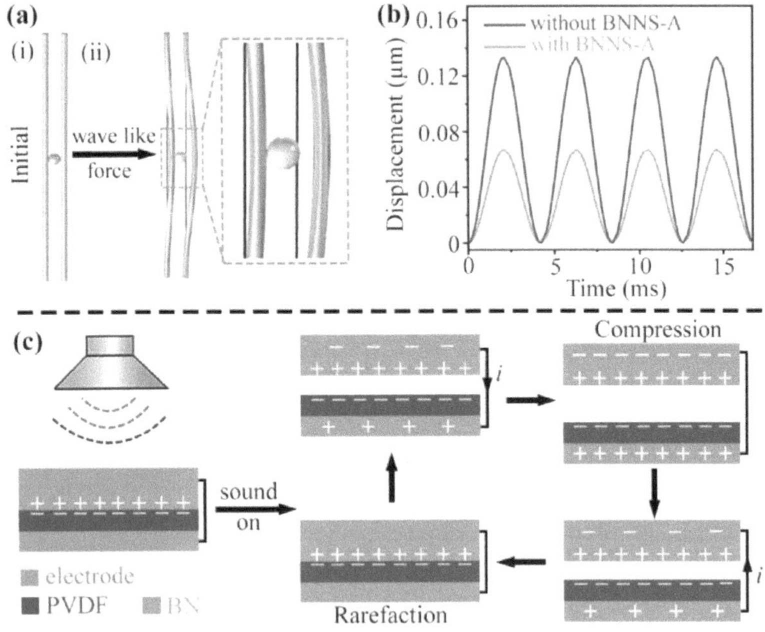

Figure 3.7 (a) COMSOL Multiphysics simulation illustrating PVDF fibers with and without a BNNS-A: (i) initial configuration where a BNNS-A is attached to a PVDF fiber (right) and in contact with a neighboring PVDF fiber (left); (ii) deformation of the two PVDF fibers under an analytical point force function resembling a wave, resulting in the separation of the BNNS-A from its neighboring PVDF fiber. (b) Comparison of displacement in the middle of PVDF fibers with (green curve) and without a BNNS-A (blue curve). (c) Schematic diagrams illustrating the endogenous triboelectric effect of +PVDF/-BN membrane in response to a sound wave [24].

3.4.2 Anti-Corrosion Coating

The BNNS-NH_2 dispersion in ethanol, with a high concentration of 46 mg mL^{-1}, can be effortlessly applied to a copper (Cu) sheet using a paintbrush, resulting in a highly uniform coating (Figure 3.8a). This coating adheres strongly to the Cu surface, providing long-lasting protection [48,49]. In contrast, BNNS dispersions with lower stable concentrations (<20 mg mL^{-1}) failed to form uniform coatings using the same method. The BNNS coatings significantly enhance both the anti-corrosion (oxidation) performance and the insulating properties of the Cu sheet. Electrochemical impedance spectroscopy analysis, presented as Nyquist plots (Figure 3.8b), shows that the semicircle diameter increased from 436.3 Ω for bare Cu to 1899.0 Ω for BNNS-coated Cu, indicating a marked improvement in charge transfer resistance between the Cu sheet and the NaCl solution. Furthermore, Tafel analysis (Figure 3.8c) revealed the corrosion potential (E_{corr}), corrosion current (I_{corr}), and corrosion rate for coated and uncoated Cu sheets. For the bare Cu sheet, E_{corr} and I_{corr} were measured at −370.9 mV and 104.9 μA, respectively. After applying the BNNS coating, E_{corr} shifted positively to −336.9 mV, and I_{corr} decreased significantly to 19.6 μA, reducing the corrosion rate from 1.22 mm per year to 0.23 mm per year. These results highlight the protective effectiveness of the BNNS coating.

The anti-oxidation performance of BNNS-coated Cu foil was further evaluated under heat treatment at 200°C, 400°C, and 600°C for 10 minutes in air (Figure 3.8d). On uncoated Cu foil,

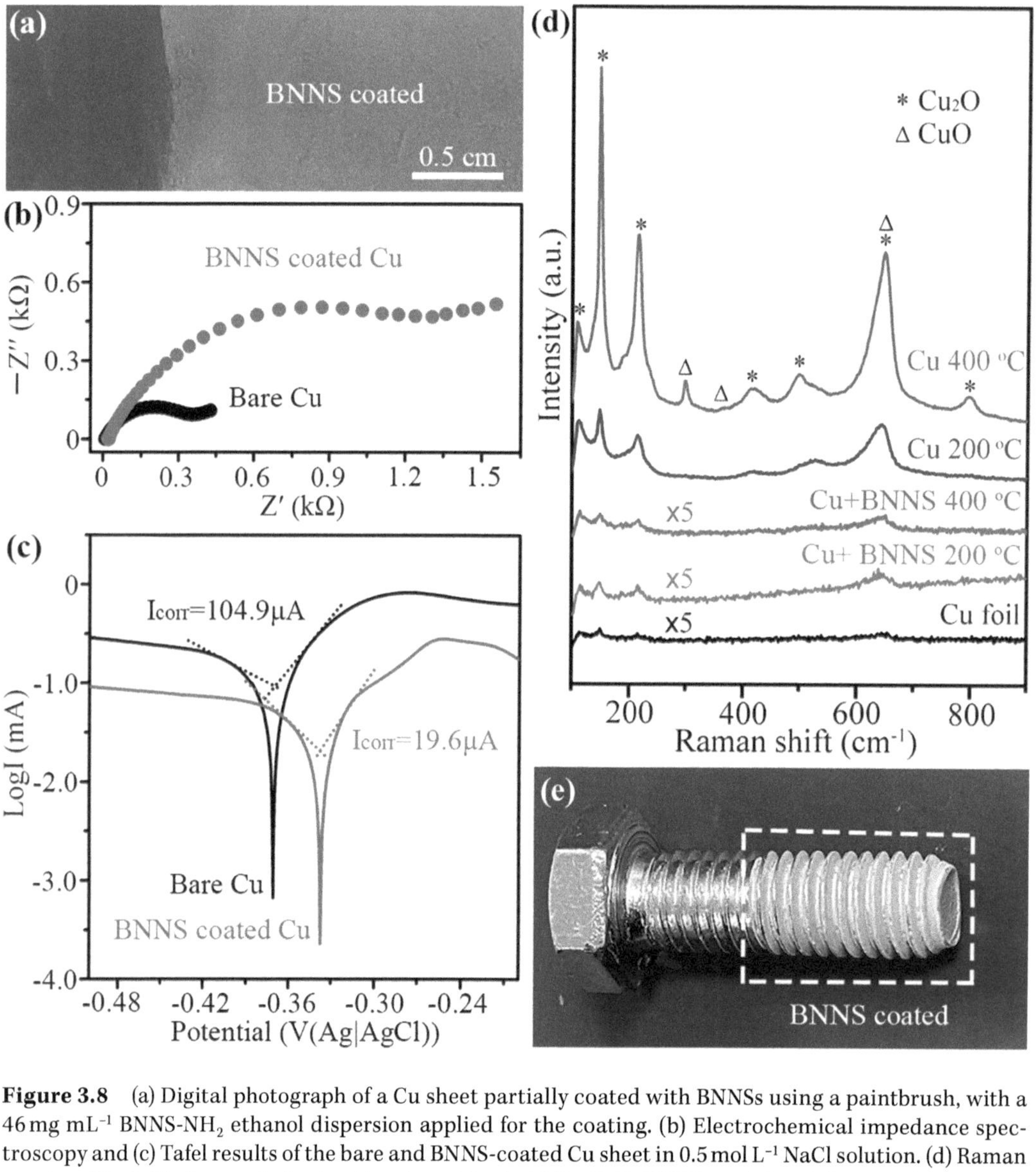

Figure 3.8 (a) Digital photograph of a Cu sheet partially coated with BNNSs using a paintbrush, with a 46 mg mL^{-1} BNNS-NH_2 ethanol dispersion applied for the coating. (b) Electrochemical impedance spectroscopy and (c) Tafel results of the bare and BNNS-coated Cu sheet in 0.5 mol L^{-1} NaCl solution. (d) Raman spectra of Cu foil before and after heat treatment in air. (e) Partially BNNS-coated stainless-steel screw with an uneven surface. (Reprinted with permission from [23]. Copyright 2021 American Chemical Society.)

a trace amount of cuprous oxide (Cu_2O) was detected (black curve), likely due to the natural formation of a thin oxide layer on the Cu surface [50]. Interestingly, BNNS-coated Cu foil showed only trace amounts of Cu_2O even after heating at 200°C and 400°C (Cu+BNNS curves), demonstrating the coating's effective oxidation resistance. In contrast, uncoated Cu samples exhibited enhanced Raman vibrational modes of Cu_2O after heating at 200°C (Cu 200 curve), and cupric oxide (CuO) appeared after heating at 400°C (Cu 400 curve) [50,51]. At 600°C, the BNNS coating was less effective against oxidation, likely due to oxygen penetration through nanochannels between the BNNS layers. This oxidation caused the Cu foil's surface to darken compared to pure copper. Unlike graphene, which accelerates the corrosion of metals through galvanic reactions during long-term contact [52,53], BNNSs offer long-term protection due to their insulating properties. Additionally, the high concentration of BNNS dispersions allows for easy application to irregular surfaces, as illustrated in Figure 3.8e. These outstanding anti-oxidation properties open up significant opportunities for BNNSs to be used as protective films for metal substrates against corrosion and oxidation [23].

3.4.3 Ion-Selective Adsorption

It has been demonstrated that the molecular separation capability of filtration membranes made from amino-functionalized BNNSs is primarily dependent on the channel size of the network [54]; *h*-BN nanoflakes are particularly efficient for ion separation, and their performance can be tuned by adjusting the solution pH, which alters the surface charge density [30]. In addition to the channel size and surface charge density, the type of functional group on the BNNSs (either negatively or positively charged) also plays a significant role in ion separation performance. To explore the effects of functional groups and minimize the influence of the BNNS network's channel size, the ion-selective adsorption behavior of BNNS-OH and BNNS-NH_2 was compared by soaking BNNSs into $FeCl_3$, K_2MnO_4, and $K_4Fe(CN)_6$ solutions. These ions were chosen for two main reasons: (1) BNNSs have low affinity for these ions, and (2) the ions differ in color and valency. Figure 3.9 shows the concentration changes of these solutions before and after adsorption by BNNS-OH or BNNS-NH_2. The results indicate that BNNS-OH was more effective at capturing cations (Fe^{3+}), as shown in Figure 3.9a, while BNNS-NH_2 adsorbed more anions ($Fe(CN)_6^{4-}$ and MnO_4^{2-}), as seen in Figure 3.9b and c. This difference is attributed to the formation of BNNS-O^- and BNNS-NH^{3+} in BNNS-OH and BNNS-NH_2 dispersions, respectively. Specifically, BNNS-OH adsorbed 34.34% more Fe^{3+}, 28.16% less MnO_4^{2-}, and 19.31% less $Fe(CN)_6^{4-}$ compared to BNNS-NH_2. Furthermore, BNNS-NH_2 showed a greater affinity for anions with larger valency. While the difference in ion-selective adsorption between negatively and positively charged functional groups was not dramatic, these findings provide valuable insights for the selective separation of ions using functionalized 2D membranes.

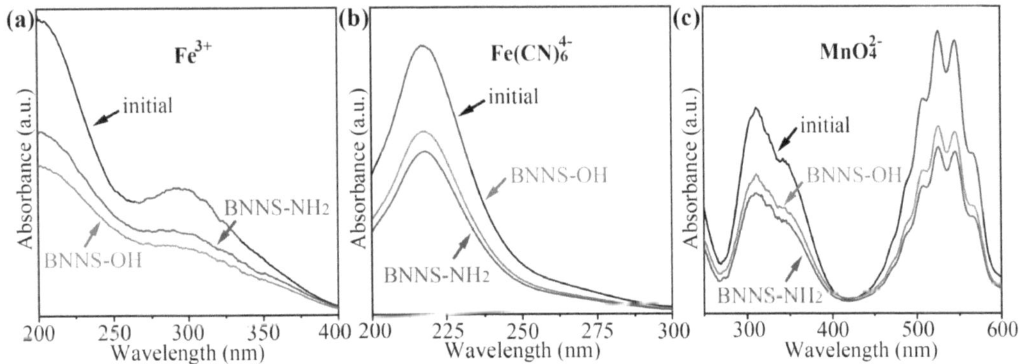

Figure 3.9 UV-vis spectra of (a) $FeCl_3$, (b) $K_4Fe(CN)_6$, and (c) K_2MnO_4 solutions at a concentration of 10^{-3} mol L^{-1}, shown before (black curves) and after adsorption by either BNNS-OH (red curves) or BNNS-NH_2 (blue curves). (Reprinted with permission from [23]. Copyright 2021 American Chemical Society.)

3.5 CONCLUSIONS

In this chapter, BNNSs functionalized with -OH or $-NH_2$ groups were successfully fabricated using a one-step ball milling method. The functionalization significantly improved the solubility of BNNSs, which were highly hydrophobic in most solvents. The resulting BNNS-NH_2 and BNNS-OH showed stable dispersion in water and common organic solvents at high concentrations for extended periods, up to several months. The dispersion mechanisms were thoroughly investigated. BNNS-NH_2 in ethanol achieved the highest stable concentration (46 ± 2 mg mL^{-1}) due to its increased hydrophilicity and strong hydrogen bonding with ethanol molecules. Additionally, the high concentration of stable BNNS dispersion ensured the uniformity of electrosprayed BNNS-A for acoustic energy harvesting. The excellent dispersibility of functionalized BNNSs also enabled the creation of BNNS coatings on various substrates, enhancing insulating and anti-oxidation properties. The positively and negatively charged functional groups improved the ion-selective separation performance of BNNS membranes through electrostatic interactions. In summary, the enhanced solubility of functionalized BNNSs in various solvents opens up a range of potential applications, including BNNS-reinforced polymer composites, ultralight materials, protective films, ion separation, and water purification.

REFERENCES

1. K. S. Novoselov et al., Electric field effect in atomically thin carbon films. *Science* **306**, 666–669 (2004).
2. K. Watanabe, T. Taniguchi, H. Kanda, Direct-bandgap properties and evidence for ultraviolet lasing of hexagonal boron nitride single crystal. *Nature Materials* **3**, 404–409 (2004).
3. D. Wickramaratne, L. Weston, C. G. Van de Walle, Monolayer to bulk properties of hexagonal boron nitride. *The Journal of Physical Chemistry C* **122**, 25524–25529 (2018).
4. L. H. Li, J. Cervenka, K. Watanabe, T. Taniguchi, Y. Chen, Strong oxidation resistance of atomically thin boron nitride nanosheets. *ACS Nano* **8**, 1457–1462 (2014).
5. Q. Cai et al., High thermal conductivity of high-quality monolayer boron nitride and its thermal expansion. *Science Advances* **5**, eaav0129 (2019).
6. L. Lindsay, D. Broido, Enhanced thermal conductivity and isotope effect in single-layer hexagonal boron nitride. *Physical Review B* **84**, 155421 (2011).
7. A. Falin et al., Mechanical properties of atomically thin boron nitride and the role of interlayer interactions. *Nature Communications* **8**, 15815 (2017).
8. Q. Cai et al., Molecule-induced conformational change in boron nitride nanosheets with enhanced surface adsorption. *Advanced Functional Materials* **26**, 8202–8210 (2016).
9. H. Zhu et al., Highly thermally conductive papers with percolative layered boron nitride nanosheets. *ACS Nano* **8**, 3606–3613 (2014).
10. W. Lei, D. Portehault, D. Liu, S. Qin, Y. Chen, Porous boron nitride nanosheets for effective water cleaning. *Nature Communications* **4**, 1777 (2013).
11. G. Giovannetti, P. A. Khomyakov, G. Brocks, P. J. Kelly, J. Van Den Brink, Substrate-induced band gap in graphene on hexagonal boron nitride: Ab initio density functional calculations. *Physical Review B* **76**, 073103 (2007).
12. X.-B. Wang et al., Biomass-directed synthesis of 20 g high-quality boron nitride nanosheets for thermoconductive polymeric composites. *ACS Nano* **8**, 9081–9088 (2014).
13. Q. Cai et al., Boron nitride nanosheets: Thickness-related properties and applications. *Advanced Functional Materials* **34**, 2403669 (2024).
14. Z. Liu et al., Challenges and solutions in surface engineering and assembly of boron nitride nanosheets. *Materials Today* **44**, 194–210 (2021).

15. X. Wang et al., "Chemical Blowing" of thin-walled bubbles: High-throughput fabrication of large-area, few-layered BN and C_x-BN nanosheets. *Advanced Materials* **35**, 4072–4076 (2011).
16. L. H. L. Deepika et al., High-efficient production of boron nitride nanosheets via an optimized ball milling process for lubrication in oil. *Scientific Reports* **4**, 7288 (2014).
17. L. H. Li et al., Large-scale mechanical peeling of boron nitride nanosheets by low-energy ball milling. *Journal of Materials Chemistry* **21**, 11862–11866 (2011).
18. Y. I. Chen, *Nanotubes and Nanosheets: Functionalization and Applications of Boron Nitride and Other Nanomaterials* (CRC Press, 2015).
19. J.-H. Choi, J. Kim, D. Seo, Y.-S. Seo, Purification of boron nitride nanotubes via polymer wrapping. *Materials Research Bulletin* **48**, 1197–1203 (2013).
20. Y.-T. R. Lau et al., Facile and mild strategy toward biopolymer-coated boron nitride nanotubes via a glycine-assisted interfacial process. *The Journal of Physical Chemistry C* **117**, 19568–19576 (2013).
21. W. Lei et al., Boron nitride colloidal solutions, ultralight aerogels and freestanding membranes through one-step exfoliation and functionalization. *Nature Communications* **6**, 8849 (2015).
22. S. Chen et al., Simultaneous production and functionalization of boron nitride nanosheets by sugar-assisted mechanochemical exfoliation. *Advanced Materials* **31**, 1804810 (2019).
23. H. Jiang et al., Boron nitride nanosheet dispersion at high concentrations. *ACS Applied Materials & Interfaces* **13**, 44751–44759 (2021).
24. H. Shao et al., Electrosprayed boron nitride nanosheet aggregates for enhanced acoustic energy harvesting with poly (vinylidene fluoride) nanofiber membranes. *Nano Energy* **121**, 109195 (2024).
25. C. Tang, Y. Bando, Y. Huang, C. Zhi, D. Golberg, Synthetic routes and formation mechanisms of spherical boron nitride nanoparticles. *Advanced Functional Materials* **18**, 3653–3661 (2008).
26. T. Shen, S. Liu, W. Yan, J. Wang, Highly efficient preparation of hexagonal boron nitride by direct microwave heating for dye removal. *Journal of Materials Science* **54**, 8852–8859 (2019).
27. Y. Xue et al., Template-free synthesis of boron nitride foam-like porous monoliths and their high-end applications in water purification. *Journal of Materials Chemistry A* **4**, 1469–1478 (2016).
28. D. Li, M. B. Müller, S. Gilje, R. B. Kaner, G. G. Wallace, Processable aqueous dispersions of graphene nanosheets. *Nature Nanotechnology* **3**, 101–105 (2008).
29. D. H. Everett, *Basic Principles of Colloid Science* (Royal Society of Chemistry, 2007).
30. A. Pendse et al., Charged layered boron nitride-nanoflake membranes for efficient ion separation and water purification. *Small* **15**, 1904590 (2019).
31. A. Siria et al., Giant osmotic energy conversion measured in a single transmembrane boron nitride nanotube. *Nature* **494**, 455–458 (2013).
32. D. F. Lewis, Hydrogen bonding in human P450-substrate interactions: A major contribution to binding affinity. *The Scientific World Journal* **4**, 1074–1082 (2004).
33. A. Vila, R. A. Mosquera, On the different origin of the stabilization of oxygen versus sulphur H-bond complexes with water. *Chemical Physics* **291**, 73–80 (2003).
34. W. S. MacGregor, The chemical and physical properties of DMSO. *Annals of the New York Academy of Sciences* **141**, 3–12 (1967).
35. A. Einstein, *Investigations on the Theory of the Brownian Movement* (Courier Corporation, 1956).
36. M. Sharma, S. Yashonath, Size dependence of solute diffusivity and Stokes-Einstein relationship: Effect of van der Waals interaction. *Diffusion Fundamentals* **7**, 11 (2007).
37. F. A. Boettcher, J. H. Mills, J. L. Swerdloff, B. L. Holley, Auditory evoked potentials in aged gerbils: Responses elicited by noises separated by a silent gap. *Hearing Research* **102**, 167–178 (1996).
38. N. Poole, Y. Mo, J. Brusey, M. Langley, R. Hazelden, Power consumption of microelectronic equipment for wireless sensor networks. *Nanotechnology NSTI* **1**, 534–537 (2009).
39. F. U. Khan, Electromagnetic based acoustic energy harvester for low power wireless autonomous sensor applications. *Sensor Review* **38**, 298–310 (2018).
40. B. Duan et al., Bioinspired PVDF piezoelectric generator for harvesting multi-frequency sound energy. *Advanced Electronic Materials* **9**, 2300348 (2023).
41. H. Zhao et al., Dual-tube helmholtz resonator-based triboelectric nanogenerator for highly efficient harvesting of acoustic energy. *Advanced Energy Materials* **9**, 1902824 (2019).

42. Z. Yang, X. Zhang, G. Xiang, 2D boron nitride nanosheets in polymer nanofibers for triboelectric nanogenerators with enhanced performance and flexibility. *ACS Applied Nano Materials* **5**, 16906–16911 (2022).
43. H. Kuang et al., Piezoelectric boron nitride nanosheets for high performance energy harvesting devices. *Nano Energy* **80**, 105561 (2021).
44. L. Pang et al., Triboelectric nanogenerator based on polyimide/boron nitride nanosheets/polyimide nanocomposite film with enhanced electrical performance. *ACS Applied Electronic Materials* **4**, 3027–3035 (2022).
45. Y. Liu, J. Ping, Y. Ying, Recent progress in 2D-nanomaterial-based triboelectric nanogenerators. *Advanced Functional Materials* **31**, 2009994 (2021).
46. J. Zhang et al., Enhancing the triboelectricity of stretchable electrospun piezoelectric polyvinylidene fluoride/boron nitride nanosheets composite nanofibers. *Composites Communications* **22**, 100535 (2020).
47. M. Basner et al., Auditory and non-auditory effects of noise on health. *The Lancet* **383**, 1325–1332 (2014).
48. Z. Liu et al., Ultrathin high-temperature oxidation-resistant coatings of hexagonal boron nitride. *Nature Communications* **4**, 2541 (2013).
49. L. H. Li, T. Xing, Y. Chen, R. Jones, Boron nitride nanosheets for metal protection. *Advanced Materials Interfaces* **1**, 1300132 (2014).
50. N. A. Mohemmed Shanid, M. Abdul Khadar, V. Sathe, Frohlich interaction and associated resonance enhancement in nanostructured copper oxide films. *Journal of Raman Spectroscopy* **42**, 1769–1773 (2011).
51. F. A. Akgul, G. Akgul, N. Yildirim, H. E. Unalan, R. Turan, Influence of thermal annealing on microstructural, morphological, optical properties and surface electronic structure of copper oxide thin films. *Materials Chemistry and Physics* **147**, 987–995 (2014).
52. F. Zhou, Z. Li, G. J. Shenoy, L. Li, H. Liu, Enhanced room-temperature corrosion of copper in the presence of graphene. *ACS Nano* **7**, 6939–6947 (2013).
53. M. Schriver et al., Graphene as a long-term metal oxidation barrier: Worse than nothing. *ACS Nano* **7**, 5763–5768 (2013).
54. C. Chen et al., Functionalized boron nitride membranes with ultrafast solvent transport performance for molecular separation. *Nature Communications* **9**, 1902 (2018).

Chapter 4

Engineering Lightweight Composites for Extreme Environments

Pranjal Nautiyal, Sohail M.A.K. Mohammed, Kazue Orikasa, and Arvind Agarwal

4.1 INTRODUCTION

There is a critical need for lightweight structural materials capable of withstanding harsh operating conditions. In particular, high-speed transportation, defense, space exploration, and energy technologies require lightweight materials for agility, fuel economy, and efficiency. Metals and alloys with high strength-to-weight ratios, *e.g.*, aluminum, titanium, and magnesium, are attractive for these applications. However, their mechanical properties (*e.g.*, strength and ductility) fall short of high-strength steels, as illustrated in the Ashby plot in Figure 4.1 [1]. This gap in mechanical properties is a major barrier to their adoption in critical load-bearing components. Reinforcing lightweight metals and alloys with stronger and harder ceramic phases is a popular strategy to enhance their mechanical properties [2]. In the last two decades, the discovery of new nanomaterials and the capability to synthesize them in large quantities have boosted efforts to develop novel nanostructured composites [3–10]. In particular, the discoveries of carbon nanotubes (CNTs) and graphene, both exhibiting exceptional mechanical

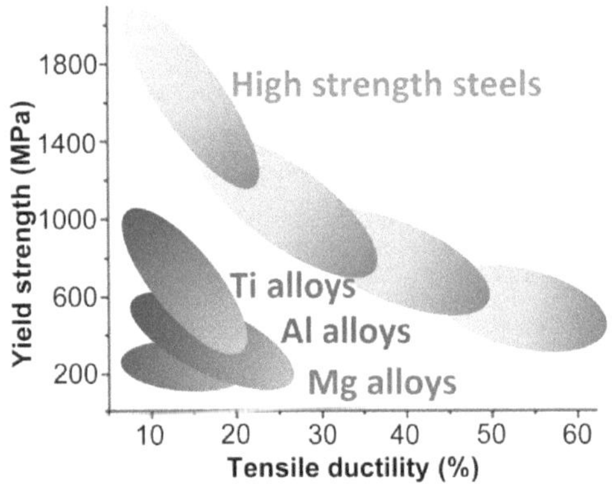

Figure 4.1 Comparison of the yield strength and ductility of lightweight alloys against high-strength structural steels shows a significant gap in load-carrying capacity. (Adapted with permission from Springer et al. [1]. Copyright (2017) The Author(s).)

DOI: 10.1201/9781003647812-5

strength and stiffness, ushered in the development of metal matrix nanocomposites with significantly improved load-carrying capacities [11,12]. However, both these carbon-based nanomaterials are susceptible to oxidation and thermal degradation at temperatures below 500°C [13]. As a result, the temperature window for fabricating CNT and graphene-reinforced composites is limited. This is a major drawback since most structural alloys melt at much higher temperatures. As a result, the processing of CNT and graphene-based composites is challenging and significantly limited.

BNNTs have emerged as an attractive alternative to CNT and graphene for engineering MMCs. BNNTs possess an excellent tensile strength (>60 GPa) and elastic modulus (>1 TPa) [14–19]. Additionally, they are thermally stable and resist oxidation at temperatures exceeding 700°C–1000°C [13,20,21]. In inert atmospheres, BNNTs are stable up to 2000°C [22]. They are also chemically less reactive than carbon-based nanomaterials. Their superior thermo-oxidative and chemical stability makes BNNTs suitable for high-temperature metal manufacturing operations.

In this chapter, we cover the latest advancements in the fabrication and characterization of BNNT-reinforced MMCs. We discuss their potential as structural materials for extreme environments. We also highlight knowledge gaps and directions for future research to advance these composites toward technological applications.

4.2 MICROSTRUCTURE FORMATION UNDER EXTREME PROCESSING ENVIRONMENTS

BNNT-reinforced MMCs have been fabricated using powder metallurgy, solidification, deformation processing, and a combination of these approaches. Here, we discuss the important microstructure formation mechanisms and their influence on the mechanical properties of the fabricated composites. Additionally, we emphasize the challenges and opportunities that these manufacturing methods present for designing high-performance nanocomposites.

4.2.1 Sintering at Elevated Temperatures

BNNT-reinforced Al and Ti composites have been fabricated by sintering at elevated temperatures. In this approach, metal/BNNT powder mixtures are pressed in a die, and the resulting green body is sintered either in a vacuum or inert gas environment for several hours. The required sintering temperature and time depend on the material properties: metal powders with higher melting points require higher sintering temperatures and longer sintering times for densification. For instance, Al-BNNT composites require sintering temperatures of 500°C–600°C and sintering duration of ~1–3 hours [23,24]. However, sintering of Ti-BNNT is performed at higher temperatures ranging from 800 to 1100°C [25]. BNNTs generally retain their structure during sintering, but interfacial reactions between BNNTs and metal matrices can form new phases. For instance, AlN and AlB_2 are formed during the consolidation of Al-BNNT composites [24]. Sintering of Ti-BNNT composites produces a TiB phase in the microstructure, as shown in Figure 4.2 [25]. These reactions are strongly temperature dependent. Ti-BNNT composites sintered at 800°C do not contain any TiB phase, whereas raising the sintering temperature to 1100°C results in a complete transformation of BNNTs to TiB. The formation of the TiB phase also results in higher compressive strength. For instance, Ti-BNNT composites sintered at 1100°C have a compressive strength of ~1421 MPa, which is ~44% higher than the compressive

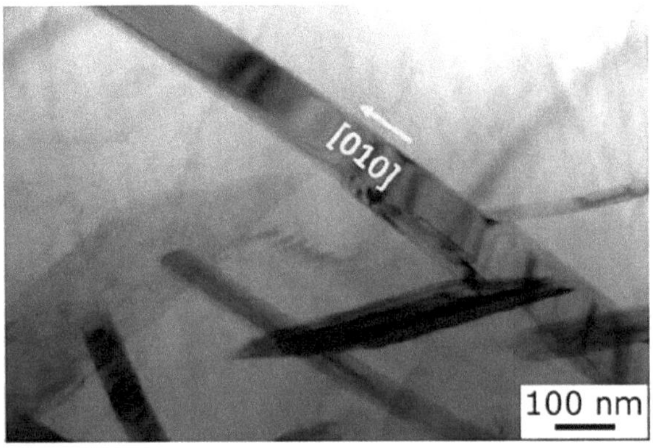

Figure 4.2 Formation of TiB needles in the composite due to interfacial reactions between Ti and BNNT. (Reprinted with permission from Bhuiyan et al. [25]. Copyright (2017) The Materials Research Society.)

strength of composites sintered at 800°C, ~984 MPa [25]. Therefore, *in-situ* interfacial reactions during sintering can be exploited to augment the load-bearing capacity of BNNT composites.

BNNT addition also produces fine-grained and homogeneous microstructures. This is due to pinning of grain boundaries by the nanotubes [23]. The microstructure and mechanical properties of sintered composites can be further optimized through post-processing operations, such as rolling and annealing. A study on Al-BNNT composites found that warm rolling created elongated grains, and subsequent annealing led to finer grains through recrystallization. These fine-grained composites exhibited a 110% improvement in tensile strength compared to unreinforced Al. The study also reported an improvement in ductility from 0.6% (after rolling) to 1% (after annealing). This was attributed to a reduction of internal stresses. These findings underscore the importance of post-processing operations for designing nanocomposites with superior strength-ductility synergy.

4.2.2 Spark Plasma Sintering (SPS): Electric Field and Pressure Assisted Processing

In SPS, powders are consolidated under the dual influence of Joule heating caused by direct current and uniaxial compressive stresses. The sintering is performed in a conductive die and vacuum environment. SPS processing of Al-BNNT composites has been conducted in a temperature window of 500°C–600°C, with compressive stresses ranging from 50 to 80 MPa [26,27]. The sintering times in these studies varied from 15 to 60 minutes, which are shorter than the conventional sintering discussed in the previous section. This is advantageous for preserving nanotubes and minimizing chemical reactions and phase transformations during composite fabrication. Multiple studies reported that agglomeration of BNNTs inhibits densification during SPS [26,27]. These agglomerates can produce micropores or holes at the grain boundaries, undermining the mechanical properties of composites. Therefore, it is crucial to prepare powders with homogeneous dispersion of nanotubes and metals for sintering dense and robust nanocomposites.

SPS has also been employed to fabricate sandwich-structured Al-BNNT and Mg-BNNT composites (Figure 4.3a) [28,29]. Nautiyal and co-workers used a BNNT mat sputter-coated with Al and placed it between Al powder in a graphite die for SPS [29]. After sintering, the pellet was cold rolled to produce thin sheets. It was discovered that the rolling process aligns BNNTs in the

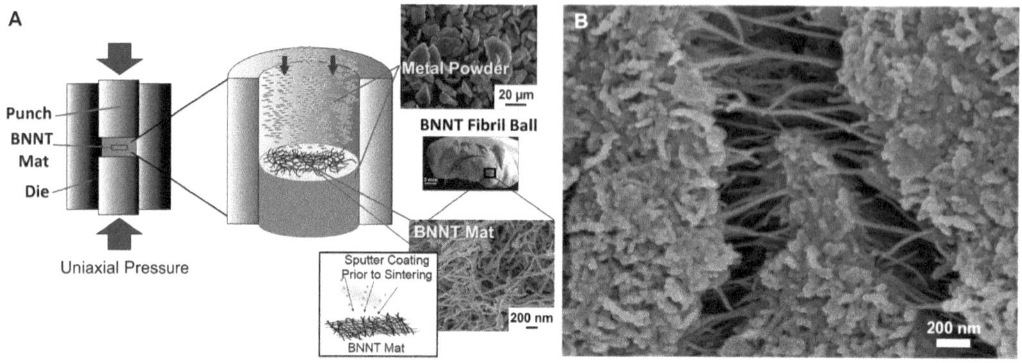

Figure 4.3 (a) Schematic representation of current- and pressure-assisted consolidation of BNNT-reinforced composites during SPS processing. (b) Scanning electron micrograph showing aligned BNNTs in a cold-rolled Al-BNNT composite. (Reprinted with permission from Nautiyal et al. [28]. Copyright (2020) WILEY-VCH Verlag GmbH & Co. KGaA, Weinheim, and Nautiyal et al. [29]. Copyright (2016) WILEY-VCH Verlag GmbH & Co. KGaA, Weinheim.)

microstructure (Figure 4.3b) [29]. These sandwich-structured Al-BNNT composites exhibited ~400% enhancement in tensile strength compared to pure Al.

Multiple studies have reported the occurrence of chemical reactions between BNNTs and metal powders during SPS. Nautiyal and co-workers observed the formation of the AlN phase at the Al/BNNT interface [29]. SPS of BNNT-reinforced AZ31 magnesium alloy produces Mg_3N_2 and AlN phases in the microstructure [28]. These reaction products were not homogeneously observed throughout the metal/nanotube interface; instead, they were scarce and scattered, which suggests slow reaction kinetics. SPS of higher melting point metals, such as Ti alloys, is performed at higher temperatures (750°C–1000°C). Therefore, interfacial reactions between BNNTs and Ti are more pronounced than in Al and Mg-based composites. Bustillos et al. examined the effect of sintering temperature on interfacial reactions in BNNT-reinforced Ti6Al4V composites [30]. The authors observed the formation of TiB and TiN phases during SPS at 950°C. To minimize the formation of these products, they explored low-temperature sintering of the composite at 750°C. The reduction in temperature was compensated by raising the sintering pressure from 60 MPa (at 950°C) to 555 MPa (at 750°C). The interfacial reactions were significantly arrested at lower temperatures. This is shown in the X-ray photoelectron spectroscopy maps in Figure 4.4. Both sintering conditions produced >97% densification.

In a subsequent study published by the same group, Nautiyal and co-workers investigated the mechanical properties of these Ti-BNNT composites by conducting indentation tests [31]. Ti-BNNT sintered at 950°C exhibited a 50% higher hardness and elastic modulus compared to the composites sintered at 750°C. This was attributed to a higher concentration of hard and stiff ceramic phases in the microstructure. The authors also performed *in-situ* indentation tests inside a scanning electron microscope to determine deformation mechanisms in the composites. It was revealed that the interfacial reaction products deflect propagating cracks, which leads to superior mechanical reinforcement. However, the study did not investigate the effect of reaction products on the ductility of composites. It is expected that excessive brittle phases will suppress ductility, which is not desirable. The interplay of pressure and temperature in SPS processing provides the ability to control interfacial reactions, which is important to optimize the mechanical properties of BNNT-reinforced composites.

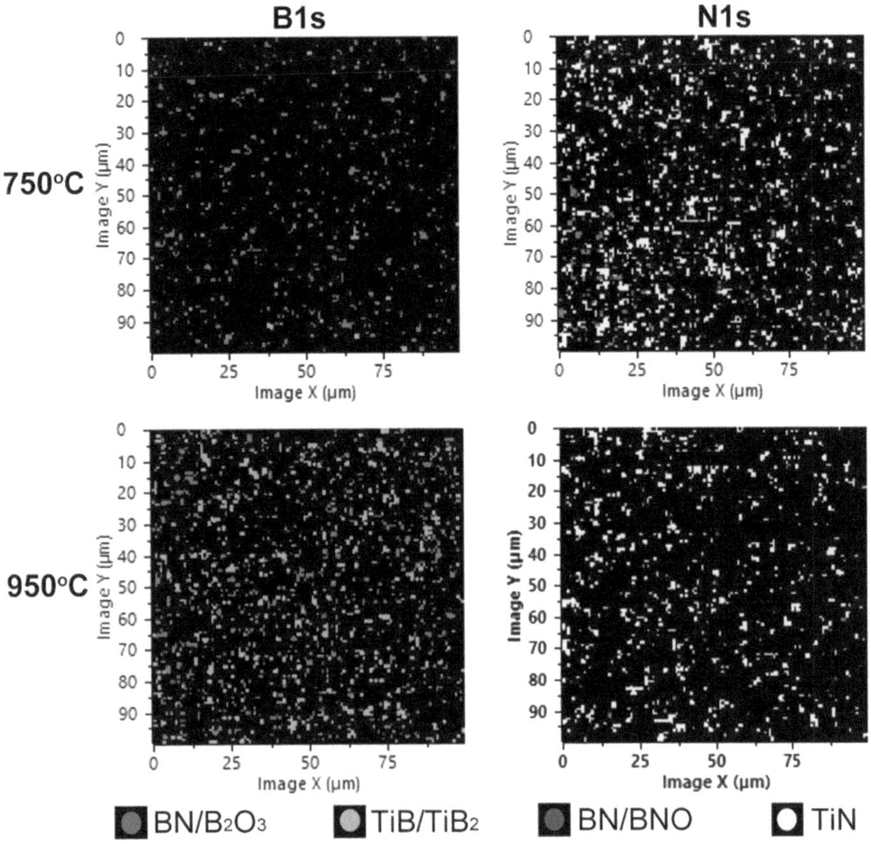

Figure 4.4 X-ray photoelectron spectroscopy mapping of Ti-BNNT composites showing arrested interfacial reactions/phase transformations during sintering at 750°C. (Reprinted with permission from Nautiyal et al. [31]. Copyright (2022) WILEY-VCH Verlag GmbH & Co. KGaA, Weinheim.)

4.2.3 High Pressure Torsion (HPT): Extreme Plastic Deformation Processing

The HPT processing involves severe plastic deformation of powders under extreme pressures and mechanical strains. In this method, the powder is compressed between two rotating anvils, as shown in Figure 4.5a [32]. HPT has been employed to fabricate Al-BNNT composites with 2.5–5 GPa applied pressure and 1 rpm rotation speed. Due to severe plastic deformation and low processing temperatures, the grain size of HPT-processed composites is smaller (50–500 nm) [32] compared to those produced by conventional high-temperature sintering and SPS approaches (2–20 μm) [23,27]. Furthermore, no interfacial reactions were reported in HPT-processed composites, likely due to the low processing temperature. However, the nanotubes were smashed, torn, and bundled (see Figure 4.5b) [27] due to severe shear deformation during processing. Nevertheless, the composites exhibited up to 70% improvement in tensile strength on adding 5 wt% BNNT to Al. Post-fabrication heat treatment of Al-BNNT composites at 350°C and 450°C was reported to produce a further 10% improvement in tensile strength. This was attributed to improved bonding between Al and BNNTs due to heat treatment.

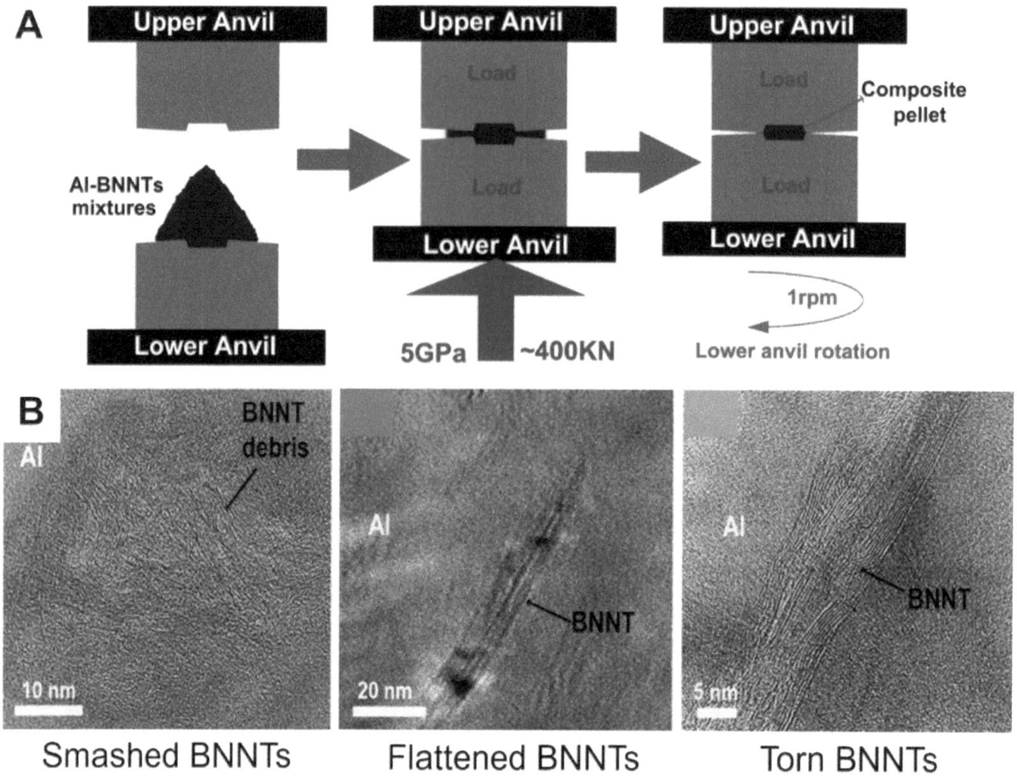

Figure 4.5 (a) Schematic representation of HPT processing. (b) Mechanisms of nanotube degradation due to extreme deformation during HPT processing. (Adapted with permission from Xue et al. [32]. Copyright (2015) Elsevier Ltd., and Yamaguchi et al. [27]. Copyright (2014) Elsevier B.V.)

A major challenge with HPT processing is the inferior ductility of fabricated composites. For instance, adding just 3 wt% BNNTs to Al resulted in a significant reduction in strain to failure from 26% (pure Al) to 6% (Al-BNNT composite). Further addition of BNNTs resulted in such dramatic embrittlement that the authors could not even machine dog-bone samples for tensile testing. Therefore, there is a need to further explore microstructure engineering strategies to balance the strength and ductility of HPT-processed BNNT MMCs.

4.2.4 Solidification Processing

BNNTs can be introduced in molten metal to fabricate MMCs via a solidification approach. Capturing BNNTs in the melt is necessary for manufacturing composites with a uniform microstructure. However, this is challenging because of repulsive interactions, such as van der Waals force, intermolecular interactions, interfacial force, and the integrated disjoining pressure [33]. These forces push away the nanoparticles in front of a moving solidification front, causing them to accumulate. On the other hand, a viscous drag force opposes the repulsion of nanoparticles, which allows the solidification front to quickly catch up with the nanofillers and capture (or engulf) them. The competition between these two phenomena governs microstructure formation during the solidification processing of nanocomposites. Xu et al. derived the critical velocity expression for the capture of nanoparticles by balancing these forces [33]:

$$V_{cr} = \frac{A_{sys}\left(D_{cr} - 2D_s\right)}{36\pi R \eta_{bulk} D_{cr}^2} \tag{4.1}$$

where A_{sys} is the Hamaker constant for the nanomaterial capture system, D_{cr} is the critical distance between the solidification front and the nanoparticles for the capture of nanoparticles, D_s is the diameter of the liquid molecule, R is the nanoparticle radius, and η_{bulk} is the viscosity of the melt. This relationship predicts that critical velocities for capturing fine nanoparticles in Al melt are on the order of 10^2–10^3 μm s^{-1}, which are significantly higher than the typical solidification front velocity in regular mold casting processes (~1 μm s^{-1}) [33]. This poses a challenge for the effective integration of nanotubes in metals via solidification.

The ultra-high critical speed can be lowered to some extent by increasing the viscosity of the melt, η_{bulk}. The viscosity of molten Al is temperature-sensitive. Regression analysis by Assael et al. gives the relationship between viscosity (η) and temperature (T) [34]:

$$\log_{10}\left(\eta/\eta^0\right) = -a_1 + \frac{a_2}{T} \tag{4.2}$$

where η_0=1 MPa s, a_1=0.7324, and a_2=803.49 K. Decreasing the melt temperature from 1000°C to 660°C causes the viscosity to drop by an order of magnitude. Therefore, choosing relatively lower processing temperatures (near melting point) during solidification will be helpful for capturing BNNTs in the Al melt.

Nautiyal et al. adopted this approach for synthesizing Al-BNNT composites via a solidification approach [35]. The authors employed a 700°C temperature, which is just above the melting point of Al. BNNTs were introduced and stirred in molten Al inside a muffle furnace. The Al/BNNT mixture was soaked at this temperature for 1 hour and then allowed to cool slowly inside the furnace until it reached the ambient temperature. The authors reported the successful integration of BNNTs in the microstructure, as shown in the transmission electron micrograph in Figure 4.6. The nanotubes retained their structure and chemical composition in the solidified composite. The study also reported the formation of AlN, AlB_2, and AlB_{10} phases due to interfacial reactions between Al and BNNT. AlN was the predominant phase, forming a 4–5 nm thick layer at the interface (see Figure 4.6). AlB_2, on the other hand, was found at fewer locations at the metal/nanotube interface.

The formation of AlN on BNNT also affects the wetting of nanotubes by molten Al during solidification processing. AlN has a surface tension of ~660 mN m^{-1} [36], which is higher than the surface tension of BNNT (~27 mN m^{-1}) [37]. As a result, the equilibrium contact angle (θ) between the molten Al and the nanotube surface will change, which is given by the relationship:

$$\cos\theta = \frac{\gamma_s - \gamma_{s/l}}{\gamma_l} \tag{4.3}$$

where γ_s is the surface energy of the solid phase being wetted, γ_l is the surface tension of the liquid, and $\gamma_{s/l}$ is the solid–liquid interfacial tension. The formation of the AlN product at the interface will enhance the wettability of nanotubes owing to AlN's higher γ_s. Previous results also show that the effective θ decreases from ~150° for the h-BN/Al interface to ~50° for the AlN/Al interface [38–41].

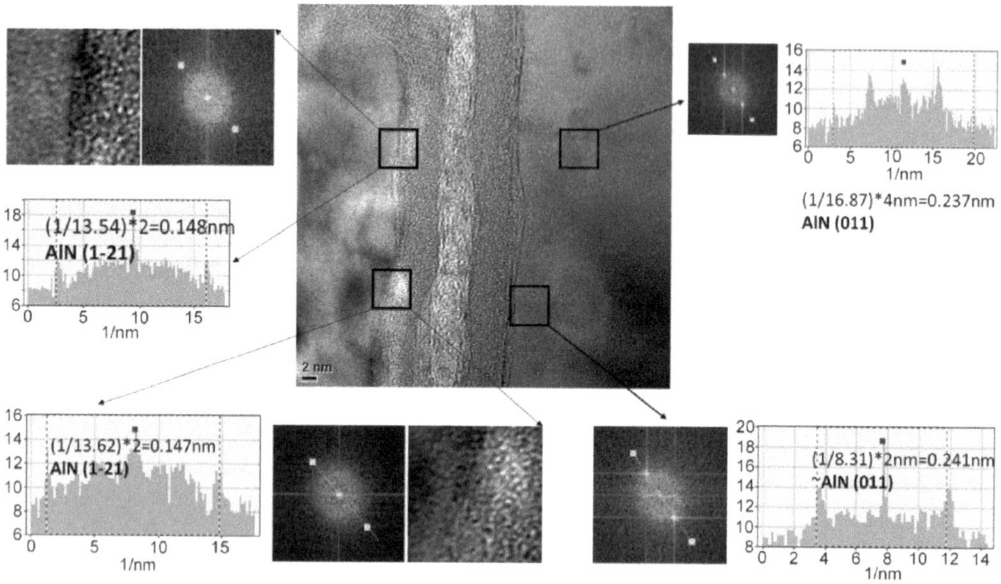

Figure 4.6 High-resolution transmission electron microscopy shows integration of BNNT in the Al matrix and the formation of interfacial reaction products at the Al/BNNT interface during solidification processing. (Reprinted with permission from Nautiyal et al. [35]. Copyright (2017) Acta Materialia Inc.)

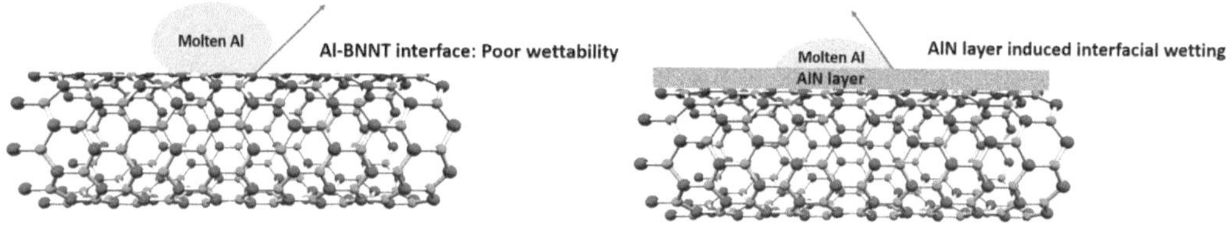

Figure 4.7 Schematic representation of reactive wetting of BNNT due to AlN formation during solidification processing. (Reprinted with permission from Nautiyal et al. [35]. Copyright (2017) Acta Materialia Inc.)

Thus, the formation of AlN assists in solidification processing. This reactive wetting mechanism is illustrated in Figure 4.7.

The formation of AlN also leads to improved work of adhesion (W_{ad}), which is given by the relationship:

$$W_{ad} = \gamma_l (1 + \cos\theta) \tag{4.4}$$

Based on this equation, the work of adhesion increases from 117 mJ m^{-2} for $\theta = 150°$ (Al/BN interface) to 1430 mJ m^{-2} for $\theta = 50°$ (Al/AlN interface). An increase in the work of adhesion is indicative of stronger interfacial bonding, which is desirable for superior load-bearing capacity.

Nautiyal and co-workers also reported filling of nanotubes by Al during solidification processing (see Figure 4.8a). This observation was ascribed to capillary action, with Laplace pressure driving the suction of molten Al into the hollow nanotubes:

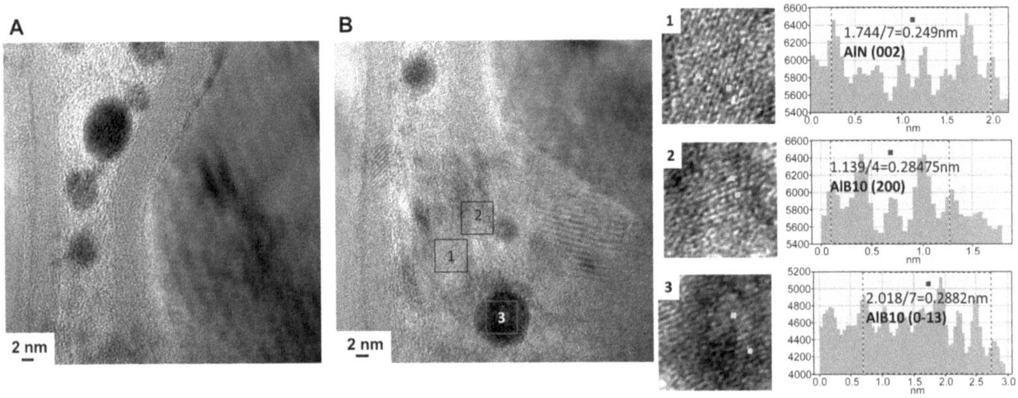

Figure 4.8 (a) Capillarity-induced filling of BNNTs by Al during solidification processing, (b) Reaction products inside BNNTs. (Reprinted with permission from Nautiyal et al. [35]. Copyright (2017) Acta Materialia Inc.)

$$\Delta P = \frac{2\gamma_l \cos\theta}{r} \tag{4.5}$$

where ΔP is the pressure difference across the liquid–vapor interface and r is the nanotube diameter. ΔP is positive if $\theta < 90°$. As mentioned above, θ for the Al/BNNT interface is 150°, and therefore, ΔP is negative (–300 MPa). Therefore, capillary suction of Al into BNNT is not expected to be spontaneous. However, the formation of AlN on the BNNT surface modifies θ to 50°. Correspondingly, ΔP becomes positive (220 MPa), which is favorable for capillarity-induced filling of BNNTs by molten Al. The authors also observed the formation of AlN and AlB_{10} phases inside the nanotubes (Figure 4.8b), suggesting that chemical reactions also took place inside the nanotubes after some Al droplets were drawn in.

Reactive wetting and filling of nanotubes can have profound consequences for the mechanical properties of nanocomposites. On one hand, the formation of hard ceramic phases is likely to enhance the elastic modulus and yield strength of composites; however, these brittle phases could also undermine the ductility of fabricated composites. Therefore, a fundamental understanding of how these microstructure formation mechanisms influence the mechanical properties of nanocomposites is vital. We discuss the mechanical behavior of MMCs in more detail in Section 4.3.

4.2.5 Plasma Spray: Rapid Solidification Processing

Plasma spray is a thermal spray technique in which feedstock powders are injected into a plasma jet and accelerated toward a substrate for layer-by-layer deposition. The particles melt *in-flight* as they are exposed to high-temperature plasma [42]. They rapidly solidify upon impacting the substrate. Nautiyal et al. deposited 300 μm thick Al-BNNT coatings by plasma spray (Figure 4.9a) [43]. BNNTs retained their structural integrity and chemical composition in the fabricated composites, despite exposure to extreme temperatures, as shown in the scanning electron micrograph in Figure 4.9b. The solidification rate (R) during plasma spray is given by the relationship [44]:

$$R = \frac{h(T_P - T_s)}{\rho L_f} \tag{4.6}$$

where h is the heat transfer coefficient, T_P is the particle melting temperature, T_S is the substrate temperature, ρ is the particle density, and L_f is the latent heat of fusion of the particle. For aluminum, the solidification rate during plasma spray is on the order of 10^5 µm s^{-1} [44], which is at least two orders of magnitude higher than the critical velocity required for capturing nanoparticles in aluminum melt (based on Equation 4.1) [33]. Therefore, plasma spray is an effective solidification approach for integrating BNNTs in the Al matrix.

Trace amounts of AlN and AlB_2 were observed in the microstructure due to interfacial reactions between BNNT and Al during plasma spray (Figure 4.9). However, these phases were not

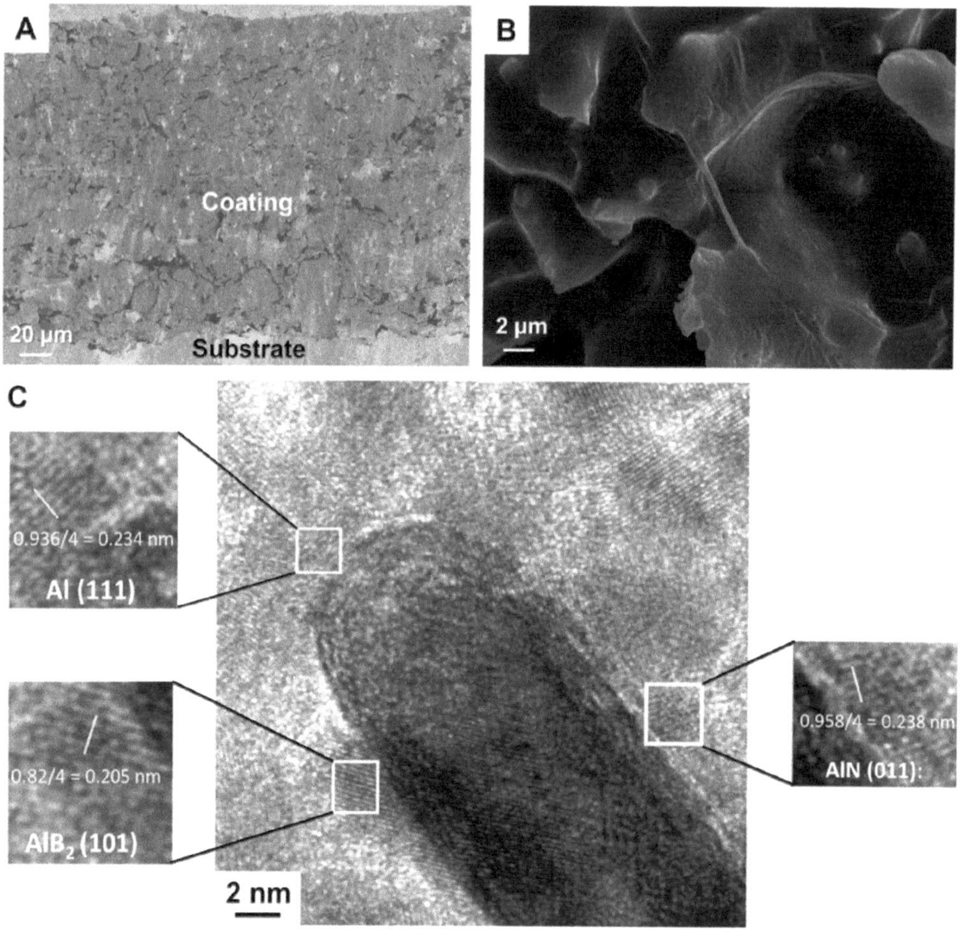

Figure 4.9 Plasma spray deposition of Al-BNNT composite coatings: (a) Scanning electron micrograph showing the cross-section of the coating. (b) Scanning electron micrograph confirming the integration of BNNTs in the composite coatings. (c) Formation of interfacial reaction products revealed by high-resolution transmission electron microscopy of the BNNT/Al interface. (Reprinted with permission from Nautiyal et al. [43]. Copyright (2018) Acta Materialia Inc.)

observed throughout the microstructure. In fact, the majority of the nanotubes did not react. This is because the particles are exposed to elevated temperatures for extremely short durations (<1 ms [45]) and the cooling rate is also very high (>10^6 K s^{-1}) [9]. As a result, there is limited time for interfacial chemical reactions. The authors also postulated that these interfacial reaction products assist in the capture of nanotubes in the moving solidification front. Thermodynamically, the feasibility of nanotube capture is predicted by the Gibbs free energy change during solidification [33]:

$$\Delta G = \pi r^2 l \left(\sigma_{iS} - \sigma_{iL}\right) \tag{4.7}$$

where σ_{iS} and σ_{iL} are the interfacial energies associated with the BNNT/solidified Al interface and the BNNT/molten Al interface, respectively. The formation of AlN on the nanotube surface lowers σ_{iS} (which was discussed in the previous section). Therefore, the interfacial reactions are expected to promote the capture of nanotubes by the molten metal front, which is crucial for engineering nanocomposites with homogeneous microstructure.

The authors also conducted nanoindentation experiments to ascertain the effect of BNNT integration on the mechanical properties of plasma-sprayed coatings. The measurements revealed a heterogeneous mechanical response. While 80% of the indents exhibited elastic moduli and hardnesses representative of pure Al, about 20% of the indents exhibited up to 45% higher elastic moduli and up to 10 times higher hardness than pure Al. These results confirm that BNNTs and the interfacial reaction products enhance the load-carrying capacity of the coatings.

4.2.6 Ultrasonic Cavitation Assisted Processing

Ultrasonic treatment (UST) utilizes very high frequencies of 18–20 kHz during melt processing to produce MMCs [46–49]. The introduction of cyclic high-intensity sound waves into the melt pool through ultrasonication results in homogeneous dispersion of BNNTs and the formation of numerous microbubbles, serving as nucleation sites for metal solidification. The cyclic process of bubble formation, growth, pulsation, and collapse creates transient 'hot spots' characterized by high pressure (1000 MPa), elevated temperatures (~5000°C), and high heating/cooling rate (10^{10} K s^{-1}) [50,51]. The significant impact from bubble collapse, along with localized high temperatures, enhances the wettability of the BNNTs and facilitates effective reinforcement into the matrix [52,53]. The microbubbles and cavitation activity induced by UST enhance the formation of heterogeneous nucleation sites and enable uniform distribution of BNNTs in the composite [50,54–56]. Therefore, UST provides a highly effective processing pathway for grain refinement, deagglomeration, and dispersion of reinforcements. These mechanisms are illustrated in Figure 4.10.

Recently, Mohammed et al. [57] employed UST for processing BNNT-reinforced aluminum matrix composites. The authors initially utilized UST for dispersing BNNTs in aluminum powder by a wet mixing approach at room temperature. The Al-BNNT composite powders were then melted, and UST was performed again during melt processing to homogeneously distribute BNNTs in the Al matrix. The solidified composites were then cold rolled. The authors reported synergistic grain refinement due to UST and BNNT reinforcement, as shown in Figure 4.11a. Despite exposure to a high temperature of 750°C, the nanotubes remained well-dispersed, embedded, and retained in the matrix after solidification (Figure 4.11b), confirming that BNNTs can withstand high processing temperatures. The resulting composites

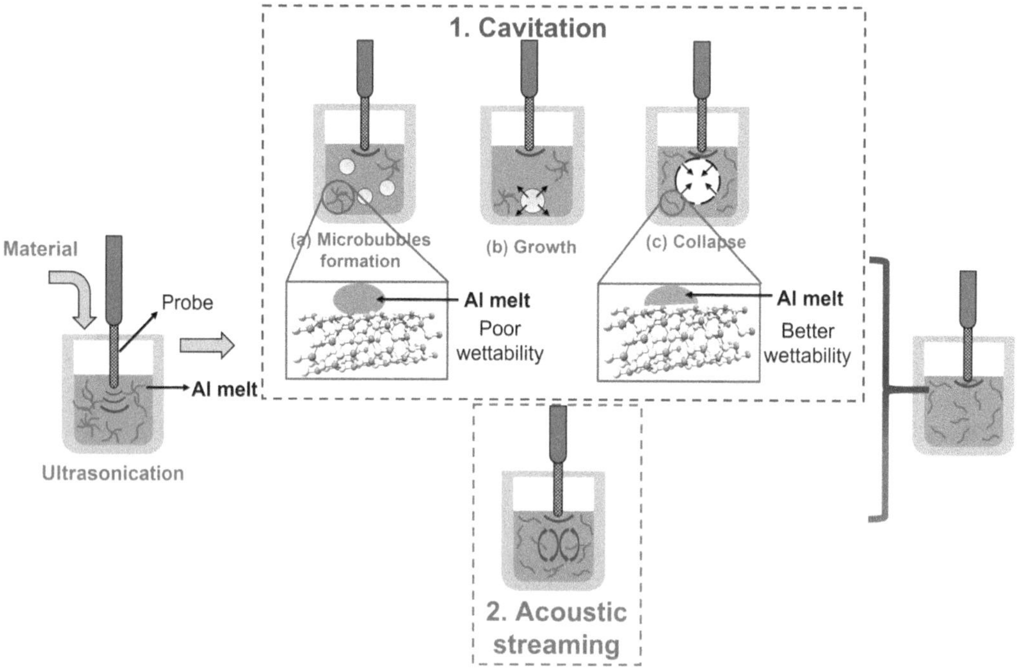

Figure 4.10 Schematic representation of microstructure formation mechanisms during ultrasonic cavitation assisted processing of MMCs. Reprinted with permission from Mohammed et al. [57]. Copyright (2023) The Authors.)

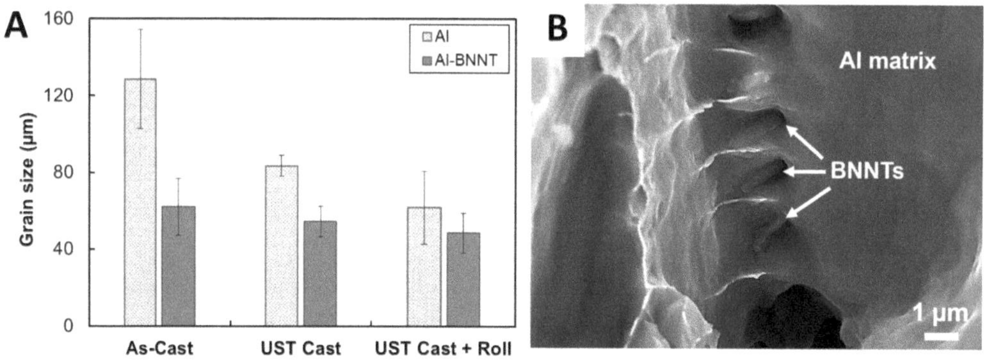

Figure 4.11 Ultrasonic cavitation-assisted casting of Al-BNNT composites: (a) Synergistic grain refinement due to BNNT reinforcement, UST processing, and cold rolling of Al-BNNT composites. (b) Scanning electron micrograph showing the fracture surface of the Al-BNNT composite. (Reprinted with permission from Mohammed et al. [57]. Copyright (2023) The Authors.)

showed a ~97% improvement in yield strength compared to unreinforced aluminum. These findings confirm that UST is a highly effective approach for engineering high-performance BNNT-reinforced composites.

4.2.7 Additive Manufacturing (AM)

In recent years, there has been a surge in interest in AM processes owing to their capability to fabricate near net-shaped structures through layer-by-layer material deposition [58]. However, AM techniques are characterized by extremely rapid cooling rates. The ultra-short processing times create non-equilibrium solidification conditions, offering a limited timeframe for the control of microstructure evolution and the incorporation of nanotubes. Chao and co-workers employed a blown powder directed energy deposition (DED) process for 3D printing Ti-BNNT composites using BNNT-decorated Ti powder [59]. The composite powders were injected into the focused beam of a 250 W laser. Addition of just 0.4 wt% BNNTs caused prior-β grain size reduction from 337 μm (for Ti-6Al-4V) to 9 μm (for 0.4 wt% BNNT-reinforced Ti-6Al-4V). Moreover, BNNT addition transformed columnar prior-β grains into near-equiaxed grains. The BNNT-reinforced alloy also exhibited 50% higher compressive yield strength than the as-built Ti-6Al-4V. These results are remarkable considering that trace amounts of BNNT were added to the alloy. It is noteworthy that all nanotubes were transformed into TiB nanowhiskers during DED, which was attributed to chemical reactions between BNNTs and the Ti matrix (Figure 4.12).

A crucial step in facilitating scalable AM, especially DED, is the successful production of precursors or electrodes. Mohammed et al. [60] fabricated Al-BNNT electrodes using ultrasonic cavitation and rolling. This composite electrode was then employed for wire-fed DED of Al-BNNT composites. The microstructure of the deposited composite consisted of a mix of fine and coarse grains (see Figure 4.13a–c), which is indicative of uneven solidification rates resulting from manual deposition. The BNNTs were integrated into the microstructure (Figure 4.13d and e) and maintained their structural integrity despite exposure to temperatures exceeding 1000°C during deposition. The Al-BNNT composites exhibited ~130% improvement in yield strength compared to pure Al, which the authors attributed to grain refinement and efficient stress transfer to BNNTs. These results demonstrate the feasibility of engineering load-bearing BNNT composites *via* far-from-equilibrium AM processes.

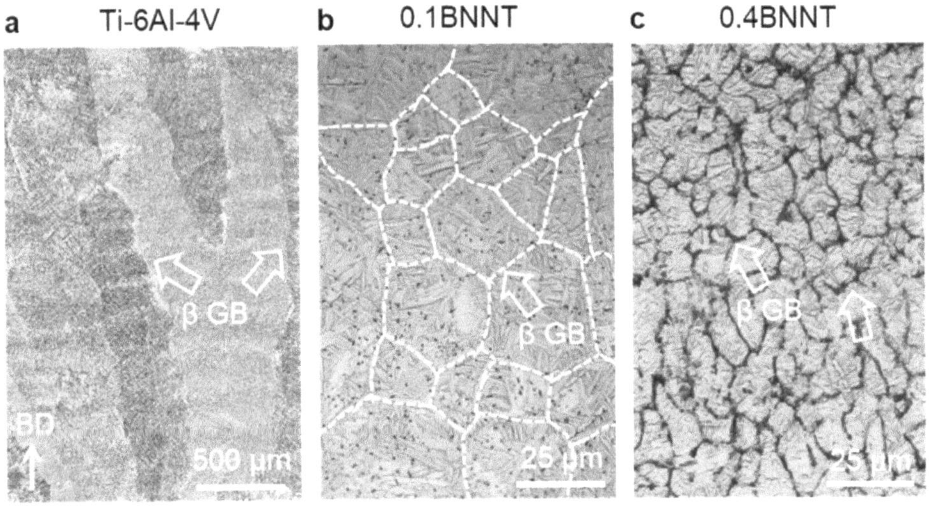

Figure 4.12 BNNT-induced grain refinement during laser DED of Ti-BNNT composites: (a) unreinforced Ti-6Al-4V alloy, (b) 0.1 wt% BNNT-reinforced composite, and (c) 0.4 wt% BNNT-reinforced composite. (Reprinted with permission from Chao et al. [59]. Copyright (2021) Elsevier B.V.)

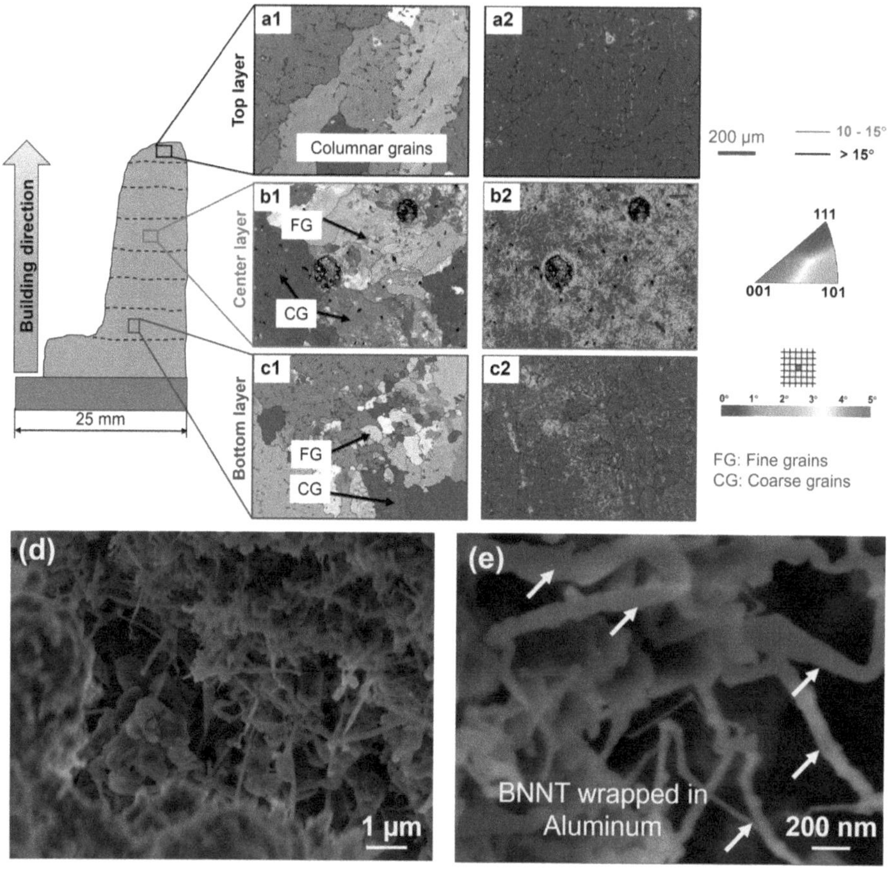

Figure 4.13 Wire-fed DED of Al-BNNT composites: electron backscatter diffraction maps showing the (a1), (b1), (c1) orientation maps and (a2), (b2), (c2) KAM. Integration of BNNTs in melt pool (d, e) BNNTs integrated in the microstructure of Al-BNNT composites. (Reprinted with permission from Mohammed et al. [60]. Copyright (2023) Wiley-VCH GmbH.)

4.3 MECHANISMS OF STRENGTHENING INDUCED BY BORON NITRIDE NANOSHEETS

The experimental results indicate that the addition of BNNTs can improve the mechanical properties of Al, Mg, and Ti matrix composites. The underlying strengthening mechanisms in nanotube-reinforced MMCs are commonly Hall–Petch strengthening, load transfer, Orowan strengthening, and thermal mismatch-induced dislocation strengthening [61–63]. Here, we briefly discuss these mechanisms:

4.3.1 Hall–Petch Strengthening

Nanotubes can act as nanoscale nucleation sites during processing. These sites can increase nucleation rates, producing fine-grained microstructures. This leads to improved yield strength (σ_y), given by the Hall–Petch relationship:

$$\sigma_y = \sigma_i + kd^{-1/2} \tag{4.8}$$

where σ_i is the intrinsic strength of the material, d is the grain size, and k is a constant.

4.3.2 Load Transfer Mechanism

The load transfer between the matrix and nanotubes occurs by shear force, assuming the nanotubes are uniformly aligned along the loading direction. The effective load transfer depends on the length of the nanotubes, which must exceed the critical length, l_c, given by [62],

$$l_c = \frac{\sigma_f d}{\sigma_m} \tag{4.9}$$

where σ_f and σ_m are the strengths of the nanotube and matrix, respectively, and d is the diameter of the nanotube.

4.3.3 Orowan Strengthening

The dislocations tend to move as the material is strained until they confront an obstacle. The typical obstacles are grain boundaries and reinforcements in the composite. Since the BNNTs are very strong and cannot be sheared by the dislocations, the dislocations are likely to pile up at the point of their interaction with the obstacle (BNNT), thereby increasing the amount of stress required to overcome the resistance to dislocation movement. At sufficiently high stresses, these dislocations are expected to bypass the BNNTs and form an Orowan loop, leading to Orowan strengthening. The activation of Orowan strengthening is highly dependent on the geometry of the BNNTs. The Orowan looping strengthening is quantified by the Orowan–Ashby equation,

$$\Delta\sigma_O = \frac{0.13 G_m b}{d_{\mathrm{BNNT}}\left[\left(\frac{1}{2\,V_{\mathrm{BNNT}}}\right)^{1/3} - 1\right]} \ln\left(\frac{d_{\mathrm{BNNT}}}{2b}\right) \tag{4.10}$$

where $\Delta\sigma_O$ is the strength improvement by Orowan looping, G_m is the shear modulus of the matrix, b is the Burgers vector, and d_{BNNT} and V_{BNNT} are the diameter and volume fraction of BNNTs, respectively.

4.3.4 Thermal Mismatch

The thermal mismatch mechanism in composites is associated with the huge variation in coefficients of thermal expansion of BNNTs and the matrix. This can produce a high dislocation density in the microstructure during cooling. The dislocation-dense composites exhibit higher yield strength (σ_y) due to strain hardening, given by the relationship:

$$\sigma_y = \sigma_i + k\rho^{1/2} \tag{4.11}$$

where ρ is the dislocation density in the microstructure, σ_i is the intrinsic strength of the material, and k is a constant. This mechanism is discussed further in a case analysis below.

4.3.5 Case Analysis

Electron backscatter diffraction analysis can be used to determine strengthening mechanisms in reinforced MMCs. Kernel Average Misorientation (KAM) is a parameter that serves as a metric indicating intragranular strain and dislocation activity. It quantifies the average misorientation value between a central point within the kernel and all surrounding points on the Kernel's perimeter, on a pixel-by-pixel basis. Typically, it is employed for qualitative assessment of stored energy resulting from point defects and dislocations formed during deformation, as evidenced by geometrically necessary dislocations (GNDs). This stored energy serves as the driving force for recrystallization processes.

In a study on casting of Al and Al-BNNT composites, the effect of BNNT reinforcement and ultrasonic processing on GND density was estimated [57]. The microstructure of as-case Al was characterized by a uniform strain distribution and a GND value of $\sim 0.69 \times 10^{14} m^{-2}$. The addition of BNNT reinforcement produced strain concentration in the microstructure, and the GND value was increased to $\sim 1.37 \times 10^{14} m^{-2}$. Ultrasonication of the MMC melt for 90 seconds resulted in a further enhancement of GND density to $\sim 1.68 \times 10^{14} m^{-2}$. This rise in GND value correlates directly with the formation of low-angle grain boundaries. Low-angle grain boundaries function as arrays of dislocations that interact with each other, consequently forming dislocation pile-up zones. This results in enhanced yield stress.

In a separate study, Mohammed et al. [60] measured the dislocation densities of DED Al-BNNT composites by KAM. They discovered that BNNTs introduce dislocations in composites during solidification processing owing to the difference in the thermal expansion coefficient of the Al matrix and BNNTs. The quantitative distribution of dislocations is captured in KAM maps in Figure 4.13 (a2), (b2), and (c2). The mutual interaction between dislocations and the restricted movement of dislocations by BNNT obstacles increases the strength and hardness of composites.

4.4 PERFORMANCE OF COMPOSITES UNDER HARSH CONDITIONS

4.4.1 Mechanical Properties at Elevated Temperatures

Lightweight MMCs are desired for high-temperature applications. Titanium composites are particularly attractive owing to their superior specific strength and stiffness compared to steel and nickel-based materials [64]. Bhuiyan et al. investigated the high-temperature mechanical properties of Ti-BNNT composites [25]. The nanocomposites in this study were fabricated by sintering over a temperature range of 800°C–1100°C. The authors investigated the effect of BNNT concentration on the compressive strength of composites at room temperature and 500°C. The study reported the highest yield strength for composites synthesized at 1100°C, as shown in Figure 4.14. This was attributed to chemical reactions between BNNT and Ti, forming TiB whiskers in the microstructure. BNNT-induced enhancement in mechanical strength was observed at both room temperature and 500°C. This suggests that BNNT is a promising nanofiller for engineering high-temperature composites.

One potential concern is that BNNTs in composite materials may react with the metal phase when in service at elevated operating temperatures. As a result, the microstructure of the composites will dynamically evolve, which may also alter their mechanical properties during operation. Unfortunately, there is no quantitative information on how the mechanical properties of

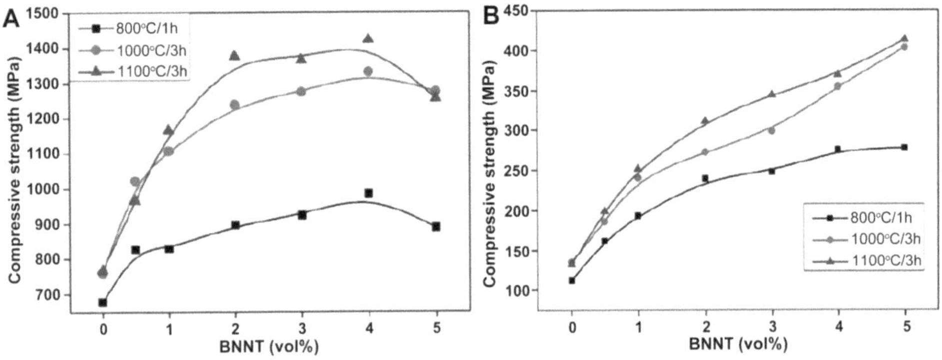

Figure 4.14 Compressive strength of sintered Ti-BNNT composites at: (a) room temperature and (b) 500°C. (Reprinted with permission from Bhuiyan et al. [25]. Copyright (2017) The Materials Research Society.)

BNNT MMCs evolve over time, after prolonged exposure to elevated temperatures. Further analysis of temperature- and time-dependent microstructure evolution and deformation mechanisms is required to enable high-temperature BNNT composites for extreme environments.

4.4.2 Tribology in Harsh Conditions

Mechanical components in contact and relative motion experience friction and wear. Under harsh conditions, unchecked friction and wear can lead to catastrophic mechanical failures. Therefore, there is a critical need for wear-resistant, self-lubricating materials for extreme operating conditions. Several studies have investigated the frictional and wear properties of BNNT-reinforced composites using ball-on-disk tribometry. In these experiments, stainless steel or ceramic spherical balls are slid against the MMC samples. The effect of BNNT reinforcement on the tribological performance of cold-sprayed aluminum deposits has been recently reported in the literature [65–67]. Kulkarni and co-workers [65] observed a reduction in the coefficient of friction from ~0.63 for pure Al coatings to below ~0.55 for BNNT-reinforced Al, which suggests that BNNTs may be acting as solid lubricants at sliding contacts. BNNT addition also enhanced the wear resistance of aluminum deposits, with a ~25% reduction in wear volume during harsh sliding experiments. The superior wear resistance of Al-BNNT composites was ascribed to a higher hardness-to-elastic modulus ratio (*H/E*) due to BNNT incorporation in the matrix. Higher values of *H/E* translate to higher elastic-strain-to-failure [68], which is desirable for redistributing the applied contact load over the contact area.

Bacca et al. [67] studied the tribological properties of BNNT-reinforced Ti composites for lunar exploration applications. The Ti-BNNT composite displayed improved wear resistance and lower friction compared to unreinforced Ti alloy. The addition of 1 wt% BNNT reduced the coefficient of friction from over 0.41 to under 0.37 and mitigated the wear volume loss by 10.2%. The improved wear resistance was attributed to superior hardness and yield strength of reinforced composites [30,31]. The authors also performed wear testing of Ti-BNNT composites with a zircon lunar regolith simulant. Zircon particles are known to be highly abrasive among lunar regolith simulants. Consequently, wear tests involving zircon on the pristine Ti sample resulted in 5.16 times greater wear volume compared to tests without it (see Table 4.1). Similarly, in Ti-BNNT tests,

TABLE 4.1 RESULTS OF TRIBOLOGICAL TESTING ON PRISTINE TI AND TI-BNNT SAMPLES [67]

Sample	Lunar regolith simulant	Coefficient of friction	Improvement in CoF from pristine Ti	Wear volume (mm^3)	Improvement in wear due to BNNT
Pristine Ti	None	0.412	–	0.049	–
Ti-BNNT	None	0.368	9.76%	0.044	10.2%
Pristine Ti	Zircon	0.448	–	0.253	–
Ti-BNNT	Zircon	0.434	3.13%	0.188	25.5%

the inclusion of zircon led to a wear volume approximately 4.27 times higher than tests without it. Additionally, zircon particles contributed to an increase in frictional resistance, with the coefficient of friction for Ti and Ti-BNNT increasing by 8.73% and 17.9%, respectively. However, despite these effects, the incorporation of BNNTs successfully enhanced the wear resistance of Ti. In comparison to the unreinforced Ti sample, the Ti-BNNT sample demonstrated a notable 25.5% reduction in wear volume. However, it's worth noting that the coefficient of friction of the Ti-BNNT sample did not exhibit as substantial an improvement as the reduction in wear volume, showing only a 3.13% difference compared to unreinforced Ti.

In addition to BNNTs, some studies have explored the potential of h-BN on the tribological properties of MMCs. Sukumaran et al. [66] examined the wear resistance of h-BN-reinforced titanium coatings for lunar structural components. The coatings were manufactured by vacuum plasma spray and atmospheric plasma spray approaches. The abrasive wear resistance of the deposited coatings was investigated in the presence of JSC-1A lunar regolith and no-regolith conditions and compared with the standard structural Al and Ti alloys. In no-regolith conditions, the atmospheric plasma-sprayed Ti-2 vol.% hBN coatings exhibited 90% and 65% improvement in wear resistance compared to Al6061 and Ti6Al4V substrates, respectively. Vacuum plasma-sprayed composite coatings also exhibited an improvement of 93% and 40%, respectively, for the same scenario. In the presence of JSC-1A lunar regolith, atmospheric plasma spray composite coatings showed an enhancement in wear resistance by 82% and 90%, compared to Al6061 and Ti6Al4V substrates, respectively. Interestingly, vacuum plasma spray composite coatings did not exhibit any material loss at all. Instead, the authors reported some mass gain due to the embedding of JSC-1A particles. The improved wear resistance of these coatings was ascribed to the synergistic effect of lubricious hBN and hard TiB phases in the microstructure. While the hard boride phase resists the penetration of abrasive particles, the hBN layers easily shear off at sliding contacts due to weak van der Waals bonds. This produces a solid lubrication effect where the ball slides over the substrate without significant contact, thus reducing friction as well as mitigating wear of coatings.

4.4.3 Shielding Radiation with BNNT Composites

Neutron radiation shielding is critical in nuclear energy and space industries to protect equipment and human life [69]. Shielding against neutron radiation is complex due to its chargeless nature, distinguishing it from other forms of radiation. Neutrons interact directly with atomic nuclei, unaffected by electric fields [70]. The extent of this interaction, quantified by cross-sections, depends on the neutron energy. Inelastic scattering, elastic scattering, and neutron absorption are the primary modes in which neutrons interact with matter. Fast neutrons (>10 MeV) undergo inelastic scattering, while those with energies between 1 and 10 MeV experience elastic scattering. Thermal neutrons, moving at slower speeds, are absorbed into nuclei without emitting particles, leading to nucleus excitation and subsequent emission of γ-rays and α-rays. BNNTs have neutron shielding capability due to the high neutron absorption cross-section of ^{10}B (up to 3800 barns), the highest among the low atomic number elements [71]. Bacca and co-workers investigated the radiation shielding capability of spark plasma sintered Ti-BNNT composites [67]. The samples were fabricated at 750°C and 950°C. The composites fabricated at the lower temperature exhibited superior improvement in linear and mass absorption coefficients, as shown in Figure 4.15. This is because of limited interfacial reactions between BNNT and Ti during low-temperature sintering. Therefore, more BNNTs are retained in the microstructure of composites sintered at 750°C [30,31], resulting in superior radiation shielding. Neutron shielding by BNNT MMCs involves both absorption and scattering. Boron in the BNNTs absorbs neutrons during nuclear reactions, releasing lithium atoms, alpha particles, and gamma radiation, significantly reducing neutron flux [72]. Additionally, when fast neutrons collide with the nuclei of metal (matrix) atoms, they lose energy through elastic scattering [73]. This energy loss slows down the neutrons, making them less harmful. These preliminary results suggest that BNNT MMCs are strong candidates for space applications where radiation protection is critical.

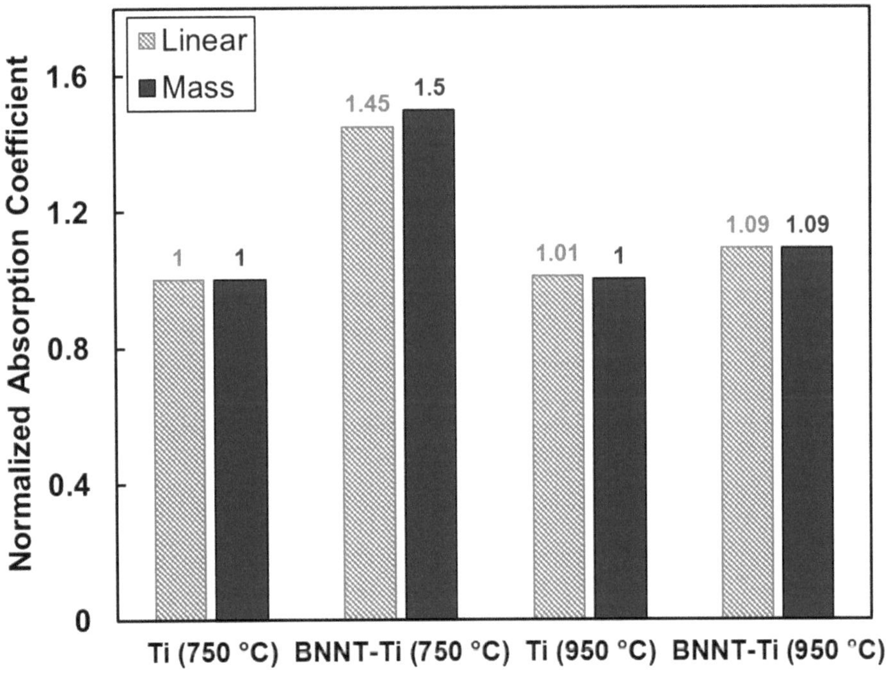

Figure 4.15 Neutron shielding results for Ti and Ti-BNNT composites sintered at different temperatures, showing normalized linear and mass absorption coefficients. (Reproduced from [67]. Reprinted with permission from Bacca et al. [67]. Copyright (2022) The Author(s).)

4.5 CONCLUSIONS AND FUTURE OUTLOOK

In this chapter, we discussed various aspects of BNNT-based MMCs, from processing to mechanical, tribological, and radiation shielding properties. We reviewed developments in composite fabrication using a wide range of conventional and advanced manufacturing techniques, such as high-temperature sintering, SPS, HPT consolidation, solidification processing, plasma spray, ultrasonic cavitation, DED, and cold rolling. We discussed microstructure formation mechanisms, such as densification, grain size evolution, and phase transformations, during the processing of composites under the influence of elevated temperatures, stresses, electric fields, and acoustic energies. BNNTs generally retain their structural and chemical integrity under far-from-equilibrium extreme processing conditions. BNNTs also react with metals during processing to form interfacial reaction products. The formation of these interphases is accelerated at elevated temperatures. These phases enhance the strength of nanocomposites. We emphasized the importance of homogeneous dispersion of nanotubes in metal matrices for effective strengthening. We also discussed various strengthening mechanisms activated by nanotube reinforcements in MMCs. Lastly, case studies on the performance of BNNT-reinforced composites at elevated temperatures, abrasive conditions, and radiation exposure were discussed.

Scalable manufacturing of BNNT composites will be crucial for their widespread industrial adoption. We highlighted case studies on laser- and electric arc-assisted AM of BNNT-reinforced composites. However, elevated temperatures in these processes can lead to unchecked chemical reactions. Moreover, these processes can generate undesirable microstructural defects, such as solidification cracks and porosity. Recently, low-temperature and solid-state processes, such as cold spray and friction stir AM, are gaining attention for 3D printing alloys and MMCs. Due to low processing temperatures, undesirable phase transformations can be avoided in these manufacturing approaches. However, these processes involve severe plastic deformation of powder feedstock [74,75]. Can BNNTs retain their structural integrity under harsh deformation conditions? How do elevated stresses and strain rates during processing influence BNNT integration and dispersion in the MMC microstructure? A fundamental understanding of microstructure evolution mechanisms under these far-from-equilibrium processing conditions is critically needed.

We discussed the importance of post-processing operations, such as rolling and annealing, to design the microstructure of nanocomposites. However, there is limited work on quantifying the kinetics of microstructure evolution during these operations. Moreover, there is very limited exploration of the processing parameter space. How do heat treatment temperature and time affect interfacial bonding between the metal matrix and nanotubes? What is the effect of cold-rolling speed and number of passes on the structure and integrity of BNNTs in the microstructure? Establishing these relationships will be crucial for predictably designing the microstructure of BNNT-reinforced composites.

Multiple studies have found that the formation of interfacial products during nanocomposite processing can be advantageous for mechanical performance. More than one product can form during composite fabrication. For instance, Al-BNNT composites can contain AlN, AlB_2, and AlB_{10} phases. However, there are no systematic studies on the mechanical characterization of these constituent phases. We currently lack a quantitative understanding of how each of these phases contributes to the load-carrying capacity of composites. Since the overall deformation behavior of composites is influenced by the stress transfer between the metal matrix, interfacial reaction products, and BNNTs, the quantification of phase-specific properties is crucial. Advanced *in-situ* nanomechanical characterization techniques can be employed to determine the mechanical response of the constituent phases in BNNT MMCs. The phase-specific properties can then feed into finite element models to predict the mechanical performance of composites. This will be a major step forward toward theory-driven design and optimization of BNNT composites.

The mechanical behavior of BNNT composites has been widely investigated under tensile, compressive, and indentation loading. However, we lack a fundamental understanding of fatigue, creep, or impact-induced deformation of these composites. These failure modes are encountered in extreme operating conditions. How do interfacial reaction products and BNNT loading fraction affect these time- and loading rate-dependent mechanical properties? What is the influence of temperature on these deformation mechanisms? High temperature, high strain rate, and long-duration mechanical tests coupled with microscopic analyses can fill this knowledge gap.

A vast majority of research efforts have focused on laboratory-scale fabrication and testing of BNNT-reinforced composites to understand their fundamental properties. In recent years, several patents have been filed and granted on the design and manufacturing of BNNT-reinforced composites using wire-based AM, thermal spray, field-assisted sintering, deformation processing, and ultrasonic casting [76–78]. Translating these innovations to products and deployable technologies will require prototyping, scaling up, optimization, and pilot-scale demonstrations and validation in relevant environments. Moreover, industry partnerships will also be crucial to integrate these materials into existing production processes to ensure reliability and cost-effectiveness.

These are some crucial and impactful directions for future research on BNNT MMCs. The recent advancements in advanced manufacturing, advanced characterization techniques, and data-driven modeling approaches are expected to drive significant scientific and technological advancements in BNNT-based high-performance structural composites.

REFERENCES

1. H. Springer, C. Baron, A. Szczepaniak, V. Uhlenwinkel, D. Raabe, Stiff, light, strong and ductile: Nano-structured high modulus steel. *Scientific Reports* 7, 2757 (2017).
2. D. B. Miracle, Metal matrix composites-From science to technological significance. *Composites Science and Technology* **65**, 2526–2540 (2005).
3. X. Fu et al., Orientation-dependent tensile behavior of nanolaminated graphene-Al composites: An in situ study. *Metallurgical and Materials Transactions A* **49**, 5229–5234 (2018).
4. K. Chu, J. Wang, Y. Liu, Z. Geng, Graphene defect engineering for optimizing the interface and mechanical properties of graphene/copper composites. *Carbon* **140**, 112–123 (2018).
5. K. T. Kim, S. I. Cha, S. H. Hong, S. H. Hong, Microstructures and tensile behavior of carbon nanotube reinforced Cu matrix nanocomposites. *Materials Science and Engineering: A* **430**, 27–33 (2006).
6. S. R. Bakshi, D. Lahiri, R. R. Patel, A. Agarwal, Nanoscratch behavior of carbon nanotube reinforced aluminum coatings. *Thin Solid Films* **518**, 1703–1711 (2010).
7. S. R. Bakshi et al., Carbon nanotube reinforced aluminum composite coating via cold spraying. *Surface and Coatings Technology* **202**, 5162–5169 (2008).
8. Y. Shimizu et al., Multi-walled carbon nanotube-reinforced magnesium alloy composites. *Scripta Materialia* **58**, 267–270 (2008).
9. T. Laha, A. Agarwal, T. McKechnie, S. Seal, Synthesis and characterization of plasma spray formed carbon nanotube reinforced aluminum composite. *Materials Science and Engineering A* **381**, 249 (2004).
10. M. W. Smith et al., Very long single and few-walled boron nitride nanotubes via the pressurized vapor/condenser method. *Nanotechnology* **20**, 505604 (2009).

11. A. Agarwal, S. R. Bakshi, D. Lahiri, *Carbon Nanotubes: Reinforced Metal Matrix Composites* (CRC Press, Taylor & Francis Group, 2011).
12. A. Nieto, A. Bisht, D. Lahiri, C. Zhang, A. Agarwal, Graphene reinforced metal and ceramic matrix composites: A review. *International Materials Reviews* **62**, 241 (2017).
13. Y. Chen, J. Zou, S. J. Campbell, G. Le Caer, Boron nitride nanotubes: Pronounced resistance to oxidation. *Applied Physics Letters* **84**, 2430–2432 (2004).
14. X. Wei, M.-S. Wang, Y. Bando, D. Golberg, Tensile tests on individual multi-walled boron nitride nanotubes. *Advanced Materials* **22**, 4895–4899 (2010).
15. Y. Huang et al., Thin boron nitride nanotubes with exceptionally high strength and toughness. *Nanoscale* **5**, 4840–4846 (2013).
16. D. Golberg et al., Direct force measurements and kinking under elastic deformation of individual multiwalled boron nitride nanotubes. *Nano Letters* **7**, 2146–2151 (2007).
17. R. Arenal, M.-S. Wang, Z. Xu, A. Loiseau, D. Golberg, Young modulus, mechanical and electrical properties of isolated individual and bundled single-walled boron nitride nanotubes. *Nanotechnology* **22**, 265704 (2011).
18. D.-M. Tang et al., Mechanical properties of bamboo-like boron nitride nanotubes by in situ TEM and MD simulations: Strengthening effect of interlocked joint interfaces. *ACS Nano* **5**, 7362–7368 (2011).
19. J. Garel et al., Ultrahigh torsional stiffness and strength of boron nitride nanotubes. *Nano Letters* **12**, 6347–6352 (2012).
20. P. Nautiyal et al., Oxidative unzipping and transformation of high aspect ratio boron nitride nanotubes into "white Graphene Oxide" platelets. *Scientific Reports* **6**, 29498 (2016).
21. D. Golberg, Y. Bando, K. Kurashima, T. Sato, Synthesis and characterization of ropes made of BN multiwalled nanotubes. *Scripta Materialia* **44**, 1561–1565 (2001).
22. M. J. Tank et al., Extreme thermal stability and dissociation mechanisms of purified boron nitride nanotubes: Implications for high-temperature nanocomposites. *ACS Applied Nano Materials* **5**, 12444–12453 (2022).
23. A. Bisht et al., Effect of warm rolling and annealing on the mechanical properties of aluminum composite reinforced with boron nitride nanotubes. *Materials Science and Engineering A* **710**, 366–373 (2018).
24. S. K. Singhal, A. K. Srivastava, R. Pasricha, R. B. Mathur, Fabrication of Al-matrix composites reinforced with amino functionalized boron nitride nanotubes. *Journal of Nanoscience and Nanotechnology* **11**, 5179–5186 (2011).
25. M. M. H. Bhuiyan, et al., Boron nitride nanotube reinforced titanium metal matrix composites with excellent high-temperature performance. *Journal of Materials Research* **32**, 3744–3752 (2017).
26. D. Lahiri, et al., Boron nitride nanotubes reinforced aluminum composites prepared by spark plasma sintering: Microstructure, mechanical properties and deformation behavior. *Materials Science & Engineering A* **574**, 149–156 (2013).
27. M. Yamaguchi, F. Meng, K. Firestein, K. Tsuchiya, D. Golberg, Powder metallurgy routes toward aluminum boron nitride nanotube composites, their morphologies, structures and mechanical properties. *Materials Science & Engineering A* **604**, 9–17 (2014).
28. P. Nautiyal et al., Interface engineering and direct observation of strengthening behavior in field-sintered boron nitride nanotube-magnesium alloy composite. *Advanced Engineering Materials* **22**, 2000170 (2020).
29. P. Nautiyal et al., Directionally aligned ultra-long boron nitride nanotube induced strengthening of aluminum-based sandwich composite. *Advanced Engineering Materials* **18**, 1747–1754 (2016).
30. J. Bustillos et al., Boron nitride nanotube–reinforced titanium composite with controlled interfacial reactions by spark plasma sintering. *Advanced Engineering Materials* **22**, 2000702 (2020).
31. P. Nautiyal, et al., In situ investigation of deformation mechanisms induced by boron nitride nanotubes and nanointerphases in Ti–6Al–4V alloy. *Advanced Engineering Materials* **24**, 2200610 (2022).
32. Y. Xue, et al., Aluminum matrix composites reinforced with multi-walled boron nitride nanotubes fabricated by a high-pressure torsion technique. *Materials & Design* **88**, 451–460 (2015).
33. J. Q. Xu, L. Y. Chen, H. Choi, X. C. Li, Theoretical study and pathways for nanoparticle capture during solidification of metal melt. *Journal of Physics Condensed Matter* **24**, 255304 (2012).

34. M. J. Assael, et al., Reference data for the density and viscosity of liquid aluminum and liquid iron. *Journal of Physical and Chemical Reference Data* **35**, 285–300 (2006).
35. P. Nautiyal, A. Gupta, S. Seal, B. Boesl, A. Agarwal, Reactive wetting and filling of boron nitride nanotubes by molten aluminum during equilibrium solidification. *Acta Materialia* **126**, 124–131 (2017).
36. W. Martienssen, H. Warlimont, Springer *Handbook of Condensed Matter and Materials Data* (Springer, 2005).
37. K. Yum, M. F. Yu, Measurement of wetting properties of individual boron nitride nanotubes with the Wilhelmy method using a nanotube-based force sensor. *Nano Letters* **6**, 329–333 (2006).
38. P. Shen, H. Fujii, K. Nogi, Effect of temperature and surface roughness on the wettability of boron nitride by molten Al. *Journal of Materials Science* **42**, 3564–3568 (2007).
39. X. M. Xue, J. T. Wang, M. X. Quan, Wettability and spreading kinetics of liquid aluminium on boron nitride. *Journal of Materials Science* **26**, 6391–6395 (1991).
40. G. Rosazza Prin, T. Baffie, M. Jeymond, N. Eustathopoulos, Contact angles and spreading kinetics of Al and Al-Cu alloys on sintered AlN. *Materials Science and Engineering A* **298**, 34–43 (2001).
41. Y. V Naidich, N. Y. Taranets, Wettability of aluminium nitride by tin-aluminium melts. *Journal of Materials Science* **33**, 3993–3997 (1998).
42. S. Bhusal, et al., A computational approach for predicting microstructure and mechanical properties of plasma sprayed ceramic coatings from powder to bulk. *Surface and Coatings Technology* **374**, 1–11 (2019).
43. P. Nautiyal, C. Zhang, B. Boesl, A. Agarwal, Non-equilibrium wetting and capture of boron nitride nanotubes in molten aluminum during plasma spray. *Scripta Materialia* **151**, 71–75 (2018).
44. S. Sampath, H. Herman, Rapid solidification and microstructure development during plasma spray deposition. *Journal of Thermal Spray Technology* **5**, 445 (1996).
45. S. R. Bakshi, V. Singh, S. Seal, A. Agarwal, Aluminum composite reinforced with multiwalled carbon nanotubes from plasma spraying of spray dried powders. *Surface and Coatings Technology* **203**, 1544–1554 (2009).
46. C. J. Todaro et al., Grain structure control during metal 3D printing by high-intensity ultrasound. *Nature Communications* **11**, 142 (2020).
47. D. G. Eskin, J. Mi, Eds., *Solidification Processing of Metallic Alloys under External Fields* (Springer International Publishing, 2018).
48. C. Ruirun et al., Effects of ultrasonic vibration on the microstructure and mechanical properties of high alloying TiAl. *Scientific Reports* **7**, 41463 (2017).
49. D. Sui, Q. Han, Ultrasound-assisted cast-on method: Obtaining high-quality metallurgical bonds between a bare steel insert and A354 aluminum alloy within a composite casting. *Journal of Materials Processing Technology* **311**, 117783 (2023).
50. G. I. Eskin, *Ultrasonic Treatment of Light Alloy Melts* (CRC Press, 1998).
51. K. S. Suslick, W. L. Nyborg, Ultrasound: Its chemical, physical and biological effects. *The Journal of the Acoustical Society of America* **87**, 919–920 (1990).
52. X. Li, Y. Yang, X. Cheng, Ultrasonic-assisted fabrication of metal matrix nanocomposites. *Journal of Materials Science* **39**, 3211–3212 (2004).
53. N. Balasubramani, et al., An overview and critical assessment of the mechanisms of microstructural refinement during ultrasonic solidification of metals. *Ultrasonics Sonochemistry* **89**, 106151 (2022).
54. S. Mohammed, P. Nautiyal, T. Paul, C. Zhang, A. Agarwal, A critique on boron nitride nanotube reinforced metal matrix composites. *Critical Reviews in Solid State and Materials Sciences* **49**, 482–519 (2023).
55. O. V. Abramov, Action of high intensity ultrasound on solidifying metal. *Ultrasonics* **25**, 73–82 (1987).
56. G. I. I. Eskin, Broad prospects for commercial application of the ultrasonic (cavitation) melt treatment of light alloys. *Ultrasonics Sonochemistry* **8**, 319–325 (2001).
57. S. M. A. K. Mohammed, T. Paul, D. John, C. Zhang, A. Agarwal, Understanding the role of ultrasonic cavitation assisted casting of boron nitride nanotube-reinforced aluminum matrix composite. *Journal of Materials Research and Technology* **25**, 2405–2418 (2023).
58. B. Palacios et al., Role of structural hierarchy on mechanics and electrochemistry of wire arc additive manufactured (WAAM) single phase titanium. *Journal of Manufacturing Processes* **93**, 239–249 (2023).

59. Q. Chao et al., Nanoparticle-mediated ultra grain refinement and reinforcement in additively manufactured titanium alloys. *Additive Manufacturing* **46**, 102173 (2021).
60. S. M. A. K. Mohammed, A. A. Aleman, D. John, T. Paul, A. Agarwal, Exploring the potential of wire fed direct energy deposition of aluminum–boron nitride nanotube composite: Microstructural evolution and mechanical properties. *Advanced Engineering Materials* **25**, 1438–1656 (2023).
61. S. M. A. K. Mohammed et al., Deformation behavior and strengthening mechanisms in a CNT-reinforced bimodal-grained aluminum matrix nanocomposite. *Materials Science and Engineering: A* **817**, 141370 (2021).
62. B. Chen et al., Length effect of carbon nanotubes on the strengthening mechanisms in metal matrix composites. *Acta Mater* **140**, 317–325 (2017).
63. R. George, K. T. T. Kashyap, R. Rahul, S. Yamdagni, Strengthening in carbon nanotube/aluminium (CNT/Al) composites. *Scripta Materialia* **53**, 1159–1163 (2005).
64. M. D. Hayat, H. Singh, Z. He, P. Cao, Titanium metal matrix composites: An overview. *Composites Part A: Applied Science and Manufacturing* **121**, 418 (2019).
65. A. Kulkarni, D. Tauber, T. Y. Ansell, Reinforcing cold-sprayed Al coatings with boron nitride nanotubes and micro-boron carbide and their effect on surface mechanical and dry sliding behavior. *Journal of Tribology* **145**, 121402 (2023).
66. A. K. Sukumaran et al., Tribological and radiation shielding response of novel titanium-boron nitride coatings for lunar structural components. *Surface and Coatings Technology* **476**, 130300 (2024).
67. N. Bacca et al., Tribological and neutron radiation properties of boron nitride nanotubes reinforced titanium composites under lunar environment. *Journal of Materials Research* **37**, 4582–4593 (2022).
68. B. D. Beake, The influence of the H/E ratio on wear resistance of coating systems – insights from small-scale testing. *Surface and Coatings Technology* **442**, 128272 (2022).
69. S. A. Thibeault, et al., Nanomaterials for radiation shielding. *MRS Bulletin* **40**, 836–841 (2015).
70. S. T. Abdulrahman, Z. Ahmad, S. Thomas, A. A. Rahman, "Introduction to neutron-shielding materials" in S. T. Abdulrahman, S. Thomas, Z. Ahmad (eds) *Micro and Nanostructured Composite Materials for Neutron Shielding Applications* (Elsevier, 2020), pp. 1–23.
71. K. Orikasa, et al., Foam with direction: Unraveling the anisotropic radiation shielding properties of 2D boron nitride nanoplatelet foams. *Npj 2D Materials and Applications* **8**, 13 (2024).
72. E. Cheraghi, S. Chen, J. T. W. Yeow, Boron nitride-based nanomaterials for radiation shielding: A review. *IEEE Nanotechnology Magazine* **15**, 8–17 (2021).
73. M. B. Stone, L. Crow, V. R. Fanelli, J. L. Niedziela, Characterization of shielding materials used in neutron scattering instrumentation. *Nuclear Instruments and Methods in Physics Research A* **946**, 162708 (2019).
74. R. S. Mishra, R. S. Haridas, P. Agrawal, Friction stir-based additive manufacturing. *Science and Technology of Welding and Joining* **27**, 141–165 (2022).
75. Y. Zou, Cold spray additive manufacturing: Microstructure evolution and bonding features. *Accounts of Materials Research* **2**, 1071–1081 (2021).
76. A. Agarwal, C. Zhang, T. Paul, S. Mohammed, D. John, Boron nitride nanotube aluminum composites and methods of manufacturing the same. U.S. Patent, US12005497B2, Jun. 11, 2024.
77. A. Agarwal, B. Boesl, P. Nautiyal, C. Rudolf, C. Zhang, Aluminum-boron nitride nanotube composites and method for making the same. U.S. Patent, US11148201B2 Oct. 19, 2021.
78. A. Agarwal, P. Nautiyal, B. Boesl, Boron nitride nanotube-magnesium alloy composites and manufacturing methods thereof. U.S. Patent, US11131007B1, Sep. 28, 2021.

Section II

Properties in Focus

A Spectrum of Possibilities and Impacts

Chapter 5

Structures and Properties of Boron Nitride Nanomaterials

Feng Liu and Zhenguo Huang

5.1 INTRODUCTION

Nanomaterials are defined by having at least one dimension ranging from 1 to 100 nm in three-dimensional space. Due to their small size, quantum effects, and surface effects, nanomaterials often exhibit unique physical and chemical properties [1]. Since the discovery of carbon nanotubes (CNTs) in 1991 and the expansion of graphene-based nanomaterials after 2004, low-dimensional nanomaterials have attracted widespread attention in both scientific and industrial communities [2]. Their applications continue to evolve in fields such as electronics, optics, energy storage and conversion, and environmental remediation, significantly enhancing human life.

Among these materials, BN stands out as a well-known ceramic material composed of equal amounts of boron (B) and nitrogen (N) atoms. As elements adjacent to carbon in the periodic table, BN is isostructural with similarly structured carbon lattices and exists in various allotropes [3]. A common and stable polymorph is hexagonal BN (*h*-BN). There are many structural and property similarities between *h*-BN and graphene, leading to *h*-BN being referred to as 'white graphene'. *h*-BN exhibits high thermal conductivity, a wide bandgap, strong ultraviolet (UV) emission, excellent oxidation resistance, good mechanical strength, and chemical inertness, making it a subject of interest in materials science, electronics, energy, and environmental research [4].

5.2 STRUCTURES OF BN NANOMATERIALS

BN crystal structures can be categorized into four types based on the hybridization modes between B and N: *h*-BN, cubic BN (*c*-BN), wurtzite BN (*w*-BN), and rhombohedral BN (*r*-BN) [5]. As illustrated in Figure 5.1a–d, *h*-BN and *r*-BN have structures similar to graphite, while *c*-BN and *w*-BN are analogous to diamond [6]. *h*-BN features a stacked lamellar structure with sp^2-hybridized B and N atoms in each layer, as shown in Figure 5.1a. It possesses a hexagonal crystalline lattice with a D_6h point group and a $P6_3/mmc$ space group [7,8]. The stacking sequence in *h*-BN follows an AA' pattern, where each N atom is positioned between two B atoms in adjacent layers and vice versa [8]. Compared to graphite's 0.334 nm atomic layer spacing, *h*-BN has

DOI: 10.1201/9781003647812-7

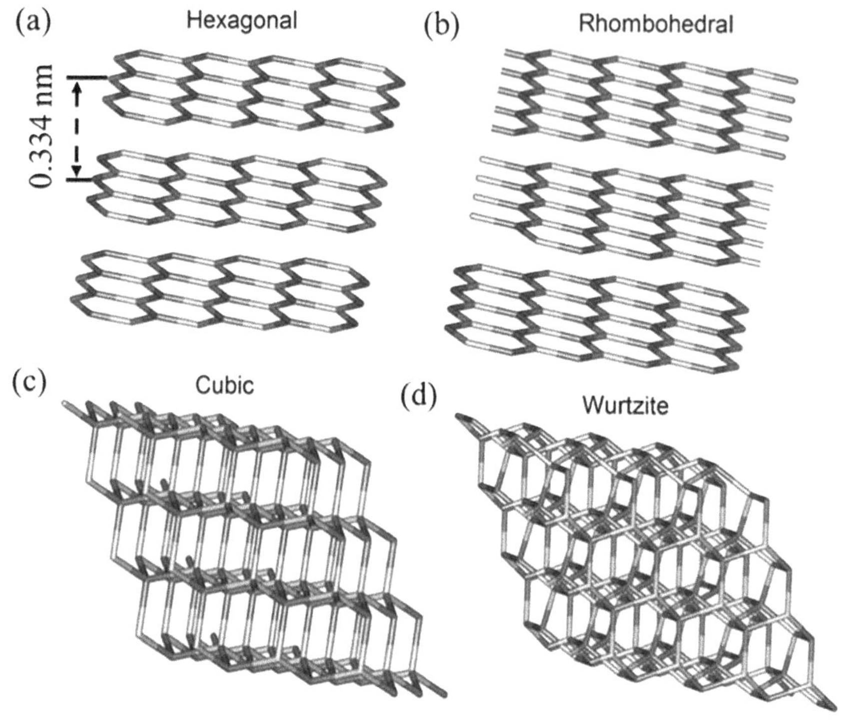

Figure 5.1 Bulk BN with different crystal structures: (a) hexagonal, (b) rhombohedral, (c) cubic, and (d) wurtzite. (Reproduced with permission Ref. [6]. Copyright 2014, The Royal Society of Chemistry.)

a slightly smaller interlayer distance of 0.333 nm [9]. *r*-BN also exhibits sp^2 hybridization, but its honeycomb structure shows regular inclined stacking, as depicted in Figure 5.1b. Figure 5.1c displays the cubic structure of *c*-BN, consisting of sp^3-hybridized B and N atoms, forming a robust 3D network due to the alternating arrangement of B and N atoms.

The alternating arrangement of B and N atoms forms a vast 3D tetrahedral network in *c*-BN, with B-N-B or N-B-N bonds exhibiting a covalent nature and a bond angle of 109° [10]. This closely packed structure gives *c*-BN its renowned super hardness. *w*-BN, depicted in Figure 5.1d, is a metastable phase of BN that can be produced by compressing *h*-BN under high pressure. Unlike *h*-BN, the B and N atoms in *w*-BN are tetrahedrally coordinated, forming an alternating arrangement of tightly packed layers in an AA-BB-AA sequence [11]. The bond length in *h*-BN is approximately 0.147 nm, whereas the B–N bond length in *w*-BN extends to about 0.157 nm with a bond angle of 109.5° [12]. Table 5.1 presents the crystal structural data for various BN polymorphs for comparative purposes. Due to their stability and abundance, research has primarily focused on *h*-BN and *c*-BN, particularly *h*-BN nanomaterials, which have demonstrated exceptional performance across various applications.

As an isoelectronic form of carbon, *h*-BN, also known as white graphite, is composed of atomically flat BN layers held together by van der Waals interactions. Monolayered *h*-BN is isostructural to graphene, featuring sp^2 hybridized, strong covalent, and polarized boron–nitrogen (B–N) bonds along the plane. Due to the difference in electronegativity between boron (2.04) and nitrogen (3.04), the B–N bond is strongly polarized with an electronegative nitrogen, leading to strong

TABLE 5.1 THE CRYSTAL STRUCTURE PARAMETERS OF BN

Crystal names	Crystal structures	Space group	Lattice constant	Hybridization modes
h-BN	Hexagonal	$P6_3/mmc$ (194)	a = b = 2.5044 Å, c = 6.6522 Å	sp^2
r-BN	Rhombohedral	R_3m (160)	a = b = 2.5042 Å, c = 9.99 Å	sp^2
c-BN	Cubic	$F\text{-}4_3m$ (216)	a = b = c = 3.615 Å	sp^3
w-BN	Wurtzite	$P6_3/mc$ (186)	a = b = 2.5505 Å, c = 4.210 Å	sp^3

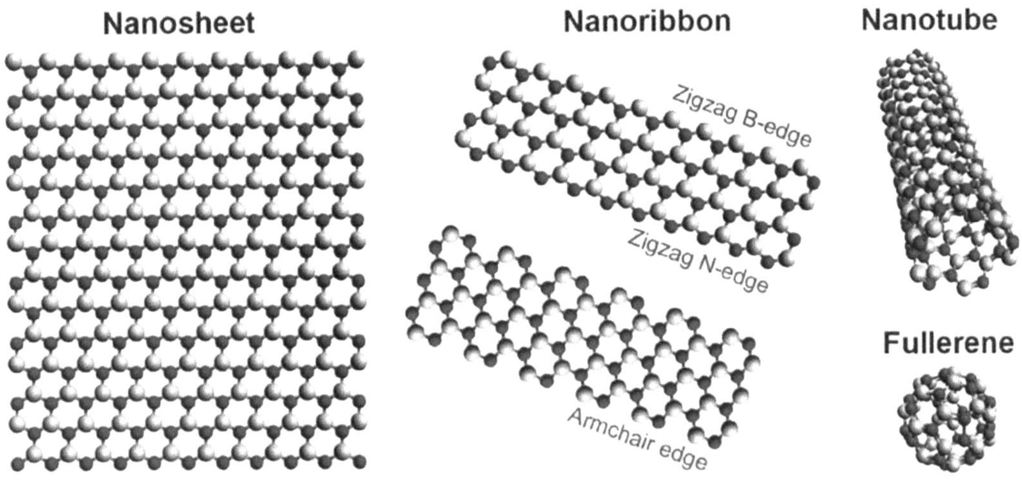

Figure 5.2 BNNS, nanoribbon (zigzag and armchair configurations), nanotube, and zero-dimensional fullerene. (Reproduced with permission Ref. [6]. Copyright 2014, The Royal Society of Chemistry.)

'lip-lip' interactions within individual BN layers and potential anisotropic properties [12]. It is noteworthy that low-dimensional BN nanomaterials, such as BNNSs, BN nanoribbons, BNNTs, and fullerenes or quantum dots, can be derived from bulk *h*-BN through further processing. Similar to graphene, *h*-BN is highly versatile and adaptable in various geometric configurations. BNNSs are single or few-layered BN sheets exfoliated from *h*-BN, and BNNTs can be viewed as a tubular structure formed by curling BNNSs [4]. BN nanoribbons are BNNSs with varying lengths but narrow widths, and depending on their orientation, the atoms can be arranged in zigzag or armchair conformations. Different morphological structure models of nanostructured BN are illustrated in Figure 5.2 [6].

5.2.1 One-Dimensional BNNTs

B–N bonds exhibit both ionic and covalent characteristics due to the disparity in electronegativity between boron and nitrogen atoms. This dual bonding nature facilitates 'lip-lip' interactions among layers, which stabilize the growth of open-ended BNNTs by forming bridges between consecutive shells, allowing them to evolve into a metastable energy minimum, similar to the growth mechanism of CNTs [13]. BNNTs, like CNTs, can be classified into single-walled nanotubes and multi-walled nanotubes based on the number of layers.

An ideal single-walled BNNT is a defect-free tubular structure formed by rolling *h*-BN layers into a seamless cylinder, typically with diameters ranging from a few to tens of nanometers [14]. Multi-walled BNNTs are composed of several or even dozens of coaxially nested single-walled BNNTs with different diameters, and the spacing between tubes is approximately 0.33–0.34 nm [15]. One-dimensional BNNTs typically exhibit a helical configuration known as chirality. However, BNNTs can be either achiral or chiral. In achiral nanotubes, the mirror image of the tube is consistent with the original configuration, as seen in armchair and zigzag BNNTs. These types can be classified based on the cross-sectional shape of the nanotube. Conversely, the mirror image of chiral tubes does not simply replicate the original configuration but rather reflects the specific helical symmetry of the tube [16].

The chiral properties and diameter of nanotubes can be determined by the chirality vector (C_h):

$$C_h = na_1 + ma_2(n,m)$$

where a_1 and a_2 are the primitive lattice vectors of the monolayered BNNS, as shown in Figure 5.3a [4]. The direction of a_1 is referred to as the sawtooth direction, and the angle between C_h and a_1 is the chiral angle θ. n and m are integers. Based on the chiral angle, nanotubes can be classified into zigzag type, armchair type, and helical tube. Figure 5.3b–d presents the models of a single-walled chiral BNNT in which its C_h values are (15, 0), (8, 8), and (8, 5), respectively. The diameter of the tube is defined as

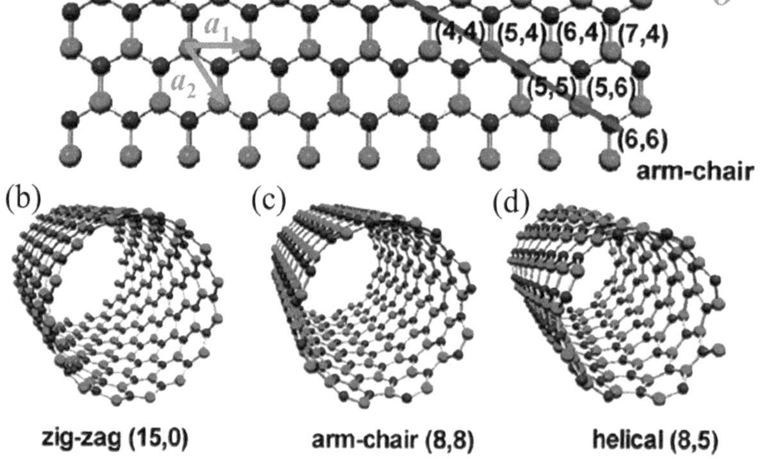

Figure 5.3 (a) Definition of BNNTs, (b) zigzag BNNT, (c) armchair BNNT, and (d) chiral BNNT. (Reproduced with permission Ref. [4]. Copyright 2010, American Chemical Society.)

$$d = \frac{C_1}{\pi} = a\frac{\sqrt{3\left(n^2 + m^2 + nm\right)}}{\pi}$$

where a is the bond length of B–N in the plane. The rotation angle θ, which is critical for representing the chirality of the tube, is defined as:

$$\cos\theta = \frac{C_h \cdot a_1}{|C_h||a_1|} = \frac{2n+m}{2\sqrt{n^2 + m^2 + nm}}$$

This equation uniquely defines the chirality of the tube.

Structurally, BNNTs differ from CNTs in their tendency to adopt specific chirality during the preparation process. While the synthesis of traditional CNTs typically results in a mix of different chiral tubes, including zigzag and armchair types, BNNTs exhibit a stronger propensity to form zigzag nanotubes rather than armchair-shaped ones. The concept of tension energy, defined as the energy difference between the nanotube and its sheet-like precursor structure [17], plays a key role in this tendency. Theoretical calculations suggest that while the magnitude of tension energy is independent of the chirality of BNNTs, the energy required to form armchair-shaped BNNTs is lower than that needed to form armchair-shaped BN nanoribbons of the same width [18]. Given that BNNTs can be considered as rolled-up BN nanoribbons and considering that tension energy does not vary with chirality, there is a natural tendency for BNNTs to adopt a zigzag configuration during their formation [19]. Furthermore, the end-cap structures of BNNTs also reflect this structural preference. Zigzag-typed BNNTs typically form flat-topped end caps, while armchair-type BNNTs tend to develop tapered end caps. Most chiral BNNTs, on the other hand, exhibit amorphous end-cap morphologies.

5.2.2 Two-Dimensional BNNSs

From a structural perspective, two-dimensional BNNSs can be envisioned as monolayer or few-layer materials exfoliated from bulk *h*-BN or obtained by unrolling single- or multi-walled BNNTs. For monolayer BNNSs, the edges may terminate in either a zigzag or armchair configuration. Theoretical calculations and experimental evidence suggest that armchair edge termination in *h*-BN nanoribbons is energetically more favorable than zigzag edge termination. However, when *h*-BN is grown on a copper substrate, nitrogen-terminated zigzag edges are more stable than boron-terminated edges, leading to the formation of equilateral BN triangles during the growth process [20,21]. Multilayer BNNSs can be viewed as single layers of *h*-BN stacked in specific configurations. Unlike the Bernal (AB) stacking of graphite, *h*-BN exhibits up to six distinct bilayer stacking configurations, as illustrated in Figure 5.4 [22,23]. The AA stacking of *h*-BN features two completely parallel atomic layers, though this arrangement is energetically unstable [22]. In AA' stacking, boron atoms are positioned directly above nitrogen atoms and vice versa [23]. In AC stacking, nitrogen atoms occupy the center of the hexagons in the second layer, differing from the AB configuration where boron atoms hold this position in *h*-BN [24]. Besides the AA' stacking of *h*-BN, Bernal stacked AB and AC configurations have been reported as thermodynamically stable structures for *h*-BN [11,25].

Identifying the number of layers in BNNSs is challenging due to the extremely thin nature of these sheets, which often results in crumpling or folding of the planes and edges. Transmission electron microscopy (TEM) is a state-of-the-art characterization tool that can directly observe

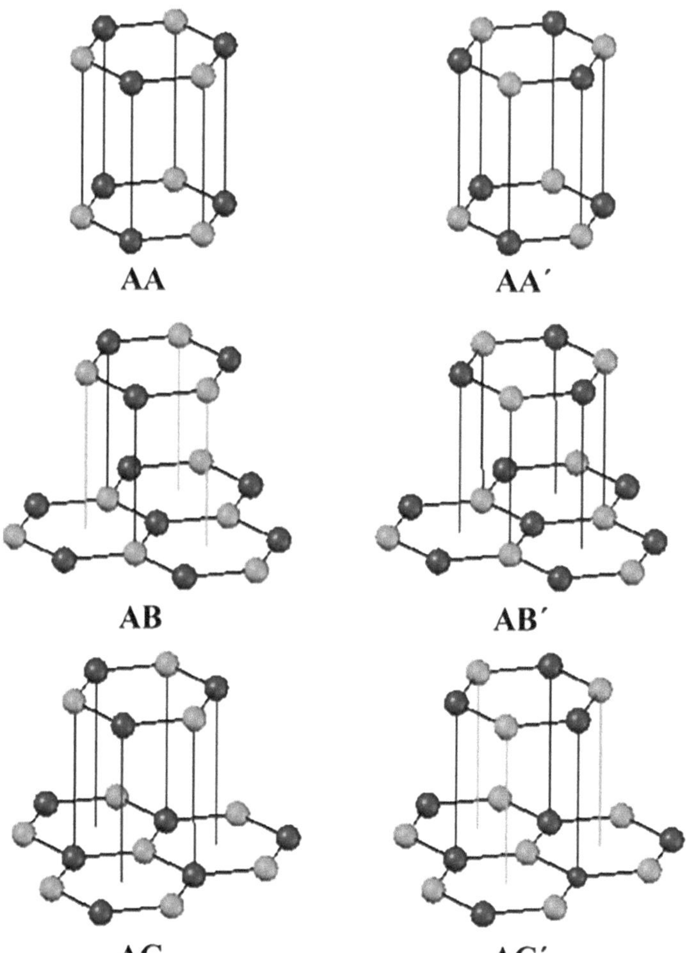

Figure 5.4 Six possible types of bilayer *h*-BN stacking [24].

the number of layers at crumpled and folded areas in high-resolution TEM mode [26]. The thickness of BNNSs can also be determined by calculating the d-spacing using TEM. However, when the number of layers exceeds 20, TEM becomes less effective for counting the layers. In such cases, atomic force microscopy (AFM) provides a more accurate measurement of thickness and roughness by constructing a topographic image of the sample surface in tapping mode [27]. Nevertheless, AFM can serve as a complementary approach, as the measured height of a monolayer often exceeds theoretical values due to the presence of trapped moisture, water, or impurities on the BNNS surfaces [28].

In addition to TEM and AFM, other techniques like optical microscopy, Raman spectroscopy, and annular dark field imaging in scanning TEM are also used to determine the number of layers in BNNSs [29–31]. Since the absorption wavelength of *h*-BN lies beyond the visible region, the layer form (mono- or few-layer) of BNNSs can be identified by their optical contrast. Raman spectroscopy is particularly useful for this purpose, as BNNSs and BNNTs share similar Raman spectral characteristics, both being derived from bulk *h*-BN. The E_{2g} Raman mode of bulk *h*-BN appears around 1366 cm^{-1}. However, reducing the thickness of BNNSs results in a shift of the E_{2g} band, blue-shifted for monolayer BN and red-shifted for few-layer BNNSs [30]. For instance, a single-walled BNNT exhibits an E_{2g} band at 1370 cm^{-1} due to its reduced size [31]. Annular dark field–scanning TEM imaging is also employed to identify individual B and N atoms and to determine the layer thickness of BNNSs based on annular dark field intensity [32].

5.3 PROPERTIES OF BN NANOMATERIALS

h-BN nanomaterials have garnered significant attention from both scientific and technological perspectives, largely due to their highly anisotropic properties. The strong covalent bonding and affinity between boron (B) and nitrogen (N) confer *h*-BN with several outstanding characteristics, including high anisotropy, mechanical strength, thermal insulation, large thermal conductivity, excellent oxidation resistance, and remarkable thermal and chemical stability [4]. For instance, *h*-BN decomposes only at 1000°C in air, 1400°C in a vacuum, and nearly 2850°C in an inert atmosphere [10]. Moreover, *h*-BN is inert to both strong acids and bases, making it difficult to functionalize or form heterostructures—unlike its counterpart, graphite, which can easily be converted to graphene oxide via Hummer's or modified Hummer's methods [33,34]. With a pore diameter of just 0.012 nm in its hexagonal structure, BNNSs have been shown to be impermeable to argon atoms [35], making *h*-BN an ideal candidate for protective layers in hazardous and corrosive environments. Additionally, *h*-BN's chemical inertness contributes to its biocompatibility, making it suitable for biomedical applications. This section will further explore the specific properties of BNNTs and BNNSs.

5.3.1 Mechanical Properties

The mechanical properties of *h*-BN nanomaterials have been extensively studied both theoretically, using density functional theory (DFT), and experimentally, through techniques like TEM and AFM. However, mechanical properties such as elastic modulus and tensile strength can vary depending on factors like material defect density, morphology, thickness, chirality, shear rate, and temperature. Theoretical studies indicate that BNNTs possess a high elastic modulus, ranging from 0.837 to 0.912 TPa, which is slightly lower than that of CNTs, which range from 1.22 to 1.25 TPa [36,37]. While the activation energy for defect creation in BNNTs is relatively low compared to CNTs, their formation energy is higher [38], making CNTs stronger at room temperature, whereas BNNTs are more stable and somewhat stronger at elevated temperatures [39].

The mechanical properties of BNNTs also differ based on the preparation method, resulting in varying defect densities and corresponding variations in experimental results. For instance, Chopra and Zettl observed and calculated the thermal vibration amplitude of a single cantilever multi-walled BNNT under TEM, yielding an axial elastic modulus of 1.22±0.24 TPa, which is higher than Hernandez's theoretical calculations [40]. In another study, Golberg et al. used a piezoelectric-driven support device combining TEM and AFM to directly measure the elastic modulus of multi-walled BNNTs, finding values ranging from 0.5 to 0.6 TPa [41]. Additionally, Golberg's experimental results also demonstrated that BNNTs exhibit good elastic-plastic deformation capabilities, alongside their high tensile strength and elastic modulus [42].

The structural similarities between BNNSs and graphene offer an excellent opportunity to systematically study and compare their mechanical properties. It has been observed that the

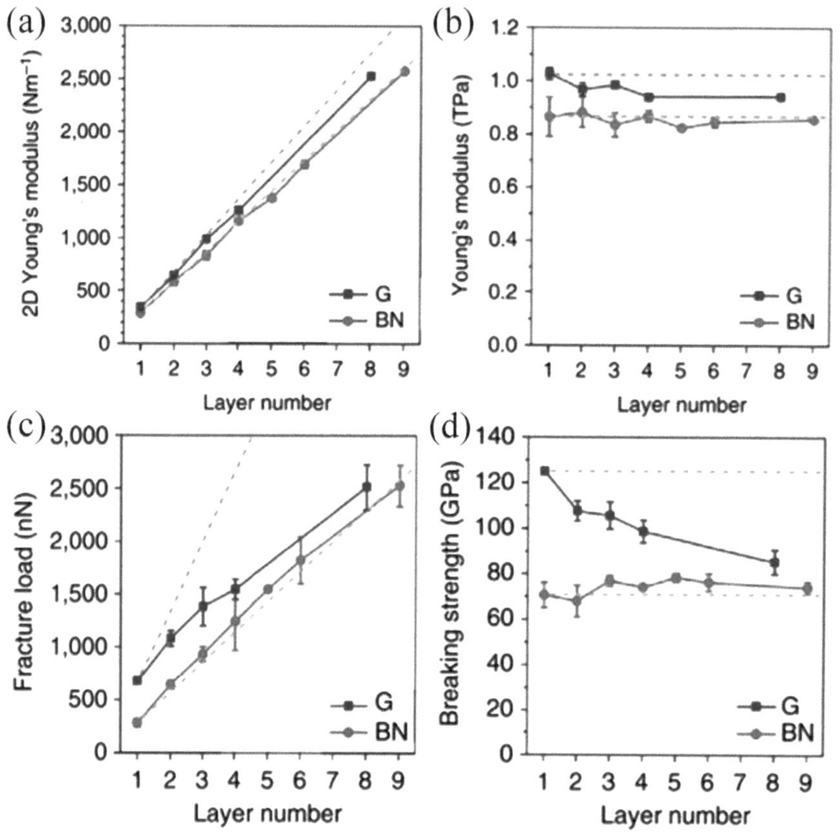

Figure 5.5 (a) 2D Young's modulus, (b) volumetric Young's modulus, (c) fracture load, and (d) breaking strength of graphene (blue line) and *h*-BN nanosheets (red line) of different thicknesses. (Reproduced with permission Ref. [43]. Copyright 2017, Springer Nature.)

thickness and defect levels significantly influence the Young's modulus of both graphene and *h*-BN. As illustrated in Figure 5.5, the Young's modulus, fracture load, and breaking strength of graphene are slightly higher than those of *h*-BN [43]. Experimental studies have established that the elastic modulus of graphene ranges between 0.5 and 2.4 TPa, with a commonly accepted value around 1 TPa (normalized Young's modulus by a thickness of 0.335 nm, $E_{2D}=342\,N\,m^{-1}$). As the number of layers in graphene increases from 1 layer (1L) to 9 layers (9L), the Young's modulus decreases from 1.026 to 0.942 TPa [43]. In contrast, *h*-BN exhibits nonlinear elastic deformation under tension [44,45].

DFT predicts that the Young's modulus of monolayer *h*-BN falls between 0.716 and 0.977 TPa ($E_{2D}=279.2\,N\,m^{-1}$). The elastic modulus of BNNSs with a thickness of 1–2 nm (2–5 layers) has been theoretically and experimentally determined to range from 184 to 510 N m^{-1} using AFM nanoindentation techniques [46–48]. However, the experimental bending modulus of 1–2 nm thick BNNSs (8.8–16 N m^{-1}) is significantly lower than the theoretical value (31.2 N m^{-1}), possibly due to stacking faults or the presence of defects in chemical vapor deposition-grown BNNSs [48]. In comparison, high-quality monolayer *h*-BN prepared by tape exfoliation, which has a lower defect density, shows a Young's modulus of 0.865 TPa in similar tests.

Kong et al. measured a Young's modulus of 1.16±0.1 TPa in a chemical vapor deposition-grown BNNS with a thickness of 15 nm (approximately 45 layers) using AFM [49]. This multilayer *h*-BN film, characterized by a large area and intact structure, exhibits high mechanical strength. At a load of 2400 N m^{-1}, with a strain of 0.35% and stress of 37 GPa, an elastic modulus of 18,000 N m^{-1} was recorded for the *h*-BN film. Although the predicted Young's modulus of the BNNS is isotropic, its fracture strength and strain are anisotropic. Due to this anisotropy, the fracture strength of the BNNS varies between 68 and 215 GPa [50]. Additionally, the relationship between thickness and fracture strain, as well as strength, is similar to that of the Young's modulus. The bending modulus of *h*-BN increases as the thickness of the thin film decreases [48].

5.3.2 Electronic Properties

Graphene is an excellent conductor of electricity, whereas *h*-BN acts as an insulator due to its wide bandgap, which ranges from approximately 5.2–7.0 eV for bulk *h*-BN [51]. However, there has been ongoing debate regarding the electronic band structure of *h*-BN. Figure 5.6a illustrates the relationship between the energy of electronic states and their wave vectors in bulk *h*-BN. The conduction band minimum is aligned with the valence band maximum, exhibiting a high symmetry transition along the K-point, which suggests that BN possesses a direct bandgap [8]. In contrast, Figure 5.6b shows a band structure of *h*-BN with indirect transitions. Here, the transition of electrons is not highly symmetric. The valence band maximum is aligned with K, while the minimum energy of the conduction band is characterized by the M momentum vector. This indicates that any transition requires changes in both momentum and energy [52].

Early theoretical work, employing the tight binding approximation, studied the electronic structure of BNNTs, establishing that BNNTs are semiconductors with either direct or indirect bandgaps. For example, local DFT calculations for (10, 0) zigzag and (10, 10) armchair single-walled BNNTs demonstrated that (10, 0) tubes are direct bandgap semiconductors, while (10, 10) tubes have indirect bandgaps, both around 4 eV, which is slightly less than the experimental values [18]. This discrepancy is mainly due to the traditional DFT method's limitations in accurately describing the unoccupied states of semiconductors. As the diameter of BNNTs decreases, the bandgap of zigzag BNNTs decreases rapidly, while the bandgap of armchair-type BNNTs remains relatively unchanged [18].

Moreover, the bandgap of *h*-BN is strongly influenced by the number of layers and stacking. A transition from an indirect to a direct bandgap has been observed when scaling bulk *h*-BN down to a monolayer of 2D *h*-BN [53]. As depicted in Figure 5.6c, a direct-to-indirect crossover occurs at a high symmetry point when the thickness exceeds a single monolayer. This phenomenon arises because the conduction band at the M point is highly sensitive to interlayer coupling. Variations in the number of layers, stacking order, and interlayer separation can significantly alter the band structure [53,54].

In addition to these factors, the magnitude of the bandgap is also related to size rather than the symmetry of transitions. Due to quantum confinement and changes in the electron effective mass, the gap between the valence band maximum along the K-Γ line and the conduction band minimum at the M point widens from 5.95 to 6.47 eV as *h*-BN is reduced from bulk to a monolayer [53]. However, this increase is only significant when the level of defects or impurities is negligible. In reality, defects and functional groups can disrupt the electronic state composition, making such increases in bandgap uncommon [55]. For example, cutting two-dimensional BN atomic layers into one-dimensional nanoribbons introduces boundary states that result in rich electrical and magnetic properties. In the case of zigzag BN nanoribbons with all boundaries saturated

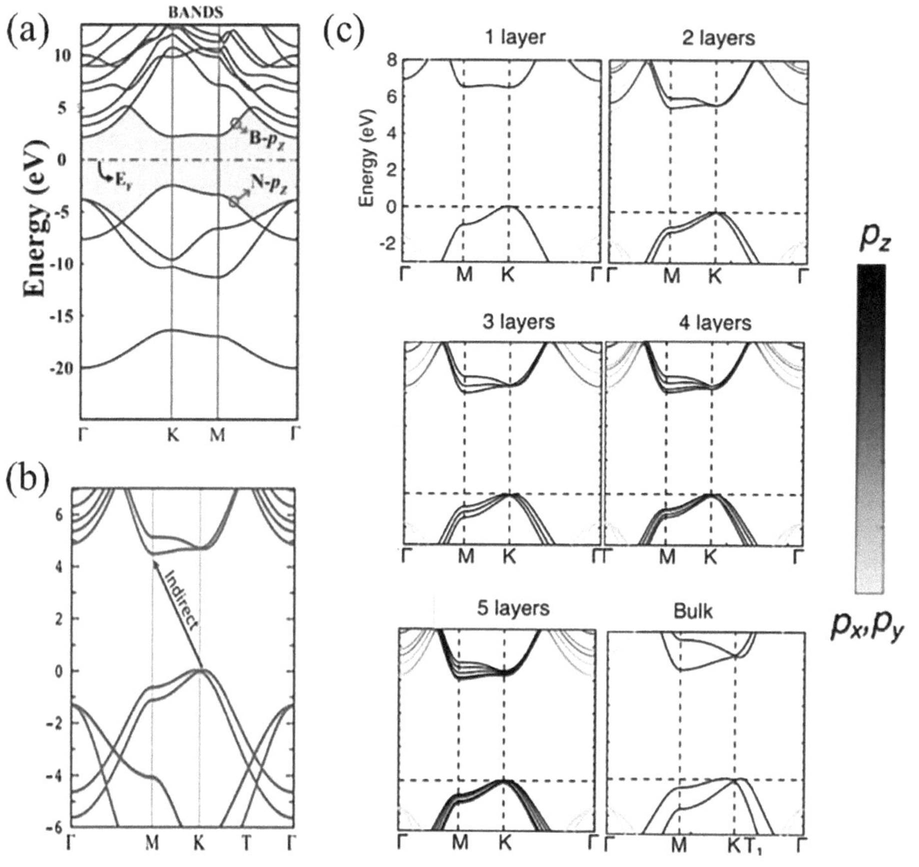

Figure 5.6 (a) Electronic band structure of bulk *h*-BN showing the direct transition along with K. (Reproduced with permission Ref. [8]. Copyright 2009, American Physical Society.) (b) Band structure of *h*-BN revealing the indirect band gap along K-M. (Reproduced with permission Ref. [52]. Copyright 2016, American Chemical Society.) (c) Band structure of *h*-BN in AA' stacking configuration. (Reproduced with permission Ref. [53]. Copyright 2018, American Chemical Society.)

with hydrogen atoms, the hydrogen-saturated zigzag BN nanoribbons are indirect bandgap semiconductors with a bandgap greater than 4.0 eV [56].

Additionally, introducing defects into single-layer BN materials is an effective way to modulate their electronic properties. Defects, which are difficult to avoid during the growth and preparation of crystal materials, play a crucial role in determining the properties of nanomaterials. Traditional materials often exhibit localized defects, such as vacancy defects and Stone–Wales defects. However, if line defect structures are introduced into single-layer BN materials, these defects generate paired defect states within the bandgap [57]. Unlike the localized electronic states caused by point defects, the impurity states resulting from line defects are extended states, where occupied and unoccupied defect states belong to different defects. This unique electronic structure enables the effective separation of electrons and holes, with conduction occurring along different line defects. Heteroatom doping and the introduction of atomic groups have also been reported to modulate surface electronic states and adjust the bandgap [58–60].

5.3.3 Optical Properties

Due to the large bandgap of bulk *h*-BN material, UV or deep UV (DUV) luminescence is observed when the sample is excited by light or an electron beam. The wavelength of luminescence depends on the state of the sample. Pure *h*-BN exhibits emission around 225 nm, attributed to bandgap exciton recombination. Figure 5.7a shows an example of *h*-BN absorption and luminescence at 8 and 83 K, respectively. In Figure 5.7b, absorption peaks at 207.5, 207.7, and 208.5 nm correspond to 5.985, 5.962, and 5.945 eV, respectively. The luminescence of *h*-BN is primarily generated by a series of s-like exciton absorption bands. The dashed line in Figure 5.7b represents the Elliot step, connecting the discrete energy levels of the band structure [61].

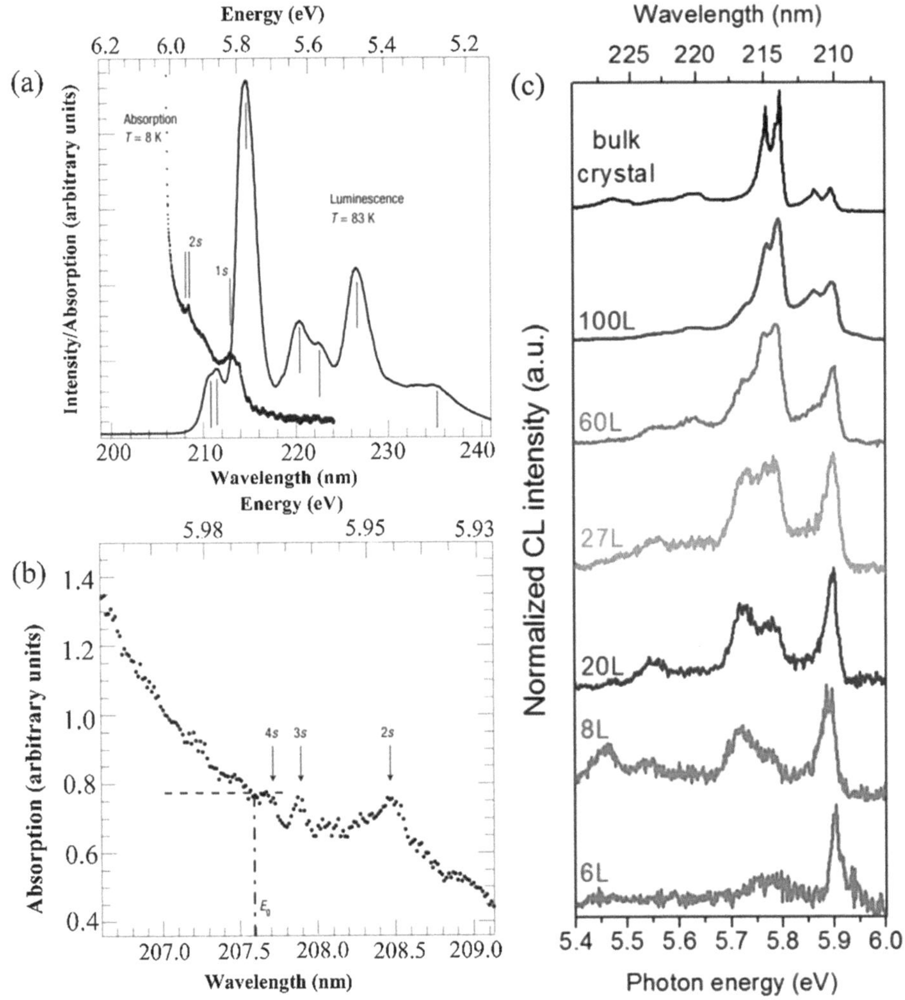

Figure 5.7 (a) Absorption and luminescence spectra of *h*-BN at 8 and 83 K, respectively. (Reproduced with permission Ref. [61]. Copyright 2004, Springer Nature.) (b) Absorption spectrum around the 2 s-like exciton structure. (Reproduced with permission Ref. [61]. Copyright 2004, Springer Nature.) (c) Cathodoluminescence spectra of bulk *h*-BN and BNNSs with different thicknesses [66].

In the visible light region (380–750 nm), *h*-BN does not exhibit any absorption peaks, resulting in a low-concentration suspension and a white appearance in the solid state [62]. A few-layered BNNS is highly transparent, with over 99% transparency from near-UV to visible light (250–900 nm) [47]. A sharp absorption for the BNNS is observed between 210–250 nm [61]. A monolayer BNNS absorbs at approximately 200 nm, while the absorption wavelength shifts slightly for a few-layered BNNS [47]. Though no precise correlation between layer thickness and UV absorption has been reported, monolayer *h*-BN demonstrates a relatively higher bandgap (6.07 eV) than multilayer or bulk *h*-BN due to the absence of interlayer interactions [63].

The strong UV emission of a BNNS makes it a promising material for DUV photodetection devices [64,65]. A few-layered BNNS has been applied in high-performance DUV photodetectors, with an on/off ratio greater than 10^3 under DUV light illumination at 212 nm [65]. In addition to photoluminescence, a BNNS also exhibits strong cathodoluminescence, which is influenced by defect centers in the BNNS. As shown in Figure 5.7c, cathodoluminescence intensity is thickness-dependent, decreasing as the thickness is reduced [66]. Furthermore, a BNNS displays nonlinear optical properties, such as two-photon absorption, making it suitable for sensitive optical components or laser protection [67].

For a BNNT, extensive theoretical and experimental research has examined its luminescence and absorption characteristics. Recent studies show that BNNT absorption is closely related to exciton effects [68,69], though much theoretical research on BNNTs' optical properties focuses on band structure calculations [70]. Luminescence and absorption spectra of a BNNT depend on its chirality and external electric field [71]. Laurent et al. studied single-wall BNNTs synthesized via laser ablation, observing two optical transitions at 4.45 and 5.5 eV [72]. These transitions likely occur between the van Hove singularities of the one-dimensional density of states, though the 4.45 eV peak may also result from exciton effects. Photoluminescence and cathodoluminescence studies suggest that BNNTs are effective UV emitters, with emission peaks at 230, 279, 338, and 460 nm, depending on exciton effects, defect levels, and functional groups [71].

5.3.4 Thermal Properties

The ionic nature of the B–N bonds in *h*-BN significantly reduces its decomposition temperature. Bulk *h*-BN begins to decompose at 3000°C in air, whereas BNNTs are stable only up to 1000°C–1700°C. Monolayer *h*-BN oxidizes at 700°C and is fully burned at 800°C [73]. However, *h*-BN nanomaterials still possess much greater antioxidation capacity than carbon-based materials. Golberg et al. found that oxidation of CNTs started from 550°C and was completed by 700°C, whereas BNNTs were oxidized at 900°C–950°C, forming boron oxide [74]. Both BNNTs and BNNSs remain stable in air up to 700°C–850°C [73]. High-crystallinity, low-defect BNNTs experienced oxidation of the tube cap at 900°C. A 5 nm thick BNNS was reported to resist oxidation up to 1100°C on a Ni substrate and up to 500°C on a Cu substrate in a 300 mTorr oxygen atmosphere for 30 minutes [75].

With strong sp^2 B–N bonds, *h*-BN exhibits in-plane thermal conductivity ranging from 100–2000 W m^{-1} K^{-1} [76,77]. However, along the longitudinal direction, the value drops significantly. Bulk *h*-BN has a slightly lower thermal conductivity than graphite, while nanostructured *h*-BN exhibits superior conductivity. For instance, BNNTs can achieve thermal conductivity up to 350 W m^{-1} K^{-1} at room temperature, outperforming CNTs [67]. Inspired by the Boltzmann theory, a few-layered BNNS was expected to have higher thermal conductivity than bulk *h*-BN [67], yet experimental results indicate otherwise (Figure 5.8). Contaminants introduced during preparation result in lower thermal conductivity values for 11-layer (360 W m^{-1} K^{-1}) and 5-layer

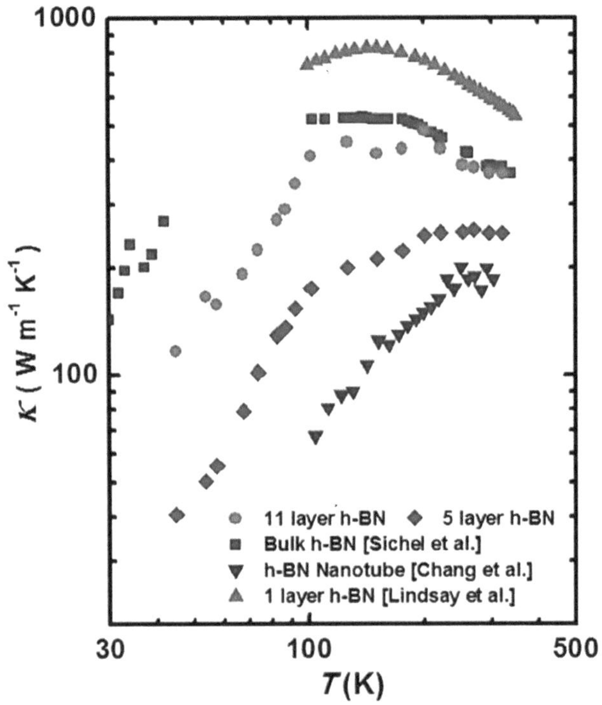

Figure 5.8 Thermal conductivity of *h*-BN with different thicknesses at different temperatures. (Reproduced with permission Ref. [78]. Copyright 2013, American Chemical Society.)

(250 W m^{-1} K^{-1}) *h*-BN compared to bulk *h*-BN (390 W m^{-1} K^{-1}) [78]. Recent studies show that monolayer BNNSs without surface contamination can achieve values up to 751 W m^{-1} K^{-1} at room temperature [79]. Chapter 7 will provide a detailed discussion on the thermal conductivity of BNNSs.

5.3.5 Toxicity

BN nanomaterials, such as BNNTs and BNNSs, exhibit lower toxicity to the human body than their carbon counterparts [12]. The biocompatibility of BNNSs and BNNTs makes *h*-BN suitable for applications in cosmetics and biomedicine [80–82]. The concentration, size, and shape of *h*-BN influence its biocompatibility, as do the types of cell cultures used in testing. Recent studies show that BNNTs do not accumulate in the body, whereas CNTs may accumulate in the lungs and kidneys [12,83]. BNNTs loaded with doxorubicin can specifically target and kill brain cancer cells in vitro without affecting healthy brain cells [83]. However, the toxicity of *h*-BN nanomaterials has also been reported. For example, Augustine et al. found that long BNNTs exhibited higher cytotoxicity than shorter ones [84]. Additionally, higher doses of BNNTs in pulmonary exposure could cause damage [85].

Impurities in BN or BNNT materials can influence their toxicity, so controlling residual pollutants during production can enhance biocompatibility. BNNSs have shown better biocompatibility than graphene [86], with studies indicating that low BN concentrations (10–100 μg mL^{-1}) stimulate cell proliferation and increase viability regardless of the size (5–200 nm) and shape (nanospheres, BNNSs, BNNTs) [87]. Low concentrations of *h*-BN (44 μg mL^{-1}) have also been shown to reduce doxorubicin-induced oxidative stress, improving cell viability [88]. In vivo experiments on albino rats suggest that large doses of BNNS (1600 μg kg^{-1}) can cause severe organ damage, whereas smaller doses (50–800 μg kg^{-1}) are non-toxic [89].

5.4 CONCLUSIONS AND CHALLENGES

This chapter has reviewed the definitions, structural characteristics, and properties of nanostructured BN materials (1D BNNTs and 2D BNNSs). The mechanical, electronic, optical, thermal, and toxicological properties of *h*-BN nanomaterials are extensively examined. BNNTs and BNNSs share atomic compositions and arrangements with BN, resulting in similar properties in terms of mechanical strength, thermal conduction, electrical insulation, chemical stability, and high-temperature oxidation. BNNTs and BNNSs each offer unique advantages and applications in the fields of materials science and nanotechnology.

REFERENCES

1. E. Roduner, Size matters: Why nanomaterials are different. *Chemical Society Reviews* **35**, 583–592 (2006).
2. K. S. Novoselov et al. Electric field effect in atomically thin carbon films. *Science* **306**, 666–669 (2004).
3. I. Itskou, A. L'Hermitte, S. Marchesini, T. Tian, C. Petit, How to tailor porous boron nitride properties for applications in interfacial processes. *Accounts of Materials Research* **4**, 143–155 (2023).
4. D. Golberg et al. Boron nitride nanotubes and nanosheets. *ACS Nano* **4**, 2979–2993 (2010).
5. A. E. Naclerio, P. R. Kidambi, A review of scalable hexagonal boron nitride (h-BN) synthesis for present and future applications. *Advanced Materials* **35**, 2207374 (2023).
6. A. Pakdel, Y. Bando, D. Golberg, Nano boron nitride flatland. *Chemical Society Reviews* **43**, 934–959 (2014).
7. G. Cassabois, P. Valvin, B. Gil, Hexagonal boron nitride is an indirect bandgap semiconductor. *Nature Photonics* **10**, 262–266 (2016).
8. M. Topsakal, E. Aktürk, S. Ciraci, First-principles study of two-and one-dimensional honeycomb structures of boron nitride. *Physical Review B—Condensed Matter and Materials Physics* **79**, 115442 (2009).
9. S. M. Gilbert et al. Alternative stacking sequences in hexagonal boron nitride. *2D Materials* **6**, 021006 (2019).
10. D. Pan et al. The properties and preparation methods of different boron nitride nanostructures and applications of related nanocomposites. *The Chemical Record* **20**, 1314–1337 (2020).
11. G. Constantinescu, A. Kuc, T. Heine, Stacking in bulk and bilayer hexagonal boron nitride. *Physical Review Letters* **111**, 036104 (2013).
12. S. Roy et al. Structure, properties and applications of two-dimensional hexagonal boron nitride. *Advanced Materials* **33**, 2101589 (2021).
13. R. Saito, M. Fujita, G. Dresselhaus, M. S. Dresselhaus, Electronic structure of chiral graphene tubules. *Applied Physics Letters* **60**, 2204–2206 (1992).
14. A. Loiseau, F. Willaime, N. Demoncy, G. Hug, H. Pascard, Boron nitride nanotubes with reduced numbers of layers synthesized by arc discharge. *Physical Review Letters* **76**, 4737 (1996).
15. A. Loiseau et al., Boron nitride nanotubes. *Carbon* **36**, 743–752 (1998).
16. L. Xiuling, W. Xiaojun, Low-dimensional boron nitride nanomaterials. *Journal of University of Science and Technology of China* **44**, 389 (2014).
17. Y. Kinoshita, N. Ohno, Electronic structures of boron nitride nanotubes subjected to tension, torsion, and flattening: A first-principles DFT study. *Physical Review B—Condensed Matter and Materials Physics* **82**, 085433 (2010).

18. H. Xiang, J. Yang, J. Hou, Q. Zhu, First-principles study of small-radius single-walled BN nanotubes. *Physical Review B* **68**, 035427 (2003).
19. R. Arenal, M. Kociak, A. Loiseau, D.-J. Miller, Determination of chiral indices of individual single-and double-walled boron nitride nanotubes by electron diffraction. *Applied Physics Letters* **89**, 073104 (2006).
20. R. Mukherjee, S. Bhowmick, Edge stabilities of hexagonal boron nitride nanoribbons: A first-principles study. *Journal of Chemical Theory and Computation* **7**, 720–724 (2011).
21. B. Huang, H. Lee, B.-L. Gu, F. Liu, W. Duan, Edge stability of boron nitride nanoribbons and its application in designing hybrid BNC structures. *Nano Research* **5**, 62–72 (2012).
22. C.-J. Kim et al., Stacking order dependent second harmonic generation and topological defects in h-BN bilayers. *Nano Letters* **13**, 5660–5665 (2013).
23. M. H. Khan et al., Few-atomic-layered hexagonal boron nitride: CVD growth, characterization, and applications. *Materials Today* **20**, 611–628 (2017).
24. M. H. Khan et al., Carbon-and crack-free growth of hexagonal boron nitride nanosheets and their uncommon stacking order. *Nanoscale* **8**, 15926–15933 (2016).
25. R. M. Ribeiro, N. M. Peres, Stability of boron nitride bilayers: Ground-state energies, interlayer distances, and tight-binding description. *Physical Review B—Condensed Matter and Materials Physics* **83**, 235312 (2011).
26. A. W. Robertson et al., Structural distortions in few-layer graphene creases. *ACS Nano* **5**, 9984–9991 (2011).
27. L. Zitzler, S. Herminghaus, F. Mugele, Capillary forces in tapping mode atomic force microscopy. *Physical Review B* **66**, 155436 (2002).
28. P. Nemes-Incze, Z. Osváth, K. Kamarás, L. Biró, Anomalies in thickness measurements of graphene and few layer graphite crystals by tapping mode atomic force microscopy. *Carbon* **46**, 1435–1442 (2008).
29. P. Hartel, H. Rose, C. Dinges, Conditions and reasons for incoherent imaging in STEM. *Ultramicroscopy* **63**, 93–114 (1996).
30. R. V. Gorbachev et al., Hunting for monolayer boron nitride: Optical and Raman signatures. *Small* **7**, 465–468 (2011).
31. S. Reich et al., Resonant Raman scattering in cubic and hexagonal boron nitride. *Physical Review B—Condensed Matter and Materials Physics* **71**, 205201 (2005).
32. O. L. Krivanek et al., Atom-by-atom structural and chemical analysis by annular dark-field electron microscopy. *Nature* **464**, 571–574 (2010).
33. J. Chen, B. Yao, C. Li, G. Shi, An improved Hummers method for eco-friendly synthesis of graphene oxide. *Carbon* **64**, 225–229 (2013).
34. W. S. Hummers Jr, R. E. Offeman, Preparation of graphitic oxide. *Journal of The American Chemical Society* **80**, 1339–1339 (1958).
35. M. H. Khan et al., Atomically thin hexagonal boron nitride nanofilm for Cu protection: The importance of film perfection. *Advanced Materials* **29**, 111359 (2017).
36. L. Vaccarini et al., Mechanical and electronic properties of carbon and boron–nitride nanotubes. *Carbon* **38**, 1681–1690 (2000).
37. M. Santosh, P. K. Maiti, A. Sood, Elastic properties of boron nitride nanotubes and their comparison with carbon nanotubes. *Journal of Nanoscience and Nanotechnology* **9**, 5425–5430 (2009).
38. G. Kim, J. Park, S. Hong, First-principles study of substitutional carbon pair and Stone–Wales defect complexes in boron nitride nanotubes. *Chemical Physics Letters* **522**, 79–82 (2012).
39. T. Li, Z. Tang, Z. Huang, J. Yu, A comparison between the mechanical and thermal properties of single-walled carbon nanotubes and boron nitride nanotubes. *Physica E: Low-dimensional Systems and Nanostructures* **85**, 137–142 (2017).
40. N. G. Chopra, A. Zettl, Measurement of the elastic modulus of a multi-wall boron nitride nanotube. *Solid State Communications* **105**, 297–300 (1998).
41. D. Golberg et al., Direct force measurements and kinking under elastic deformation of individual multiwalled boron nitride nanotubes. *Nano Letters* **7**, 2146–2151 (2007).

42. J. Garel et al., Ultrahigh torsional stiffness and strength of boron nitride nanotubes. *Nano Letters* **12**, 6347–6352 (2012).
43. A. Falin et al., Mechanical properties of atomically thin boron nitride and the role of interlayer interactions. *Nature Communications* **8**, 15815 (2017).
44. W. Gao, A. Tkatchenko, Sliding mechanisms in multilayered hexagonal boron nitride and graphene: The effects of directionality, thickness, and sliding constraints. *Physical Review Letters* **114**, 096101 (2015).
45. N. Marom et al., Stacking and registry effects in layered materials: The case of hexagonal boron nitride. *Physical Review Letters* **105**, 046801 (2010).
46. Q. Peng, W. Ji, S. De, Mechanical properties of the hexagonal boron nitride monolayer: Ab initio study. *Computational Materials Science* **56**, 11–17 (2012).
47. L. Song et al., Large scale growth and characterization of atomic hexagonal boron nitride layers. *Nano Letters* **10**, 3209–3215 (2010).
48. C. Li, Y. Bando, C. Zhi, Y. Huang, D. Golberg, Thickness-dependent bending modulus of hexagonal boron nitride nanosheets. *Nanotechnology* **20**, 385707 (2009).
49. S. M. Kim et al., Synthesis of large-area multilayer hexagonal boron nitride for high material performance. *Nature Communications* **6**, 8662 (2015).
50. J. Wu, B. Wang, Y. Wei, R. Yang, M. Dresselhaus, Mechanics and mechanically tunable band gap in single-layer hexagonal boron-nitride. *Materials Research Letters* **1**, 200–206 (2013).
51. J. Wang, F. Ma, W. Liang, M. Sun, Electrical properties and applications of graphene, hexagonal boron nitride (h-BN), and graphene/h-BN heterostructures. *Materials Today Physics* **2**, 6–34 (2017).
52. J. Kang, L. Zhang, S.-H. Wei, A unified understanding of the thickness-dependent bandgap transition in hexagonal two-dimensional semiconductors. *Journal of Physical Chemistry Letters* **7**, 597–602 (2016).
53. D. Wickramaratne, L. Weston, C. G. Van de Walle, Monolayer to bulk properties of hexagonal boron nitride. *The Journal of Physical Chemistry C* **122**, 25524–25529 (2018).
54. Y. Ji et al. Chemical vapor deposition growth of large single-crystal mono-, bi-, tri-layer hexagonal boron nitride and their interlayer stacking. *ACS Nano* **11**, 12057–12066 (2017).
55. S. Tang, J. Yu, L. Liu, Tunable doping and band gap of graphene on functionalized hexagonal boron nitride with hydrogen and fluorine. *Physical Chemistry Chemical Physics* **15**, 5067–5077 (2013).
56. X.-J. Wu, M.-H. Wu, X. C. Zeng, Chemically decorated boron-nitride nanoribbons. *Frontiers of Physics in China* **4**, 367–372 (2009).
57. X. Li, X. Wu, X. C. Zeng, J. Yang, Band-gap engineering via tailored line defects in boron-nitride nanoribbons, sheets, and nanotubes. *ACS Nano* **6**, 4104–4112 (2012).
58. T. Sainsbury et al., Oxygen radical functionalization of boron nitride nanosheets. *Journal of the American Chemical Society* **134**, 18758–18771 (2012).
59. F. Xiao et al., Edge-hydroxylated boron nitride nanosheets as an effective additive to improve the thermal response of hydrogels. *Advanced Materials,* **27**, 7196–7203 (2015).
60. F. Liu, R. Han, A. Nattestad, X. Sun, Z. Huang, Carbon-and oxygen-doped hexagonal boron nitride for degradation of organic pollutants. *Surface Innovations* **9**, 222–230 (2020).
61. K. Watanabe, T. Taniguchi, H. Kanda, Direct-bandgap properties and evidence for ultraviolet lasing of hexagonal boron nitride single crystal. *Nature Materials* **3**, 404–409 (2004).
62. Y. Lin, T. V. Williams, J. W. Connell, Soluble, exfoliated hexagonal boron nitride nanosheets. *Journal of Physical Chemistry Letters* **1**, 277–283 (2010).
63. K. K. Kim et al., Synthesis of monolayer hexagonal boron nitride on Cu foil using chemical vapor deposition. *Nano Letters* **12**, 161–166 (2012).
64. M. Gao et al., Catalyst-free growth of two-dimensional hexagonal boron nitride few-layers on sapphire for deep ultraviolet photodetectors. *Journal of Materials Chemistry C* **7**, 14999–15006 (2019).
65. H. Liu et al., High-performance deep ultraviolet photodetectors based on few-layer hexagonal boron nitride. *Nanoscale* **10**, 5559–5565 (2018).
66. L. Schué et al., Dimensionality effects on the luminescence properties of hBN. *Nanoscale* **8**, 6986–6993 (2016).

67. S. Angizi et al., A comprehensive review on planar boron nitride nanomaterials: From 2D nanosheets towards 0D quantum dots. *Progress in Materials Science* **124**, 100884 (2022).
68. L. Wirtz, A. Marini, A. Rubio, Excitons in boron nitride nanotubes: Dimensionality effects. *Physical Review Letters* **96**, 126104 (2006).
69. S. Konabe, Exciton effect on shift current in single-walled boron-nitride nanotubes. *Physical Review B* **103**, 075402 (2021).
70. Y. S. Itas et al., Computational studies of the excitonic and optical properties of armchair SWCNT and SWBNNT for optoelectronics applications. *Crystals* **12**, 870 (2022).
71. M. Silly et al., Luminescence properties of hexagonal boron nitride: Cathodoluminescence and photoluminescence spectroscopy measurements. *Physical Review B—Condensed Matter and Materials Physics* **75**, 085205 (2007).
72. J. Lauret et al., Optical transitions in single-wall boron nitride nanotubes. *Physical Review Letters* **94**, 037405 (2005).
73. L. H. Li, J. Cervenka, K. Watanabe, T. Taniguchi, Y. Chen, Strong oxidation resistance of atomically thin boron nitride nanosheets. *ACS Nano* **8**, 1457–1462 (2014).
74. D. Golberg, Y. Bando, C. Tang, C. Zhi, Boron nitride nanotubes. *Advanced Materials* **19**, 2413–2432 (2007).
75. Z. Liu et al., Ultrathin high-temperature oxidation-resistant coatings of hexagonal boron nitride. *Nature Communications* **4**, 2541 (2013).
76. C. Sevik, A. Kinaci, J. B. Haskins, T. Çağın, Characterization of thermal transport in low-dimensional boron nitride nanostructures. *Physical Review B—Condensed Matter and Materials Physics* **84**, 085409 (2011).
77. T. Ouyang et al., Thermal transport in hexagonal boron nitride nanoribbons. *Nanotechnology* **21**, 245701 (2010).
78. I. Jo et al., Thermal conductivity and phonon transport in suspended few-layer hexagonal boron nitride. *Nano Letters* **13**, 550–554 (2013).
79. Q. Cai et al., High thermal conductivity of high-quality monolayer boron nitride and its thermal expansion. *Science Advances* **5**, eaav0129 (2019).
80. M. M. Fiume et al., Safety assessment of boron nitride as used in cosmetics. *International Journal of Toxicology* **34**, 53S–60S (2015).
81. N. Wang et al., Toxicity evaluation of boron nitride nanospheres and water-soluble boron nitride in Caenorhabditis elegans. *International Journal of Nanomedicine*, **12**, 5941–5957 (2017).
82. A. Merlo, V. Mokkapati, S. Pandit, I. Mijakovic, Boron nitride nanomaterials: Biocompatibility and bio-applications. *Biomaterials Science* **6**, 2298–2311 (2018).
83. D. De Pasquale et al., Homotypic targeting and drug delivery in glioblastoma cells through cell membrane-coated boron nitride nanotubes. *Materials & Design* **192**, 108742 (2020).
84. J. Augustine et al., Assessing size-dependent cytotoxicity of boron nitride nanotubes using a novel cardiomyocyte AFM assay. *Nanoscale Advances* **1**, 1914–1923 (2019).
85. X. Xin et al., Toxicity evaluation following pulmonary exposure to an as-manufactured dispersed boron nitride nanotube (BNNT) material in vivo. *NanoImpact* **19**, 100235 (2020).
86. D. Chimene, D. L. Alge, A. K. Gaharwar, Two-dimensional nanomaterials for biomedical applications: Emerging trends and future prospects. *Advanced Materials* **27**, 7261–7284 (2015).
87. D. V. Shtansky et al., Recent progress in fabrication and application of BN nanostructures and BN-based nanohybrids. *Nanomaterials* **12**, 2810 (2022).
88. I. C. Taskin, O. Sen, M. Emanet, M. Culha, B. Yilmaz, Hexagonal boron nitrides reduce the oxidative stress on cells. *Nanotechnology* **31**, 215101 (2020).
89. F. Kar et al., Hexagonal boron nitride nanoparticles trigger oxidative stress by modulating thiol/disulfide homeostasis. *Human & Experimental Toxicology* **40**, 1572–1583 (2021).

Chapter 6

Unique Adsorption Capability of Boron Nitride Nanosheets for Enhanced Sensing

Qiran Cai, Wei Gan, Lu Hua Li, Srikanth Mateti, and Ying Ian Chen

6.1 INTRODUCTION

Surface enhanced Raman spectroscopy (SERS) is a powerful and non-destructive analytical technique capable of detecting chemical and biological molecules at extremely low concentrations [1,2]. This makes it widely used in various fields, including chemical analysis, environmental protection, food processing, the pharmaceutical industry, and diagnostic biology [3–5]. There are two main mechanisms that explain the SERS effect: the electromagnetic enhancement (EM) mechanism and the chemical enhancement (CM) mechanism [5,6]. The EM mechanism involves local electromagnetic fields that are amplified by localized surface plasmon resonance. This amplification can enhance the Raman signal by factors of approximately 10^6–10^{10} [5,7–9]. Noble metal nanostructures, such as gold (Au), silver (Ag), and copper (Cu), are commonly employed in this context because their plasmon resonance frequencies are optimal in the visible and near-infrared ranges [10,11]. The CM mechanism, on the other hand, relates to chemical interactions between analytes and the substrate, including charge transfer and the resonant Raman effect. This interaction increases molecular polarizability and, consequently, the Raman cross-section, with EFs ranging from 10 to 1000 [5,7,12].

The high cost of noble metals and their fabrication processes make it highly desirable to develop reusable SERS substrates that exhibit high sensitivity, reproducibility, and homogeneity. Research has shown that heating the substrates in an oxygen-containing atmosphere is the most effective method for completely removing adsorbed analyte molecules, thus enhancing reusability [13,14]. However, this approach requires SERS substrates to withstand oxidation at high temperatures, such as 400°C. Unfortunately, plasmonic metals, *e.g.*, Ag and Cu, are prone to oxidation even under ambient conditions, making them particularly vulnerable at elevated temperatures in oxygen-rich environments. For instance, Ag nanoparticles (NPs) can lose over 50% of their plasmon enhancement after just 28 days of exposure to the ambient environment [15]. To tackle this issue, thin oxide films, including alumina (Al_2O_3), silica (SiO_2), and zinc oxide (ZnO), have been applied to Ag NPs. These coatings help reduce oxidation of the plasmonic metals and minimize substrate-induced disturbances, thereby increasing stability [16–19]. However, thicker coatings are often needed for effective oxidation prevention, which can significantly diminish SERS enhancement [18,20,21]. Additionally, these ceramic coatings tend to have low affinity for probe molecules, leading to reduced SERS sensitivity.

DOI: 10.1201/9781003647812-8

2D materials, with their unique characteristics and properties, offer new opportunities for SERS applications. For instance, graphene can enhance Raman signals through CM [12,22,23] and reduce the fluorescence background of probe molecules [24]. Additionally, graphene can be combined with plasmonic metal particles to further enhance SERS *via* EM [25–27]. Due to its flexibility and gas impermeability, graphene also provides some protection for metal NPs against oxidation [28]. However, this protective effect is only short-term [29,30] and can actually accelerate oxidation over time due to galvanic reactions between graphene and the underlying metal [31,32]. Furthermore, graphene does not address the reusability challenge, as it begins to react with oxygen at temperatures as low as 250°C [33]. In contrast, BN nanosheets, which are structural analogues of graphene, possess comparable properties. More importantly, BN nanosheets are wide bandgap insulators and maintain thermal stability beyond 800°C in air [34]. This makes them suitable as a barrier to protect metals from oxidation and corrosion at high temperatures without causing galvanic corrosion [35,36]. Additionally, BN nanosheets exhibit strong adsorption capabilities for organic molecules, significantly enhancing SERS sensitivity [14]. In this chapter, we will discuss the unique adsorption behavior of BN nanosheets, the underlying mechanisms [37], and their applications in reusable SERS with high sensitivity [38–41] and MEF [42].

6.2 UNIQUE ADSORPTION CAPABILITY OF BN NANOSHEETS

Compared to bulk materials, nanomaterials generally offer superior adsorption capabilities, traditionally attributed to their larger surface areas. However, a study comparing the adsorption behavior of atomically thin BN and bulk *h*-BN presents a different perspective [37]. Specifically, monolayer (1L) and bilayer (2L) BN were mechanically exfoliated from high-quality bulk *h*-BN single crystals. These atomically thin BN sheets, along with bulk *h*-BN, were immersed in a 10^{-3} mol L^{-1} aqueous rhodamine 6G (R6G) solution and then washed with Milli-Q water to remove any non-adsorbed R6G molecules. As shown in Figure 6.1a–c, after 300 seconds of immersion, the surfaces of 1L and

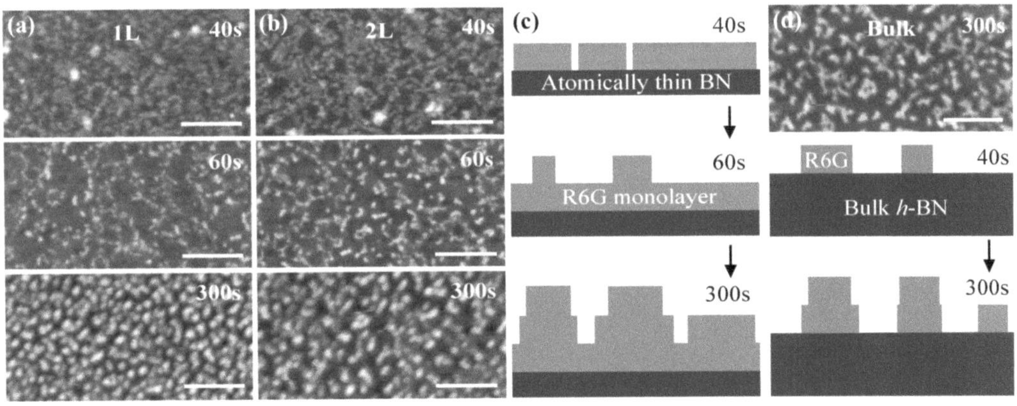

Figure 6.1 AFM images of 1L (a) and 2L (b) BN after the immersion in aqueous solution of R6G (10^{-3} mol L^{-1}) for 40, 60, and 300 seconds, with a height range of 5 nm. (c) Schematic diagrams of the adsorption process of R6G on atomically thin BN. (d) AFM image of R6G adsorbed on a bulk *h*-BN crystal with a thickness of approximately 1 μm under the same condition for 300 seconds and the corresponding schematic diagrams. Scale bars in (a, b, and d) are 250 nm. (Reprinted with permission from Ref. [37]. Copyright 2016, John Wiley & Sons, Inc.)

2L BN were fully covered by a 2–3-layer thick coating of R6G molecules, while only sparse, discontinuous R6G islands were observed on the surface of bulk *h*-BN (Figure 6.1d). These results indicate that, despite having the same surface area, atomically thin BN exhibits superior adsorption capability compared to bulk *h*-BN.

Density functional theory calculations were used to investigate the adsorption of R6G on atomically thin BN and bulk *h*-BN, focusing on π–π interactions. Figure 6.2 illustrates the adsorption of an R6G molecule on a 1L BN. The results showed that R6G prefers a 'lying-down' orientation, with the xanthene ring parallel to the BN plane and positioned approximately 0.8 nm above it (Figure 6.2a). However, this configuration is not stable due to the stress caused by the adsorption. The high flexibility of 1L BN allows it to curve or distort to alleviate this stress. This

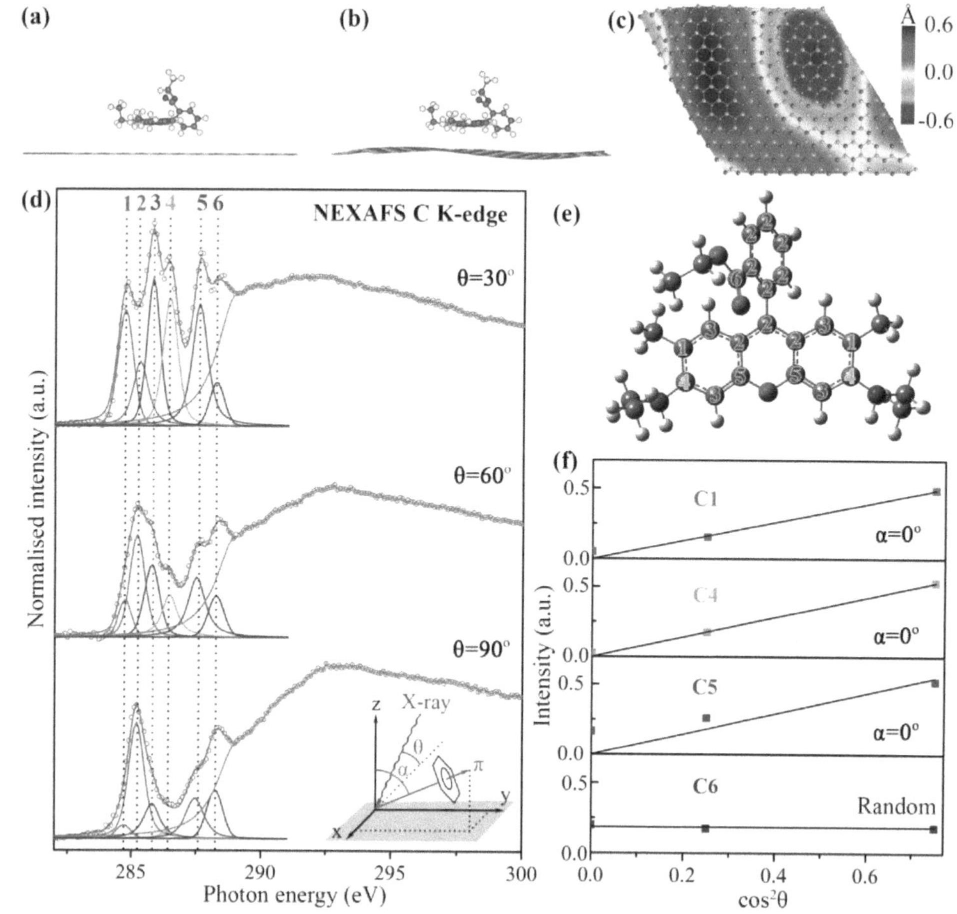

Figure 6.2 (a) Side view of an R6G molecule on a BNNS in its initial state (no deformation). (b) Side view of the same R6G molecule on the BNNS after a conformational change (equilibrium state), accompanied by the corresponding height mapping shown in (c). (d) Fits to the π^* resonance of the angularly dependent NEXAFS spectra for the 9 Å R6G deposited on the BNNS, with the incidence geometry illustrated in the inset. (e) Carbon atoms corresponding to the six fitted π^* transition subpeaks in (d). (f) Cosine-squared dependence of the intensities for the fitted C1, C4, C5, and C6 1s-π^* transitions. (Reprinted with permission from Ref. [37]. Copyright 2016, John Wiley & Sons, Inc.)

deformation, known as a conformation change, induces an average compressive strain of −0.16% in the 1L BN and results in increased adsorption energy and efficiency (Figure 6.2b and c). In contrast, bulk *h*-BN, which was modeled as a fixed 1L BN in the calculations, remains rigid and non-deformable upon R6G adsorption (Figure 6.2a), leading to lower adsorption energy and efficiency. To validate these theoretical findings, angular-dependent near-edge X-ray absorption fine structure (NEXAFS) spectroscopy was employed to investigate the configuration of R6G molecules on 1L BN. The 1L BN was initially cleaned at 425°C in an ultrahigh vacuum to remove carbon contamination, and a monolayer (~0.9 nm) of R6G molecules was adsorbed on the surface. The C K-edge NEXAFS spectra of R6G were then collected at varying X-ray incident angles (θ). The NEXAFS intensity of carbon atoms in the xanthene ring (*i.e.*, C1, C4, and C5) exhibited a strong dependence on $\cos^2\theta$ (Figure 6.2d–f), confirming that the xanthene ring of the R6G molecule is parallel to the BN plane [43].

Although visualizing the subtle morphological changes in atomically thin BN due to R6G molecule adsorption is challenging, the resulting strain can be detected using Raman spectroscopy [44]. Figure 6.3 compares the Raman E_{2g} mode (G band) of suspended 2L BN before and after adsorption of R6G molecules, achieved by immersion in a 10^{-3} mol L^{-1} R6G aqueous solution for 60 seconds. The insets show the corresponding optical microscopy images of the 2L BN. Prior to R6G adsorption, the G band frequency of the suspended 2L BN was approximately 1366.6 cm^{-1}, which is close to that of bulk *h*-BN. After R6G adsorption, the G band frequency increased to 1368.3 cm^{-1}. The average frequency upshift of nine pieces of suspended 2L BN due to R6G adsorption was 1.6 cm^{-1}, as shown in Figure 6.3b. In contrast, a downshift of −1.3 cm^{-1} was observed in SiO_2/Si substrate-supported 2L BN after R6G adsorption, and no change in the G band frequency was detected for bulk *h*-BN after R6G adsorption. This frequency shift is about an order of magnitude larger than the system error of Raman measurements, indicating that it is not due to measurement error. The Raman frequency shift in BN nanosheets can generally be attributed to strain, doping, or temperature changes [44]. Unlike graphene, BN nanosheets are insulators and are minimally affected by doping from molecule adsorption [45,46]. Additionally, the observed opposite frequency shifts in suspended versus substrate-supported BN nanosheets cannot be explained by doping effects. Since Raman measurements were conducted at room temperature, temperature effects are also not relevant. Thus, the frequency upshift (in suspended BN) and downshift (in substrate-supported BN) after R6G adsorption are attributed to compressive and tensile strain, respectively.

The Raman frequency shift ($\Delta\omega$) in BN nanosheets due to biaxial strain change ($\Delta\varepsilon$) can be expressed as [47]:

$$\Delta\omega = -2\omega_0\gamma\Delta\varepsilon \tag{6.1}$$

where $\Delta\omega_0$ is the Raman frequency of strain-free BN nanosheets, and γ is the Grüneisen parameter of *h*-BN. Initially, suspended BN nanosheets are strain-free [48]. The adsorption of R6G induces deformation and compressive strain (negative $\Delta\varepsilon$) in these nanosheets due to structural mismatch, resulting in an upshift in the Raman frequency (Figure 6.3c). In contrast, substrate-supported BN nanosheets are initially under compressive strain because of the uneven SiO_2/Si substrate [44]. This initial strain is greater than the compressive strain induced by R6G adsorption. Consequently, the adsorption of R6G can partially lift BN nanosheets from the SiO_2/Si substrate, reducing their deformation or compressive strain and leading to a downshift in the Raman frequency (Figure 6.3d). From the energy point of view, this is plausible. Although the interaction between BN nanosheets and SiO_2/Si substrates has not been directly measured,

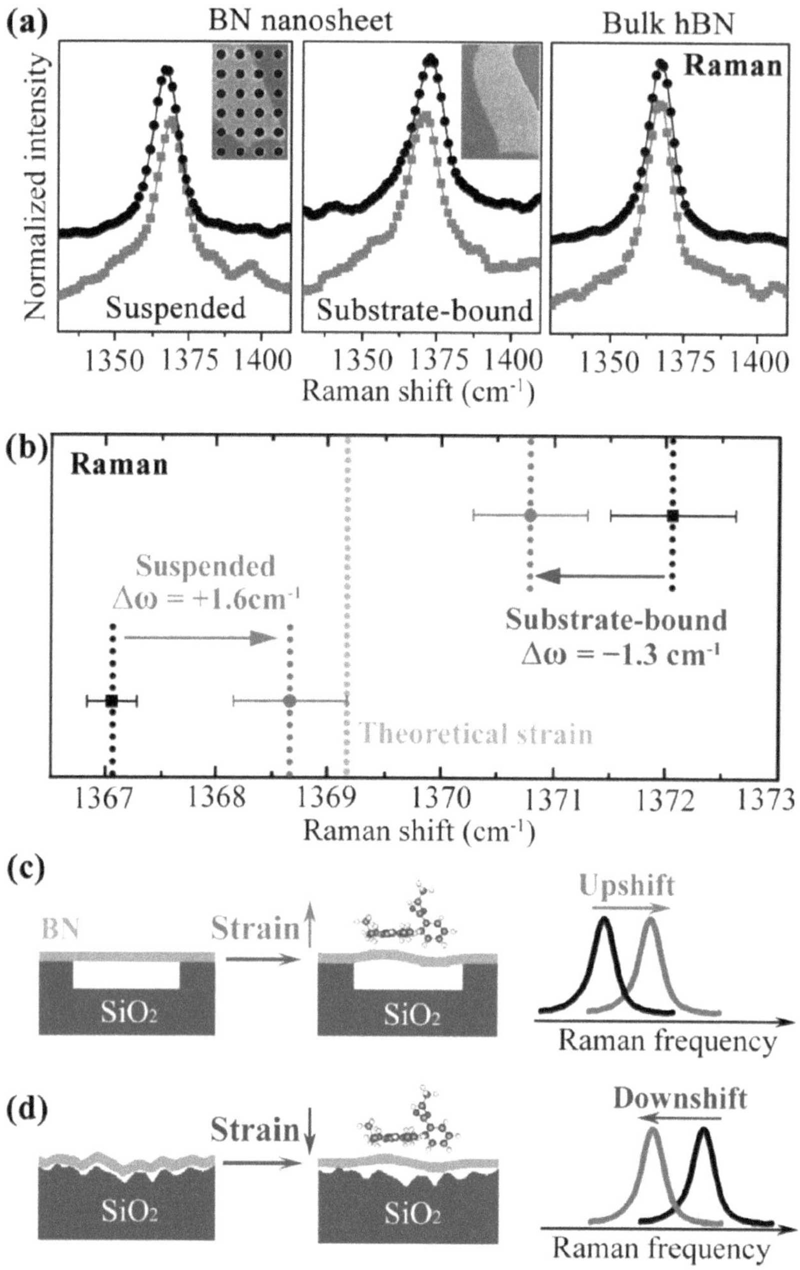

Figure 6.3 (a) Comparison of the Raman spectra for suspended and substrate-bound BN nanosheets, as well as bulk h-BN, before (black) and after (red) the adsorption of R6G molecules. Optical images are provided as insets. (b) Summary of the average frequency shifts of the Raman G band for suspended and substrate-bound BN nanosheets, shown before (black) and after (red) R6G adsorption. The theoretically predicted strain is indicated in gold. (c) and (d) Diagrams illustrating strain changes in suspended and substrate-bound BN nanosheets, respectively, following R6G adsorption, along with the corresponding Raman shifts. (Reprinted with permission from Ref. [37]. Copyright 2016, John Wiley & Sons, Inc.)

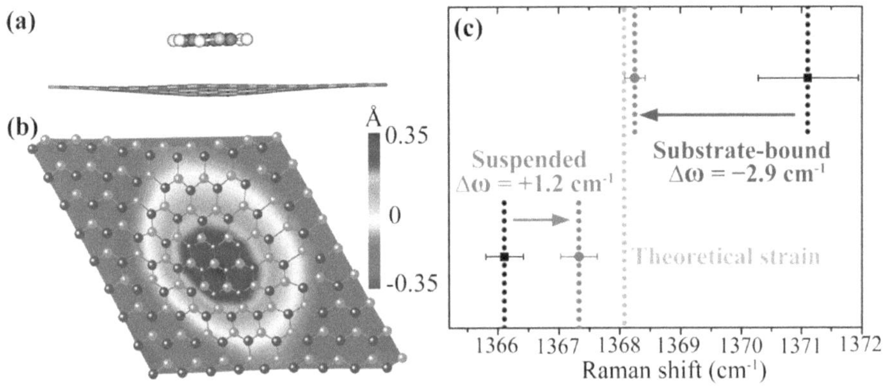

Figure 6.4 (a) Top view of a 4-MBA molecule on a BNNS after a conformational change (equilibrium state), along with the corresponding height mapping shown in (b). (c) Frequency shifts of the Raman G band before (black) and after (red) the adsorption of 4-MBA molecules on both suspended and substrate-bound BN nanosheets, with the theoretically predicted strain indicated in gold. (Reprinted with permission from Ref. [37]. Copyright 2016, John Wiley & Sons, Inc.)

previous studies report that the adhesion energy between graphene and SiO_2/Si ranges from 0.24 to 0.31 J m^{-2} [49–51]. Additionally, the adsorption energy of R6G on a BN nanosheet has been calculated to be 0.26 J m^{-2}, which is comparable to the substrate's adhesion energy [37].

The conformational change in BN nanosheets is not limited to the adsorption of R6G. Adsorption of 4-mercaptobenzoic acid (4-MBA) molecules also induces deformation in BN nanosheets. Figure 6.4a and b illustrates the conformational change and resulting strain distribution in a BN nanosheet after 4-MBA adsorption. Density functional theory calculations show that this conformational change results in an average strain of −0.087% in the BN nanosheet. Similarly, different G band frequency shifts were observed in suspended versus substrate-supported BN nanosheets after 4-MBA adsorption (Figure 6.4c). Specifically, a suspended 2L BN nanosheet exhibited an average upshift of $1.2 \pm 0.4\, cm^{-1}$ in the G band frequency, while its substrate-supported counterpart showed a downshift of $2.9 \pm 0.6\, cm^{-1}$. This indicates that conformational changes due to molecular adsorption are common phenomena in interactions between molecules and 2D materials with low rigidity or high flexibility. The exceptional adsorption capability of BN nanosheets makes them promising for applications such as water purification with high efficiency, SERS, and MEF, providing high sensitivity (as discussed in Sections 6.3 and 6.4).

6.3 BN NANOSHEETS FOR SERS

As aforementioned, the advent of 2D materials, especially graphene, has provided new possibilities for SERS. For example, graphene has proven to be an excellent SERS substrate, not only enhancing Raman signals of analytes *via* CM [12,45], but also capable of capturing probe molecules in a more controllable way and quenching fluorescence [52–54]. However, the strong Raman signal of graphene itself, including G and 2D bands, and possibly the D band, will interfere with Raman spectroscopy of analytes. Furthermore, graphene has a relatively low thermal stability, starting to be oxidized in air at around 250°C [33], which cannot offer reusability to the SERS substrate because the substrate needs to be heated at around 400°C in air or oxygen-containing gas to remove analytes for reuse [14]. In addition to comparable properties with graphene, such as aforementioned special adsorption capability [37], BN nanosheets can sustain in air over 800°C without oxidation [34], and only have a Raman G band of low intensity, which is too weak to interfere with the SERS spectrum of the analyte [55]. Therefore, this section will discuss the application of BN nanosheets for reusable SERS with enhanced sensitivity. It will explore three specific configurations: BN nanosheet-supported metal nanostructures, BN nanosheet-veiled metal nanostructures, and BN/graphene van der Waals heterostructure-veiled metal nanostructures.

6.3.1 BN Nanosheet-Supported Metal Nanostructure

Although 2D materials are ideal for studying the mechanisms of CM [12,45], their enhancement factors (EFs) are generally too low for practical applications. Noble metal nanostructures, known for providing much higher EFs, remain dominant in highly sensitive SERS applications, particularly for single-molecule detection. While the strength of the electromagnetic field is crucial, SERS sensitivity also significantly depends on the amount of analyte captured on the substrate. Commonly used plasmonic metals (*e.g.*, Au, Ag, Cu) have low affinity for non-thiol and non-amino molecules, which limits their sensitivity. Moreover, developing a reusable SERS substrate is a viable solution to mitigate the high costs associated with noble metals. The incorporation of BN nanosheets can enhance the capture of probe molecules, leading to increased sensitivity. Additionally, their excellent thermal stability ensures the effectiveness of the SERS substrate [39].

BN nanosheets were initially mechanically exfoliated onto a SiO_2/Si substrate from high-quality *h*-BN single crystals and located using optical microscopy with a DP71 camera (Figure 6.5). Their thickness was measured with an atomic force microscope (AFM). Next, a layer of Au film was sputtered onto the BN nanosheets, followed by annealing in argon (Ar) at 600°C for 1 hour. This process allowed the Au film to transform into Au particles. The size and distribution of the Au particles can be controlled by adjusting the sputtering time, current, and annealing temperature. Most Au particles exhibit a hexagonal shape with a flat top, indicating high crystallinity

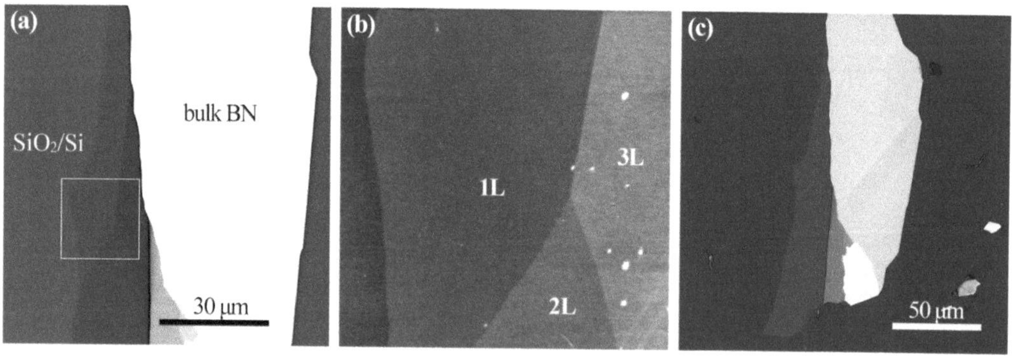

Figure 6.5 (a) Optical image showing a 1–3L BN alongside bulk BN on a SiO_2/Si substrate. (b) AFM image of the 1–3L BN, highlighted within the white square in panel (a). (c) Optical image of the few-layer BN coated with a thin film of sputtered Au. (Reproduced from Ref. [39] with permission from the Royal Society of Chemistry.)

and a preferential orientation of the (111) planes parallel to the substrate surface (Figure 6.6). This suggests that the Au film partially melted at 600°C (which is below Au's melting point) and recrystallized during cooling. Interestingly, under the same sputtering and annealing conditions, the size distribution of Au particles vary slightly between 1L BN, bulk *h*-BN, and SiO_2/Si wafers. Generally, Au particles on bulk *h*-BN are larger than those on 1L BN, which in turn are larger than those on SiO_2/Si wafers. This variation is attributed to the different surface diffusion coefficients of these materials. During annealing, the Au film breaks into small clusters, which then migrate and coalesce with surrounding clusters to form larger particles. A higher surface diffusion coefficient results in a higher migration rate and a greater likelihood of forming larger Au particles [56]. Since Au has a higher diffusion coefficient on BN than on SiO_2/Si wafers [57], it results in larger particles on BN. Additionally, the high flexibility of 1L BN causes it to conform more closely to the profile of the SiO_2/Si wafer, leading to higher roughness compared to bulk *h*-BN [58,59]. This increased roughness reduces the diffusion coefficient of Au, resulting in smaller particles on 1L BN compared to bulk *h*-BN [39].

To compare the SERS performance of the three different substrates, various sputtering times were used to produce Au particles with similar sizes and distributions on each substrate, as shown in Figure 6.6a–c. The Raman enhancement of the three Au-decorated substrates was evaluated using a 10^{-6}mol L^{-1} aqueous solution of R6G (Figure 6.6d). Interestingly, the SERS enhancement on the BN substrates (1L and bulk) was significantly greater than that on the SiO_2/Si substrate (black curve). This increased enhancement is attributed to the stronger adsorption capability of BN toward R6G molecules *via* π–π interactions, compared to the SiO_2/Si wafer. As illustrated in Figure 6.6e, more R6G molecules are adsorbed on the BN, particularly in the gaps between the Au particles—known as 'hot spots'—leading to a significant increase in Raman signal. Furthermore, the 1L BN substrate exhibited a stronger Raman signal than bulk *h*-BN. This is

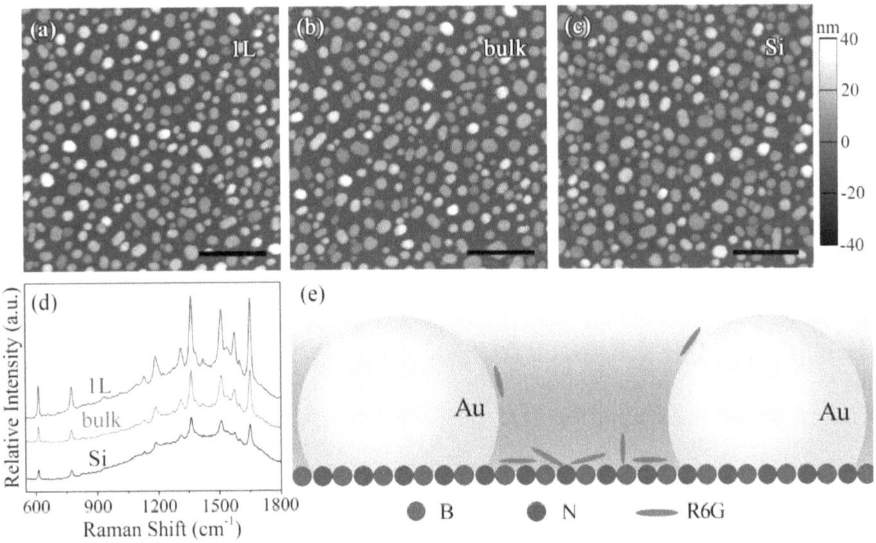

Figure 6.6 AFM images of Au particles on (a) 1L BN, (b) bulk BN, and (c) SiO_2/Si substrates. The scale bars represent 500 nm. (d) Raman spectra of R6G on the three SERS substrates. (e) Schematic illustrating the adsorption of R6G molecules on 1L BN, highlighting the enhancement of SERS signals. (Reproduced from Ref. [39] with permission from the Royal Society of Chemistry.)

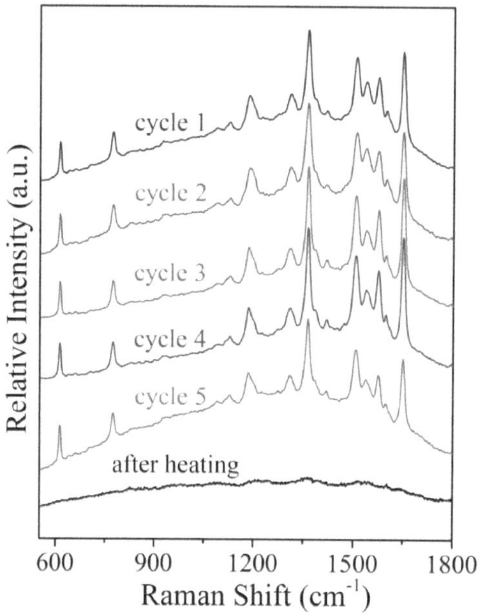

Figure 6.7 Raman spectra of R6G at a concentration of 10^{-6} mol L^{-1} on a SERS substrate of Au-decorated 1L BN after five cycles of a reusability test. For comparison, a typical Raman spectrum obtained after heating the substrate at 400°C to remove adsorbed R6G molecules is also included. All spectra were acquired under identical conditions. (Reproduced from Ref. [39] with permission from the Royal Society of Chemistry.)

due to the greater adsorption capability of 1L BN, which allows it to capture more R6G molecules and consequently enhance the SERS signal [37]. Importantly, unlike graphene, which exhibits strong Raman signals, the Raman G band of 1L BN is too weak to be detected. This lack of interference makes 1L BN an ideal substrate for Raman spectroscopy of analytes.

To regenerate the SERS substrate, the adsorbed R6G molecules were completely removed by heating the substrate at 400°C in air for 5 minutes, after which no R6G Raman signals were detected. The substrate was then re-immersed in a solution of R6G with the same concentration. This heating and re-adsorption process was repeated five times on the same 1L BN-Au SERS substrate. Figure 6.7 presents the Raman spectra of R6G obtained from each of the five cycles of reuse. Notably, no significant loss of SERS enhancement was observed after five cycles. The excellent reusability of the SERS substrate is attributed to the high thermal stability of BN nanosheets, which allows for the complete removal of analytes through heat treatment while maintaining their strong adsorption capabilities.

6.3.2 BN Nanosheet-Veiled Metal Nanostructure

Although the aforementioned structure where plasmonic metals are deposited on top of BNNSs can enhance both the sensitivity and reusability of SERS substrates to some extent, it has at least two significant disadvantages: (1) Metal particles cover most of the surface area of BNNSs, which limits the available area for molecule adsorption and thus reduces sensitivity; (2) metal particles are exposed to air during heat treatment without protection, leading to oxidation, especially for metals that are prone to it, such as Ag particles [60]. To fully utilize the exceptional properties of BN nanosheets, it has been proposed to cover metal particles with BN nanosheets [38,61]. This approach not only preserves more surface area for probe molecule adsorption but also protects the metal particles from oxidation, thereby significantly enhancing both sensitivity and reusability.

To fabricate highly sensitive and reusable SERS substrates, a 10 nm thick silver (Ag) film was sputter-coated onto a SiO_2/Si wafer. This was followed by the mechanical exfoliation of BN nanosheets from high-quality BN single crystals onto the Ag film. The BN-coated substrates were then annealed in an Ar atmosphere at 450°C for 1 hour. This annealing process transformed the Ag film into particles with suitable sizes and distributions. Figure 6.8a–d presents AFM images of SERS substrates covered with 2L and 17L BN nanosheets. After annealing, the flexible 2L BN wrinkled to conform to the underlying Ag particles, while the 17L BN remained relatively rigid with less deformation. Consequently, the 2L BN better preserves the plasmonic hot spots, resulting in stronger Raman enhancement due to the exponential decrease in plasmon-induced electromagnetic fields with distance. The BN nanosheets have minimal impact on the size and distribution of the Ag particles. The SERS sensitivity of bare Ag substrates and those covered with BN nanosheets of varying thicknesses was tested by immersing them in a 10^{-8} mol L^{-1} aqueous solution of R6G for 1 hour (Figure 6.8e). As anticipated, the Raman spectrum of R6G was strongest for the 2L BN-covered Ag particles (2L/Ag), with the SERS intensity decreasing as BN thickness increased. In contrast, the bare Ag particles on SiO_2/Si (Ag/SiO_2) substrates showed no Raman signal for R6G at concentrations of 10^{-8} or 10^{-7} mol L^{-1}, demonstrating that 2L BN enhances the sensitivity of SERS substrates by over two orders of magnitude. This enhancement is attributed to two main factors: (1) The atomically thin BN more efficiently captures analyte molecules, and (2) the BN nanosheets' outstanding flexibility better preserves Raman hot spots

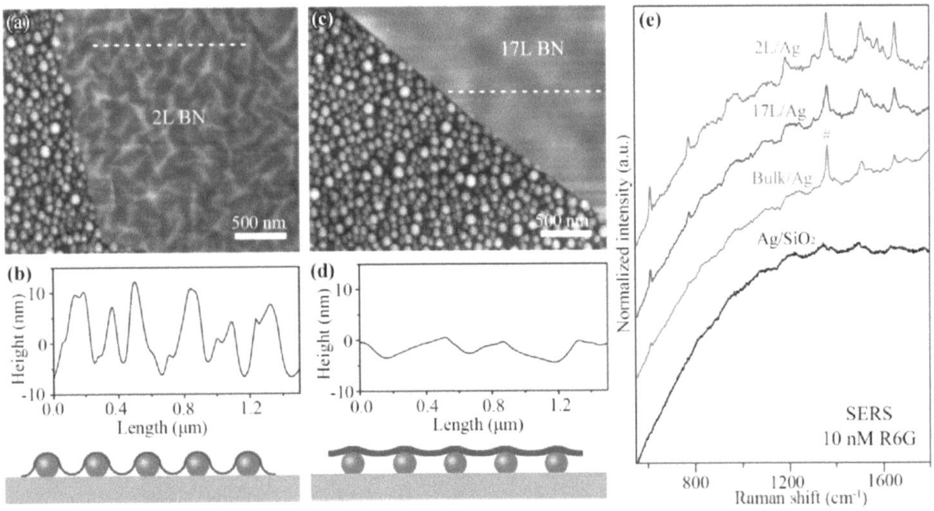

Figure 6.8 AFM images of Ag NPs covered by (a) 2L and (c) 17L BN. Panels (b) and (d) show height traces and schematic diagrams corresponding to the dotted lines in (a) and (c), respectively. (e) SERS spectra of R6G at a concentration of 10 nmol L^{-1} from the 2L and 17L BN, as well as bulk BN and the Ag/SiO_2 substrate obtained *via* the adsorption method. The symbol # indicates the Raman G band (E_{2g} mode) of bulk *h*-BN. (Adapted with permission from Ref. [38]. Copyright 2016, John Wiley & Sons, Inc.)

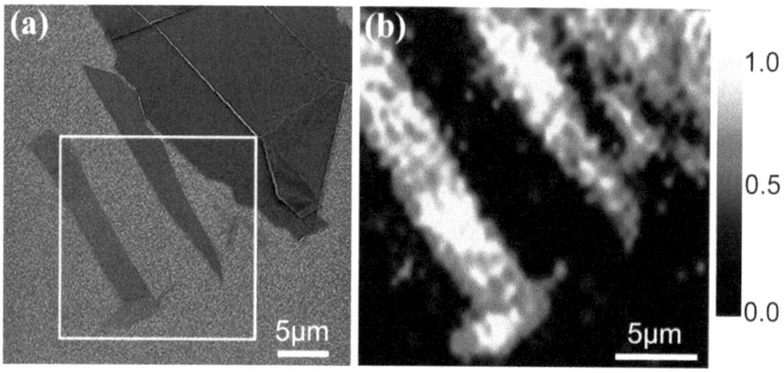

Figure 6.9 (a) Scanning electron microscope image showing Ag particles covered with BN of varying thicknesses. (b) Corresponding Raman mapping of the squared area in (a) after immersion in a 10^{-6} mol L^{-1} R6G solution for 1 hour. In the mapping, a value of 0.0 indicates no detected intensity, while 1.0 represents the strongest measured Raman peak at 612 cm^{-1}. (Adapted with permission from Ref. [38]. Copyright 2016, John Wiley & Sons, Inc.)

[38]. Given that the Ag/SiO_2 substrate, despite having the strongest electromagnetic field, did not produce a detectable Raman signal due to its low affinity for aromatic molecule adsorption, while the bulk BN-covered Ag particle (Bulk/Ag) substrate with weaker electromagnetic fields still showed a weak Raman signal, it is believed that the former factor (adsorption capability) plays a decisive role in the observed enhancement [38].

Raman mapping was used to assess the homogeneity and reproducibility of the SERS substrates, as illustrated in Figure 6.9. The contrast between areas with and without BN coverage is striking. The Ag/SiO_2 regions appear predominantly black, indicating significantly weaker Raman sensitivity, while the BN/Ag areas exhibit Raman signals that are several times stronger. Additionally, the Raman enhancement in the BN-covered areas is relatively uniform. The weaker Raman signals observed around the edges of the BN nanosheets are likely due to a mixing of signals from both the BN/Ag and Ag/SiO_2 regions. As previously mentioned, BN nanosheets are promising barriers for long-term protecting metals from oxidation due to their excellent thermal stability, high impermeability, and resistance to galvanic corrosion. This makes them ideal for regenerating SERS substrates by heating them in air to remove adsorbed organic molecules. The reusability of 2L/Ag, Bulk/Ag, and Ag/SiO_2 substrates over up to 30 cycles is shown in Figure 6.10. In each cycle, the substrates were heated at 360°C in air and then reused by immersing them in a R6G solution (10^{-6} mol L^{-1}), since the Raman signal of the Ag/SiO_2 substrate was too weak for the reusability test at concentrations lower than 10^{-6} mol L^{-1}. After the heating/cleaning treatment, no Raman signals from R6G were detected on the 2L/Ag and Bulk/Ag samples, except for a Si band around 1000 cm^{-1} and/or the G band of h-BN at 1366 cm^{-1} (black spectra in Figure 6.10a and b), indicating effective removal of the analyte. After 30 cycles of reuse, the Raman intensity of R6G from the 2L/Ag substrate reduced by only about 10%, indicating that repeated heating in air has a minimal effect on the plasmonic enhancement of Ag NPs. Raman enhancement from the Bulk/Ag substrate slightly weakened with the increase of cycle numbers, dropping by 40% after 30 cycles of reuse. The difference was attributed to the different morphologies of BN nanosheets and bulk *h*-BN on Ag particles. BN nanosheets are flexible, allowing them to be wrinkled to better 'seal' the underlying Ag particles from air, while thick h-BN is not able to deform as much as BN nanosheets, and between thick h-BN and SiO_2 wafer, there exists a gap where air can penetrate

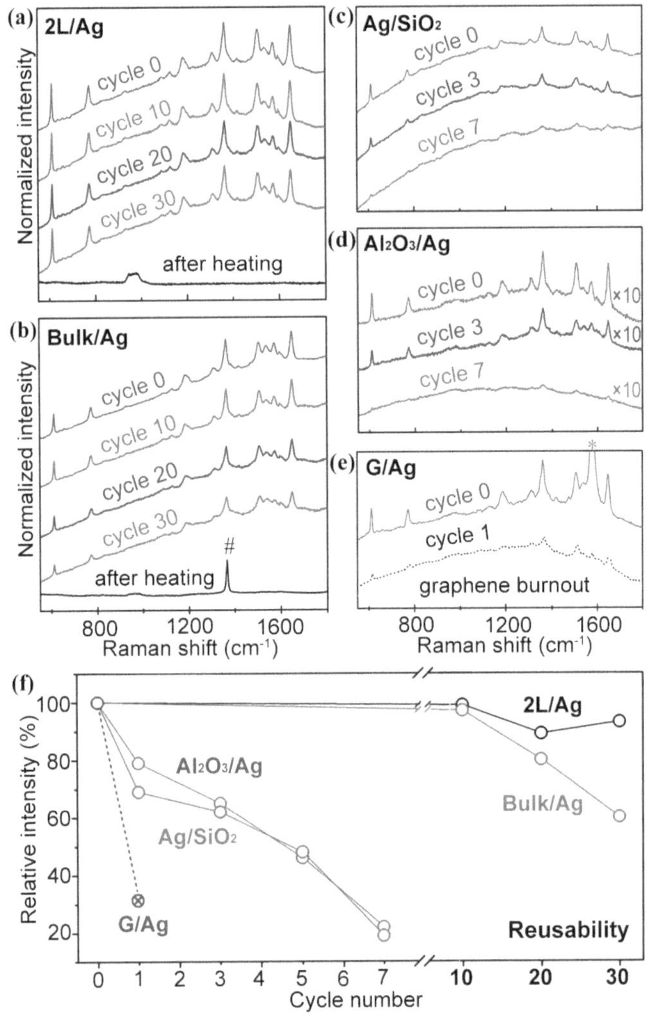

Figure 6.10 Reusability tests of various SERS substrates. (a) 2L/Ag tested over 30 cycles. (b) Bulk/Ag tested for 30 cycles; the # symbol indicates the G band of bulk h-BN. The black spectra in panels (a) and (b) show results after heating without re-immersion. (c) Ag particles on SiO_2 without BN protection (Ag/SiO_2) tested for seven cycles. (d) Al_2O_3/Ag tested for seven cycles, and (e) G/Ag tested for one cycle, where * denotes the G band of graphite. (f) Summary of the changes in Raman enhancement across these SERS substrates, based on the intensity of the 612 cm^{-1} peak. (Adapted with permission from Ref. [38]. Copyright 2016, John Wiley & Sons, Inc.)

to oxidize Ag particles. In contrast, bare Ag substrates experienced an approximately 80% loss in SERS enhancement after seven cycles of reuse [38].

For comparison, the reusability of graphene- and Al_2O_3-protected SERS substrates was investigated, since both have been used as barriers to protect Ag particles against oxidation [15,18]. A 1.07 nm thick Al_2O_3 film, comparable in thickness to 2L BN, was deposited on Ag/SiO_2 (Al_2O_3/Ag) by atomic layer deposition. The Raman intensity of R6G from the Al_2O_3/Ag substrate was about 10 times weaker than that from 2L/Ag. There are two reasons contributing to the reduced SERS enhancement: (1) Al_2O_3 has low affinity toward R6G molecules, resulting in less R6G captured on the substrate; and (2) Al_2O_3 reduces the strength of the electromagnetic field [62]. Furthermore, Al_2O_3 offered almost no protection to Ag particles, and its SERS enhancement dropped by 80% after seven cycles of reuse, similar to its bare Ag counterpart (Figure 6.10d and f). In addition, a 10 nm thick graphene layer was mechanically exfoliated on an Ag film, and the same procedure was used to prepare the graphene-covered Ag particle-enabled SERS substrate (G/Ag). The G/Ag exhibited comparable SERS enhancement to Bulk/Ag due to the strong adsorption capability of graphene, but with a lower fluorescence background resulting from charge transfer between graphene and R6G molecules. However, the G/Ag substrate is not reusable due to the inability of graphene against oxidation at 360°C in air, and the enhancement dropped to 30% of its initial intensity after one cycle of reuse (Figure 6.10e and f).

X-ray photoelectron spectroscopy was used to analyze the oxidation levels of Ag particles on these substrates after reusability tests (Figure 6.11). It is evident that Ag particles protected by BN nanosheets exhibited a dominating silver peak at 368.2 eV, indicating that almost no obvious oxidation happened after 30 cycles of reuse. In contrast, both Al_2O_3-protected Ag particles and bare Ag particles showed a strong silver oxide peak at 367.6 eV, suggesting a serious oxidation on

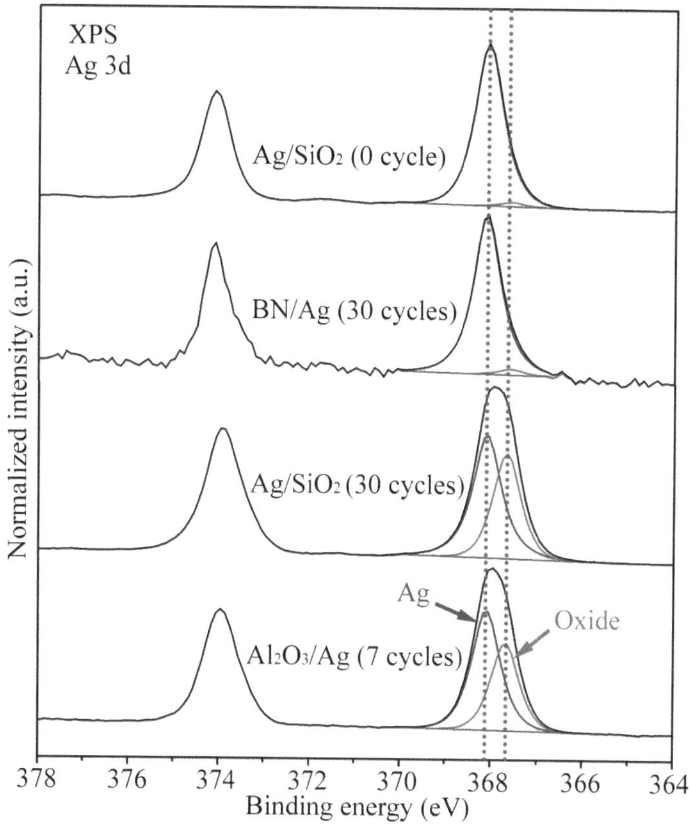

Figure 6.11 X-ray photoelectron spectroscopy spectra of Ag particles with and without BN protection, both before and after reusability tests. Dotted lines indicate the positions of the signals corresponding to silver and silver oxide. (Adapted with permission from Ref. [38]. Copyright 2016, John Wiley & Sons, Inc.)

Ag particles after reusability testing. These findings further confirmed that BN nanosheets are an excellent material for reusable SERS devices.

6.3.3 BNNS/Graphene Van Der Waals Heterostructure-Veiled Metal Nanostructures

As previously mentioned, BN nanosheets can significantly enhance the sensitivity, reusability, and homogeneity of conventional SERS substrates [38]. However, fluorescence background is another factor that affects the accuracy and sensitivity of SERS, particularly when sensing fluorescent and biological molecules. Fluorescence has a cross-section approximately six orders of magnitude larger than that of Raman scattering [63], leading to strong background signals that increase noise levels and obscure weak Raman signals [41]. Although Raman scattering can be enhanced by factors of 10^2 through CM and 10^3–10^8 through EM, fluorescence is simultaneously enhanced by up to 10^3. As a result, fluorescence remains at least 10 times stronger than Raman signals [41]. Electrically conductive substrates can help quench the fluorescence background by conducting away excited electrons. Using an excitation light source in the near-infrared region can also mitigate this issue, but some important wavelength-dependent Raman information may be lost [41,64]. Graphene is a well-known fluorescence quencher because of its conductivity and can protect Ag particles in ambient conditions due to its high impermeability [63,65], though it lacks reusability. Therefore, combining graphene with BN nanosheets, such as BN and graphene van der Waals heterostructure, could offer a promising solution to achieve highly sensitive, reusable, and homogeneous SERS while effectively suppressing the fluorescence background [60].

The BN nanosheet/graphene van der Waals heterostructure (BN/G) was fabricated by manually stacking [60], as shown in Figure 6.12. Specifically, BN nanosheets and graphene were exfoliated onto SiO_2/Si wafers. The BN nanosheet was then transferred onto the graphene layer using a PMMA-assisted transfer method. Next, a 10 nm thick layer of silver (Ag) film was sputter-coated onto the SiO_2/Si substrate, followed by the transfer of the BN/G structure on top. The substrate was subsequently annealed in an Ar atmosphere at 450°C for 1 hour to convert the Ag film into Ag particles, resulting in the BN/G/Ag SERS substrate (Figure 6.12). Figure 6.13 illustrates the optical and AFM images of the 1L BN and graphene, as well as the BN/G and BN/G/Ag SERS substrates. The AFM images show that the BN/G structure closely follows the profile of the underlying Ag particles, which have an average size of 45 nm and are approximately 10 nm apart. This structure effectively preserves plasmonic 'hot spots' and provides protection against the oxidation of Ag particles [60].

To create a monolayer of R6G molecules on the BN/G/Ag SERS substrate, 20 μL of a 10^{-6} mol L^{-1} R6G ethanol solution was drop-cast onto each 15×15 mm substrate and then dried using a fan. The SERS performance of the BN/G/Ag substrate, including sensitivity/EF, fluorescence background quenching, and reusability, was evaluated and compared with that of Ag particles covered by graphene or BN alone (*i.e.*, G/Ag and BN/Ag) and bare Ag, as shown in Figure 6.14a. The bare Ag substrate exhibited a strong fluorescence background (blue-shaded area) and reduced SERS sensitivity. This fluorescence occurred because the π electrons in the carbon atoms ($2p_z$ orbitals) and the *n* electrons in the oxygen atoms (non-bonding electrons) in the xanthene structure were excited to higher energy states (π*) by the laser. These π* electrons then relaxed back to the ground state by emitting photons [66]. While the fluorescence from R6G molecules directly in contact with Ag particles could be quenched, fluorescence from R6G molecules situated between Ag particles was not effectively quenched, leading to a strong fluorescence background. In contrast, the G/Ag substrate exhibited significantly enhanced Raman peaks

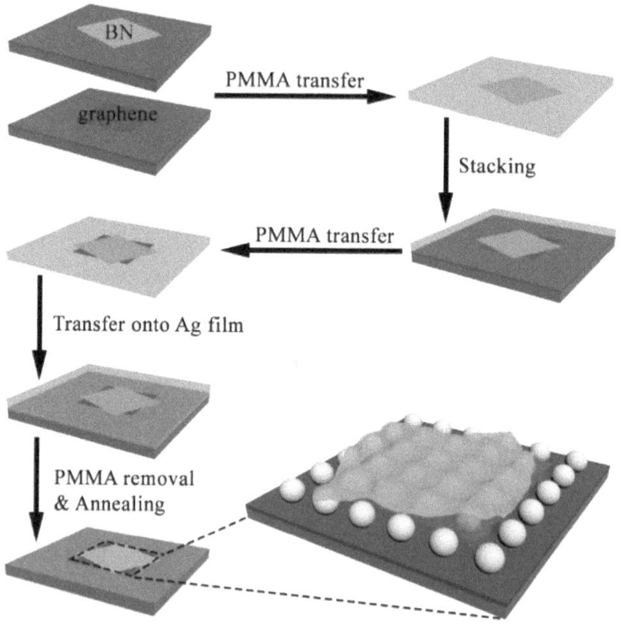

Figure 6.12 Schematic diagram illustrating the preparation of the BN/G/Ag substrate. (Reprinted with permission from Ref. [41]. Copyright 2021 American Chemical Society.)

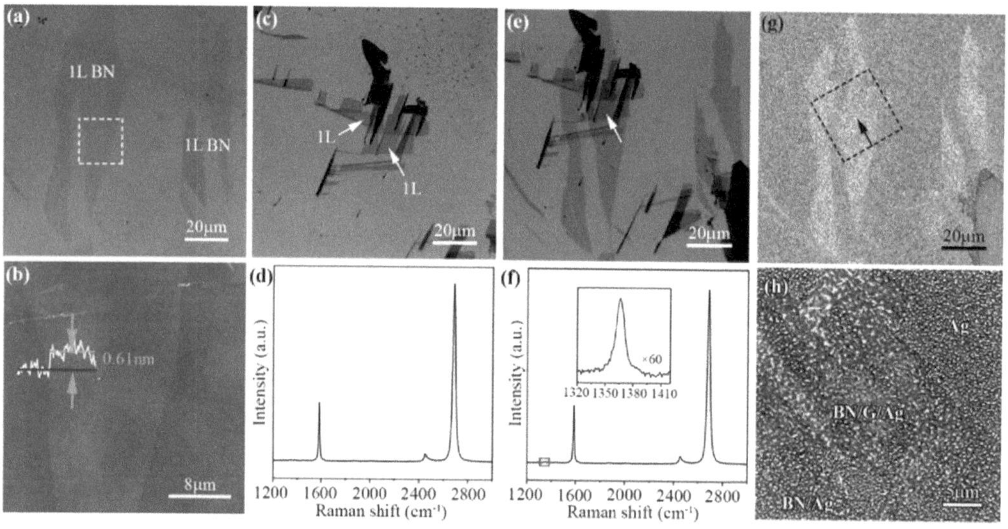

Figure 6.13 Optical (a) and AFM (b) images of 1L BN samples on SiO_2/Si. (c) Optical and Raman spectra (d) of graphene at various thicknesses on SiO2/Si. (e) Optical image and (f) Raman spectrum (f) of a BN/G heterostructure, including an enlarged view of the Raman G band for 1L BN within the red rectangle. (g) Optical and (h) AFM images of the BN/G/Ag SERS substrate after converting the Ag film into NPs by annealing at 450°C in Ar for 1 hour. (Reprinted with permission from Ref. [41]. Copyright 2021 American Chemical Society.)

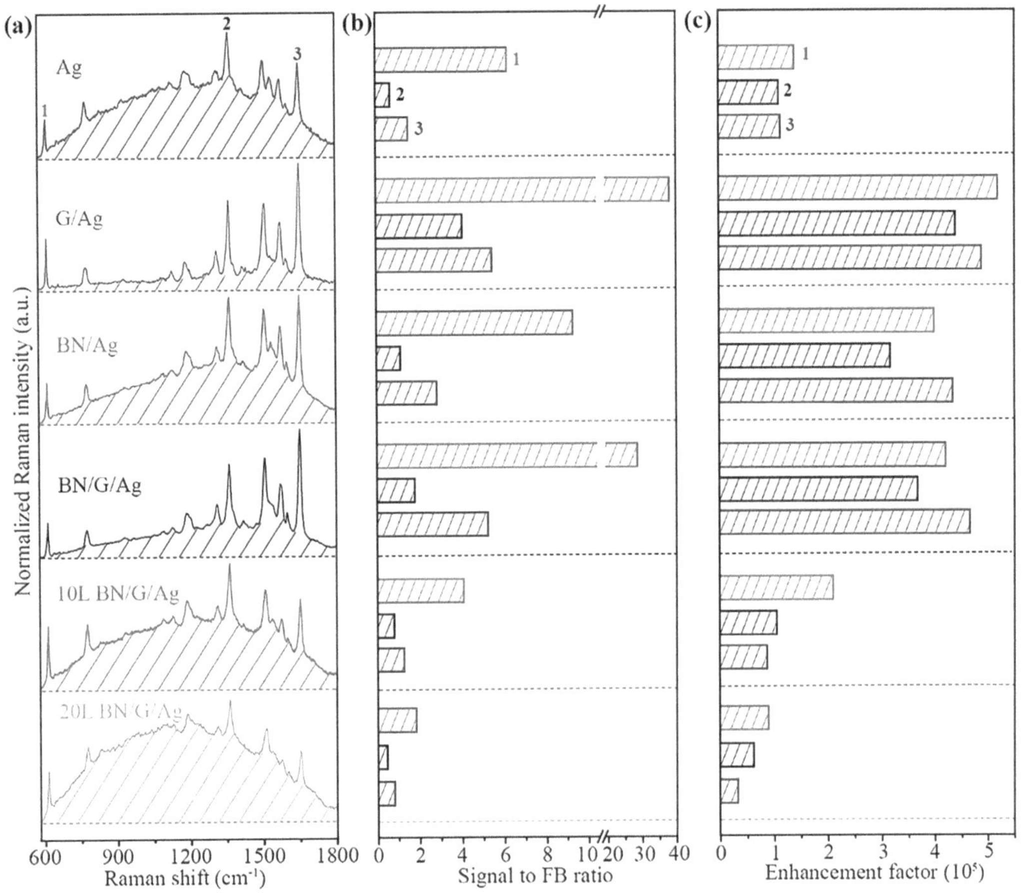

Figure 6.14 (a) Raman spectra of 10^{-6}mol L^{-1} R6G on various substrates: Ag NPs, G/Ag, BN/Ag, BN/G/Ag, 10L BN/G/Ag, and 20L BN/G/Ag. (b) Comparison of the S/FB ratio and (c) EF for the 613, 1363, and 1650 cm^{-1} peaks of R6G across these SERS substrates. (Reprinted with permission from Ref. [41]. Copyright 2021 American Chemical Society.)

with a much weaker fluorescence background, attributed to the quenching effect of graphene (Figure 6.14a). The BN/Ag substrate also increased the Raman intensity of R6G, but with a moderate fluorescence background. Although the π* electrons of R6G molecules directly above the Ag particles can tunnel through the 1L BN and transfer to the underlying Ag particles, resulting in quenched fluorescence, this charge transfer is less effective for R6G molecules situated between Ag particles, similar to the bare Ag substrate [41].

In comparison, the BN/G/Ag substrate demonstrated strong Raman intensity with a substantially suppressed fluorescence background. To illustrate the fluorescence quenching effect of BN/G/Ag, the signal-to-fluorescence background ratio (S/FB) and EF for the 613, 1363, and 1650 cm^{-1} Raman peaks of R6G were compared across different SERS substrates, as shown in Figure 6.14b and c. For the 1363 cm^{-1} peak, the Ag substrate had an S/FB ratio and EF of 0.67 and 1.11×10^5, respectively. The G/Ag substrate achieved the highest S/FB ratio (4.16) and EF (4.40×10^5) due to graphene's additional CM and reduced fluorescence background. The BN/Ag substrate had an S/FB ratio of 1.10 and EF of 3.15×10^5. Compared to the Ag and BN/Ag substrates, the BN/G/Ag substrate improved the S/FB ratio by 168% and 64%, respectively, and the EF by 229% and 16%, respectively. The improvements of the BN/G/Ag substrate are primarily due to the following factors: (1) The excited electrons of R6G molecules can tunnel through the 1L BN, facilitating charge transfer between R6G molecules and graphene, leading to both fluorescence quenching and CM in addition to EM. (2) The high flexibility of BN/G allows the heterostructure to closely follow the profile of the underlying Ag NPs, maximizing the preservation of plasmonic hot spots. Furthermore, BN nanosheets exhibit excellent adsorption capabilities for aromatic molecules due to π–π interactions and conformational changes, further enhancing Raman sensitivity [41]. When thicker BN (*e.g.*, 10L and 20L) was used in the van der Waals heterostructure, the fluorescence background noticeably increased with the thickness of BN (as shown in pink and orange in Figure 6.14a), while the Raman enhancement decreased. This was attributed to reduced electron tunneling through the thicker BN and a decreased strength of the electromagnetic field. These findings further support the previous explanation. Similar to the previously discussed 2L/Ag SERS substrate, the BN/G/Ag SERS substrate is also highly reusable, with only a slight loss of SERS enhancement after multiple reuse cycles. This high reusability is due to the protective role of BN nanosheets, which shield the underlying graphene and Ag particles from oxidation during heat treatments used to regenerate the SERS substrate.

6.4 BN NANOSHEETS FOR MEF

Fluorescence is highly useful in various fields, such as chemical detection, biomedical diagnosis, and forensics [67–70]. However, its limited intensity hinders wider applications. MEF addresses this issue by amplifying the luminescence intensity through localized surface plasmon resonance [71–73]. As aforementioned, quenching normally occurs when fluorophores are in close proximity to metals *via* non-radiative decay. To mitigate this quenching effect and maximize luminescence, a dielectric spacer is necessary between the fluorophore and the plasmonic metal. Traditional spacer materials, such as polymers, alumina (Al_2O_3), and silicon dioxide (SiO_2), tend to have considerable thickness and inherent defects, resulting in unsatisfactory sensitivity and intensity of MEF [74–77]. In contrast, BN nanosheets could be an ideal spacer material for MEF due to their atomically thin thickness, high impermeability, and excellent dielectric properties.

The fabrication process of the MEF substrate is shown in Figure 6.15. Specifically, all Ag films were deposited at a pressure of $\sim5\times10^{-2}$ mbar and current of 20 mA using a Kaoduen sputtering system. Then, BN and graphene nanosheets were mechanically exfoliated on the Ag/SiO_2/Si substrates. Suitable nanosheets with specific thickness were located by optical microscopy, and their thickness was determined using an AFM. Finally, the Ag films were heated under an argon (Ar) gas atmosphere at different temperatures to convert to Ag NPs, forming the MEF substrate. The rhodamine 6G (R6G) water solution (10^{-6} mol L^{-1}) was spin-coated onto different MEF substrates at 2000 rpm for 1 minute using a spin-coater. For the reusability test, the substrates with R6G were annealed at 350°C in air for 5 minutes and re-spin-coated at the same parameters up to 30 times.

To clarify the distinct effects of BN and graphene nanosheets on MEF, high-quality BN and graphene nanosheets with thickness from 4~20 nm were mechanically exfoliated on Ag films (70 nm) covered silicon (SiO_2/Si) wafers, shown in Figure 16.6a and b. The R6G molecules were spin-coated on top of BN/Ag/SiO_2/Si and graphene/Ag/SiO_2/Si substrates under the same conditions.

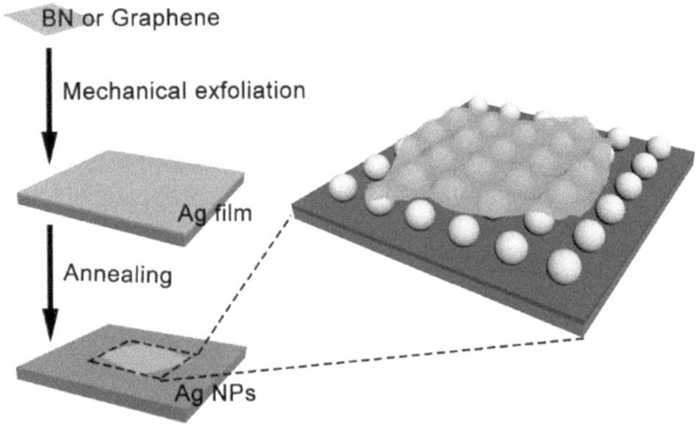

Figure 6.15 Schematic diagram illustrating the preparation of the BN/Ag or G/Ag substrate.

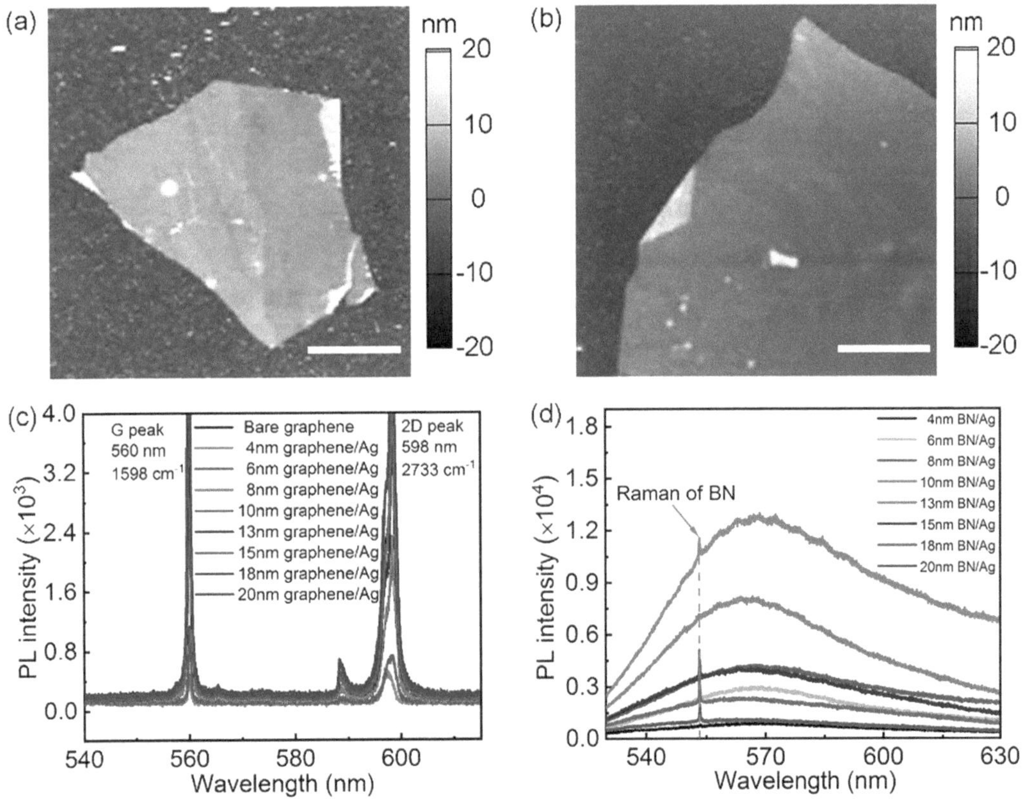

Figure 6.16 Schematic diagram illustrating the preparation of the BN/Ag or G/Ag substrate. Typical AFM images of the exfoliated (a) graphene and (b) BN on 70 nm Ag films produced by sputtering (both scale bars 5 μm). PL spectra of R6G on the 70 nm Ag film covered by (c) graphene and (d) BN sheets of different thicknesses. The sharp peaks in (c) and (d) were Raman signals of graphene and BN, respectively. (Reprinted with permission from Ref. [42]. Copyright 2019 American Chemical Society.)

As expected, the photoluminescence (PL) signal from the graphene/Ag/SiO_2/Si substrate was negligible regardless of the graphene thickness, while the PL signal from the BN/Ag/SiO_2/Si substrate was significantly stronger and varied with the BN thickness, shown in Figure 16.6c and d, respectively. This discrepancy is attributed to the synergistic effect between plasmon-enhanced excitation and antenna effect [78–81]. As can be seen in Figure 16.6d, the optimized thickness of the BN spacer is determined to be between 10~14 nm. To maximize the EF, we focused on the BN nanosheets with thickness within this range while converting the Ag films into more plasmonically effective Ag NPs by annealing. As the MEF largely depends on the overlap between the localized surface plasmon resonance spectrum and the excitation spectrum of the fluorescence, we optimized the localized surface plasmon resonance band of Ag NPs by adjusting both the Ag film thickness and annealing temperatures. Figure 6.17a and b illustrates the AFM images of the Ag/SiO_2/Si MEF substrates before and after annealing, with the BN nanosheets of 12 nm. As shown in Figure 6.17c, the 12 nm BN provided the best enhancement with a PL intensity of 5.5×10^6, while the signal of bare Ag NPs without a spacer layer was approximately 6 orders of magnitude weaker due to the quenching effect. As shown in Figure 6.17d, the EF of 12 nm BN/Ag/SiO_2/Si MEF substrates was calculated to be ~95±5, which is the highest among those previously reported results [82–84]. Besides, the PL mapping from the box in Figure 6.17e is shown in Figure 6.17f, indicating that the PL signal from our MEF substrate was homogeneous.

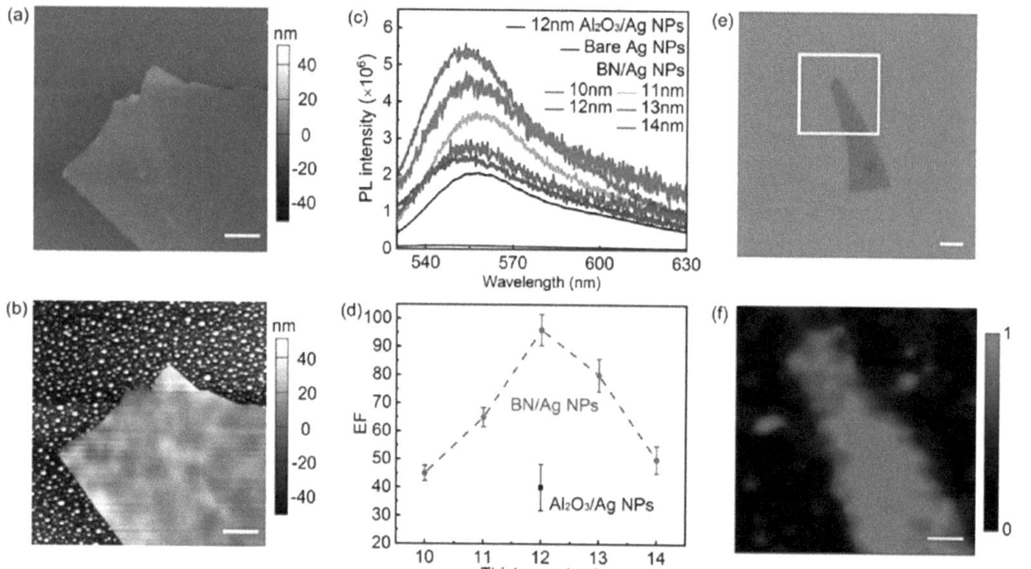

Figure 6.17 MEF from the thin BN-sheet-covered Ag NPs. AFM images of the 12 nm thick BN sheet on the 12 nm Ag film on SiO_2/Si (a) before and (b) after the annealing at 400°C (scale bar: 2 μm). (c) Representative PL spectra of R6G on the Ag NPs covered by BN sheets having different thicknesses (10–14 nm) and 12 nm Al_2O_3 and that of the structure without a spacer (bare Ag NPs). (d) Fluorescence EFs as a function of the BN thickness. (e) Optical image of the 12 nm BN-covered Ag NPs (scale bar: 5 μm). (f) Corresponding fluorescence map of the spin-coated R6G, within the white square area shown in (e), where the color scale bar between 0.0 and 1.0 represents the weakest to the strongest fluorescence, respectively (scale bar: 2 μm). (Reprinted with permission from Ref. [42]. Copyright 2019 American Chemical Society.)

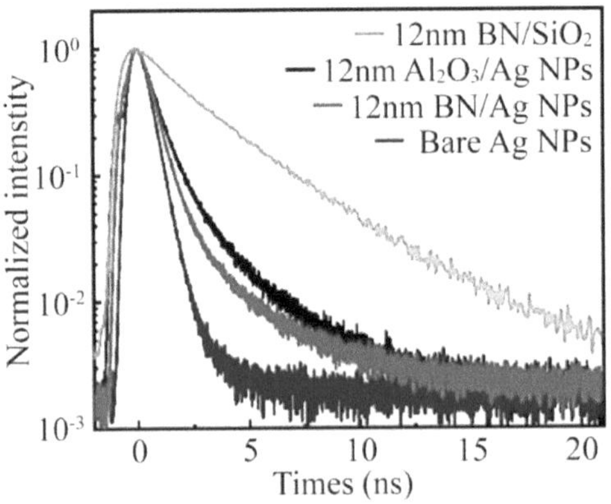

Figure 6.18 Fluorescence lifetime. The lifetimes of R6G on the bare Ag NPs (0.46 ns), 12 nm BN/Ag NPs (1.12 ns), 12 nm Al_2O_3/Ag NPs (2.06 ns), and 12 nm BN/SiO_2 (3.83 ns) are compared. (Reprinted with permission from Ref. [42]. Copyright 2019 American Chemical Society.)

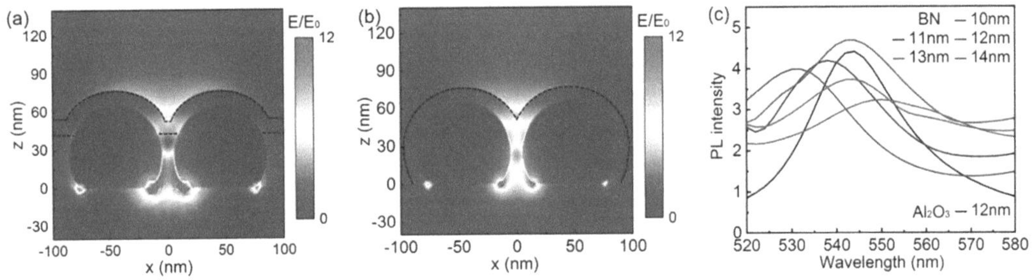

Figure 6.19 Theoretical simulations. Calculated electric field enhancements in the *x*–*z* plane for the (a) 12 nm BN/Ag dimers and (b) 12 nm Al_2O_3/Ag dimers; (c) simulated fluorescence spectra for the different MEF substrates. (Reprinted with permission from Ref. [42]. Copyright 2019 American Chemical Society.)

To elucidate the enhancement mechanism, we measured the fluorescence lifetime of R6G on various substrates, including bare Ag NPs (0.46 ns), 12 nm BN/Ag NPs (1.12 ns), 12 nm Al_2O_3/Ag NPs (2.06 ns), and 12 nm BN/SiO_2 (3.83 ns), as shown in Figure 6.18. The bare Ag NPs exhibited the shortest lifetime and weakest fluorescence, which is due to the quenching effect (molecules sit directly on Ag NPs) dominating the Purcell factor [81]. As for the other three MEF substrates, the shorter lifetime means stronger coupling between R6G and the plasmonic Ag NPs, leading to stronger PL intensity [85–87].

Finite-element calculations were also conducted to further understand the physical origin of the MEF from Ag NPs spaced by Al_2O_3 and BN, shown in Figure 16.9a and b. The plasmon-enhanced electric field was polarized along the *x* axis, which also corresponds to the plasmonic dimer axis. The electric field on the BN spacer was found to be stronger than that of Al_2O_3, which agrees well with the previous lifetime results. As the *x* component of the electric field was stronger, it should excite dipoles with the same orientation more efficiently. Theoretically and experimentally, results indicated that the R6G physisorbed on BN preferred a 'lying-down' orientation, where the xanthene group is parallel to the BN surface [37]. As the orientation of R6G on Al_2O_3 has not been reported, we had to assume that the R6G molecules on top of Ag NPs were oriented along the x axis (similar to the case of BN), while R6G in the valley between two Ag NPs was oriented along the z axis. The improved excitation efficiency is due to the aligned orientation of molecules and plasmonic electric field, leading to the best MEF performance of R6G on BN/Ag NP substrates, shown in Figure 16.9c. Our calculations indicated that both plasmon-enhanced excitation and antenna effect contributed to the large PL intensity, and the plasmonic geometry of our Ag NPs was well-suited for MEF.

6.5 CONCLUSIONS

In conclusion, our research demonstrates that BN nanosheets possess a strong surface adsorption capability, as confirmed by both experimental and theoretical analyses. Despite having the same surface area, atomically thin BN nanosheets adsorb organic molecules more efficiently than bulk h-BN crystals. Theoretical calculations and Raman spectroscopy reveal that this phenomenon arises from conformational changes in the nanosheets, which increase both adsorption energy and efficiency. In contrast, bulk *h*-BN crystals do not experience these conformational changes due to their significantly higher rigidity. When combined with their exceptional properties—such as high thermal stability, flexibility, and gas impermeability—this strong adsorption capability makes BN nanosheets a promising material for reusable SERS and MEF applications, offering enhanced sensitivity and homogeneity. This study not only provides a fresh perspective on the adsorption behavior of nanomaterials but also expands the applications of reusable SERS and MEF while reducing fabrication costs.

REFERENCES

1. D. Jeanmaire, R. Van Duyne, Heterocyclic, aromatic, and aliphatic amines adsorbed on the anodized silver electrode. *Journal of Electroanalytical Chemistry* **84**, 1–20 (1977).
2. M. G. Albrecht, J. A. Creighton, Anomalously intense Raman spectra of pyridine at a silver electrode. *Journal of the American Chemical Society* **99**, 5215–5217 (1977).
3. S. Nie, S. R. Emory, Probing single molecules and single nanoparticles by surface-enhanced Raman scattering. *Science* **275**, 1102–1106 (1997).
4. L. Yang et al., A novel ultra-sensitive semiconductor SERS substrate boosted by the coupled resonance effect. *Advanced Science* **6**, 1900310 (2019).
5. J. Langer et al., Present and future of surface-enhanced Raman scattering. *ACS Nano* **14**, 28–117 (2019).
6. A. Campion, P. Kambhampati, Surface-enhanced Raman scattering. *Chemical Society Reviews* **27**, 241–250 (1998).
7. S. Schlücker, Surface-enhanced Raman spectroscopy: Concepts and chemical applications. *Angewandte Chemie International Edition* **53**, 4756–4795 (2014).
8. K. N. Kanipe, P. P. Chidester, G. D. Stucky, M. Moskovits, Large format surface-enhanced Raman spectroscopy substrate optimized for enhancement and uniformity. *ACS Nano* **10**, 7566–7571 (2016).
9. W. J. Cho, Y. Kim, J. K. Kim, Ultrahigh-density array of silver nanoclusters for SERS substrate with high sensitivity and excellent reproducibility. *ACS Nano* **6**, 249–255 (2012).
10. L. Brus, Noble metal nanocrystals: Plasmon electron transfer photochemistry and single-molecule Raman spectroscopy. *Accounts of Chemical Research* **41**, 1742–1749 (2008).

11. Y. Yang et al., Controlled fabrication of silver nanoneedles array for SERS and their application in rapid detection of narcotics. *Nanoscale* **4**, 2663–2669 (2012).
12. X. Ling et al., Can graphene be used as a substrate for Raman enhancement? *Nano Letters* **10**, 553–561 (2010).
13. P. Aldeanueva-Potel, E. Faoucher, R. A. Alvarez-Puebla, L. M. Liz-Marzan, M. Brust, Recyclable molecular trapping and SERS detection in silver-loaded agarose gels with dynamic hot spots. *Analytical Chemistry* **81**, 9233–9238 (2009).
14. Y. Lin, C. E. Bunker, K. S. Fernando, J. W. Connell, Aqueously dispersed silver nanoparticle-decorated boron nitride nanosheets for reusable, thermal oxidation-resistant surface enhanced Raman spectroscopy (SERS) devices. *ACS Applied Materials & Interfaces* **4**, 1110–1117 (2012).
15. X. Li et al., Silver nanoparticles protected by monolayer graphene as a stabilized substrate for surface enhanced Raman spectroscopy. *Carbon* **66**, 713–719 (2014).
16. X. Y. Zhang, J. Zhao, A. V. Whitney, J. W. Elam, R. P. Van Duyne, Ultrastable substrates for surface-enhanced Raman spectroscopy: Al_2O_3 overlayers fabricated by atomic layer deposition yield improved anthrax biomarker detection. *Journal of the American Chemical Society* **128**, 10304–10309 (2006).
17. J. F. Li et al., Shell-isolated nanoparticle-enhanced Raman spectroscopy. *Nature* **464**, 392–395 (2010).
18. S. M. Mahurin, J. John, M. J. Sepaniak, S. Dai, A reusable surface-enhanced Raman scattering (SERS) substrate prepared by atomic layer deposition of alumina on a multi-layer gold and silver film. *Applied Spectroscopy* **65**, 417–422 (2011).
19. L. W. Ma, Y. Huang, M. J. Hou, Z. Xie, Z. J. Zhang, Silver nanorods wrapped with ultrathin Al_2O_3 layers exhibiting excellent SERS sensitivity and outstanding SERS stability. *Scientific Reports* **5**, 12890 (2015).
20. J. A. Dieringer et al., Surface enhanced Raman spectroscopy: New materials, concepts, characterization tools, and applications. *Faraday Discussions* **132**, 9–26 (2006).
21. J. R. Anema, J. F. Li, Z. L. Yang, B. Ren, Z. Q. Tian, Shell-isolated nanoparticle-enhanced Raman spectroscopy: Expanding the versatility of surface-enhanced Raman scattering. *Annual Review of Analytical Chemistry* **4**, 129–150 (2011).
22. X. Ling, J. Zhang, First-layer effect in graphene-enhanced Raman scattering. *Small* **6**, 2020–2025 (2010).
23. H. H. Cheng et al., Graphene-quantum-dot assembled nanotubes: A new platform for efficient Raman enhancement. *ACS Nano* **6**, 2237–2244 (2012).
24. E. S. Thrall, A. C. Crowther, Z. H. Yu, L. E. Brus, R6G on graphene: High raman detection sensitivity, yet decreased Raman cross-section. *Nano Letters* **12**, 1571–1577 (2012).
25. W. Xu et al., Graphene-veiled gold substrate for surface-enhanced Raman spectroscopy. *Advanced Materials* **25**, 928–933 (2013).
26. W. Xu et al., Surface enhanced Raman spectroscopy on a flat graphene surface. *Proceeding of the National Academy of Science of the United States of America* **109**, 9281–9286 (2012).
27. Z. Zhang et al., A facile one-pot method to high-quality Ag-graphene composite nanosheets for efficient surface-enhanced Raman scattering. *Chemical Communications* **47**, 6440–6442 (2011).
28. L. H. Li et al., Dielectric screening in atomically thin boron nitride nanosheets. *Nano Letters* **15**, 218–223 (2014).
29. D. Prasai, J. C. Tuberquia, R. R. Harl, G. K. Jennings, K. I. Bolotin, Graphene: Corrosion-inhibiting coating. *ACS Nano* **6**, 1102–1108 (2012).
30. S. Chen et al., Oxidation resistance of graphene-coated Cu and Cu/Ni alloy. *ACS Nano* **5**, 1321–1327 (2011).
31. F. Zhou, Z. Li, G. J. Shenoy, L. Li, H. Liu, Enhanced room-temperature corrosion of copper in the presence of graphene. *ACS Nano* **7**, 6939–6947 (2013).
32. M. Schriver et al., Graphene as a long-term metal oxidation barrier: Worse than nothing. *ACS Nano* **7**, 5763–5768 (2013).
33. L. Liu et al., Graphene oxidation: Thickness-dependent etching and strong chemical doping. *Nano Letters* **8**, 1965–1970 (2008).

34. L. H. Li, J. Cervenka, K. Watanabe, T. Taniguchi, Y. Chen, Strong oxidation resistance of atomically thin boron nitride nanosheets. *ACS Nano* **8**, 1457–1462 (2014).
35. L. H. Li, T. Xing, Y. Chen, R. Jones, Boron nitride nanosheets for metal protection. *Advanced Materials Interfaces* **1**, 1300132 (2014).
36. Z. Liu et al., Ultrathin high-temperature oxidation-resistant coatings of hexagonal boron nitride. *Nature Communications* **4**, 2541 (2013).
37. Q. Cai et al., Molecule-induced conformational change in boron nitride nanosheets with enhanced surface adsorption. *Advanced Functional Materials* **26**, 8202–8210 (2016).
38. Q. Cai et al., Boron nitride nanosheets improve sensitivity and reusability of surface-enhanced Raman spectroscopy. *Angewandte Chemie* **128**, 8545–8549 (2016).
39. Q. Cai et al., Boron nitride nanosheets as improved and reusable substrates for gold nanoparticles enabled surface enhanced Raman spectroscopy. *Physical Chemistry Chemical Physics* **17**, 7761–7766 (2015).
40. Q. Cai et al., Boron nitride nanosheet-veiled gold nanoparticles for surface-enhanced Raman scattering. *ACS Applied Materials & Interfaces* **8**, 15630–15636 (2016).
41. Q. Cai et al., Two-dimensional Van der Waals heterostructures for synergistically improved surface-enhanced Raman spectroscopy. *ACS Applied Materials & Interfaces* **12**, 21985–21991 (2020).
42. W. Gan et al., Atomically thin boron nitride as an ideal spacer for metal-enhanced fluorescence. *ACS Nano* **13**, 12184–12191 (2019).
43. L. Hua Li et al., High-resolution x-ray absorption studies of core excitons in hexagonal boron nitride. *Applied Physics Letters* **101**, 191604 (2012).
44. Q. Cai et al., Raman signature and phonon dispersion of atomically thin boron nitride. *Nanoscale* **9**, 3059–3067 (2017).
45. X. Ling et al., Raman enhancement effect on two-dimensional layered materials: Graphene, h-BN and MoS_2. *Nano Letters* **14**, 3033–3040 (2014).
46. Z. Chen et al., Physical adsorption and charge transfer of molecular Br_2 on graphene. *ACS Nano* **8**, 2943–2950 (2014).
47. T. Mohiuddin et al., Uniaxial strain in graphene by Raman spectroscopy: G peak splitting, Grüneisen parameters, and sample orientation. *Physical Review B—Condensed Matter and Materials Physics* **79**, 205433 (2009).
48. Q. Cai et al., High thermal conductivity of high-quality monolayer boron nitride and its thermal expansion. *Science Advances* **5**, eaav0129 (2019).
49. Z. Zong, C.-L. Chen, M. R. Dokmeci, K.-T. Wan, Direct measurement of graphene adhesion on silicon surface by intercalation of nanoparticles. *Journal of Applied Physics* **107**, 026104 (2010).
50. S. P. Koenig, N. G. Boddeti, M. L. Dunn, J. S. Bunch, Ultrastrong adhesion of graphene membranes. *Nature Nanotechnology* **6**, 543–546 (2011).
51. N. G. Boddeti et al., Mechanics of adhered, pressurized graphene blisters. *Journal of Applied Mechanics* **80**, 040909 (2013).
52. M. Chen et al., 2D materials: Excellent substrates for surface-enhanced Raman scattering (SERS) in chemical sensing and biosensing. *TrAC Trends in Analytical Chemistry* **130**, 115983 (2020).
53. H. Lai, F. Xu, Y. Zhang, L. Wang, Recent progress on graphene-based substrates for surface-enhanced Raman scattering applications. *Journal of Materials Chemistry B* **6**, 4008–4028 (2018).
54. Z. Yin et al., Recent progress on two-dimensional layered materials for surface enhanced Raman spectroscopy and their applications. *Materials Today Physics* **18**, 100378 (2021).
55. R. V. Gorbachev et al., Hunting for monolayer boron nitride: Optical and Raman signatures. *Small* **7**, 465–468 (2011).
56. H. Zhou et al., Thickness-dependent morphologies of gold on N-layer graphenes. *Journal of the American Chemical Society* **132**, 944–946 (2010).
57. F. P. Chiaramonte, B. N. Rosenthal, Wettability of pyrolytic boron nitride by aluminum. *Journal of the American Ceramic Society* **74**, 658–661 (1991).
58 M. Ishigami, J.-H. Chen, W. G. Cullen, M. S. Fuhrer, E. D. Williams, Atomic structure of graphene on SiO_2. *Nano Letters* **7**, 1643–1648 (2007).

59. U. Stöberl, U. Wurstbauer, W. Wegscheider, D. Weiss, J. Eroms, Morphology and flexibility of graphene and few-layer graphene on various substrates. *Applied Physics Letters* **93**, 051906 (2008).
60. Q. Cai et al., Boron nitride nanosheets for surface-enhanced Raman spectroscopy. *Materials Today Physics* **22**, 100575 (2022).
61. G. Kim et al., Hexagonal boron nitride/Au substrate for manipulating surface plasmon and enhancing capability of surface-enhanced Raman spectroscopy. *ACS Nano* **10**, 11156–11162 (2016).
62. J. A. Dieringer et al., Introductory lecture surface enhanced Raman spectroscopy: New materials, concepts, characterization tools, and applications. *Faraday Discussions* **132**, 9–26 (2006).
63. L. Xie, X. Ling, Y. Fang, J. Zhang, Z. Liu, Graphene as a substrate to suppress fluorescence in resonance Raman spectroscopy. *Journal of the American Chemical Society* **131**, 9890–9891 (2009).
64. H. Sato, H. Chiba, H. Tashiro, Y. Ozaki, Excitation wavelength-dependent changes in Raman spectra of whole blood and hemoglobin: Comparison of the spectra with 514.5-, 720-, and 1064-nm excitation. *Journal of Biomedical Optics* **6**, 366–370 (2001).
65. W. Lin et al., Graphene-based fluorescence-quenching-related Fermi level elevation and electron-concentration surge. *Nano Letters* **16**, 5737–5741 (2016).
66. J. R. Lakowicz, Radiative decay engineering: Biophysical and biomedical applications. *Analytical Biochemistry* **298**, 1–24 (2001).
67. Y. Zhang et al., Real-time tracking of fluorescent magnetic spore–based microrobots for remote detection of *C. diff* toxins. *Science Advances* **5**, eaau9650 (2019).
68. A. Camposeo et al., Metal-enhanced near-infrared fluorescence by micropatterned gold nanocages. *ACS Nano* **9**, 10047–10054 (2015).
69. D. Kim et al., Dual-color fluorescent nanoparticles showing perfect color-specific photoswitching for bioimaging and super-resolution microscopy. *Nature Communications* **10**, 3089 (2019).
70. F. Cheng et al., Enhanced photoluminescence of monolayer WS_2 on Ag films and nanowire–WS_2–film composites. *ACS Photonics* **4**, 1421–1430 (2017).
71. J. R. Lakowicz, Radiative decay engineering 5: Metal-enhanced fluorescence and plasmon emission. *Analytical Biochemistry* **337**, 171–194 (2005).
72. A. J. Haes, S. Zou, J. Zhao, G. C. Schatz, R. P. Van Duyne, Localized surface plasmon resonance spectroscopy near molecular resonances. *Journal of the American Chemical Society* **128**, 10905–10914 (2006).
73. P. Pompa et al., Metal-enhanced fluorescence of colloidal nanocrystals with nanoscale control. *Nature Nanotechnology* **1**, 126–130 (2006).
74. M. Yi et al., Fluorescence enhancement caused by plasmonics coupling between silver nano-cubes and silver film. *Plasmonics* **6**, 213–217 (2011).
75. T. Murakami, Y. Arima, M. Toda, H. Takiguchi, H. Iwata, Effect of dielectric spacer thickness on signal intensity of surface plasmon field-enhanced fluorescence spectroscopy. *Analytical Biochemistry* **421**, 632–639 (2012).
76. M. Saboktakin et al., Metal-enhanced upconversion luminescence tunable through metal nanoparticle–nanophosphor separation. *ACS Nano* **6**, 8758–8766 (2012).
77. P. J. Tarcha, J. DeSaja-Gonzalez, S. Rodriguez-Llorente, R. Aroca, Surface-enhanced fluorescence on SiO_2-coated silver island films. *Applied Spectroscopy* **53**, 43–48 (1999).
78. M. Bauch, K. Toma, M. Toma, Q. Zhang, J. Dostalek, Plasmon-enhanced fluorescence biosensors: A review. *Plasmonics* **9**, 781–799 (2014).
79. J.-F. Li, C.-Y. Li, R. F. Aroca, Plasmon-enhanced fluorescence spectroscopy. *Chemical Society Reviews* **46**, 3962–3979 (2017).
80. P. Anger, P. Bharadwaj, L. Novotny, Enhancement and quenching of single-molecule fluorescence. *Physical Review Letters* **96**, 113002 (2006).
81. H. Kuhn, Classical aspects of energy transfer in molecular systems. *The Journal of Chemical Physics* **53**, 101–108 (1970).
82. F. Xie et al., Nanoscale control of Ag nanostructures for plasmonic fluorescence enhancement of near-infrared dyes. *Nano Research* **6**, 496–510 (2013).
83. M. R. Gartia, A. Hsiao, M. Sivaguru, Y. Chen, G. L. Liu, Enhanced 3D fluorescence live cell imaging on nanoplasmonic substrate. *Nanotechnology* **22**, 365203 (2011).

84. A. J. Wood et al., Influence of silver grain size, roughness, and profile on the extraordinary fluorescence enhancement capabilities of grating coupled surface plasmon resonance. *RSC Advances* **5**, 78534–78544 (2015).
85. G. W. Ford, W. H. Weber, Electromagnetic interactions of molecules with metal surfaces. *Physics Reports* **113**, 195–287 (1984).
86. C. Ciracì, A. Rose, C. Argyropoulos, D. R. Smith, Numerical studies of the modification of photodynamic processes by film-coupled plasmonic nanoparticles. *Journal of the Optical Society of America B* **31**, 2601–2607 (2014).
87. C. Tserkezis, N. Stefanou, M. Wubs, N. A. Mortensen, Molecular fluorescence enhancement in plasmonic environments: Exploring the role of nonlocal effects. *Nanoscale* **8**, 17532–17541 (2016).

Chapter 7

Thickness-Dependent Thermal Properties of Boron Nitride Nanosheets

Qiran Cai, Lu Hua Li, Srikanth Mateti, and Ying Ian Chen

7.1 INTRODUCTION

With the growing demand for miniaturization in advanced electronics, effective thermal dissipation has become critical. Inefficient heat transfer can lead to significant local temperature increases, known as 'hot spots', which can severely impact performance, reliability, longevity, and safety of electronics [1,2]. Consequently, considerable efforts have been devoted to finding materials with exceptional thermal conductivity to address these overheating challenges. Traditional materials like graphite and diamond offer high thermal conductivity, exceeding 1000 W m^{-1} K^{-1} at room temperature. Emerging materials, such as carbon nanotubes [3], graphene [4], cubic boron arsenide (*c*-BAs) [5–7], and monoisotopic cubic boron nitride [8], also demonstrate remarkable thermal conductivities. However, graphene, graphite, and carbon nanotubes are electrically conductive, which limits their use in certain electronics that require insulating thermal conductors to prevent short circuits. Although diamond, *c*-BAs, and cubic boron nitride are electrical insulators suitable for high-power electronics, they are costly to produce due to the high-temperature and pressure processes required, and their brittleness makes them challenging to incorporate into flexible devices [9].

As discussed in Chapter 6, boron nitride nanosheets (BNNSs), a new form of *h*-BN, hold promise for thermal management in advanced electronics due to their exceptional properties. BNNSs possess a wide bandgap of approximately 6 eV, which is thickness independent [10], and are among the strongest electrically insulating materials while being highly flexible and stretchable [11]. Additionally, BNNSs exhibit excellent thermal stability, withstanding temperatures over 800°C in air without oxidation [12]. Monolayer (1L) BN, in particular, has demonstrated high thermal conductivity of 751 W m^{-1} K^{-1} [13], which can be enhanced by about 30% through isotope engineering [9]. Various factors, such as length, thickness, defects, isotope disorder, temperature, and strain, influence the thermal conductivity of BNNSs [14]. This chapter will focus on the thickness-dependent thermal conductivity of naturally occurring and monoisotopic BNNSs.

DOI: 10.1201/9781003647812-9

7.2 IN-PLANE THERMAL CONDUCTIVITY OF NATURALLY OCCURRING BN NANOSHEETS

Compared to their low cross-plane thermal conductivity, which results from weak interlayer van der Waals interactions, BNNSs exhibit exceptional in-plane thermal conductivity due to the strong in-plane B–N covalent bonds [13,15]. In BNNSs, thermal transport is primarily governed by phonons due to the absence of free electrons resulting from their large bandgap [16]. Two main techniques have been developed to measure the thermal conductivity of BNNSs: the microbridge technique [17–19] and the optothermal Raman method [9,13,20,21]. The microbridge technique generally requires complex device fabrication and a polymer-involved transfer process, which can introduce contamination. In contrast, the optothermal Raman method offers a noncontact approach for characterizing energy transport in 2D materials. It provides advantages such as minimal sample preparation, high spatial resolution, and material specificity. Therefore, this study employed the optothermal Raman technique to experimentally measure the thermal conductivity of monolayer (1L), bilayer (2L), and trilayer (3L) BN and used DFT calculations to investigate the associated thermal transport mechanisms [9,13].

7.2.1 Basic of the Optothermal Raman Technique

For measuring thermal conductivity, the BNNSs are suspended over an 80 nm gold (Au) film on a silicon wafer with perforated microwells of radius R. The Au film acts as a heat sink. A laser beam with radius r_0 is focused on the center of the suspended BNNSs, causing a local temperature increase due to laser absorption. Heat then transfers radially toward the edge of the microwell (the heat sink). The temperature distribution in the suspended BNNSs can be determined using the following heat diffusion equation in cylindrical coordinates, assuming diffusive phonon transport [22]:

$$\kappa \frac{1}{r}\frac{d}{dr}\left(r\frac{dT}{dr}\right)+q'''=0 \tag{7.1}$$

This simplifies to a Bessel's equation:

$$\frac{d^2T}{dr^2}+\frac{1}{r}\frac{dT}{dr}+\frac{q'''}{\kappa}=0 \tag{7.2}$$

where κ is the thermal conductivity of the suspended BNNS, r is the radial distance from the center of the laser beam, and q''' is the volumetric optical heating, given by:

$$q'''=\frac{q''}{t}\exp\left(\frac{-r^2}{r_0^2}\right) \tag{7.3}$$

Here, t is the thickness of the BNNS, and q'' is the peak absorbed laser power per unit area at the center of the beam spot. The total absorbed laser power Q is:

$$Q=\int_0^{\infty} q''\exp\left(\frac{-r^2}{r_0^2}\right)2\pi r\,dr=q''\pi r_0^2 \tag{7.4}$$

With the boundary conditions $T(r=R)=T_a$ (ambient temperature), and $Q\left(1-\exp\left(-R^2/r_0^2\right)\right)$ $=-2\pi R\kappa t\left(\frac{dT}{dr}\right)|_{r=R}$, the resulting temperature distribution in the suspended BNNS is:

$$T(r)=T_a+\frac{Q}{2\pi\kappa t}\ln\left(\frac{R}{r}\right)\beta(r) \tag{7.5}$$

where:

$$\beta(r)=\left(1+\frac{Ei\left(-r^2/r_0^2\right)-Ei\left(-R^2/r_0^2\right)}{2\ln\left(\frac{R}{r}\right)}\right) \tag{7.6}$$

The temperature T_m of the BNNS measured by the Raman technique is approximated as:

$$T_m \approx \frac{\int_0^R T(r)\exp\left(-\frac{r^2}{r_0^2}\right)rdr}{\int_0^R \exp\left(-\frac{r^2}{r_0^2}\right)rdr} \tag{7.7}$$

For diffusive transport analysis, the equivalent thermal resistance R_g of the suspended BNNS is:

$$R_g \equiv \frac{T_m-T_a}{Q}=\alpha\frac{\ln\left(\frac{R}{r_0}\right)}{2\pi\kappa t} \tag{7.8}$$

Thus, the thermal conductivity κ is:

$$\kappa=\frac{\alpha Q\ln\left(\frac{R}{r_0}\right)}{2\pi t\left(T_m-T_a\right)} \tag{7.9}$$

where α is the Gaussian profile factor of the laser beam:

$$\alpha=\frac{T_m-T_a}{T_0-T_a}\beta(r_0) \tag{7.10}$$

Here, T_0 is the temperature of the suspended BNNS at a radial distance of r_0. In our experiment setup, α is approximately 0.97. When accounting for heat loss Q_{air} to surrounding air and *via* radiation:

$$\kappa=\frac{\alpha\left(Q-Q_{\text{air}}\right)\ln\left(\frac{R}{r_0}\right)}{2\pi t\left(T_m-T_a\right)} \tag{7.11}$$

where:

$$Q_{\text{air}}=\int_{r_0}^R 2\pi h\left(T-T_a\right)r\,dr+\pi r_0^2 h\left(T_m-T_a\right) \tag{7.12}$$

Here, h is the heat transfer coefficient of h-BN. For small temperature differences between the object and the ambient, the quadratic expression for radiation can be linearized, resulting in the total heat transfer coefficient as the sum of convective (h_c) and radiative components. Therefore, $h=h_c+\varepsilon\sigma 4T^3$, where $h_c=3475$ W m^{-2} K^{-1} for BN sheets, ε is the emissivity (0.8 for h-BN), and $\sigma=5.670373\times10^{-8}$ W m^{-2} K^{-4} is the Stefan–Boltzmann constant [18]. Therefore, to accurately measure the thermal conductivity of BNNSs, it is essential to calculate the following factors: the first-order temperature coefficient, laser absorbance, and laser beam radius. These factors will be discussed in detail in the following sections.

7.2.2 First-Order Temperature Coefficient

BNNSs were mechanically exfoliated from high-quality h-BN single crystals using Scotch tape onto two different substrates: a SiO_2/Si wafer and an 80 nm Au-coated SiO_2/Si wafer (Au/Si). Both substrates had prefabricated microwells with a diameter of 3.8 μm and connecting trenches with a width of 200 nm. The trenches act as vents to prevent air trapped in the BN-covered microwells from expanding due to heating. The prepared BNNSs were free of defects, grain boundaries, and surface contamination. An optical microscope was used to locate the BNNSs, and their thickness was measured using atomic force microscopy (AFM). Figure 7.1a and b shows optical and AFM images of a 1L BN with a thickness of 0.48 nm on the SiO_2/Si wafer. To determine the temperature coefficients, the temperature-dependent Raman spectra of three different 1–3L BN samples were investigated: suspended over SiO_2/Si, bound to SiO_2/Si, and suspended over Au/Si. A relatively low laser power of 0.84–1.63 mW was used to minimize the laser heating effects. Figure 7.1c and d presents typical Raman spectra of 1L BN suspended over and bound to SiO_2/Si across a temperature range from 293 to 403 K. The Raman frequency shifts for 1–3L and bulk h-BN are summarized in Figure 7.1e. Linear fits of the form $\omega-\omega_0=\chi T$ were applied to estimate the first-order temperature coefficients (χ), where $\omega-\omega_0$ represents the change in G band frequency due to temperature variation and T is the temperature. The suspended 1–3L BN exhibited similar temperature coefficients: -0.0223 ± 0.0012, -0.0214 ± 0.0010, and -0.0215 ± 0.0007 cm^{-1} K^{-1}, respectively, which are close to those of bulk h-BN single crystals (-0.0191 ± 0.0005 cm^{-1} K^{-1}). In contrast, the temperature coefficients for the substrate-bound 1–3L BN were significantly different: -0.0558 ± 0.0011, -0.0480 ± 0.0022, and -0.0380 ± 0.0011 cm^{-1} K^{-1}, respectively [13].

Three factors contribute to the downshift in G band frequency with increasing temperature: the thermal expansion of the BN lattice, anharmonic phonon-phonon interactions, and the mismatch in the TECs between BNNSs and the SiO_2/Si wafer. Theoretically, these factors should be consistent for both suspended and SiO_2/Si-supported BNNSs [13]. However, as shown in Figure 7.1e, BNNSs bound to SiO_2/Si exhibit larger temperature coefficients. This discrepancy is attributed to the 'hanging down' effect of suspended BNNSs, where the BNNSs can be positioned several nanometers below the surface of the substrate (Figure 7.1f). A similar phenomenon has been observed in graphene [23]. The 'hanging down' effect allows suspended BNNSs to mitigate the impact of TEC mismatch during heating (the third factor). Specifically, suspended BNNSs can either peel off or adhere more effectively to the microwell walls with minimal energy dissipation, thereby accommodating strain caused by TEC mismatch (Figure 7.1g and h). This hypothesis is supported by the similar fitting slopes observed for 1–3L BN suspended over both Au/Si and SiO_2/Si substrates. Therefore, the temperature coefficients of 1–3L BN suspended over SiO_2/Si were used for calculating the thermal conductivity in this study [13]. Conversely, the differing temperature-dependent Raman shifts of 1–3L BN bound to SiO_2/Si suggest variations in the TEC of BNNSs with varying thickness, which will be discussed in detail in Section 7.4.

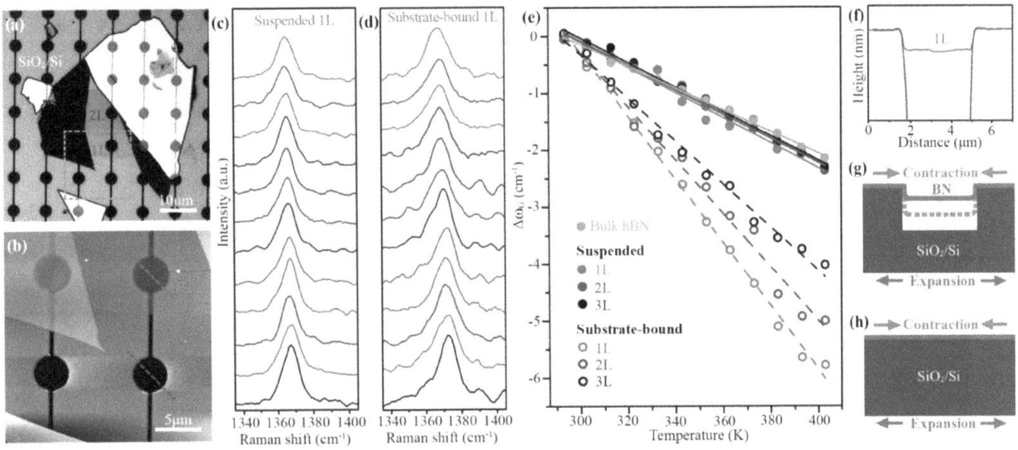

Figure 7.1 (a) Optical image of a 1–2L BN on the SiO_2/Si substrate with prefabricated microwells. (b) AFM image of the area outlined in (a). (c) Raman G band spectra of 1L BN suspended over SiO_2/Si, and (d) Raman G band spectra of 1L BN bound to SiO_2/Si, at various temperatures ranging from 293 to 403 K in 10 K intervals. Data are shown in arbitrary units (a.u.). (e) Summary of G band frequency shifts for suspended and substrate-bound 1–3L BN as a function of temperature, including corresponding linear fits. (f) AFM height profiles corresponding to the dashed lines in (b). (g) and (h) Schematic diagrams illustrating the thermal expansion of suspended and substrate-bound BN nanosheets. Licensed under a Creative Commons Attribution (CC BY) license [13].

7.2.3 Laser Absorbance

The total laser power absorbed by a BNNS (Q) equals the product of the laser absorbance (I_{ab}) and the laser power (P). As indicated by Equation (7.11), the accuracy of thermal conductivity measurements using the optothermal Raman technique is significantly influenced by the laser absorbance [4,22]. To accurately determine the light absorbance of 1–3L BN at 514 nm, BNNSs were first mechanically exfoliated onto a SiO_2/Si wafer and then transferred to silicon nitride (SiN_x) transmission electron microscopy (TEM) grids with patterned 2 μm holes using polymethyl-methacrylate (PMMA) as a transfer medium (Figure 7.2a–c). The PMMA was then completely removed by heating the sample at 550°C in air. The optical absorption of the BNNSs was measured by comparing the laser power in empty holes to that in holes covered by BN, giving $I_{ab} = (P_{empty} - P_{BN})/P_{empty}$. The light absorbance of 1–3L BN at 514 nm was found to be 0.35±0.14%, 0.62±0.19%, and 1.04±0.10%, respectively (Figure 7.2d) [13]. These values are approximately 15% of those measured for graphene.

7.2.4 Laser Beam Radius

The laser beam radius, r_0 (using a 100× objective lens), was determined by performing a micro-Raman scan across the boundary of a partially Au-coated Si wafer. Figure 7.3a and b displays an optical image of the boundary with and without Au coating, along with the corresponding Raman mapping of the Si frequency at 520.5 cm^{-1} across the boundary. The measured intensity (I) was proportional to the total laser power incident on the sample. A step size of 0.1 μm was used for the mapping. Figure 7.3c shows the intensity distribution as a function of the

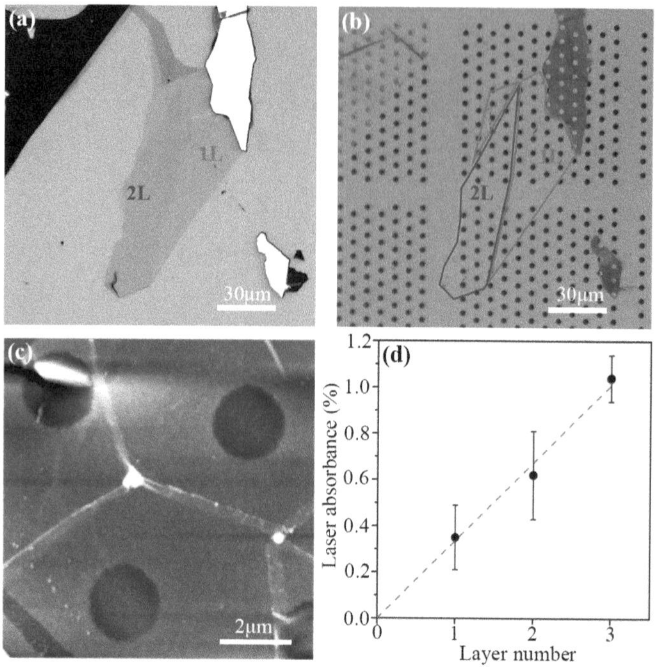

Figure 7.2 Optical images of 1–2L BN: (a) as exfoliated on a SiO_2/Si substrate and (b) after transfer to a SiN_x TEM grid. (c) AFM image showing the BN suspended over the TEM grid. (d) Laser absorbance data for 1–3L BN with corresponding linear fit. Licensed under a Creative Commons Attribution (CC BY) license [13].

distance (x) from the boundary. A Gaussian function, $\exp\left(-x^2/r_0^2\right)$, was fitted to the slope dI/dx to calculate r_0, which was found to be $0.31 \pm 0.01\,\mu m$ (Figure 7.3d). Here, the laser beam radius is defined as the position where the laser intensity decreases to $1/e$ of its maximum intensity. Whether r_0 is defined at $1/e$ or $1/e^2$ intensity does not affect the resulting value of thermal conductivity [9].

7.2.5 Thermal Conductivity

To measure thermal conductivity, the effect of laser heating on the Raman frequency of 1–3L BN suspended over Au/Si was examined [13]. The Au film, with its significantly higher thermal conductivity compared to SiO_2, served as a heat sink to maintain the edge of the suspended BNNSs at room temperature during measurements. Figure 7.4a and b displays optical and AFM images of a 1L BN with a thickness of 0.52 nm covering four microwells on the Au/Si substrate. Figure 7.4c–e presents Raman spectra of the G bands for suspended 1–3L BN under varying laser powers. A downshift in the Raman frequency was observed for all samples with increasing laser power, indicating a rise in local temperature. However, this temperature increase was modest, approximately 25 K. The change in Raman frequency induced by the laser was related to the thermal conduction capabilities of the suspended BNNSs to the edge of the microwells, *i.e.*, the heat sink [13].

According to Equation (7.11), the temperature-dependent thermal conductivity of suspended 1–3L BN was calculated and is compared in Figure 7.5a. The thermal conductivity of 1L BN did

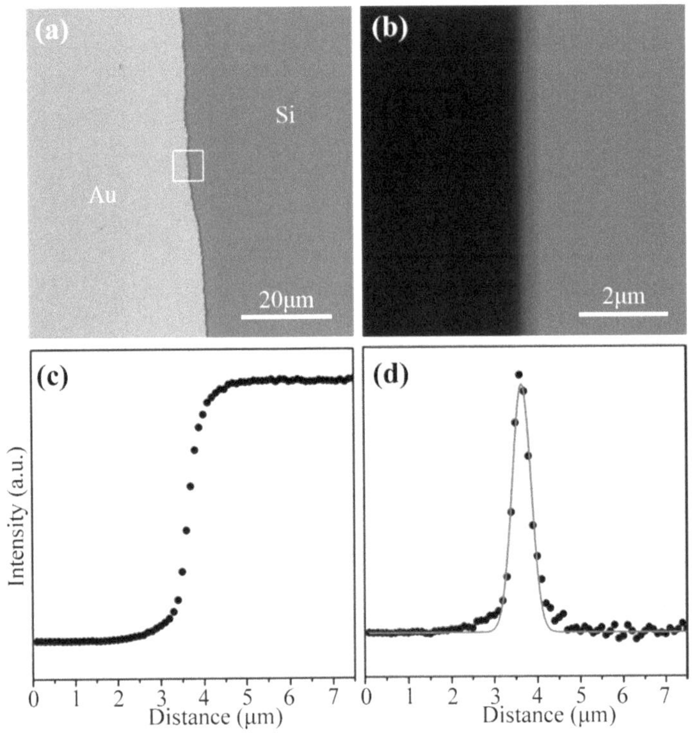

Figure 7.3 (a) Optical microscopy image showing the boundary with and without Au coating on a Si wafer, used to estimate the laser beam radius. (b) Raman mapping of the Si band at 520.5 cm^{-1} within the area outlined in (a). (c) Distribution of the intensity of the Si Raman band across the boundary. (d) Gaussian fit of the slope dI/dx in (c). Licensed under a Creative Commons Attribution (CC BY) license [13].

not exhibit the expected inverse proportionality to measured temperature (T_m). This deviation is attributed to uncertainties in the optothermal measurements, particularly due to 1L BN's extremely low optical absorption and thus small temperature change. For a better comparison, the thermal conductivity values were averaged, yielding values close to room temperature: 751±340, 646±242, and 602±247 W m^{-1} K^{-1} for 1–3L BN, respectively. The optothermal method may underestimate thermal conductivity by neglecting nonequilibrium effects among different phonon polarizations [13]. For 1L BN, heat loss to air (Q_{air}) accounts for only about 2.6% of the total heat dissipation during laser heating, with even smaller percentages for 2L and 3L BN (1.4% and 1.0%, respectively). To provide context, Figure 7.5c compares the thermal conductivity of BNNSs with common semiconductors and insulators as a function of their bandgaps. 1L BN ranks as the third most thermally conductive material among semiconductors and insulators, following diamond and *c*-BAs. When considering density, 1L BN has the second highest thermal conductivity per unit weight, after diamond, due to its density of 2.2 g cm^{-3} compared to *c*-BAs's density of 5.2 g cm^{-3} [13].

The thermal conductivities of 1–3L BN were also assessed using molecular dynamics simulations, yielding values of 849, 740, and 634 W m^{-1} K^{-1} for 1L, 2L, and 3L BN, respectively, at 300 K (Figure 7.5a, open rhombus). These simulated values align with experimental results and confirm the trend that thermal conductivity decreases with increasing thickness. To elucidate the effect

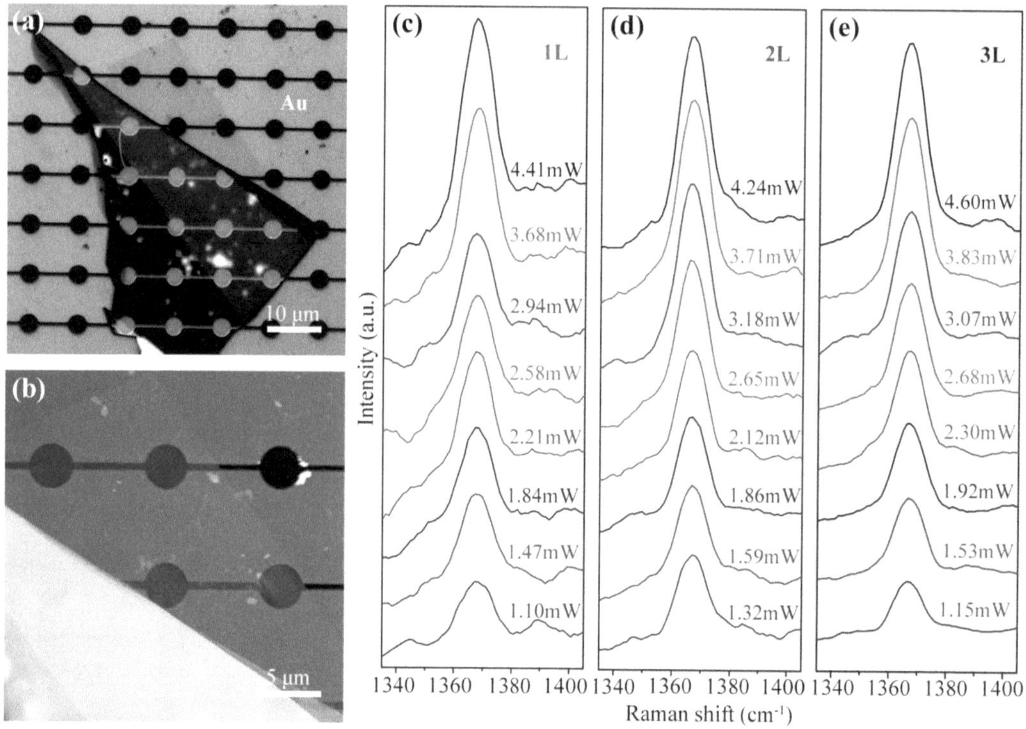

Figure 7.4 Optical image (a) and AFM image (b) of a 1L BN on an Au/Si. (c–e) Raman G bands of suspended 1L, 2L, and 3L BN, respectively, under varying laser powers. The G band intensity is displayed in arbitrary units (a.u.). Licensed under a Creative Commons Attribution (CC BY) license [13].

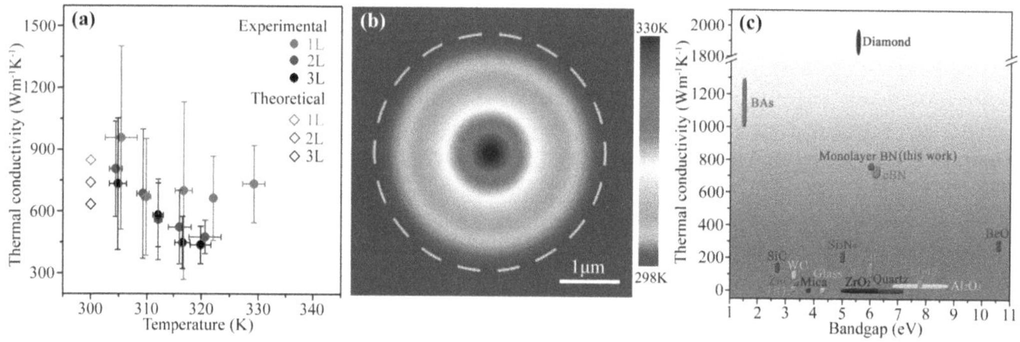

Figure 7.5 (a) Experimental thermal conductivity of suspended 1–3L BN as a function of temperature (filled circles), compared with theoretical values at 300 K (open rhombus). (b) Temperature distribution of a suspended 1L BN over 3.8 µm microwells under laser heating, with temperatures reaching up to 330 K and the heat sink maintained at 298 K. The dashed circle marks the edge of the suspended BN. (c) Comparison of thermal conductivity values for various common semiconductors and insulators. Licensed under a Creative Commons Attribution (CC BY) license [13].

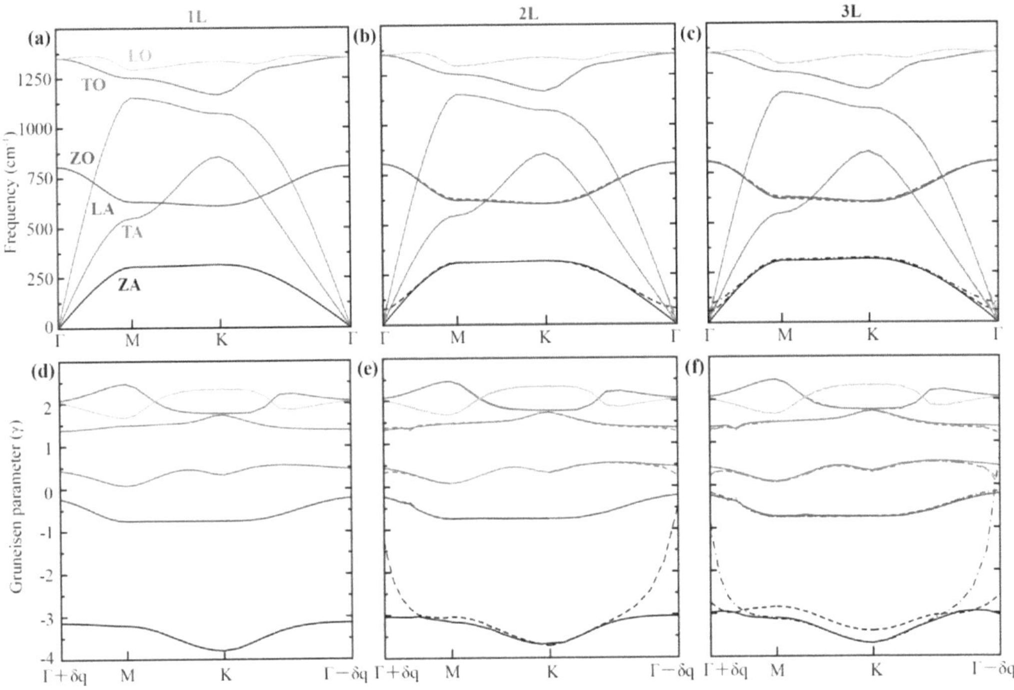

Figure 7.6 (a–c) Phonon dispersion for 1–3L BN, calculated using DFT. (d–f) Grüneisen parameters for 1–3L BN, also calculated by DFT. Dashed curves indicate additional phonon branches and corresponding Grüneisen parameters introduced by additional BN layers. Licensed under a Creative Commons Attribution (CC BY) license [13].

of thickness, phonon dispersion and Grüneisen parameters for 1–3L BN were analyzed using DFT. Figure 7.6 illustrates that BNNSs have three optical phonon branches—longitudinal optical, transverse optical, and ZO—and three acoustic phonon branches—LA, TA, and out-of-plane acoustic (ZA). Similar to graphene, optical phonon branches contribute minimally to thermal transport in BNNSs due to their low group velocities. Instead, acoustic phonons, particularly ZA phonons, play a significant role in thermal conductivity [24,25]. At near-room temperatures, Umklapp scattering is the primary factor limiting the thermal conductivity of BNNSs that are free of defects, grain boundaries, and contamination. The addition of BN layers introduces more ZA phonon states, which enhances Umklapp scattering and reduces thermal conductivity. Additionally, adding BN layers increases the ZA phonon frequency and decreases the absolute value of the Grüneisen parameter for the ZA branch. According to the equation [26,27]:

$$\frac{1}{\tau} = 2\gamma^2 \frac{k_B T \omega^2}{M v^2 \omega_m} \tag{7.13}$$

where τ is the intrinsic phonon relaxation time related to phonon-phonon Umklapp scattering, γ is the Grüneisen parameter, M is the atomic mass, ω_m is the Debye frequency, T is the temperature, k_B is the Boltzmann constant, and v is the averaged sound velocity. The increase in ZA frequency and decrease in the Grüneisen parameter lead to shorter phonon relaxation times and enhanced Umklapp scattering. Furthermore, Lindsay et al. demonstrated that ZA phonons contribute approximately 75% to the total thermal conductivity of 1L BN at room temperature, with this contribution decreasing to around 60% as the thickness increases to 5L [28]. This reduction occurs because ZA phonons, which vibrate perpendicular to the BN plane, are significantly affected by interlayer coupling. In contrast, TA and LA phonons, which vibrate within the BN plane, are less affected by increasing thickness. A similar thickness-dependent behavior in thermal conductivity has also been observed in graphene and MoS_2 [29,30].

7.3 IN-PLANE THERMAL CONDUCTIVITY OF ISOTOPE-MODIFIED BN NANOSHEETS

Isotope engineering influences many fundamental properties of solids, such as lattice parameters, disorder, elastic constants, vibrations, band structure, exciton and polariton dispersion, and scattering. These changes can lead to a variety of notable phenomena and applications, including the enhancement of superconducting transition temperatures, extended lifetimes for organic light-emitting diodes, faster optical fibers, precise quantification of proteomes, and ultratrace environmental analysis [31–33]. In comparison to carbon, which consists of 98.9% ^{12}C and 1.1% ^{13}C, naturally occurring boron nitride (^{Nat}BN) exhibits greater isotope disorder. Boron contains two stable boron isotopes: 19.9% ^{10}B and 80.1% ^{11}B. Naturally occurring nitrogen is more than 99.6% ^{14}N, making it effectively isotopically pure. Isotope engineering can alter various properties, including phonon energy, electronic bandgap, electronic density distribution, and thermal conductivity [9]. For instance, ^{12}C purified graphene exhibits an in-plane thermal conductivity that is 36% higher than that of natural graphene [34]. Additionally, bulk h-^{10}BN has a thermal conductivity of 585 W m^{-1} K^{-1}, which is approximately 39% higher than that of bulk h-^{Nat}BN [35]. This section will explore the effect of isotopes on the thermal conductivity of 1–3L BN and investigate the underlying mechanisms [9].

7.3.1 Preparation and Characterization of Monoisotopic BN Nanosheets

High-quality, contamination-free isotopically pure BNNSs were mechanically exfoliated from monoisotopic bulk h-BN crystals grown using the nickel-chromium solvent method [36,37]. Secondary ion mass spectrometry measurements revealed that the bulk h-^{10}BN crystals contain 99.2% ^{10}B, while bulk h-^{11}BN crystals consist of 99.9% ^{11}B. As previously described, the $^{10}BNNSs$ and $^{11}BNNSs$ were suspended over prefabricated microwells (3.8 μm in diameter) connected by 200 nm-wide trenches on an 80 nm Au film-coated SiO_2/Si wafer, with the Au film serving as the heat sink. Figure 7.7a and b shows an AFM image and height profile of a suspended 1L ^{10}BN, with a thickness of 0.50 nm. The chemical composition, crystal structure, and quality of the isotopically pure samples were analyzed using near-edge X-ray absorption fine structure spectroscopy and compared to those of a single crystal h-^{Nat}BN synthesized by the high-pressure Ba-BN solvent method (Figure 7.7c) [38]. All samples exhibited sharp π^* resonances at 192.0 eV, corresponding to *1s* core electron transitions to the unoccupied antibonding orbitals of B atoms bonded to three N atoms in sp^2 configurations, confirming their hexagonal crystal structure. The absence of satellite peaks indicates that the isotopically pure samples are of high chemical purity [39,40]. TEM was used to assess the crystalline quality of the monoisotopic BNNSs. As shown in Figure 7.7d and e, the 1L ^{10}BN displays a clear diffraction pattern with single sets of diffraction spots arranged with sixfold symmetry. The dark-field TEM image further confirms the monocrystalline nature of the nanosheets, with no visible grain boundaries (Figure 7.7f).

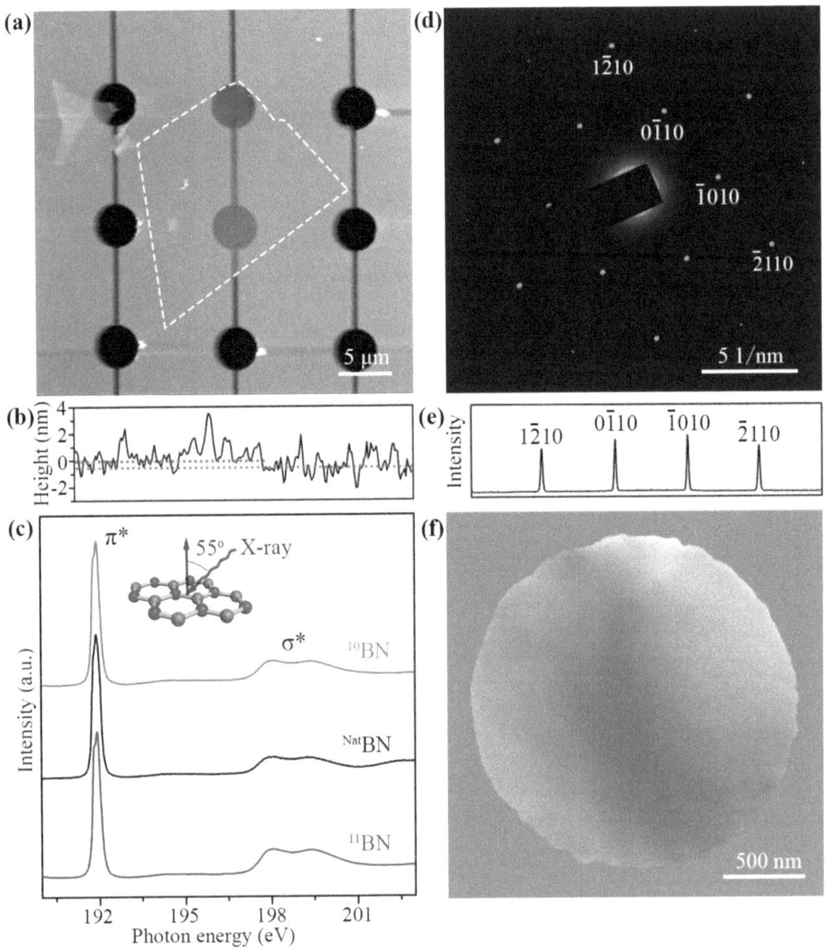

Figure 7.7 (a) AFM image of a 1L ^{10}BN sheet suspended over an Au/Si substrate. (b) Height profile of the 1L ^{10}BN sheet shown in (a). (c) Near-edge X-ray absorption fine structure spectra of the B K-edge for bulk ^{10}BN, ^{Nat}BN, and ^{11}BN crystals, taken at an incidence angle of 55°. (d) Selected area electron diffraction pattern of a suspended 1L ^{10}BN sheet. (e) Corresponding diffraction spot intensities. (f) Dark-field TEM image of the 1L ^{10}BN. Licensed under a Creative Commons Attribution (CC BY) license [9].

Figure 7.8 presents a comparison of the Raman spectra for suspended 1–3L and bulk ^{10}BN, ^{Nat}BN, and ^{11}BN. The G bands of bulk ^{10}BN, ^{Nat}BN, and ^{11}BN crystals were centered at 1392.0 cm^{-1}, 1366.9 cm^{-1}, and 1358.0 cm^{-1}, respectively. This variation is due to the fact that Raman frequency is inversely proportional to the square root of the mean atomic mass. The Raman G band frequencies of the 1–3L suspended BN are close to those of their bulk counterparts but exhibit lower peak intensities. Additionally, isotope disorder affects the full width at half maximum of the G bands, with values of 5.9, 9.4, and 5.6 cm^{-1} for bulk ^{10}BN, ^{Nat}BN, and ^{11}BN crystals, respectively [41]. The broader bandwidth observed in the 1–3L BN compared to the bulk *h*-BN is attributed to stronger surface scattering, which impacts the vibrational excitation lifetime [12,42].

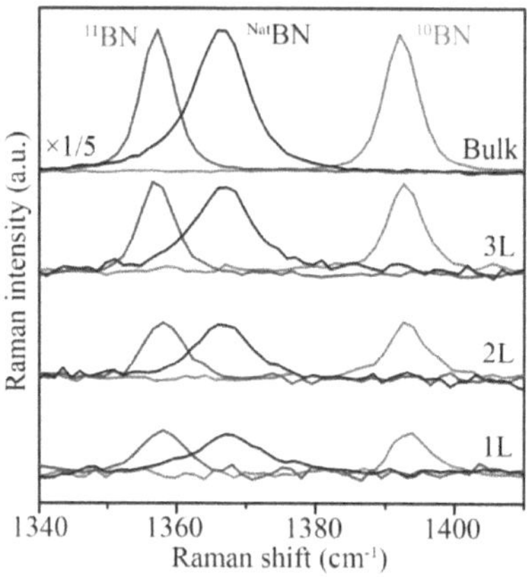

Figure 7.8 Figure 7.8. Comparison of Raman spectra for 1L, 2L, 3L, and bulk samples of ^{10}BN, ^{Nat}BN, and ^{11}BN. Licensed under a Creative Commons Attribution (CC BY) license [9].

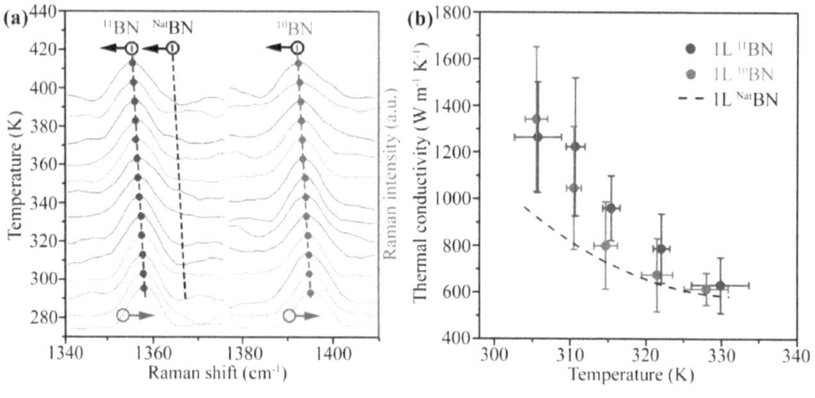

Figure 7.9 (a) Temperature dependence of the Raman G band frequency for 1L ^{10}BN, ^{Nat}BN, and ^{11}BN sheets, along with the corresponding fitting curves. (b) Experimental thermal conductivity of 1L ^{10}BN (red) and ^{11}BN (blue), including standard deviations, plotted as a function of temperature. The thermal conductivity of 1L ^{Nat}BN is shown for comparison (black, dashed line). Licensed under a Creative Commons Attribution (CC BY) license [9].

7.3.2 Thermal Conductivity Measurements

As previously discussed, the first-order temperature coefficient of suspended monoisotopic BNNSs on SiO_2/Si wafers was investigated using a hot plate with precise temperature control. To minimize the laser heating effect, a small laser power of approximately 1.5 mW was utilized. Figure 7.9a illustrates the G band frequency of 1L ^{10}BN and ^{11}BN as a function of temperature, ranging from 293 to 413 K in 10 K intervals. The corresponding linear fits are represented by the equation $\omega - \omega_0 = \chi T$, where $\omega - \omega_0$ indicates the change in G band frequency due to temperature, and χ represents the first-order temperature coefficient. The temperature coefficients for 1L ^{10}BN and ^{11}BN are -0.0223±0.0018 cm^{-1} K^{-1} (right dashed line) and −0.0220±0.0022 cm^{-1} K^{-1} (left dashed line), respectively. These values are comparable to that of ^{Nat}BN, which is −0.0223±0.0012 cm^{-1} K^{-1} (black dashed line in the middle). As noted in Section 7.2.2, the volumetric thermal expansion of the substrate during heating has a minimal impact on the temperature coefficient due to the 'hang-down' nature of the suspended BNNSs [13].

Additionally, the laser absorbance of 1L ^{10}BN and ^{11}BN was measured using the procedure outlined in Section 7.2.3, resulting in a value of (0.32±0.13)%, similar to that of 1L ^{Nat}BN. The beam size of the 488 nm laser was also measured according to the method described in Section 7.2.4, yielding a value of (0.32±0.01) μm. Subsequently, the effects of laser heating were investigated for 1L ^{10}BN and ^{11}BN. Thermal conductivity values were calculated using Equation (7.12) and are illustrated in Figure 7.9b. Due to the narrow temperature range and the inherent uncertainty of the optothermal technique, the thermal conductivity was averaged, resulting in values of 958±355 W m^{-1} K^{-1} for 1L ^{10}BN and 1009±313 W m^{-1} K^{-1} for 1L ^{11}BN. The heat loss to air during laser heating was minimal, accounting for only about 1.8% of the total heat dissipation [9]. The results indicate that the thermal conductivity of 1L monoisotopic BN is approximately 34% higher than that of 1L ^{Nat}BN and 140% higher than that of bulk *h*-^{Nat}BN [13].

To gain a comprehensive understanding of the isotope effects on the thermal conductivity of BNNSs, *ab initio* calculations were performed to analyze phonon-phonon, phonon-isotope, and phonon-boundary scatterings. It is clear from Figure 7.10a that the cumulative thermal conductivity of 1L ^{10}BN, ^{11}BN, and ^{Nat}BN increases with sample length from 0 to approximately 4 μm, after which it plateaus. Consistent with experimental results, 1L ^{11}BN (blue curve) exhibits slightly higher thermal conductivity than 1L ^{10}BN (red curve), which is attributed to the higher isotope purity in 1L ^{11}BN. Figure 7.10b and c shows the cumulative thermal conductivity as a function of phonon frequency and phonon dispersion, respectively. In all BN sheets, ZA phonons contribute the most to thermal conductivity. There is minimal difference in cumulative thermal conductivity at phonon frequencies below 320 cm^{-1}, indicating that isotope disorder has a negligible impact on ZA phonons. However, in the frequency range of 320–600 cm^{-1}, differences in cumulative thermal conductivity between 1L ^{Nat}BN and monoisotopic BN begin to appear. In the 600–800 cm^{-1} range, corresponding to the ZO phonon branch (dark gray in Figure 7.10b and c), the deviation becomes more pronounced. This suggests that ZO phonon-isotope scattering significantly limits the thermal conductivity of 1L ^{Nat}BN. While optical phonon branches generally contribute little to the thermal conductivity of BNNSs, isotope mixing may enhance acoustic-optical phonon scattering between ZO and TA/LA phonons, thus reducing the contribution of TA and LA phonons to the thermal conductivity of ^{Nat}BN [9].

In addition, the thermal conductivity of few-layer monoisotopic BN was measured using the same procedure, yielding values of 880±174 W m^{-1} K^{-1} and 833±209 W m^{-1} K^{-1} for 2L and 3L ^{10}BN,

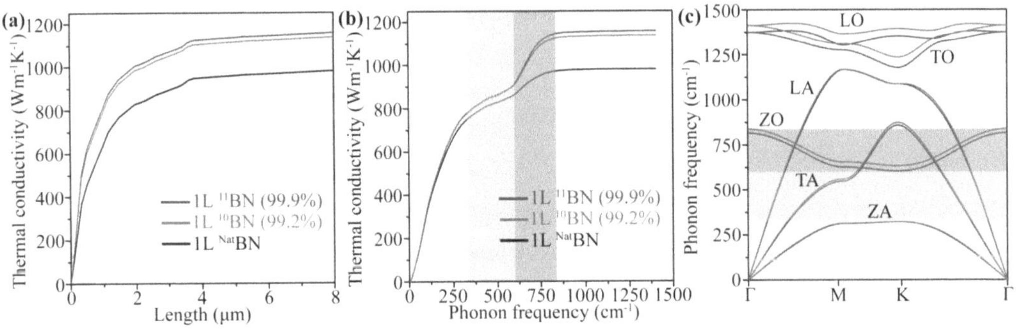

Figure 7.10 (a) Theoretical cumulative thermal conductivity of 1L ^{10}BN (99.2%), ^{11}BN (99.9%), and ^{Nat}BN as a function of sample length. (b) Theoretical cumulative thermal conductivity of the same materials as a function of phonon frequency. (c) Phonon dispersion for 1L ^{10}BN (red) and ^{11}BN (blue). Licensed under a Creative Commons Attribution (CC BY) license [9].

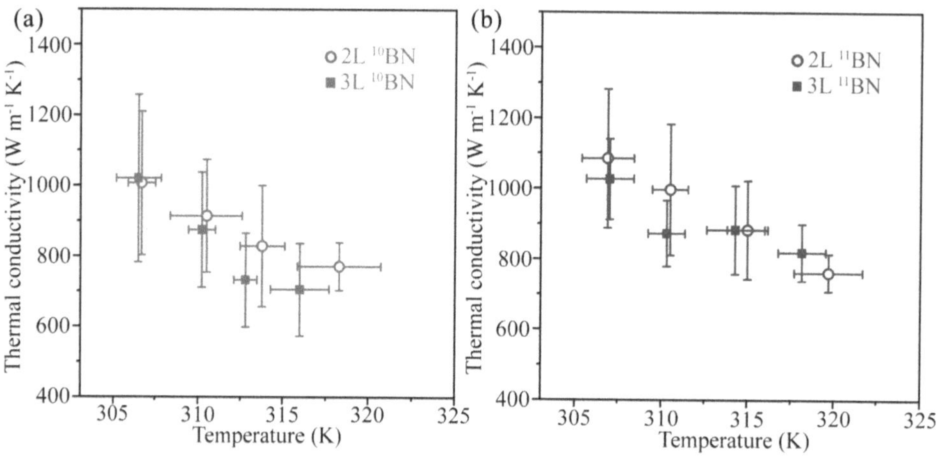

Figure 7.11 Experimental thermal conductivity of 2L and 3L ^{10}BN (a) and ^{11}BN (b), including standard deviations, as a function of temperature. Licensed under a Creative Commons Attribution (CC BY) license [9].

respectively, and 930±204 W m^{-1} K^{-1} and 902±193 W m^{-1} K^{-1} for 2L and 3L ^{11}BN, respectively, as shown in Figure 7.11. As previously discussed, the decrease in thermal conductivity with increasing thickness is attributed to enhanced Umklapp scattering of ZA phonons (Section 7.2.5). Interestingly, the differences in thermal conductivity among 1L–3L monoisotopic BN are smaller compared to those among 1L–3L ^{Nat}BN. Specifically, for ^{10}BN, the ratios are 1:0.92:0.87, and for ^{11}BN, the ratios are 1:0.92:0.89, whereas for ^{Nat}BN, the ratios are 1:0.86:0.80 [13]. This smaller variation in thermal conductivity for monoisotopic BN is due to the larger contribution of ZA phonons to the overall thermal conductivity in ^{Nat}BN compared to monoisotopic BN. As previously mentioned, reducing isotope disorder has minimal impact on ZA phonons but increases the contribution of TA and LA phonons to thermal conductivity. Adding BN layers to 1L BN significantly affects ZA phonons. Consequently, the enhanced Umklapp scattering of ZA phonons with additional BN layers leads to a more pronounced decrease in thermal conductivity in few-layer ^{Nat}BN compared to monoisotopic BN [9].

7.4 NEGATIVE TEC

With their outstanding thermal conductivity, high electrical resistivity, and other exceptional properties, BNNSs are emerging as promising materials for thermal management in next-generation flexible and wearable devices [28,43]. However, the TEC is a critical property for minimizing strain and preventing failure due to temperature changes in electronic packaging and van der Waals 2D heterostructures. Graphene, the most extensively studied two-dimensional material, exhibits a negative TEC over a wide temperature range [44–52]. For example, Zakharchenko et al. reported an average TEC of graphene to be -4.8×10^{-6} K^{-1} in the range of 0–300 K, though it becomes positive at temperatures above 900 K [45]. As a structural analog to graphene, 1L BN is also theoretically predicted to have a negative TEC. For example, Sevik's first-principles calculations using density functional perturbation theory predicted a room-temperature TEC of -5.7×10^{-6} K^{-1} based on the generalized gradient approximation and -6.2×10^{-6} K^{-1} based on the local density approximation. These values remain negative between 0 and 1500 K [46]. As discussed in Section 7.2.2, BNNSs bound to SiO_2/Si wafers exhibit varying temperature-dependent Raman shifts. This variation is attributed to the mismatch in TECs between BNNSs and SiO_2/Si, as well as changes in the TEC with BNNS thickness. By analyzing the Raman frequency shift, the TEC of BNNSs can be experimentally determined. Therefore, this section will discuss the TEC of both isotope-purified and naturally occurring BNNSs using the Raman method [13,53].

7.4.1 Basics of Measuring TEC Using Raman Spectroscopy

The temperature-dependent Raman frequency shift of free-standing BNNSs can be attributed to two main factors: lattice expansion ($\Delta\omega_G^E$) and anharmonic effect ($\Delta\omega_G^A$) [13,50]:

$$\Delta\omega_G = \Delta\omega_G^E(T_m) + \Delta\omega_G^A(T_m) \tag{7.14}$$

where T_m is the sample temperature. As discussed in Section 7.2.2, suspended BNNSs can be considered 'free-standing' because of their 'hanging down' nature; thus, their temperature-dependent G band frequency shift can also be described by Equation (7.14). For substrate-supported BNNSs, in addition to $\Delta\omega_G^E$ and $\Delta\omega_G^A$, the thermal expansion mismatch between the SiO_2/Si substrate and BNNSs induces biaxial strain when temperature changes, which results in an additional G band frequency shift ($\Delta\omega_G^S(T_m)$). Therefore, the Raman frequency shift for substrate-supported BNNSs could be expressed as:

$$\Delta\omega_G = \Delta\omega_G^E(T_m) + \Delta\omega_G^A(T_m) + \Delta\omega_G^S(T_m) \tag{7.15}$$

where:

$$\Delta\omega_G^S(T_m) = \beta\varepsilon(T_m) = \beta\int_{297}^{T_m}\left(\alpha_{SiO_2/Si}(T) - \alpha_{BN}(T)\right)dT \tag{7.16}$$

Here, β is the biaxial strain coefficient for the G band of BNNSs, given by $\beta = -2\gamma\omega_0$, where γ is the Grüneisen parameter and ω_0 is the strain-free G band frequency [50]. $\alpha_{SiO_2/Si}$ and α_{BN} represent the temperature-dependent TEC of SiO_2/Si and BNNSs, respectively. Consequently, $\Delta\omega_G^S(T_m)$ can be determined by analyzing the difference in G band frequency shifts caused by temperature variations between suspended and substrate-supported BNNSs, allowing for the calculation of the BNNS TEC using Equation (7.16). Note that this equation is valid only if there is no slippage of BNNSs during the temperature increase.

7.4.2 TEC of Monoisotopic BN Nanosheets

To investigate the temperature-dependent G band frequency shift of substrate-supported samples, isotope-enriched BNNSs were mechanically exfoliated onto Si wafers with a 75 nm thick SiO_2 layer (SiO_2/Si) from bulk monoisotopic *h*-BN single crystals [53]. As mentioned earlier, ^{10}BN and ^{11}BN have isotope purities of 99.2% and 99.9%, respectively. The resulting BNNSs were identified using an optical microscope, and their thicknesses were measured with AFM, as shown in Figure 7.12. Figure 7.12c and f compares the Raman spectra of 1L, 2L, 3L, and bulk ^{10}BN and ^{11}BN. The G band of bulk *h*-^{10}BN is centered at 1392.8 cm^{-1}, while that of bulk *h*-^{11}BN is centered

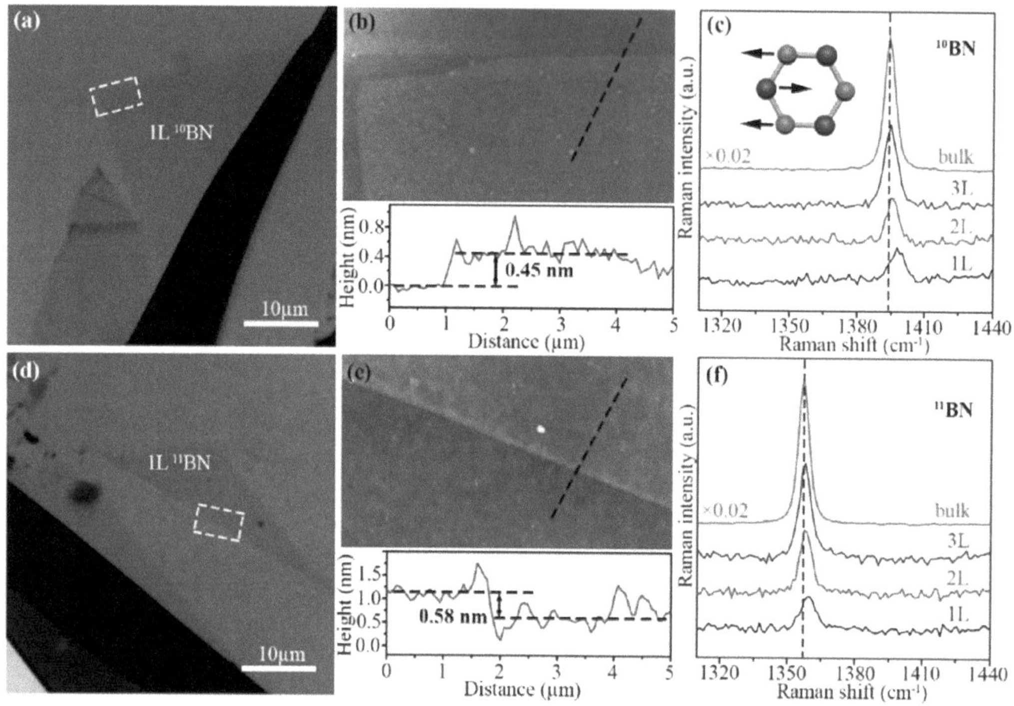

Figure 7.12 (a) Optical image of a 1L ^{10}BN sheet. (b) AFM image and corresponding height profile of the white rectangular area indicated in (a). (c) Raman spectra of 1L, 2L, 3L, and bulk ^{10}BN. (d–f) Optical image, AFM image, and Raman spectra of ^{11}B-enriched BN, respectively. (Used with permission of IOP Publishing, from [53]; permission conveyed through Copyright Clearance Center, Inc.)

at 1357.5 cm^{-1}, due to the Raman frequency being inversely proportional to the square root of the mean atomic mass. Additionally, as the thickness decreases, the G band frequency shows a slight upshift, and the bandwidth slightly increases. These frequency shifts are attributed to substrate-induced in-plane strain and reduced interlayer interactions in thinner BNNSs. The broadening of the G band is linked to in-plane strain from substrate interactions and increased surface scattering in atomically thin sheets, which affects the vibrational excitation lifetime [12,42].

Figure 7.13 displays the temperature-dependent Raman spectra of substrate-supported 1L ^{10}BN and ^{11}BN at temperatures ranging from 293 to 333 K, as well as those for 1L suspended ^{10}BN and ^{11}BN. It is evident that the G band frequency of all BN sheets exhibits a redshift with increasing temperature. Substrate-supported samples show a larger shift compared to their suspended counterparts due to the additional contribution of $\Delta\omega_G^S(T_m)$. Figure 7.14a illustrates $\Delta\omega_G$ as a function of temperature, with corresponding fits based on Equations (7.15) and (7.16). The TEC of the top surface of the SiO_2/Si wafer ($\alpha_{SiO_2/Si}$) was calculated using the finite element method due to a lack of experimental data [53]. The calculation showed that $\alpha_{SiO_2/Si}$ is approximately 80% of the TEC of Si, rather than that of SiO_2. The Grüneisen parameters for ^{10}BN and ^{11}BN were also calculated, resulting in β values of −56.79 and −55.41 cm^{-1}/% for 1L ^{10}BN and ^{11}BN, respectively. Consequently, the TECs at 300 K were found to be $(-5.1\pm0.8)\times10^{-6}$ K^{-1} for 1L ^{10}BN and $(-4.6\pm0.8)\times10^{-6}$ K^{-1} for 1L ^{11}BN (Figure 7.14b and c).

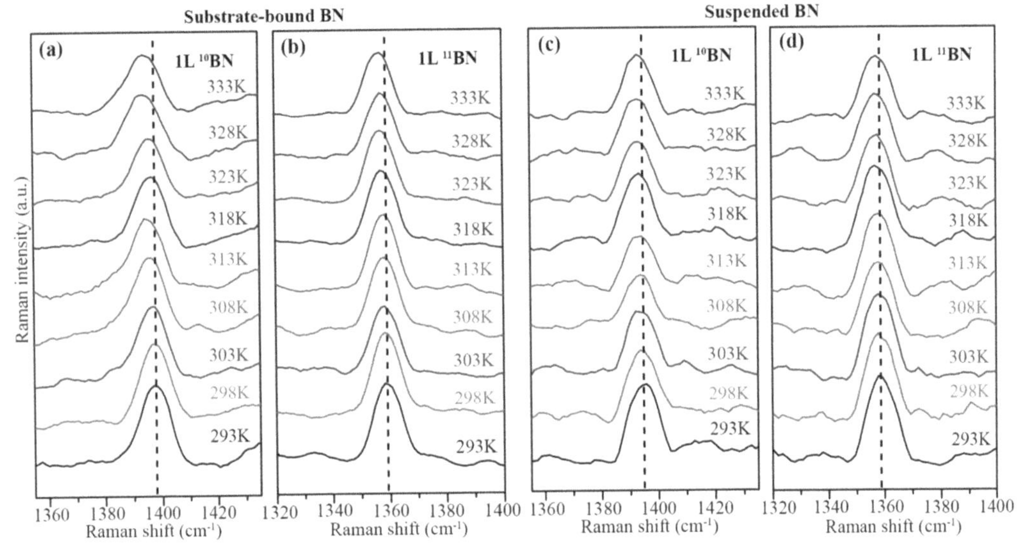

Figure 7.13 Temperature-dependent Raman spectra of substrate-supported (a) 1L ^{10}BN and (b) 1L ^{11}BN on SiO_2/Si, recorded at temperatures ranging from 293 to 333 K. (c) and (d) Temperature-dependent Raman spectra of suspended 1L ^{10}BN and 1L ^{11}BN, respectively, over the same temperature range. (Used with permission of IOP Publishing, from [53]; permission conveyed through Copyright Clearance Center, Inc.)

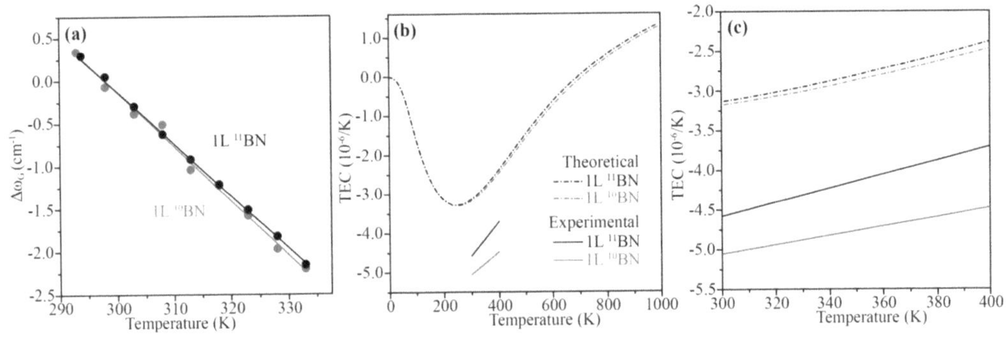

Figure 7.14 (a) G band frequency shifts as a function of temperature for 1L monoisotopic BN bound to SiO_2/Si, with corresponding fits using the TEC as a fitting parameter. (b) Experimental (solid lines) and theoretical (short-dashed-dotted lines) curves of the TEC for 1L ^{10}BN (red) and 1L ^{11}BN (black). (c) Enlarged view of the data shown in (b), focusing on the temperature range from 300 to 400 K. (Used with permission of IOP Publishing, from [53]; permission conveyed through Copyright Clearance Center, Inc.)

The TECs of 1L monoisotopic BN were also calculated using DFT, as indicated by the dashed lines in Figure 7.14b and c. At room temperature, the theoretical values are -3.17×10^{-6} K^{-1} for 1L ^{10}BN and -3.13×10^{-6} K^{-1} for 1L ^{11}BN. These theoretical values are consistent with the experimental measurements. The small deviation between theoretical and experimental results is attributed to limitations in the exchange–correlation functional used to model fundamental vibrational modes [13,42]. Additionally, the theoretical calculations did not account for the underlying substrate, which may also contribute to the observed deviation [53]. According to the DFT results, both 1L ^{10}BN and ^{11}BN exhibit a negative TEC below 700 K. This negative thermal expansion is due to competitive anharmonic couplings between in-plane stretching modes and out-of-plane bending modes in BN [53]. Specifically, in-plane anharmonicities contribute to positive thermal expansion, while bending modes lead to negative thermal expansion. The TEC α can be expressed as:

$$\alpha=\left(\beta_T(\perp)+\beta_T(Z)\right)B_{2D}^{-1} \tag{7.17}$$

where $\beta_T(\perp)$ represents the sum of in-plane polarization, $\beta_T(Z)$ represents the out-of-plane modes, and B_{2D} is the bulk modulus.

$$\beta_T(\perp)=\frac{\gamma(\perp)h^2}{2\pi\hat{c}^2k_BT^2}\int_{\omega_t(\perp)}^{\omega_s(\perp)}d\omega\frac{\omega^3e^{h\omega/k_BT}}{(e^{h\omega/k_BT}-1)^2} \tag{7.18}$$

$$\beta_T(Z)=\frac{\gamma(Z)h^2q^2}{4\pi k_BT^2}\int_{\omega_t(Z)}^{\omega_s(Z)}d\omega\frac{\omega e^{h\omega/k_BT}}{(e^{h\omega/k_BT}-1)^2} \tag{7.19}$$

According to these equations [49], $\beta_T(\perp)$ is positive, while the negative out-of-plane Grüneisen parameter $\gamma(Z)$ results in a negative $\beta_T(Z)$. Additionally, $\omega_t(\perp)>\omega_t(Z)$ leads to $|\beta_T(Z)|>\beta_T(\perp)$ at relatively low temperatures, resulting in a negative TEC α. Compared to 1L ^{11}BN, lighter ^{10}BN has a larger out-of-plane vibration amplitude, leading to a slightly larger absolute value for the TEC [53].

The TECs of 2L and 3L ^{10}BN and ^{11}BN were also measured using the same procedure. The values obtained are $(-3.6\pm0.7)\times10^{-6}$ K^{-1} for 2L ^{10}BN, $(-3.4\pm0.6)\times10^{-6}$ K^{-1} for 3L ^{10}BN, $(-3.3\pm0.4)\times10^{-6}$ K^{-1} for 2L ^{11}BN, and $(-3.0\pm0.3)\times10^{-6}$ K^{-1} for 3L ^{11}BN, as shown in Figure 7.15. As anticipated, the TEC of BNNSs decreases with increasing thickness, for example, from $(-5.1\pm0.8)\times10^{-6}$ K^{-1} for 1L to $(-3.4\pm0.6)\times10^{-6}$ K^{-1} for 3L ^{10}BN. This decrease is attributed to the reduced contribution of $\beta_T(Z)$ due to the increased bending rigidity of thicker BN layers [53].

7.4.3 TEC of Naturally Occurring BN Nanosheets

The TECs of naturally occurring BNNSs were measured using the same procedure [13]. BNNSs of 1L, 2L, and 3L were exfoliated on a Si wafer with a 90 nm thick SiO_2 layer. The measured β values for the 1L, 2L, and 3L BN were −56.07, −56.03, and −55.99 cm^{-1}/%, respectively. The TECs of the SiO_2/Si substrate were also calculated using the finite element method over a temperature range of 300–400 K. The estimated TECs for the 1L, 2L, and 3L BN at room temperature were $(-3.58\pm0.18)\times10^{-6}$ K^{-1}, $(-2.55\pm0.28)\times10^{-6}$ K^{-1}, and $(-1.67\pm0.20)\times10^{-6}$ K^{-1}, respectively, as illustrated in Figure 7.16. The small variation between the TECs of natural BN and monoisotopic BN can be attributed to differences in synthesis techniques and impurity concentrations. Similar to monoisotopic BN, the absolute value of the TEC for natural BNNSs decreases with increasing layer thickness due to the greater bending rigidity of thicker BN layers.

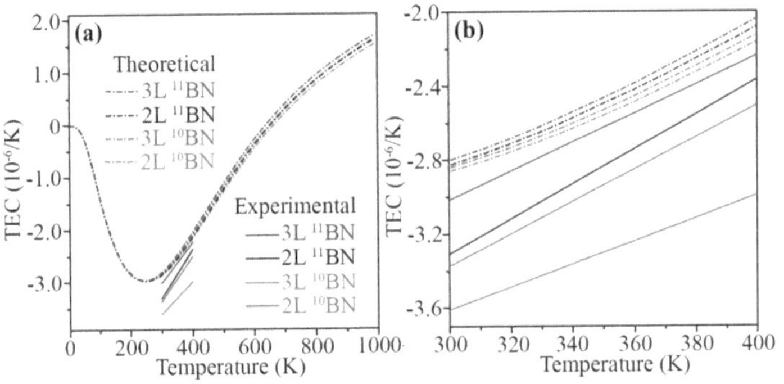

Figure 7.15 (a) Experimental (solid lines) and theoretical (short-dashed-dot lines) curves of the TEC for 2L ^{10}BN (red), 3L ^{10}BN (pink), 2L ^{11}BN (black), and 3L ^{11}BN (blue). Enlarged view of (a) focusing on the temperature range from 300 to 400 K. (Used with permission of IOP Publishing, from [53]; permission conveyed through Copyright Clearance Center, Inc.)

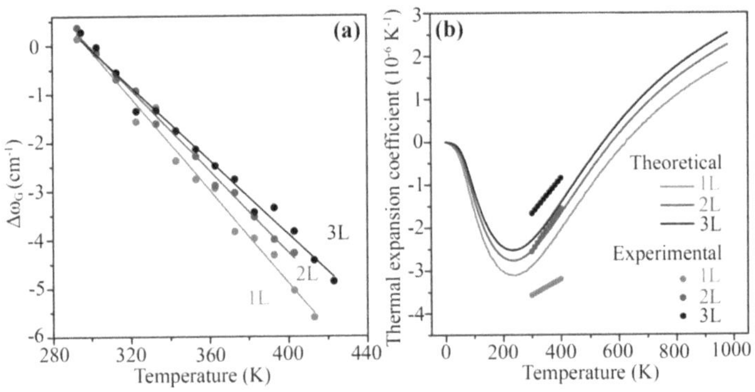

Figure 7.16 (a) G band frequency shifts as a function of temperature for 1L, 2L, and 3L ^{Nat}BN bound to SiO_2/Si. The corresponding fits are shown using the TEC as a fitting parameter. (b) Comparison of the experimental (dots) and theoretical (solid lines) TECs for 1L, 2L, and 3L ^{Nat}BN. Licensed under a Creative Commons Attribution (CC BY) license [13].

7.5 CONCLUSIONS

In summary, high-quality, surface-clean monolayer and few-layer BN nanosheets were successfully produced through mechanical exfoliation. The intrinsic in-plane thermal conductivity of these nanosheets was measured using the optothermal Raman technique. The average thermal conductivity values obtained for 1–3L ^{Nat}BN were 751±340, 646±242, and 602±247 W m^{-1} K^{-1} at room temperature, respectively. Notably, a decrease in thermal conductivity with increasing thickness was observed, attributed to interlayer interactions that create more phonon branches and available states for Umklapp scattering in few-layer BN. Furthermore, isotope purification enhanced the thermal conductivity of 1L BN by approximately 30%, yielding values of 958±355 and 1009±313 W m^{-1} K^{-1} for 1L ^{10}BN and ^{11}BN, respectively, at near-room temperature. This enhancement is linked to reduced ZO phonon scattering due to isotope engineering, which subsequently lowers the acoustic-optical scattering between ZO and transverse/LA phonons. The TEC of BN nanosheets was also investigated. The findings indicate a negative TEC within the temperature range of 0–700 K, which is attributed to anharmonic couplings between in-plane stretching modes and out-of-plane bending modes. Additionally, the absolute values of the TEC decrease with increasing atomic thickness, reflecting a diminished contribution from the out-of-plane bending modes. This study enhances the understanding of the thermal properties of 2D materials and highlights the potential of BN nanosheets for thermal management in next-generation electronics.

REFERENCES

1. Q. Wang, J. Tao, Z. Cui, T. Zhang, G. Chen, Passive enhanced heat transfer, hotspot management and temperature uniformity enhancement of electronic devices by micro heat sinks: A review. *International Journal of Heat and Fluid Flow* **107**, 109368 (2024).
2. J. Lee, I. Mudawar, Low-temperature two-phase microchannel cooling for high-heat-flux thermal management of defense electronics. *IEEE Transactions on Components and Packaging Technologies* **32**, 453–465 (2009).
3. Y.-K. Kwon, P. Kim, "Unusually high thermal conductivity in carbon nanotubes" in Shindé, S.L., Goela, J.S. (eds) *High Thermal Conductivity Materials*, Springer, New York (2006) pp. 227–265.
4. A. A. Balandin et al., Superior thermal conductivity of single-layer graphene. *Nano Letters* **8**, 902–907 (2008).
5. S. Li et al., High thermal conductivity in cubic boron arsenide crystals. *Science* **361**, 579–581 (2018).
6. F. Tian et al., Unusual high thermal conductivity in boron arsenide bulk crystals. *Science* **361**, 582–585 (2018).
7. J. S. Kang, M. Li, H. Wu, H. Nguyen, Y. Hu, Experimental observation of high thermal conductivity in boron arsenide. *Science* **361**, 575–578 (2018).
8. K. Chen et al., Ultrahigh thermal conductivity in isotope-enriched cubic boron nitride. *Science* **367**, 555–559 (2020).
9. Q. Cai et al., Outstanding thermal conductivity of single atomic layer isotope-modified boron nitride. *Physical Review Letters* **125**, 085902 (2020).
10. L. H. Li, Y. Chen, Atomically thin boron nitride: Unique properties and applications. *Advanced Functional Materials* **26**, 2594–2608 (2016).
11. A. Falin et al., Mechanical properties of atomically thin boron nitride and the role of interlayer interactions. *Nature Communications* **8**, 15815 (2017).
12. L. H. Li, J. Cervenka, K. Watanabe, T. Taniguchi, Y. Chen, Strong oxidation resistance of atomically thin boron nitride nanosheets. *ACS Nano* **8**, 1457–1462 (2014).
13. Q. Cai et al., High thermal conductivity of high-quality monolayer boron nitride and its thermal expansion. *Science Advances* **5**, eaav0129 (2019).
14. L. An et al., Hexagonal boron nitride nanosheets: Preparation, heat transport property and application as thermally conductive fillers. *Progress in Materials Science* **138**, 101154 (2023).
15. P. Jiang, X. Qian, R. Yang, L. Lindsay, Anisotropic thermal transport in bulk hexagonal boron nitride. *Physical Review Materials* **2**, 064005 (2018).
16. X. Gu, R. Yang, Phonon transport and thermal conductivity in two-dimensional materials. *Annual Review of Heat Transfer* **19**, 1–65 (2016).
17. I. Jo et al., Thermal conductivity and phonon transport in suspended few-layer hexagonal boron nitride. *Nano Letters* **13**, 550–554 (2013).
18. M. Alam, M. Bresnehan, J. Robinson, M. Haque, Thermal conductivity of ultra-thin chemical vapor deposited hexagonal boron nitride films. *Applied Physics Letters* **104**, 013113 (2014).
19. C. Wang et al., Superior thermal conductivity in suspended bilayer hexagonal boron nitride. *Scientific Reports* **6**, 25334 (2016).

20. H. Zhou et al., High thermal conductivity of suspended few-layer hexagonal boron nitride sheets. *Nano Research* **7**, 1232–1240 (2014).
21. Z. Lin, C. Liu, Y. Chai, High thermally conductive and electrically insulating 2D boron nitride nanosheet for efficient heat dissipation of high-power transistors. *2D Materials* **3**, 041009 (2016).
22. W. Cai et al., Thermal transport in suspended and supported monolayer graphene grown by chemical vapor deposition. *Nano Letters* **10**, 1645–1651 (2010).
23. L. Zhou, Y. Wang, G. Cao, Boundary condition and pre-strain effects on the free standing indentation response of graphene monolayer. *Journal of Physics: Condensed Matter* **25**, 475303 (2013).
24. S. Ghosh et al., Dimensional crossover of thermal transport in few-layer graphene. *Nature Materials* **9**, 555–558 (2010).
25. L. Lindsay, D. Broido, N. Mingo, Flexural phonons and thermal transport in multilayer graphene and graphite. *Physical Review B—Condensed Matter and Materials Physics* **83**, 235428 (2011).
26. G. Zhang, Y.-W. Zhang, Thermoelectric properties of two-dimensional transition metal dichalcogenides. *Journal of Materials Chemistry C* **5**, 7684–7698 (2017).
27. P. Klemens, D. Pedraza, Thermal conductivity of graphite in the basal plane. *Carbon* **32**, 735–741 (1994).
28. L. Lindsay, D. Broido, Theory of thermal transport in multilayer hexagonal boron nitride and nanotubes. *Physical Review B—Condensed Matter and Materials Physics* **85**, 035436 (2012).
29. A. A. Balandin, Thermal properties of graphene and nanostructured carbon materials. *Nature Materials* **10**, 569–581 (2011).
30. D. L. Nika, A. A. Balandin, Phonons and thermal transport in graphene and graphene-based materials. *Reports on Progress in Physics* **80**, 036502 (2017).
31. C. C. Tong, K. C. Hwang, Enhancement of OLED efficiencies and high-voltage stabilities of light-emitting materials by deuteration. *The Journal of Physical Chemistry C* **111**, 3490–3494 (2007).
32. S. Konegger-Kappel, T. Prohaska, Spatially resolved analysis of plutonium isotopic signatures in environmental particle samples by laser ablation-MC-ICP-MS. *Analytical and Bioanalytical Chemistry* **408**, 431–440 (2016).
33. X. Shen et al., IonStar enables high-precision, low-missing-data proteomics quantification in large biological cohorts. *Proceedings of the National Academy of Sciences* **115**, E4767–E4776 (2018).
34. S. Chen et al., Thermal conductivity of isotopically modified graphene. *Nature Materials* **11**, 203–207 (2012).
35. C. Yuan et al., Modulating the thermal conductivity in hexagonal boron nitride via controlled boron isotope concentration. *Communications Physics* **2**, 43 (2019).
36. T. Vuong et al., Isotope engineering of van der Waals interactions in hexagonal boron nitride. *Nature Materials* **17**, 152–158 (2018).
37. S. Liu et al., Single crystal growth of millimeter-sized monoisotopic hexagonal boron nitride. *Chemistry of Materials* **30**, 6222–6225 (2018).
38. T. Taniguchi, K. Watanabe, Synthesis of high-purity boron nitride single crystals under high pressure by using Ba–BN solvent. *Journal of Crystal Growth* **303**, 525–529 (2007).
39. I. Caretti, I. Jiménez, Point defects in hexagonal BN, BC_3 and BC_xN compounds studied by x-ray absorption near-edge structure. *Journal of Applied Physics* **110**, 023511 (2011).
40. L. H. Li et al., High-resolution x-ray absorption studies of core excitons in hexagonal boron nitride. *Applied Physics Letters* **101**, 191604 (2012).
41. B. R. Carvalho et al., Probing carbon isotope effects on the Raman spectra of graphene with different ^{13}C concentrations. *Physical Review B* **92**, 125406 (2015).
42. Q. Cai et al., Raman signature and phonon dispersion of atomically thin boron nitride. *Nanoscale* **9**, 3059–3067 (2017).
43. C. Sevik, A. Kinaci, J. B. Haskins, T. Çağın, Characterization of thermal transport in low-dimensional boron nitride nanostructures. *Physical Review B—Condensed Matter and Materials Physics* **84**, 085409 (2011).
44. N. Mounet, N. Marzari, First-principles determination of the structural, vibrational and thermodynamic properties of diamond, graphite, and derivatives. *Physical Review B—Condensed Mutter and Materials Physics* **71**, 205214 (2005).

45. K. Zakharchenko, M. Katsnelson, A. Fasolino, Finite temperature lattice properties of graphene beyond the quasiharmonic approximation. *Physical Review Letters* **102**, 046808 (2009).
46. C. Sevik, Assessment on lattice thermal properties of two-dimensional honeycomb structures: Graphene, h-BN, h-MoS_2, and h-$MoSe_2$. *Physical Review B* **89**, 035422 (2014).
47. C.-W. Kim, S.-H. Kang, Y.-K. Kwon, Rigid unit modes in sp-sp^2 hybridized carbon systems: Origin of negative thermal expansion. *Physical Review B* **92**, 245434 (2015).
48. J.-W. Jiang, B.-S. Wang, J.-S. Wang, H. S. Park, A review on the flexural mode of graphene: Lattice dynamics, thermal conduction, thermal expansion, elasticity and nanomechanical resonance. *Journal of Physics: Condensed Matter* **27**, 083001 (2015).
49. K. Michel, S. Costamagna, F. Peeters, Theory of thermal expansion in 2D crystals. *Physica Status Solidi (B)* **252**, 2433–2437 (2015).
50. D. Yoon, Y.-W. Son, H. Cheong, Negative thermal expansion coefficient of graphene measured by Raman spectroscopy. *Nano Letters* **11**, 3227–3231 (2011).
51. X. Hu et al., Mapping thermal expansion coefficients in freestanding 2D materials at the nanometer scale. *Physical Review Letters* **120**, 055902 (2018).
52. S. Tian et al., Temperature-dependent Raman investigation on suspended graphene: Contribution from thermal expansion coefficient mismatch between graphene and substrate. *Carbon* **104**, 27–32 (2016).
53. Q. Cai et al., Isotope effect on the thermal expansion coefficient of atomically thin boron nitride. *2D Materials* **8**, 034006 (2021).

Chapter 8

Mechanical Properties of Boron Nitride Nanosheets

The Role of Interlayer Interactions and Isotope Effects

Qiran Cai, Srikanth Mateti, Alexey Falin, Lu Hua Li, and Ying Ian Chen

8.1 INTRODUCTION

Two-dimensional (2D) nanomaterials, for example, graphene [1], boron nitride nanosheets (BNNSs) [2], transition metal dichalcogenides [3], and several others, have attracted significant attention due to their enhanced mechanical properties and potential applications in composites, nanoelectromechanical systems, and sensing devices. Understanding the mechanical behavior of these materials is the key to realizing their full potential. While traditional bulk materials usually have surface defects and internal defects that are often the major cause of inaccuracies in measuring their mechanical properties, 2D materials benefit from atomic-scale uniformity, low disorder, and lack of surface defects. These qualities enable precise, defect-free measurements at small scales, typically performed using an atomic force microscopy (AFM)-based indentation method. Consequently, they provide opportunities to study very fine mechanical phenomena that have previously proven difficult to investigate in conventional materials, which is quite exciting for both fundamental and applied research.

Among 2D materials, atomically thin BN exhibits several unique properties that distinguish it from graphene. While important, the mechanical properties of BNNSs have not been thoroughly examined. Notably, the interlayer interactions in few-layer BN differ from those in graphene, significantly influencing their mechanical behavior. For example, while more than a 30% drop in the mechanical strength of graphene occurs with the increase of thickness from one to eight layers, the strength of BNNSs practically does not change with additional layers [2]. These different characteristics make investigations of the mechanical properties of BNNSs relevant for unlocking their potential in applications where strength, durability, and stability are required. Adding to this uniqueness, BNNSs exhibit a subtle but intriguing isotope effect on their elasticity and strength. In the conventional isotope effect, differences in nuclear mass typically cause lighter isotopes to exhibit higher vibrational energy and longer bond lengths, resulting in weaker bonds at low-energy states [4]. However, BNNSs with the lighter ^{10}B isotope exhibit greater elasticity and strength compared to those with heavier ^{11}B, primarily due to the dominant influence of nuclear charge distribution rather than nuclear mass [5]. Despite advancements in techniques and the availability of isotopically pure crystals, experimentally measuring isotope effects on mechanical properties remains a significant challenge.

DOI: 10.1201/9781003647812-10

This chapter focuses on the mechanical properties of high-quality mono- and few-layer BN, which turn out to be among the strongest insulators. Few-layer BN remains similarly strong to the case of monolayer BN for up to nine layers, while such an increase in the number of layers in graphene significantly degrades its strength. The reason for this difference comes from unique interlayer interactions in BN that provide integrity under high stress and strain. Coupled with their excellent thermal stability, BNNSs become ideal candidates for mechanical reinforcement.

8.2 BASICS OF MECHANICAL PROPERTIES OF TWO-DIMENSIONAL MATERIALS

Mechanical properties of bulk crystals are often lower than their theoretical predictions due to the presence of defects that initiate fractures [6]. In contrast, the mechanical properties of reduced-dimensional materials, such as 2D materials, differ significantly from those of their bulk counterparts. While defects also impair the mechanical properties of 2D materials, reducing dimensionality typically decreases the number of defects [7]. As a result, 2D materials can often exhibit their intrinsic, theoretically predicted strength [8]. This subsection explores the fundamentals of Young's modulus, breaking strength, and fracture behavior of 2D materials studied using nanoindentation techniques.

8.2.1 Young's Modulus

Young's modulus, also known as the tensile or elastic modulus, is a key property for assessing the stiffness of elastic materials. At the macroscopic scale, it is defined as the ratio of stress to strain, where strain is the elongation relative to the object's initial length, and stress is the applied force per unit area. This relationship remains linear up to the point of plastic deformation or brittle failure [9]. On the nanoscale, however, elastic behavior is more complex, involving both linear and non-linear components, as detailed below. These differences between macro- and nanoscale mechanics necessitate new methods for measuring the Young's modulus of 2D materials. Young's modulus at the nanoscale is primarily measured using two methods: in-plane tensile testing and out-of-plane indentation. Although theoretical models for in-plane tensile tests are well-established, implementing these tests for 2D materials is challenging due to the practical difficulty of stretching such materials [10]. Conversely, indentation methods benefit from well-developed experimental procedures, but their theoretical basis remains less robust, and they face challenges such as substrate effects. To minimize substrate interference, 2D materials must often be suspended over micro-wells. Elastic properties of suspended nanosheets can also be studied using bulge tests, though this method requires large nanosheets and may underestimate stiffness [11–13].

A key challenge in determining the volumetric/conventional Young's modulus for 2D materials lies in defining their active thickness, an issue known as the 'Yakobson paradox' [14,15]. While theoretical estimates rely on covalent or van der Waals radii and π-bond systems, no direct experimental method exists for precisely determining monolayer thickness. The interlayer distance values are often used as an approximation for the thickness of 2D materials, as the interlayer distance is comparable to the thickness of the monolayer [1,9,16]. By specifying the assumed thickness, results can be compared across studies. Alternatively, the use of 2D Young's modulus (E^{2D}), which normalizes elastic energy to the material's surface area rather than volume, eliminates the ambiguity of thickness determination. For comparison, E^{2D} can be converted to a volumetric Young's modulus by dividing it by the chosen layer thickness [1,17,18].

To measure elastic-plastic behavior, materials are subjected to deformation, where they first undergo elastic (reversible) deformation and, beyond a certain stress threshold, transition to plastic (irreversible) deformation. 2D materials exhibit exceptional elasticity under significant loads, reverting to their original structure upon unloading, as seen in matching stress–strain curves during indentation experiments. However, exceeding a critical indentation depth causes lattice damage, leading to plastic deformation and eventual fracture [19–21]. Indentation of suspended 2D nanosheets involves both in-plane elastic stretching and out-of-plane bending. Using techniques like AFM, the process begins with a linear load-deformation relationship due to pre-tension $\left(\sigma_0{}^{2D}\right)$ in the membrane. Pre-tension depends on sample preparation factors like temperature, moisture, or crystallization and can affect fracture onset and dimensional stability [20,22–24]. As the applied load increases, the stiffness of the membrane becomes the dominant factor, resulting in a cubic load-deformation response that is governed by Young's modulus and the material's geometric parameters [17]. At larger deformations (>5%), nonlinear behavior emerges, associated with the breaking of lattice symmetry. Less than 1% of the membrane area undergoes such large deformation, minimizing its impact on the evaluation of mechanical properties [25,26]. Therefore, the total load–displacement relationship $F(\delta)$ combines pre-tension and membrane stretching effects:

$$F(\delta) = \pi\sigma_0{}^{2D}\delta + \left(\frac{q^3 E^{2D}}{a^2}\right) * \delta^3 \tag{8.1}$$

where q is a Poisson's ratio-dependent constant, and a is the suspended membrane radius [27].

Several factors influence measurements, including pre-tension, crystal orientation, temperature, strain rate, and nanosheet size. Deflection during indentation may be underestimated due to the effect of sliding between nanosheets because of weak interlayer interaction, leading to overestimation of Young's modulus [10]. Additionally, elastic properties are highly dependent on nanosheet size, as weak adhesion between smaller sheets and substrates complicates measurements. Proper fixation or larger nanosheets can mitigate these issues. Adhesion of nanosheets to substrates further complicates analysis due to its effect on pre-tension [28,29]. Defects, intrinsic to 2D materials, significantly affect elastic properties. The Poisson's ratio, typically strain-dependent and difficult to measure experimentally, affects calculations of pre-tension and Young's modulus [16,30,31]. Some 2D materials may exhibit negative Poisson's ratios (auxetic behavior) due to broken lattice symmetry or bond rotation mechanisms under strain [32–37]. Despite these challenges, 2D materials often exhibit mechanical properties similar to theoretically predicted properties of defect-free bulk crystals [8]. Understanding the elastic properties of 2D materials is crucial for advancing fundamental research and enabling future technological applications. By carefully accounting for variables like thickness, pre-tension, and substrate effects, researchers can obtain reliable measurements for these promising materials.

8.2.2 Breaking Strength and Fracture Behavior

In indentation experiments, mechanical properties beyond Young's modulus, such as breaking (fracture or ultimate) strength and strain, can also be determined. Breaking strength refers to the maximum stress a material can endure before structural failure. In bulk materials, it is typically lower than the theoretical (ideal) strength due to defects. However, 2D materials, which can be nearly defect-free, often exhibit breaking strengths close to the theoretical values calculated from atomic bond strengths [1,38,39]. The defect-free nature of 2D materials has been confirmed in graphene using tunneling microscopy [40]. The fracture behavior of 2D materials can be attributed to elastic instability at the contact point under breaking stress [41]. These materials generally remain isotropic under strain up to 15%, though nonlinear behavior starts at 5%. Under larger strains, softening occurs around the indenter, leading to elastic instability and eventual fracture [41]. Experimentally, breaking strength can be effectively estimated *via* AFM nanoindentation [1,3,42], and its value depends on Young's modulus, the maximum applied force, and the indenter's radius [1,21]:

$$\sigma_m{}^{2D} = \left(\frac{F_{\max} E^{2D}}{4\pi r} \right)^{\frac{1}{2}} \tag{8.2}$$

However, this value is often overestimated in experiments, as nonlinear deformation near the fracture is not accounted for. Accurate corrections can be made using finite element methods based on force–displacement curves [1,3,42].

Breaking strength is an intrinsic property governed by a material's nature and experimental conditions. For instance, graphene, with its strong in-plane covalent bonds and short bond lengths, exhibits the highest breaking strength among 2D materials. In contrast, monolayer BN, despite its structural similarity to graphene, has weaker polar B–N bonds, resulting in lower strength. Defects significantly reduce breaking strength [8,43–45], for example, with even a single vacancy defect in graphene decreasing strength by more than 5%. Chemical vapor deposition (CVD)-grown graphene, whether mono- or polycrystalline, often has more defects, leading to a substantial reduction in strength compared to pristine samples [46,47]. The fracture mechanism varies between mono- and polycrystalline CVD-grown graphene. For monocrystalline nanosheets, the fracture typically initiates at defective sites under stress and often propagates along zigzag directions. Polycrystalline nanosheets fail at grain boundaries containing defect arrays, with fractures propagating between adjacent defects [41,42,45]. Interestingly, grain boundaries with larger angular mismatches tend to have higher breaking strengths. Bilayer nanosheets display wide variations in strength due to differences in stacking, with some cases showing rapid fracture (~4 fs), suggesting nearly instantaneous failure [43,48]. While extensive research has been conducted on these processes, a comprehensive understanding remains elusive.

Experimentally measured breaking strength depends heavily on the indenter tip geometry. Smaller tips require less force to penetrate membranes. However, the size of the suspended membrane (*e.g.*, the micro-well size) usually has minimal impact [30]. Strain rate also affects measured strength, with higher rates producing higher values. Beyond a critical loading rate, the structure has insufficient time to respond, causing higher measured strength than theoretical predictions [21]. Chirality also influences breaking strength; for example, graphene's armchair direction has higher strength compared to other orientations [49]. Breaking strength is a key mechanical property of 2D materials. It reflects their ability to withstand stress and defines fracture limits, making it essential for practical applications and understanding nanosheet mechanical behavior.

8.3 MECHANICAL PROPERTIES OF BNNSS

2D materials such as graphene, BNNSs, and molybdenum disulfide (MoS_2) exhibit superior mechanical properties compared to their bulk counterparts. These nanosheets are highly flexible and can withstand significant stretching without failure, despite behaving as brittle as glass under extreme loads [1,50]. Graphene monolayers, for instance, are among the strongest materials known [1]. While graphene is brittle and fractures suddenly under high strain [10], it demonstrates anisotropic and nonlinear elastic behavior under such conditions. Its elastic modulus ranges from 0.5 to 2.4 TPa [22,26,38,51,52], with AFM measurements frequently reporting a Young's modulus of ~1.02 TPa or a 2D Young's modulus of 342 N m^{-1}, assuming an effective monolayer thickness of 0.335 nm [1]. This value decreases as the number of layers increases. Few-layer MoS_2 exhibits Young's modulus of 170–370 GPa and a breaking strength of ~23 GPa [3,19,27]. Similarly, its mechanical properties degrade significantly with increasing layer numbers, primarily due to defects and interlayer sliding [3]. This thickness-dependent mechanical behavior has also been observed in other 2D materials, including tungsten dichalcogenides and chromium trihalides [53,54]. Mechanical properties vary considerably among different 2D materials. This section discusses the mechanical properties of BNNSs, along with factors such as thickness and isotope effects that influence their mechanical behavior [2,5].

8.3.1 Thickness Effect

The mechanical properties of monolayer (1L) graphene have been extensively studied, revealing it as one of the strongest known materials, with a Young's modulus of approximately 1 TPa [1]. For few-layer graphene, both Young's modulus and strength have been observed to decrease with increasing thickness [55–60]. This reduction is attributed to the strong in-plane covalent bonds within each layer and weak van der Waals interactions between layers [58,61]. In contrast to graphene, BN nanosheets have partly ionic interlayer interactions, resulting in distinct mechanical behaviors between graphene and BNNSs. This subsection examines in detail the role of interlayer interactions in influencing the mechanical properties of few-layer BNNSs.

8.3.1.1 Sample Preparation and Characterization

BNNSs were mechanically exfoliated using Scotch tape [62,63] from high-quality *h*-BN single crystals [64] and transferred onto silicon substrates covered with 90 nm-thick silicon oxide (SiO_2/Si). These substrates had pre-fabricated micro-wells with a radius of 650 nm. Figure 8.1a shows an optical microscopy image of 1L BN covering seven micro-wells, with the corresponding AFM image displayed in Figure 8.1b. Height trace measurements determined the thickness of the 1L BN to be 0.48 nm (Figure 8.1c). Figure 8.1d presents the Raman spectrum of the suspended 1L BN, showing a G band frequency centered at 1366.5 cm^{-1}, which is close to that of bulk *h*-BN (1366.4 cm^{-1}) [65]. Bilayer (2L) and trilayer (3L) BN, with average thicknesses of 0.85 and 1.02 nm, respectively, were produced using the same procedure.

8.3.1.2 Mechanical Tests by Indentation

The mechanical properties of BNNSs and graphene were investigated through indentation tests at the centers of their suspended regions using a Cypher AFM. Diamond-tipped cantilevers were employed due to the high strength of BNNSs and graphene. The spring constants of the cantilevers were determined using a combination of the thermal noise method [66] and the Sader method [67], while the radii of the tips were measured *via* TEM. Indentation experiments were conducted under ambient conditions on relatively large nanosheets to minimize inaccuracies caused by slippage on the substrate. A constant loading/unloading velocity of 0.5 μm s^{-1} was maintained throughout. Only loading/unloading curves with no significant hysteresis were used for fitting and subsequent calculations. The AFM displacements were converted into the deflection (δ) of the nanosheets to obtain load–displacement curves, as follows:

$$\delta = \Delta Z_{\text{piezo}} - \delta_{\text{tip}} \tag{8.3}$$

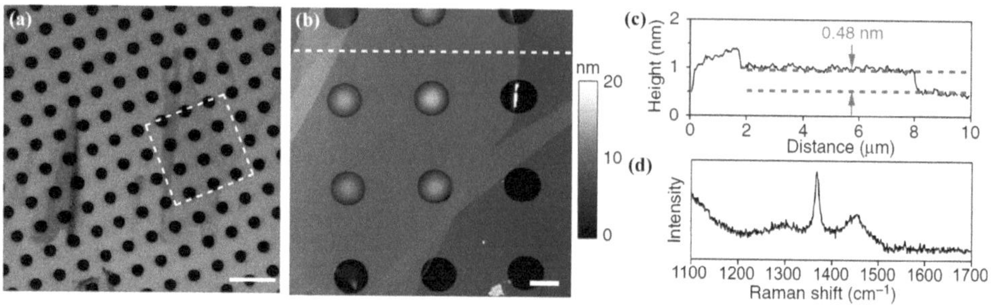

Figure 8.1 (a) Optical image of a 1L BN on a SiO_2/Si substrate featuring micro-wells with a radius of 650 nm. (b) AFM image of the BNNS highlighted within the square in (a). (c) Height profile corresponding to the dashed line in (b). (d) Raman spectrum of the suspended portion of the 1L BN. Scale bars: 5 μm in (a) and 1 μm in (b). Licensed under a Creative Commons Attribution (CC BY) license [2].

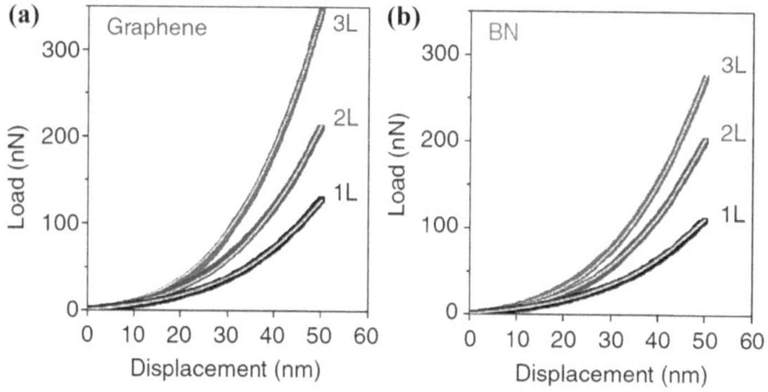

Figure 8.2 Load–displacement curves with corresponding fittings for (a) 1–3L graphene and (b) 1–3L BN. Licensed under a Creative Commons Attribution (CC BY) license [2].

Here, δ_{tip} is the deflection of the AFM tip, and ΔZ_{piezo} is the z-displacement of the AFM piezo/sample [22]. The deflection behavior of 2D materials during indentation consists of two regions: under small uniaxial loads, the elastic response is linear, while larger loads and deformations result in a cubic load–displacement relationship [1,55]. The total load–displacement relationship, $F(\delta)$, can be calculated using Equation (8.1).

The 2D effective Young's modulus (E^{2D}) of the nanosheet can be converted to the conventional bulk (volumetric) modulus (E) by dividing it by the nanosheet's thickness. For 1L BN, an effective thickness of 0.334 nm and a Poisson's ratio of 0.211 were used [49,68], while for graphene, the effective thickness was 0.335 nm, with a Poisson's ratio of 0.165 [1]. The elastic moduli of BNNSs and graphene were determined by fitting the loading curves using Equation (8.1) [1,3,69]. Typical loading curves for 1–3L graphene and BN up to a displacement of ~50 nm, along with their corresponding fittings, are shown in Figure 8.2.

8.3.1.3 Young's Modulus and Breaking Strength of BNNSs

Figure 8.3 compares the Young's moduli of BNNSs of varying thicknesses to those of graphene. The average E^{2D} for 1L BN is 289±24 N m^{-1}, aligning well with theoretical predictions [30,49,51,68,70]. For 2L and 3L BN, the E^{2D} values are 590±38 and 822±44 N·m^{-1}, respectively.

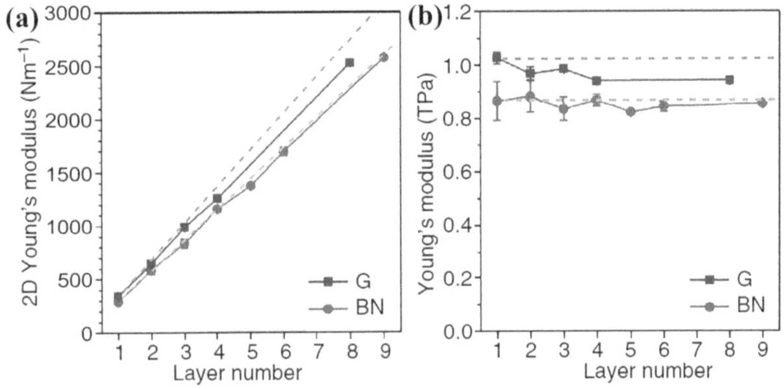

Figure 8.3 2D Young's modulus (E^{2D}) of BNNSs (solid circles) and graphene (solid squares) across different thicknesses, with dashed lines representing projections calculated as the product of monolayer E^{2D} and layer number. (b) Volumetric Young's modulus (E) of BNNSs and graphene for various thicknesses, with dashed lines indicating the E values of 1L BN and 1L graphene. Licensed under a Creative Commons Attribution (CC BY) license [2].

In comparison, the E^{2D} values for 1–3L graphene are 342±8, 645±16 and 985±10 N·m^{-1} respectively, consistent with previous AFM-based studies [1,55]. Dashed lines in Figure 8.3a represent projections of E^{2D} with increasing thickness, calculated by multiplying the E^{2D} of monolayers by their layer number. The deviation between the experimental data and the projections indicates how E^{2D} changes with thickness for both materials. Graphene shows a greater deviation compared to BNNSs as the thickness increases. This trend is more apparent in Figure 8.3b, which shows the (volumetric) Young's modulus E for varying thicknesses. For 1L graphene, E is 1.026±0.022 TPa, but it decreases to 0.942±0.003 TPa for 8L graphene. In contrast, the E of 1L BN (0.865±0.073 TPa) remains nearly unchanged for 9L BN (0.856±0.003 TPa).

The strengths of BNNSs and graphene with varying thicknesses were determined from load–displacement curves and fracture loads using finite element simulation. Figure 8.4 presents the fracture loads for BNNSs and graphene of different thicknesses. As observed with Young's modulus (Figure 8.3), the dashed lines represent projections obtained by multiplying the fracture load of 1L BN or graphene by the layer number. Notably, the fracture loads of BNNSs closely followed the red dashed line, whereas multilayer graphene deviated increasingly from the blue dashed line as the thickness increased. For instance, the fracture load of 8L graphene was 53.7% lower than eight times the fracture load of 1L graphene. These trends are also evident in the mechanical strengths (Figure 8.4b). The strengths of 1–3L BN were measured at 70.5±5.5, 68.0±6.8, and 76.9±2.3 GPa, respectively. In comparison, the breaking strengths of graphene are 125.0±0, 107.7±4.3, 105.6±6.0, and 85.3±5.4 GPa for 1L, 2L, 3L, and 8L, respectively.

8.3.1.4 The Role of Interlayer Interaction

The thickness-dependent strength of graphene is attributed to interlayer slippage in few-layer graphene during indentation [58]. As noted earlier, a similar phenomenon has been observed in many other 2D materials [53,54,71]. In contrast, BNNSs exhibit thickness-independent Young's modulus and strength. To explore this distinct mechanical behavior, *ab initio* DFT calculations, including van der Waals interactions *via* the VASP code [72,73], were used to analyze the sliding

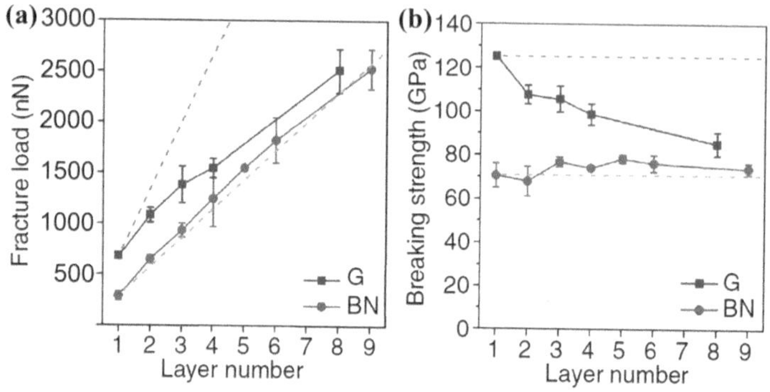

Figure 8.4 (a) Fracture load and (b) breaking strength of BNNSs (solid circles) and graphene (solid squares) with varying thicknesses. The dashed lines represent projections calculated by multiplying the fracture load (breaking strength) of 1L BN or graphene by the layer number. Licensed under a Creative Commons Attribution (CC BY) license [2].

energy in 2L BN and graphene as a function of radial distance from the indentation center. Finite element simulations reveal that the majority of the suspended nanosheet regions experience minimal in-plane strain and no out-of-plane compression, even under fracture loads. Thus, the interlayer interaction in the low-strained regions can be approximated using the sliding energy in standard, unstrained 2L BN and graphene. However, regions closer to the indentation center experience significant in-plane tensile strain and out-of-plane compression, altering their interlayer sliding behavior. Figure 8.5a shows strain distribution (solid lines) and out-of-plane pressure (dashed lines) calculated for 2L BN and graphene within 10 nm of the indentation center under their respective fracture loads.

For van der Waals-corrected DFT calculations, four combinations of strain and hydraulic pressure (compression) were examined to capture interlayer interactions at radial distances of 10, 4, 2, and 0 nm from the indentation center. For 2L BN, the conditions are as follows: 10 nm, 5.7% + 0 GPa; 4 nm, 9.8% + 5.3 GPa; 2 nm, 12.4% + 14.2 GPa; 0 nm, 14.5% strain + 14.1 GPa. For 2L graphene, these correspond to 10 nm: 7.2% + 0 GPa; 4 nm: 12.4% + 8.3 GPa; 2 nm, 16.8% + 17.9 GPa; 0 nm, 21.7% strain + 16.9 GPa pressure. In equilibrium (no strain and compression), the sliding energy for 2L BN was only slightly larger than that for 2L graphene (7.22 *vs.* 5.12 meV per unit cell). With strain alone (*e.g.*, at 10 nm), the sliding energies increased to 11.64 meV for 2L graphene and 21.57 meV for 2L BN (Figure 8.5c). As strain and pressure increased, their behaviors diverged. At 4 nm, 2L graphene's sliding energy nearly vanished (0.92 meV/cell), while 2L BN's increased significantly (69.56 meV/cell, Figure 8.5d). At the center (0 nm), 2L graphene exhibited negative sliding energy (–112.26 meV/cell), whereas that of BN reached 582.84 meV/cell (Figure 8.5f).

The stark difference in sliding energy between 2L BN and graphene under various in-plane tensile strain and out-of-plane compression conditions arises from their distinct electronic properties. In graphene, the semi-metallic nature leads to orbital overlap under high strain and pressure, facilitating interlayer slippage. Conversely, the insulating nature and polar 2pz orbitals of BN localize electronic density, enhancing interlayer adhesion. As a result, graphene layers slide more readily during indentation (Figure 8.6a), while BN layers near the indentation center remain firmly bonded, resisting sliding (Figure 8.6b), and contributing to their distinct mechanical behaviors.

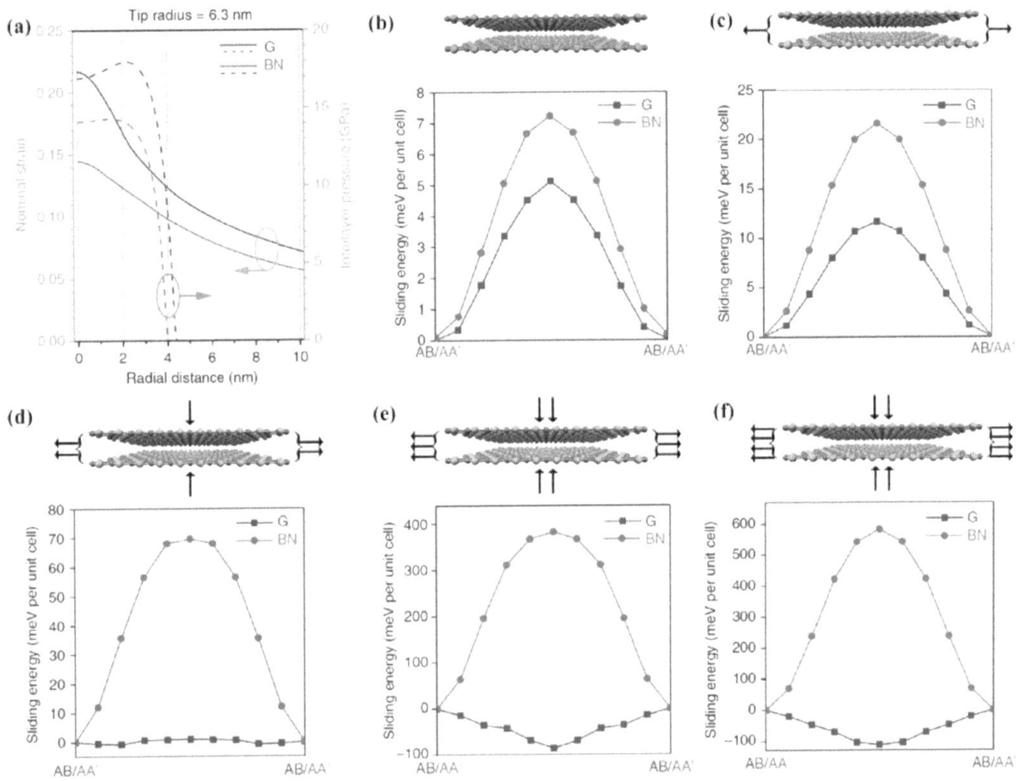

Figure 8.5 (a) Finite element calculations showing in-plane strain (solid lines) and out-of-plane pressure (dashed lines) in 2L BN and graphene within a 10 nm radial distance from the indentation center. Sliding energies of 2L BN (solid circles) and graphene (solid squares) are shown at the equilibrium state (b), and at 10 nm (c), 4 nm (d), 2 nm (e), and 0 nm (f) from the indentation center. Licensed under a Creative Commons Attribution (CC BY) license [2].

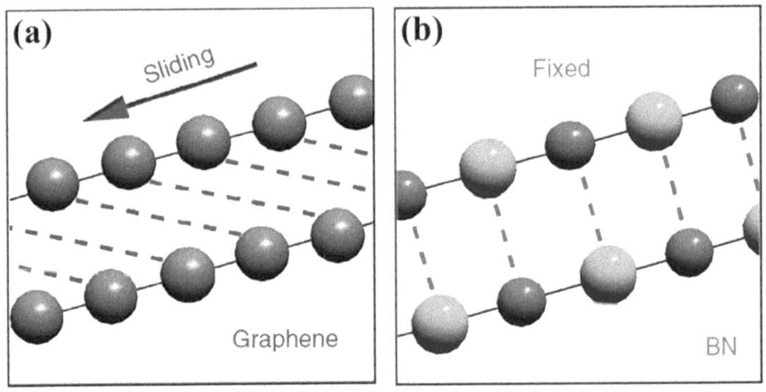

Figure 8.6 (a) Diagram illustrating the ease of slippage in graphene. (b) Diagram showing the resistance to interlayer sliding in BN. Licensed under a Creative Commons Attribution (CC BY) license [2].

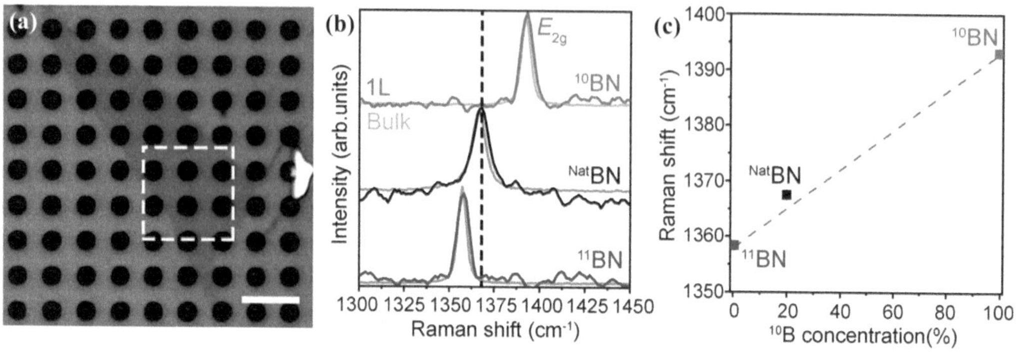

Figure 8.7 (a) Optical microscopy image of a 1L ^{11}BN on a SiO_2/Si substrate with micro-wells of 1.6 μm in diameter. (b) Comparison of the Raman spectra of suspended ^{10}BN, ^{Nat}BN, and ^{11}BN monolayers (darker color) and their bulk crystals (lighter color and at ~1/50 intensity). (c) Raman frequency dependence on ^{10}B concentration. Licensed under a Creative Commons Attribution (CC BY) license [5].

8.3.2 Isotope Effect

An isotope is a variation of a chemical element that has the same number of protons but a different number of neutrons, resulting in slightly different atomic masses. The mechanical properties of BN can be significantly influenced by its isotopic composition. ^{11}BN and ^{10}BN differ in neutron number, affecting their atomic mass and interactions within the material. ^{Nat}BN, the most abundant form of BN, is a mixture of both isotopes, with a higher proportion of ^{11}BN. These isotopic differences lead to variations in bond strength, stiffness, and fracture resistance, with lighter isotopes like ^{10}BN generally exhibiting stronger mechanical properties due to changes in atomic interactions [5]. Studying these isotope effects helps deepen our understanding of how atomic mass influences material behavior.

Figure 8.7a shows optical microscopy images of a 1L ^{11}BN covering ten micro-wells. The differences in isotope composition are observed *via* Raman spectroscopy (Figure 8.7b). The spectra compare the Raman E_{2g} peaks of ^{10}BN, natural boron nitride (^{Nat}BN), and ^{11}BN, along with the corresponding bulk crystals (at ~1/50 intensity). In the suspended form, 1L ^{10}BN and ^{11}BN exhibited Raman frequencies similar to their bulk counterparts, indicating the absence of substrate-induced strain [74], as observed in the case of ^{Nat}BN [75]. The average Raman frequencies for ^{10}BN, ^{Nat}BN, and ^{11}BN were 1393.2±0.3, 1367.5±0.4, and 1358.4±0.3 cm^{-1}, respectively. These values showed a roughly linear relation with the change in atomic mass, as shown by the dashed line in Figure 8.7c [76]. The small deviation for 1L ^{Nat}BN was due to isotopic mass disorder [51,77].

Using AFM-based nanoindentation, the mechanical properties of 1L ^{10}BN, ^{Nat}BN, and ^{11}BN were measured. Typical loading displacement curves for 1L ^{10}BN, ^{Nat}BN, and ^{11}BN are compared above, along with the corresponding fitting results in Figure 8.8a. All curves were fitted up to the cubic regime to meet the requirements for correct Young's modulus calculation [78]. The isotope effect is depicted more clearly in Figure 8.8b, which shows that the effective Young's moduli (E^{2D}) of 1L ^{10}BN, ^{Nat}BN, and ^{11}BN closely follow a linear trend (dashed line). In Figure 8.8c, breaking strength also increases with decreasing atomic mass, following the same trend as Young's modulus (Figure 8.8b). The characteristic peculiar isotope effect arises from the difference in nuclear charge distribution, rather than differences in atomic mass. Compared to the ^{11}B atom, the ^{10}B atom has a smaller radius of effective nuclear charge. The nuclei of these isotopes are structured

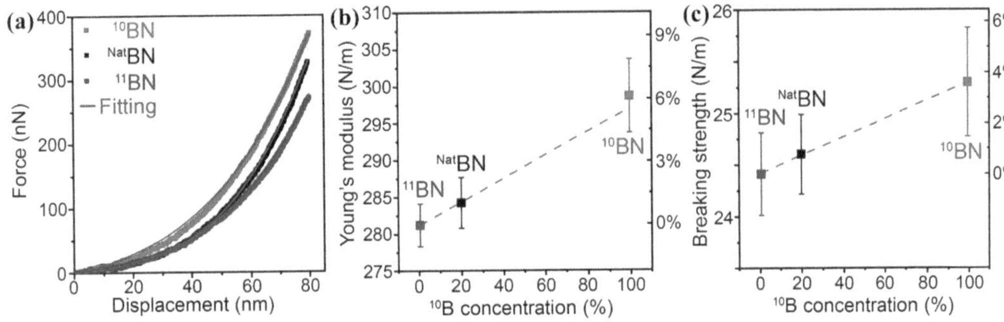

Figure 8.8 (a) Load–displacement curves and the corresponding fittings. (b) Optical microscopy image of a 1L ^{11}BN on a SiO_2/Si substrate with micro-wells of 1.6 μm in diameter. Young's modulus (b) and breaking strength (c) dependence on ^{10}B contents with standard errors. Licensed under a Creative Commons Attribution (CC BY) license [5].

differently. In ^{10}B, a deuteron binds to two alpha-like clusters less strongly than that of a triton in the stronger binding in ^{11}B. Hence, the ^{10}B nucleus pulls electrons closer, which causes a smaller atomic radius and shorter covalent bond with nitrogen atoms [79].

The ^{10}B has a larger effective nuclear charge, which makes the interaction potential with the nitrogen atom narrower (B–N bond) and enhances the rigidity of the bond [80]. Compared to most light-element isotopes, except for carbon (C), this nuclear charge effect in ^{10}B is unique. The nuclear charge effect is often counterbalanced by the mass effect, which predicts that elements with lighter isotopes generally have weaker interactions with neighboring atoms and reduced mechanical properties. The bond dissociation energy (BDE) results indicate that bond strength is typically influenced by the isotopic mass effect, as variations in isotopic mass directly affect the BDE. Since ^{10}B is lighter than ^{11}B, it has a higher zero-point vibrational energy level and therefore needs less energy to break the B–N bond. However, due to the nuclear charge effect, the bond length of ^{10}B–N is shorter than that of ^{11}B–N, resulting in a higher BDE per unit cell value compared to ^{11}B–N. Therefore, ^{10}B in its single-layer form is expected to have a higher fracture strength as compared to ^{11}BN.

Temperature brings a lot of confusion because it changes the vibrational states and the bonding energies of isotopes having different masses. At maximum temperature, heavier isotopes undergo a lot more changes in their states compared to lighter isotopes [81,82]. This means that the bonding energies of heavier isotopes decrease faster than those of lighter isotopes. This aspect of isotopic BN becomes even more important concerning room temperature (300 K), directly from the findings of our experimental and theoretical studies, as the bonding of isotopic BN changes at higher temperatures. DFT results at 0 K and room temperature indicate that the BDE of ^{11}B–N decreases more rapidly with increasing temperature. The difference in bond lengths (lattice constant) between the isotopes shrinks, resulting in smaller differences in unit cell area and BDE per unit cell. So, it becomes very hard to detect differences in fracture strength between the isotopes at room temperature compared to cryogenic ones, but variations in Young's modulus are more apparent.

The B isotope effects on the mechanical properties of 1L BN were found to match well with the DFT calculations. The effects that are the result of the strong nuclear charge apparent in B isotopes may also be observed in neighboring elements like carbon (C), nitrogen (N), and oxygen (O).

For example, in the cluster structure model of the nucleus, ^{12}C has three tightly bound alpha-like clusters, whereas ^{13}C has the same three clusters plus an extra neutron [83]. This additional neutron slightly increases the effective nuclear charge radius of ^{13}C, hence the longer bonds found in ^{13}C as opposed to ^{12}C [84]. The nuclei of ^{14}N and ^{15}N also appear to have three alpha-like clusters, and ^{14}N features a deuteron, while ^{15}N carries a triton, concerning the two stable isotopes of nitrogen: ^{14}N and ^{15}N. A framework of nucleons results in ^{16}O getting four alpha-like clusters, while ^{17}O and ^{18}O incorporate one or two additional neutrons. These nuclear structures give an indication that similar anomalous isotope effects could be possible in carbon, nitrogen, and oxygen, where the nuclear charge effect might outweigh the mass effect; however, the effect is expected to be weaker than in boron. For example, molecular dynamics (MD) simulations have been able to predict the Young's modulus [85,86] and fracture strength reduction in graphene with the heavier isotope of carbon [86].

To investigate this further, it is worth synthesizing and exploring *h*-BN single crystals with ^{14}N and ^{15}N isotopes, which will allow us to study one-atom-thin $^{10}B^{14}N$, $^{10}B^{15}N$, $^{11}B^{14}N$, and $^{11}B^{15}N$, leading to more generalized conclusions. However, for elements much lighter or heavier than boron, the anomalous isotope effect is unlikely to be observed as the mass effect dominates in lighter elements, and the nuclear size effect becomes more significant in heavier elements.

8.4 CONCLUSIONS

This chapter highlights the remarkable mechanical properties and unique features of BNNSs, emphasizing their potential for both fundamental research and practical applications. BNNSs exhibit strength and stiffness comparable to graphene; experimental characterizations have demonstrated exceptional mechanical properties, with a recorded Young's modulus of 0.865 ± 0.073 TPa and a fracture strength of 70.5 ± 5.5 GPa for monolayer BN, ranking it among the strongest insulating materials. Unlike graphene, whose strength diminishes significantly with added layers, few-layer BN retains the strength of its monolayer counterpart due to enhanced interlayer interactions under strain and compression. Additionally, both experimental results and theoretical calculations reveal a subtle yet significant isotope effect, where nanosheets made from lighter isotopes (^{10}BN) show greater elasticity and strength than those made from heavier isotopes (^{11}BN), driven by nuclear charge rather than mass. These findings position BNNSs as an ideal platform for studying mechanical phenomena while also showcasing their potential in engineering and technological applications under extreme performance and temperature conditions.

REFERENCES

1. C. Lee, X. Wei, J. W. Kysar, J. Hone, Measurement of the elastic properties and intrinsic strength of monolayer graphene. *Science* **321**, 385–388 (2008).
2. A. Falin et al., Mechanical properties of atomically thin boron nitride and the role of interlayer interactions. *Nature Communications* **8**, 15815 (2017).
3. S. Bertolazzi, J. Brivio, A. Kis, Stretching and breaking of ultrathin MoS_2. *ACS Nano* **5**, 9703–9709 (2011).
4. W. M. White, "Chapter 9: Stable isotopes geochemistry" in W. M. White (ed) *Geochemistry* (Wiley-Blackwell, 2013), pp. 388–396.
5. A. Falin et al., Anomalous isotope effect on mechanical properties of single atomic layer Boron Nitride. *Nature Communications* **14**, 5331 (2023).

6. M. A. Meyers, K. K. Chawla, *Mechanical Behavior of Materials,* 2nd Edition (Cambridge University Press, 2009), pp. 1–856.
7. G. Stan, S. Krylyuk, A. V. Davydov, I. Levin, R. F. Cook, Ultimate bending strength of Si nanowires. *Nano Letters* **12**, 2599–2604 (2012).
8. G. H. Lee et al., High-strength chemical-vapor-deposited graphene and grain boundaries. *Science* **340**, 1073–1076 (2013).
9. W. Philipp, V. I. Viktoria, J. R. Mark, R. B. Patrick, P. E. Christopher, Mechanical properties of nanosheets and nanotubes investigated using a new geometry independent volume definition. *Journal of Physics: Condensed Matter* **25**, 155302 (2013).
10. G. Cao, Atomistic studies of mechanical properties of graphene. *Polymers* **6**, 2404–2432 (2014).
11. D. Lloyd et al., Band gap engineering with ultralarge biaxial strains in suspended monolayer MoS_2. *Nano Letters* **16**, 5836–5841 (2016).
12. L. B. Freund, S. Suresh, *Thin Film Materials: Stress, Defect Formation, and Surface Evolution* (Cambridge University Press, 2003).
13. R. J. T. Nicholl et al., The effect of intrinsic crumpling on the mechanics of free-standing graphene. *Nature Communications* **6**, 8789 (2015).
14. O. A. Shenderova, V. V. Zhirnov, D. W. Brenner, Carbon nanostructures. *Critical Reviews in Solid State and Materials Sciences* **27**, 227–356 (2002).
15. B. I. Yakobson, C. J. Brabec, J. Bernholc, Nanomechanics of carbon tubes: Instabilities beyond linear response. *Physical Review Letters* **76**, 2511–2514 (1996).
16. L. Boldrin, F. Scarpa, R. Chowdhury, S. Adhikari, Effective mechanical properties of hexagonal boron nitride nanosheets. *Nanotechnology* **22**, 505702 (2011).
17. K. Liu, J. Q. Wu, Mechanical properties of two-dimensional materials and heterostructures. *Journal of Materials Research* **31**, 832–844 (2016).
18. R. Al-Jishi, G. Dresselhaus, Lattice-dynamical model for graphite. *Physical Review B* **26**, 4514–4522 (1982).
19. A. Castellanos-Gomez et al., Mechanical properties of freely suspended semiconducting graphene-like layers based on MoS_2. *Nanoscale Research Letters* **7**, 233 (2012).
20. K.-T. Wan, S. Guo, D. A. Dillard, A theoretical and numerical study of a thin clamped circular film under an external load in the presence of a tensile residual stress. *Thin Solid Films* **425**, 150–162 (2003).
21. W. Wang et al., Nanoindentation experiments for single-layer rectangular graphene films: A molecular dynamics study. *Nanoscale Research Letters* **9**, 41 (2014).
22. I. W. Frank, D. M. Tanenbaum, A. M. van der Zande, P. L. McEuen, Mechanical properties of suspended graphene sheets. *Journal of Vacuum Science and Technology B* **25**, 2558–2561 (2007).
23. J. S. Bunch et al., Electromechanical resonators from graphene sheets. *Science* **315**, 490–493 (2007).
24. M. W. Pruessner et al., Mechanical property measurement of InP-based MEMS for optical communications. *Sensors and Actuators A: Physical* **105**, 190–200 (2003).
25. Q. Lu, R. Huang, Nonlinear mechanics of single-atomic-layer graphene sheets. *International Journal of Applied Mechanics* **01**, 443–467 (2009).
26. X. Wei, B. Fragneaud, C. Marianetti, J. Kysar, Nonlinear elastic behavior of graphene: Ab initio calculations to continuum description. *Physical Review B* **80**, 205407 (2009).
27. A. Castellanos-Gomez et al., Elastic properties of freely suspended MoS_2 nanosheets. *Advanced Materials* **24**, 772–775 (2012).
28. J. R. Withers, D. E. Aston, Nanomechanical measurements with AFM in the elastic limit. *Advances in Colloid and Interface Science* **120**, 57–67 (2006).
29. B. J. Robinson et al., Nanomechanical mapping of graphene layers and interfaces in suspended graphene nanostructures grown via carbon diffusion. *Thin Solid Films* **550**, 472–479 (2014).
30. A. Bosak et al., Elasticity of hexagonal boron nitride: Inelastic x-ray scattering measurements. *Physical Review B* **73**, 041402 (2006).
31. A. Sakhaee-Pour, Elastic properties of single-layered graphene sheet. *Solid State Communications* **149**, 91–95 (2009).
32. K. E. Evans, M. A. Nkansah, I. J. Hutchinson, S. C. Rogers, Molecular network design. *Nature* **353**, 124–124 (1991).

33. V. H. Ho, D. T. Ho, S. Y. Kwon, S. Y. Kim, Negative Poisson's ratio in periodic porous graphene structures. *Physica Status Solidi B-Basic Solid State Physics* **253**, 1303–1309 (2016).
34. H. F. Wang et al., Strain effects on borophene: Ideal strength, negative Possion's ratio and phonon instability. *New Journal of Physics* **18**, 073016 (2016).
35. J. W. Jiang, H. S. Park, Negative Poisson's ratio in single-layer graphene ribbons. *Nano Letters* **16**, 2657–2662 (2016).
36. J. N. Grima et al., Tailoring graphene to achieve negative Poisson's ratio properties. *Advanced Materials* **27**, 1455 (2015).
37. J. W. Jiang, T. C. Chang, X. M. Guo, H. S. Park, Intrinsic negative Poisson's ratio for single-layer graphene. *Nano Letters* **16**, 5286–5290 (2016).
38. F. Liu, P. Ming, J. Li, Ab initio calculation of ideal strength and phonon instability of graphene under tension. *Physical Review B* **76**, 064120 (2007).
39. R. Khare et al., Coupled quantum mechanical/molecular mechanical modeling of the fracture of defective carbon nanotubes and graphene sheets. *Physical Review B* **75**, 075412 (2007).
40. E. Stolyarova et al., High-resolution scanning tunneling microscopy imaging of mesoscopic graphene sheets on an insulating surface. *Proceedings of the National Academy of Sciences* **104**, 9209–9212 (2007).
41. X. D. Wei, J. W. Kysar, Experimental validation of multiscale modeling of indentation of suspended circular graphene membranes. *International Journal of Solids and Structures* **49**, 3201–3209 (2012).
42. K. Kim et al., Ripping graphene: Preferred directions. *Nano Letters* **12**, 293–297 (2012).
43. H. I. Rasool, C. Ophus, W. S. Klug, A. Zettl, J. K. Gimzewski, Measurement of the intrinsic strength of crystalline and polycrystalline graphene. *Nature Communications* **4**, 2811 (2013).
44. C. S. Ruiz-Vargas et al., Softened elastic response and unzipping in chemical vapor deposition graphene membranes. *Nano Letters* **11**, 2259–2263 (2011).
45. Y. Wei et al., The nature of strength enhancement and weakening by pentagon–heptagon defects in graphene. *Nature Materials* **11**, 759–763 (2012).
46. P. Y. Huang et al., Grains and grain boundaries in single-layer graphene atomic patchwork quilts. *Nature* **469**, 389–392 (2011).
47. Q. Yu et al., Control and characterization of individual grains and grain boundaries in graphene grown by chemical vapour deposition. *Nature Materials* **10**, 443–449 (2011).
48. B. Bhushan, D. Luo, S. R. Schricker, W. Sigmund, S. Zauscher, *Handbook of Nanomaterials Properties* (Springer, 2014).
49. H. Tongwei, L. Ying, W. Chengyuan, Effects of temperature and strain rate on the mechanical properties of hexagonal boron nitride nanosheets. *Journal of Physics D: Applied Physics* **47**, 025303 (2014).
50. R. Mas-Balleste, C. Gomez-Navarro, J. Gomez-Herrero, F. Zamora, 2D materials: To graphene and beyond. *Nanoscale* **3**, 20–30 (2011).
51. K. N. Kudin, G. E. Scuseria, B. I. Yakobson, C_2F, BN, and C nanoshell elasticity from ab initio computations. *Physical Review B* **64**, 235406 (2001).
52. J.-U. Lee, D. Yoon, H. Cheong, Estimation of Young's modulus of graphene by Raman spectroscopy. *Nano Letters* **12**, 4444–4448 (2012).
53. A. Falin et al., Mechanical properties of atomically thin tungsten dichalcogenides: WS_2, WSe_2, and WTe_2. *ACS Nano* **15**, 2600–2610 (2021).
54. F. Cantos-Prieto et al., Layer-dependent mechanical properties and enhanced plasticity in the van der Waals chromium trihalide magnets. *Nano Letters* **21**, 3379–3385 (2021).
55. C. Lee et al., Elastic and frictional properties of graphene. *Physica Status Solidi (B)* **246**, 2562–2567 (2009).
56. Y. Zhang, C. Pan, Measurements of mechanical properties and number of layers of graphene from nano-indentation. *Diamond and Related Materials* **24**, 1–5 (2012).
57. M. Neek-Amal, F. M. Peeters, Nanoindentation of a circular sheet of bilayer graphene. *Physical Review B* **81**, 235421 (2010).
58. X. D. Wei et al., Recoverable slippage mechanism in multilayer graphene leads to repeatable energy dissipation. *ACS Nano* **10**, 1820–1828 (2016).
59. S. Scharfenberg et al., Probing the mechanical properties of graphene using a corrugated elastic substrate. *Applied Physics Letters* **98**, 091908 (2011).

60. Q. Y. Lin et al., Step-by-step fracture of two-layer stacked graphene membranes. *ACS Nano* **8**, 10246–10251 (2014).
61. K. Behfar, R. Naghdabadi, Nanoscale vibrational analysis of a multi-layered graphene sheet embedded in an elastic medium. *Composites Science and Technology* **65**, 1159–1164 (2005).
62. L. H. Li et al., Dielectric screening in atomically thin boron nitride nanosheets. *Nano Letters* **15**, 218–223 (2015).
63. L. H. Li, J. Cervenka, K. Watanabe, T. Taniguchi, Y. Chen, Strong oxidation resistance of atomically thin boron nitride nanosheets. *ACS Nano* **8**, 1457–1462 (2014).
64. T. Taniguchi, K. Watanabe, Synthesis of high-purity boron nitride single crystals under high pressure by using Ba-BN solvent. *Journal of Crystal Growth* **303**, 525–529 (2007).
65. Q. R. Cai et al., Raman signature and phonon dispersion of atomically thin boron nitride. *Nanoscale* **9**, 3059–3067 (2017).
66. J. L. Hutter, J. Bechhoefer, Calibration of atomic-force microscope tips. *Review of Scientific Instruments* **64**, 1868–1873 (1993).
67. J. E. Sader, J. W. M. Chon, P. Mulvaney, Calibration of rectangular atomic force microscope cantilevers. *Review of Scientific Instruments* **70**, 3967–3969 (1999).
68. Q. Peng, W. Ji, S. De, Mechanical properties of the hexagonal boron nitride monolayer: Ab initio study. *Computational Materials Science* **56**, 11–17 (2012).
69. L. Song et al., Large scale growth and characterization of atomic hexagonal boron nitride layers. *Nano Letters* **10**, 3209–3215 (2010).
70. M. Mirnezhad, R. Ansari, H. Rouhi, Mechanical properties of multilayer boron nitride with different stacking orders. *Superlattices and Microstructures* **53**, 223–231 (2013).
71. K. Liu et al., Elastic properties of chemical-vapor-deposited monolayer MoS_2, WS_2, and their bilayer heterostructures. *Nano Letters* **14**, 5097–5103 (2014).
72. G. Kresse, J. Hafner, Ab-initio molecular-dynamics for open-shell transition-metals. *Physical Review B* **48**, 13115–13118 (1993).
73. G. Kresse, J. Furthmuller, Efficient iterative schemes for ab initio total-energy calculations using a plane-wave basis set. *Physical Review B* **54**, 11169–11186 (1996).
74. Q. Cai et al., Raman signature and phonon dispersion of atomically thin boron nitride. *Nanoscale* **9**, 3059–3067 (2017).
75. Q. Cai et al., High thermal conductivity of high-quality monolayer boron nitride and its thermal expansion. *Science Advances* **5**, eaav0129 (2019).
76. R. Cuscó et al., Isotopic effects on phonon anharmonicity in layered van der Waals crystals: Isotopically pure hexagonal boron nitride. *Physical Review B* **97**, 155435 (2018).
77. R. Cuscó, J. H. Edgar, S. Liu, J. Li, L. Artús, Isotopic disorder: The prevailing mechanism in limiting the phonon lifetime in hexagonal BN. *Physical Review Letters* **124**, 167402 (2020).
78. D. Vella, B. Davidovitch, Indentation metrology of clamped, ultra-thin elastic sheets. *Soft Matter* **13**, 2264–2278 (2017).
79. B. Maaß et al., Nuclear charge radii of $^{10,11}B$. *Physical Review Letters* **122**, 182501 (2019).
80. W. A. Steele, Vapor pressure of nearly classical liquids. *The Journal of Chemical Physics* **34**, 802–803 (1961).
81. V. Kogan, Isotope effects in the structural properties of solids. *Soviet Physics Uspekhi* **5**, 951 (1963).
82. A. J. Giles et al., Ultralow-loss polaritons in isotopically pure boron nitride. *Nature Materials* **17**, 134–139 (2018).
83. R. Bijker, F. Iachello, Cluster structure of light nuclei. *Progress in Particle and Nuclear Physics* **110**, 103735 (2020).
84. L. Schaller et al., Nuclear charge radii of the carbon isotopes ^{12}C, ^{13}C and ^{14}C. *Nuclear Physics A* **379**, 523–535 (1982).
85. J.-W. Jiang, J.-S. Wang, B. Li, Young's modulus of graphene: A molecular dynamics study. *Physical Review B—Condensed Matter and Materials Physics* **80**, 113405 (2009).
86. Y. Y. Zhang, Y. Gu, Mechanical properties of graphene: Effects of layer number, temperature and isotope. *Computational Materials Science* **71**, 197–200 (2013).

Section III

Keeping It Cool

Advanced Thermal Management with BN Nanosheets

Chapter 9

Nanoscale Boron Nitride-Polymer Composites in Thermal Management of Electronic Devices

Jiangfeng Fan, Xiaoliang Zeng, and Rong Sun

9.1 INTRODUCTION

The rapid development of computer and network communication technologies has enhanced the connectivity between devices and individuals. With the widespread adoption of smart devices such as smartphones and smart cars, along with the growing demand for large data centers, efficient heat dissipation is crucial for the performance and reliability of electronic components [1,2]. However, this presents significant challenges in effectively managing the heat generated by electronic devices [3–5].

Polymers, which are characterized by low density, easy processing, and high insulation, are the main materials for the thermal management of electronic devices. The TC of conventional polymers, typically below 0.5 W $m^{-1}{\cdot}K^{-1}$, fails to satisfy consumer demands [5,6]. To overcome this limitation, researchers have begun integrating high-thermal-conductivity additives into polymer matrixes, resulting in composites that maintain both flexibility and enhanced TC [7–11]. Commonly used thermally conductive additives include metal particles, carbon-based materials, and ceramics. Notably, hexagonal BNNSs possess a two-dimensional structure akin to graphene and can achieve TC levels of up to 600 W $m^{-1}{\cdot}K^{-1}$ [12,13], making them competitive with nanostructured carbon materials [14]. In addition, BNNSs exhibit remarkable chemical stability, heat resistance, and insulation properties, surpassing those of conventional carbon materials. This makes them an outstanding choice as fillers for thermally conductive composites [15,16]. Nonetheless, the TCs of polymers filled with ceramics tend to be relatively low, with most systems showing values below 3 W $m^{-1}{\cdot}K^{-1}$ [17–19]. In *h*-BN-polymer composites, the conductivity is notably hindered by excessive phonon scattering in the polymer matrix and considerable resistance at the polymer-*h*-BN interface. This issue is even more pronounced in composites with a low concentration of fillers, where individual particles are surrounded by large amounts of polymer, which obstructs efficient lattice vibrations through the thermally insulating medium. To enhance TC, it is essential to increase the filler volume past a critical percolation threshold—typically around 30 vol% [20–26]. Achieving a percolated microstructure allows for direct contact between *h*-BN platelets, which strengthens interactions among them, reduces phonon scattering at the polymer-*h*-BN interface, and establishes effective

DOI: 10.1201/9781003647812-12

thermal conduction pathways throughout the matrix. Furthermore, aligning these particles strategically can boost this effect, facilitating percolation along the covalently bonded basal plane of *h*-BN. Thus, a well-thought-out design of the filler structure is crucial for forming efficient thermal pathways within the polymer matrix, lowering interfacial thermal resistance, and enhancing the TC of the composite materials overall [27,28]. To achieve these objectives, various techniques have been adopted to achieve these goals, such as filler orientation control [29], filler agglomeration [30], 3D network construction [31,32], and filler self-assembly [33]. Lattice vibrations occurring within the basal plane of one *h*-BN platelet can effectively transfer to the basal plane of an adjacent platelet at the *h*-BN-*h*-BN interface, greatly enhancing phonon transfer throughout the material. The creation of composites that achieve high TCs around 10 W $m^{-1}\cdot K^{-1}$, combined with ease of processing, is essential for supporting increased power densities in electronic devices.

9.2 OVERVIEW OF NANOSCALE BN AND POLYMER COMPOSITES

9.2.1 Characteristics and Fabrication Methods of Nanoscale BN

BN-reinforced polymer nanocomposites with enhanced TC are highly desirable for a wide range of applications. These nanocomposites can be produced using various methods, allowing for both isotropic and anisotropic thermal properties. To achieve isotropic TC, one approach is to blend BN fillers with isotropic TC directly into a polymer matrix. This method ensures a homogeneous dispersion of BN throughout the composite, facilitating efficient heat transfer in all directions. Another method involves fabricating 3D BN structures and subsequently filling the voids or gaps with a polymer matrix. This approach creates a network of thermally conductive pathways, enabling heat to be effectively transferred in multiple directions within the composite. Furthermore, by controlling the distribution of *h*-BN flakes or BNNSs within a polymer matrix, it is possible to establish thermally conductive pathways in specific directions, resulting in anisotropic thermal properties. This approach allows for tailoring the composite's TC along preferred directions, which can be advantageous in certain applications.

9.2.1.1 Isotropic BN Filler-Enhanced Polymer Nanocomposites

The synthesis of BN spheres (BNSs) incorporating randomly distributed *h*-BN flakes or BNNSs provides a means to isotropically enhance the TC of polymers. In the study by Zhang et al., similar BNSs were employed as fillers in an epoxy matrix (Figure 9.1a and b), achieving a TC of 1.66 W $m^{-1}\cdot K^{-1}$ with a 40wt% loading of BN [34]. The spherical configuration of the BN contributed to the epoxy nanocomposite exhibiting similar TCs in both in-plane and cross-plane orientations (Figure 9.1c). However, it was noted that these BNSs were fragile and prone to breaking during manufacturing. Furthermore, due to the high viscosity of the pre-cured epoxy nanocomposite, the BN loading was capped at around 40wt% [35]. To tackle the issues of high viscosity and inadequate mechanical strength, Ren et al. developed new spherical BN using spray drying and vacuum sintering methods (Figure 9.1d). The spray drying process enabled the production of larger BNSs, which helped lower the viscosity of the BN/polymer mixture. Following this, vacuum sintering at 1800°C for 2 hours significantly improved the connectivity between the BN platelets, resulting in more durable and robust BNSs. The final product, spherical BN-enhanced polydimethylsiloxane (PDMS), achieved a TC of 2.3 W $m^{-1}\cdot K^{-1}$ at a BN content of 50wt%, roughly three times greater than that of *h*-BN platelets.

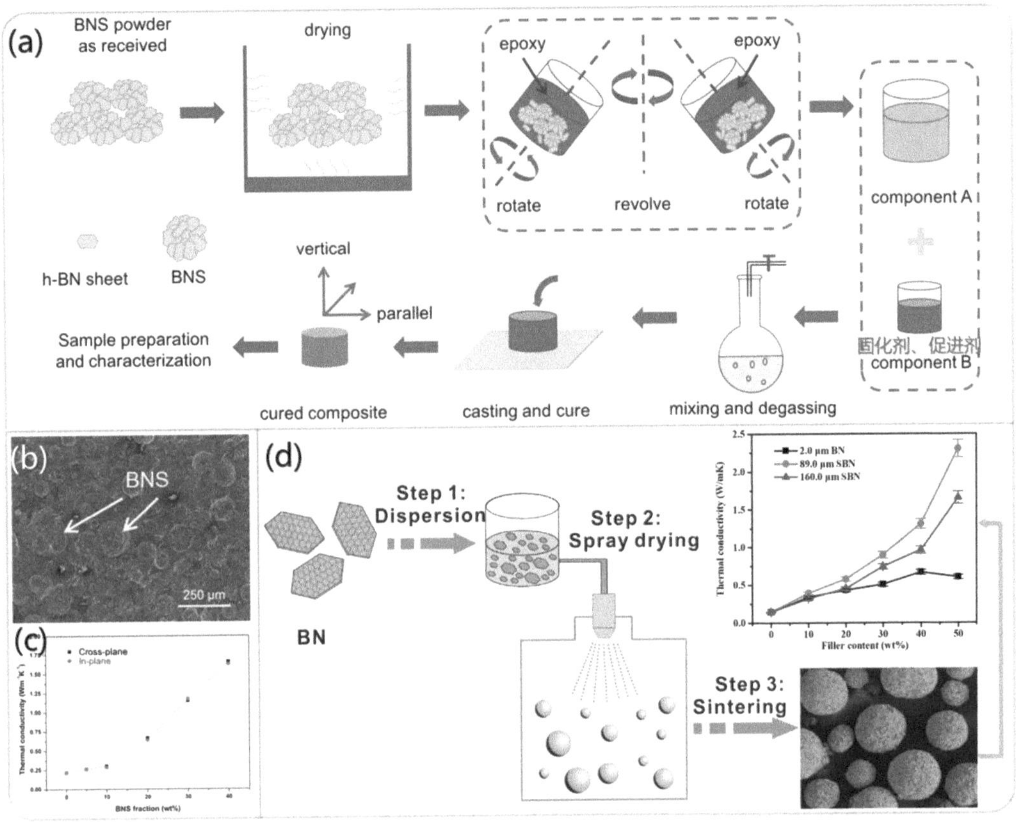

Figure 9.1 (a) The fabrication process of epoxy/BNS composites. (b) SEM image of the BNS/epoxy nanocomposite. (c) Cross-plane and in-plane TC of BNS/epoxy. (Reprinted with permission from Ref. [34]. Copyright 2019, Polymers.) (d) Schematic illustration of the spherical BN fabrication procedure and TC of PDMS with spherical BN fillers of varying sizes as a function of filler concentration [35]. (Reprinted with permission from Ref. [35]. Copyright 2019, *Chemical Engineering Journal.*)

9.2.1.2 3D BN Frame-Enhanced Polymer Nanocomposites

To improve the TC of polymer nanocomposites, one straightforward method is to increase the quantity of thermally conductive fillers. However, a higher filler content can negatively affect other polymer properties, such as mechanical strength and optical performance [36,37]. A more efficient approach to enhance TC while keeping the filler content low is to create 3D thermally conductive pathways within the polymer matrix. This 3D filler architecture includes branches oriented horizontally, vertically, or randomly, thereby forming isotropic thermal conduction routes throughout the polymer. A variety of techniques can be used to produce these 3D fillers, such as foaming, freeze-drying, and chemical vapor deposition, among others [36,38–40].

9.2.1.2.1 Ice Template-Assisted Strategy

Zeng et al. have introduced an advanced epoxy resin integrated with a 3D BNNS network, resulting in exceptional thermal properties. This composite achieves a TC of 2.85 W m^{-1}·K^{-1} while maintaining a low thermal expansion coefficient of $(24–32)\times10^{-6}$ K^{-1}, all at a BNNS content of

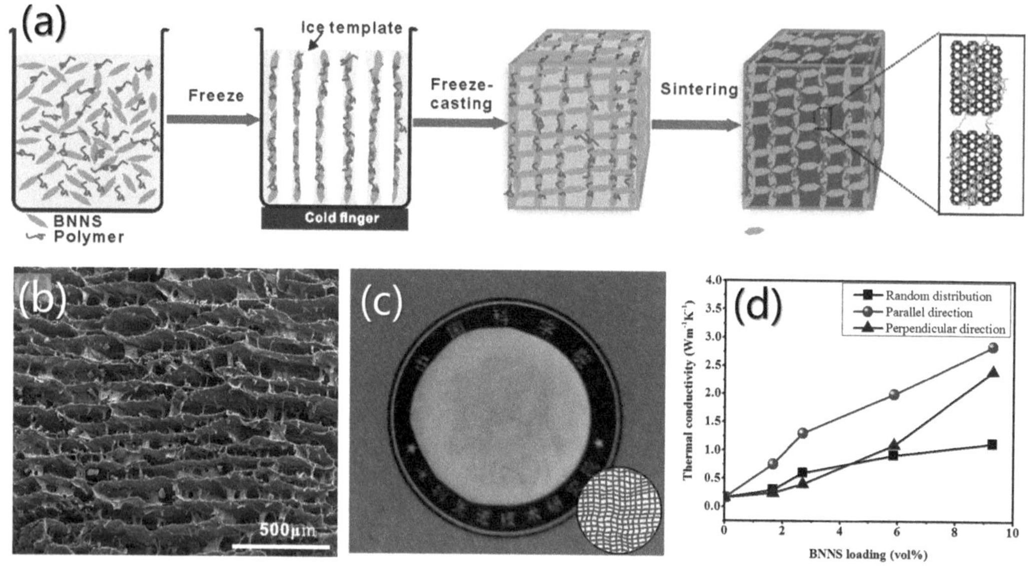

Figure 9.2 (a) Schematic representation illustrating the preparation process of 3D-BNNS aerogels using the ice template technique. (b) SEM image of the cross-section of the 3D-BNNS aerogel. (c) Optical image of the 3D-BNNS aerogel nanocomposite at 9.29 vol% of BNNSs. (d) TC of 3D-BNNS aerogel nanocomposites as a function of BNNS fractions [36]. (Reprinted with permission from Ref. [36]. Copyright 2015, *Small.*)

only 9.3 vol% [36]. The production process, illustrated in Figure 9.2a, begins with the creation of a 3D BNNS aerogel using an ice template method. Polyvinyl alcohol (PVA) is employed as a binder, and the aerogel undergoes sintering to eliminate organic impurities. Subsequently, liquid epoxy resin is infused into the porous aerogel. Upon curing, the final 3D-BNNS/epoxy composite retains the shape and dimensions of the original aerogel without any deformation or shrinkage during the filtration process (Figure 9.2b and c). This innovative nanocomposite exhibits highly uniform TC, measuring 2.85 W $m^{-1}{\cdot}K^{-1}$ in-plane and 2.40 W $m^{-1}{\cdot}K^{-1}$ out-of-plane, even with a modest filler concentration of 9.3 vol% (Figure 9.2d). The enhancement in thermal performance and isotropy is primarily due to its distinct layered structure, which facilitates the formation of thermal conduction pathways throughout the polymer matrix (Figure 9.5b and c). In contrast, an epoxy nanocomposite with randomly oriented BNNSs shows a significantly lower TC of just 1.13 W $m^{-1}{\cdot}K^{-1}$ at the same filler level (Figure 9.2d).

Bai et al. developed a BN/epoxy composite featuring a 3D conductive network that resembles nacre, utilizing a bidirectional freezing method [41] (Figure 9.3). In this process, BNNSs were arranged into an aerogel characterized by long-range aligned layers, which were then infused with epoxy resin. This meticulously structured 3D conductive network facilitates extended phonon pathways, resulting in significantly enhanced TC of 6.07 W $m^{-1}{\cdot}K^{-1}$ at a relatively modest BN content of 15 vol%, outperforming similar composites reported in existing studies. Additionally, the composite exhibits an outstanding electrical resistivity of 2×10^{12} Ω·cm and impressive thermal stability, with a glass transition temperature of 120°C. These properties position the composite for diverse applications, such as thermal interface materials (TIMs), in advanced electronic packaging technologies.

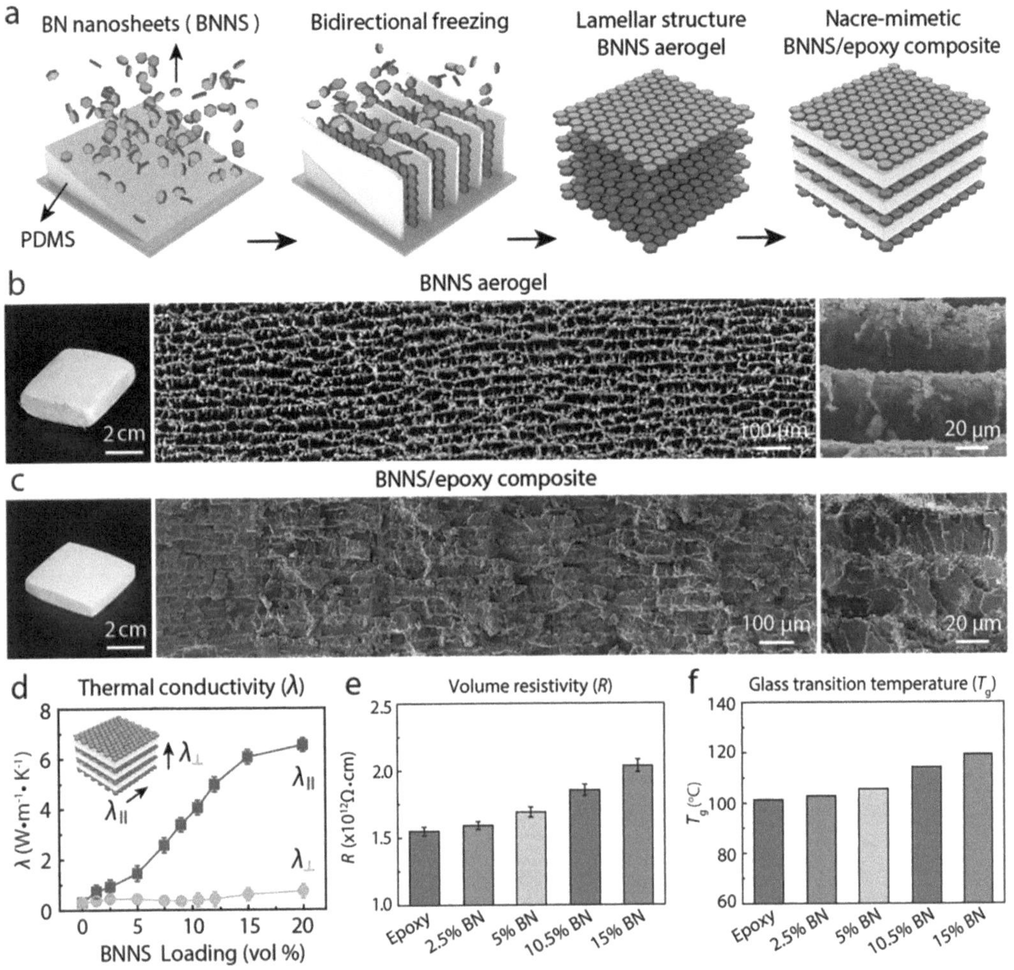

Figure 9.3 Fabrication and characterization of the BNNS/epoxy composite. (a) Schematic illustration of the fabrication route using a bidirectional freezing technique together with epoxy resin infiltration. Optical and SEM images of (b) a BNNS aerogel and (c) a BNNS/epoxy composite. The cross-sectional SEM images show that the long-range aligned lamellar structure of the BNNS aerogel was well maintained in the composite after the infiltration and curing of epoxy resin. (d) TC of our BNNS/epoxy composites in the parallel ($\lambda\|$) and perpendicular ($\lambda\perp$) directions with respect to the lamellar layers. (e) Volume resistivity (R) and (f) glass transition temperature (T_g) as a function of BNNS loading in the composite [41]. (Reprinted with permission from Ref. [41]. Copyright 2019, *Advanced Functional Materials.*)

Furthermore, Zhao et al. developed a modified bidirectional freezing and hot-pressing process to produce BN/polyurethane (BN/PU) nanocomposites (Figure 9.4) [42]. By applying horizontal and vertical temperature gradients, both horizontally and vertically aligned BNNSs were created during the ice crystal formation process. The resulting nanocomposites demonstrated exceptional TC at a BN content of 80 vol%, achieving values of 39.0 W m^{-1}·K^{-1} in the in-plane direction and 11.5 W m^{-1}·K^{-1} in the out-of-plane direction.

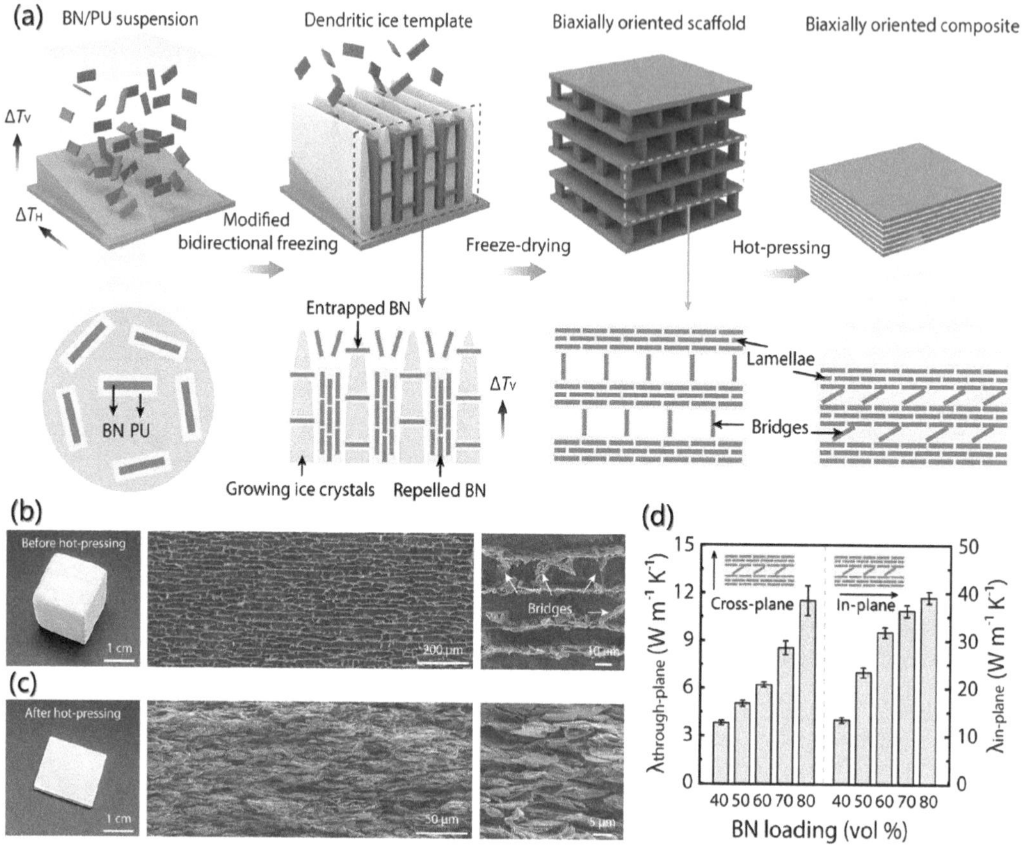

Figure 9.4 (a) Schematic for fabrication of the BN/PU nanocomposite. (b) Structure of the BN/PU porous scaffold, showing a lamellar architecture with interlayer bridges. (c) Structure of the BN/PU bulk composite, maintaining the lamellar structure after hot-pressing. (d) Through-plane and in-plane TC of the BN/PU bulk composites at different BN loadings [42]. (Reprinted with permission from Ref. [42]. Copyright 2022, *ACS Nano*.)

9.2.1.2.2 Expansion-Flow-Assisted Alignment

Bai et al. developed a universal and scalable method for constructing vertically aligned structures of anisotropic fillers in composites assisted by the expansion flow (using 2D BN platelets as a proof-of-concept) [43] (Figure 9.5). In a 3D printed expansion mold, BN/SG inks were transferred from a narrow channel to a wider channel. The mold's design significantly affects the orientation of the BN platelets, thereby influencing the TC of the BN/SG strips. Within the silicone gel strips, the BN platelets exhibit a curved orientation, with vertical alignment in the center and horizontal alignment near the surfaces. This vertical arrangement of BN in the central region results in an impressive through-plane TC of up to 5.65 W $m^{-1}{\cdot}K^{-1}$.

9.2.1.2.3 Compression Molding

Gan et al. introduced an innovative method for creating elastic porous materials by utilizing PMMA microspheres as sacrificial components. This technique draws inspiration from the directional assembly of *h*-BN through foaming and salt template methods, as well as the unique

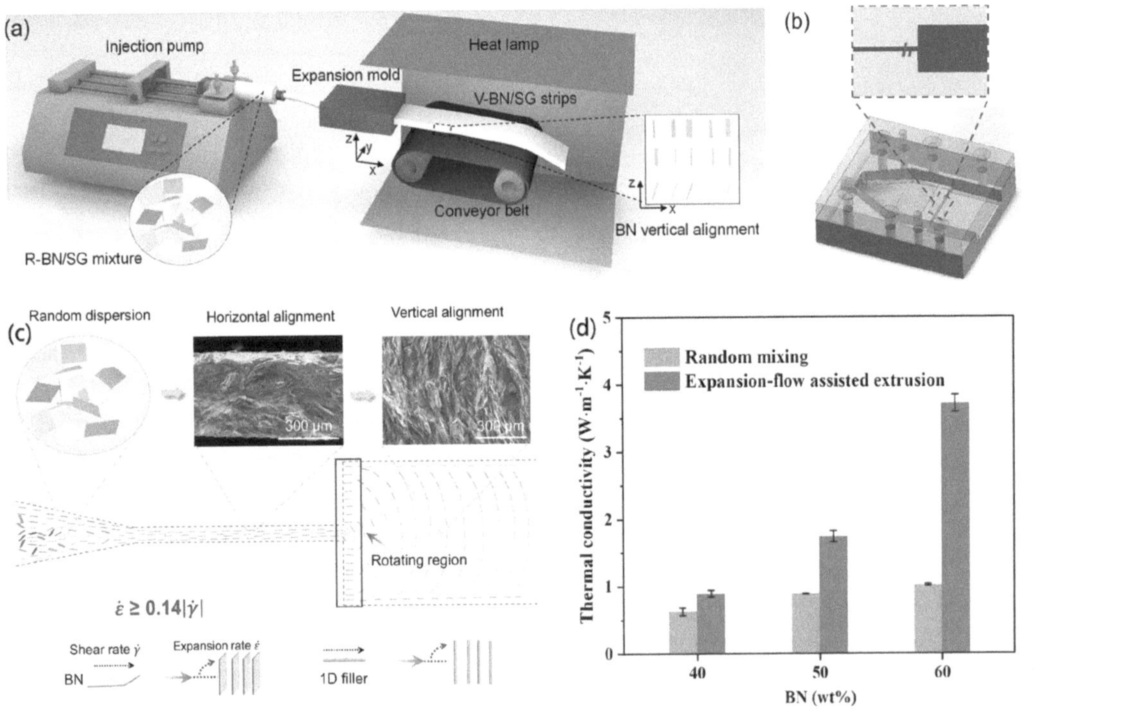

Figure 9.5 (a) Schematic of the continuous fabrication of V-BN/SG strips. (b) The expansion mold designed by the Solidworks software and the inset showing the changed height of microchannels from small h to large H. (c) Illustration of the rotating process of BN platelets from parallel to perpendicular to the flow direction. The inset shows the alignment rotation of BN in the horizontal region and the vertical region. (d) Through-plane TC of BN/SG strips as a function of filler loading [43]. (Reprinted with permission from Ref. [43]. Copyright 2022, *Nano-Micro Letter.*)

attributes of the *h*-BN platelet structure [44] (Figure 9.6a). Unlike traditional foaming and salt template methods, this hard-template technique allows for precise control over the pore sizes in the resulting 3D network and enhances the effectiveness of fillers in thermal conduction. Consequently, it is particularly suitable for developing 3D thermal conductive networks using various high TC fillers. Characterization analyses and 3D finite element simulations reveal that the presence of 3D BN is crucial for enhancing the TC of the composites. In comparison to nanoscale *h*-BN, the micron-scale *h*-BN exhibits a more organized arrangement along the pore walls in 3D-BN, optimizing the in-plane TC of *h*-BN while minimizing interfacial thermal resistance. The resulting lightweight composite demonstrates an impressive TC of 1.868 W m^{-1}·K^{-1} (Figure 9.6b) and an exceptional volume resistivity of 3.66×10^{13} Ω·m (Figure 9.6c) with an mBN loading of 18.33 vol%.

Fu et al. described an easy method for producing polymer composites that exhibit high TC and insulation by establishing a continuous network of BN (*h*-BN) fillers at the interface and incorporating graphene nanoplatelets (GNPs) within polyamide 6 (PA6) particles [45] (Figure 9.6d and e). Initially, PA6/GNPs binary composites were created through melt mixing PA6 with GNPs. The resulting composites were then ground into particles. A small amount of vinyl acetate-ethylene

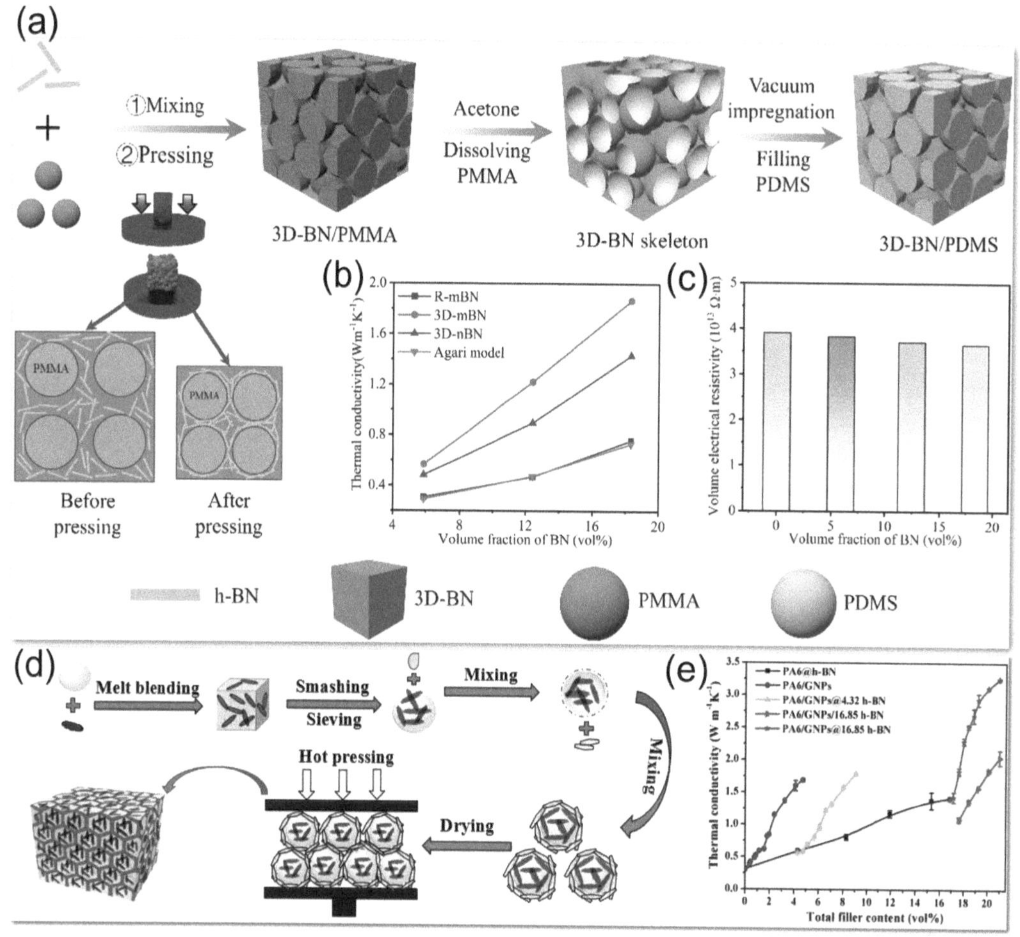

Figure 9.6 (a) Schematic diagram of preparation procedures of 3D-BN/PDMS. (b) The TCs and (c) Volume of composites with different *h*-BN contents [44]. (Reprinted with permission from Ref. [44]. Copyright 2022, Polymer.) (d) Schematic illustration of preparation procedures of PA6/GNPs@*h*-BN composites with the segregated double network. (e) TC of PA6 composites prepared via different fabricating methods with different total filler loadings [45]. (Reprinted with permission from Ref. [45]. Copyright 2019, *Composites Science & Technology*.)

resin was used as a binder to apply varying amounts of *h*-BN onto the surfaces of these particles. By employing simple compression molding, the final composites featuring a continuous segregated dual network were achieved, positioning *h*-BN at the interface of the PA6/GNPs binary particles. Consequently, by adjusting the GNP content within the PA6 particles and the quantity of *h*-BN coated on the surfaces, both thermal and electrical conductivity of the composites could be effectively managed. For example, with a total filler content of 18.82 vol%, including 1.97 vol% GNPs inside the PA particles and 16.85 vol% *h*-BN on the surface, the TC saw an almost tenfold increase (from 0.29 to 2.69 W m^{-1}·K^{-1}), while still maintaining excellent electrical insulation (as low as 4.9×10^{-9} S m^{-1}). Furthermore, the insulating shell formed by *h*-BN can obstruct the conductive pathways of the GNP network, allowing the PA6/GNPs@*h*-BN ternary composites to be tailored for specific thermal and electrical properties based on the filler ratios. Additionally,

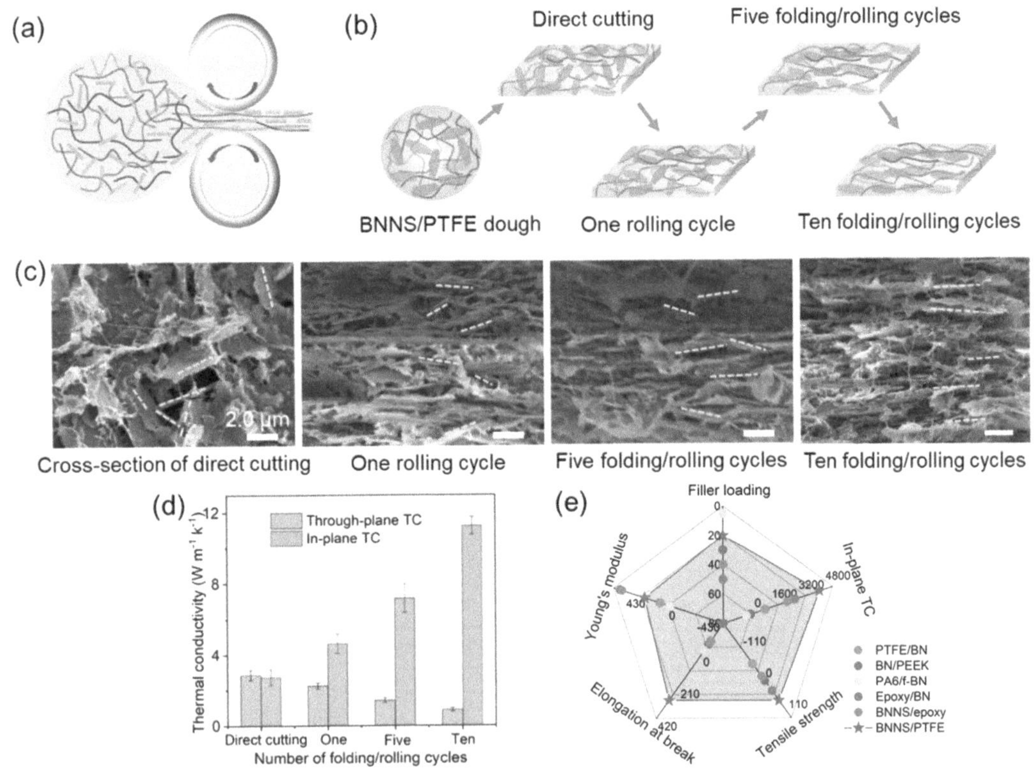

Figure 9.7 (a) Schematic of the roller pressing process. (b) Production of four films from the same dough. (c) Cross-sectional SEM images of the four films. (d) TC of the four films. (e) Radar diagram of the integrated performance improvement of the BNNS/PTFE composite film compared to results for other reported BNNS composites [46]. (Reprinted with permission from Ref. [46]. Copyright 2023, *ACS Nano*.)

anisotropic thermal composites were produced by further compression molding to decrease sample thickness, significantly enhancing TC in the desired direction. In summary, a customized distribution of the filler network within the composite offers a promising strategy for fabricating polymers that combine high TC with outstanding electrical insulation.

9.2.1.2.4 Roller Pressing

Qiu et al. introduced a mechanical interlocking method that effectively integrates BNNSs at high concentrations (up to 20 wt%) into a polytetrafluoroethylene matrix. This results in a ductile, easily moldable, and reusable BNNS/polytetrafluoroethylene composite dough [46] (Figure 9.7). By carefully adjusting the solvent levels, the dough achieves excellent cohesion and flexibility, allowing it to be shaped through techniques such as roller pressing, kneading, and cutting. Additionally, the orientation of the BNNSs can be modified via repeated roller pressing. A composite film with parallel BNNS orientation exhibits a remarkable 4408% increase in in-plane TC, alongside improvements of 334.0%, 69%, 266%, and 302% in tensile modulus, tensile strength, toughness, and elongation at break, respectively. Its low dielectric constant (2.44) and loss (0.001) make it suitable for thermal management applications in complex surfaces and high-speed/high-frequency signal transmission devices.

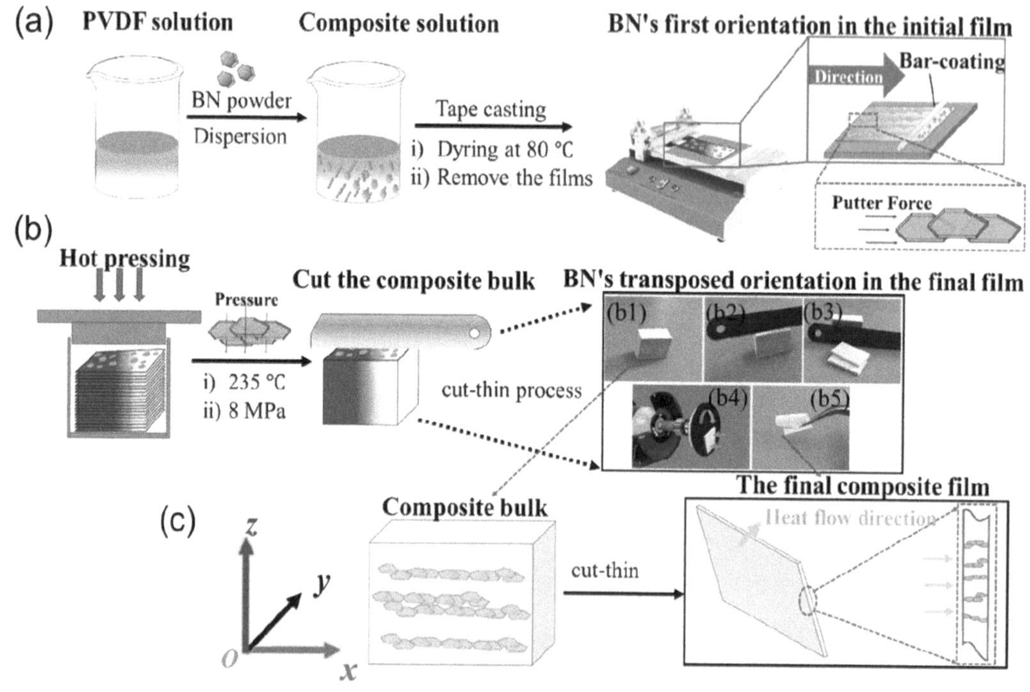

Figure 9.8 The schematic diagram of the step-by-step orientation method used to prepare x-BN/PVDF composite films [47]. (Reprinted with permission from Ref. [47]. Copyright 2019, *Composites Part A*.)

Song et al. reported a step-by-step orientation method (including bar-coating, hot-pressing, cutting, and thinning) for the fabrication of a through-plane BN path for excellent TIM application in poly(vinylidene fluoride) composite films [47] (Figure 9.8). The effects of BN content on TC and electrical insulation properties were studied. At a filler loading of 30 wt%, the composite achieved an impressive through-plane TC of 3.50 W m^{-1}·K^{-1}. Additionally, the composites demonstrated excellent electrical insulation ($>10^{13}$ Ω·cm) and effective performance at high frequencies.

9.2.2 Structural and Morphological Characterization for Nanoscale BN-Polymer Composites

Various characterization techniques have been employed in numerous studies to investigate the distribution of fillers. Among these techniques, scanning electron microscopy (SEM) and X-ray diffraction (XRD) have been extensively utilized to provide valuable insights into the arrangement and alignment of BNs within the matrix, shedding light on their effective dispersion and orientation.

9.2.2.1 SEM

There are many other research reports that utilize SEM to investigate fillers' orientation structures. Bai et al. conducted SEM characterization in longitudinal (x–z plane) and transverse direction (y–z plane) of samples to confirm the alignment structure and rotating process of BN platelets [43] (Figure 9.9a). The longitudinal SEM image of the 60 V-BN/SG (0.2–2) strips

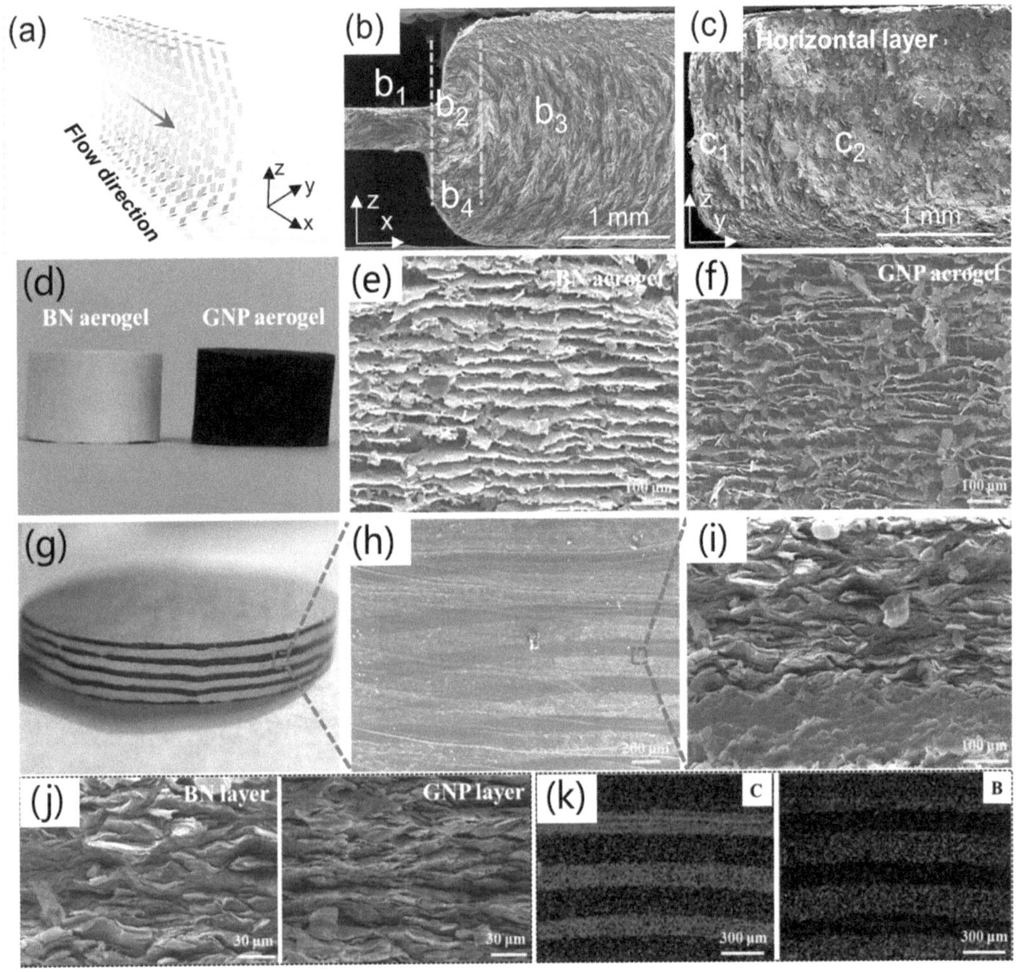

Figure 9.9 (a) Schematic of BN alignment in the expansion mold. SEM characterization in longitudinal (x–z plane, (b) and transverse direction (y–z plane, (c) of samples [43]. (Reprinted with permission from Ref. [43]. Copyright 2022, *Nano-Micro Letter.*) (d)–(k) Morphology characterization of the bidirectional BN (GNP) aerogels and alternatively multilayered composites: (d) Digital photo of BN and GNP aerogels; (e) SEM of the BN aerogel indicating oriented lamellar structure; (f) SEM of the GNP aerogel indicating oriented lamellar structure; (g) digital photograph of the BN/GNP layered sheet; (h)–(i) SEM observation of the BN/GNP layered sheet under different magnifications; (j) SEM describing the orientation state of BN (left) and GNP (right) sheets in respective layers; (k) EDX mapping of the C and B element in multilayered structure [49]. (Reprinted with permission from Ref. [49]. Copyright 2023, *Journal of Alloys and Compounds.*)

(Figure 9.9b) exhibits three distinct regions: the horizontal region (b_1), the rotating region (b_2), and the vertical region (b_3). In the horizontal region (b_1), BN platelets align parallel to the narrow channel wall due to strong shear stress. In the rotating region (b_2), BN undergoes a sudden rotation influenced by expansion stress, occurring within about 500 μm from the wide channel entrance. Close to the centerline (b_3), the BN platelets exhibit a highly vertical alignment, gradually rotating toward the horizontal direction near the channel wall (b4), which confirms a curved orientation (Figure 9.9b). The transverse SEM image (Figure 9.9c) indicates that in the central area of the channel, the basal plane of BN is perpendicular to the flow direction and aligned vertically along the thickness (c_2). Conversely, near the channel's side wall, the BN basal plane is parallel to the flow direction but remains vertically aligned due to shear stress (c_1). The transverse SEM images also confirm the horizontally aligned layer close to the top or bottom walls (Figure 9.9c, b_4). Huang et al. also observed the cross-section of 10-BNNS-M/Poly(p-phenylene-2,6-benzobisoxazole) (PBO) paper by SEM, showing the nacre-like multilayered internal microstructure of the BNNS-M/PBO insulating paper [48].

Zhao et al. prepared a dense and layered BN/GNP compositing film by alternatively stacking and pressing at room temperature along the transverse direction [49] (Figure 9.9d–k). The cross-sectional SEM observation clearly reveals the formation of alternating structure, and the conductive GNP layers are sandwiched between insulating BN layers (Figure 9.9g–i). The alignment of BN and GNPs remains well maintained even under compression, as evidenced by the network of interconnected BN branches (or GNPs) linking adjacent hierarchical BN platelets (or GNPs) in Figure 9.9j. Additionally, the Energy Dispersive X-ray Spectroscopy (EDS) mapping in Figure 9.9k shows that B and C elements are distinctly positioned in the BN/CS and GNP/CS layers, respectively, creating a multilayered elemental distribution with clear separations. This further confirms the oriented structure of the fillers in the composite materials.

9.2.2.2 XRD

XRD measurements were performed to ascertain the orientation of BNNSs in the composites. As reported previously, the horizontally and vertically oriented BN are responsible for the (0 0 2) and (1 0 0) peaks, respectively [50,51]. Figure 9.10a presents a schematic diagram confirming the orientation characteristics of BN aligned with the through-plane of the composite. This is evident from the intensity ratios at $2\theta=26.76°$ and $2\theta=41.60°$, which correspond to the diffraction from the (0 0 2) and (1 0 0) planes of BN in the XRD results [47,52].

The XRD results for the x-BN/PVDF composites are shown in Figure 9.10b. These composites display characteristic diffraction peaks at $2\theta=26.7°$, 41.6°, 43.7°, 50.1°, 54.9°, and 59.6°, attributed to the (0 0 2), (1 0 0), (1 0 1), (1 0 2), (0 0 4), and (1 0 3) crystal planes of BN (JCPDS 45–0894). For comparison, the in-plane XRD of the BN/PVDF composites was also analyzed (Figure 9.10c). The results indicate a notable difference in peak intensities between the two measurement directions. The through-plane test reveals both (0 0 2) and (1 0 0) peaks, while the in-plane test of BN/PVDF shows only the (0 0 2) peak, suggesting a well-oriented arrangement of BN platelets within the composites.

$$\delta=\frac{I_{100}}{I_{100}+I_{002}} \tag{9.1}$$

The peak intensity ratio (δ) is defined as the relative intensity of the (1 0 0) peak (I_{100}) compared to the sum of the relative intensities of the (1 0 0) peak (I_{100}) and the (0 0 2) peak (I_{002}).

Figure 9.10d indicates that Hu et al. found that epoxy resin/oriented BN composites have a significantly higher I_{002}/I_{100} ratio of 29, compared to just six for epoxy resin/random BN composites at the same loading of 40 wt% [53]. This finding confirms the orientation of BN platelets within the composites, as illustrated in the two insets. Furthermore, the average orientation angle f of BN crystallites is estimated from the XRD diffraction peaks in the 2q range of 10°–60° by using the equation reported by Uchikoshi [54–56] to numerically evaluate the improvement of the BN crystalline orientation.

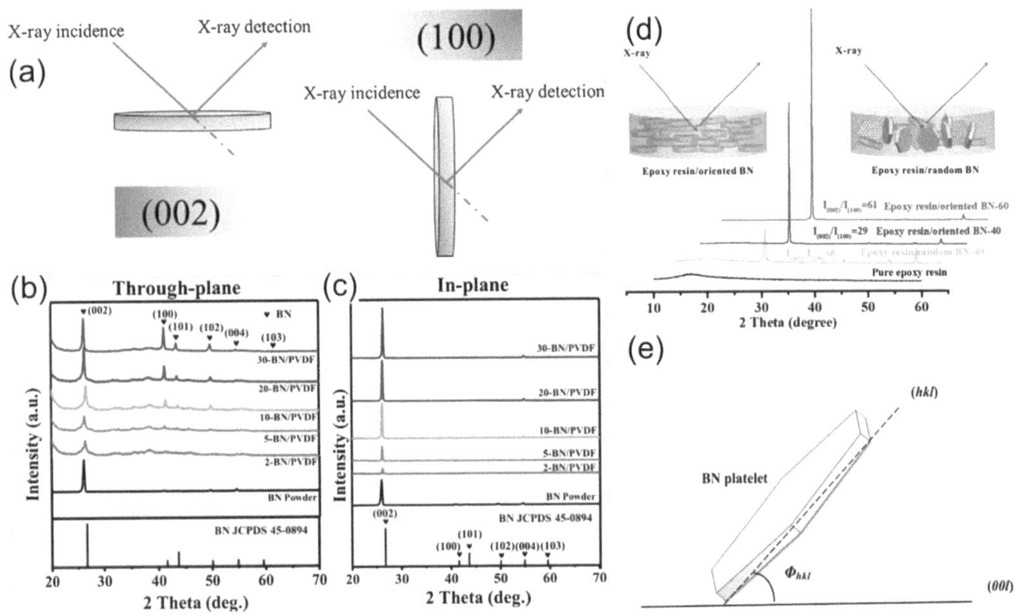

Figure 9.10 (a) Illustration of the effect of the orientation of BN platelets on the XRD pattern. (b, c) XRD patterns of x-BN/PVDF in through-plane (b) and in-plane (c) [47]. (Reprinted with permission from Ref. [52]. Copyright 2018, *Composites Part A-Applied Science and Manufacturing*.) (d) XRD spectra for pure epoxy resin, epoxy resin/oriented BN composites with 40 and 60 wt% (referred to as epoxy resin/oriented BN-40 and epoxy resin/oriented BN-60), and epoxy resin/random BN composites with 40 wt% (denoted as epoxy resin/random BN-40). The two insets illustrate these composites schematically. (e) The interplanar angle between the crystal plane (hkl) and the base plane (00l) is defined as F_{hkl} [53]. (Reprinted with permission from Ref. [53]. Copyright 2018, *Composites Science & Technology*.)

$$\bar{\phi} = \frac{\sum (I_{\text{hkl}} + \phi_{\text{hkl}})}{\sum I_{\text{hkl}}} \tag{9.2}$$

where I_{hkl} is the intensity of the diffraction peaks and F_{hkl} is the angle between the crystal plane (hkl) and the base plane (00l) (Figure 9.10e). ϕ_{hkl} is calculated by the following equation:

$$\cos\phi_{\text{hkl}} = \frac{\frac{\sqrt{3}a}{2c} l}{\sqrt{h^2 + k^2 + \text{hk} + \frac{3a^2}{4c^2} l^2}} \tag{9.3}$$

where a and c denote the crystalline parameters of the BN crystal (a=0.2173, c=0.6657). Consequently, the average orientation degree of BN in the epoxy resin/oriented BN composite is 3.0°. This represents a significant decrease compared to the 16.5° orientation in the composite with randomly distributed BN at the same loading of 40 wt%. In this, the corresponding improvement of BN orientation degree is up to 81.8%, which quantitatively indicates the enhanced orientation of BN for epoxy resin/oriented BN composites.

9.2.2.3 Transmission Electron Microscopy (TEM) and Micro-Computed Tomography (CT)

TEM and micro-CT are also important techniques for characterizing filler network structures, although their applications in BN-polymer composites are less reported. Li et al. reported that in the TEM of the oriented BNNS-supported carbon nanotubes (CNTs)/epoxy resin (Figure 9.11b and c), the stacked oriented layer of BN/ZC still existed, and linkage of CNTs between BNNSs was also shown, consistent with the result of SEM (Figure 9.11a) [57]. In the work of Zhang et al., the strong hydrogen bond interaction induced the tannic acid (TA)-tungsten disulfide (WS2) nanosheets to self-assemble into an interwoven network around the PU latex microsphere [58]. TEM images of TAWS2/PU nanocomposites show that there is an ordered interwoven network of WS2 nanosheets in the PU matrix (Figure 9.11d).

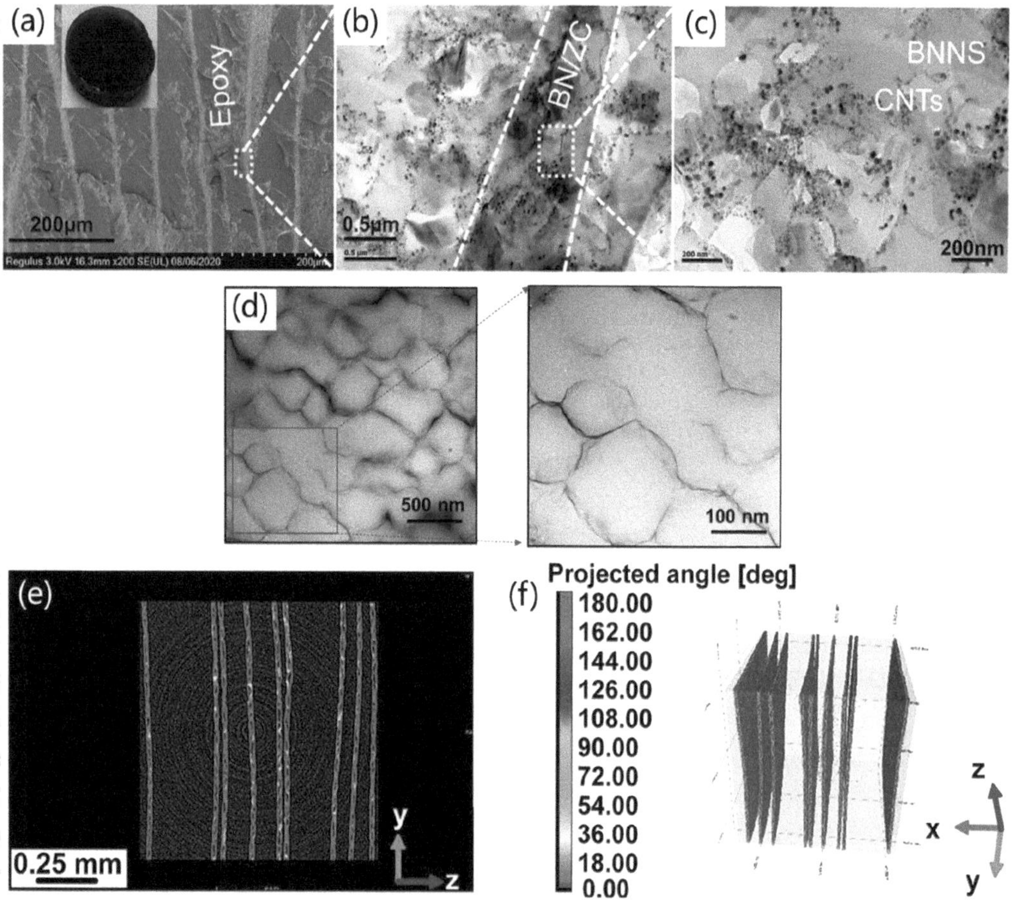

Figure 9.11 (a) SEM images of the fractured surface of o-BN/ZC/ER (175 mg mL^{-1}) composites. (b,c) TEM images of o-BN/ZC-ER composites [57]. (Reprinted with permission from Ref. [57]. Copyright 2021, Advanced Materials Interfaces.) (d) TEM images of 16 wt% TA-WS2/PU composites with a 3D conductive network [58]. (Reprinted with permission from Ref. [58]. Copyright 2021, *Nature Communication*.) (e,f) Micro-CT cross-section and 3D structure images of the V-GP/PB TIM [59]. (Reprinted with permission from Ref. [59]. Copyright 2021, *Chemistry of Materials*.)

To further explore the distribution of the graphite film in the polybutadiene matrix, Zeng et al. used micro-CT to characterize the microstructure of the vertically oriented graphite/polybutadiene composites. The graphite films are fully vertically through in the polybutadiene matrix (Figure 9.11e and f) [59].

9.3 MECHANISMS FOR ENHANCING TC IN NANOSCALE BN-POLYMER COMPOSITES

9.3.1 Factors Influencing the Enhancement of TC

Both theoretical and experimental findings suggest that establishing a continuous filler network is essential, as it significantly facilitates the transport of phonons and electrons. The intrinsic TC, shape, size, and aspect ratio of fillers, filler loading, fillers' hybrid and dispersion, and interfaces between BNNS and polymeric matrixes are the main parameters that largely influence the TC of composites. Thus, it is necessary for us to take a comprehensive view of these factors.

9.3.1.1 Intrinsic TC of Thermally Conductive Fillers

Thermally conductive fillers play a crucial role in determining the λ values of polymer composites. The three main types of these fillers are metals, ceramics, and carbon materials [60,61]. The characteristics of these fillers are especially critical during the preparation of thermally conductive polymer composites. Factors such as size, loading amount, surface morphology, and overall performance significantly affect the λ values of these composites [62]. Therefore, gaining a comprehensive understanding of the physical and chemical properties of thermally conductive fillers is essential for producing high-quality composites.

Table 9.1 provides a summary of the λ values for commonly used thermally conductive fillers [63–70]. In general, the λ values of polymer composites correlate positively with those of the thermally conductive fillers [71,72]. However, when the λ values of the fillers are substantially higher than those of the polymer matrix, improvements in the λ values of the composites under the same loading conditions are minimal [73]. Table 9.2 presents the λ values for several representative polymer composites [7–11,74–94].

9.3.1.2 Filler Loading

The λ values of polymer composites are generally positively correlated with the loading of thermally conductive fillers. However, there is nearly no percolation threshold for TC that can cause a significant increase in the λ values of these composites, except for some extremely high λ value fillers like graphene and CNTs [99–102]. In the initial phases of developing thermally conductive polymer composites, researchers typically incorporate a large amount of fillers (usually over 30 vol%) into the polymer matrix to achieve higher final λ values (>4 W m^{-1} K^{-1}) [64]. When the loading of thermally conductive fillers is low, they struggle to make contact and establish effective thermal conduction pathways and networks, leading to limited improvements in λ values. Therefore, a higher loading of fillers is necessary for achieving elevated λ values. However, excessive filler addition can result in reduced mechanical properties, poor processability, and increased costs. Additionally, there are limitations to the addition of thermally conductive fillers; for spherical rigid fillers of equal radius, the theoretical maximum volume fraction is 0.637 [103].

TABLE 9.1 λ VALUES OF COMMONLY USED THERMALLY CONDUCTIVE FILLERS [70]

Fillers	λ (W m^{-1}·K^{-1})	Fillers	λ (W m^{-1}·K^{-1})
Silver (Ag)	429	Copper (Cu)	401
Gold (Au)	317	Aluminum (Al)	237
Zinc (Zn)	121	Nickel (Ni)	90
Iron (Fe)	80	Tin (Sn)	67
Alumina (Al_2O_3)	35	Magnesium oxide (MgO)	40
Zinc oxide (ZnO)	30	Silica (SiO_2)	1.5
Cerium dioxide (CeO_2)	260	Aluminum nitride (AlN)	320
Silicon carbide (SiC)	80	Silicon nitride (SiN)	180
BN	180	BN nanosheets (BNNS)	751 (λ‖)[a]
Boron arsenide (BAs)	1000	Black phosphorus (BP)	70
Carbon black (CB)	6–174	Carbon fiber (CF)	100 (Axial)
CNTs	3000 (Axial)	Diamond	2000
Graphite	100–400(λ‖)	Graphene	5300 (λ‖)

[a] λ‖ represents the in-plane TC.

TABLE 9.2 λ VALUES OF THE THERMALLY CONDUCTIVE POLYMER COMPOSITES [70]

Fillers and Loading	Matrix	λ (W m^{-1}·K^{-1})	Enhancement	Testing Method	References
1 wt% graphene	Nylon	0.79	93%	CTHERM TCI	[7]
20 wt% CNT/MoS_2/graphene	EP	4.60 (λ‖), 0.48 (λ⊥)	2200% (λ‖), 140% (λ⊥)	Laser flash	[8]
9 wt% rGO/Fe_3O_4	EP	1.21 (λ‖), 0.08 (λ?)	Not reported	Laser flash	[9]
2.4 vol% tellurium NWs	EP	1.63 (λ‖), 0.38 (λ⊥)	675% (λ‖), 189% (λ⊥)	Laser flash	[10]
3 wt% Ag + 12 wt% rGO	PI	2.12	783%	Hot disk	[11]
10 wt% s-MWCNT	PVDF	1.55	762%	Transient hot-wire	[74]
20 wt% *h*-BN	PA66	26.13 (λ‖)[a], 3.68 (λ⊥)[a]	Not reported	Hot disk	[75]
10 wt% f-MWCNT-g-rGO	PI	1.60	493%	Hot disk	[76]
0.1 wt% rGO + 20.4 wt% GNPs	EP	1.56	680%	Hot disk	[77]
33 wt% BNNS	PVDF	16.3 (λ‖)	Not reported	Laser flash	[78]

(*Continued*)

TABLE 9.2 (*CONTINUED*) λ VALUES OF THE THERMALLY CONDUCTIVE POLYMER COMPOSITES [70]

Fillers and Loading	Matrix	λ (W m^{-1}·K^{-1})	Enhancement	Testing Method	References.
20 wt% Te/MoS_2/Ag	EP	10.4 (λ‖), 0.49 (λ⊥)	4160% (λ‖), 101% (λ?)	Laser flash	[79]
5 wt% AlN	PBS	0.32	67%	Laser flash	[80]
2 wt% SiO_2@GNPs	PDMS	0.50	155%	Steady-state	[81]
1 wt% graphene	PVDF	0.48	240%	PPMS-9	[82]
0.25 wt% graphene	PA6	0.69	188%	TC 3000E	[83]
20 vol% ZnOs/ZnOw	SR	1.31	550%	Hot disk	[84]
30 wt% TA-GNPs	PLA	0.77	Not reported	Hot disk	[85]
50 wt% BT@C + 10 wt% SiC	PVDF-HFP	0.92	Not reported	CTHERM TCI	[86]
40 vol% *h*-BN	PLGA	2.10	Not reported	Laser flash	[87]
5 wt% CMG	PI	1.05	275%	Hot disk	[88]
20 wt% Fe_3O_4@MG	SR	0.64 (λ⊥)	433% (λ⊥)	DRL-III	[89]
23 vol% f-Al_2O_3	EP	2.58	1000%	Hot disk	[90]
30 vol% BN-PDA-KH560	NBR	0.41	160%	Laser flash	[91]
62 vol% Si_3N_4	PTFE	1.30	Not reported	Laser flash	[92]
4.4 vol% 3D-BNNS	NFC	1.56	734%	Laser flash	[93]
2.39 vol% rGO + 2.39 vol% AgNPs	PS	0.54	238%	Laser flash	[94]
40 wt% AlN/AgNPs	EP	4.72	1715%	LW-9389	[95]
10 wt% s-MWCNT	PVDF	1.46	711%	Hot disk	[96]
35 wt% GNPs	PS	4.72 (λ‖)	2260% (λ‖)	Hot disk	[97]
38 wt% Al_2O_3 + 2 wt% AlN	PLA	0.72	150%	Laser flash	[98]

[a] λ‖ is the in-plane TC and λ⊥ is the through-plane TC.

Most studies indicate that the λ values of polymer composites improve as the loading of thermally conductive fillers increases [71]. However, some research has shown that the λ values initially rise and then fall with increasing filler content. This decline is mainly due to the agglomeration of highly conductive fillers, which disrupts thermal conduction pathways. Zhang et al. created modified BN/polyvinylidene fluoride (m-BN/PVDF) thermally conductive films using electrospinning. The λ value of the m-BN/PVDF film containing 30 wt% BN fillers reached a peak of 7.29 W m^{-1}·K^{-1}. However, as the BN loading increased to 40 wt%, the λ value dropped to 6.50 W m^{-1}·K^{-1} [104]. A similar trend was observed in the work of Guiney and colleagues, who fabricated BN/poly(lactic-co-glycolic acid) composites via 3D printing. Initially, the λ values increased with

BN loading, but beyond a critical threshold of 40 wt%, the λ values began to decline despite further increases in BN content [87].

9.3.1.3 Shape, Size, and Aspect Ratio

Thermally conductive fillers come in various shapes, including spherical, polyhedral, linear, and sheet forms, with sizes ranging from nanometers to millimeters. These characteristics can significantly affect the λ values of polymer composites [105]. Research indicates that larger fillers can decrease the contact area between the polymer matrix and the fillers, leading to reduced interfacial thermal resistance and increased λ values. Conversely, some studies suggest that smaller fillers may be more effective in enhancing λ values [106–108]. However, the impact of filler loading is often overlooked. It is widely recognized that smaller fillers, at low loadings, result in a much greater number of contact points compared to larger fillers. This makes smaller fillers more effective at establishing thermal conduction pathways and networks, which can enhance λ values. Yet, once an appropriate loading is achieved, these conduction pathways and networks are fully developed. At this point, larger fillers create fewer interfaces with the polymer matrix, which can further enhance λ values [109,110]. Ren et al. incorporated BN fillers of different particle sizes into PDMS. Their findings showed that when the BN loading was below 10 wt%, the λ value of the BN/PDMS composite with 2.0 μm fillers exceeded that of the composite with 160.0 μm fillers. However, when the BN loading surpassed 10 wt%, the λ value of the BN/PDMS composite with 160.0 μm fillers increased more rapidly than the others [35]. Similarly, Park et al. observed that in epoxy composites, the λ value of those containing long CNTs (55 W $m^{-1}·K^{-1}$) was higher than that of composites with short CNTs (20 W $m^{-1}·K^{-1}$) at the same loading level (60 wt%) [111].

Moreover, the shape of thermally conductive fillers influences the type of contact between them, which can be point, line, or surface contact, with the contact area hierarchy being 'point contact < line contact < surface contact'. These contact areas directly impact the thermal resistance among the fillers, thereby affecting the λ values of polymer composites [112]. Spherical fillers create point contacts, linear fillers establish line contacts, and layered fillers result in surface contacts. Research by Fu et al. indicated that lamellar fillers are more effective than spherical fillers in enhancing the λ values of epoxy composites, and they outperform polyhedral fillers with numerous angular edges [71]. Additionally, linear fillers (such as CNTs, CFs, and nanowires) and lamellar fillers (like expanded graphite, graphene, and BNNSs) exhibit significant anisotropy in their λ values due to their structural characteristics, whereas spherical fillers typically demonstrate isotropy. Consequently, the λ values of polymer composites are also influenced by the orientation of the thermally conductive fillers.

Given the complexity of the shapes and sizes of thermally conductive fillers, using the aspect ratio (the ratio of diameter to thickness or length to diameter) provides a more coherent way to illustrate their impact on the λ values of polymer composites [113,114]. Assuming that the fillers are uniformly dispersed in the polymer matrix, linear or lamellar fillers are significantly longer than spherical fillers at the same loading. This allows phonons to propagate more easily along linear or lamellar fillers, enhancing thermal conduction in the composites. Finite element analyses support this conclusion [115]. Additionally, high aspect ratios enable the fillers to overlap, forming aggregates known as 'grape-like' structures, which can establish thermal conduction paths with lower thermal resistance (see Figure 9.12) [70]. However, at a macroscopic scale, the aspect ratio does not significantly influence the thermal conduction mechanism [114]. Research by Yu et al. demonstrated that the λ values of GNPs/epoxy composites increased with the aspect

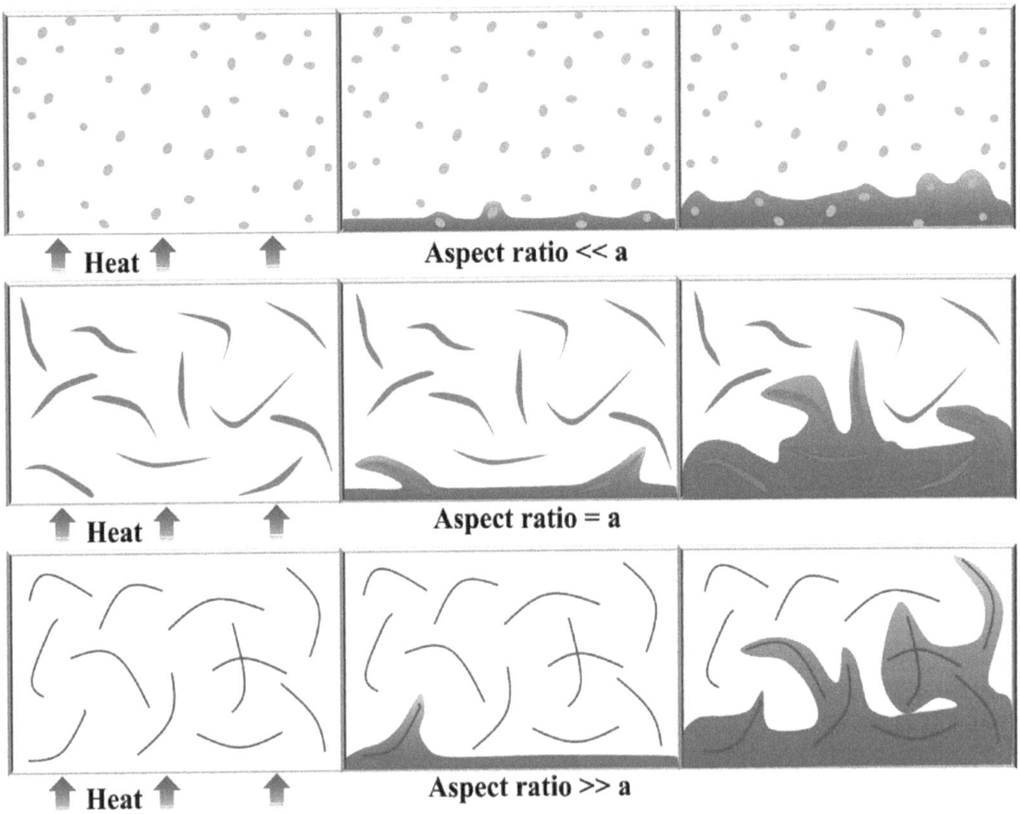

Figure 9.12 Heat transfer in polymer composites vs aspect ratio fillers [70]. (Reprinted with permission from Ref. [70]. Copyright 2006, *Composites Part A: Applied Science and Manufacturing.*)

ratio of GNPs. At the highest aspect ratio (200), the λ value of the composite containing 5.4 vol% GNPs reached 1.45 W $m^{-1}{\cdot}K^{-1}$, while the composite with the lowest aspect ratio (30) had a λ value of only 1.08 W $m^{-1}{\cdot}K^{-1}$ [116]. Given the benefits of high aspect ratios for thermal conduction, many researchers are focused on studying such fillers. For instance, exfoliating BN into BNNS can raise its λ value to 300 W $m^{-1}{\cdot}K^{-1}$, significantly enhancing the TC of the composites. The λ value of the BNNS/PVDF composite film developed by Chen et al. reached 16.3 W $m^{-1}{\cdot}K^{-1}$ with a 33 wt% BNNS content, which is 43 times greater than that of pure PVDF films [78]. Furthermore, a high-quality monolayer BNNS produced via chemical vapor deposition achieved an exceptional λ value of 751 W $m^{-1}{\cdot}K^{-1}$ [65].

However, an excessively high aspect ratio of thermally conductive fillers can lead to processing challenges in polymer composites and negatively impact their λ values. This is because a high aspect ratio indicates that the fillers are oversized in one dimension. Furthermore, larger sizes can heighten the likelihood of defects within the composites. These factors collectively hinder the enhancement of λ values. Research by Kim and Roy has shown that the λ values of polymer composites rise with increasing aspect ratio until a certain threshold is reached. Beyond this point, further increases in the aspect ratio of the fillers result in minimal changes to the λ values of the composites [117,118].

9.3.1.4 Hybrid

Researchers have determined that the use of a single type of thermally conductive filler is insufficient to achieve the theoretically high λ values in polymer composites [119,120]. This limitation arises mainly from phonon scattering due to defects and interfaces, along with processing difficulties caused by the excessive addition of fillers. Incorporating hybrid thermally conductive fillers (of various shapes or types) into the polymer matrix typically results in better TC compared to composites using only one type of filler. This improvement is largely attributed to the ability of hybrid fillers to enhance the dispersion of conductive fillers within the matrix, more effectively bridge adjacent fillers to create thermal conduction pathways, or help reduce voids in the polymer matrix [98,121,122]. Common hybrid thermally conductive fillers include binary hybrids (like SiCw/SiCp [123], MWCNT/Al_2O_3 [124], Ag/graphene [94], Al_2O_3/Ag [125], graphene/CNT [126,127], micro-BN/nano-BN [21]) and multi-hybrids (such as BN/GNPs/CF [128], Cu/CNT/CF [129], CF/*h*-BN/Cu [130]).

The preparation of hybrid thermally conductive fillers typically involves methods such as direct mixing, physical adsorption, and chemical bonding. The direct mixing approach is suitable for fillers of various sizes, relying on the spatial compatibility among them. Smaller thermally conductive fillers can occupy the voids between larger fillers, thus enhancing the likelihood of establishing thermal conduction pathways [131]. In physical adsorption, common techniques include electrostatic adsorption and π–π interactions, which facilitate the direct formation of hybrid fillers with specific structures under certain conditions or chemical reactions [132]. Guo et al. leveraged the interaction between the negative charge of graphene oxide (GO) and the positive charge of Ag+ to create Ag/rGO hybrid fillers with a 'point-plane' structure. The Ag+ ions were adsorbed onto the GO sheets and subsequently reduced using glucose (see Figure 9.13) [11]. Liang et al. developed 3D GNPs/rGO hybrid fillers, effectively addressing the aggregation issue of GNPs and constructing complete thermal conduction pathways. The λ value of GNPs/rGO/epoxy composites containing 20.4 wt% GNPs reached 1.56 W m^{-1}·K^{-1}, while the λ value for standard GNPs/epoxy composites with the same GNPs content was only 1.01 W m^{-1}·K^{-1} [77].

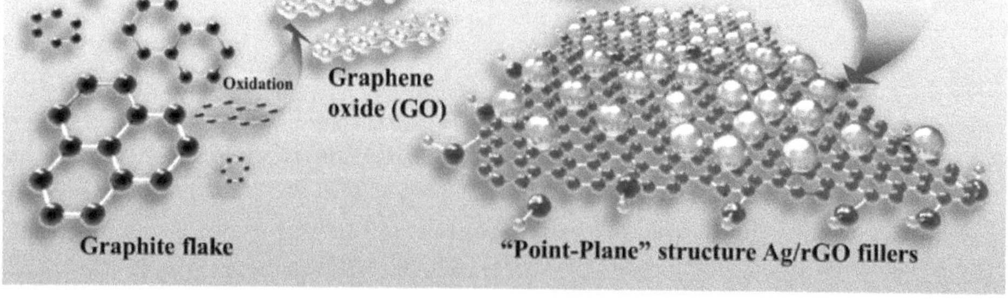

Figure 9.13 Schematic diagram for fabrication of Ag/rGO hybrid fillers by physical adsorption [11]. (Reprinted with permission from Ref. [11]. Copyright 2019, *ACS Applied Materials & Interfaces*.)

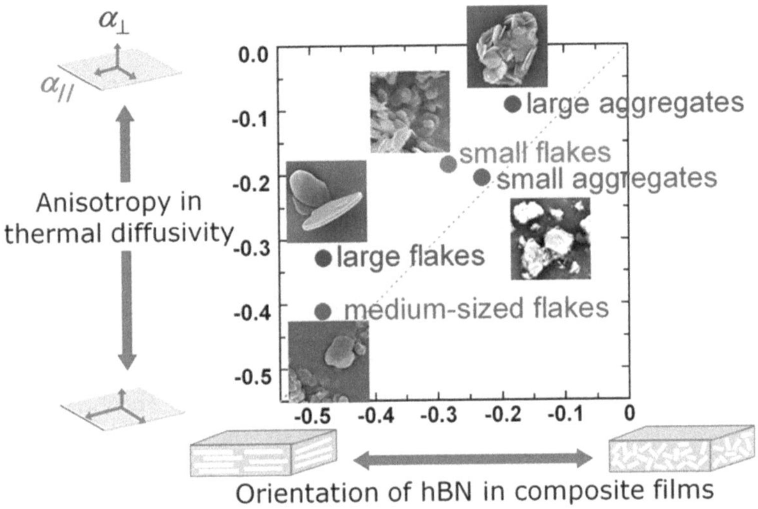

Figure 9.14 Dispersion of *h*-BN particles affecting the anisotropic thermal diffusion [134]. (Reprinted with permission from Ref. [134]. Copyright 2013, ACS Applied Materials & Interfaces.)

9.3.1.5 Dispersion

Based on current understanding, uniform dispersion of thermally conductive fillers may not significantly enhance the λ values of polymer composites. When these fillers are evenly distributed throughout the polymer matrix, low-λ polymer molecules encase them, which hampers effective thermal conduction. Conversely, appropriate agglomeration of the fillers can create thermal conduction pathways in the clustered regions, potentially improving the λ values of the composites [133]. Tanimoto et al. observed that agglomerated fillers have a more pronounced effect on the anisotropic thermal diffusivity in BN/PI composite films compared to uniformly dispersed fillers (Figure 9.14) [134]. Burger et al. examined how different dispersion states affect λ values and confirmed that λ values in a uniform dispersion state were relatively low [16]. Particularly in nanofluids, it has been noted that as the size of the aggregates increases, the λ values initially rise sharply before stabilizing [135–137].

9.3.1.6 Interfaces between BNNS and Polymeric Matrixes

Interfaces play a crucial role in the TC (κ) of composites, as heat flow (phonon transport) is often obstructed at these boundaries. The thermal coupling characteristics at the interfaces between different materials, like inorganic crystals and organic polymers, frequently lead to interfacial thermal resistance due to mismatched phonon spectra. Smaller filler particles create a higher density of interfaces, making interfacial thermal resistance a more pressing concern in nanocomposites. To gain a comprehensive understanding of the thermal transport properties of polymer composites, it is essential to investigate the thermal coupling at the filler/polymer interfaces. Thermal energy transfer across interfaces has been a subject of interest for many years [138–140]. Additionally, poor affinity between the surfaces of thermally conductive fillers and the matrix polymers can lead to defects (such as voids and pores), which significantly hinder the enhancement of κ in the composites. This lack of compatibility may also result in unintentional agglomeration and uneven distribution of the fillers during the composite manufacturing process [141]. Uncontrolled structural changes can obstruct the formation of thermal conduction pathways within the composites. Therefore, interfacial properties remain critical for assembling desired functional structures in these materials [142–144].

The thermal conductance at the interface between BNNS and matrix polymers is a critical factor influencing the enhancement of κ in BNNS-based composite materials. Interfacial thermal conductance (h) is defined as the inverse of interfacial thermal resistance (Ω), with common units being watts per square meter per kelvin (W m^{-2} K^{-1}). Reports on experimental measurements of h between h-BN and organic materials are limited. Recently, several research groups have utilized molecular dynamics simulations to report h values for h-BN basal planes interacting with organic substances. Ma et al. found that the h value between monolayer h-BN and hydrocarbon molecules with six carbon atoms ranges from 90 to 150 MW m^{-2} K^{-1} [145]. Liu and Rao reported an h value of 36–40 MW m^{-2} K^{-1} for the interaction between BNNS and amorphous n-octadecane [146]. Their findings indicated that as the number of h-BN layers increases, the h value decreases. Yang et al. derived an h value of 365 MW m^{-2} K^{-1} for the interface between BNNS and epoxy resin [147]. Wu et al. systematically altered the number of layers of BNNSs encapsulated in paraffin [148], obtaining h values of 18.2 MW m^{-2} K^{-1} for monolayer h-BN and 280 MW m^{-2} K^{-1} for 20 layers of BNNSs. Li et al. reported an h value of 240 MW m^{-2} K^{-1} for the interface between monolayer h-BN and amorphous cellulose [149]. Despite the considerable variation in calculated h values, the aforementioned simulation results are significant. Based on these simulations, the h value for h-BN fillers with optimized layer numbers and amorphous organic polymers is expected to range between 20 and 300 MW m^{-2} K^{-1}.

In the area of chemical bonding, researchers have grafted reactive groups onto various thermally conductive fillers using a pretreatment process, subsequently creating hybrid conductive fillers through chemical reactions. This modification allows for the introduction of polar functional groups on the surfaces of the fillers, providing the benefit of strong bonding (see Figure 9.15) [150]. A theory posits that these polar functional groups can attract organic molecules closer to the interfaces [151,152]. An increase in polymer density near the interfaces enhances the vibrational coupling among the components at the interface. However, excessive pretreatment can compromise the inherent properties of the thermally conductive fillers, which may lead to a decrease in the λ value. Song et al. employed reversible addition-fragmentation chain transfer polymerization to create polymer-functionalized carbon nanotubes (PCNTs), which were then reacted with GO. By reducing PCNT/GO, they produced PCNT@rGO hybrid conductive fillers, effectively addressing the agglomeration issues of CNTs and graphene. The λ value of the PCNT@ rGO/SBR composites containing 3 wt% PCNT@rGO was measured to be 0.45 W m^{-1}·K^{-1}, whereas the λ values for PCNT/SBR and rGO/SBR were only 0.29 and 0.26 W m^{-1}·K^{-1}, respectively [153].

9.3.2 Measurement Methods for TC

The most widely used instruments for characterizing TIMs are those that adhere to ASTM D5470, the 'Standard Test Method for Thermal Transmission Properties of Thermally Conductive Electrical Insulation Materials' [154]. This method employs opposing meter bars with known TC—one heated and the other cooled—to sandwich the TIM, effectively simulating the heat transfer surfaces in real-world applications. As discussed in recent studies [155–158], these meter bars are designed to ensure one-dimensional heat conduction along their length. This setup allows embedded thermal probes (such as thermocouples, resistance temperature detectors (RTDs), and thermistors) to measure the temperature gradient and derive the temperatures of the hot and cold surfaces. The heat flux through the thermal joint containing the TIM can be calculated by multiplying the temperature gradient by the TC of the meter bars. When combined

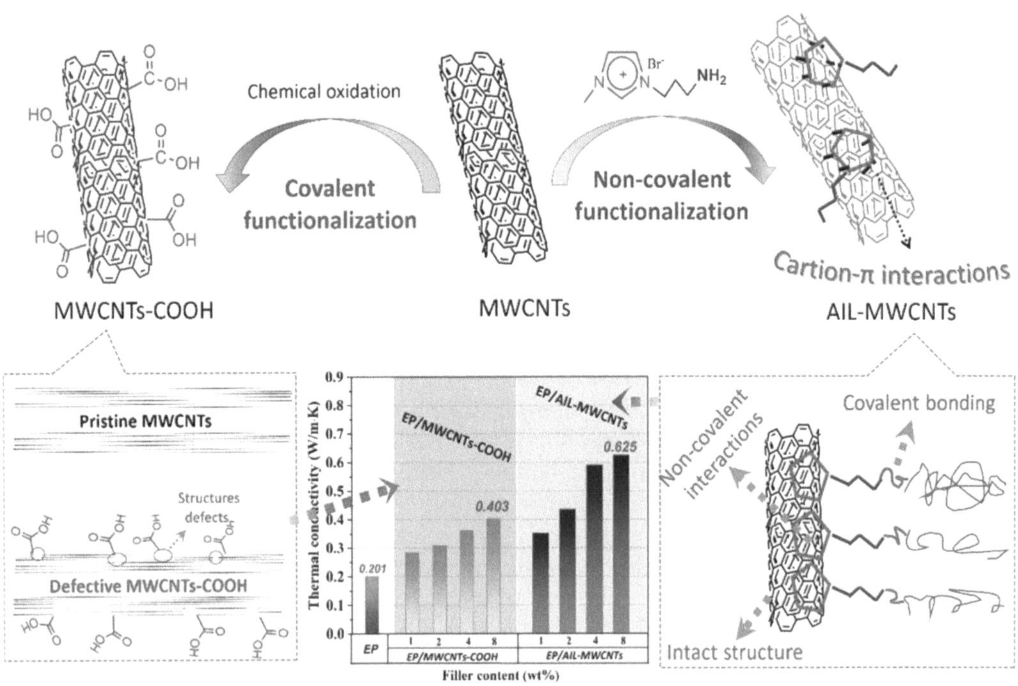

Figure 9.15 Schematic illustration of chemical bonding improving λ values of epoxy composites [150]. (Reprinted with permission from Ref. [150]. Copyright 2019, *Composites Part A: Applied Science and Manufacturing.*)

with the measured temperature difference between the hot and cold surfaces, this is sufficient to determine the net thermal resistance. Key advantages of the ASTM D5470 method include accounting for the thermal contact resistance between the TIM and the heat transfer surfaces, the ability to easily vary and measure the applied load, adjust operating temperatures, control surface roughness, and measure the bond line thickness, which aids in estimating effective TC. However, high-performance TIMs can lead to lower thermal resistance at the joint, necessitating careful attention to ensure accurate temperature measurements while minimizing excessive heat transfer from the meter bars to the environment [155,156].

A recently developed technique involves the application of infrared microscopy in conjunction with embedded temperature sensors within ASTM D5470-type assemblies [159–162]. This method employs high-resolution infrared thermography to assess thermal gradients in focused regions of thermal joints, effectively addressing the resolution limitations of standard ASTM D470 instruments. Smith et al. [159] successfully demonstrated this technique, measuring solder joint thermal resistances as low as 0.01 ± 0.001 cm^2 KW^{-1}, which is quite remarkable. Additionally, this approach allows for the measurement of thermal gradients across different layers in a TIM stack assembly, aiding in the evaluation of each layer's contribution—including that of semiconductor dies and interface layers—to the overall thermal resistance of the TIM [160].

However, one significant limitation of ASTM methods is that they may not accurately replicate the thermomechanical conditions found in real integrated circuit packages. Thermal test vehicles can measure TIM thermal resistance in situ, providing a more accurate representation of performance in practical applications [161]. For instance, Lee et al. [162] illustrated this by employing a thermal test vehicle that follows the JESD51–2 standard, using a Flip Chip Ball Grid Array package as the test device. Taylor and Garimella [163] created a novel thermal test vehicle embedded with capacitive sensors to measure bond line thickness and detect voids in thin dielectric TIMs, such as thermal grease, thereby providing a new method for quantifying the degradation of TIM quality over extended use. Na et al. [164] utilized a transient method where a semiconductor power chip served as both the heat source and temperature sensor. Using a Mentor Graphics T3Ster1.1C, they initiated a heat pulse at the die and monitored the test chip's temperature over specified intervals. The resulting temperature-time history was then analyzed with structural functions, which break down the signal into components along the thermal pathway, enabling the estimation of thermal resistance across each thermal impedance zone, including the TIM.

A commonly used transient method for assessing the thermal properties of TIM is the flash method [165]. This technique typically employs infrared detectors to track the temperature rise of a sample, which is subjected to energy bursts from a laser or xenon flash lamp on the opposite side, allowing for the determination of the sample's thermal diffusivity. To calculate TC, separate measurements of specific heat and density are required. While this method is generally effective for free-standing thin, solid, and homogeneous samples, it has also been adapted for layered TIM assemblies to replicate the realistic thermomechanical load conditions found in integrated circuit packages [166].

The TC (k) of thin films and the interfacial thermal conductance GGG of different materials are crucial for the performance and reliability of micro- and nanomaterials and devices [167]. Measuring these properties poses significant challenges [167–169]. Currently, the predominant techniques for measuring the TC of thin films can be categorized into frequency-domain and time-domain methods. The electrically heated 3-ω method [170–173] and time-domain thermoreflectance (TDTR) using ultrafast lasers [167,168,174] are key examples of these approaches. Figure 9.16 illustrates the setup of the ultrafast thermoreflectance apparatus in Ronggui Yang's laboratory, which is designed for both TDTR and frequency-domain thermoreflectance (FDTR) measurements [175].

The 3ω technique employs a metal microbridge on thin film samples, serving as both an electrical heater and a temperature sensor. This approach has proven to be highly effective for measuring the thermophysical properties of dielectric films [170]. However, applying the 3ω method to conductive or semiconducting materials poses challenges in sample preparation due to the requirement for electrical isolation between the microbridge and the sample [176]. This necessary layer of insulation introduces additional thermal resistance, which can compromise the technique's sensitivity and accuracy [172]. Typically, the frequency range used for the 3ω method spans from 5 Hz to 200 kHz [173,177]. Using a typical thermal diffusivity value $\alpha=3\times10^{-6}\,m^2s^{-1}$, the thermal penetration depth d can be several to hundreds of microns from the relationship of $L=\sqrt{\alpha/(\pi f)}$ [173]. Thus, this quasisteady-state method is not sensitive to interfacial thermal conductance or resistance, making it ineffective for precise measurements of TC in nanoscale films.

Ultrafast thermoreflectance techniques typically employ subpicosecond lasers. The laser pulses are divided into two beams: one pump beam excites the sample, while the other probe beam measures changes in reflectivity or diffraction related to temperature. In the TDTR method, the probe beam reaches the sample surface at various time intervals after the pump beam through a mechanical delay stage. The time decay of the measured signals is analyzed to derive TC (k) and interfacial thermal conductance (G) using a thermal transport model. Since its

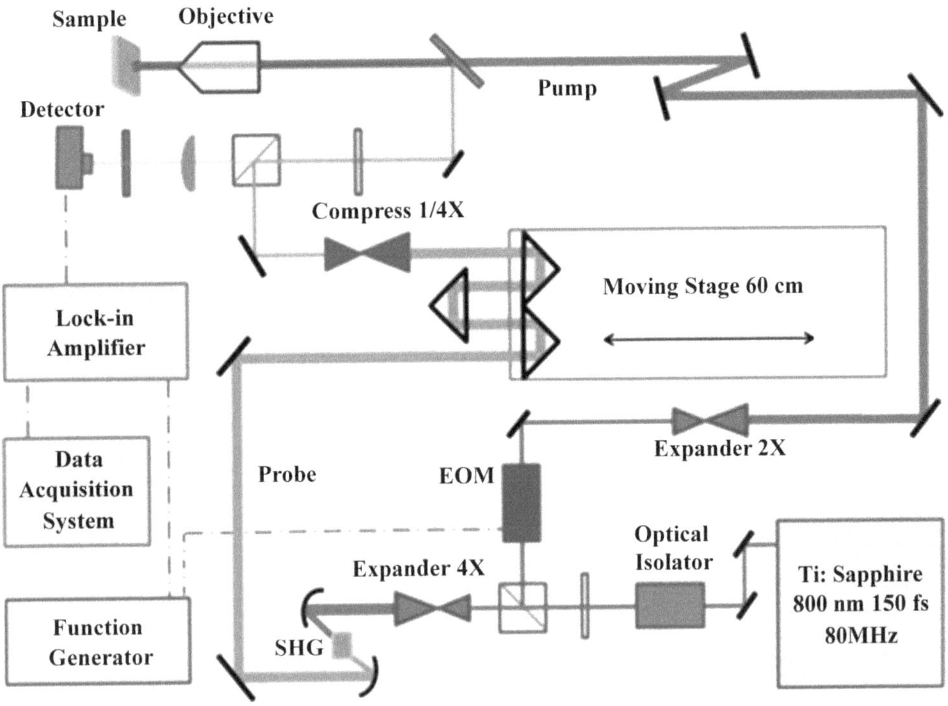

Figure 9.16 Schematic diagram of the optical paths of our ultrafast pump-and-probe thermoreflectance system [175]. (Reprinted with permission from Ref. [175]. Copyright 2010, *Journal of Applied Physics*.)

introduction in the 1980s [174,178,179], TDTR has been widely utilized to measure the TC of both solids [180] and liquids [181], covering a range from the lowest TC observed in fully dense solids [182] to the high conductivities of metals and diamond [179,183], and it can also evaluate the interfacial conductance of different materials [168,174,184–187]. Measurement errors can arise from imperfections in the stage motion, overlap of pump and probe pulses, and divergence of the delayed optical beam [167,175,179]. In recent years, several improvements have been made to mitigate these issues [167,175,179], yet optical alignment involving a moving platform remains relatively cumbersome.

The recently proposed FDTR technique has the potential to eliminate the need for a moving stage, enabling the measurement of TC and heat capacity in bulk materials [188] as well as in thin metal films that function as transducer layers in pump-probe setups [189]. FDTR can measure TC across a broad range, from 1 to 250 W $m^{-1}\cdot K^{-1}$. In terms of implementation, the FDTR setup is similar to that of the TDTR method, with pump pulses modulated at variable frequencies and thermoreflectance signals recorded at a fixed delay between the probe and pump pulses. This setup avoids imperfections that could arise from the motion of the pump-probe delay stage. In addition to the three methods mentioned earlier, another frequency-domain technique based on continuous wave lasers, known as phase-sensitive transient thermoreflectance, is also employed for TC measurements [190]. Compared to the FDTR method using ultrafast lasers, phase-sensitive transient thermoreflectance operates in a lower frequency range and serves as an optical 3-ω method, which is less sensitive to interfacial thermal conductance.

TABLE 9.3 COMPARISONS AMONG 3ω, TDTR, AND FDTR METHODS [175]

	3ω	TDTR	FDTR
Probing method	Metal bridge	Noncontact	Noncontact
Sensitive thermal properties	K	k, G	k, G
Mechanical stage problem	No	Yes	No
Modulation frequency	$0.05 < f < 200$ kHz	f~1–10 MHz	$0.1 < f < 20$ MHz
Thermal penetration depth	$400 > L > 2$ µm	L~1–0.3 µm	$3 > L > 0.2$ µm

Table 9.3 provides a comparison of the 3-ω, TDTR, and FDTR techniques. The FDTR method operates within a frequency range of 0.1–20 MHz, achieving thermal penetration depths typically between 200 nm and several micrometers, making it well-suited for measuring the TC of nanoscale thin films. Furthermore, FDTR signals include transient heat transfer data that can be utilized to assess interfacial thermal conductance and resistance. Therefore, we believe that the FDTR method retains all the advantages of the TDTR approach compared to the 3-ω method, such as sensitivity to TC in both thin films and interfaces, as well as being a noncontact technique that circumvents the issues related to moving mechanical stages in TDTR measurements.

9.3.3 Contribution of Nanoscale BN Fillers to Thermal Conduction

The performance of TIMs is best assessed by their area-averaged thermal resistance when compressed between two heat transfer surfaces. A key principle of TIMs is to address the issue of contact area. They mitigate air-filled microscopic roughness voids by deforming under acceptable assembly pressures. This requirement for deformation introduces a critical engineering trade-off: low mechanical shear strength facilitates shape changes and conformal contact, which typically results in TC approximately two orders of magnitude lower than that of the associated packaging and thermal management components. Furthermore, the overall geometry and mechanics of thermal interface formation are inefficient in generating the shear stress required for flow processes. The high aspect ratio of TIM deformation, referred to as 'poker-chip' geometry, is evident in various applications such as tack testing [191] and nanoimprinting [192,193]. In viscoelastic solids—materials with a permanent shear modulus and viscous characteristics—this geometry leads to a rapidly increasing hydrostatic pressure component of the total stress with rising Poisson's ratio, resulting in purely hydrostatic deformation under long-term incompressible conditions. Although this sharp local geometry does not hinder the filling of microscale cavities, it creates issues related to bond line thickness. This parasitic global effect occurs because the inability to thin TIMs allows them to occupy significant gaps between mating surfaces, even if they can fill local voids. Consequently, while TIMs are only two orders of magnitude less effective as thermal conductors than metals, their ability to achieve conductance levels of 1×10^{-4} cm^2 KW^{-1} at pristine interfaces is compromised by the presence of bond lines. For high-performance pastes rated at 5 W m^{-1}·K^{-1}, a typical bond line thickness of 50 µm results in an effective interface resistance of 1 cm^2 KW^{-1} due to the bulk properties of the TIM.

The advancements in the current generation of deformable commercial TIMs primarily arise from reducing bond line thickness, largely defined by deformability, and/or enhancing the bulk TC of the TIM. The significance of interfacial thermal resistance between TIMs and mating surfaces seems to manifest primarily in high-performance TIMs, such as CNT forests. Regarding

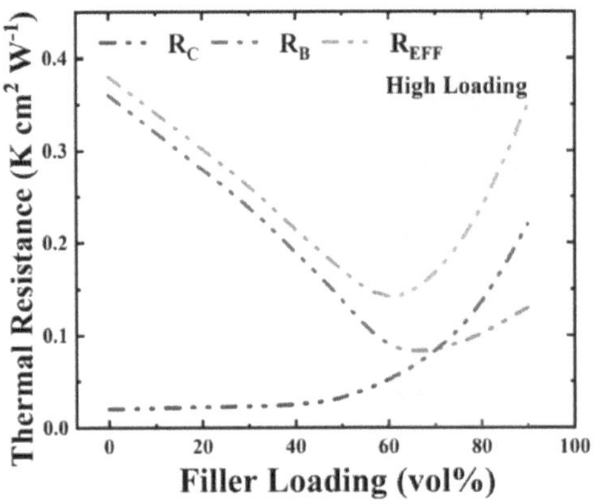

Figure 9.17 Effect of particle fillers on the thermal resistance of TIMs [199].

total thermal contact resistance, Prasher [194] and Luo et al. [195] noted that most studies focus on enhancing k_{TIM} by incorporating improved materials and/or fillers within existing polymer matrixes. However, research aimed at understanding and enhancing bond line thickness and solid-liquid contact resistances, using more realistic models of loaded paste flow and capillary physics, is relatively sparse [196–200]. Techniques that modify TIM deformation during the assembly flow through surface patterning have demonstrated the potential to reduce bond line thickness from tens of micrometers to single micrometers [201–203]. Additionally, optimizing the distribution of loaded particles can be achieved by designing hierarchical channels and increasing the rate of TIM compression [204,205]. These innovative effects are summarized in Figure 9.17 [199].

The emergence of larger surface roughness, waviness, damage, and contamination typically arises as by-products of machining, polishing, casting, and handling methods during the packaging of devices and the integration of thermal hardware, such as heat pipes, heat spreaders, and heat sinks. These imperfections significantly impair thermal interface performance compared to the aforementioned values. For instance, Dou et al. [206] recently reviewed the mating performance of bare machined metal surfaces, and their summary is presented in Figure 9.18 (which also includes the performance characteristics of commercial TIMs and the limits of smooth silicon membranes regarding surface-force adhesion pressure influenced by atomic-scale roughness for comparison). At relevant pressures for thermal interface assemblies, typically around 100 kPa, the thermal contact resistance of machined surfaces varies between 1 and 10 $cm^2\,KW^{-1}$, with roughness (measured by the standard Ra parameter) ranging from sub-micrometer to several micrometers. Overall, these actual interfaces exhibit contact resistance that is five orders of magnitude higher than that of pristine interfaces. The influence of roughness-induced air gaps on thermal conductance degradation is strongly indicated by the observed relationship between resistance and surface roughness: a one-order drop in resistance corresponds to a similar drop in roughness Ra from 2 to 0.2 μm. Assuming a rough square relationship between contact area and linear roughness, further reducing the roughness of bonded semiconductor surfaces to the nanoscale leads to a four-order reduction in contact resistance, approximately $1 \times 10^{-4}\,cm^2\,KW^{-1}$, which is consistent with the ranges observed for wafer-bonded and other pristine interfaces. Additionally, applying clamping pressures of 10 MPa or more can decrease resistance by a factor of three to four, although there is considerable variability in the collected data. This observation aligns with the rapid increase in contact area due to deformation, and the relative reduction

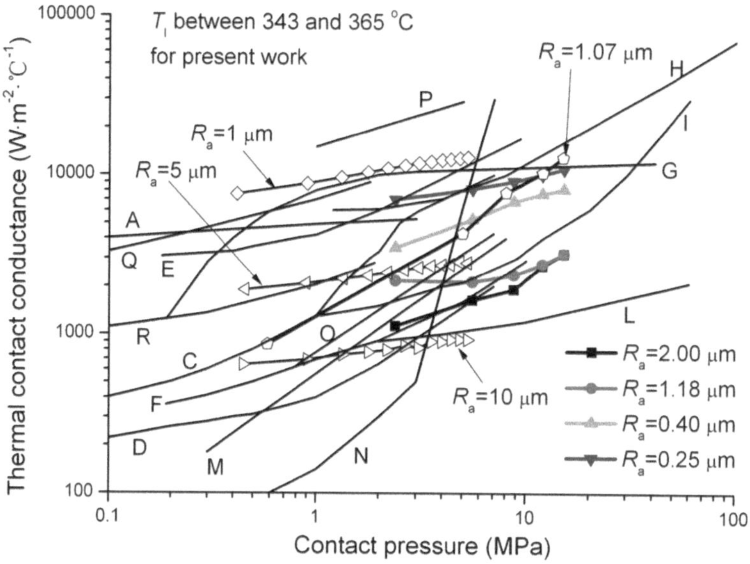

Figure 9.18 Range of contact thermal resistance from bare machined metal contacts showing the effect of surface roughness and contact pressure to general effect of all commercial TIMs [206]. (Reprinted with permission from Ref. [206]. Copyright 2016, *International Journal of Heat and Mass Transfer.*)

in resistance is significantly greater than that seen in completely void-free contacts, despite smaller changes in pressure [207].

Therefore, for the BN/polymer composites used as thermal pads by filler network orientation and slicing techniques, these methods objectively preserve the high flexibility and toughness properties of the polymer. Optimizing the slicing techniques employed in thermal pad manufacturing allows for precise control of the thickness of the composite material. Modulating polymer properties enables the attainment of a lower modulus. Optimizing the filler arrangement and orientation achieves high TC at lower filler content. All these methods contribute to reducing thermal resistance and improving heat dissipation. Alternatively, incorporating high content fillers into a soft elastomer, even exceeding 50 vol%, is a straightforward approach to achieving high TC. However, the simple blending method presents challenges when incorporating such large amounts of nanoscale BN fillers. On the other hand, excessive entanglements during the blending process not only reduce filler content and compromise TC but also lead to an undesired increase in the material's modulus and thickness, resulting in elevated interfacial contact thermal resistance. These drawbacks hinder the effective utilization of BN fillers and limit the overall enhancement of thermal conduction.

9.4 APPLICATION OF NANOSCALE BN-POLYMER COMPOSITES IN THERMAL MANAGEMENT OF ELECTRONIC DEVICES

9.4.1 Thermal Management Challenges in Electronic Devices

In the absence of TIMs, when two nominally flat and smooth solid surfaces come together to form a dry joint [208], the roughness of both surfaces limits the actual contact area to a small fraction (approximately 1%–2%) of the apparent contact area [209], as illustrated in Figure 9.19a.

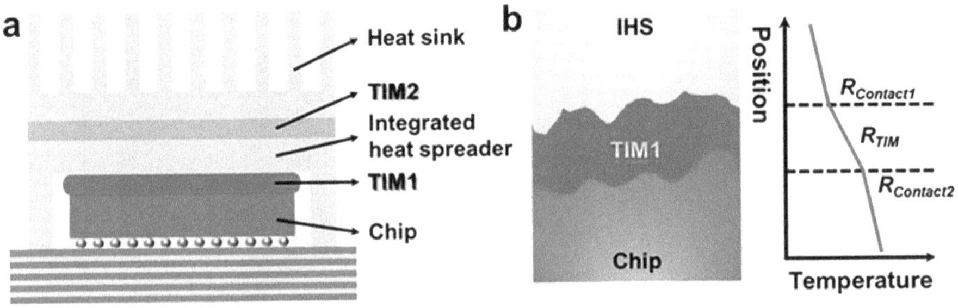

Figure 9.19 (a) A typical ball grid array electronics package with TIM1 and TIM2. (b) Working principle of the TIM [210].

If we disregard radiative heat transfer between the surfaces, heat flow through this interface occurs via two primary conductive paths: solid-to-solid conduction at the contact points and conduction through the air in the noncontact regions, which is a relatively poor thermal conductor compared to solid materials. These two conductive paths restrict the transfer of heat from one surface to another, a phenomenon termed thermal contact resistance. To minimize the temperature drop at the junction, it is essential to reduce this thermal contact resistance. This necessitates replacing the air gap between the mating surfaces with a material that has a TC higher than that of air, referred to as TIM. Figure 9.19b illustrates the total thermal resistance (θ_{TIM}) resulting from the insertion of a TIM between the two mating surfaces, which can be expressed as follows [210]:

$$\theta_{\mathrm{TIM}} = \frac{\mathrm{BLT}}{k_{\mathrm{TIM}}} + \theta_{c1} + \theta_{c2} \tag{9.4}$$

where k_{TIM} refers to TC, bond line thickness indicates the bond line thickness, and θ_{c1} and θ_{c2} denote the contact resistances of the TIM with the two bonding surfaces. The primary goal of thermal management in electronic packaging is to efficiently transfer heat from the semiconductor device to the surrounding environment; thus, minimizing the total thermal resistance from the junction to the ambient (θ_{ja}) is essential. A reduction in thermal resistance results in a smaller temperature drop from the heat-generating device to the environment.

Figure 9.20 illustrates the schematics of two microprocessor package assemblies utilized in practical electronic cooling applications, along with their associated thermal resistances. Additionally, numerous other variants are available [211]. The figure indicates that the heat transfer throughout the entire assembly comprises three primary stages (with corresponding thermal resistances) [212]:

- Heat transfer within the device package (θ_{jc}).
- Heat transfer from the package to the heat sink (θ_{cs}).
- Heat transfer from the heat sink to the surrounding environment (θ_{sa}).

To minimize total thermal resistance, the primary bottlenecks are the interfacial resistances between the package and heat sink [194]. These resistances arise from the mating surfaces, which, in practical scenarios, are often neither fully conforming nor sufficiently smooth, resulting in a notable increase in thermal contact resistance. Consequently, thermal management in the chip area is a significant concern.

Like the surface roughness of heat sinks and heat-generating devices, the compliance of TIMs under specific pressures is crucial for minimizing thermal contact resistance. In this context,

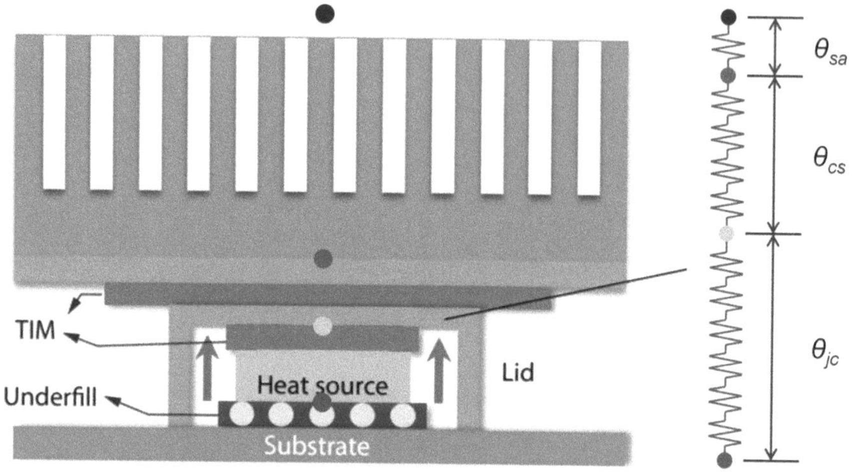

Figure 9.20 The superimposition of different layers of materials in order to build the assembly leads to a creation of a thermal resistance network [212].

using a soft material as the interface is advantageous, as it can adapt to the roughness of the two mating surfaces under appropriate pressures (approximately 200 kPa). Besides contact thermal resistance, the bulk thermal resistance of the interface material is also significant in assessing the material's overall performance, which can be characterized by its intrinsic TC.

9.4.2 Application of Thermally Conductive Polymer Nanocomposites

9.4.2.1 Thermal Engineering of Wearable Electronics

Wearable electronics have spurred advancements across various fields, including alternative energy harvesting, personal health management, and the Internet of Things [213–219]. To enhance comfort and convenience, these devices are typically designed to be compact, which can lead to heat buildup during high-frequency operation. Because they are in direct contact with the skin, their temperature must not significantly exceed that of the human body. This necessitates that the encapsulation materials used in wearable electronics possess high TC for effective heat dissipation and safety. Research into thermally conductive polymer nanocomposites for managing heat in flexible wearable electronics has made notable progress. Figure 9.21 displays a simplified diagram and an enlarged view of a skin-like thermohaptic device. In this design, a copper (Cu) serpentine electrode is embedded in a thermally conductive elastomer nanocomposite. This electrode is engineered to endure substantial bidirectional strain while maintaining electrical connectivity, thus preventing discomfort from elevated temperatures (i.e., significantly above 37 °C) for users. The elastomer not only provides flexibility to the Cu electrode but also facilitates rapid heat dissipation due to its high TC [220]. Furthermore, thanks to the thermally conductive elastomer, the device's surface temperature can swiftly respond to feedback from artificial thermal sensations, offering both heating and cooling effects on the skin [220]. Figure 9.21 also depicts the design of an advanced thermal skin aimed at preventing skin burns and device failures. The core innovation lies in a hybrid structure that combines high TC with thermal energy storage capability. The combination of a thermally conductive PDMS/Ag

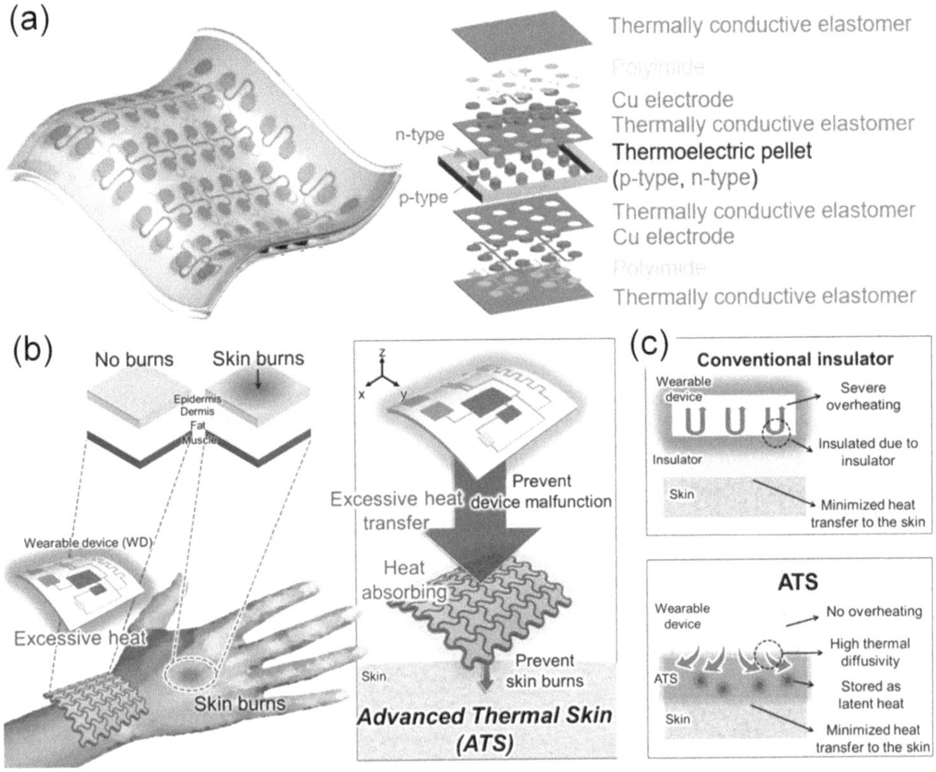

Figure 9.21 (a) Simplified illustration and an exploded view of the skin-like thermohaptic device. (Reprinted with permission from Ref. [220]. Copyright 2020, *Advanced Functional Materials*.) (b) Conceptual overview of the advanced thermal skin and its operational mechanism of preventing heat transfer to the skin. (c) Schematic illustration of comparison between the conventional insulator and advanced thermal skin [221]. (Reprinted with permission from Ref. [221]. Copyright 2023, *Nano Energy*.)

serpentine electrode structure and a heat-storing sodium acetate hydrogel serves as the interface layer between the wearable device and the skin. In this configuration, the high TC PDMS/Ag serpentine electrode effectively transfers excess heat, while the high-heat-storing hydrogel absorbs heat without causing a significant temperature increase, thereby preventing both skin burns and device malfunctions by absorbing heat released from the device [221].

Thermally conductive polymer nanocomposite fabrics exhibit considerable potential for application in wearable electronics. For instance, when combined with a thermoelectric generator, thermally conductive cellulose/BNNS nanocomposite fibers can effectively harvest energy from body heat, enabling continuous self-powering of wearable devices and ensuring reliable performance [222]. Anisotropic heat dissipation is more advantageous than isotropic dissipation but presents challenges for wearable electronics. Additionally, maintaining high TC during deformation is crucial, as significant heat can be generated due to changes in resistance. An effective assembly method has been proposed to address these challenges [223]. In particular, a continuous tetrahedral BNNS array was developed to achieve anisotropic thermal conduction, resulting in a through-plane TC of 1.15 W $m^{-1}{\cdot}K^{-1}$ and an in-plane TC of 11.05 W $m^{-1}{\cdot}K^{-1}$ at 16 wt % BNNS. The tetrahedral structures feature a wavy cross-section, allowing for flexibility and stretchability without compromising TC during the deformation of the nanocomposite film. A high in-plane TC facilitates rapid dissipation of heat generated by localized hotspots. Therefore, the controlled fabrication of polymer nanocomposites with thoughtfully designed thermal conduction network structures is significant for advancing the next generation of flexible wearable electronics.

9.4.2.2 TIMs

Heat accumulation in electronic devices can elevate operating temperatures, negatively impacting stability, reliability, and lifespan. As miniaturization, integration, and power density continue to advance rapidly, modern electronic devices impose increasingly urgent and higher demands for effective heat dissipation [224]. Typically, efficient heat dissipation requires that devices be connected to heat sinks via materials with low thermal resistance. However, achieving perfect surface contact between the device and the heat sink is often challenging due to their high modulus. This situation results in numerous voids and gaps at the contact surfaces, which are filled with air, leading to increased thermal resistance and reduced heat dissipation efficiency [225]. To mitigate thermal resistance, thermally conductive materials, such as TIMs, are commonly employed to fill these voids and grooves. Enhancing heat dissipation can be achieved by increasing the TC and decreasing the contact thermal resistance of TIMs [226]. Thermally conductive phase change materials (PCMs) represent a promising option for TIMs in device cooling applications. As illustrated in Figure 9.22b, phase change greases can serve as TIMs to enhance the cooling of central processing units (CPUs). When compared to CPUs without TIMs, those coated with thermally conductive greases demonstrate significantly lower operating temperatures. The heat capacity is a critical factor influencing the performance of phase change TIMs. As shown in Figure 9.22c–e, while TIM-1 (80% Ga grease) and TIM-2 (75% Ga0.40In0.28Sn0.32 grease) exhibit comparable TCs (approximately 5.5 W $m^{-1}{\cdot}K^{-1}$), TIM-2, with its sustained latent heat, allows for a slower temperature rise in the CPU and lower peak temperatures during transient operation. This highlights the importance of continuous latent heat in safeguarding electronic components from thermal damage [227].

Ideal TIMs should possess high through-plane TC and low modulus [4]. In some instances, electrical insulation is also crucial. A high through-plane TC facilitates efficient heat transfer between the heat source and the heat sink [228,229]. The low modulus of TIMs allows for better surface contact between the device and the heat sink, thus reducing contact thermal resistance [230]. Additionally, the electrical insulation provided by TIMs can prevent short circuits in devices under an electric field, ensuring high operational stability of the entire system in certain applications [48]. In reality, there is a trade-off regarding TIM performance: flexible TIMs with a lower modulus can achieve improved surface contact, which decreases thermal resistance at the interface, but such materials generally exhibit lower breakdown strength. Moreover, a high TC is typically attained only by incorporating a significant amount of reinforcing materials, which inevitably leads to an increased modulus [231]. Therefore, the development of highly thermally conductive TIMs with low modulus and electrical insulation properties remains a significant challenge.

Aligning nanoscale reinforcement materials within a polymer matrix significantly enhances TC, with common techniques including the ice-templated method, chemical vapor deposition, and 3D printing [232–234]. Current research on TIMs primarily emphasizes improving the TC of these nanocomposites, while the influence of material properties on the heat dissipation efficiency of devices has been somewhat overlooked. Future studies should prioritize the impact of contact thermal resistance between these nanocomposites and devices, as it is a crucial factor

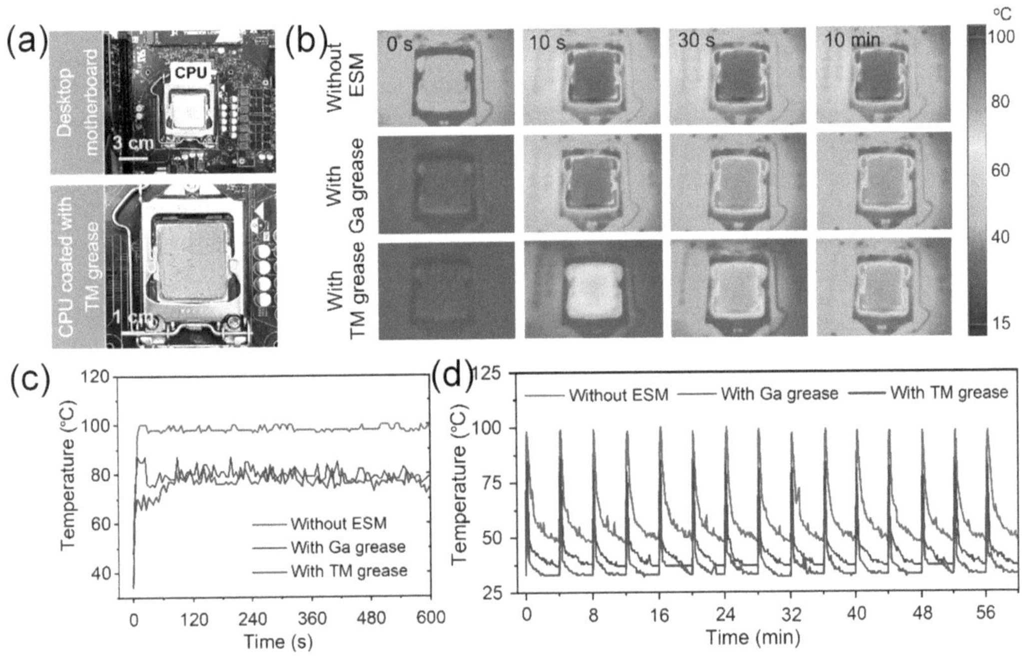

Figure 9.22 (a) An optical image of a desktop computer CPU on the motherboard (top) and the same CPU after TIM application (bottom). (b) Thermal infrared images showing the CPU without TIM (top), with TIM-1 (middle, 80 vol% Ga grease), and with TIM-2 (bottom, 75 vol% Ga0.40In0.28Sn0.32 grease) while operating at full load. The TIMs have a thickness of approximately 0.5 mm, and both possess a TC of about 5.5 W $m^{-1}{\cdot}K^{-1}$. (c) Real-time temperature curves for the CPU. (d) Real-time temperature responses of the CPU operating without TIM, with TIM-1, and with TIM-2 during 15 cycles of instantaneous thermal shock [227]. (Reprinted with permission from Ref. [227]. Copyright 2022, *Proceedings of the National Academy of Sciences of the United States of America.*)

influencing the heat dissipation performance of advanced devices [235–237]. Thus, the development of advanced TIMs aimed at reducing thermal resistance between heat sinks and devices remains a vital focus in thermal management.

9.4.2.3 Battery Thermal Management

Batteries serve as critical electrochemical energy storage devices, advancing toward higher energy density, power density, and enhanced safety [238]. Both charging and discharging processes generate heat due to charge transfer and chemical reactions. If the heat is not dissipated promptly, a significant temperature increase can occur, leading to capacity loss, shortened cycle life, and thermal runaway, which may result in accidents such as explosions and fires [239]. Additionally, the internal temperature gradient can create uneven current density distribution, accelerating battery aging and failure [240]. Consequently, effective thermal management that addresses both internal and external heat dissipation is essential for ensuring the safe, stable, and reliable operation of batteries. The separator plays a key role in limiting internal heat dissipation. As an inert component (depicted in the battery configuration with a separator in Figure 9.23a), it does not engage in electrochemical reactions but significantly influences the electrochemical performance and safety of batteries [86]. Currently, the most widely used separators in lithium-ion batteries are porous membranes made from polyolefins, which typically have low TC [241]. Research has shown that thermally conductive separators can facilitate uniform thermal and lithium-ion flux distribution, thereby mitigating thermal shrinkage of the separator, preventing the formation of 'hot spots', reducing cracking at the solid electrolyte interface, and inhibiting lithium dendrite growth.

A finite element simulation (illustrated in Figure 9.23b) demonstrates that the through-plane TC of the separator significantly impacts the temperature rise in 18,650 batteries. A separator with higher TC promotes a more uniform temperature distribution within the battery [242]. Experimental results indicate that the thermally conductive BNNS nanocomposite separator yields a more consistent temperature distribution in lithium-ion batteries compared to the pure poly(vinylidene fluoride-hexafluoropropylene) separator, thereby enhancing the cycling stability and safety of the batteries [243]. All-solid-state batteries featuring high-capacity electrode materials are viewed as the next-generation power sources, offering high energy density along with improved safety features. However, the low TC of polymer electrolytes restricts the thermal

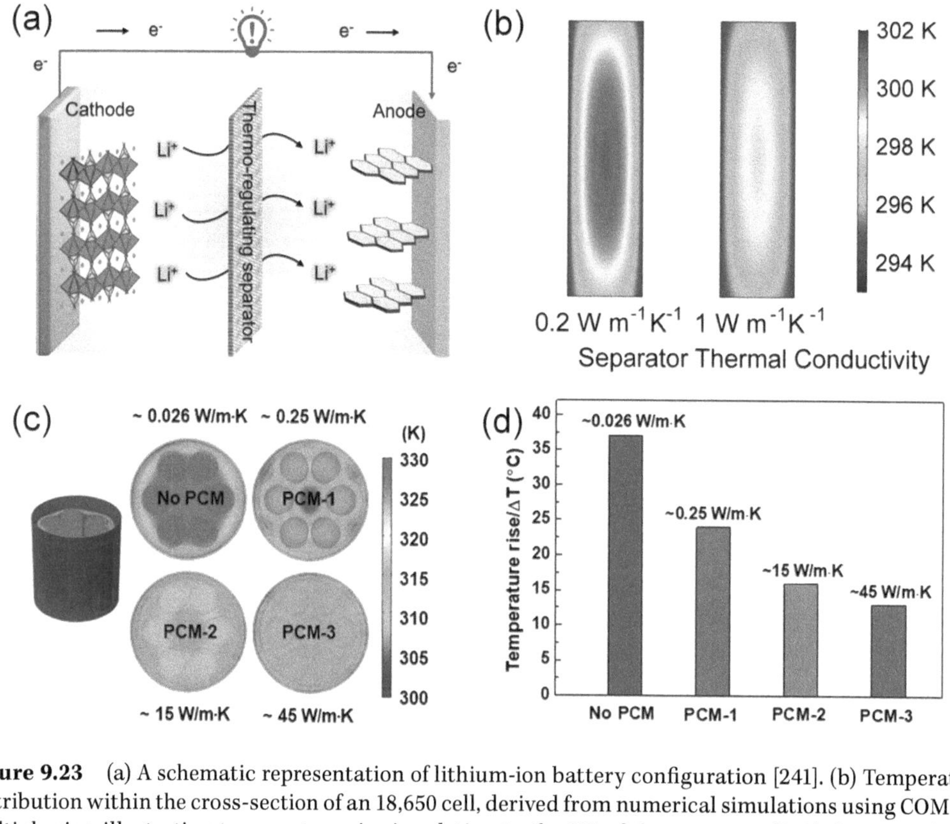

Figure 9.23 (a) A schematic representation of lithium-ion battery configuration [241]. (b) Temperature distribution within the cross-section of an 18,650 cell, derived from numerical simulations using COMSOL Multiphysics, illustrating temperature rise in relation to the TC of the separator [242]. (Reprinted with permission from Ref. [242]. Copyright 2016, Nano Energy.) (c) Numerical simulation results of temperature profiles inside battery packs, based on experimentally measured specific heat and TC values [249,250]. (Reprinted with permission from Ref. [249]. Copyright 2014, *Journal of Power Sources.*) (d) Temperature increase within battery packs after ten charge–discharge cycles, comparing packs wrapped without PCMs (exposed to air) and those with PCM-1, PCM-2, and PCM-3 [249,250]. (Reprinted with permission from Ref. [249]. Copyright 2014, *Journal of Power Sources.*)

performance of these batteries [244]. Recent research has highlighted the significant potential of thermally conductive fillers, such as BNNS and Al_2O_3, which possess low electrical conductivity and excellent electrochemical stability, for use in solid-state composite electrolytes [242–244].

Polymer nanocomposites with elevated TC values play a crucial role in the external thermal management of batteries. Research indicates that a 3D-printed graphene/TPU nanocomposite can effectively lower the temperature rise of operating batteries by approximately 5°C [245]. Importantly, these thermally conductive nanocomposites typically serve as auxiliary materials for battery heat dissipation. Currently, the thermal management of battery packs primarily relies on external liquid cooling systems. However, passive thermal management utilizing PCMs is viewed as a sophisticated and effective approach for battery applications [246]. Recent studies have shown that PCMs with high TC can significantly reduce temperature rise in batteries during operation [247,248]. The temperature decrease can be attributed to two main factors: (1) PCMs can absorb substantial heat while maintaining a stable temperature during the phase change, and (2) their high TC facilitates rapid thermal response. Experimental and simulation results presented in Figure 9.23c and d indicate that, compared to batteries without PCMs and those containing pure PCMs, the graphene-filled PCMs (PCM-2, PCM-3) exhibit significantly lower temperature rises in the battery cylinders, even after ten charge–discharge cycles [249,250]. Future research should focus on developing thermally conductive PCMs with high latent heat capacity for long-term thermal management in batteries.

9.4.2.4 Polymer-Based Dielectric Capacitors

Recently, the energy storage capabilities of dielectric polymer capacitors at elevated temperatures (greater than 150°C) have garnered significant interest, particularly for miniaturized and highly integrated electric vehicle circuits [251]. The decline in electrical performance of capacitors under high voltage, typically caused by Joule heating during operation, remains a substantial challenge. Dielectric polymer capacitors are primarily composed of wound polymer films, where the Joule heat generated from conduction losses is transferred from the core to the exterior (see Figure 9.24a) [252]. In this context, the temperature at the center of the material reaches its peak, known as the highest internal temperature (HIT) during capacitor operation [253,254]. If this HIT surpasses the glass transition temperature of the dielectric, the capacitor will fail to function effectively due to the relaxation of polymer chain segments. Introducing thermally conductive reinforcements like BNNSs can significantly lower the HIT of polymer dielectric capacitors (illustrated in Figure 9.24b). The addition of 10 vol% BNNSs to cross-linked divinyltetramethyldisiloxane-bis(benzocyclobutene) raises its TC from 0.3 to 1.8 W $m^{-1}{\cdot}K^{-1}$ [255]. Calculations indicate that the HIT of BNNS/divinyltetramethyldisiloxane-bis(benzocyclobutene) nanocomposite film capacitors approaches ambient temperature (152°C vs. 150°C), while c-BCB film capacitors show a significantly higher HIT (267°C–269°C) under the same conditions. Moreover, the enhanced TC of dielectrics plays a crucial role in reducing the HIT of larger devices. While high TC in polymer dielectrics is vital for effective high-temperature energy storage, the incorporation of materials like BNNSs may introduce challenges in processing and other aspects [256,257]. In some instances, improving TC might compromise electrical insulation. Therefore, future research should concentrate on two main areas: (1) developing advanced dielectric polymers with high intrinsic TC and (2) investigating innovative synthesis methods for nanocomposites. Both approaches must aim to enhance TC without negatively impacting the thermal and processing characteristics of dielectric materials.

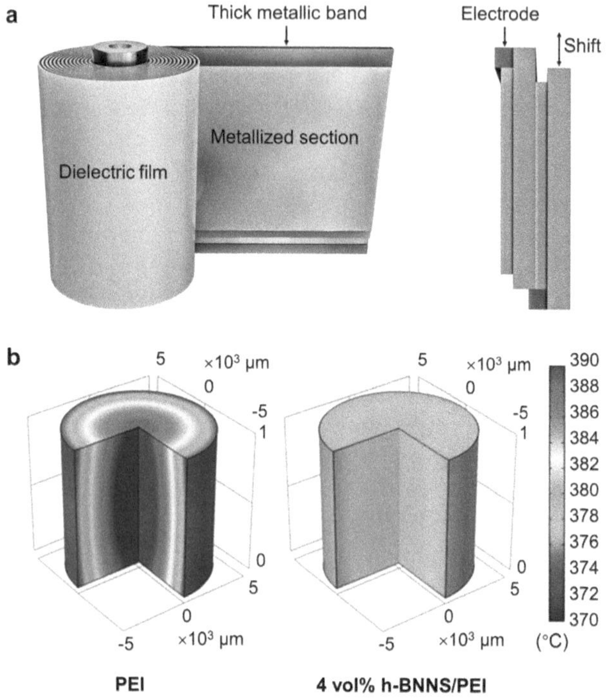

Figure 9.24 (a) Single-section metallized capacitor. (b) Internal temperature distribution of pure poly(ether imide) and 4 vol% *h*-BNNS/poly(ether imide) nanocomposite [102,250]. (Reprinted with permission from Ref. [102]. Copyright 2019, *Advanced Materials*.)

9.4.2.5 Thermal Management of Electrical Equipment

The aim of achieving Net Zero Carbon by 2050 has led to a heightened demand for clean energy, complicating the existing power systems. Ideally, the next generation of power systems should be more compact, with higher power density and operational voltage compared to current systems. However, the low TC of existing insulating materials hinders the advancement of power equipment. Insulating materials with low TC restrict heat dissipation within power devices, causing significant heat buildup in the system [258]. This accumulation results in reduced equipment capacity, shorter lifespan of electrical insulation, and decreased grid stability. Thus, future power systems will necessitate insulating materials with high TC.

For instance, in power cables, increased capacity requirements lead to enhanced Joule heating in the conductors, raising the temperature gradient in the insulation and causing substantial electric field distortion within it. This scenario heightens the risk of cable failure and accelerates the aging process of the insulation. Utilizing high TC insulating layers allows for rapid dissipation of the Joule heat generated by the cable conductors, thereby lowering the temperature gradient in the insulation and enhancing operational stability. Simulations indicate that increasing the insulation's TC from 0.29 to 0.95 $W\ m^{-1}{\cdot}K^{-1}$ can reduce the cable core temperature by over 8°C. Additionally, improved cooling efficiency can boost current-carrying capacity by 14.6% [259,260].

Generators are the primary source of electricity for industrial activities and daily use. While air-cooled generators offer advantages like simple design, high reliability, and easy maintenance compared to water- and hydrogen-cooled counterparts, their low heat dissipation rate restricts capacity improvements. As conductors cut through magnetic fields to generate electricity, they produce considerable heat. This heat must be transferred through the main insulation of the stator coil for dissipation into the surrounding air. Although reducing the thickness of the main insulation can lower thermal resistance, it also compromises overall dielectric strength and operational safety. Therefore, enhancing the TC of the main insulation while preserving its high breakdown strength is a viable solution. Tari et al. demonstrated that raising the TC of the main insulation from 0.25 to 0.5 W $m^{-1}{\cdot}K^{-1}$ can decrease the temperature of the main insulation in a 250 MVA turbo generator by 10°C [261]. This improvement could potentially double the generator capacity from 250 to 500 MVA with high TC insulation. A significant challenge is that nanoscale reinforcements used to increase the TC of impregnating resins (like epoxy) complicate generator manufacturing. These reinforcements can significantly elevate resin viscosity and may adversely affect the dielectric properties of the stator coil's main insulation. Researchers aim to improve the TC of impregnating resins such as epoxy without compromising their processing and insulating capabilities; however, this goal remains elusive. Innovative strategies are necessary to tackle this issue.

In addition to power cables and generators, other electrical equipment is also pursuing polymer nanocomposites that offer high TC while maintaining electrical insulation. This includes applications in solid-insulated switchgear [262] and saturable reactors (illustrated in Figure 9.25) [263–265]. Despite significant advancements over the past decade, numerous challenges persist. The main goal is to achieve a synergistic enhancement of TC along with other attributes, such as electrical insulation, mechanical strength, and processability of polymer nanocomposites. Furthermore, ensuring compatibility with existing generator manufacturing processes is crucial.

9.4.2.6 Solar Thermal Energy Storage

Solar energy represents an optimal focus within the renewable energy sector. The energy delivered by sunlight to the Earth in just one hour surpasses the total electricity consumed by humanity over an entire year. In comparison to solar photovoltaic cells, thermoelectric technology—which

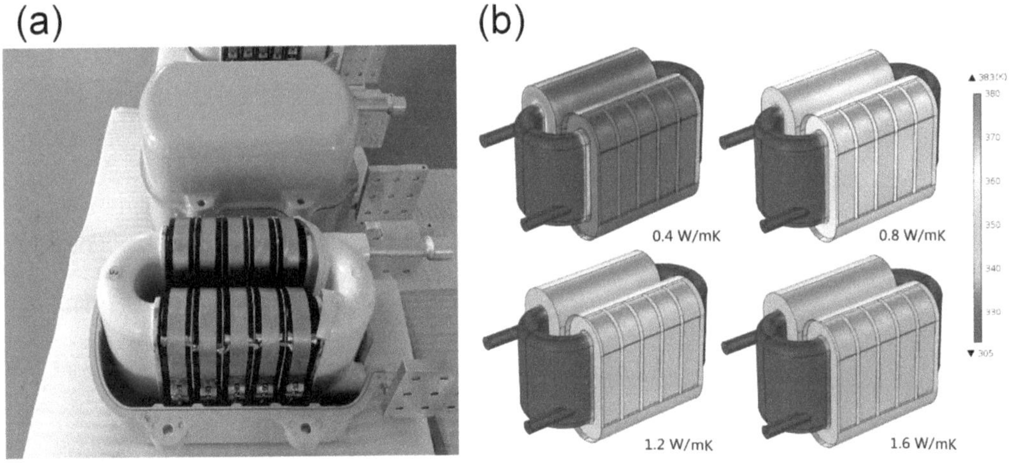

Figure 9.25 (a) Optical photo of saturable reactors. (b) Effect of TC of epoxy resin insulation on the steady-state temperature distribution of a saturable reactor [265]. (Reprinted with permission from Ref. [265]. Copyright 2020, *High Voltage*.)

utilizes the Seebeck effect to convert temperature differences into electrical potential—offers a more efficient approach to harnessing solar energy. This method consists of five key processes: energy concentration, absorption, conduction, storage, and power generation [266]. The thermal storage system is crucial for ensuring that solar thermoelectric generators can operate continuously, even during nighttime, overcast, or rainy conditions. PCMs, as latent heat storage mediums, have found extensive application in solar energy storage (see Figure 9.26a) [267]. However, the low TC (approximately 0.2 W m^{-1}·K^{-1}) of PCMs makes them susceptible to hotspot formation under concentrated solar irradiation. This uneven heat distribution can lead to suboptimal energy conversion efficiency and may damage material structures due to localized overheating [268]. Thus, improving the TC of PCMs is vital for advancing solar thermoelectric technology. Liu et al. demonstrated that incorporating vertically and radially aligned BN (11.65 vol%) as a reinforcement resulted in a 7.5-fold enhancement in through-plane TC and a 13.7-fold enhancement in in-plane TC for polyethylene glycol (PEG). With rapid and uniform heat transfer capabilities, the PCM can effectively manage concentrated solar beams. The solar thermoelectric generator, integrated with a high TC PCM, achieved a remarkable real-world power density of 40.28 W m^{-2} and a photothermal energy conversion efficiency of 85.1%.

Carbonaceous nanoscale reinforcements, including CNTs and graphene, exhibit significant promise for solar thermal conversion due to their exceptional TC and distinctive optical characteristics, such as strong absorption of sunlight, a wide absorption spectrum, and high photothermal conversion efficiency. Research indicates that incorporating 4% CNTs into PEG-based PCMs raises the TC from 0.26 to 0.42 W m^{-1}·K^{-1}, while maintaining an energy storage density exceeding 120 J g^{-1}. This enhancement leads to a marked improvement in energy conversion and storage efficiency of the phase change nanocomposites, increasing from 44.8% to 89.3% [269]. Similarly, the TC of PEG nanocomposites containing 4% silver-functionalized graphene increases from approximately 0.21 to 0.41 W m^{-1}·K^{-1}. Consequently, the photothermal conversion efficiency of

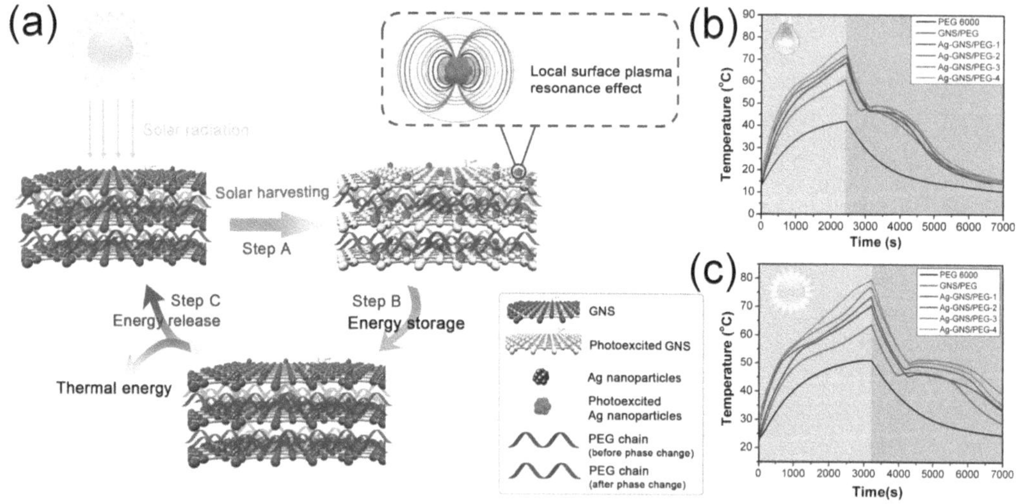

Figure 9.26 (a) Schematic illustration of visible-light-thermal energy conversion and storage. (b) Light-to-thermal energy conversion curves of PCMs with different TCs under solar simulator radiation. (c) Solar-to-thermal energy conversion curves of PCMs with different TCs under sunlight radiation [267]. (Reprinted with permission from Ref. [267]. Copyright 2019, *Applied Energy*.)

nanocomposites with 4% silver-functionalized graphene improves by 92% compared to pure PEG, attributed to the enhanced TC (see Figure 9.26b and c) [267].

The increase in TC contributes to improved energy storage efficiency of PCMs. However, the addition of reinforcements tends to lower the thermal energy storage density of these materials [270,271]. Therefore, it is essential to develop polymer nanocomposites that possess both high TC and latent heat for effective solar energy conversion and storage. Significant advancements have been made in this area. For example, a phase change composite featuring a high TC of 8.87 W $m^{-1}{\cdot}K^{-1}$ and an outstanding latent heat retention of 98.7% was successfully created by integrating lightweight, high-quality graphene aerogels into paraffin [272]. Other materials, such as metal–organic frameworks, titanium dioxide foams, and highly graphitized carbon networks, also demonstrate considerable TC enhancement and high latent heat retention, which significantly boosts solar energy conversion and storage capabilities [248,273,274].

9.4.2.7 Biomedical Applications

Bone cement is a medical material used in orthopedic surgeries primarily for securing joint prostheses and filling the space between the prosthesis and the human bone (see Figure 9.27a) [275]. Commonly, a combination of powdered PMMA (Polymethyl Methacrylate) and liquid MMA (Methyl Methacrylate) serves as bone cement. However, the polymerization reaction of MMA generates significant heat, leading to local temperatures reaching 80°C–120°C, which may cause necrosis in adjacent tissues [275,276]. While pre-cooling bone cement can help lower the peak reaction temperature, it does not entirely eliminate the overheating issue. CNTs serve as effective reinforcements for enhancing the TC and thermal stability of bone cement. Studies indicate that incorporating 5 vol% CNTs into bone cement can boost the TC of the nanocomposites by 232%, decrease the peak temperature by 48%, and extend the operational time by 80.8%. Moreover, the mechanical properties, including compressive strength, fracture toughness, and fatigue life, are also improved [277,278]. The distribution of reinforcements and the interface between the reinforcement and matrix greatly affect the TC and mechanical characteristics. Thus, the surface chemistry of nanoscale reinforcements should be thoughtfully designed.

Gutta-percha nanocomposites, composed mainly of gutta-percha, zinc oxide, and barium sulfate, are widely used as filling materials in root canal therapy [279,280]. During the procedure, gutta-percha is heated and melted to fill the root canal and its minor branches under pressure. Once the root canal is entirely filled, the gutta-percha solidifies, providing effective sealing (Figure 9.27b) [281,282]. Typically, the TC of dental adhesives correlates positively with filling depth; however, gutta-percha has a relatively low TC of around 0.19 W $m^{-1}{\cdot}K^{-1}$) [281]. This necessitates high-temperature heating for rapid melting during root canal procedures, which can risk burns and tissue necrosis. Increasing the TC can effectively lower the operating temperature of gutta-percha, thereby minimizing damage to surrounding tissues during treatment [283]. To further enhance the TC of gutta-percha and reduce the heating temperature, integrating high TC reinforcements is a viable approach. CNTs can connect with zinc oxide in gutta-percha, creating a synergistic effect on TC. A small amount of CNTs can also decrease the shear viscosity of gutta-percha, improving filling effectiveness; however, excessive CNTs may increase shear viscosity and reduce fluidity, consequently affecting filling depth (see Figure 9.27c) [281]. Besides CNTs, nanodiamonds are another option for enhancing the TC of gutta-percha [284]. Additionally, loading antibiotics, such as amoxicillin, onto the surface of the reinforcements used in gutta-percha can improve both the mechanical stability and TC, while also reducing the risk of bacterial infection, thus enhancing therapeutic outcomes.

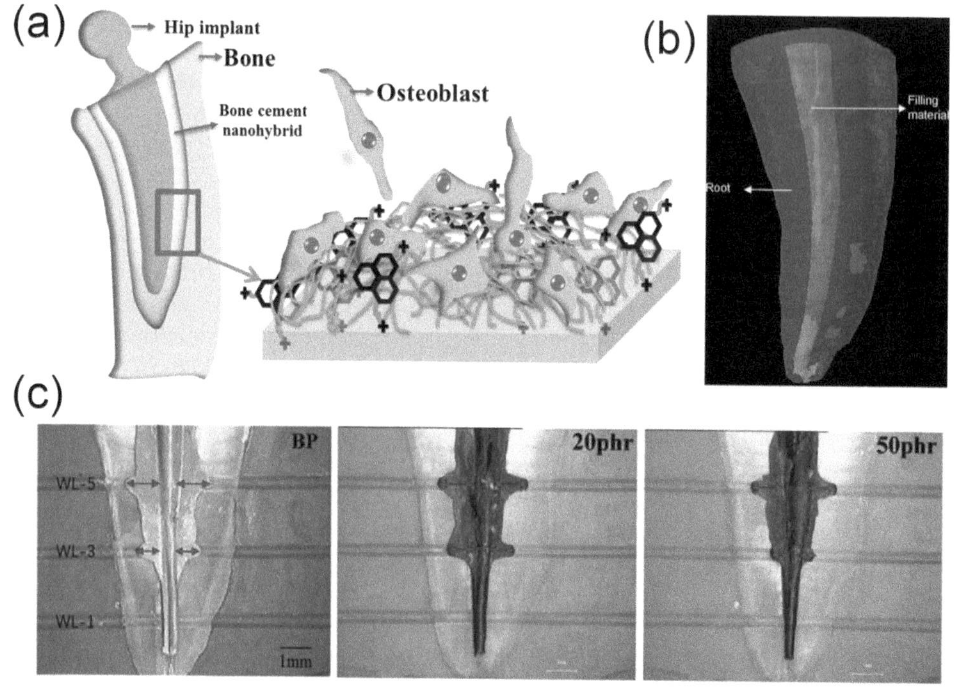

Figure 9.27 (a) Schematic illustration of the application of bone cement in joint replacement [275]. (Reprinted with permission from Ref. [275]. Copyright 2017, *Bioconjugate Chemistry*.) (b) 3D reconstructed model of an obturated root [280]. (Reprinted with permission from Ref. [280]. Copyright 2009, *Journal of Endodontics*.) (c) Comparison of filling depth of various gutta-percha composites with different TCs [281]. (Reprinted with permission from Ref. [281]. Copyright 2019, *Composites Part B: Engineering*.)

In conclusion, a high TC is essential for the safe use of gutta-percha and can improve the success rate of root canal treatments. However, there are limited reports on reinforcements that enhance the TC of gutta-percha. In biomedical applications, the biocompatibility of these reinforcements must be taken into account. Furthermore, the amount of reinforcement should be minimized to maintain the low shear viscosity of gutta-percha nanocomposites. Therefore, the development of gutta-percha should concentrate on achieving both high TC and low shear viscosity, while also exploring additional reinforcements that exhibit good biocompatibility and high TC.

9.4.2.8 Carbon Dioxide Capture

The ongoing rise in global warming has led to a range of issues, including the melting of glaciers and extreme climatic shifts, which have drawn widespread societal attention to the impact of human-generated carbon dioxide (CO_2) emissions. This concern has spurred the advancement of materials and technologies for carbon capture and storage in recent years [285–287]. A solid amine-based adsorbent has emerged as a promising candidate for carbon capture due to its high selectivity for CO_2, low heat requirement, and minimal energy needed for regeneration. It features numerous chemical adsorption sites, ensuring effective CO_2 adsorption even at low partial pressures of CO_2 [131,288]. The evolution of these adsorbents has gone through three phases: amine-impregnated porous materials, amine-grafted porous materials, and amine-polymerized porous materials. Compared to the initial amine-based adsorbents, the latter two generally exhibit superior long-term stability, attributed to the robust bonding between organic amines and porous supports. However, both the organic amines, such as polyethylenimine (e.g., PEII), and the porous supports (e.g., silica and clay) of these adsorbents have low TC. In contrast, chemisorption releases substantial exothermic heat during CO_2 capture, making the adsorbents prone to overheating, which can lead to partial degradation of organic amines and a decrease in adsorption capacity. Therefore, enhancing TC is crucial for preventing overheating and degradation of amine-based adsorbents.

Recently, there has been a focus on carbon-based nanomaterials for developing high-performance adsorbents for CO_2 capture. A 3D carbon scaffold made of hydroxylated graphene and hydroxylated CNTs provides a high TC of 5.1 W $m^{-1}{\cdot}K^{-1}$ for PEII nanocomposites, ensuring good cyclic stability of the nanocomposite adsorbent even at a harsh desorption temperature of 135°C [289]. Similarly, a covalently grafted PEII nanocomposite with 3D graphene exhibits large surface areas, strong covalent bonds between PEII and graphene, and high TC, ensuring high adsorption capacity and excellent thermal stability at a high desorption temperature of 135°C [290]. Notably, an ideal support for an amine-based adsorbent should also have a low heat capacity, good durability, and high TC. In addition to carbon-based nanomaterials, other materials such as MXene have also been used in amine-based adsorbents for CO_2 capture. A recent study indicates that the TC of the MXene-supported PEII nanocomposite adsorbent increases from 0.17 to 1.60 W $m^{-1}{\cdot}K^{-1}$ of polyethylenimine. Consequently, the nanocomposite adsorbent demonstrates a large working capacity of 2.6 mmol g^{-1}, as well as stable cycling stability and high mechanical stability against attrition [291]. Despite the significant enhancement in TC, the large-scale exfoliation of MXene remains challenging, which limits its commercial applications.

9.4.2.9 Radiative Cooling

Radiative cooling is a passive technique for reducing temperature by altering the spectral properties of materials. In contrast to conventional active cooling systems that depend on compressors, radiative cooling is advantageous due to its zero-energy consumption and zero emissions, which are crucial for conserving energy. The fundamental principle is to cover an enclosed outdoor area with materials that exhibit high infrared emissivity, enabling the heat within the area to be radiated out into the coldness of outer space (3 K) via thermal radiation. The Earth's atmosphere has a transparent window between 8 and 13 μm in wavelength when there is no cloud or fog cover, which matches the wavelength range of the blackbody radiation of objects at approximately 300 K. As a result, thermal radiation can pass directly through the atmosphere and reach outer space, with minimal absorption, leading to a reduction in the object's temperature.

For effective daytime radiative cooling, materials need to have not only high infrared emissivity within the atmospheric window but also high reflectivity to sunlight. In addition to porous natural materials and polymers that are suitable for daytime radiative cooling, other materials such as metamaterials, photonic crystal materials, and nanocomposites are also used to achieve this effect. However, to achieve high solar reflectivity, traditional daytime radiative cooling materials often include low-TC optical scatterers like micropores or SiO_2 particles, which results in decreased TC of the materials. This is not ideal for transferring the cooling power acquired by the materials to the object that requires cooling (Figure 9.28a and b). This is because, in scenarios with internal heat sources, the thermal conduction and radiation of radiative cooling materials are interconnected, and the heat inside the space must be conducted to the material's surface before it can be radiated away. Therefore, it is essential to develop radiative cooling materials with high TC to address the limitations of the heat dissipation process.

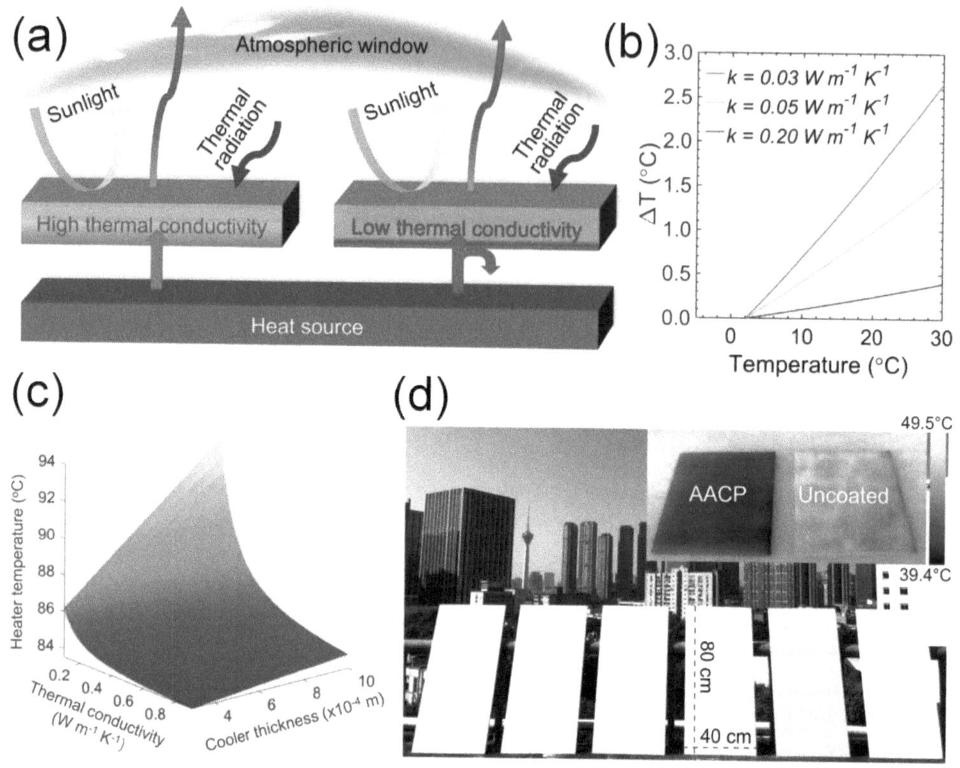

Figure 9.28 (a) Schematic illustrating the radiative and conductive heat transfer of radiative cooling materials with high and low TC. (b) Temperature difference between the top and bottom surfaces with different TCs as a function of the temperatures of the materials [293]. (Reprinted with permission from Ref. [293]. Copyright 2022, *Nano Energy*.) (c) Simulated temperature of the underneath heat source as a function of thermal resistance of the radiative cooler [292]. (Reprinted with permission from Ref. [292]. Copyright 2022, *Advanced Functional Materials*.) (d) Radiative cooling material coated outdoor wall tiles. The inset is an infrared image showing the temperature drop of the coated tile [294]. (Reprinted with permission from Ref. [294]. Copyright 2022, *Nature Communication*.)

Based on theoretical calculations, optical simulations, and experimental studies on various composite systems, Li et al. established design principles for high TC radiative cooling materials. (1) The optical scatterer should be paired with a polymer matrix that has an energy bandgap higher than that of sunlight, preventing the material from absorbing solar heat. (2) There should be a substantial difference in refractive index between the optical scatterer and the polymer matrix to increase the efficiency of sunlight scattering. (3) The optical scatterer should possess high TC to effectively enhance the TC of the composite material. (4) An optical scatterer with a 2D structure is preferable to improve the backscattering efficiency of sunlight. Following these guidelines, BN was selected as the optical scatterer, and PDMS, which has high infrared emissivity, was chosen as the matrix. The resulting composites demonstrated a sunlight reflectance of over 98% and an infrared emissivity of 90.3%, achieving a continuous 24-hour radiative cooling effect. Moreover, the composite has a significantly higher TC (1.4 $W\ m^{-1}\cdot K^{-1}$) compared to traditional radiative cooling materials, greatly improving the zero-energy dissipation efficiency of spaces that generate heat internally. When the heat generation rate is 700 $W\ m^{-2}$, it achieves an 18°C cooling effect compared to PDMS, and this experimental outcome is well supported by theoretical calculations (Figure 9.28c) [292]. While the advancement of high TC radiative cooling materials has considerably improved the heat dissipation efficiency of power equipment and electronic devices, the cooling performance is still not sufficient to meet the growing demands due to the inherent limitations of radiative heat transfer. Therefore, a critical future research direction is to further increase the heat dissipation capability of radiative cooling materials.

9.4.3 Challenges and Considerations in the Practical Implementation of Nanoscale BN-Polymer Composites

Elastomer composites filled with additives have garnered considerable interest for their capability to undergo significant reversible deformation with minimal applied force. However, attaining high TC typically necessitates exceedingly high filler loadings (exceeding 50%), which unfortunately can detrimentally affect the composites' softness and toughness due to the increased prevalence of entanglements within the polymer chains. To tackle this issue, Zeng et al. [295] introduced a straightforward solvent-thermal approach to fine-tune the trade-off between Young's modulus, ductility, and toughness in highly filled elastomer composites, exemplified by the C-PDMS/60 wt% BN elastomer composite. This optimization is achieved through the formation of a curled conformation resulting from the disentanglement of excessively entangled polymer chains, as well as enhanced mixing between the BN filler and the polymer matrix. The alleviation of trapped entanglements effectively lowers the Young's modulus (2.80 MPa) of the C-PDMS/60 wt% BN elastomer composite, while the robust unfolding and stretching capacity of the curled conformation confers exceptional ductility (~492%), thereby achieving high toughness (~2.80 MJ m^{-3}). Moreover, the improved mixing facilitates the incorporation of a high BN filler loading (60 wt%) in the C-PDMS/60 wt% BN elastomer composite, resulting in elevated TC (1.65 W $m^{-1}{\cdot}K^{-1}$). The overall performance of the C-PDMS/60 wt% BN composite represents a notable advancement in the realm of highly filled elastomer composites. Capitalizing on these advantageous attributes, the curled PDMS/BN elastomer composites are poised to function as effective TIMs for efficient heat dissipation and hold substantial promise for applications within the flexible electronics sector (Figure 9.29).

The TC of BN/polymer nanocomposites typically remains below 20 W $m^{-1}{\cdot}K^{-1}$ in both in-plane and cross-plane orientations, which is inadequate for efficient heat dissipation in advanced electronic applications. Enhancing the TC of *h*-BN/polymer nanocomposites presents several significant challenges for future research. Firstly, there is a necessity to develop isotropic BN fillers and 3D-BN structures with improved crystalline quality and continuity to achieve superior TC. Secondly, it is imperative to enhance the dispersibility and interaction between the BN structure and the polymer matrix to facilitate the formation of an interconnected BN network, thereby reducing interfacial thermal resistance and improving phonon conduction. Lastly, optimizing the structural design of the nanocomposites is crucial for establishing more effective thermal transport pathways. Addressing these challenges necessitates innovative methodologies in material synthesis and nanocomposite processing. Techniques such as chemical vapor deposition, atomic layer deposition, and ball milling can produce high-quality BNNSs as foundational materials. Advanced fabrication methods, including spray drying and spark plasma sintering, can yield isotropic BN fillers with enhanced TC. For the fabrication of 3D structures, nanofabrication and additive manufacturing techniques can be employed to create 3D-BN structures with improved crystalline quality and connectivity, thereby enhancing thermal dissipation pathways

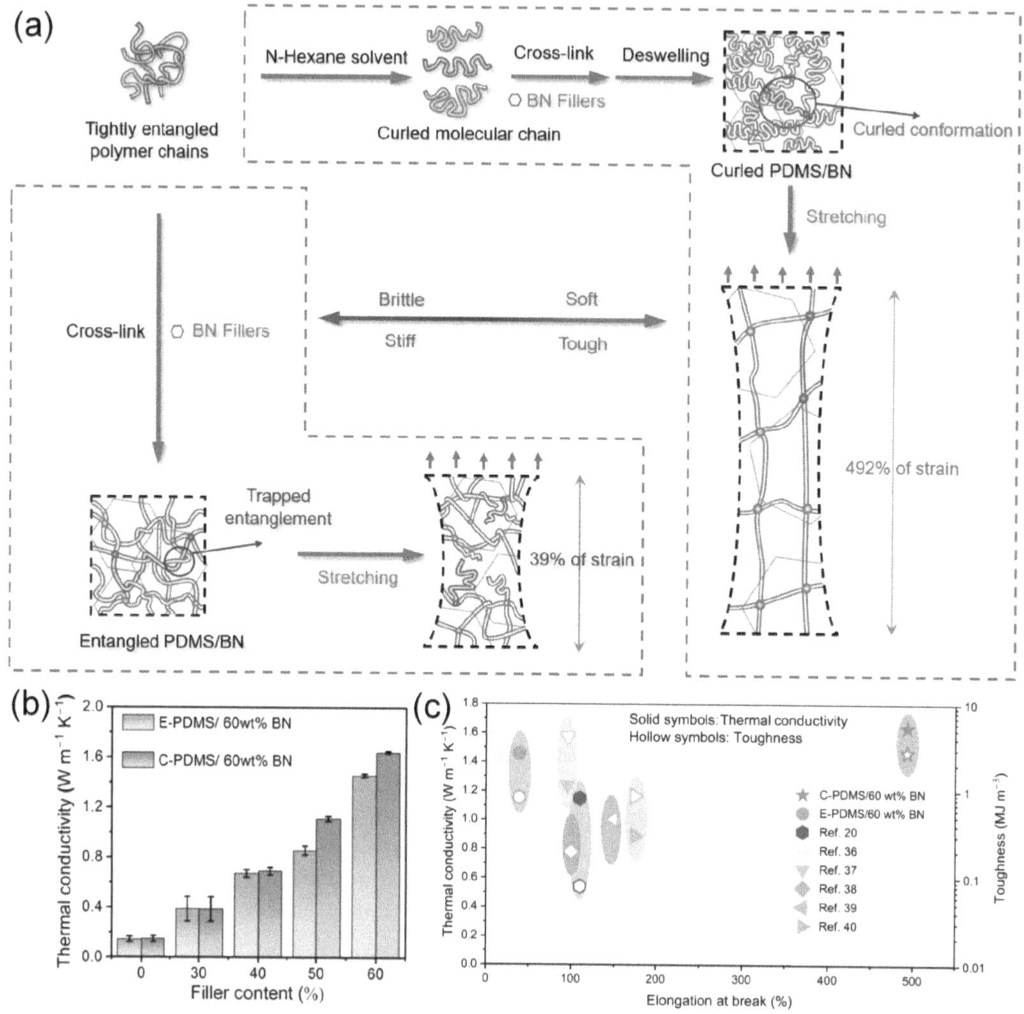

Figure 9.29 (a) Schematic diagram of the synthesis process of the curled PDMS/BN elastomer composites (C-PDMS/BN) and entangled PDMS/BN elastomer composites (E-PDMS/BN) (dimensions are not to scale) and the contribution of the curled conformation to the characteristics of the elastomer composites. (b) TC of the C-PDMS/60 wt% BN and E-PDMS/60 wt% BN at room temperature with different filler content. (c) Summary of TC, elongation at break, and toughness of thermally conductive elastomer composites recently reported in the literature [295]. (Reprinted with permission from Ref. [295]. Copyright 2023, *Chemistry of Materials.*)

within the nanocomposites. Interfacial engineering plays a critical role in improving the dispersibility and interaction between BN fillers and the polymer matrix. Surface modification and functionalization of BN fillers can enhance compatibility and dispersion, thereby strengthening interfacial bonding with the polymer matrix and minimizing interfacial thermal resistance. Furthermore, the design and optimization of nanocomposites can be significantly enhanced through computational modeling and simulation, which can optimize the distribution and orientation of BN fillers within the polymer matrix, thereby establishing a uniform thermally conductive network throughout the composite.

Collaborative research and development initiatives involving chemists, materials scientists, engineers, and industry partners are essential for comprehensively addressing these challenges. By leveraging multidisciplinary expertise and innovative strategies, these collaborative efforts can facilitate the translation of laboratory findings into scalable and practical solutions. Collectively, these strategies represent a holistic approach to overcoming the current limitations in the TC of BN/polymer nanocomposites, thereby enabling their effective application in the thermal management of advanced electronic systems.

9.4.4 Case Studies and Applications of Nanoscale BN-Polymer Composites in Electronic Devices

The BNNSs/epoxy resin composite, containing 15vol% BNNSs, has been synthesized using the freeze-drying technique with the objective of creating a nacre-mimetic 3D thermally conductive network characterized by robust thermal stability. The reported TC of 6.07 W $m^{-1}.K^{-1}$ for the BNNSs/epoxy composite is nearly 32 times greater than that of the unmodified resin, underscoring the significant role of *h*-BN. To evaluate the commercial viability of this composite, Han et al. [41] conducted a comparative analysis between a commercially available silicone sheet and the BNNSs/epoxy composite as TIMs, which were employed in conjunction with a 20W light-emitting device (LED) chip and a copper heat sink (Figure 9.30a and b). An infrared camera was utilized to monitor the variations in surface temperature of the LED chips. Figure 9.30c illustrates the temperature change map for both types of chips over different time intervals following activation. The results indicate that the silicon chip experiences a rapid increase in temperature compared to the BNNSs/epoxy composite. The notable temperature differential of 10°C (Figure 9.30d) highlights the superior performance of the BNNSs/epoxy composite in thermal management applications. Furthermore, the thermal stability of this composite was assessed in Figure 9.30e by recording the chip temperature during 'on' (4 minutes) and 'off' (2 minutes) phases, demonstrating exceptional thermal stability [41].

Bai et al. investigated the application of their synthesized V-BN/SG polymer composites within a battery pack, highlighting their potential for use in electronic cooling systems [43]. As illustrated in Figure 9.31a, two lithium-ion batteries (3.7 V, 3000 mAh) were connected in series and encased in iron foil. A 60 V-BN/SG (0.2–2) strip, serving as a TIM with a thickness of 1 mm, was positioned in the gap between the battery and the iron foil, followed by compression to ensure optimal contact between the TIMs and both the battery and the iron foil. A thermocouple was placed between the battery and the TIMs to monitor the temperature of the battery's surface. To maintain the battery temperature below the critical threshold of 60°C, a bottom graphitic heat sink and TIMs were employed to enhance heat dissipation from the battery pack. The discharging process was executed using electrical resistances of 1 Ω, 0.5 Ω, or by connecting two resistors in parallel, resulting in calculated discharging rates of approximately 2.74°C, 4.94°C, and 6.16°C, respectively. Infrared imaging revealed that the temperatures of the top surface of the battery pack equipped with TIMs consistently remained lower than those without TIMs during the discharging process at 4.94°C, indicating a more efficient heat transfer from the battery to the metal housing (Figure 9.31b). The surface temperature was recorded from the central red square area of the battery pack. As the discharging rate increased, the temperature differential of the battery's surface rose from 0.9°C (2.47°C), to 3.2°C (4.94°C), and ultimately to 5.9°C (6.16°C), attributable to

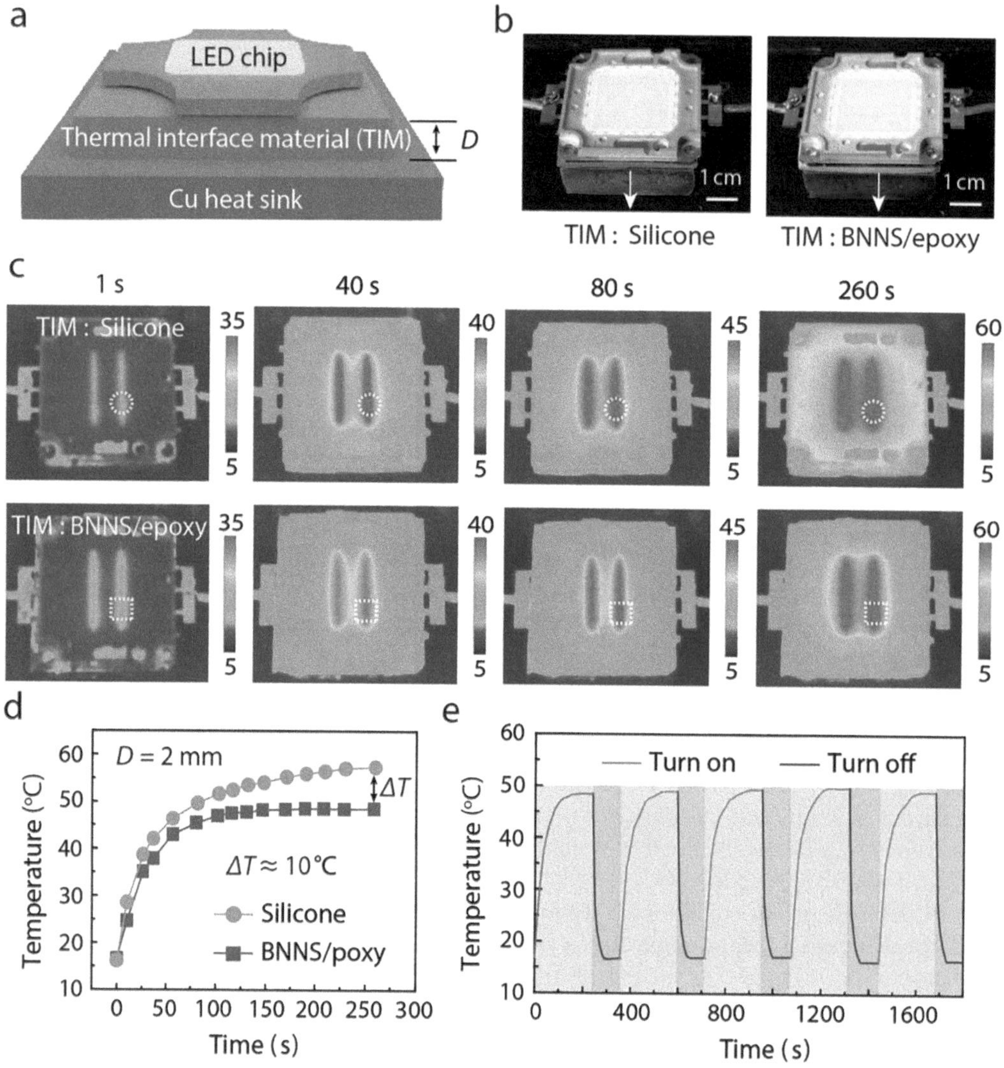

Figure 9.30 Demonstration of the as-mentioned BNNSs/epoxy composite as a TIM: (a) Schematic diagram of a LED chip integrated with a TIM and a Cu heat sink; (b) optical images showing two LED chips integrated with commercial silicone and our BNNSs/epoxy composite as TIMs (both TIMs are 40×40×2 mm in size); (c) sequential top-view infrared images of the two LED chips, indicating more efficient thermal dissipation using the present composite as a TIM; (d) comparison between the temperature increase of the same spot on the two chips, depicted by the white dotted circle and square, respectively, in (c); (e) the chip temperature monitored during cyclic 'on' (4 minutes) and 'off' (2 minutes) states, indicating the thermal stability of the BNNSs/epoxy composite [41]. (Reprinted with permission from Ref. [41]. Copyright 2019, *Adv. Funct. Mater.*)

the superior TC of the TIMs compared to that of air (0.02 W m^{-1}·K^{-1}) (Figure 9.31c and d). During the discharging (4.94°C) and charging (1.4°C) cycles (Figure 9.31e), the temperature exhibited an increase during discharging and a decrease during charging, reflecting the lower charging rate.

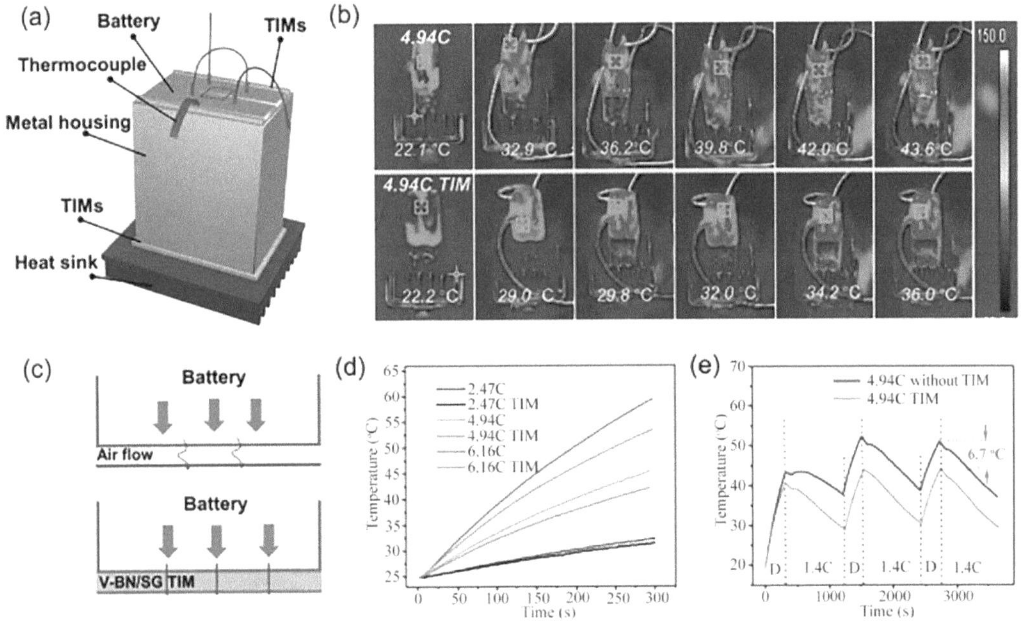

Figure 9.31 Battery thermal dissipation performance of 60 V-BN/SG (0.2–2) strips. (a) Schematic illustration of the TIM temperature measurement system in the battery setup. (b) IR images of the top surface of batteries with and without V-BN/SG strips as TIMs under a 4.94C discharging rate at room temperature. (c) Schematic of TIMs used to bridge the battery and metal housing. Temperature revolution of the battery surface measured by a thermocouple as a function of time (d) under 2.74C, 4.94C, 6.16C discharging rate, and (e) during three charging (1.4C)-discharging (4.94C) cycles [43]. (Reprinted with permission from Ref. [43]. Copyright 2022, Nano-Micro Letters.)

Notably, the maximum temperature recorded in the second cycle exceeded that of the first cycle due to heat accumulation. The temperature differential observed in the third cycle was approximately 6.7°C, underscoring the enhanced thermal dissipation capabilities of the battery pack when utilizing the 60 V-BN/SG (0.2–2) TIMs. Additionally, the compression testing of the V-BN/SG strips indicated a compressive stress of approximately 1.53 MPa at 50% compressive strain, demonstrating the significant flexibility and resilience of the TIMs.

Polytetrafluoroethylene (PTFE)-based composites exhibiting high TC are critical materials for thermal management in communication devices. Their low dielectric constant and loss tangent minimize interference with signal transmission, in contrast to traditional heat dissipation films, such as conductive graphene or copper films, which can significantly disrupt signals. Qiu et al. developed a BNNS/PTFE composite characterized by high TC, a low dielectric constant of 2.44 at 15 GHz, and a dielectric loss tangent of 0.001 at the same frequency. Notably, the dielectric constant remains relatively stable across a broad temperature range of −40°C to +80°C, which is essential for ensuring the stable long-term operation of telecommunication devices at high speeds and frequencies. The researchers assessed the signal interference of the BNNS/PTFE composite on an antenna within a 5G mobile device, revealing that the composite film exhibited negligible signal interference in the N78 (3.3–4.2 GHz) and N79 (4.4–5.0 GHz) frequency bands when compared to a bare antenna. Furthermore, the volume resistivity of the 20 wt% BNNS film was measured at 3.4×10^{10} Ω cm, significantly exceeding the electrical insulation standard of

10^9 Ω cm for electronic applications. The low dielectric properties and high TC of the BNNS/PTFE composites substantially enhance the thermal dissipation capacity of communication devices, thereby providing greater design flexibility for future telecommunications technologies. Additionally, the BNNS/PTFE composite is poised to play a vital role in the development of next-generation high-speed and high-frequency circuit boards. The 20 wt% BNNS film also demonstrates considerable folding strength and thermal reliability for prolonged use, making it suitable for thermal management in foldable flexible electronics. The roller-pressed film retains a minimal amount of solvent, which contributes to its excellent malleability and adhesion properties. Moreover, the composite film can be easily conformed to complex surfaces and exhibits a very low thermal expansion coefficient. As illustrated, the 20 wt% BNNS composite film can adhere tightly to a heating element through natural drying without the need for binders, facilitating effective heat dissipation with an equilibrium temperature difference of 15.7°C. Infrared imaging indicates that the 20 wt% BNNS composite film, utilized as a heat dissipation medium, demonstrates rapid thermal diffusion while maintaining a low temperature on intricate surfaces. For specific thermal management applications that necessitate a high through-plane TC, a BNNS/PTFE film with favorable processability can be employed to create a perpendicularly oriented film by rolling and slicing a thin section from a BNNS/PTFE composite roll. The original 20 wt% BNNS film, containing approximately 0.8 wt% solvent prior to sintering, exhibited strong adhesion between layers during the rolling process. By cutting a thin slice in the radial direction (x–z plane) and subsequently thinning it, a film was produced in which the BNNSs were oriented perpendicular to the surface (Figure 9.32).

In contemporary society, transformers are widely utilized, ranging from microtransformers embedded in printed circuit boards (PCBs) to large-scale ultra-high-voltage power apparatus. The accumulation of heat significantly affects the longevity and reliability of transformers. In a study conducted by Huang et al., a commercially available control transformer was selected as a representative example to evaluate the efficacy of BNNS-M/PBO paper as a thermally conductive insulating material for transformers. The BNNS-M/PBO paper was applied around the winding (refer to Figure 9.33a). As illustrated in Figure 9.33b and c, the application of BNNS-M/PBO paper resulted in a more uniform heat distribution and lower surface temperatures throughout the operational process of the transformer. This improvement can be attributed to the high TC of the BNNS-M/PBO paper, which mitigates heat accumulation and enhances heat dissipation, functioning effectively as a non-combustible thermal spreading film. To further emphasize the advantages of BNNS-M/PBO paper as an insulating material, a radar plot was constructed to compare its properties with those of conventional insulating materials, such as Nomex 410, glass mica tape, mica paper, and polyimide film, which are commonly employed in large-scale electrical equipment (see Figure 9.33d). The results clearly indicate that the overall performance of BNNS-M/PBO paper surpasses that of the other insulating materials in terms of ultimate strength, strain-to-failure, breakdown strength, TC, and heat resistance.

Bai et al. utilized a computer cooling system to assess the heat dissipation capabilities of V-BNF/SR TIM by contrasting their performance with that of a commercial thermal pad, which has a TC of 7.5 W m^{-1}·K^{-1} [296]. As illustrated in Figure 9.34a, the V–BNF/SR TIM, measuring $25 \times 25 \times 1$ mm^3, was positioned between a CPU (i5–4570) and a heat sink, with the applied packaging force monitored by a pressure sensor. The temperature changes were recorded using a thermocouple placed at the center of the CPU. The CPU was operated at its maximum power output of 84 W, facilitated by professional detection software (AIDA64). Figure 9.34b demonstrates that the equilibrium temperature of the CPU utilizing V–BNF/SR TIMs is reduced by 28.7°C in

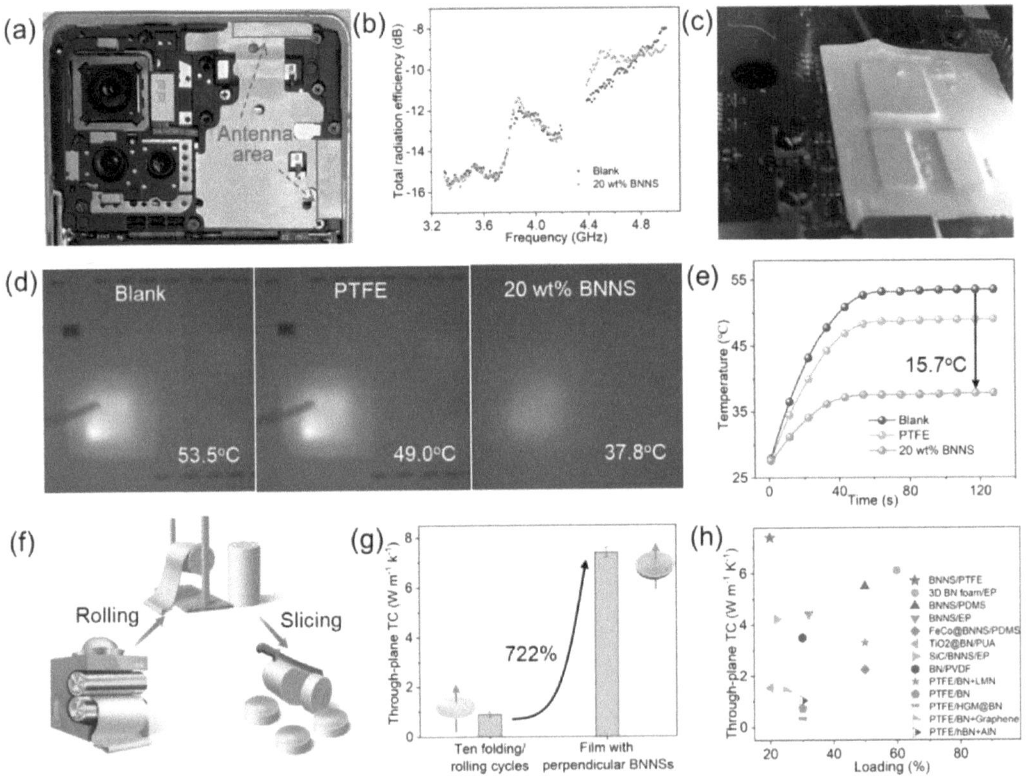

Figure 9.32 (a) Photo of a 20 wt% BNNS composite film attached to a mobile phone antenna. (b) Total radiation efficiency of antennas in different frequency bands. (c) Photo of a 20 wt% BNNS composite film attached to a heating element. (d) Infrared images of comparative surface temperature and (e) corresponding temperature changes as a function of time. (f) Diagram of the rolling and slicing process used to prepare films in which the BNNSs are oriented perpendicular to the surface. (g) Comparison of through-plane TC between the original film and the one with perpendicularly oriented BNNSs. (h) Comparison of the through-plane TC of our BNNS/PTFE film with other reported BNNS composites [46]. (Reprinted with permission from Ref. [46]. Copyright 2023, *ACS Nano*.)

comparison to the uncoated CPU. Additionally, the application of SG/V–BNF/SR TIMs results in a temperature decrease of 47.3°C, achieving a lower equilibrium temperature of 43.9°C, which is comparable to that of the thermal pad. Notably, finite element simulations (Figure 9.34c and d) indicate that the temperature of the CPU with SG/V–BNF-SR TIMs (1 mm thickness) can be 4.87°C lower than that with the thermal pad of equivalent thickness, attributable to the reduced total thermal resistance of SG/V–BNF/SR TIMs. This finding underscores the superior heat dissipation performance of these TIMs in electronic cooling applications. The temperature profile depicted in Figure 9.34e illustrates the temperature fluctuations of the CPU subjected to SG/V–BNF/SR TIMs during a ten-cycle test. In each cycle, the CPU was operated at maximum power until it reached a steady-state temperature, after which the software was deactivated. The minimal variation in the maximum CPU temperature, ranging from 43.4°C to 45.6°C, reflects the stable thermal conduction properties of SG/V–BNF/SR TIMs, thereby highlighting their potential for application in electronic cooling systems.

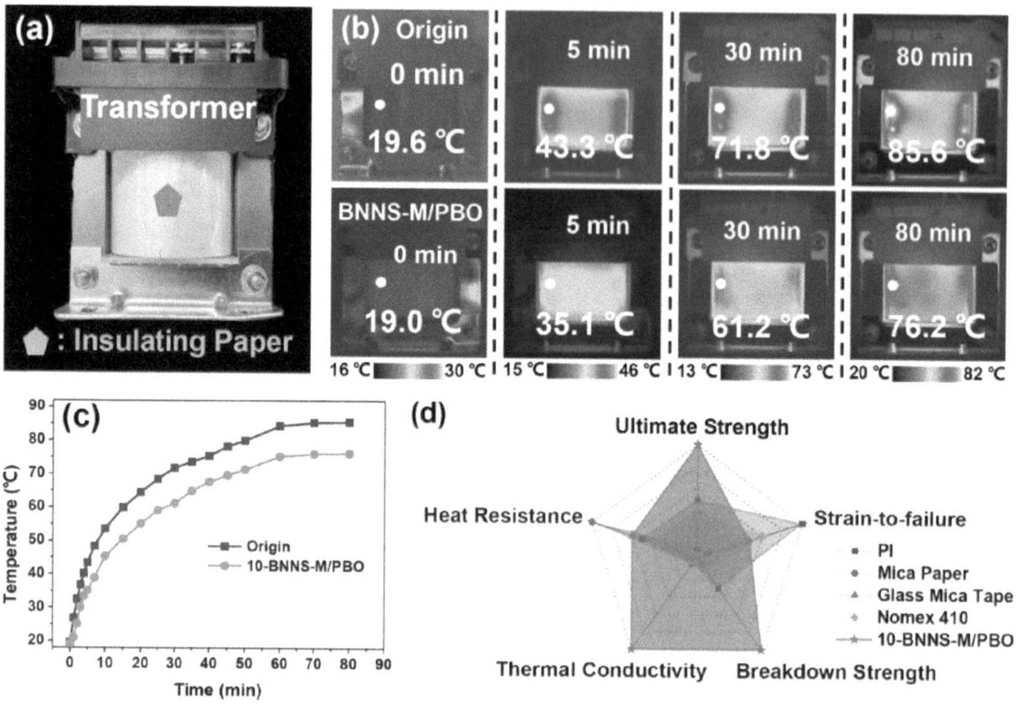

Figure 9.33 (a) Optical photograph of the control transformer. (b) Infrared thermal images of the control transformer integrated with wrapped with 10-BNNS-M/PBO paper. (c) Corresponding temperature variations. (d) Radar plot showing a comparison among 10-BNNS-M/PBO paper, Nomex 410, glass mica tape, mica paper, and PI film. The results are normalized by the maximum value of each term [48]. (Reprinted with permission from Ref. [48]. Copyright 2022, *ACS Nano*.)

9.5 CONCLUSION AND OUTLOOK

9.5.1 Conclusions

h-BN and BNNSs are distinguished by their exceptional TC, electrical insulation, and minimal thermal expansion, rendering them highly suitable for thermal management in the forthcoming generation of electronic devices. These devices are characterized by their multifunctionality, compact design, and high power density. In recent decades, extensive research has significantly advanced the status of polymer nanocomposites, making them among the most prevalent materials utilized in various applications. This perspective offers a comprehensive review of the conventional preparation methods for thermally conductive polymer nanocomposites and underscores the critical role of TC in the thermal management of electronics, power equipment, solar energy conversion and storage, biomedicine, and environmental protection. While TC is a key property parameter of materials, other factors such as thermal resistance, as well as mechanical, processing, and electrical properties, are also vital for the diverse applications of polymer nanocomposites.

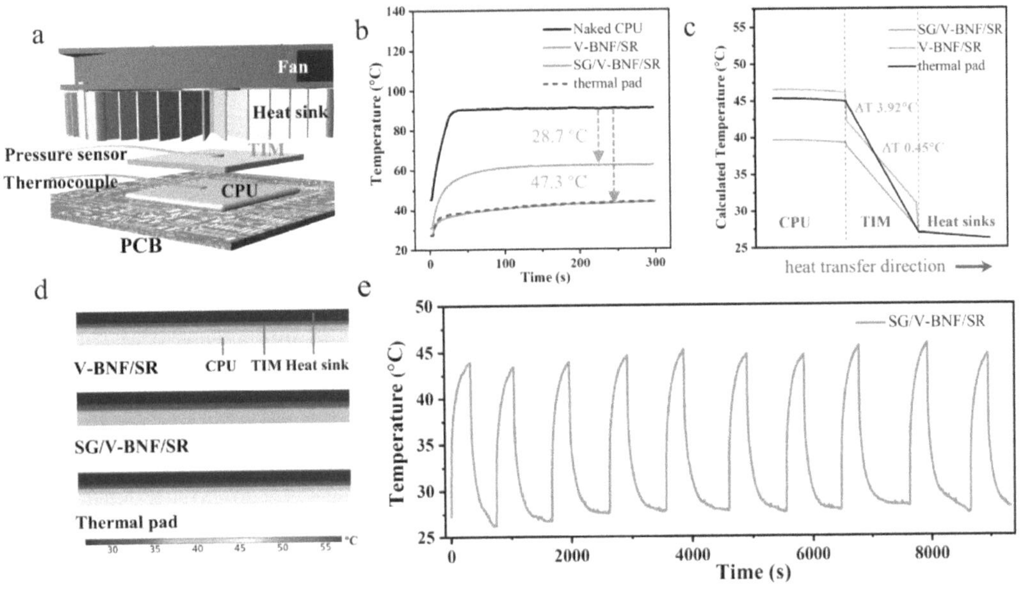

Figure 9.34 (a) Schematic diagram of the test system. (b) The plot of temperature evolution of the CPU equipped with different TIMs at the maximum running power (84 W). (c) Calculated temperature distribution along the heat transfer from the CPU to the heat sink. (d) Calculated temperature profiles of the cooling system loaded with V–BNF/SR, SG/V–BNF/SR, and commercial thermal pad. (e) Thermal cycling stability of the SG/V–BNF/SR TIM in the cooling system [296]. (Reprinted with permission from Ref. [296]. Copyright 2024, *Composites Part B*.)

9.5.2 Outlook

The sustainable development of energy, the environment, and society has put forward higher requirements for the TC of emerging materials. Nevertheless, emerging applications of thermally conductive polymer nanocomposites are based on application-driven, in-depth fundamental research. More systematic studies are required to address the following issues regarding thermally conductive polymer nanocomposites.

1. Over the past decade, significant progress has been made in enhancing the isotropic properties of these materials. Key strategies include the following: (1) the development of BN fillers exhibiting isotropic TC, (2) the fabrication of 3D BN structures, and (3) the establishment of thermally conductive pathways through the controlled distribution of *h*-BN flakes or BNNSs within a polymer matrix. These methodologies have successfully achieved a two-order magnitude increase in the isotropic TC of polymers. Nonetheless, the TC of the resulting nanocomposites typically remains below 20 W $m^{-1}{\cdot}K^{-1}$ in both in-plane and cross-plane directions, which is inadequate for effective heat dissipation in advanced electronic applications.
2. Considerable advancements have been made in enhancing the performance of thermally conductive polymer nanocomposites. However, high filler loading does not

consistently yield the anticipated high TC. It is important to recognize that TC is merely one of several material property parameters; other characteristics, including mechanical, processing, and electrical properties, are equally significant. Therefore, it is imperative to synergistically enhance TC alongside these other properties, ensuring that improvements in TC do not compromise the overall material performance.

3. While substantial focus has been directed toward enhancing TC in thermally conductive polymer nanocomposites, the aspect of thermal resistance—defined as the material thickness divided by the TC and heat transfer area—has received comparatively little attention. From an engineering perspective, thermal resistance is often more critical than TC itself. For instance, in TIMs, thermal resistance is influenced not only by TC but also by material thickness and interface contact quality. Consequently, the benefits of improved TC may be mitigated or negated by inadequate interface contact and increased TIM thickness in practical applications. Thus, efforts to reduce the thermal resistance of thicker materials should be prioritized in future research endeavors.
4. The recyclability of polymer nanocomposites has garnered significant attention and is vital for the sustainable development of materials. Future efforts should focus on developing environmentally friendly recycling technologies for the high-value components of thermally conductive polymer nanocomposites, while also striving to minimize recycling costs to ensure sustainability. Currently, enhancements in material recyclability predominantly rely on dynamic covalent reactions, necessitating the maintenance of thermal stability to meet application requirements.
5. Materials that are thermally conductive yet electrically insulating have broader applications than those that are both thermally and electrically conductive. For example, polymer nanocomposites with high TC and low dielectric constants are essential in the realms of 5G and 6G technologies. BN has been widely recognized for its ability to enhance the TC of polymer nanocomposites while preserving their electrical insulating properties. Future research should explore alternative thermally conductive yet electrically insulating reinforcements to meet the demands of relevant industrial applications.
6. Finally, it is essential to consider the practical aspects of developing low-cost, scalable, and environmentally sustainable fabrication processes for thermally conductive polymer nanocomposites. In the context of large-scale applications, factors such as scalability, applicability, and cost-effectiveness are paramount. Prioritizing the affordability of raw materials and the efficiency of manufacturing processes, while ensuring the environmental sustainability of the final products, is crucial.

Collaborative research and development initiatives involving chemists, materials scientists, engineers, and industry stakeholders are vital for comprehensively addressing these challenges. By harnessing multidisciplinary expertise and innovative methodologies, these collaborative efforts can facilitate the translation of laboratory findings into scalable and practical solutions. Collectively, these strategies represent a holistic approach to overcoming the current limitations in the TC of BN/polymer nanocomposites, thereby enabling their effective application in the thermal management of advanced electronic systems.

REFERENCES

1. A. Giri, P. E. Hopkins, A review of experimental and computational advances in thermal boundary conductance and nanoscale thermal transport across solid interfaces, *Adv. Funct. Mater.* 2020, 30(8), 1903857.
2. H. Chen, V. V. Ginzburg, J. Yang, Y. Yang, W. Liu, Y. Huang, Thermal conductivity of polymer-based composites: fundamentals and applications, *Prog. Polym. Sci.* 2016, 59, 41–85.
3. Z. Yan, G. Liu, J. M. Khan, A. A. Balandin, Graphene quilts for thermal management of high-power GaN transistors, *Nat. Commun.* 2012, 3(1), 827.
4. Y. Cui, Z. Qin, H. Wu, M. Li, Y. Hu, Flexible thermal interface based on self-assembled boron arsenide for high-performance thermal management, *Nat. Commun.* 2021, 12(1), 1284.
5. Y. Xu, D. Kraemer, B. Song, Z. Jiang, J. Zhou, J. Loomis, Nanostructured polymer films with metal-like thermal conductivity, *Nat. Commun.* 2019, 10(1), 1771.
6. S. Shen, A. Henry, J. Tong, R. Zheng, G. Chen, Polyethylene nanofibres with very high thermal conductivities, *Nat. Nanotechnol.* 2010, 5(4), 251–255.
7. Y. F. Zhuang, X. Y. Cao, J. N. Zhang, Y. Y. Ma, X. X. Shang, J. X. Lu, Monomer casting nylon/graphene nanocomposite with both improved thermal conductivity and mechanical performance, *Compos. Part A* 2019, 120, 49–55.
8. C. Ji, C. Yan, Y. Wang, S. Xiong, F. Zhou, Y. Li, Thermal conductivity enhancement of CNT/MoS_2/graphene–epoxy nanocomposites based on structural synergistic effects and interpenetrating network, *Compos. Part B* 2019, 163, 363–370.
9. Y. Liu, M. Lu, K. Wu, S. Yao, X. Du, G. Chen, Anisotropic thermal conductivity and electromagnetic interference shielding of epoxy nanocomposites based on magnetic driving reduced graphene oxide@Fe_3O_4, *Compos. Sci. Technol.* 2019, 174, 1–10.
10. C. Yan, T. Yu, C. Ji, X. Zeng, J. Lu, R. Sun, 3D interconnected high aspect ratio tellurium nanowires in epoxy nanocomposites: serving as thermal conductive expressway, *J. Appl. Polym. Sci.* 2019, 136(6), 47054.
11. Y. Guo, X. Yang, K. Ruan, J. Kong, M. Dong, J. Zhang, Reduced graphene oxide heterostructured silver nanoparticles significantly enhanced thermal conductivities in hot-pressed electrospun polyimide nanocomposites, *ACS Appl. Mater. Interfaces* 2019, 11(28), 25465–25473.
12. L. Lindsay, D. A. Broido, Enhanced thermal conductivity and isotope effect in single-layer hexagonal boron nitride, *Phys. Rev. B Condens. Matter* 2011, 84(15), 155421.
13. L. Lindsay, D. A. Broido, Theory of thermal transport in multilayer hexagonal boron nitride and nanotubes, *Phys. Rev. B Condens. Matter* 2012, 85(3), 035436.
14. A. A. Balandin, Thermal properties of graphene and nanostructured carbon materials, *Nat. Mater.* 2011, 10(8), 569–581.
15. V. Guerra, C. Wan, T. McNally, Thermal conductivity of 2D nano-structured boron nitride (BN) and its composites with polymers, *Prog. Mater. Sci.* 2019, 100, 170–186.
16. N. Burger, A. Laachachi, M. Ferriol, M. Lutz, V. Toniazzo, D. Ruch, Review of thermal conductivity in composites: mechanisms, parameters and theory, *Prog. Polym. Sci.* 2016, 61, 1–28.
17. Z. Lin, Y. Liu, S. Raghavan, K. S. Moon, S. K. Sitaraman, C. P. Wong, Magnetic alignment of hexagonal boron nitride platelets in polymer matrix: toward high performance anisotropic polymer composites for electronic encapsulation, *ACS Appl. Mater. Interfaces* 2013, 5, 7633–7640.
18. K. Kim, J. Kim, Vertical filler alignment of boron nitride/epoxy composite for thermal conductivity enhancement via external magnetic field, *Int. J. Therm. Sci.* 2016, 100, 29–36.
19. X. Yang, C. Liang, T. Ma, Y. Guo, J. Kong, J. Gu, M. Chen, J. Zhu, A review on thermally conductive polymeric composites: classification, measurement, model and equations, mechanism and fabrication methods, *Adv. Compos. Hybr. Mater.* 2018, 1, 207–230.
20. K. Sato, H. Horibe, T. Shirai, Y. Hotta, H. Nakano, H. Nagai, K. Mitsuishi, K. Watari, Thermally conductive composite films of hexagonal boron nitride and polyimide with affinity-enhanced interfaces, *J. Mater. Chem.* 2010, 20, 2749–2752.

21. T. L. Li, S. L. C. Hsu, Enhanced thermal conductivity of polyimide films via a hybrid of micro- and nano-sized boron nitride, *J. Phys. Chem. B* 2010, 114, 6825–6829.
22. A. Devpura, P. E. Phelan, R. S. Prasher, Percolation theory applied to the analysis of thermal interface materials in flipchip technology, *ITHERM 2000. The Seventh Intersociety Conference on Thermal and Thermomechanical Phenomena in Electronic Systems* (Cat. No. 00CH37069), 23–26 May 2000, 21–28.
23. X. Huang, P. Jiang, T. Tanaka, A review of dielectric polymer composites with high thermal conductivity, *IEEE Electr. Insul. Mag.* 2011, 27, 8–16.
24. C. Yu, W. Gong, W. Tian, Q. Zhang, Y. Xu, Z. Lin, M. Hu, X. Fan, Y. Yao, Hotpressing induced alignment of boron nitride in polyurethane for composite films with thermal conductivity over 50 Wm^{-1} K^{-1}, *Compos. Sci. Technol.* 2018, 160, 199–207.
25. J. Zhang, X. Wang, C. Yu, Q. Li, Z. Li, C. Li, H. Lu, Q. Zhang, J. Zhao, M. Hu, Y. Yao, A facile method to prepare flexible boron nitride/poly(vinyl alcohol) composites with enhanced thermal conductivity, *Compos. Sci. Technol.* 2017, 149, 41–47.
26. H.-B. Cho, T. Nakayama, H. Suematsu, T. Suzuki, W. Jiang, K. Niihara, E. Song, N.S.A. Eom, S. Kim, Y.-H. Choa, Insulating polymer nanocomposites with highthermal-conduction routes via linear densely packed boron nitride nanosheets, *Compos. Sci. Technol.* 2016, 129, 205–213.
27. Q. Weng, X. Wang, X. Wang, Y. Bando, D. Golberg, Functionalized hexagonal boron nitride nanomaterials: emerging properties and applications, *Chem. Soc. Rev.* 2016, 45(14), 3989–4012.
28. X. He, Y. Wang, Recent advances in the rational design of thermal conductive polymer composites, *Ind. Eng. Chem. Res.* 2021, 60(3), 1137–1154.
29. H. Wang, A. S. Tazebay, G. Yang, H. T. Lin, W. Choi, C. Yu, Highly deformable thermal interface materials enabled by covalently-bonded carbon nanotubes, *Carbon* 2016, 106, 152–157.
30. J. Zhang, G. Shi, C. Jiang, S. Ju, D. Jiang, 3D bridged carbon nanoring/graphene hybrid paper as a high-performance lateral heat spreader, *Small* 2015, 11(46), 6197–6204.
31. Z. Liang, Y. Pei, C. Chen, B. Jiang, Y. Yao, H. Xie, General, vertical, three-dimensional printing of two-dimensional materials with multiscale alignment, *ACS Nano* 2019, 13(11), 12653–12661.
32. C. Ji, Y. Wang, Z. Ye, L. Tan, D. Mao, W. Zhao, Ice-templated MXene/Ag-epoxy nanocomposites as high-performance thermal management materials, *ACS Appl. Mater. Interfaces* 2020, 12(21), 24298–24307.
33. Y. Yao, X. Zeng, R. Sun, J. Xu, C. Wong, Highly thermally conductive composite papers prepared based on the thought of bioinspired engineering, *ACS Appl. Mater. Interfaces* 2016, 8(24), 15645–15653.
34. H. Zhang, R. Huang, Y. Li, H. Li, Z. Wu, J. Huang, B. Yu, X. Gao, J. Li, L. Li, Optimization of boron nitride sphere loading in epoxy: enhanced thermal conductivity and excellent electrical insulation, *Polymers* 2019, 11, 1335.
35. L. Ren, X. Zeng, R. Sun, J.-B. Xu, C.-P. Wong, Spray-assisted assembled spherical boron nitride as fillers for polymers with enhanced thermally conductivity, *Chem. Eng. J.* 2019, 370, 166–175.
36. X. Zeng, Y. Yao, Z. Gong, F. Wang, R. Sun, J. Xu, C.P. Wong, Ice-templated assembly strategy to construct 3D boron nitride nanosheet networks in polymer composites for thermal conductivity improvement, *Small* 2015, 11, 6205–6213. https://doi.org/10.1002/smll.201502173.
37. J. F. Fan, S. C. Ding, X. L. Zeng, S. Gao, Z. B. Wen, X. L. Zeng, R. Sun, L. L. Ren, Gecko-inspired adhesive structures enable efficiently thermal conductance and vibration dissipation in a highly mismatched system. *Chem. Eng. J.* 2022, 445, 136754.
38. H. J. Hong, S. M. Kwan, D. S. Lee, S. M. Kim, Y. H. Kim, J. S. Lim, J. Y. Hwang, H. S. Jeong, Highly flexible and stretchable thermally conductive composite film by polyurethane supported 3D networks of boron nitride, *Compos. Sci. Technol.* 2017, 152, 94–100.
39. W. Liang, X. Ge, J. Ge, T. Li, T. Zhao, X. Chen, M. Zhang, J. Ji, X. Pang, R. Liu, Three-dimensional heterostructured reduced graphene oxide-hexagonal boron nitride-stacking material for silicone thermal grease with enhanced thermally conductive properties, *Nanomaterials* 2019, 9, 938.
40. H. Zhao, X. Song, H. Zeng, 3D white graphene foam scavengers: vesicant-assisted foaming boosts the gram-level yield and forms hierarchical pores for superstrong pollutant removal applications, *NPG Asia Mater.* 2015, 7, e168.
41. J. K. Han, G. L. Du, W. W. Gao, H. Bai, An anisotropically high thermal conductive boron nitride/epoxy composite based on nacre-mimetic 3D network, *Adv. Funct. Mater.* 2019, 29, 1900412.

42. N. Zhao, J. Li, W. Wang, W. Gao, H. Bai, Isotropically ultrahigh thermal conductive polymer composites by assembling anisotropic boron nitride nanosheets into a biaxially oriented network, *ACS Nano* 2022, 16, 18959–18967.
43. H. Y. Niu, H. C. Guo, L. Kang, L. C. Ren, R. C. Lv, S. L. Bai, Vertical alignment of anisotropic fillers assisted by expansion flow in polymer composites, *Nano-Micro Lett.* 2022, 14, 153.
44. L. Liu, Y. Li, X. L. Yu, J. H. Du, J. M. Zhang, J. P. Li, G. Y. Gan, Three-dimensional network of hexagonal boron nitride filled with polydimethylsiloxane with high thermal conductivity and good insulating properties for thermal management applications, *Polymer* 2022, 262, 125440.
45. X. Zhang, K. Wu, Y. H. Liu, B. W. Yu, Q. Zhang, F. Chen, Q. Fu, Preparation of highly thermally conductive but electrically insulating composites by constructing a segregated double network in polymer composites, *Compos. Sci. Technol.* 2019, 175, 135–142.
46. L. S. Xu, K. Zhan, S. Y. Ding, J. Y. Zhu, M. S. Liu, W. R. Fan, P. Duan, K. Luo, B. F. Ding, B. L. Liu, Y. L. Liu, H. M. Cheng, L. Qiu, A malleable composite dough with well-dispersed and high-content boron nitride nanosheets, *ACS Nano* 2023, 17, 4886–4895.
47. Q. S. Song, W. Zhu, Y. Deng, F. X. Hai, Y. L. Wang, Z. P. Guo, Enhanced through-plane thermal conductivity and high electrical insulation of flexible composite films with aligned boron nitride for thermal interface material, *Compos. Part A* 2019, 127, 105654.
48. Y. Chen, H. G. Zhang, J. Chen, Y. T. Guo, P. K. Jiang, F. Gao, H. Bao, X. Y. Huang, Thermally conductive but electrically insulating polybenzazole nanofiber/boron nitride nanosheets nanocomposite paper for heat dissipation of 5G base stations and transformers, *ACS Nano* 2022, 16, 14323–14333.
49. H. Zhao, L. X. Ran, Y. J. Xie, F. R. Sun, L. F. Yi, L. J. Zhao, J. R. Wu, Achieving highly anisotropic thermal and electrical conductivities via synergistic distribution of boron nitride and graphene nanosheets in multilayered composites, *J. Alloys Compound.* 2023, 935, 167971.
50. C. Yuan, B. Duan, L. Li, B. Xie, M. Huang, X. Luo, Thermal conductivity of polymerbased composites with magnetic aligned hexagonal boron nitride platelets, *ACS Appl. Mater. Interfaces* 2015, 7(23), 13000–13006.
51. C. Zhi, Y. Bando, C. Tan, D. Golberg, Effective precursor for high yield synthesis of pure BN nanotubes, *Solid State Commun.* 2005, 135(1–2), 67–70.
52. N. Sun, J. Sun, X. Zeng, P. Chen, J. Qian, R. Xia, R. Sun, Hot-pressing induced orientation of boron nitride in polycarbonate composites with enhanced thermal conductivity, *Compos. A: Appl. Sci. Manuf.* 2018, 110, 45–52.
53. J. T. Hu, Y. Huang, X. L. Zeng, Q. Li, L. L. Ren, R. Sun, J. B. Xu, C. P. Wong, Polymer composite with enhanced thermal conductivity and mechanical strength through orientation manipulating of BN, *Compos. Sci. Technol.* 2018, 160, 127–137.
54. Z. Kuang, Y. Chen, Y. Lu, L. Liu, S. Hu, S. Wen, Y. Mao, L. Zhang, Fabrication of highly oriented hexagonal boron nitride nanosheet/elastomer nanocomposites with high thermal conductivity, *Small* 2015, 11(14), 1655–1659.
55. T. Uchikoshi, T. S. Suzuki, Y. Sakka, Crystalline orientation of alumina ceramics prepared by electrophoretic deposition under a high magnetic field, *J. Mater. Sci.* 2006, 41(24), 8074–8078.
56. T. Uchikoshi, T. S. Suzuki, H. Okuyama, Y. Sakka, Fabrication of textured alumina by electrophoretic deposition in a strong magnetic field, *J. Mater. Sci.* 2004, 39(3), 861–865.
57. X. Li, B. Wu, Y. Li, M. M. Alam, P. Chen, R. Xia, C. T. Lin, J. S. Qian, Construction of oriented interconnected BNNS skeleton by self-growing CNTs leading high thermal conductivity, *Adv. Mater. Interfaces* 2021, 8, 2001910.
58. Y. Y. Wang, X. Huang, X. X. Zhang, Ultrarobust, tough and highly stretchable self-healing materials based on cartilage-inspired noncovalent assembly nanostructure. *Nat. Commun.* 2021, 12, 1291.
59. J. W. Li, Y. X. Zhang, T. Liang, X. Bai, Y. S. Pang, X. L. Zeng, Q. H. Hu, W. D. Tu, Z. Q. Ye, G. P. Du, R. Sun, X. L. Zeng, Thermal interface materials with both high through-plane thermal conductivity and excellent elastic compliance, *Chem. Mater.* 2021, 33, 8926–8937.
60. R. Zhang, X. Shi, L. Tang, Z. Liu, J. Zhang, Y. Guo, J. Gu, Thermally conductive and insulating epoxy composites by synchronously incorporating Si-sol functionalized glass fibers and BN fillers, *Chin. J. Polym. Sci.* 2020, 38, 730–739.

61. Z. Li, L. Wang, Y. Li, Y. Feng, W. Feng, Carbon-based functional nanomaterials: preparation, properties and applications, *Compos. Sci. Technol.* 2019, 179, 10–40.
62. A. Tessema, D. Zhao, J. Moll, S. Xu, R. Yang, C. Li, S. K. Kumar, A. Kidane, Effect of filler loading, geometry, dispersion and temperature on thermal conductivity of polymer nanocomposites, *Polym. Test.* 2017, 57, 101–106.
63. N. Mehra, L.W. Mu, T. Ji, X.T. Yang, J. Kong, J.W. Gu, J.H. Zhu, Thermal transport in polymeric materials and across composite interfaces, *Appl. Mater. Today* 2018, 12, 92–130.
64. Z. Han, A. Fina, Thermal conductivity of carbon nanotubes and their polymer nanocomposites: a review, *Prog. Polym. Sci.* 2011, 36(7), 914–944.
65. Q. Cai, D. Scullion, W. Gan, A. Falin, S. Zhang, K. Watanabe, T. Taniguchi, Y. Chen, E.J.G. Santos, L.H. Li, High thermal conductivity of high-quality monolayer boron nitride and its thermal expansion, *Sci. Adv.* 2019, 5(6), eaav0129.
66. S. Yang, W. Li, S. Bai, Q. Wang, Fabrication of morphologically controlled composites with high thermal conductivity and dielectric performance from aluminum nanoflake and recycled plastic package, *ACS Appl. Mater. Interfaces* 2019, 11(3), 3388–3399.
67. H. M. Fang, S. L. Bai, C. P. Wong, Microstructure engineering of graphene towards highly thermal conductive composites, *Compos. A: Appl. Sci. Manuf.* 2018, 112, 216–238.
68. C. Xiao, Y. Tang, L. Chen, X. Zhang, K. Zheng, X. Tian, Preparation of highly thermally conductive epoxy resin composites via hollow boron nitride microbeads with segregated structure, *Compos. A: Appl. Sci. Manuf.* 2019, 121, 330–340.
69. L. Qiu, H. Y. Zou, X. T. Wang, Y. H. Feng, X. X. Zhang, J. N. Zhao, X. H. Zhang, Q. W. Li, Enhancing the interfacial interaction of carbon nanotubes fibers by Au nanoparticles with improved performance of the electrical and thermal conductivity, *Carbon* 2019, 141, 497–505.
70. Y. Guo, K. Ruan, X. Shi, X. Yang, J. Gu, Factors affecting thermal conductivities of the polymers and polymer composites: a review. *Compos. Sci. Technol.* 2020, 193, 108134.
71. Y. X. Fu, Z. X. He, D. C. Mo, S. S. Lu, Thermal conductivity enhancement with different fillers for epoxy resin adhesives, *Appl. Therm. Eng.* 2014, 66(1), 493–498.
72. J. W. Gu, N. Li, L. D. Tian, Z. Y. Lv, Q. Y. Zhang, High thermal conductivity graphite nanoplatelet/UHMWPE nanocomposites, *RSC Adv.* 2015, 5(46), 36334–36339.
73. G. W. Lee, M. Park, J. Kim, J. I. Lee, H. G. Yoon, Enhanced thermal conductivity of polymer composites filled with hybrid filler, *Compos. A: Appl. Sci. Manuf.* 2006, 37(5), 727–734.
74. H. Guo, Q. Wang, J. Liu, C. Du, B. Li, Improved interfacial properties for largely enhanced thermal conductivity of poly(vinylidene fluoride)-based nanocomposites via functionalized multi-wall carbon nanotubes, *Appl. Surf. Sci.* 2019, 487, 379–388.
75. J. You, H. H. Choi, Y. M. Lee, J. Cho, M. Park, S. S. Lee, J. H. Park, Plasma-assisted mechanochemistry to produce polyamide/boron nitride nanocomposites with high thermal conductivities and mechanical properties, *Compos. B Eng.* 2019, 164, 710–719.
76. Y. Guo, K. Ruan, X. Yang, T. Ma, J. Kong, N. Wu, J. Zhang, J. Gu, Z. Guo, Constructing fully carbon-based fillers with a hierarchical structure to fabricate highly thermally conductive polyimide nanocomposites, *J. Mater. Chem. C* 2019, 7(23), 7035–7044.
77. C. B. Liang, H. Qiu, Y. Y. Han, H. B. Gu, P. Song, L. Wang, J. Kong, D. P. Cao, J. W. Gu, Superior electromagnetic interference shielding 3D graphene nanoplatelets/reduced graphene oxide foam/epoxy nanocomposites with high thermal conductivity, *J. Mater. Chem. C* 2019, 7(9), 2725–2733.
78. J. Chen, X. Huang, B. Sun, P. Jiang, Highly thermally conductive yet electrically insulating polymer/boron nitride nanosheets nanocomposite films for improved thermal management capability, *ACS Nano* 2019, 13(1), 337–345.
79. C. Yan, T. Yu, C. Ji, D. J. Kang, N. Wang, R. Sun, C. P. Wong, Tailoring highly thermal conductive properties of Te/MoS_2/Ag heterostructure nanocomposites using a bottom-up approach, *Adv. Electron. Mater.* 2019, 5(1), 1800548.
80. Z. Lule, J. Kim, Surface modification of aluminum nitride to fabricate thermally conductive poly(butylene succinate) nanocomposite, *Polymers* 2019, 11(1), 148.

81. C. Shen, H. Wang, T. Zhang, Y. Zeng, Silica coating onto graphene for improving thermal conductivity and electrical insulation of graphene/polydimethylsiloxane nanocomposites, *J. Mater. Sci. Technol.* 2019, 35(1), 36–43.
82. J. Tong, H. X. Huang, M. Wu, Simultaneously facilitating dispersion and thermal reduction of graphene oxide to enhance thermal conductivity of poly(vinylidene fluoride)/graphene nanocomposites by water in continuous extrusion, *Chem. Eng. J.* 2018, 348, 693–703.
83. R. Wang, L. Wu, D. Zhuo, J. Zhang, Y. Zheng, Fabrication of polyamide 6 nanocomposite with improved thermal conductivity and mechanical properties via incorporation of low graphene content, *Ind. Eng. Chem. Res.* 2018, 57(32), 10967–10976.
84. C. Li, B. Liu, Z. Gao, H. Wang, M. Liu, S. Wang, C. Xiong, Electrically insulating ZnOs/ZnOw/silicone rubber nanocomposites with enhanced thermal conductivity and mechanical properties, *J. Appl. Polym. Sci.* 2018, 135(27), 46454.
85. H. Lin, L. X. Pei, L. Z. Zhang, Enhanced thermal conductivity of PLA-based nanocomposites by incorporation of graphite nanoplatelets functionalized by tannic acid, *J. Appl. Polym. Sci.* 2018, 135(26), 46397.
86. X. H. Zhang, C. Tan, Y. H. Ma, F. Wang, W. T. Yang, $BaTiO_3$@carbon/silicon carbide/poly(vinylidene fluoride-hexafluoropropylene) three-component nanocomposites with high dielectric constant and high thermal conductivity, *Compos. Sci. Technol.* 2018, 162, 180–187.
87. L. M. Guiney, N. D. Mansukhani, A. E. Jakus, S. G. Wallace, R. N. Shah, M. C. Hersam, Three-dimensional printing of cytocompatible, thermally conductive hexagonal boron nitride nanocomposites, *Nano Lett.* 2018, 18(6), 3488–3493.
88. Y. Q. Guo, G. J. Xu, X. T. Yang, K. P. Ruan, T. B. Ma, Q. Y. Zhang, J. W. Gu, Y. L. Wu, H. Liu, Z. Guo, Significantly enhanced and precisely modeled thermal conductivity in polyimide nanocomposites with chemically modified graphene via in situ polymerization and electrospinning-hot press technology, *J. Mater. Chem. C* 2018, 6(12), 3004–3015.
89. Y. Shi, W. Ma, L. Wu, D. Hu, J. Mo, B. Yang, S. Zhang, Z. Zhang, Magnetically aligning multilayer graphene to enhance thermal conductivity of silicone rubber composites, *J. Appl. Polym. Sci.* 2019, 136(37), 47951.
90. C. Xiao, L. Chen, Y. Tang, X. Zhang, K. Zheng, X. Tian, Three dimensional porous alumina network for polymer composites with enhanced thermal conductivity, *Compos. A: Appl. Sci. Manuf.* 2019, 124, 105511.
91. D. Yang, X. Kong, Y. Ni, D. Gao, B. Yang, Y. Zhu, L. Zhang, Novel nitrile-butadiene rubber composites with enhanced thermal conductivity and high dielectric constant, *Compos. A: Appl. Sci. Manuf.* 2019, 124, 105447.
92. Y. Yuan, Z. Li, L. Cao, B. Tang, S. Zhang, Modification of Si3N4 ceramic powders and fabrication of Si3N4/PTFE composite substrate with high thermal conductivity, *Ceram. Int.* 2019, 45(13), 16569–16576.
93. X. W. Wang, P. Y. Wu, 3D vertically aligned BNNS network with long-range continuous channels for achieving a highly thermally conductive composite, *ACS Appl. Mater. Interfaces* 2019, 11(32), 28943–28952.
94. J. Huang, W. Yang, J. Zhu, L. Fu, D. Li, L. Zhou, Silver nanoparticles decorated 3D reduced graphene oxides as hybrid filler for enhancing thermal conductivity of polystyrene composites, *Compos. A: Appl. Sci. Manuf.* 2019, 123, 79–85.
95. C. L. Li, H. L. Guo, X. Tian, X. G. Tian, Transient response for a half-space with variable thermal conductivity and diffusivity under thermal and chemical shock, *J. Therm. Stresses* 2017, 40(3), 389–401.
96. H. Guo, J. Liu, Q. Wang, M. Liu, C. Du, B. Li, L. Feng, High thermal conductive poly(vinylidene fluoride)-based composites with well-dispersed carbon nanotubes/graphene three-dimensional network structure via reduced interfacial thermal resistance, *Compos. Sci. Technol.* 2019, 181, 107713.
97. G. H. Kim, D. Lee, A. Shanker, L. Shao, M. S. Kwon, D. Gidley, J. Kim, K. P. Pipe, High thermal conductivity in amorphous polymer blends by engineered interchain interactions, *Nat. Mater.* 2014, 14, 295.
98. Z. Lule, J. Kim, Thermally conductive and highly rigid polylactic acid (PLA) hybrid composite filled with surface treated alumina/nano-sized aluminum nitride, *Compos. A: Appl. Sci. Manuf.* 2019, 124, 105506.

99. A. Oluwalowo, N. Nguyen, S. L. Zhang, J. G. Park, R. Liang, Electrical and thermal conductivity improvement of carbon nanotube and silver composites, *Carbon* 2019, 146, 224–231.
100. R. Aradhana, S. Mohanty, S. K. Nayak, Novel electrically conductive epoxy/reduced graphite oxide/silica hollow microspheres adhesives with enhanced lap shear strength and thermal conductivity, *Compos. Sci. Technol.* 2019, 169, 86–94.
101. Y. Su, J. J. Li, G. J. Weng, Theory of thermal conductivity of graphene-polymer nanocomposites with interfacial Kapitza resistance and graphene-graphene contact resistance, *Carbon* 2018, 137, 222–233.
102. H. X. Chen, Z. B. Pan, W. L. Wang, Y. Y. Chen, S. Xing, Y. Cheng, X. P. Ding, J. J. Liu, J. W. Zhai, J. H. Yu, Ultrahigh discharge efficiency and improved energy density in polymer-based nanocomposite for high-temperature capacitors application. *Compos. A: Appl. Sci. Manuf.* 2021, 142,106266.
103. I. L. Ngo, S. V. P. Vattikuti, C. Byon, A modified Hashin-Shtrikman model for predicting the thermal conductivity of polymer composites reinforced with randomly distributed hybrid fillers, *Int. J. Heat Mass Transf.* 2017, 114, 727–734.
104. D. L. Zhang, J. W. Zha, W. K. Li, C. Q. Li, S. J. Wang, Y. Wen, Z. M. Dang, Enhanced thermal conductivity and mechanical property through boron nitride hot string in polyvinylidene fluoride fibers by electrospinning, *Compos. Sci. Technol.* 2018, 156, 1–7.
105. Y. Sohn, T. Han, J. H. Han, Effects of shape and alignment of reinforcing graphite phases on the thermal conductivity and the coefficient of thermal expansion of graphite/copper composites, *Carbon* 2019, 149, 152–164.
106. S. Moradi, Y. Calventus, F. Roman, J. M. Hutchinson, Achieving high thermal conductivity in epoxy composites: effect of boron nitride particle size and matrixfiller interface, *Polymers* 2019, 11(7), 1156.
107. W. Yu, D. M. France, J. L. Routbort, S. U. S. Choi, Review and comparison of nanofluid thermal conductivity and heat transfer enhancements, *Heat Transf. Eng.* 2008, 29(5), 432–460.
108. W. Zhou, J. Zuo, W. Ren, Thermal conductivity and dielectric properties of Al/PVDF composites, *Compos. A: Appl. Sci. Manuf.* 2012, 43(4), 658–664.
109. B. L. Zhu, J. Ma, J. Wu, K. C. Yung, C. S. Xie, Study on the properties of the epoxymatrix composites filled with thermally conductive AlN and BN ceramic particles, *J. Appl. Polym. Sci.* 2010, 118(5), 2754–2764.
110. H. Y. Ng, X. Lu, S. K. Lau, Thermal conductivity of boron nitride-filled thermoplastics: effect of filler characteristics and composite processing conditions, *Polym. Compos.* 2005, 26(6), 778–790.
111. J. G. Park, Q. Cheng, J. Lu, J. Bao, S. Li, Y. Tian, Z. Liang, C. Zhang, B. Wang, Thermal conductivity of MWCNT/epoxy composites: the effects of length, alignment and functionalization, *Carbon* 2012, 50(6), 2083–2090.
112. A. Yu, P. Ramesh, X. Sun, E. Bekyarova, M. E. Itkis, R. C. Haddon, Enhanced thermal conductivity in a hybrid graphite nanoplatelet - carbon nanotube filler for epoxy composites, *Adv. Mater.* 2008, 20(24), 4740–4744.
113. A. Rai, A. L. Moore, Enhanced thermal conduction and influence of interfacial resistance within flexible high aspect ratio copper nanowire/polymer composites, *Compos. Sci. Technol.* 2017, 144, 70–78.
114. L. Riviere, A. Lonjon, E. Dantras, C. Lacabanne, P. Olivier, N. R. Gleizes, Silver fillers aspect ratio influence on electrical and thermal conductivity in PEEK/Ag nanocomposites, *Eur. Polym. J.* 2016, 85, 115–125.
115. B. Mortazavi, M. Baniassadi, J. Bardon, S. Ahzi, Modeling of two-phase random composite materials by finite element, Mori–Tanaka and strong contrast methods, *Compos. B Eng.* 2013, 45(1), 1117–1125.
116. A. Yu, P. Ramesh, M. E. Itkis, E. Bekyarova, R. C. Haddon, Graphite nanoplateletepoxy composite thermal interface materials, *J. Phys. Chem. C* 2007, 111(21), 7565–7569.
117. X. Shen, Z. Wang, Y. Wu, X. Liu, Y. B. He, J. K. Kim, Multilayer graphene enables higher efficiency in improving thermal conductivities of graphene/epoxy composites, *Nano Lett.* 2016, 16(6), 3585–3593.
118. V. Varshney, J. Lee, J. S. Brown, B. L. Farmer, A. A. Voevodin, A. K. Roy, Effect of length, diameter, chirality, deformation, and strain on contact thermal conductance between single-wall carbon nanotubes, *Front. Mater.* 2018, 5(17), UNSP 17.
119. J. W. Gu, Q. Y. Zhang, J. Dang, C. Xie, Thermal conductivity epoxy resin composites filled with boron nitride, *Polym. Adv. Technol.* 2012, 23(6), 1025–1028.

120. J. W. Gu, J. J. Du, J. Dang, W. C. Geng, S. H. Hu, Q. Y. Zhang, Thermal conductivities, mechanical and thermal properties of graphite nanoplatelets/polyphenylene sulfide composites, *RSC Adv.* 2014, 4(42), 22101–22105.
121. J. W. Gu, Z. Y. Lv, X. T. Yang, G. E. Wang, Q. Y. Zhang, Fabrication and properties of thermally conductive epoxy resin nanocomposites filled with f GNPs/PNBRs hybrid fillers, *Sci. Adv. Mater.* 2016, 8(5), 972–979.
122. S. C. Chen, Y. Y. Feng, M. M. Qin, T. X. Ji, W. Feng, Improving thermal conductivity in the through-thickness direction of carbon fibre/SiC composites by growing vertically aligned carbon nanotubes, *Carbon* 2017, 116, 84–93.
123. J. Gu, Q. Zhang, J. Dang, C. Yin, S. Chen, Preparation and properties of polystyrene/SiCw/SiCp thermal conductivity composites, *J. Appl. Polym. Sci.* 2012, 124(1), 132–137.
124. A. A. A. Arani, F. Pourmoghadam, Experimental investigation of thermal conductivity behavior of MWCNTS-Al_2O_3/ethylene glycol hybrid nanofluid: providing new thermal conductivity correlation, *Heat Mass Trans.* 2019, 55(8), 2329–2339.
125. L. Ren, X. Zeng, X. Zhang, R. Sun, X. Tian, Y. Zeng, J. B. Xu, C. P. Wong, Silver nanoparticle-modified alumina microsphere hybrid composites for enhanced energy density and thermal conductivity, *Compos. A: Appl. Sci. Manuf.* 2019, 119, 299–309.
126. Z. F. Liu, Z. H. Chen, F. Yu, Enhanced thermal conductivity of microencapsulated phase change materials based on graphene oxide and carbon nanotube hybrid filler, *Sol. Energy Mater. Sol. Cells* 2019, 192, 72–80.
127. F. Lv, M. Qin, F. Zhang, H. Yu, L. Gao, P. Lv, W. Wei, Y. Feng, W. Feng, High crossplane thermally conductive hierarchical composite using graphene-coated vertically aligned carbon nanotubes/graphite, *Carbon* 2019, 149, 281–289.
128. M. Owais, J. Zhao, A. Imani, G. R. Wang, H. Zhang, Z. Zhang, Synergetic effect of hybrid fillers of boron nitride, graphene nanoplatelets, and short carbon fibers for enhanced thermal conductivity and electrical resistivity of epoxy nanocomposites, *Compos. A: Appl. Sci. Manuf.* 2019, 117, 11–22.
129. F. Yan, L. Liu, M. Li, M. J. Zhang, L. Shang, L. H. Xiao, Y. H. Ao, One-step electrodeposition of Cu/CNT/CF multiscale reinforcement with substantially improved thermal/electrical conductivity and interfacial for properties of epoxy composites, *Compos. A: Appl. Sci. Manuf.* 2019, 125, 105530.
130. X. R. Zheng, S. Kim, C. W. Park, Enhancement of thermal conductivity of carbon fiber-reinforced polymer composite with copper and boron nitride particles, *Compos. A: Appl. Sci. Manuf.* 2019, 121, 449–456.
131. J. Gu, Y. Guo, X. Yang, C. Liang, W. Geng, L. Tang, N. Li, Q. Zhang, Synergistic improvement of thermal conductivities of polyphenylene sulfide composites filled with boron nitride hybrid fillers, *Compos. A: Appl. Sci. Manuf.* 2017, 95, 267–273.
132. Y. Wei, Y. P. Shi, Z. Y. Jiang, X. F. Zhang, H. H. Chen, Y. H. Zhang, J. W. Zhang, C. H. Gong, High performance and lightweight electromagnetic wave absorbers based on TiN/RGO flakes, *J. Alloys Compd.* 2019, 810, 151950.
133. J. M. Wernik, S. A. Meguid, Recent developments in multifunctional nanocomposites using carbon nanotubes, *Appl. Mech. Rev.* 2011, 63(5), 050801.
134. M. Tanimoto, T. Yamagata, K. Miyata, S. Ando, Anisotropic thermal diffusivity of hexagonal boron nitride-filled polyimide films: effects of filler particle size, aggregation, orientation, and polymer chain rigidity, *ACS Appl. Mater. Interfaces* 2013, 5(10), 4374–4382.
135. W. Evans, R. Prasher, J. Fish, P. Meakin, P. Phelan, P. Keblinski, Effect of aggregation and interfacial thermal resistance on thermal conductivity of nanocomposites and colloidal nanofluids, *Int. J. Heat Mass Transf.* 2008, 51(5), 1431–1438.
136. Y. Feng, B. Yu, P. Xu, M. Zou, The effective thermal conductivity of nanofluids based on the nanolayer and the aggregation of nanoparticles, *J. Phys. D Appl. Phys.* 2007, 40(10), 3164–3171.
137. J. Wensel, B. Wright, D. Thomas, W. Douglas, B. Mannhalter, W. Cross, H. P. Hong, J. Kellar, P. Smith, W. Roy, Enhanced thermal conductivity by aggregation in heat transfer nanofluids containing metal oxide nanoparticles and carbon nanotubes, *Appl. Phys. Lett.* 2008, 92(2), 023110.
138. X. Wei, T. Zhang, T. Luo, Thermal energy transport across hard-soft interfaces, *ACS Energy Lett.* 2017, 2, 2283–2292.

139. X. Zhang, G. Yang, B. Cao, Bonding-enhanced interfacial thermal transport: mechanisms, materials, and applications, *Adv. Mater. Interfaces* 2022, 9, 2200078.
140. D. Ma, Y. Xing, L. Zhang, Reducing interfacial thermal resistance by interlayer, *J. Phys. Condens. Matter* 2023, 35, 053001.
141. W. R. Caseri, Nanocomposites of polymers and inorganic particles: preparation, structure and properties, *Mater. Sci. Technol.* 2006, 22, 807–817.
142. D. Chen, S. Torquato, Designing disordered hyperuniform two-phase materials with novel physical properties, *Acta Mater.* 2018, 142, 152–161.
143. K. Saito, K. Sato, Y. Tominaga, Y. Imai, Computational prediction of microstructures in α-alumina/PMMA composites and its experimental verification, *Polym. Compos.* 2022, 43, 339–346.
144. Y. Han, K. Li, Z. Li, J. Liu, S. Hu, S. Wen, L. Liu, L. Zhang, Molecular dynamics simulation of the formation mechanism of the thermal conductive filler network of polymer nanocomposites, *Phys. Chem. Chem. Phys.* 2022, 24, 4334–4347.
145. R. Ma, X. Wan, T. Zhang, N. Yang, T. Luo, Role of molecular polarity in thermal transport of boron nitride-organic molecule composites, *ACS Omega* 2018, 3, 12530–12534.
146. X. Liu, Z. Rao, Interfacial thermal conductance across hexagonal boron nitride & paraffin based thermal energy storage materials, *J. Energy Storage* 2020, 32, 101860.
147. X. Yang, X. Wang, W. Wang, Y. Fu, Q. Xie, Atomic-scale insights into interface thermal resistance between epoxy and boron nitride in nanocomposites, *Int. J. Heat Mass Transf.* 2020, 159, 120105.
148. S. Wu, Q. Chen, D. Chen, D. Peng, Y. Ma, Multiscale study of thermal conductivity of boron nitride nanosheets/paraffin thermal energy storage materials, *J. Energy Storage* 2021, 41, 102931.
149. M. Li, J. Liu, W. Yu, Y. Zhang, H. Zhou, Atomistic molecular dynamic simulations of the thermal transport across h-BN/cellulose nanocrystal interface, *Int. J. Heat Mass Transf.* 2021, 171, 121043.
150. C. Chen, X. J. Li, Y. F. Wen, J. W. Liu, X. W. Li, H. X. Zeng, Z. G. Xue, X. P. Zhou, X. L. Xie, Noncovalent engineering of carbon nanotube surface by imidazolium ionic liquids: a promising strategy for enhancing thermal conductivity of epoxy composites, *Compos. A: Appl. Sci. Manuf.* 2019, 125, 105517.
151. T. Luo, J.R. Lloyd, Enhancement of thermal energy transport across graphene/graphite and polymer interfaces: a molecular dynamics study, *Adv. Funct. Mater.* 2012, 22, 2495–2502.
152. T. Zhang, A. R. Gans-Forrest, E. Lee, X. Zhang, C. Qu, Y. Pang, F. Sun, T. Luo, Role of hydrogen bonds in thermal transport across hard/soft material interfaces, *ACS Appl. Mater. Interfaces* 2016, 8, 33326–33334.
153. S. Song, Y. Zhang, Carbon nanotube/reduced graphene oxide hybrid for simultaneously enhancing the thermal conductivity and mechanical properties of styrene -butadiene rubber, *Carbon* 2017, 123, 158–167.
154. ASTM. ASTM D5470-12, Standard test method for thermal transmission properties of thermally conductive electrical insulation materials. ASTM International, West Conshohocken, PA, 2012. https://doi.org/10.1520/D5470-12.
155. R. Kempers, P. Kolodner, A. Lyons, A. J. Robinson, A high-precision apparatus for the characterization of thermal interface materials, *Rev. Sci. Instrum.* 2009, 80(9), 95111.
156. B. R. Burg, M. Kolly, N. Blasakis, D. Gschwend, J. Zürcher, T. Brunschwiler, Steady-state low thermal resistance characterization apparatus: the bulk thermal tester, *Rev. Sci. Instrum.* 2015, 86(12), 124903.
157. B. Sponagle, D. Groulx, Measurement of thermal interface conductance at variable clamping pressures using a steady state method, *Appl. Therm. Eng.* 2016, 96, 671–681.
158. P. Zhang, Y. Xuan, Q. Li, A high-precision instrumentation of measuring thermal contact resistance using reversible heat flux, *Exp. Therm. Fluid. Sci.* 2014, 54, 204–211.
159. A. N. Smith, N. R. Jankowski, L. M. Boteler, Measurement of high-performance thermal interfaces using a reduced scale steady-state tester and infrared microscopy, *J Heat Transfer* 2016, 138(4), 41301.
160. A. M. Marconnet, N. Yamamoto, M. A. Panzer, B. L. Wardle, K. E. Goodson, Thermal conduction in aligned carbon nanotube–polymer nanocomposites with high packing density, *ACS Nano* 2011, 5(6), 4818–4825.
161. J. Due, A. J. Robinson, Reliability of thermal interface materials: a review, *Appl. Therm. Eng.* 2013, 50(1), 455–463.

162. Y. T. Lee, S. Shanmugan, D. Mutharasu, Thermal resistance of CNTs-based thermal interface material for high power solid state device packages, *Appl. Phys. A* 2014, 114(4), 1145–1152.
163. S. H. Taylor, S. V. Garimella, Capacitive sensing of local bond layer thickness and coverage in thermal interface materials. *Int. J. Heat Mass Transf.* 2016, 97, 26–31.
164. N. R. Na, K. Hasegawa, X. S. Zhou, M. Nihei, S. Noda, Denser and taller carbon nanotube arrays on Cu foils useable as thermal interface materials, *Jpn. J. Appl. Phys.* 2015, 54(9), 95102.
165. W. J. Parker, R. J. Jenkins, C. P. Butler, G. L. Abbott, Flash method of determining thermal diffusivity, heat capacity, and thermal conductivity, *J. Appl. Phys.* 1961, 32(9), 1679–1684.
166. V. Khuu, M. Osterman, A. Bar-Cohen, M. Pecht, Effects of temperature cycling and elevated temperature/humidity on the thermal performance of thermal interface materials, *IEEE Trans. Device Mater. Reliab.* 2009, 9(3), 379–391.
167. D. G. Cahill, W. K. Ford, K. E. Goodson, G. D. Mahan, A. Majumdar, H. J. Maris, R. Merlin, S. R. Phillpot, Nanoscale thermal transport, *J. Appl. Phys.* 2003, 93, 793.
168. D. G. Cahill, K. Goodson, A. Majumdar, Thermometry and thermal transport in micro/nanoscale solid-state devices and structures, *ASME J. Heat Transf.* 2002, 124, 223.
169. G. Chen, D. Borca-Tascuic, R. Yang, Nanoscale heat transfer, in *Encyclopedia of Nanoscience and Nanotechnology*, edited by H. S. Nalwa, American Scientific Publishers, Valencia, CA, 2004, Vol. 7, 429–459.
170. D. G. Cahill, R. O. Pohl, Thermal conductivity of amorphous solids above the plateau, *Phys. Rev. B* 1987, 35, 4067.
171. D. G. Cahill, Thermal conductivity measurement from 30 to 750 K: the 3ω method, *Rev. Sci. Instrum.* 1990, 61, 802.
172. T. Borca-Tasciuc, A. R. Kumar, G. Chen, Data reduction in 3ω method for thin-film thermal conductivity determination, *Rev. Sci. Instrum.* 2001, 72, 2139.
173. Y. K. Koh, S. L. Singer, W. Kim, J. O. Zide, H. Lu, D. G. Cahill, A. Majumdar, A. C. Gossard, Comparison of the 3ω method and time-domain thermoreflectance for measurements of the cross-plane thermal conductivity of epitaxial semiconductors, *J. Appl. Phys.* 2009, 105, 054303.
174. R. J. Stoner, H. J. Maris, Kapitza conductance and heat flow between solids at temperatures from 50 to 300 K, *Phys. Rev. B* 1993, 48, 16373.
175. J. Zhu, D. W. Tang, W. Wang, J. Liu, K. W. Holub, R. G. Yang, Ultrafast thermoreflectance techniques for measuring thermal conductivity and interface thermal conductance of thin films, *J. Appl. Phys.* 2010, 108, 094315.
176. S. M. Lee, D. G. Cahill, Heat transport in thin dielectric films, *J. Appl. Phys.* 1997, 81, 2590.
177. C. E. Raudzis, F. Schatz, D. Wharam, Extending the 3ω method for thin-film analysis to high frequencies, *J. Appl. Phys.* 2003, 93, 6050.
178. C. A. Paddock, G. L. Eesley, Transient thermoreflectance from thin metal films, *J. Appl. Phys.* 1986, 60, 285.
179. W. S. Capinski, H. J. Maris, Improved apparatus for picosecond pump-and-probe optical measurements, *Rev. Sci. Instrum.* 1996, 67, 2720.
180. R. M. Costescu, M. A. Wall, D. G. Cahill, Thermal conductance of epitaxial interfaces, *Phys. Rev. B* 2003, 67, 054302.
181. A. J. Schmidt, M. Chiesa, X. Chen, G. Chen, An optical pump-probe technique for measuring the thermal conductivity of liquids, *Rev. Sci. Instrum.* 2008, 79, 064902.
182. C. Chiritescu, D. G. Cahill, N. Nguyen, D. Johnson, A. Bodapati, P. Keblinski, P. Zschack, Ultralow thermal conductivity in disordered, layered WSe_2 crystals, *Science* 2007, 315, 351.
183. J. C. Zhao, X. Zheng, D. G. Cahill, High-throughput diffusion multiples, *Mater. Today* 2005, 8, 28.
184. R. J. Stevens, A. N. Smith, P. M. Norris, Measurement of thermal boundary conductance of a series of metal-dielectric interfaces by the transient thermoreflectance technique, *J. Heat Transf.* 2005, 127, 315.
185. P. E. Hopkins, R. N. Salaway, R. J. Stevens, P. M. Norris, Temperature-dependent thermal boundary conductance at Al/Al_2O_3 and Pt/Al_2O_3 interfaces, *Int. J. Thermophys.* 2007, 28, 947.
186. P. M. Norris, P. E. Hopkins, Examining interfacial diffuse phonon scattering through transient thermoreflectance measurements of thermal boundary conductance, *J. Heat Transf.* 2009, 131, 043207.

187. D. G. Cahill, A. Bullen, S. M. Lee, Interface thermal conductance and the thermal conductivity of multilayer thin films, *High Temp. - High Press.* 2000, 32, 135.
188. A. J. Schmidt, R. Cheaito, M. Chiesa, A frequency-domain thermoreflectance method for the characterization of thermal properties, *Rev. Sci. Instrum.* 2009, 80, 094901.
189. A. J. Schmidt, R. Cheaito, M. Chiesa, Characterization of thin metal films via frequency-domain thermoreflectance, *J. Appl. Phys.* 2010, 107, 024908.
190. T. Tong, Y. Zhao, L. Delzeit, A. Kashani, M. Meyyappan, A. Majumdar, Dense vertically aligned multiwalled carbon nanotube arrays as thermal interface materials, *IEEE Trans. Compon. Packag. Technol.* 2007, 30, 92.
191. Y. Y. Lin, C. Y. Hui, H. D. Conway, A detailed elastic analysis of the flat punch (Tack) test for pressure-sensitive adhesives, *J. Polym. Sci. B* 2000, 38(21), 2769–2784.
192. G. L. Cross, B. S. O'Connell, J. B. Pethica, Influence of elastic strains on the mask ratio in glassy polymer nanoimprint, *Appl. Phys. Lett.* 2005, 86(8), 81902.
193. G. L. Cross, The production of nanostructures by mechanical forming, *J. Phys. D: Appl. Phys.* 2006, 39 (20), R363.
194. R. Prasher, Thermal interface materials: historical perspective, status, and future directions, *Proc. IEEE.* 2006, 94(8), 1571–1586.
195. X. B. Luo, R. Hu, S. Liu, K. Wang, Heat and fluid flow in highpower LED packaging and applications, *Prog. Energ. Combust.* 2016, 56, 1–32.
196. R. S. Prasher, J. Shipley, S. Prstic, P. Koning, J. L. Wang, Thermal resistance of particle laden polymeric thermal interface materials, *J. Heat Transf.* 2003, 125(6), 1170–1177.
197. R. S. Prasher, Surface chemistry and characteristics based model for the thermal contact resistance of fluidic interstitial thermal interface materials, *J. Heat Transf.* 2001, 123(5), 969–975.
198. A. Hamasaiid, M. S. Dargusch, T. Loulou, G. Dour, A predictive model for the thermal contact resistance at liquid–solid interfaces: analytical developments and validation, *Int. J. Therm. Sci.* 2011, 50(8), 1445–1459.
199. A. Hamasaiid, G. Dour, T. Loulou, M. S. Dargusch, A predictive model for the evolution of the thermal conductance at the casting–die interfaces in high pressure die casting, *Int. J. Therm. Sci.* 2010, 49(2), 365–372.
200. C. Yuan, B. Duan, L. Li, B. F. Shang, X. B. Luo, An improved model for predicting thermal contact resistance at liquid–solid interface, *Int. J. Heat Mass. Transfer* 2015, 80, 398–406.
201. B. Smith, W. Glatz, B. Michel, *Electronics Cooling Magazine*, Plymouth, Pennsylvania: ITEM Publications, 2009.
202. T. Brunschwiler, U. Kloter, R. J. Linderman, H. Rothuizen, B. Michel, Hierarchically nested channels for fast squeezing interfaces with reduced thermal resistance, *IEEE Trans. Comp. Packag. Technol.* 2007, 30(2), 226–234.
203 N. Bajaj, G. Subbarayan, S. V. Garimella, Topological design of channels for squeeze flow optimization of thermal interface materials, *Int. J. Heat Mass Transf.* 2012, 55(13), 3560–3575.
204. R. J. Linderman, T. Brunschwiler, U. Kloter, H. Toy, B. Michel, Hierarchical nested surface channels for reduced particle stacking and low-resistance thermal interfaces, *Twenty-Third Annual IEEE Semiconductor Thermal Measurement and Management Symposium*, San Jose, CA: IEEE, 2007, 87–94.
205. M. R. S. Shirazy, S. Allard, M. Beaumier, L. G. Fréchette, Effect of squeezing conditions on the particle distribution and bond line thickness of particle filled polymeric thermal interface materials. *Thermal and Thermomechanical Phenomena in Electronic Systems (ITherm), 2014 IEEE Intersociety Conference*, Orlando, FL: IEEE, 2014, 251–259.
206. R. F. Dou, T. R. Ge, X. L. Liu, Z. Wen, Effects of contact pressure, interface temperature, and surface roughness on thermal contact conductance between stainless steel surfaces under atmosphere condition, *Int. J. Heat Mass Transf.* 2016, 94,156–163.
207. G. T. Hohensee, R. Wilson, D. G. Cahill, Thermal conductance of metal–diamond interfaces at high pressure, *Nat. Commun.* 2015, 6, 6578.
208. G. Xu, *Thermal Management for Electronic Packaging*, San Diego, CA: University of California, 2006.
209. J. A. Greenwood, J. B. P. Williamson, Contact of nominally flat surfaces, *Proc. R. Soc. A: Math. Phys. Eng. Sci.* 1966, 295(1442), 300–319.

210. J. F. Fan, Z. Q. Ye; X. L. Zeng, J. B. Xu, L. L. Ren, R. Sun, C. P. Wong, Predicting bond line thickness of polymeric thermal interface materials based on the rheological properties. *J. Appl. Phys.* 2024, 136, 025301.
211. B. H. Oh, E. K. Lee, H. Y. Loo, P. T. Oh, The evolution of CPU packaging technology and future challenges. *2006 International Conference on Electronic Materials and Packaging*, Kowloon, China, 2006, 11–14, 1–6.
212. B. Zhang, Z. L. Dou, Y. Z. Zhang, Q. Fu, K. Wu, Exploring trade-offs in thermal interface materials: the impact of polymer-filler interfaces on thermal conductivity and thixotropy, *Chin. J. Polym. Sci.* 2024, 42, 916–925.
213. W. Zeng, L. Shu, Q. Li, S. Chen, F. Wang, X. M. Tao, Fiberbased wearable electronics: a review of materials, fabrication, devices, and applications, *Adv. Mater.* 2014, 26(31), 5310–5336.
214. B. Wang, S. Ma, S. Yan, J. Zhu, Readily recyclable carbon fiber reinforced composites based on degradable thermosets: a review, *Green Chem.* 2019, 21(21), 5781–5796.
215. M. Khatib, O. Zohar, H. Haick, Self-healing soft sensors: from material design to implementation, *Adv. Mater.* 2021, 33(11), 2004190.
216. Q. Li, G. Chen, Y. Cui, S. Ji, Z. Liu, C. Wan, Y. Liu, Y. Lu, C. Wang, N. Zhang, Y. Cheng, K. Q. Zhang, X. Chen, Highly thermal-wet comfortable and conformal silk-based electrodes for onskin sensors with sweat tolerance, *ACS Nano* 2021, 15(6), 9955–9966.
217. Y. Luo, M. R. Abidian, J. H. Ahn, D. Akinwande, A. M. Andrews, M. Antonietti, Z. Bao, M. Berggren, C. A. Berkey, C. J. Bettinger, J. Chen, P. Chen, W. Cheng, X. Cheng, S. J. Choi, A. Chortos, C. Dagdeviren, R. H. Dauskardt, C. A. Di, M. D. Dickey, X. Duan, A. Facchetti, Z. Fan, Y. Fang, J. Feng, X. Feng, H. Gao, W. Gao, X. Gong, C. F. Guo, X. Guo, M. C. Hartel, Z. He, J. S. Ho, Y. Hu, Q. Huang, Y. Huang, F. Huo, M. M. Hussain, A. Javey, U. Jeong, C. Jiang, X. Jiang, J. Kang, D. Karnaushenko, A. Khademhosseini, D. H. Kim, I. D. Kim, D. Kireev, L. Kong, C. Lee, N. E. Lee, P. S. Lee, T. W. Lee, F. Li, J. Li, C. Liang, C. T. Lim, Y. Lin, D. J. Lipomi, J. Liu, K. Liu, N. Liu, R. Liu, Y. Liu, Y. Liu, Z. Liu, Z. Liu, X. J. Loh, N. Lu, Z. Lv, S. Magdassi, G. G. Malliaras, N. Matsuhisa, A. Nathan, S. Niu, J. Pan, C. Pang, Q. Pei, H. Peng, D. Qi, H. Ren, J. A. Rogers, A. Rowe, O. G. Schmidt, T. Sekitani, D. G. Seo, G. Shen, X. Sheng, Q. Shi, T. Someya, Y. Song, E. Stavrinidou, M. Su, X. Sun, K. Takei, X. M. Tao, B. C. K. Tee, A. V. Thean, T. Q. Trung, C. Wan, H. Wang, J. Wang, M. Wang, S. Wang, T. Wang, Z. L. Wang, P. S. Weiss, H. Wen, S. Xu, T. Xu, H. Yan, X. Yan, H. Yang, L. Yang, S. Yang, L. Yin, C. Yu, G. Yu, J. Yu, S. H. Yu, X. Yu, E. Zamburg, H. Zhang, X. Zhang, X. Zhang, X. Zhang, Y. Zhang, Y. Zhang, S. Zhao, X. Zhao, Y. Zheng, Y. Q. Zheng, Z. Zheng, T. Zhou, B. Zhu, M. Zhu, R. Zhu, Y. Zhu, Y. Zhu, G. Zou, X. Chen, Technology roadmap for flexible sensors, *ACS Nano* 2023, 17(6), 5211–5295.
218. S. Ji, X. Chen, Enhancing the interfacial binding strength between modular stretchable electronic components, *Natl. Sci. Rev.* 2023, 10(1), nwac172.
219. K. Wu, P. Liao, R. Du, Q. Zhang, F. Chen, Q. Fu, Preparation of a thermally conductive biodegradable cellulose nanofiber/hydroxylated boron nitride nanosheet film: the critical role of edge-hydroxylation, *J. Mater. Chem. A* 2018, 6(25), 11863–11873.
220. J. Lee, H. Sul, W. Lee, K. R. Pyun, I. Ha, D. Kim, H. Park, H. Eom, Y. Yoon, J. Jung, D. Lee, S. H. Ko, Stretchable skin-like cooling/heating device for reconstruction of artificial thermal sensation in virtual reality, *Adv. Funct. Mater.* 2020, 30(29), 1909171.
221. Y. Jung, I. Ha, M. Kim, J. Ahn, J. Lee, S. H. Ko, High heat storing and thermally diffusive artificial skin for wearable thermal management, *Nano Energy* 2023, 105, 107979.
222. K. Wu, Y. Zhang, F. Gong, D. Liu, C. Lei, Q. Fu, Highly thermo-conductive but electrically insulating filament via a volumeconfinement self-assembled strategy for thermoelectric wearables, *Chem. Eng. J.* 2021, 421, 127764.
223. H. Hong, Y. H. Jung, J. S. Lee, C. Jeong, J. U. Kim, S. Lee, H. Ryu, H. Kim, Z. Ma, T. I. Kim, Anisotropic thermal conductive composite by the guided assembly of boron nitride nanosheets for flexible and stretchable electronics. *Adv. Funct. Mater.* 2019, 29(37), 1902575.
224. J. Khan, S. A. Momin, M. Mariatti, A review on advanced carbon-based thermal interface materials for electronic devices, *Carbon* 2020, 168, 65–112.
225. D. Suh, C. M. Moon, D. Kim, S. Baik, Ultrahigh thermal conductivity of interface materials by silver-functionalized carbon nanotube phonon conduits, *Adv. Mater.* 2016, 28(33), 7220–7227.

226. K. M. Shahil, A. A. Balandin, Graphene-multilayer graphene nanocomposites as highly efficient thermal interface materials, *Nano Lett.* 2012, 12(2), 861–867.
227. H. Wang, Y. Peng, H. Peng, J. Zhang, Fluidic phase-change materials with continuous latent heat from theoretically tunable ternary metals for efficient thermal management, *Proc. Natl. Acad. Sci. U. S. A.* 2022, 119(31), e2200223119.
228. W. Dai, T. Ma, Q. Yan, J. Gao, X. Tan, L. Lv, H. Hou, Q. Wei, J. Yu, J. Wu, Y. Yao, S. Du, R. Sun, N. Jiang, Y. Wang, J. Kong, C. Wong, S. Maruyama, C. T. Lin, Metal-level thermally conductive yet soft graphene thermal interface materials, *ACS Nano* 2019, 13(10), 11561–11571.
229. W. Dai, L. Lv, J. Lu, H. Hou, Q. Yan, F. E. Alam, Y. Li, X. Zeng, J. Yu, Q. Wei, X. Xu, J. Wu, N. Jiang, S. Du, R. Sun, J. Xu, C. P. Wong, C. T. Lin, A paper-like inorganic thermal interface material composed of hierarchically structured graphene/silicon carbide nanorods, *ACS Nano* 2019, 13(2), 1547–1554.
230. J. Hansson, T. M. J. Nilsson, L. Ye, J. Liu, Novel nanostructured thermal interface materials: a review, *Int. Mater. Rev.* 2018, 63(1), 22–45.
231. Y. Lin, J. Chen, P. Jiang, X. Huang, Wood annual ring structured elastomer composites with high thermal conduction enhancement efficiency, *Chem. Eng. J.* 2020, 389, 123467.
232. Y. Lin, J. Chen, S. Dong, G. Wu, P. Jiang, X. Huang, Wetresilient graphene aerogel for thermal conductivity enhancement in polymer nanocomposites, *J. Mater. Sci. Technol.* 2021, 83, 219–227.
233. S. Xu, S. Wang, Z. Chen, Y. Sun, Z. Gao, H. Zhang, J. Zhang, Electric-field-assisted growth of vertical graphene arrays and the application in thermal interface materials, *Adv. Funct. Mater.* 2020, 30(34), 2003302.
234. Z. Zhang, M. Li, Y. Wang, W. Dai, L. Li, Y. Chen, X. Kong, K. Xu, R. Yang, P. Gong, J. Zhang, T. Cai, C.-T. Lin, Nishimura, K. Li, H. N. Jiang, N. Yu, J. Ultrahigh thermal conductive polymer composites by the 3D printing induced vertical alignment of carbon fiber, *J. Mater. Chem. A* 2023, 11, 10971–10983.
235. F. Sun, T. Zhang, M. M. Jobbins, Z. Guo, X. Zhang, Z. Zheng, D. Tang, S. Ptasinska, T. Luo, Molecular bridge enables anomalous enhancement in thermal transport across hard-soft material interfaces, *Adv. Mater.* 2014, 26(35), 6093–6099.
236. K. M. Razeeb, E. Dalton, G. L. W. Cross, A. J. Robinson, Present and future thermal interface materials for electronic devices, *Int. Mater. Rev.* 2018, 63(1), 1–21.
237. D. D. L. Chung, Performance of thermal interface materials, *Small* 2022, 18(16), 2200693.
238. B. Koo, P. Goli, A. V. Sumant, P. C. d. S. Claro, T. Rajh, C. S. Johnson, A. A. Balandin, E. V. Shevchenko, Toward lithium ion batteries with enhanced thermal conductivity, *ACS Nano* 2014, 8, 7202–7207.
239. X. Feng, D. Ren, X. He, M. Ouyang, Mitigating thermal runaway of lithium-ion batteries, *Joule* 2020, 4(4), 743–770.
240. X. Du, Q. Wu, Y.-N. Wang, T.-S. Pan, Y.-M. Wei, H.-S. Chen, W.-L. Song, D.-N. Fang, Visualizing two-dimensional internal temperature distribution in cylindrical Li-ion cells, *J. Power Sources* 2020, 446, 227343.
241. Z. Liu, Q. Hu, S. Guo, L. Yu, X. Hu, Thermoregulating separators based on phase-change materials for safe lithium-ion batteries, *Adv. Mater.* 2021, 33(15), 2008088.
242. Y. Yang, X. Huang, Z. Cao, G. Chen, Thermally conductive separator with hierarchical nano/microstructures for improving thermal management of batteries, *Nano Energy* 2016, 22, 301–309.
243. Y. Liu, Y. Qiao, Y. Zhang, Z. Yang, T. Gao, D. Kirsch, B. Liu, J. Song, B. Yang, L. Hu, 3D printed separator for the thermal management of high-performance Li metal anodes, *Energy Storage Mater.* 2018, 12, 197–203.
244. X. Yin, L. Wang, Y. Kim, N. Ding, J. Kong, D. Safanama, Y. Zheng, J. Xu, D. V. M. Repaka, K. Hippalgaonkar, S. W. Lee, S. Adams, G. W. Zheng, Thermal conductive 2D boron nitride for high-performance all-solid-state lithium-sulfur batteries, *Adv. Sci.* 2020, 7(19), 2001303.
245. H. Guo, H. Zhao, H. Niu, Y. Ren, H. Fang, X. Fang, R. Lv, M. Maqbool, S. Bai, Highly thermally conductive 3d printed graphene filled polymer composites for scalable thermal management applications, *ACS Nano* 2021, 15(4), 6917–6928.
246. G. Murali, G. S. N. Sravya, J. Jaya, V. N. Vamsi, A review on hybrid thermal management of battery packs and it's cooling performance by enhanced PCM, *Renew. Sust. Energy Rev.* 2021, 150, 111513.

247. G. Cheng, Z. Wang, X. Wang, Y. He, All-climate thermal management structure for batteries based on expanded graphite/polymer composite phase change material with a high thermal and electrical conductivity, *Appl. Energy* 2022, 322, 119509.
248. S. Wu, T. Li, Z. Tong, J. Chao, T. Zhai, J. Xu, T. Yan, M. Wu, Z. Xu, H. Bao, T. Deng, R. Wang, High-performance thermally conductive phase change composites by large-size oriented graphite sheets for scalable thermal energy harvesting, *Adv. Mater.* 2019, 31(49), e1905099.
249. P. Goli, S. Legedza, A. Dhar, R. Salgado, J. Renteria, A. A. Balandin, Graphene-enhanced hybrid phase change materials for thermal management of Li-ion batteries, *J. Power Sources* 2014, 248, 37–43.
250. Y. Lin, P. L. Li, W. J. Liu, J. Chen, X. Y. Liu, P. K. Jiang, X. Y. Huang, Application-driven high-thermal-conductivity polymer nanocomposites. *ACS Nano* 2024, 18, 3851–3870.
251. Z. H. Shen, J. J. Wang, J. Y. Jiang, Y. H. Lin, C. W. Nan, L. Q. Chen, Y. Shen, Phase-field model of electro-thermal breakdown in flexible high-temperature nanocomposites under extreme conditions, *Adv. Energy Mater.* 2018, 8(20), 1800509.
252. O. G. Gnonhoue, A. Velazquez-Salazar, É. David, I. Preda, Review of technologies and materials used in high-voltage film capacitors, *Polymers* 2021, 13(5), 766.
253. J. Chen, Z. Shen, Q. Kang, X. Qian, S. Li, P. Jiang, X. Huang, Chemical adsorption on 2D dielectric nanosheets for matrix free nanocomposites with ultrahigh electrical energy storage, *Sci. Bull.* 2022, 67(6), 609–618.
254. J. Lao, H. Xie, Z. Shi, G. Li, B. Li, G.-H. Hu, Q. Yang, C. Xiong, Flexible regenerated cellulose/boron nitride nanosheet high-temperature dielectric nanocomposite films with high energy density and breakdown strength, *ACS Sustain. Chem. Eng.* 2018, 6(5), 7151–7158.
255. Q. Li, L. Chen, M. R. Gadinski, S. Zhang, G. Zhang, U. Li, E. Iagodkine, A. Haque, L. Q. Chen, N. Jackson, Q. Wang, Flexible high-temperature dielectric materials from polymer nanocomposites, *Nature* 2015, 523(7562), 576–579.
256. Z. Li, L. An, S. Khuje, J. Tan, Y. Hu, Y. Huang, D. Petit, D. Faghihi, J. Yu, S. Ren, Solution-shearing of dielectric polymer with high thermal conductivity and electric insulation, *Sci. Adv.* 2021, 7(40), eabi7410.
257. Y. Zhu, Z. Shen, Y. Li, B. Chai, J. Chen, P. Jiang, X. Huang, High conduction band inorganic layers for distinct enhancement of electrical energy storage in polymer nanocomposites, *Nano-Micro Lett.* 2022, 14(1), 151.
258. P. Li, X. Huang, Thermally conductive polymer composites for high voltage insulation, *IET Conference Proceedings*, 2021, 30–37.
259. B. X. Du, X. X. Kong, B. Cui, J. Li, Improved ampacity of buried HVDC cable with high thermal conductivity LDPE/BN insulation, *IEEE Trans. Dielectr. Electr. Insul.* 2017, 24(5), 2667–2676.
260. X. Huang, B. Sun, C. Yu, J. Wu, J. Zhang, P. Jiang, Highly conductive polymer nanocomposites for emerging high voltage power cable shields: experiment, simulation and applications, *High Volt.* 2020, 5(4), 387–396.
261. M. Tari, K. Yoshida, S. Sekito, J. Allison, R. Brutsch, A. Lutz, N. Frost, A high voltage insulating system with increased thermal conductivity for turbo generators, *Proceedings: Electrical Insulation Conference and Electrical Manufacturing and Coil Winding Technology Conference* (Cat. No. 03CH37480), IEEE, Indiana Convention Center, Indianapolis, Indiana, 2003, 613–617.
262. G. Komiya, T. Imai, Y. Miyauchi, Development of high thermal conductive epoxy composite for large current switchgear, *Electr. Eng. Jpn.* 2018, 205(4), 9–15.
263. X. Y. Li, J. W. Zha, S. J. Wang, S. L. Zhong, C. Zhang, Z. M. Dang, Effect of high-thermal conductivity epoxy resin on heat dissipation performance of saturated reactor, *IEEE Trans. Dielectr. Electr. Insul.* 2017, 24(6), 3898–3905.
264. F. Zhang, K. Fan, F. Saba, J. Yu, Graphene reinforcedgraphitized nanodiamonds matrix composites: fabrication, microstructure, mechanical properties, thermal and electrical conductivity, *Carbon* 2020, 169, 416–428.
265. W. Yang, W. J. Chen, Y. Zhou, C. Zhang, Z. Zhang, X. Chen, Y. S. Zhao, X. M. Bian, Heat dissipation characteristics of anode saturable reactors with high thermal conductivity epoxy resin used for ultra-high-voltage direct current converter valves, *High Voltage*, 2020, 5(5), 598–604.

266. L. A. Weinstein, J. Loomis, B. Bhatia, D. M. Bierman, E. N. Wang, G. Chen, Concentrating solar power, *Chem. Rev.* 2015, 115(23), 12797–12838.
267. Y. Zhang, J. Wang, J. Qiu, X. Jin, M. M. Umair, R. Lu, S. Zhang, B. Tang, Ag-graphene/PEG composite phase change materials for enhancing solar-thermal energy conversion and storage capacity, *Appl. Energy* 2019, 237, 83–90.
268. Y. Chen, Q. Zhang, X. Wen, H. Yin, J. Liu, A novel CNT encapsulated phase change material with enhanced thermal conductivity and photo-thermal conversion performance, *Sol. Energy Mater. Sol. Cells* 2018, 184, 82–90.
269. X. Du, J. Xu, S. Deng, Z. Du, X. Cheng, H. Wang, Amino-functionalized single-walled carbon nanotubes-integrated polyurethane phase change composites with superior photothermal conversion efficiency and thermal conductivity, *ACS Sustain. Chem. Eng.* 2019, 7(21), 17682–17690.
270. Y. Lin, Q. Kang, H. Wei, H. Bao, P. Jiang, Y. W. Mai, X. Huang, Spider web-inspired graphene skeleton-based high thermal conductivity phase change nanocomposites for battery thermal management, *Nano-Micro Lett.* 2021, 13(1), 180.
271. T. Wang, Y. Lin, P. Li, P. Jiang, C. Zhang, H. Xu, H. Xie, X. Huang, Unidirectional thermal conduction in electrically insulating phase change composites for superior power output of thermoelectric generators, *Compos. Sci. Technol.* 2022, 225, 109500.
272. P. Min, J. Liu, X. Li, F. An, P. Liu, Y. Shen, N. Koratkar, Z.-Z. Yu, Thermally conductive phase change composites featuring anisotropic graphene aerogels for real-time and fast-charging solarthermal energy conversion, *Adv. Funct. Mater.* 2018, 28(51), 1805365.
273. J. Yang, Y. C. Zhou, L. Y. Yang, C. P. Feng, L. Bai, M. B. Yang, W. Yang, Exploring next-generation functional organic phase change composites, *Adv. Funct. Mater.* 2022, 32(28), 2200792.
274. J. Yang, P. Yu, L. S. Tang, R. Y. Bao, Z. Y. Liu, M. B. Yang, W. Yang, Hierarchically interconnected porous scaffolds for phase change materials with improved thermal conductivity and efficient solar-to-electric energy conversion, *Nanoscale* 2017, 9(45), 17704–17709.
275. R. Sharma, G. Kapusetti, S. Y. Bhong, P. Roy, S. K. Singh, S. Singh, C. K. Balavigneswaran, K. K. Mahato, B. Ray, P. Maiti, N. Misra, Osteoconductive amine-functionalized graphene–poly-(methyl methacrylate) bone cement composite with controlled exothermic polymerization, *Bioconjugate Chem.* 2017, 28(9), 2254–2265.
276. Y. Lv, A. Li, F. Zhou, X. Pan, F. Liang, X. Qu, D. Qiu, Z. Yang, A novel composite PMMA-based bone cement with reduced potential for thermal necrosis, *ACS Appl. Mater. Interfaces* 2015, 7(21), 11280–11285.
277. Y. Wang, Y. Xu, F. Wang, Detecting the thermal properties of bone cement by temperature sensor, *Sens. Transducers* 2013, 154, 183–187.
278. Y. Xu, Y. Wang, L. Cui, An efficient functionalization method for the multiwalled carbon nanotubes and their applications in PMMA bone cement, *Sens. Transducers* 2013, 21, 36–41.
279. E. D. Gurgel-Filho, J. P. A. Feitosa, F. B. Teixeira, R. C. Monteiro de Paula, J. B. Araújo Silva, F. J. Souza-Filho, Chemical and X-ray analyses of five brands of dental gutta-percha cone, *Int. Endod. J.* 2003, 36, 302–307.
280. M. Hammad, A. Qualtrough, N. Silikas, Evaluation of root canal obturation: a three-dimensional in vitro study, *J. Endod.* 2009, 35(4), 541–544.
281. M. Dong, J. Zhang, L. Liu, G. Hou, Y. Yu, C. Yuan, X. Wang, New gutta percha composite with high thermal conductivity and low shear viscosity contributed by the bridging fillers containing ZnO and CNTs, *Compos. B Eng.* 2019, 173, 106903.
282. P. B. Kang, K. Vogt, S. E. Gruninger, M. Marshall, C. Siew, D. M. Meyer, The immuno cross-reactivity of gutta percha points, *Dent. Mater.* 2007, 23(3), 380–384.
283. M. Dong, J. Zhang, G. Hou, L. Liu, X. Qu, Y. Yu, C. Yuan, X. Wang, Thermal conductivity of GP/ZnO@CNTs nanocomposites improved greatly by orientation of CNTs under shear field, *Compos. Commun.* 2020, 17, 61–65.
284. D.-K. Lee, S. V. Kim, A. N. Limansubroto, A. Yen, A. Soundia, C.-Y. Wang, W. Shi, C. Hong, S. Tetradis, Y. Kim, N.-H. Park, M. K. Kang, D. Ho, Nanodiamond gutta percha composite biomaterials for root canal therapy, *ACS Nano* 2015, 9, 11490–11501.

285. F. Q. Liu, L. L. Wang, G. H. Li, W. Li, C. Q. Li, Hierarchically structured graphene coupled microporous organic polymers for superior CO_2 capture, *ACS Appl. Mater. Interfaces* 2017, 9(39), 33997–34004.
286. R. Balasubramanian, S. Chowdhury, Recent advances and progress in the development of graphene-based adsorbents for CO_2 capture, *J. Mater. Chem. A* 2015, 3(44), 21968–21989.
287. J. W. Lee, S. Kim, I. Torres Pineda, Y. T. Kang, Review of nanoabsorbents for capture enhancement of CO_2 and its industrial applications with design criteria, *Renew. Sust. Energy Rev.* 2021, 138, 110524.
288. S. Yang, L. Zhan, X. Xu, Y. Wang, L. Ling, X. Feng, Graphene-based porous silica sheets impregnated with polyethyleneimine for superior CO_2 capture, *Adv. Mater.* 2013, 25(15), 2130–2134.
289. B. Wu, F.-Q. Liu, S.-W. Luo, L.-Q. Zhang, F. Zou, Carbonaceous materials-supported polyethylenimine with high thermal conductivity: a promising adsorbent for CO_2 capture, *Compos. Sci. Technol.* 2021, 208, 108781.
290. F.-Q. Liu, W. Li, J. Zhao, W.-H. Li, D.-M. Chen, L.-S. Sun, L. Wang, R.-X. Li, Covalent grafting of polyethyleneimine on hydroxylated three-dimensional graphene for superior CO_2 capture, *J. Mater. Chem. A* 2015, 3(23), 12252–12258.
291. F.-Q. Liu, X. Liu, L. Sun, R. Li, C.-X. Yin, B. Wu, MXenesupported stable adsorbents for superior CO_2 capture, *J. Mater. Chem. A* 2021, 9(21), 12763–12771.
292. P. Li, A. Wang, J. Fan, Q. Kang, P. Jiang, H. Bao, X. Huang, Thermo-optically designed scalable photonic films with high thermal conductivity for subambient and above-ambient radiative cooling, *Adv. Funct. Mater.* 2022, 32(5), 2109542.
293. Y. Tian, X. Liu, Z. Wang, J. Li, Y. Mu, S. Zhou, F. Chen, M. L. Minus, G. Xiao, Y. Zheng, Subambient daytime cooling enabled by hierarchically architected all-inorganic metapaper with enhanced thermal dissipation, *Nano Energy* 2022, 96, 107085.
294. J. Song, W. Zhang, Z. Sun, M. Pan, F. Tian, X. Li, M. Ye, X. Deng, Durable radiative cooling against environmental aging, *Nat. Commun.* 2022, 13(1), 4805.
295. W. J. Wu, J. F. Fan, Y. S. Pang, C. Zhong, R. Sun, X. L. Zeng, Z. F. Hao, Soft, tough, and thermally conductive elastomer composites by constructing a curled conformation, *Chem. Mater.* 2023, 35, 7500–7510.
296. H. Y. Niu, H. C. Guo, L. Kang, L. C. Ren, R. C. Lv, L. Liu, A. Bashir, S. L. Bai, Highly thermally conductive and soft thermal interface materials based on vertically oriented boron nitride film, *Compos. Part B* 2024, 272, 111219.

Chapter 10

Isotropic Thermal Conductivity of Boron Nitride Nanosheets

Srikanth Mateti, Ying Ian Chen, Keqin Yang, Qiran Cai, Shaoming Huang, and Jiangting Wang

10.1 INTRODUCTION

Due to rapid advancements in the miniaturization of electronic devices, there is an urgent demand for efficient heat dissipation techniques. Traditional materials such as aluminum and copper have thermal conductivities of 205 and 385 W m^{-1} K^{-1}, respectively, at room temperature [1]. However, due to the metallic nature of these materials, there is still a chance of short-circuit when used over active layers of electronics. The electrically insulating nature of BN, at the same time being highly thermally conductive, makes it an ideal and alternative candidate (when compared to the metal-based heat sinks) for thermal management applications. However, BN possesses an anisotropic structure, which means it has very high thermal conduction in in-plane directions, whereas it is very low in cross-plane direction, which limits to use of BN in real-time applications [2,3].

A significant approach has been made to address the anisotropic thermal conductivity of BN, but with limited success. High-temperature sintering is considered one of the techniques to contact the BN material, thereby tuning its structure. BN has a similar honeycomb structure to graphite but possesses unique properties, such as being highly thermally conductive, but at the same time, BN is an electric insulator. This special characteristic makes it an ideal and alternative candidate (when compared to metal-based heat sinks) for thermal management applications. However, BN possesses an anisotropic structure, which means it has very high thermal conduction in in-plane directions, whereas it is very low in cross-plane direction, which limits to use of BN in real-time applications.

This chapter presents the novel production of bulk BN pellets ($10 \times 10 \times 2\,mm^3$) featuring a quasi-isotropic crystalline structure and thermal conductivity. This achievement marks the first instance of such pellets being manufactured from BNNSs utilizing SPS technology. Bulk BN with quasi-isotropic thermal properties holds promise as an effective insulating heat spreader and heat sink for extensive applications [4].

DOI: 10.1201/9781003647812-13

Figure 10.1 (a) SEM image of BNNSs. (b) Photograph of the spark plasma sintered BN pellet at 2200°C. (c) XRD patterns of the BNNS powder and sintered BN pellet at 2200°C. (Reprinted with permission from Ref. [4] Copyright 2018 Wiley-VCH.)

10.2 THERMAL CONDUCTIVITY OF BN PELLETS

10.2.1 SPS Sintering of BNNSs

BNNSs were synthesized using the ball milling method as described in Chapter 2 [5]. The scanning electron microscope (SEM) image in Figure 10.1a shows typical BNNS morphology and 2D structure. The distribution ranges of the BNNS are 600 nm in diameter and 20 nm in thickness. The BNNSs were sintered using a SPS technique on a Labox315 system (SINTER LAND INC., Japan). During the SPS process, approximately 1 g of BNNSs was carefully loaded into a square graphite die with dimensions of 1 cm × 1 cm and placed under vacuum conditions. To ensure consistent sintering, a constant pressure of 50 MPa was maintained across all experiments, while the current density was adjusted to achieve various sintering temperatures. The sintering process was carried out at temperatures ranging from 1800°C to 2300°C, with each sample being sintered for 5 minutes. A representative SPS-sintered bulk BN pellet is shown in Figure 10.1b. X-ray diffraction (XRD) patterns of the BN powder and the pellet sintered at 2200°C are presented in Figure 10.1c. The XRD pattern of the BNNSs exhibits broader, less-defined diffraction peaks due to their nanoscale structure, whereas the pattern of the sintered pellet displays distinct, sharp peaks, confirming the formation of a crystalline hexagonal BN structure. These findings highlight the structural transformation that occurs as a result of the sintering process.

10.2.2 Thermal Conductivity Measurement In-Plane and Out-Plane

The thermal conductivity of the SPS pellets was evaluated using a Netzsch LFA 457 spectrometer. In Figure 10.2a, the thermal conductivities of the pellets are depicted across a temperature range of 298–623 K, both parallel (//) to the SPS pressing direction and perpendicular to the pellet surface, as indicated by the arrow in the inset. A corresponding decline in conductivity is observed as the measurement temperature rises for pellets sintered at elevated temperatures (>2000°C). This behavior aligns with typical thermal characteristics of insulating crystals, where heat transfer occurs via phonons, and the Umklapp scattering process becomes more prominent at high temperatures.

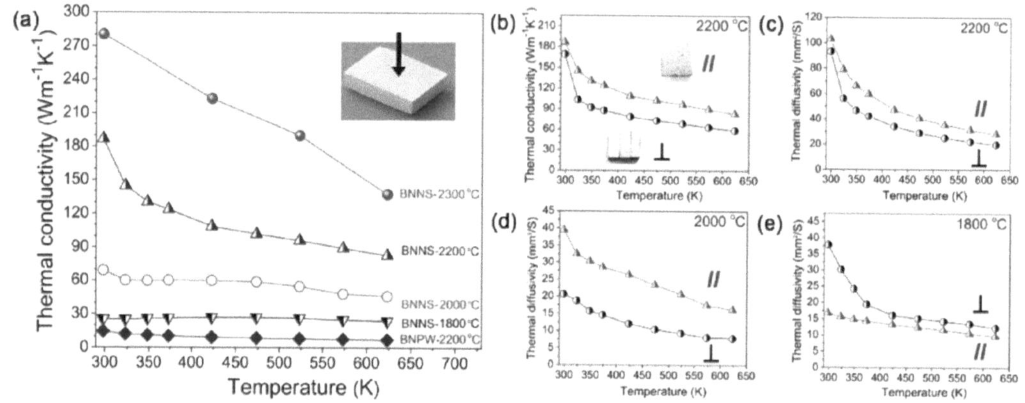

Figure 10.2 Thermal conductivity and diffusivities of BNNS pellets: (a) Thermal conductivity of sintered BNNS and BNPW pellets at various measurement temperatures under 50 MPa pressure—measurement direction parallel to pressing direction as indicated by an arrow (inset). (b) Thermal conductivity of the pellet sintered at 2200°C measured parallel (//) and perpendicular (⊥) to the SPS pressing direction. Thermal diffusivities for pellets sintered at different temperatures: (c) 2200°C, (d) 2000°C, and (e) 1800°C. (Reprinted with permission from Ref. [4], Copyright 2018 Wiley-VCH.)

Figure 10.2a also illustrates that pellets sintered at higher temperatures exhibit increased thermal conductivity. The pellet sintered at 2300°C, the highest temperature utilized, achieves the highest thermal conductivity of 280 W m^{-1} K^{-1}. While this value is lower than the reported in-plane conductivity of the hexagonal boron nitride (h-BN) crystal (390 W m^{-1} K^{-1}) [2,6], it surpasses cross-plane conductivity by a significant margin. In many cases involving hot-pressed or spark plasma-sintered samples with BN powders as starting materials, the alignment of (002) basal planes perpendicular to pressing directions typically results in an anisotropic structure [7]. In conventionally sintered pellets, the perpendicular (⊥) direction typically exhibits high thermal conductivity, while the parallel (//) direction usually demonstrates very low conductivity, often less than 20 W m^{-1} K^{-1}. This disparity is attributed to cross-plane conductivity resulting from weak van der Waals interactions between the (002) basal planes. Remarkably, our pellets sintered using a high-temperature spark plasma process showcase significantly elevated thermal conductivity in both the parallel and cross-sectional directions, a departure from previously reported normally sintered BN compacts. This observation holds great promise for enhanced performance in heat dissipation applications. For comparison, a pellet (BNPW) was fabricated from BN powders with diameters of several micrometers at 2200°C under identical SPS conditions. The thermal conductivity of the BNPW pellet in the parallel direction is approximately 14.3 W m^{-1} K^{-1}, much lower than that of pellets sintered from BNNSs at high temperatures, underscoring the pivotal role of BNNSs as starting materials.

Thermal conductivities were also assessed in the perpendicular (⊥) direction relative to the compression direction. In the case of anisotropic structures, this corresponds to the in-plane direction, where higher thermal conductivity is anticipated. However, due to the thinness of the BNNS pellets, approximately 2 mm, the cross-sectional area is too small for measurement via the laser flash system. To address this issue, the square pellets were first divided into three equal bars, each of which was then rotated 90° clockwise. These individual bars were bonded together using a thin layer of JB Weld, a high-temperature adhesive, to create a new, consolidated pellet structure (Figure 10.2b) [8]. Following this, the thermal conductivity and diffusivity of the sintered pellet at 2200°C were assessed across a temperature range from 298 to 623 K, measured in both the perpendicular (⊥) and parallel (//) directions. The results, as shown in Figure 10.2b and c, reveal that while the thermal diffusivity and conductivity in the parallel direction are slightly greater than those in the perpendicular direction, the values for both orientations are relatively close, indicating that the sintered material exhibits fairly uniform thermal properties across the two directions. This analysis provides insight into the thermal performance of the sintered pellet, demonstrating its efficiency in heat transfer across different orientations.

In the parallel and perpendicular directions, thermal conductivities are measured at 186 and 169 W m^{-1} K^{-1}, respectively. This observation is intriguing considering the highly anisotropic crystal structure of BN, where conventionally sintered BN compacts under pressure typically exhibit significant anisotropy. The specific SPS conditions used during the sintering process have clearly influenced the microstructure of the pellets in a distinct manner. Figure 10.2c–e shows the raw thermal diffusivity data, which were directly obtained through a laser flash system without any post-processing. The data reveal that for pellets sintered at 2000°C and 2200°C, thermal diffusivities measured along the parallel direction are higher than those in the perpendicular direction. In contrast, the pellet sintered at 1800°C exhibits lower thermal diffusivity in the parallel direction and higher diffusivity in the perpendicular direction, which can be attributed to the anisotropic nature of its structure. This behavior is consistent with previous studies on SPS or hot-pressed BN compacts, where the sintering conditions play a critical role in determining the alignment of grains and the resulting thermal properties. These findings highlight how variations in sintering temperature can lead to significant differences in the thermal performance of the material, emphasizing the importance of carefully selecting processing parameters to optimize the material's thermal conductivity [9,10].

10.2.3 Quasi-Isotropic Structure

The pellets underwent XRD analysis in two orientations, revealing full sets of diffraction peaks characteristic of a hexagonal structure for all BN pellets. The intensity ratio (I_{100}/I_{002}) of the (100) and (002) diffraction peaks serves as an orientation factor, indicating the degree of isotropy or anisotropy within a BN structure [8]. The starting h-BN powder exhibited a typical anisotropic structure with a low orientation factor of 0.03, primarily due to the parallel orientation of most (002) planes. As the sintering temperature increased, the intensity ratios (I_{100}/I_{002}) of BN pellets shifted from 0.17 to 0.28 (Figure 10.3a) in the direction parallel to the pressing direction. Under compression pressure, the (002) planes became more disordered in pellets sintered at higher temperatures. Similarly, in the direction perpendicular to the pressing direction, intensity ratios increased from 0.24 to 0.31 (Figure 10.3b), indicating a similar increase in disordered structure in each pressing. The 1800°C sintered pellet displayed an anisotropic structure, with more (002) basal planes perpendicular to the pressing direction, confirmed by the SEM image in Figure 10.4a. Conversely, the 2300°C sintered pellet showed nearly equal intensity ratios in both directions, suggesting a random structure.

Electron backscatter diffraction (EBSD) mapping of the 2200°C sintered pellet's surface and cross-section indicated randomly compacted BN grains, further confirming the quasi-isotropic structure as illustrated in Figure 10.4a–d. Figure 10.4a and c displays EBSD maps, where colors correspond to the misorientation angles between the pressing axis and the [0001] crystallographic direction. These maps illustrate the surfaces and cross-sections of the BNNS pellets, respectively. Regions marked in black were not indexed, which can be attributed to either the presence of grain boundaries or insufficient signal quality for reliable analysis. The dotted areas on the maps indicate regions where the [0001] direction is nearly aligned (within 0°–15°) with the compression axis, while stripped areas suggest a nearly perpendicular alignment (75°–90°) to the compression axis. The other colors correspond to intermediate misorientation angles. Analyzing

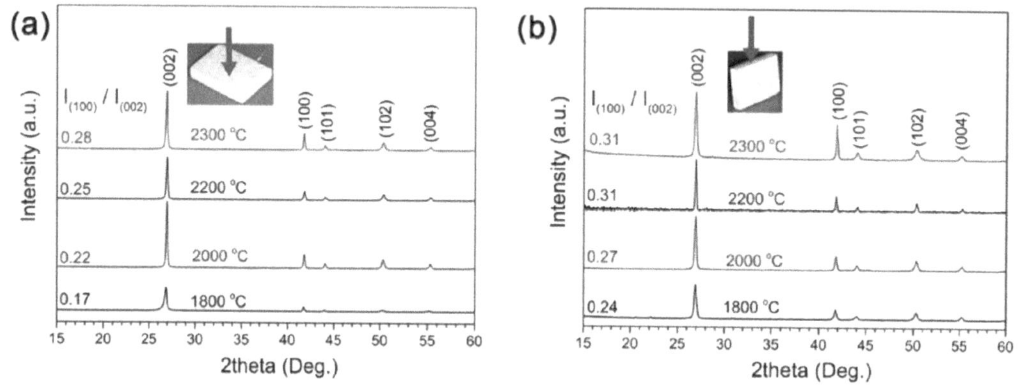

Figure 10.3 (a) XRD patterns captured from the surface of the BNNS pellets. (b) XRD patterns captured from the cross-section of the BNNS pellets. (Reprinted with permission from Ref. [4], Copyright 2018 Wiley-VCH.)

both surface and cross-sectional views, it is evident that the BN grains were compacted in a random orientation within the SPS pellet. The corresponding pole figures (Figure 10.4b and d) further support this observation, showing that no significant texture was developed during the SPS compression process. This suggests that the sintering conditions did not favor a preferred orientation of the BN grains within the pellet.

SEM analysis further confirms the varying orientations of grains within the pellets. The SEM image in Figure 10.5a, captured from the surface of the pellet sintered at 1800°C, depicts BN plates aligned parallel to each other and facing the pressing direction, aligning with the low orientation factor of 0.17. In contrast, the SEM image shown in Figure 10.5b, taken from the surface of the pellet sintered at 2200°C, reveals the presence of large grains that are randomly oriented. This observation is consistent with a relatively high orientation factor of 0.25, suggesting a shift toward a quasi-isotropic microstructure. These results highlight the anisotropic structure observed in BNNS pellets sintered at 1800°C, whereas those sintered at elevated temperatures of 2200°C–2300°C exhibit a more isotropic arrangement. Similar observations have been reported by Yuan et al., who found the development of large micrometer-sized crystallized domains (up to 30 μm) in the SPS load direction, even when subjected to higher pressures of 90 MPa and lower sintering temperatures around 1800°C. These findings further reinforce the impact of sintering temperature on the structural characteristics of BNNS pellets [11]. Discrepancies in SPS conditions may arise from differences in starting materials, such as h-BN powder.

The findings suggest an anisotropic structure in BNNS pellets sintered at 1800°C but a quasi-isotropic structure in those sintered at higher temperatures of 2200°C–2300°C. The growth mechanism of BN crystals along the pressing direction warrants further investigation. In summary, both XRD patterns and EBSD maps suggest a certain degree of randomly oriented (002) planes in both the surface and cross-sections of the pellets sintered at 2200°C. The formation of this unusual quasi-isotropic structure is discussed in subsequent sections.

10.2.4 Formation Mechanism

At elevated SPS temperatures exceeding 2000°C, significant BN crystal growth occurred, primarily contributing to the formation of the quasi-isotropic structure. In Figure 10.6a, the sample displacement along the pressing direction during SPS, correlated with heating temperature,

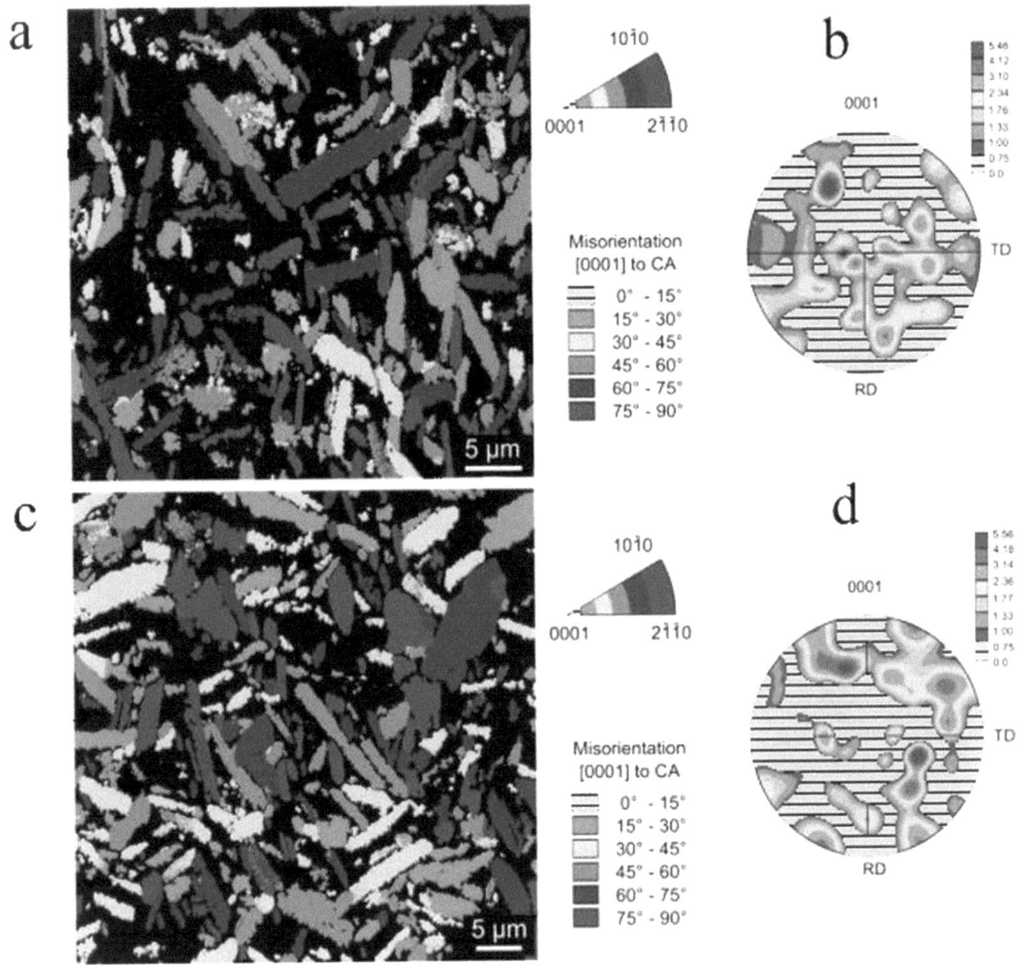

Figure 10.4 In-depth structural analysis of BNNS pellets: (a, b) EBSD map and [0001] pole figure from the surface of the pellet, respectively. (c, d) EBSD map and [0001] pole figure from the cross-section of the pellet, respectively. The EBSD maps in panels (a) and (c) are based on the misorientation angle between the compression axis and [0001] direction. (Reprinted with permission from Ref. [4], Copyright 2018 Wiley-VCH.)

is depicted. Negative displacement reflects pellet size reduction with increasing temperature under 50 MPa pressure, indicating densification and void removal. The subsequent displacement increase from 1800°C to 2000°C signifies pellet expansion due to considerable crystal growth, as corroborated by XRD and SEM analyses.

The (002) layer thickness in pellets was determined from the (002) diffraction peak width in XRD patterns (Figure 10.3a), increasing substantially from 62.4±0.5 nm at 1800°C to 534±0.5 nm at 2300°C. This aligns with SEM images showing grain size evolution (Figure 10.5a and b). Starting BNNS thickness averaged about 25.4±0.5 nm, evolving into approximately 62.4±0.5 nm thick plates post-sintering at 1800°C. Substantial crystallization from 1800°C led to micrometer-sized crystals, evident in the SEM image of pellets produced at 2200°C. This rapid crystal size increase under high sintering temperatures correlates well with the orientation factor change (Figure 10.6b), likely contributing to quasi-isotropic structure formation.

The anisotropic structure of h-BN crystals typically forms in hot-pressing and SPS processes under high compression pressure at intermediate temperatures (<1800°C) [11]. During high-temperature sintering (1800°C–2300°C), significant electric charges were applied to BNNSs via high-current pulses. Discharge between BNNS edges possibly formed links, illustrated schematically in Figure 10.6c. The electric field, parallel to the coaxial direction between electrodes, elevated charged nanosheets while compression forces pressed them down, resulting in larger (002) sheets formed in all directions [12]. TEM images confirm larger crystal sizes and improved crystalline structure of BN (002) layers in pellets sintered at higher temperatures. Enhanced BN crystal sizes and crystalline structures lead to higher thermal conductivity, as previously reported [13,14]. Thus, starting BNNSs serve as another critical factor in forming the isotropic structure. For comparison, micrometer-sized BN powders were sintered at 2200°C.

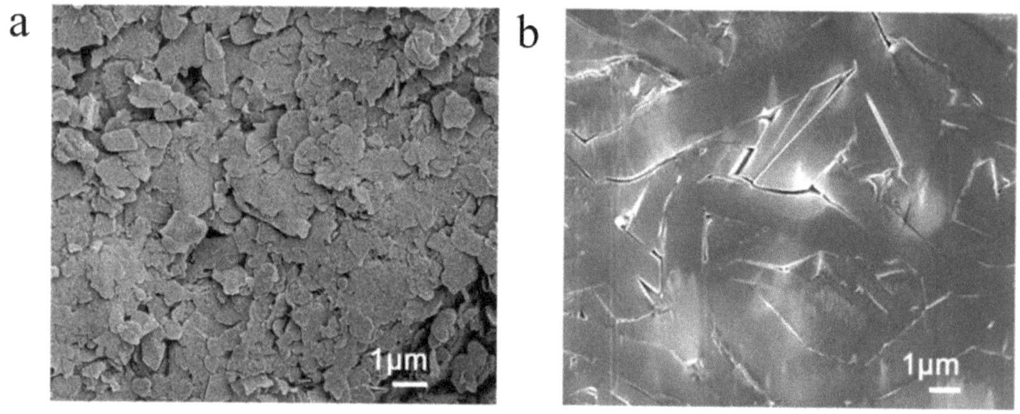

Figure 10.5 SEM images illustrating the surface of BNNS pellets sintered at 1800°C (a) and 2200°C (b). (Reprinted with permission from Ref. [4], Copyright 2018 Wiley-VCH.)

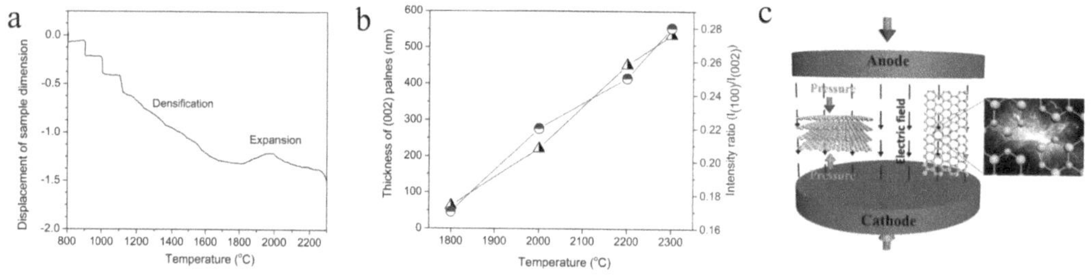

Figure 10.6 Crystal growth during SPS sintering is depicted through various analyses: (a) Displacement of the sample along the pressure direction concerning heating temperature during SPS. (b) Variations in the thickness of the (002) planes and orientation factor as a function of SPS temperature. (c) Explanation of the formation mechanism behind the quasi-isotropic structure. (Reprinted with permission from Ref. [4], Copyright 2018 Wiley-VCH.)

10.2.5 Heat Dissipation Behavior

To test the real-time applications of these BN pellets, the performance of heat dissipation behavior is crucial to evaluate. The hot plate was preset to 60°C and 100°C. Then, the sintered BNs were kept on top of the hotplate (Figure 10.7a). The variation in temperature of the BN pellets was monitored at the top of the BN pellet. The BN pellets sintered at 2200°C reached 57°C and 94°C in 30 seconds, whereas at 2300°C sintered BN pellets reached 58°C and 99°C at the same time.

Further, the heat dissipation effect of these pellets was evaluated using a thermal infrared camera, and a 100 Ω thermistor (1 x 1 mm) on a printed circuit board was used as a spot heater (Figure 10.7b). The temperature of the thermistor was controlled using a DC power supply. To monitor the temperature of both the hot spot and the pellets placed on it, an infrared thermal camera (FLIR ETS320 model) was employed. Temperature distribution images were captured once stable temperatures were reached. To enhance thermal contact between the BN pellet samples and the hot plate surface, a silver thermal compound (Formula 5) from ANTEC was applied as the thermal interface material. This setup ensured accurate and consistent temperature readings during the testing process.

A BN pellet was placed on top of the hot spot when the temperature reached 81°C and 101°C, respectively. Once the pellet's temperature stabilized, a thermal image was captured and is presented in Figure 10.7c and d. For the BNNS pellet sintered at 2200°C, a consistent temperature of 41.1°C was maintained, indicating a temperature decrease of 40°C. At a hot-spot temperature of 101°C, the stable temperature for the BNNS pellet was recorded as 49°C. Larger temperature decreases were observed with the BNNS pellet sintered at 2300°C. The stable temperatures for the pellet were 36.8°C (Figure 10.7e) and 43.0°C (Figure 10.7f), corresponding to temperature reductions of 44.2°C and 58°C for hot-spot temperatures of 81°C and 101°C, respectively.

These findings also indicate that the effectiveness of temperature reduction or heat dissipation is influenced by the thermal conductivity of the pellets. These BNNS pellets, which act as insulators, demonstrate remarkable thermal conductivity and superior heat dissipation efficiency in all three dimensions. They have the potential to substitute Al/Cu as a heat spreader or sink, offering efficient and secure heat management solutions.

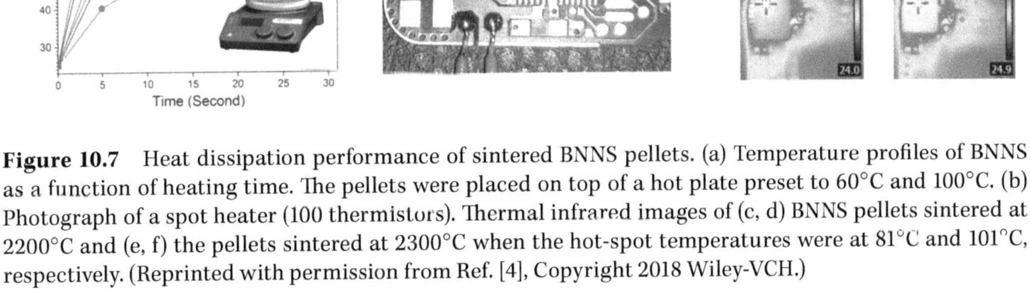

Figure 10.7 Heat dissipation performance of sintered BNNS pellets. (a) Temperature profiles of BNNS as a function of heating time. The pellets were placed on top of a hot plate preset to 60°C and 100°C. (b) Photograph of a spot heater (100 thermistors). Thermal infrared images of (c, d) BNNS pellets sintered at 2200°C and (e, f) the pellets sintered at 2300°C when the hot-spot temperatures were at 81°C and 101°C, respectively. (Reprinted with permission from Ref. [4], Copyright 2018 Wiley-VCH.)

10.3 CONCLUSIONS

Highly isotropic and thermally conductive BN pellets were fabricated through SPS of BNNSs at elevated temperatures. Pellets sintered at temperatures above 2000°C exhibited significantly enhanced thermal conductivities in both parallel and perpendicular directions relative to the SPS pressing axis. Notably, the BNNS pellet sintered at 2300°C under a pressure of 50 MPa displayed a remarkable thermal conductivity of 280 W m^{-1} K^{-1} in the parallel direction, which is typically associated with lower conductivity across the plane. This result underscores the effectiveness of optimizing SPS conditions for improving the thermal performance of BNNS-based materials. Analysis via XRD and EBSD indicated randomly oriented crystallites within the pellets post-SPS at high temperatures. At temperatures surpassing 2000°C, high-current pulses facilitated BNNS connection, forming expansive sheets under the electric field, thereby generating micrometer-sized belts. Notably, some (002) planes aligned parallel to the pressing direction, augmenting thermal conductivity in both parallel and cross-sectional directions. The BNNS pellets demonstrated efficient heat dissipation capabilities, presenting them as promising candidates for heat spreaders and heat sinks across various applications.

REFERENCES

1. H. D. Young, R. A. Freedman, A. L. Ford, *Sears and Zemansky's University Physics* (Pearson Education, 2008), vol. 1.
2. I. Jo et al., Thermal conductivity and phonon transport in suspended few-layer hexagonal boron nitride. *Nano Letters* **13**, 550–554 (2013).
3. H. Zhou et al., High thermal conductivity of suspended few-layer hexagonal boron nitride sheets. *Nano Research* **7**, 1232–1240 (2014).
4. S. Mateti et al., Bulk hexagonal boron nitride with a quasi-isotropic thermal conductivity. *Advanced Functional Materials* **28**, 1707556 (2018).
5. T. Xing et al., Gas protection of two-dimensional nanomaterials from high-energy impacts. *Scientific Reports* **6**, 35532 (2016).
6. E. Sichel, R. Miller, M. Abrahams, C. Buiocchi, Heat capacity and thermal conductivity of hexagonal pyrolytic boron nitride. *Physical Review B* **13**, 4607 (1976).
7. A. Simpson, A. Stuckes, The thermal conductivity of highly oriented pyrolytic boron nitride. *Journal of Physics C: Solid State Physics* **4**, 1710 (1971).
8. W. Xie et al., Investigation of the sintering pressure and thermal conductivity anisotropy of melt-spun spark-plasma-sintered (Bi, $Sb)_2Te_3$ thermoelectric materials. *Journal of Materials Research* **26**, 1791–1799 (2011).
9. M. Tanimoto, T. Yamagata, K. Miyata, S. Ando, Anisotropic thermal diffusivity of hexagonal boron nitride-filled polyimide films: effects of filler particle size, aggregation, orientation, and polymer chain rigidity. *ACS Applied Materials & Interfaces* **5**, 4374–4382 (2013).
10. T. Kusunose, T. Sekino, Thermal conductivity of hot-pressed hexagonal boron nitride. *Scripta Materialia* **124**, 138–141 (2016).
11. S. Yuan et al., Pure & crystallized 2D boron nitride sheets synthesized via a novel process coupling both PDCs and SPS methods. *Scientific Reports* **6**, 20388 (2016).
12. K. Yang et al., Tuning electrical and thermal connectivity in multiwalled carbon nanotube buckypaper. *Journal of Physics: Condensed Matter* **22**, 334215 (2010).
13. X. Xu et al., Length-dependent thermal conductivity in suspended single-layer graphene. *Nature Communications* **5**, 3689 (2014).
14. X. Xu, J. Chen, B. Li, Phonon thermal conduction in novel 2D materials. *Journal of Physics: Condensed Matter* **28**, 483001 (2016).

Chapter 11

Boron Nitride Spheres for Enhanced Isotropic Thermal Management

Qiran Cai, Hongbo Jiang, Srikanth Mateti, and Ying Ian Chen

11.1 INTRODUCTION

As advanced electronics evolve toward higher integration, miniaturization, lightweight designs, and increased power, effective heat dissipation has become critical for ensuring the performance, reliability, longevity, and safety of electronic components and devices [1–3]. To manage waste heat from hot spots, thermal interface materials (TIMs) are commonly employed to bridge air gaps between heat sources and heat sinks, reducing the risk of overheating [4,5]. Polymers are considered ideal materials for TIMs due to their high flexibility, low cost, lightweight nature, and ease of processing [6,7]. However, their inherently low thermal conductivity significantly limits heat transfer efficiency [8]. To address this limitation, BNNSs are widely used as filler materials, offering outstanding thermal conductivity and electrical insulation to enhance polymer-based heat dissipation systems [8,9].

The platelet structure of Boron nitride nanosheets (BNNSs) causes them to preferentially 'lie down' within the polymer matrix during fabrication, leading to highly anisotropic thermal conductivity. Combined with the anisotropic thermal conductivity of BNNSs, this leads to composites with excellent in-plane thermal conductivity but relatively low out-of-plane thermal conductivity [8,10–13]. For instance, Yan et al. reported a BNNS-reinforced polyvinyl alcohol (PVA) composite film with an impressive in-plane thermal conductivity of up to 67.6 W m^{-1} K^{-1}, but an out-of-plane thermal conductivity of only 3.6 W m^{-1} K^{-1} [8]. However, practical applications often require polymer composites with isotropic thermal conductivity. High in-plane thermal conductivity disperses localized heat laterally across a larger area, while out-of-plane thermal conductivity facilitates heat transfer from heat sources to sinks. Both are crucial for maximizing heat dissipation efficiency [14].

To achieve isotropic thermal conductivity in BNNS-reinforced polymer composites, various strategies have been explored [15–19]. One promising approach involves fabricating boron nitride spheres (BNSs), which inherently exhibit isotropic thermal conductivity and avoid complex fabrication procedures. BN microspheres have been used to isotropically enhance the thermal conductivity of polymers such as polydimethylsiloxane or epoxy [19]. However, the use of organic binder materials in their production often reduces the thermal conductivity of the resulting BNSs, leading to only moderate improvements in composite performance. This chapter

DOI: 10.1201/9781003647812-14

focuses on the fabrication and thermal conductivity measurements of 'binder-free' BNSs and their application in PVA composite films to enhance the isotropy of thermal conductivity [20,21].

11.2 FABRICATION OF BNSS

To fabricate 'binder-free' BNSs, BNNSs were first produced *via* high-energy ball milling using *h*-BN powders as the starting material and ammonia as a protective gas [22,23]. The scanning electron microscopy (SEM) image in Figure 11.1a and b reveals that the resulting BNNSs had an average lateral size of approximately 580 nm, while AFM measurements (Figure 11.1c and d) estimated their average thickness to be 3.0 nm. Transmission electron microscopy images, selected-area electron diffraction patterns, and X-ray diffraction (XRD) patterns confirmed the high crystallinity of the BNNSs. The BNSs were then fabricated through a spray-drying technique using the BNNS-water dispersion as a precursor [23]. During spray drying, the dispersion was heated and atomized into fine droplets, which were dried in a chamber with cyclone-assisted argon gas flow. This process formed larger BNSs through water evaporation. However, the absence of an organic binder led to small, mechanically weak BNSs due to insufficient interactions between BNNSs (Figure 11.2a).

To address this issue, small boron nitride nanoparticles (BNNPs) with an average size of ~100 nm were added to the BNNS-water dispersion as a secondary precursor. The BNNPs were synthesized using a ball milling method similar to that for BNNSs, but with argon (Ar) as the protective gas [22]. The incorporation of BNNPs significantly enhanced the mechanical strength

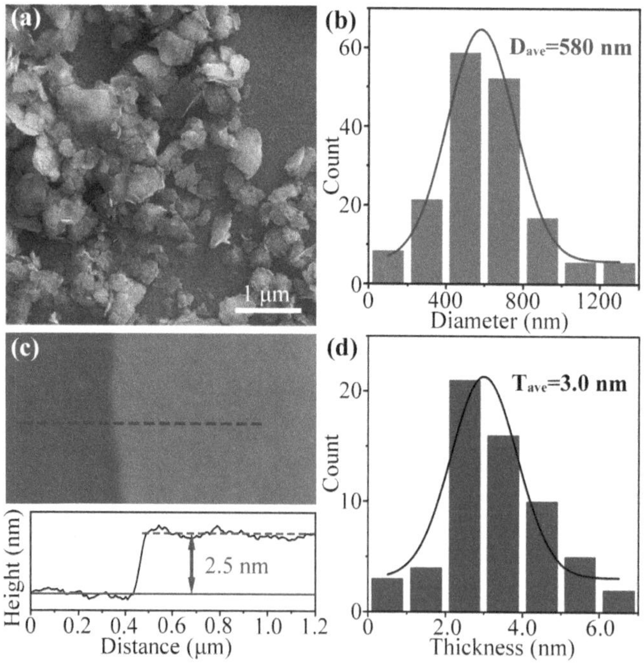

Figure 11.1 SEM image of BNNSs (a) and their size distribution (b). (c) Atomic force microscopy (AFM) image of a BNNS with the corresponding height trace. (d) Thickness distribution of BNNSs and the corresponding Gaussian fit. (Reprinted from Ref. [20], Copyright (2022), with permission from Elsevier.)

Figure 11.2 (a) SEM image of BNSs synthesized using BNNSs as the sole precursor. (b) and (c) SEM images of BNSs produced using NS70 as the precursor. (d) Size distribution corresponding to the BNSs in (b). (Reprinted from Ref. [20], Copyright (2022), with permission from Elsevier.)

Figure 11.3 SEM images of BNSs before sintering (a) and after sintering at (b) 1800°C, (c) 2000°C, and (d) 2200°C. (e) XRD patterns of BNSs before and after sintering at various temperatures. (Reprinted from Ref. [20], Copyright (2022), with permission from Elsevier.)

of the BNSs. When a precursor comprising 70 wt% BNNSs and 30 wt% BNNPs (denoted as NS-70) was used, the resulting BNSs exhibited a well-defined spherical structure with an average diameter of 7.2 μm (Figure 11.2b–d). The BNNS-to-BNNP weight ratio significantly influenced the size and thermal conductivity of the BNSs, as discussed in Section 11.3. In this process, BNNPs acted as 'binders', filling the gaps between BNNSs and strengthening their interactions [20].

To further improve the interactions between BNNSs and BNNPs and to heal structural defects for enhanced thermal conductivity, the BNSs underwent high-temperature sintering at 1800–2200°C. Morphological comparisons in Figure 11.3 show that, before sintering, BNNPs were visible on the BNS surfaces, while the BNNSs had rough textures and sharp edges. As the sintering temperature increased, the BNNPs diffused into the BNNSs, leading to larger BNNS sizes and smoother surfaces, indicative of defect healing. XRD analysis showed an increase in the intensity of the (002) peak with higher sintering temperatures, further confirming the improved crystallinity of the BNSs [20].

11.3 THERMAL CONDUCTIVITY OF BNSS

Directly measuring the thermal conductivity of a single BNS with a micrometer-scale diameter is challenging. However, this can be estimated by scaling up the structure of BNSs to a centimeter size. To achieve this, micro-sized BNSs were pressed into centimeter-sized pellets with a diameter of 12.7 mm. When the internal structure of the BNSs—randomly distributed BNNSs—remains consistent in the pellets, the thermal conductivity of the BNSs can be approximated from that of the pellets. For this method to be effective, the spherical structure of the BNSs must be preserved during the pressing process, and the interactions between BNNSs on the surfaces of adjacent BNSs must be significantly enhanced to minimize interfacial thermal resistance. The out-of-plane thermal conductivity of BNS pellets is more suitable for estimating the thermal conductivity of individual BNSs than the in-plane value, as the out-of-plane measurement reaches its maximum when the internal structure of the pellets perfectly replicates that of the spheres. Damage to the spherical structure or excessive interfacial thermal resistance between BNSs reduces the out-of-plane thermal conductivity. To address these factors, the out-of-plane thermal conductivity of BNS pellets was investigated under varying pressing pressures, pressing times, and sintering temperatures [21].

11.3.1 Influence of Pressing Pressure and Time on the Thermal Conductivity of BNS Pellets

The NS70 material was pressed at three different loads—78, 150, and 310 MPa—to determine the optimal conditions for pellet formation. Results showed that 150 MPa was the ideal loading. Lower load failed to form pellets, while higher load caused damage to the spherical structure of BNSs within the pellets. Figure 11.4 compares the room-temperature thermal conductivity of NS70 pellets pressed at 150 MPa for varying durations. Pellets pressed for 20 seconds exhibited a thermal conductivity of 31.5±2.8 W m^{-1} K^{-1}, which increased to a peak value of 37.2±2.5 W m^{-1} K^{-1} after 1 minute of pressing. However, prolonged pressing times led to reduced out-of-plane thermal conductivity, attributed to the 'lying down' of BNNSs. This structural change was confirmed by a decrease in the intensity ratio of the XRD (100) and (002) peaks (I_{100}/I_{002}), indicating the reorientation of BNNSs. SEM images of the cross-section of NS70 pellets (Figure 11.5) provide further insight. After 20 seconds of pressing, wide spaces remained between NS70 spheres, suggesting insufficient compaction. Pressing for 1 minute resulted in tighter packing and reduced void spaces within the pellets. However, extending the pressing time to 5 minutes caused the collapse of some NS70 spheres and the reorientation of BNNSs. These observations align with the trends in thermal conductivity and the I_{100}/I_{002} ratio [21].

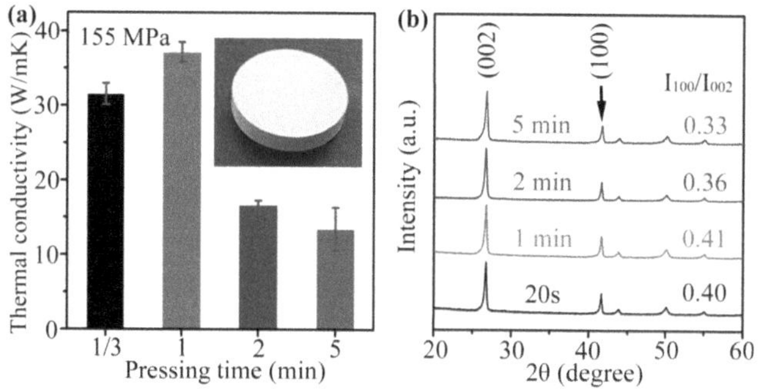

Figure 11.4 (a) Thermal conductivities and (b) XRD results of NS70 pellets with varying pressing times, under a load of 155 MPa. The inset in (a) shows a digital photograph of an NS70 pellet. (Reprinted from Ref. [21], Copyright 2024, with permission from John Wiley & Sons, Inc.)

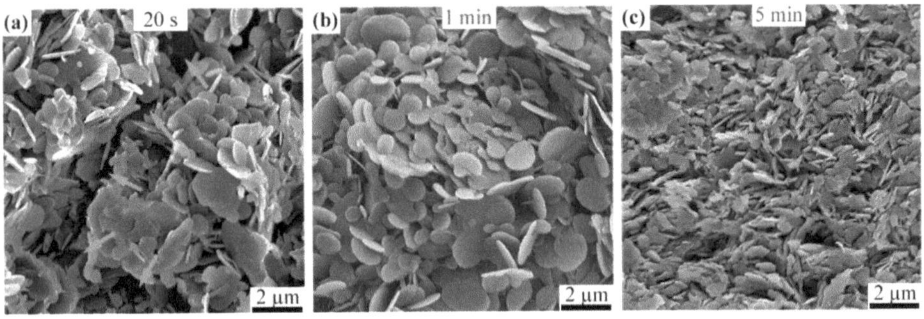

Figure 11.5 SEM images of the cross-section of NS70 pellets pressed at 155 MPa for different durations: (a) 20 seconds, (b) 1 minute, and (c) 5 minutes. (Reprinted from Ref. [21], Copyright 2024, with permission from John Wiley & Sons, Inc.)

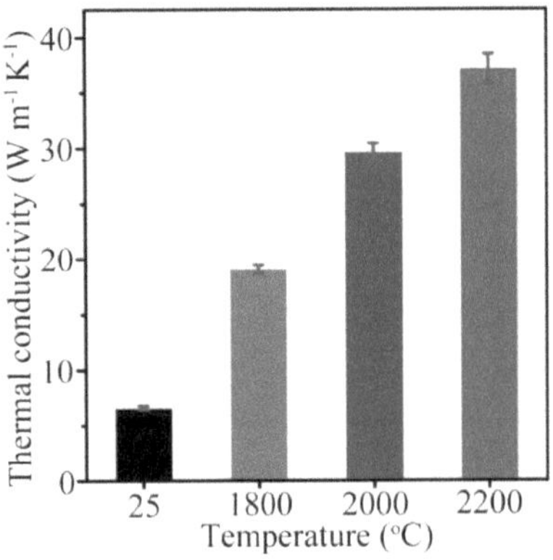

Figure 11.6 Thermal conductivity of NS70 pellets sintered at different sintering temperatures. (Reprinted from Ref. [21], Copyright 2024, with permission from John Wiley & Sons, Inc.)

11.3.2 Influence of Sintering Temperature on the Thermal Conductivity of BNS Pellets

As discussed in Section 11.2, high-temperature sintering was employed to reduce the interfacial thermal resistance between BNNSs. While pressing enhances interactions between BNNSs on the surfaces of adjacent NS70 spheres, this method is less effective than high-temperature sintering. Consequently, higher interfacial thermal resistance between BNSs within NS70 pellets leads to an underestimation of the thermal conductivity of individual BNSs. To address this, the effect of sintering temperature on the thermal conductivity of BNS pellets was investigated. Figure 11.6 shows the out-of-plane thermal conductivity of BNS pellets at different sintering temperatures. The thermal conductivity was 6.6 ± 0.8 W m^{-1} K^{-1} for non-sintered samples and increased with rising sintering temperatures, reaching 37.2 ± 2.5 W m^{-1} K^{-1} for samples sintered at 2200°C. This improvement is attributed to reduced interfacial thermal resistance between BNSs, as sintering enhances the interactions between BNNSs on the surfaces of adjacent NS70 spheres. As previously noted, NS70 pellets share a similar internal structure to NS70 spheres. By minimizing the

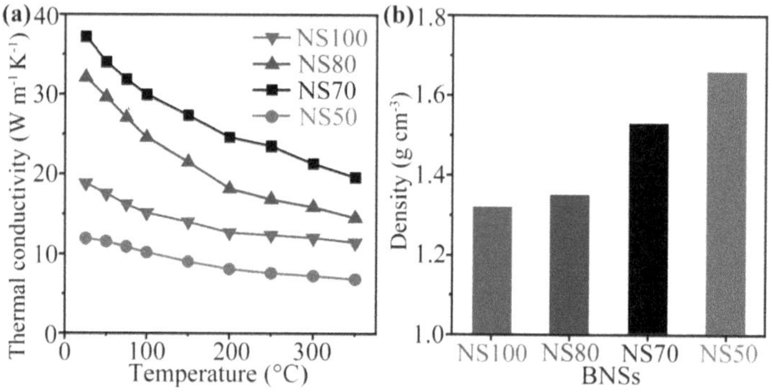

Figure 11.7 (a) Thermal conductivity and (b) density of BNSs with varying BNNS/BNNP ratios. (Reprinted from Ref. [21], Copyright 2024, with permission from John Wiley & Sons, Inc.)

interfacial thermal resistance between NS70 spheres, the average thermal conductivity of NS70 spheres is approximately equivalent to the out-of-plane thermal conductivity of NS70 pellets pressed at 155 MPa for 1 minute and subsequently sintered at 2200°C [21].

11.3.3 Factors Affecting the Thermal Conductivity of BNSs

Based on the aforementioned effective technique, the factors affecting the thermal conductivity of BNSs, including the BNNS/BNNP ratio, the use of polymer binders, and sintering temperature, were systematically investigated.

11.3.3.1 BNNS/BNNP Ratio

BNSs with varying BNNS/BNNP ratios (100:0, 80:20, 70:30, and 50:50), referred to as NS100, NS80, NS70, and NS50, respectively, are compared in Figure 11.7. NS100 exhibited a thermal conductivity of 18.8±4.2 W m^{-1} K^{-1}, largely due to gaps between randomly distributed BNNSs, as evidenced by its low density of approximately 1.32 g cm^{-3}. The addition of BNNPs filled these gaps, and high-temperature sintering improved interactions between BNNSs and BNNPs while enhancing their crystalline structure. Consequently, NS80 demonstrated an improved thermal conductivity of 32.1±3.8 W m^{-1} K^{-1}. NS70, with a BNNS/BNNP ratio of 70:30, achieved the highest thermal conductivity of 37.2±2.5 W m^{-1} K^{-1} and a density of 1.53 g cm^{-3}. Further increasing the BNNP content to 50% (NS50) increased the density to 1.66 g cm^{-3} but significantly reduced the thermal conductivity to 12.5±0.8 W m^{-1} K^{-1}. This reduction was attributed to the excessive BNNPs, which, despite enhancing density, increased thermal interface resistance and phonon scattering due to their small size (100 nm lateral dimensions). Moderately adding BNNPs improves BNNS interactions and thermal conductivity, but excessive amounts introduce detrimental boundary effects. Since the density of BNSs remains relatively low (~1.5 g cm^{-3}), applying higher pressure during high-temperature sintering is expected to further enhance their thermal conductivity [24].

11.3.3.2 Polymer Binders

Polymer binders are commonly used to fabricate BNSs [19,25–27]; however, their inclusion increases interfacial thermal resistance between BNNSs, significantly reducing thermal conductivity. As shown in Figure 11.8, 'binder-free' NS70 exhibited a thermal conductivity of 37.2±2.5 W m^{-1} K^{-1}, whereas PVA-bound BNSs (P-BNSs) had a much lower value of 2.0±0.1 W m^{-1} K^{-1} at room

Figure 11.8 (a) SEM image of P-BNSs. (b) Thermal conductivity of P-BNSs, PR-BNSs, and NS70 as a function of temperature. (Reprinted from Ref. [21], Copyright 2024, with permission from John Wiley & Sons, Inc.)

temperature. This drop is attributed to the low thermal conductivity of PVA and the separation of BNNSs by the binder, which disrupts thermal pathways. Heating the PVA-bound spheres in air at 800°C, followed by sintering at 2000°C for 10 minutes in an Ar atmosphere (PR-BNSs), improved thermal conductivity to 10.1±0.9 W m^{-1} K^{-1}. Low-temperature heating removed the binder, and high-temperature sintering enhanced BNNS interactions and reduced interfacial resistance [20,24]. However, the resulting thermal conductivity remained below that of NS70 due to voids left by the burned-out PVA, which lowered the density. Compared to the 'binder-free' NS70, the use of polymer binders unavoidably reduces the thermal conductivity of BNSs, and even post-heating cannot fully mitigate this effect [21].

11.3.3.3 Sintering Temperature

As noted in Section 11.2, high-temperature sintering effectively enhances BNNS interactions, reducing interfacial thermal resistance. Figure 11.9 compares the thermal conductivity of NS70 sintered at different temperatures. Without sintering, NS70 exhibited a low thermal conductivity of 1.4 W m^{-1} K^{-1} at room temperature, due to weak BNNS interactions and the poor crystalline structure of BNNPs, leading to significant phonon scattering. Thermal conductivity increased with higher sintering temperatures, reaching 37.2 W m^{-1} K^{-1} after sintering at 2200°C. The high thermal conductivity of sintered BNSs is attributed to (1) improved BNNS interactions and crystalline structure, and (2) the absence of organic binders, which would otherwise increase thermal resistance [20,21,24]. Additionally, the thermal conductivity of BNSs is considered isotropic due to their composition of randomly distributed BNNSs.

11.3.4 Application of BNSs in Thermal Management

The heat dissipation performance of BN pellets made from NS70, *h*-BN powders, and P-BNSs was compared as TIMs for a high-power 10 W LED light. To assess the effectiveness of these TIMs in heat management, the LED's temperature was monitored using an infrared camera. As shown in Figure 11.10a, the LED light was placed on top of a BN pellet, which was positioned atop a heat sink. Figure 11.11b–f displays temperature comparisons of the LED when different BN pellets were used as heat spreaders. Without a TIM or heat sink, the LED's temperature reached up to 130°C. However, with the NS70 pellet, which has high thermal conductivity—especially in the

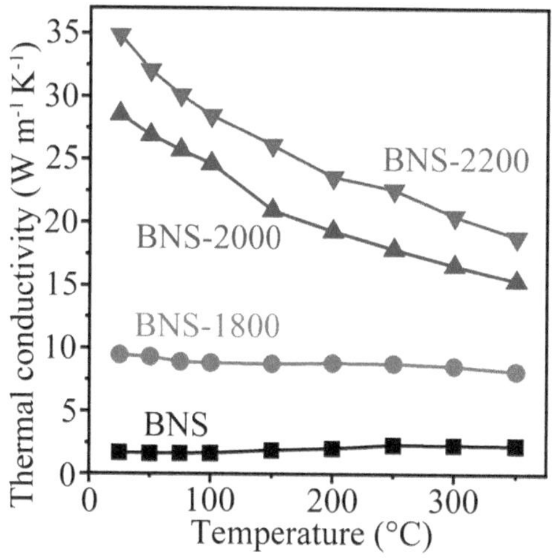

Figure 11.9 Thermal conductivity of NS70 spheres before and after sintering at 1800°C, 2000°C, and 2200°C. (Reprinted from Ref. [20], Copyright (2022), with permission from Elsevier.)

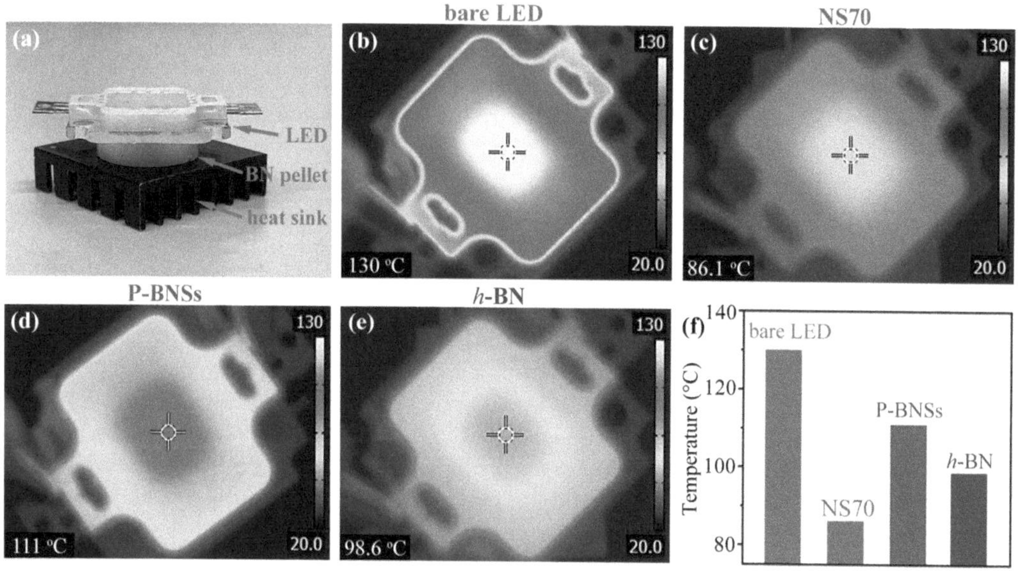

Figure 11.10 (a) Digital photograph of the heat dissipation test system. (b–e) Infrared images and corresponding temperatures (f) of the LED light with different BN TIMs. (Reprinted from Ref. [21], Copyright 2024, with permission from John Wiley & Sons, Inc.)

out-of-plane direction—heat was efficiently transferred from the LED to the heat sink, keeping the temperature below 86.1°C. This was nearly 40°C lower than the bare LED temperature, potentially extending the LED's lifespan by up to four times [28]. In contrast, the P-BNS TIM resulted in a higher LED temperature of 111°C, due to its lower thermal conductivity. Although bulk *h*-BN crystals have a high in-plane thermal conductivity of 420 W m^{-1} K^{-1} [29], the temperature of the

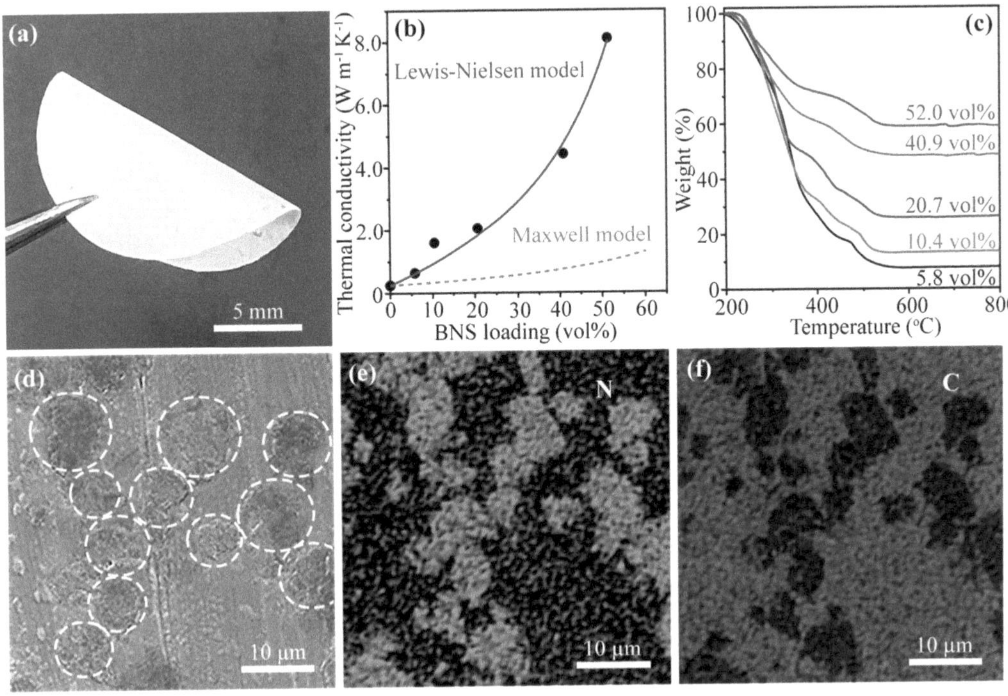

Figure 11.11 (a) Digital photograph of a BNS/PVA film with 52.0 vol% BNS loading. (b) Out-of-plane thermal conductivity of BNS/PVA as a function of BNS loading (black dots) with the corresponding Lewis-Nielsen model fit (blue curve). (c) Thermogravimetric analysis (TGA) results of BNS/PVA samples with varying BNS loadings. (d) SEM image of the cross-section of a BNS/PVA film, with white dashed circles highlighting the BNSs. Energy-dispersive X-ray spectroscopy mapping for (e) N and (f) C corresponding to (d). (Reprinted from Ref. [20], Copyright (2022), with permission from Elsevier.)

h-BN sample still reached 98.6°C. This was due to the low out-of-plane thermal conductivity (~5.7 W m^{-1} K^{-1}) of the *h*-BN platelets [29,30], which hindered efficient heat transfer from the LED to the heat sink in the cross-plane direction.

11.4 THERMAL CONDUCTIVITY AND APPLICATION OF BNS REINFORCED PVA

The 2200°C-sintered NS70 was used as a filler to enhance the thermal conductivity of PVA films (BNS/PVA), fabricated using the vacuum filtration technique (Figure 11.11a). Figure 11.11b illustrates the out-of-plane thermal conductivity of BNS/PVA as a function of BNS content, which was determined by thermogravimetric analysis (Figure 11.11c). The room-temperature thermal conductivity increased gradually with low BNS loadings (<20 vol%) and showed a significant enhancement when the BNS content increased from 40.9 vol% to 52.0 vol%, reaching 8.1 W m^{-1} K^{-1}. The semi-empirical Lewis-Nielsen model, which accounts for the shape, packing, and orientation of particles relative to the heat flux, was used to fit the experimental results (solid curve in Figure 11.11b). In this model, the thermal conductivity of the PVA composite film is expressed as [31]:

$$\frac{\kappa_c}{\kappa_p}=\frac{1+A\phi\dfrac{\kappa_f-\kappa_p}{\kappa_f+A\kappa_p}}{1-\dfrac{(\kappa_f-\kappa_p)\phi}{\kappa_f+A\kappa_p}\left[1+\dfrac{(1-\phi_m)\phi}{\phi_m^2}\right]} \tag{11.1}$$

In the model, κ_c, κ_p, and κ_f are the thermal conductivities of the composite, polymer matrix, and filler, respectively, and ϕ is the volume fraction of the filler. The constant A relates to the generalized Einstein coefficient (k_E) and depends on the filler shape, aspect ratio, and orientation relative to the heat flux. The parameter ϕ_m represents the maximum packing fraction of fillers. Using A and ϕ_m as fitting parameters, the model closely matched the experimental data across the full range of BNS loadings. The fitted ϕ_m value was 0.66, slightly higher than the theoretical 0.64 for random close packing of spheres [31]. This increase likely results from the non-uniform size distribution of BNSs, with smaller particles filling gaps between larger ones. The fitted A value was 28.5, significantly higher than the theoretical value of 1.50 for spheres. This discrepancy is attributed to the influence of the shape and aspect ratio (length/diameter). For instance, A increases from 1.58 to 8.38 for randomly oriented rods as the aspect ratio rises from 2 to 15 [31]. The large A value in the BNS/PVA sample suggests the formation of thermally conductive pathways with a rod-like structure and high aspect ratio [32]. SEM images of the BNS/PVA composite cross-section (52.0 vol% BNS) confirm the formation of such pathways within the polymer matrix (Figure 11.11d–f).

The thermal conductivities of BNS/PVA films (both in-plane and out-of-plane) were compared with those of pure PVA, BNNS-enhanced PVA (BNNS/PVA), and *h*-BN platelet-reinforced PVA (*h*-BN/PVA), all containing similar BN filler contents (~60.0 wt%). As shown in Figure 11.12a, pure PVA exhibited low isotropic thermal conductivity (~0.25 W m^{-1} K^{-1}). The BNNS/PVA composite achieved moderate enhancement, with an in-plane thermal conductivity of ~3 W m^{-1} K^{-1} and an out-of-plane value of ~1.8 W m^{-1} K^{-1}. However, these values were insufficient for efficient heat dissipation due to limited conductive pathway formation and increased phonon scattering from the small BNNS size. In contrast, the BNS/PVA films exhibited significantly higher in-plane (10.6 W m^{-1} K^{-1}) and out-of-plane (8.1 W m^{-1} K^{-1}) thermal conductivities, yielding an isotropic thermal conductivity factor $f_{(O/I)}$ of 0.76, indicative of a random BNNS distribution in the composite. Meanwhile, *h*-BN/PVA films showed a high in-plane conductivity of 12.5 W m^{-1} K^{-1} but a much lower out-of-plane value (3.5 W m^{-1} K^{-1}), resulting in $f_{(O/I)}=0.28$. This difference is attributed to the 'laying down' and anisotropic thermal conductivity of *h*-BN platelets [20]. XRD analysis confirmed the random distribution of BNNSs in BNS/PVA and the preferential orientation in *h*-BN/PVA through the intensity ratio of the (100) and (002) peaks (I_{100}/I_{002}), as shown in Figure 11.12b.

Finally, the heat dissipation performance of the composite films was evaluated using a custom-made hot spot and infrared thermal imaging [20]. The 0.25 mm-thick films—close to the thickness of TIMs [33,34]—were heated from 25°C to 100°C, and their surface temperatures were monitored by an infrared camera. Figure 11.13 shows the optical images and temperature evolution curves of the samples. BNS/PVA exhibited the fastest heat dissipation (circles) due to its superior in-plane and out-of-plane thermal conductivities. *h*-BN/PVA followed, with its lower out-of-plane conductivity (3.5 W m^{-1} K^{-1}) slowing heat transfer across the plane (squares), despite its high in-plane thermal conductivity of 12.5 W m^{-1} K^{-1}. BNNS/PVA (rhombuses) and pure PVA (pentagons) showed much slower heat dissipation due to their poor thermal conductivities in both directions [20].

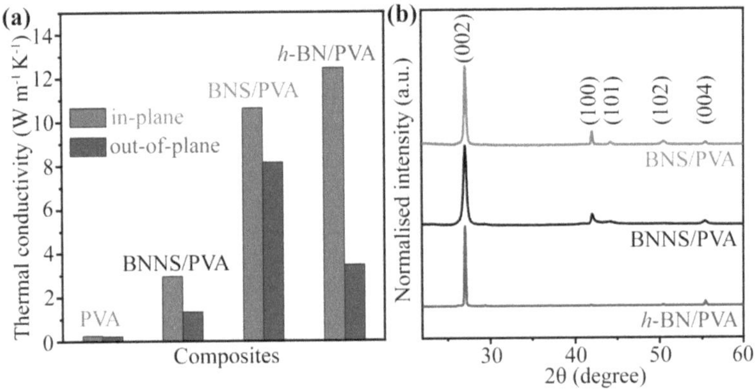

Figure 11.12 (a) Comparison of room-temperature in-plane and out-of-plane thermal conductivities for pure PVA, BNNS/PVA, BNS/PVA, and *h*-BN/PVA. (b) XRD patterns of PVA composites enhanced with BNSs, BNNSs, and *h*-BN, respectively. (Reprinted from Ref. [20], Copyright (2022), with permission from Elsevier.)

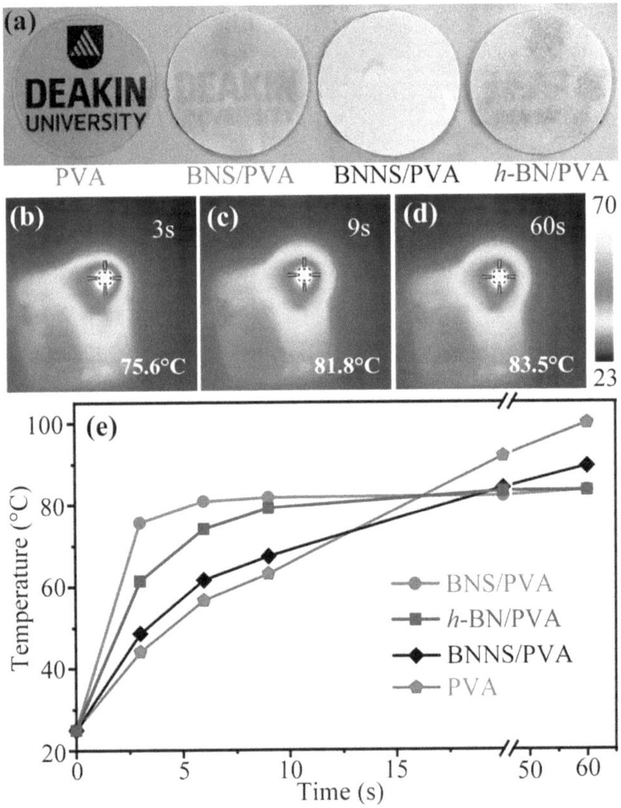

Figure 11.13 (a) Digital photographs of PVA composites. (b–d) Infrared images of BNS/PVA taken at 3, 9 and 60 seconds after being heated by a customer-made hot plate. (e) Temperature variations of PVA composites as a function of time. (Reprinted from Ref. [20], Copyright (2022), with permission from Elsevier.)

11.5 CONCLUSIONS

In summary, a spray drying technique has been developed to fabricate binder-free BNSs using BNNSs and BNNP dispersion as precursors. BNNPs act as a 'binder', filling the gaps between BNNSs and enhancing their interactions. High-temperature sintering up to 2200°C significantly improved the thermal conductivity of BNSs to 37.2 W m^{-1} K^{-1} by strengthening the interactions between BNNSs and improving their crystalline structure. The factors influencing the thermal conductivity of BNSs, including the BNNS/BNNP precursor ratio, polymer binder, and sintering temperature, were systematically examined. The BNSs were then used as fillers to enhance the thermal conductivity of PVA. Due to the random dispersion of BNNSs within the BNSs and their spherical structure, the BNS-enhanced PVA film exhibited quasi-isotropic thermal conductivity, with an in-plane value of 10.6 W m^{-1} K^{-1} and an out-of-plane value of 8.1 W m^{-1} K^{-1} at 52.0 vol% BNS loading. This high thermal conductivity is attributed to the formation of thermally conductive paths in the PVA matrix, as demonstrated both experimentally and theoretically. The two-step 'spray drying and sintering' process is easily scalable, and the BNS/PVA films, with their fast heat dissipation properties, hold promise for thermal management in advanced electronics.

REFERENCES

1. X. Huang et al., Polyhedral oligosilsesquioxane-modified boron nitride nanotube based epoxy nanocomposites: An ideal dielectric material with high thermal conductivity. *Advanced Functional Materials* **23**, 1824–1831 (2013).
2. J. Chen, X. Huang, Y. Zhu, P. Jiang, Cellulose nanofiber supported 3D interconnected BN nanosheets for epoxy nanocomposites with ultrahigh thermal management capability. *Advanced Functional Materials* **27**, 1604754 (2017).
3. K. Kang et al., High-mobility three-atom-thick semiconducting films with wafer-scale homogeneity. *Nature* **520**, 656–660 (2015).
4. R. Bahru, M. F. M. A. Zamri, A. H. Shamsuddin, N. Shaari, M. A. Mohamed, A review of thermal interface material fabrication method toward enhancing heat dissipation. *International Journal of Energy Research* **45**, 3548–3568 (2021).
5. R. Prasher, Thermal interface materials: Historical perspective, status, and future directions. *Proceedings of the IEEE* **94**, 1571–1586 (2006).
6. Y. Yao et al., Construction of 3D skeleton for polymer composites achieving a high thermal conductivity. *Small* **14**, 1704044 (2018).
7. Y. Guo, K. Ruan, X. Shi, X. Yang, J. Gu, Factors affecting thermal conductivities of the polymers and polymer composites: A review. *Composites Science and Technology* **193**, 108134 (2020).
8. Q. Yan et al., Ultrahigh-aspect-ratio boron nitride nanosheets leading to superhigh in-plane thermal conductivity of foldable heat spreader. *ACS Nano* **15**, 6489–6498 (2021).
9. L. An et al., Hexagonal boron nitride nanosheets: Preparation, heat transport property and application as thermally conductive fillers. *Progress in Materials Science* **138**, 101154 (2023).
10. J. Chen, X. Huang, B. Sun, P. Jiang, Highly thermally conductive yet electrically insulating polymer/boron nitride nanosheets nanocomposite films for improved thermal management capability. *ACS Nano* **13**, 337–345 (2018).
11. J. Han, G. Du, W. Gao, H. Bai, An anisotropically high thermal conductive boron nitride/epoxy composite based on nacre-mimetic 3D network. *Advanced Functional Materials* **29**, 1900412 (2019).
12. G. Yang et al., Highly thermal conductive poly (vinyl alcohol) composites with oriented hybrid networks: Silver nanowire bridged boron nitride nanoplatelets. *ACS Applied Materials & Interfaces* **13**, 32286–32294 (2021).

13. T. Xu et al., Polyamide composites with improved thermal conductivity for effective thermal management: The three-dimensional vertically aligned carbon network. *Composites Part B: Engineering* **224**, 109205 (2021).
14. A. L. Moore, L. Shi, Emerging challenges and materials for thermal management of electronics. *Materials Today* **17**, 163–174 (2014).
15. T. Kusunose, Y. Uno, Y. Tanaka, T. Sekino, Isotropic enhancement of the thermal conductivity of polymer composites by dispersion of equiaxed polyhedral boron nitride fillers. *Composites Science and Technology* **208**, 108770 (2021).
16. Y. Xue et al., Densely interconnected porous BN frameworks for multifunctional and isotropically thermoconductive polymer composites. *Advanced Functional Materials* **28**, 1801205 (2018).
17. K. Sato, Y. Tominaga, Y. Hotta, Y. Imai, A facile method to prepare layered solid fillers-based polymer composites with isotropic thermal conductivity. *Composites Part A: Applied Science and Manufacturing* **154**, 106776 (2022).
18. L. An et al., Quasi-isotropically thermal conductive, highly transparent, insulating and super-flexible polymer films achieved by cross linked 2D hexagonal boron nitride nanosheets. *Small* **17**, 2101409 (2021).
19. L. Ren, X. Zeng, R. Sun, J.-B. Xu, C.-P. Wong, Spray-assisted assembled spherical boron nitride as fillers for polymers with enhanced thermally conductivity. *Chemical Engineering Journal* **370**, 166–175 (2019).
20. H. Jiang et al., Quasi-isotropic thermal conductivity of polymer films enhanced by binder-free boron nitride spheres. *Composites Science and Technology* **230**, 109769 (2022).
21. H. Jiang et al., Unleashing the potential of boron nitride spheres for high-performance thermal management. *ChemNanoMat*, e202300601 (2024).
22. T. Xing et al., Gas protection of two-dimensional nanomaterials from high-energy impacts. *Scientific Reports* **6**, 35532 (2016).
23. H. Jiang et al., Boron nitride nanosheet dispersion at high concentrations. *ACS Applied Materials & Interfaces* **13**, 44751–44759 (2021).
24. S. Mateti et al., Bulk hexagonal boron nitride with a quasi-isotropic thermal conductivity. *Advanced Functional Materials* **28**, 1707556 (2018).
25. X. Zeng et al., Ice-templated assembly strategy to construct 3D boron nitride nanosheet networks in polymer composites for thermal conductivity improvement. *Small* **11**, 6205–6213 (2015).
26. F. Jiang, S. Zhou, T. Xu, N. Song, P. Ding, Enhanced thermal conductive and mechanical properties of thermoresponsive polymeric composites: Influence of 3D interconnected boron nitride network supported by polyurethane@polydopamine skeleton. *Composites Science and Technology* **208**, 108779 (2021).
27. H.-J. Hong et al., Highly flexible and stretchable thermally conductive composite film by polyurethane supported 3D networks of boron nitride. *Composites Science and Technology* **152**, 94–100 (2017).
28. J. Huang et al., Optical degradation mechanisms of mid-power white-light LEDs in LM-80-08 tests. *Microelectronics Reliability* **55**, 2654–2662 (2015).
29. P. Jiang, X. Qian, R. Yang, L. Lindsay, Anisotropic thermal transport in bulk hexagonal boron nitride. *Physical Review Materials* **2**, 064005 (2018).
30. P. Jiang, X. Qian, R. Yang, Time-domain thermoreflectance (TDTR) measurements of anisotropic thermal conductivity using a variable spot size approach. *Review of Scientific Instruments* **88**, 074901 (2017).
31. L. E. Nielsen, The thermal and electrical conductivity of two-phase systems. *Industrial & Engineering Chemistry Fundamentals* **13**, 17–20 (1974).
32. F. Kargar et al., Thermal percolation threshold and thermal properties of composites with high loading of graphene and boron nitride fillers. *ACS Applied Materials & Interfaces* **10**, 37555–37565 (2018).
33. D. Chung, Thermal interface materials. *Journal of Materials Engineering and Performance* **10**, 56–59 (2001).
34. H. Guo, X. Li, B. Li, J. Wang, S. Wang, Thermal conductivity of graphene/poly (vinylidene fluoride) nanocomposite membrane. *Materials & Design* **114**, 355–363 (2017).

Chapter 12

Boron Nitride-Based Composites for Efficient Heat Dissipation

Siyuan Ding and Ling Qiu

12.1 INTRODUCTION

Hexagonal BN (hBN) is well known for its attractive physical properties, including insulation and high thermal conductivity, which is promising for the synthesis of assemblies with improved performance. The exfoliation of hBN into BN nanosheets (BNNSs) produces a nanosheet structure that has atomic thickness but a large in-plane diameter. This 2D material nature makes it easy to assemble, significantly improving its processability and allowing for the fabrication of unique nanostructures from 0D to 3D. With good performance and the ability to fabricate unique structures, BNNS nanostructures have demonstrated their potential in realizing high-performance thermal management in different practical scenarios. In this chapter, the BN nanostructures from 0D to 3D for efficient heat dissipation will be reviewed. Since the structure achieved is highly related to the processing technique, the review will apply the fabrication method as a starting point and introduce the structures produced and the corresponding uses.

12.2 0D BN NANOSTRUCTURES

0D BN nanostructures are fabricated by granulating several small hBN (including BNNS) into sphere-like particles. These particles are produced for thermal management materials as thermally conductive fillers. hBN, due to its layered structure and highly anisotropic thermally conductive characteristics, usually requires special processing techniques to achieve effective filling so that the composites produced can effectively dissipate heat. However, 0D BN nanostructures have a sphere-like structure, which makes them easier to construct isotropic thermal conductive pathways in composites by random packing. Furthermore, these particles usually do not require specific processing techniques, but manufacturing technologies that are well established in industry. Many studies have been conducted by adopting several strategies, including template methods [1–7] and sintering methods [8,9]. However, the synthesis of strong 0D BN nanostructures with a high thermal conductivity is a long-standing problem, limiting their practical applications.

DOI: 10.1201/9781003647812-15

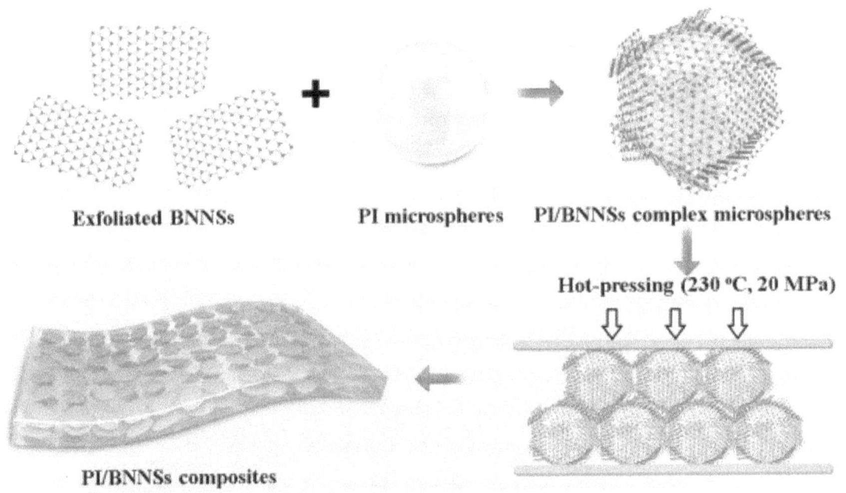

Figure 12.1 Typical processing of 0D BN nanostructures using a polymer substrate [6].

12.2.1 Fabrication of 0D BN Nanostructures

Template methods require a spherical substrate to fabricate hBN onto its surface (Figure 12.1). In this process, different materials are adopted as substrates, including polymers and inorganic materials. For polymer substrates, hBN flakes are fabricated onto the surface by adding polymer adhesives [4] or interface assembly [2,6,10]. These particles, with a high content of polymer that softens at elevated temperatures, can be directly hot-pressed to form thermal management materials (Figure 12.1). However, this polymer substrate-based strategy suffers from limited improvement of thermal conductivities. Because hBN flakes are only fabricated at the surface of the particles, the content of hBN in the particles is limited, which further limits the content of hBN in the composite prepared by these particles. Therefore, although the formation of thermal conductive pathways in the fabricated composites has been reported, due to a low filler content, the performance is still limited.

For 0D hBN particles fabricated using inorganic substrates [1,3,5,7], (e.g., glass, spherical graphite, AlN), thermal conductivity significantly surpasses that of polymer substrates, which typically range up to 1 $W{\cdot}m^{-1}{\cdot}K^{-1}$. For example, glass substrates offer around 1 $W{\cdot}m^{-1}{\cdot}K^{-1}$ [11], while spherical graphite and AlN exhibit approximately 300 and 320 $W{\cdot}m^{-1}{\cdot}K^{-1}$, respectively [12,13] However, due to different surface properties, the fabrication of BNNSs onto the surface of substrates is more complicated and usually requires a multistep chemical processing to functionalize the inorganic surface. Therefore, the absorption and the in-situ growth of hBN can be achieved. The use of inorganic sphere substrates improves the thermal conductivity of the synthesized particles; however, existing studies primarily focus on using pressing techniques to assemble these particles into thermal management materials.

The sintering method involves granulating hBN flakes into a spherical structure, followed by high-temperature sintering [8,9] (Figure 12.2). In the former step, hBN flakes are granulated into spherical agglomerates using spray drying, which have analogous shapes to the final products but are weakly bonded. In the latter step, these agglomerates are sintered at high temperatures, and the originally separated hBN bonded together by diffusion. This method produces inorganic

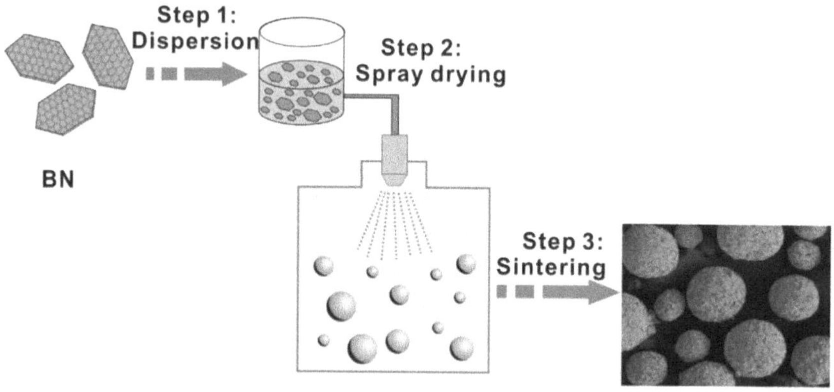

Figure 12.2 Typical processing of 0D BN nanostructures using a sintering method [8].

particles consisting mainly of hBN, which therefore share similar properties to hBN, including high thermal conductivity, chemical stability, and thermal stability. These particles are mixed with polymers to fabricate thermal management materials, and their ceramic nature allows the use of different polymer and solvent systems, resulting in wider application than those produced by above mentioned template methods. However, due to the inertness in sintering of hBN, the particles have a loose structure and tend to break under mixing in high-viscosity conditions. This leads to several problems, including limited processing efficiency, low maximum filler content, and low homogeneity. Consequently, although these particles have also been produced by industry for decades, their commercial uses are still limited.

12.2.2 Use of 0D BN Nanostructures

0D BN nanostructures have long been used in the industry because they can achieve easy fabrication of isotropic thermally conductive thermal management materials. As an insulating and yet thermally conductive filler, hBN shows great advantages over other thermal conductive fillers (Table 12.1). However, its anisotropic thermal conductive behavior (i.e., a high in-plane conductivity $\approx$750 $W{\cdot}m^{-1}{\cdot}K^{-1}$ but a low through-plane thermal conductivity $\approx$5 $W{\cdot}m^{-1}{\cdot}K^{-1}$) [14,15] requires special processing to achieve the effective filling and the construction of thermal conductive pathways, which however challenges the manufacturing methods and limits application. Granulating hBN into spherical-like particles provides a clear solution to address this problem. Besides, spherical-like particles have better flowability than flake/nanosheet particles; therefore, they are easier to incorporate into polymer composites based on proven technologies like high-speed mixing. In addition, the spherical-like particles are free to rotate in the filling of composites, leading to isotropic thermal conductivity, which is preferable for some applications. Therefore, the 0D BN nanostructures are expected to be directly mixed with polymer using industry manufacturing methods, which, however, has not been achieved mainly due to low particle strength.

12.3 1D BN NANOSTRUCTURES

1D BN nanostructures are fibers. They are fabricated by spinning the dispersion of hBN or BNNSs and polymer matrixes into 1D structures with the length $\gg$ the diameter. These fibers are further woven into fabrics used in wearable devices or incorporated into polymer matrixes to improve

TABLE 12.1 PROPERTIES OF COMMERCIALLY AVAILABLE THERMALLY CONDUCTIVE AND INSULATING FILLERS

	Thermal conductivity ($W{\cdot}m^{-1}K^{-1}$)	Volume resistivity ($\Omega{\cdot}cm$)	Chemical stability	Thermal stability	Nontoxicity	Price
hBN	700	$>10^{14}$	✓	✓	✓	✓
Al_2O_3	30	$>10^{14}$	✓	✓	✓	✓
ZnO	26	$>10^{8}$	×	×	Toxic	✓
MgO	36	$>10^{16}$	Corrosion	×	✓	✓
Si_3N_4	100	$>10^{14}$	✓	✓	✓	Expensive
BeO	300	$>10^{14}$	×	✓	Toxic	Expensive
AlN	285	$>10^{9}$	Hydrolysis	✓	✓	Expensive

mechanical and thermal performance. Therefore, the mechanical performance, especially strength and flexibility, is very important to them.

12.3.1 Fabrication of 1D BN Nanostructures

To date, many studies have been conducted to achieve the synthesis of 1D BN nanostructures, aiming at reaching better performance or some unique function. Their strategies are classified into the polymer matrix method and the pyrolysis method.

The polymer matrix method produces fibers with hBN/BNNS embedded in the polymer matrix. These fibers are developed to combine the properties of polymer fibers (stretchability, flexibility, mechanical strength) and hBN/BNNS (high thermal conductivity, electric insulation, low dielectric constant). They can be further woven into textiles with different structures, which can better adapt to various stress conditions, including curved surfaces, therefore achieving various thermal management functions. Compared to other fillers, BNNSs offer better flexibility due to their atomic thickness [16,17] This allows them to composite with the polymer matrix in fibers. As a result, significant efforts have been made to develop BNNS–polymer composite fibers, primarily using the wet spinning method.

The wet spinning method of a BNNS–polymer composite involves injecting colloidal dispersions of a BNNS and polymer matrix into a coagulation bath (Figure 12.3). This bath contains a solvent that does not dissolve the polymer but is miscible with the solvent in the spinning dispersion. Upon contact with the coagulation bath, the polymer precipitates [18], forming solid fibers with an embedded BNNS. Proper component design and a well-ordered structure are crucial to improving the properties of fibers. Strong polymers are used as the matrix, including polyvinyl alcohol [19], cellulose [20], polyimide [21], and aramid nanofiber [22,23]. Adjusting BNNS content is also essential. The arrangement of BNNSs in the fiber structure is important in reaching higher thermal conductivity and mechanical strength, which can be controlled through different techniques. Lu et. al. reported a coaxial spinning method to produce BNNS/aramid nanofiber with a high BNNS content as the shell and pure aramid nanofiber as the core [23] and used the compressive forces at the interface between the core and shell during drawing to align the BNNS in the axis direction. Another strategy aims at improving the attachment of BNNSs to the polymer matrix. Fang et. al. modified a BNNS with polydopamine, which was then covalently bonded to the polyimide matrix through in-situ polymerization [21]. Besides, due to the

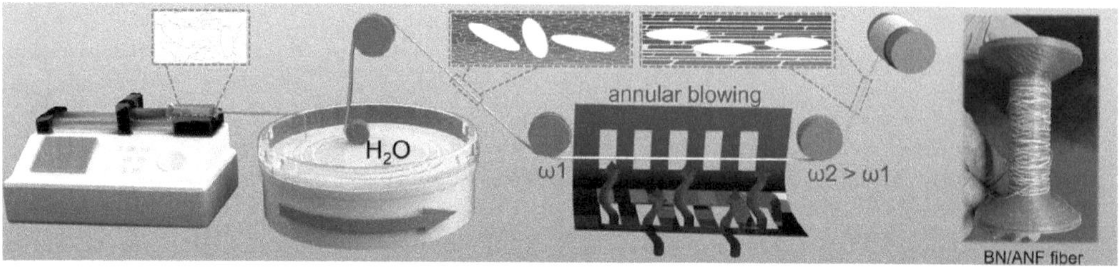

Figure 12.3 Typical process of synthesizing 1D BN nanostructures using wet spinning in the polymer matrix method [22].

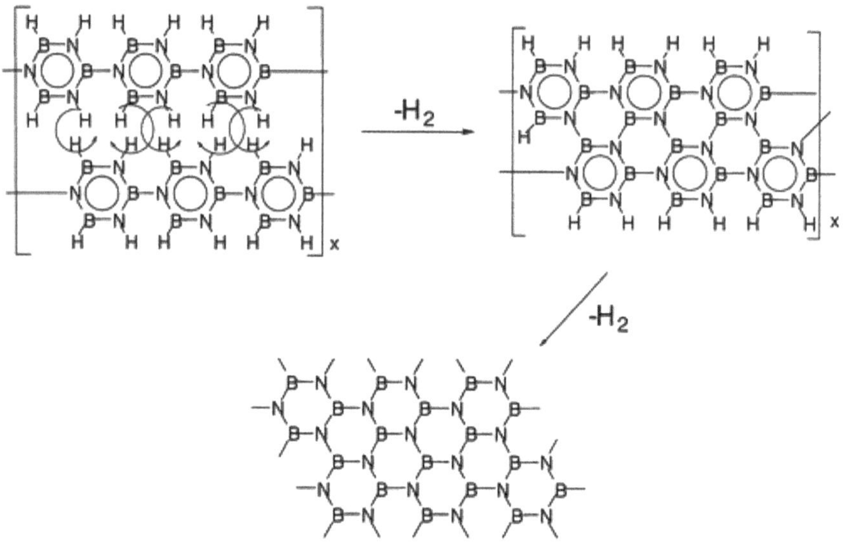

Figure 12.4 Crosslinking process for the ceramic conversion reaction of polyborazylene in the pyrolysis method of synthesizing 1D BN nanostructures [30].

similar 2D structure to graphene, the techniques used in spinning other 2D material fibers are also applicable. For example, Eom et al. revealed that the shear between the spinning needle and the nanosheets influences the arrangement of nanosheets in the fiber. By cooperatively controlling such effect, fibers with highly arranged fibers were produced, demonstrating high torsional strength [24]. These fibers, showing a good combination of BNNSs and polymers, have demonstrated applications in weaving textiles for personal thermal management [25].

Noticeably, there are some other methods synthesizing polymer matrix hBN/BNNS fibers, including coating on fabrics [26] and electrospinning, whereas they directly produced fibrous composites rather than fibers and therefore will later be discussed in the 3D BN nanostructures.

The pyrolysis method produces ceramic BN fibers based on the pyrolysis and sintering of different precursors. During the former process, the precursor fibers are transformed into BN fibers based on the polycondensation effect of borazine and its derivatives [27] (Figure 12.4), while the latter process refines the structure formed. To date, different precursors have been applied as precursors, including polyborazine [28,29], polyborazylene [30], polyalkylaminoborazine [31–33], and borylborazine [34,35]. In a typical synthesis process, these precursors, which are meltable and have low melting points, are melt-extruded to form fibers [36]. Then, during pyrolysis, the polymer components are converted into ceramics and crosslinked by intermolecular removal of small molecules, which finally produces a BN matrix. The fiber produced with this method shows strong thermal stability and chemical inertness and therefore holds great potential in applications under extreme conditions. However, due to difficulty in controlling the polycondensation process, this method produces small fibers (diameters mostly $< 10\,\mu m$) with low densities, low crystallinity, and impurities generated during pyrolysis. Consequently, their applications are mainly limited to 1D fillers to reinforce composites.

12.3.2 Use of 1D BN Nanostructures

The application of 1D BN nanostructures generally falls into two categories: weaving textiles and producing fiber reinforced composites. The former is an emerging concept driven by the needs of wearable electronics and healthcare systems, which require close contact with human skin (Figure 12.5). Therefore, these applications demand high standards for thermal management systems, including lightweight, flexibility, thermal conductivity, and skin comfort. Among these requirements, the ability to withstand complicated stress conditions while wearing is the most important, which is skillfully met by weaving fibers into textiles, a traditional technique for making clothes. Therefore, the key requirements in developing fibers for textiles are the mechanical properties to meet the requirements of weaving techniques, including flexibility, stretchability, tensile strength, and torsional strength. However, although the filling of 2D materials has demonstrated exceptional potential in improving functionality such as thermal conductivity, the ability to weave textiles using industrial techniques is still seldom discussed.

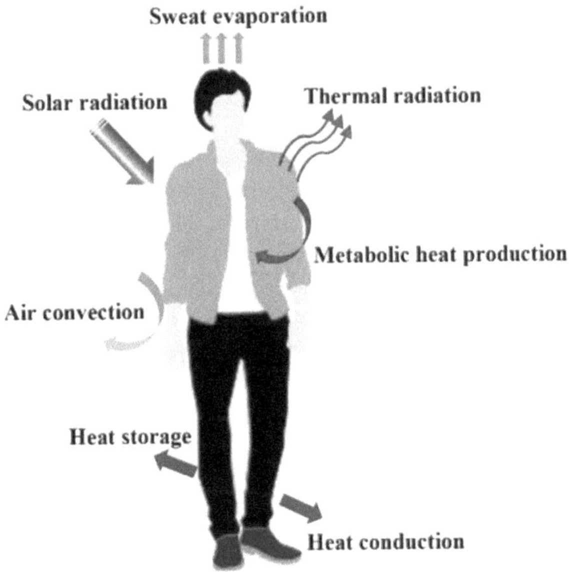

Figure 12.5 Personal thermal management based on textiles weaved by fibers[25].

12.4 2D BN NANOSTRUCTURES

2D BN nanostructures refer to films with BNNSs differently arranged in structure. Over the past few years, they have been widely investigated since the 2D structure of films has great compatibility with the layer structure of 2D materials. To date, various structures and synthesizing methods of hBN films have been extensively developed to achieve different thermal management functions. In this section, the synthesis and application of hBN/BNNS-based films will be discussed according to the structural classification of in-plane oriented films and structured films.

12.4.1 In-Plane Oriented 2D BN Nanostructures

In-plane oriented BN nanostructures have been a well-developed structure during the past years. Aligning hBN along the direction of the film plane enables a high in-plane thermal conductivity but a low through-plane thermal conductivity. This film is usually used as a thermal dissipating film, which dissipates heat from a hot spot to the whole film plane, therefore achieving cooling by enlarging the cooling area. The low through-plane thermal conductivity can also prevent the transfer of heat to the other side of the film, thus protecting the component on the other side from being overheated, which is called preventing thermal shoot-through. To synthesize these films with a high performance, several guidelines must be considered. BNNSs must be spatially densely packed and well oriented to effectively build thermally conductive pathways in the polymer matrix. In addition, interface resistance in the film structure must be reduced because the interfaces (BNNS/BNNS, BNNS/polymer, and polymer/polymer) cause severe phonon scattering, which decreases thermal conductivities. Many synthesis methods have been developed according to these guidelines, which are discussed as follows. Noticeably, these methods can be used together in pursuit of better performance, which is common in the latest studies.

The vacuum infiltration method is the mainstream lab-scale method in fabricating BNNS films, which features easy manipulation and high quality of assembly. This method involves filtrating a dispersion of BNNSs and polymers through a film driven by the negative pressure of vacuum [37–40] (Figure 12.6a). During the vacuum infiltration process, the solvent in the dispersion is removed by infiltrating through the membranes, and the polymer matrix and BNNS are assembled on the membrane to form a film. During this process, the BNNS, compared to hBN, tends to arrange parallel to the plane of the membrane (Figure 12.6b and c) due to its ultrathin structure. The assembly at the membrane interface is determined by whether BNNSs are free to move in the dispersion during infiltration, which is further affected by the dispersion condition and the flowability of the dispersion. A good dispersion of BNNSs prevents the agglomeration effect, which pre-assembles BNNSs into irregular agglomerates and thus hinders the assembly at the membrane interface. This can be achieved through functionalization [41,42] and the addition of surfactants [43,44]. The concentration of BNNSs and the viscosity of the dispersion are usually set very low to achieve a free flow [45]. The film produced by vacuum infiltration usually has a multilayer structure where BNNSs highly align along the direction of the plane [40], and the polymer matrix interconnects them and fills the interspace [46,47]. Vacuum infiltration is easy to use and produces films with fine structure; however, it is usually considered as a lab-scale method, which is mainly limited by the scalability. The low-concentration dispersion used results in low efficiency, and the area of the membrane also limits the area of the film assembled. Further, vacuum infiltration also has difficulties in assembling thick films because the thick film on the membrane impedes the flow of dispersion, leading to a decrease in assembling efficiency and worsened orientation of BNNSs.

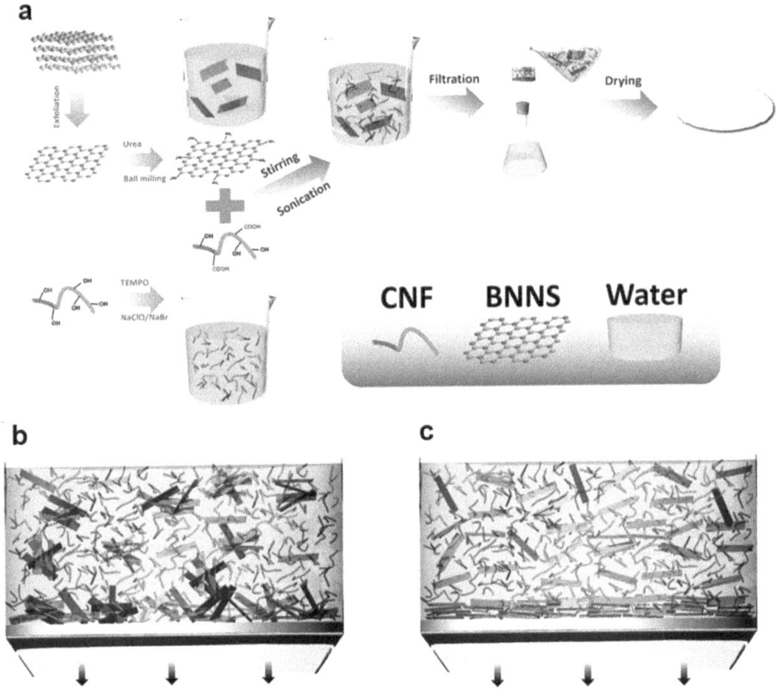

Figure 12.6 Typical process of the synthesis of 2D BN nanostructures using vacuum infiltration [48]. (a) Schematic of the synthesis process. (b,c) Assembly and orientation of hBN (b) and BNNS (c) during the vacuum infiltration.

The doctor blading method is a technique commonly used in industry to continuously produce film from a dispersion [49]. It involves spreading a viscous dispersion across a substrate and then using a blade to scrape off the excess, leaving a thin, uniform layer of the material on the substrate [50,51] (Figure 12.7a). During blading the solution, BNNSs, due to their 2D structure, tend to orient along the direction of blading under the mechanical shear [52]. This shear is applied to the dispersion by the movement of the blade and transferred to the BNNS in the dispersion [53] (Figure 12.7b). The movement of the blade along a certain direction forces the fluid and moves the top plate of the fluid, which creates a separation of the top and the bottom plates. Because the fluid has resistance to such movement of neighboring plates, which is characterized by the viscosity of the fluid, the shear force is then transferred inside the dispersion [54]. The BNNSs in the dispersion are then aligned along the direction of the shear force, which is also the direction of the movement of the blade and the flow of the fluid. The doctor blading method is controlled by the rheological properties, fluidic dynamics during shearing, and the stability of the dispersion. The former two factors determine how shear force is applied to orient BNNSs, which is further controlled by several parameters, including viscosity, viscosity behaviors, and equipment parameters. The stability of dispersion controls the drying process, and a well-dispersed system prevents the segregation, agglomeration, and pre-condensation of BNNSs. To eliminate defects such as pores and wrinkles, there are many limitations in the selection of polymer matrixes, including proper viscosity, and many additives to tune the rheological

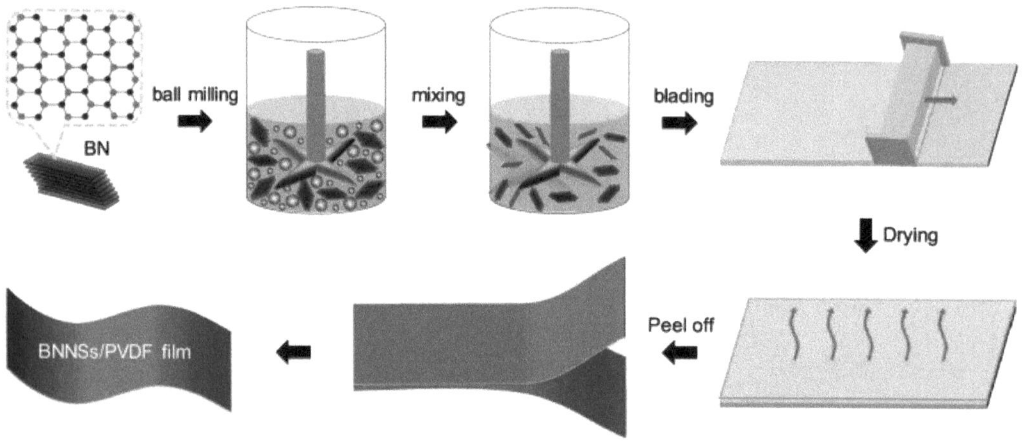

Figure 12.7 Typical process of the synthesis of 2D BN nanostructures using blading [51].

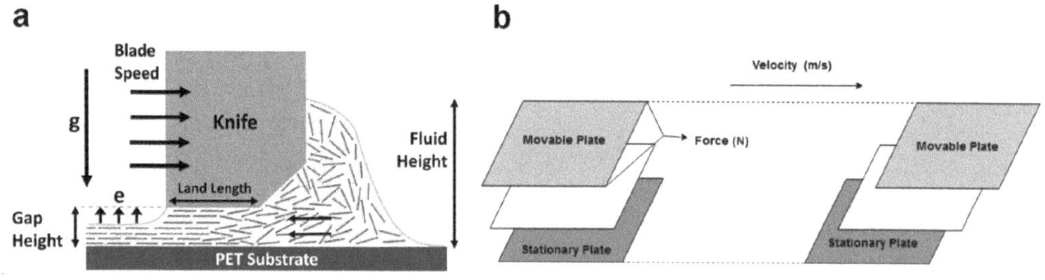

Figure 12.8 Mechanism of the orientation of 2D materials during doctor blading. (a) Schematic of the flow and orientation of 2D materials during blading [53]. (b) Simple viscosity model [54].

properties and the dispersibility are also necessary. Besides, the alignment and packing of BNNSs in the film can be affected. The doctor blading method can apply a high BNNS concentration [55], allowing for a high assembling efficiency. It also has the ability to produce continuous films with controllable thickness. However, its difficulty in controlling parameters still challenges the quality of the products (Figure 12.8).

The solid mechanical methods, including hot pressing [56,57], stretching [58–60], and roller casting [61], are usually used as a post-process to refine the structure of the assembly [62]. Hot pressing applies stress on the assembly, during which the polymer matrix softens and flows at high temperatures and high applied pressure, thus densifying the structure (Figure 12.9a). In the case of a BNNS, which has a layer structure, such a flow of polymer is coordinated with the alignment of the BNNS, promoting the orientation of BNNSs along the direction perpendicular to the plane of pressing [57]. The hot pressing process also eliminates the interspace between a polymer matrix and BNNSs and expels the pores in the structure [63], thus densifying the structure and promoting the contact between the polymer matrix and fillers. These effects are very important in improving the thermal conductivities of the assemblies because pores introduce air with a thermal conductivity of <0.3 $W{\cdot}m^{-1}\ K^{-1}$ into the structure, while the incomplete contact between

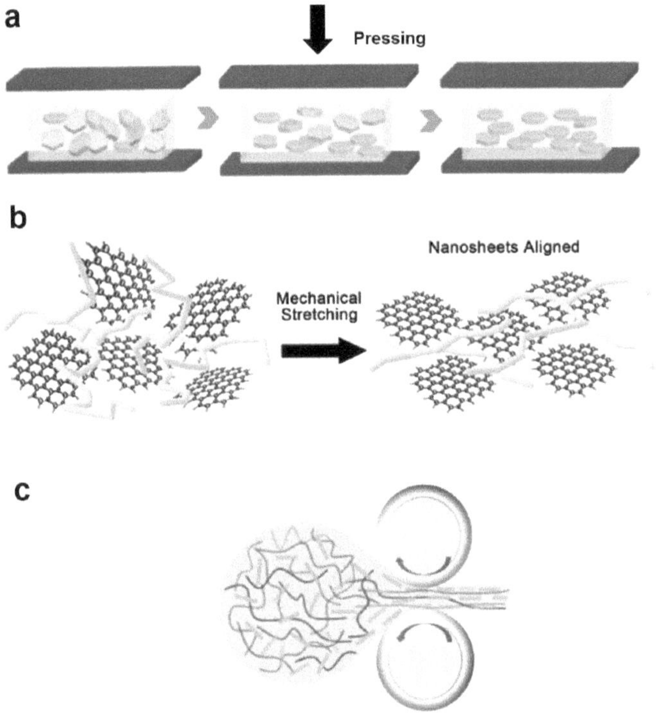

Figure 12.9 Solid mechanical method used in synthesizing 2D BN nanostructures. (a) Hot pressing [57]. (b) Stretching [58]. (c) Roller casting [61].

fillers and polymers produces severe phonon scattering. Stretching is another important strategy in achieving densification of structures. It applies in-plane tensile stress to the assemblies without breaking them (Figure 12.9b). During this process, the deformation of polymers also causes the re-arrangement of the fillers, making them orient along the direction of stretching [64]. This method can also be used for assemblies with low BNNS content since the stretching direction is the same as the orientation direction of the BNNS. The roller casting method, also known as the roll-to-roll method, involves using two opposing rollers to press and shape solid or semi-solid materials into the desired thickness and width. The two opposing rollers apply pressure on the materials, densifying them and promoting the orientation of the BNNS inside them. This method is adopted in industry to achieve continuous processing in the production line.

12.4.2 Structured 2D BN Nanostructures

Structured films are developed to achieve heat transfer behaviors or thermal management functions different from in-plane oriented films, which spread heat in the plane but hinder the heat transfer through the plane (Figure 12.10a). The main goal in this domain is to increase the through-plane thermal conductivity of the film. These films are mainly used as heat sinks (Figure 12.11a) or thermal interface materials (Figure 12.11b) in electronic devices, which require

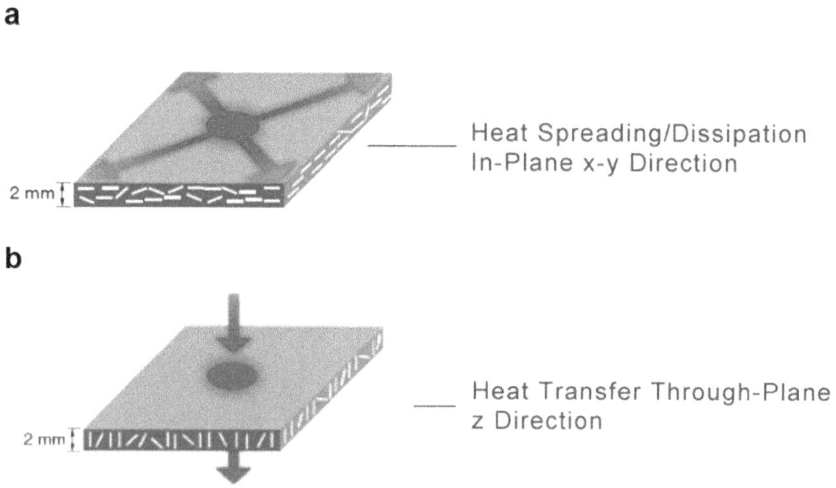

Figure 12.10 Heat transfer mode of 2D assemblies with BNNS/hBN in-plane oriented (a) and vertically aligned (b) [65].

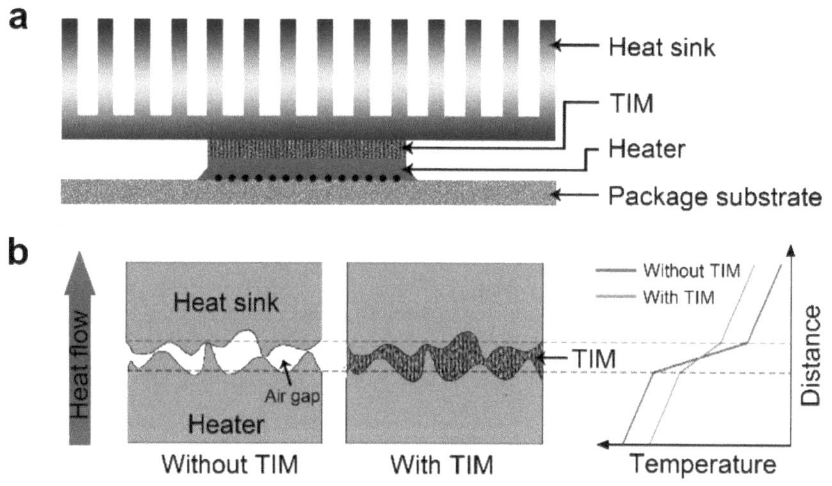

Figure 12.11 A typical thermal management system in electronics. (a) Schematics. (b) Working principle of thermal interface materials [66].

the ability to transfer heat vertically through the film. However, because of the layer structure and anisotropic thermal conductivity of hBN/BNNS, special processing and structure design are required. In the past years, several studies have been conducted to align hBN/BNNS along the through-plane direction by adopting different synthesizing methods. However, this topic has been widely investigated but still has not translated into industry. The synthesizing methods, together with their corresponding structures, mainly include stack-cutting, filler hybridization, electrospinning, and template methods.

The stack-cutting method is a top-down method, which involves assembling a bulk precursor by stacking in-plane oriented hBN/BNNS, followed by cutting it into thin films [67]. Due to the tendency of hBN/BNNSs to orientate in the direction of the plane, they are easier to fabricate in-plane assemblies like films based on the methods introduced above. By stacking these assemblies layer by layer, a bulk with BNNS highly aligned can thus be produced. Then, by cutting it perpendicular to the plane of the BNNS, films with the BNNS oriented through the plane are produced. Several methods can be applied to achieve the stacking of the stacked bulk, including vacuum infiltration [68], doctor blading [69], pressing [70], and 3D printing [71]. The stack-cutting method is easy to imagine because it is based on the combination of several proven manufacturing methods and therefore considered as one of the most scalable production methods in industry. However, it is still limited due to several problems. The accuracy and efficiency of stacking are crucial to the quality of alignment. Besides, strong cohesion in the stacked bulk is required to combine the structure; otherwise, the structure would collapse during cutting. In the cutting step, the movement of the knife deforms the bulk, resulting in microdefects or even breaking the bulk. Also, limited by current cutting technology, the thickness of the film produced cannot be very low, while the high thickness of the thermal interface material actually lowers performance. By addressing these problems, the stack-cutting method would hold great promise in manufacturing through-plane oriented BNNS films for thermal interface materials (Figure 12.12).

The filler hybridization method utilizes the interaction between fillers to construct through-plane thermally conductive pathways. Although hBN/BNNS tend to orientate parallel to the plane when using conventional processing methods in fabricating films, their arrangement can be modified by adding other fillers due to the spatial interaction of fillers with different morphological characteristics. The particles with the ability to redirect BNNSs are generally bigger. When packing them in a fluid (including dispersion and softened polymer), they first interlock and form the scaffold, which hinders the free movement of the BNNSs. Consequently, BNNSs are condensed and fabricated along this scaffold. AlN, which has a sphere-like structure and a high isotropic thermal conductivity, is reported to redirect BNNSs and therefore produce a 3D structured network with enhanced through-plane thermal conductivity [10,72]. Fillers of other shapes are also applied to redirect BNNSs, such as tetraneedle structure [73]. Except redirecting BNNSs, some smaller fillers, despite having little effect on the spatial arrangement of BNNSs, can spontaneously construct through-plane channels, connecting neighboring layers of BNNSs and enhancing through-plane thermal conductivity. These fillers are usually smaller in size and have a high thermal conductivity, including metal nanoparticles/nanowires [74,75], carbon nanotubes [76,77], and structured BN [78,79]. This method has little requirements on the processing method because the abovementioned interactions of fillers naturally occur; therefore, this method also promises the potential in manufacturing thermal management materials

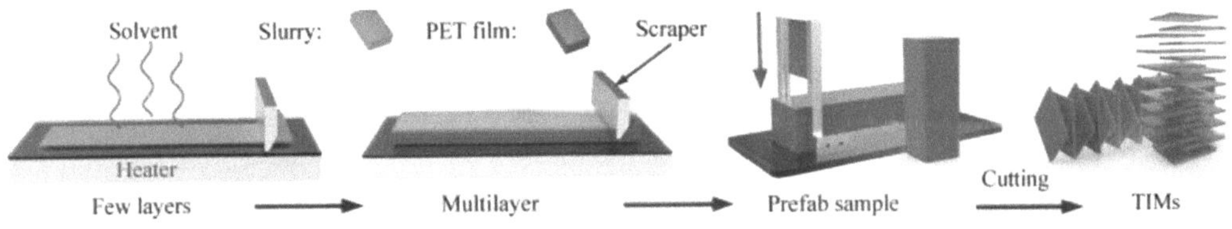

Figure 12.12 Typical process of the stack-cut method in synthesizing films with BNNSs aligned through the plane [69].

in industry. However, the limitations lie in the difficulty in controlling such interactions. The packing behaviors of fillers and their effect on the orientation of BNNSs mainly rely on the design of filler component, shape, and size, which is difficult to understand and control. Therefore, large quantities of trials are required to reach usable materials. Besides, the redirection of BNNS may also lead to deterioration of other properties, such as decreased tensile strength. In addition, most of the particles that contribute to through-plane thermal conductivity are electrically conductive, reducing the insulating and dielectric properties and therefore limiting the use in applications demanding high insulation or microwave transmission. In addition, the enhancement in thermal conductivity using this method is limited due to the limited volume fraction of thermal conductive pathways in this direction. Therefore, this method holds the potential of scalable production but mainly produces in-plane thermally conductive films with increased through-plane thermal conductivity.

The electrospinning method is a versatile and efficient technique for fabricating fibrous films, where the BNNS arrangement in the fiber and the texture structure contribute to the through-plane thermal conductivity of the film [80,81]. To achieve this, BNNSs are first dispersed with a polymer matrix and solvent. This dispersion is then loaded into the spinneret. Upon application of an electric field, the dispersion is subjected to electrostatic forces at the tip of a spinneret, typically a metallic needle. The interplay of electrostatic forces with the surface tension of the liquid leads to the formation of a phenomenon known as the Taylor cone. As the intensity of the electric field surpasses a threshold, the repulsion within the charged liquid overcomes its surface tension, resulting in the ejection of a fine jet from the apex of the Taylor cone. As the jet travels toward a grounded collector, rapid solvent evaporation occurs, culminating in the formation of solid fibers. These fibers, composed of the polymer matrix embedded with BNNSs, are deposited randomly on the collector, forming a fibrous film [82]. The BNNS was arranged in fibers to increase the thermal conductivity of the fibers, which was further woven and crosslinked to form a thermally conductive network in the film. The weaving of fibers also brings some unique mechanical behavior to the film. However, the maximum loading of BNNSs in this film is limited (typically $\leq$35 wt%), which further limits the film in achieving a high thermal conductivity. This is because a considerable amount of polymer matrix is required to achieve the formation of BNNS fibers, limiting the BNNS in every single fiber. Besides, since the fibers are woven into the film, the interspace between fibers is inevitable, which further decreases the maximum loading of BNNSs. In addition, the quality of weaving, particularly its density and uniformity, is important. Once the fibrous film is formed, it needs to be demolded to separate it from the collector, which is probably followed by a hot pressing or polymer infiltration, further limiting maximum BNNS content, to densify the structure. Controlling the electrospinning process to produce thick films presents several challenges, primarily due to the inherent complexities of the technique and the delicate balance of multiple interacting parameters. The dispersion, spinning process, and collecting/assembling process must be controlled simultaneously, which increases the difficulty in using this technique. In conclusion, the electrospinning method has the potential for automatic production, but is limited by the difficulties in manipulation and the BNNS loadings in products, which results in limited thermal conductivities (Figure 12.13).

The template method involves the use of a template for BNNSs to assemble on, thus forming a 3D BNNS network and corresponding 3D thermally conductive pathways [83]. However, this method mainly fabricates a 3D bulk composite, and the synthesis of 2D BN nanostructures is achieved by pressing them or by other post-treatments. Therefore, it was also challenged by similar problems to the stacking-cutting method, including limited minimum

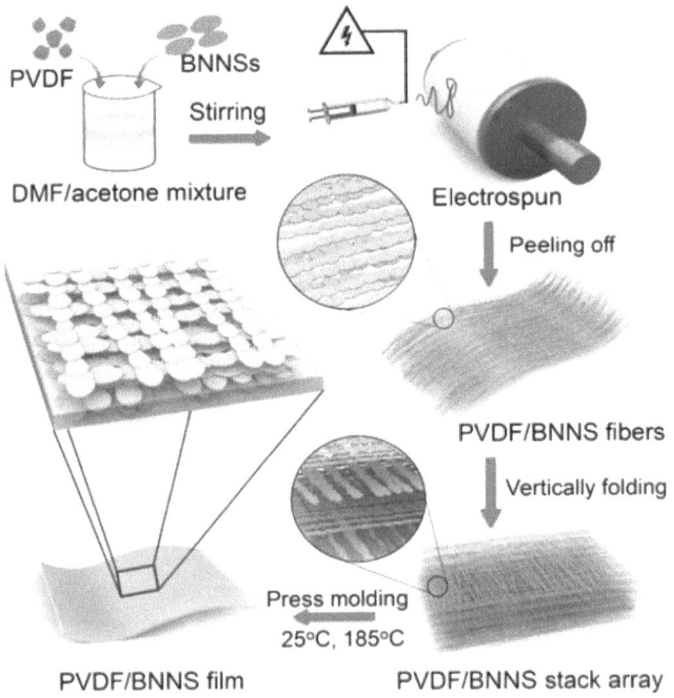

Figure 12.13 Typical process of the electrospinning method in synthesizing structured films [80].

thickness, thickness control, and limited high filler content. This method will be reviewed in the 3D BN nanostructures.

The 2D BN nanostructures, including in-plane oriented films and structured films, have been widely investigated for thermal management applications. They are subjected to development on different scales. For the in-plane oriented films, the corresponding relationship between their properties (thermal conductivity) and structures has been investigated. They have been made into products which have been used in industry, and current research mainly aims at improving the composite manufacturing factors, including product quality, processing efficiency, cost, and sustainability. However, for the structured films aiming at achieving high through-plane thermal conductivity, their development still faces several challenges. Unlike the in-plane oriented film, which requires a highly oriented structure, the realization of through-plane thermal conductivity has been divided into several approaches, which put forward different structure designs. To date, many synthesizing strategies, with their structures and processing technologies, have been constantly reported. Growing concern has been put on the requirements for practical use. Thermal resistance, for example, is defined as a comprehensive evaluation of materials' ability to fill the thermal interface. It is affected by many factors other than the through-plane thermal conductivity, which include flexibility and hardness. These properties are affected by the structure design of the structured film. The alignment of BNNSs along a certain direction can significantly increase the Young's modulus in that direction, which makes the materials hard to deform. Therefore, when it is applied as a thermal interface material, its ability to fill gaps is limited, leaving pores and poor contact interfaces in the setup. In this case, the vertically aligned structure, which can be achieved through stacking-cutting, by providing a high through-plane structure, may still have poorer performance than commercially available products, while other structures, such as 3D nanostructures, may lead to higher performance.

12.5 3D BN NANOSTRUCTURES

3D BN nanostructures involve a large variety of assemblies from porous structures to dense bulk. Based on the design of the 3D structure in them, they can achieve unique thermally conductive behaviors or other functions such as thermal insulation and heat storage.

12.5.1 Fabrication of 3D BN Nanostructures

To assemble 2D hBN into a 3D structure, many strategies have been developed over the past years, including freeze casting, template, field aligning, and additive manufacturing methods.

12.5.1.1 Freeze Casting Method

The freeze casting method, the mainstream method in fabricating 3D networks from 2D materials, uses ice as the template to assemble a 3D BNNS structure along with the phase transformation from water to ice (Figure 12.14) [83,84]. This method is a versatile technique that has been widely used to fabricate well-controlled biomimetic porous materials based on ceramics, metals, polymers, biomacromolecules, and carbon nanomaterials, endowing them with novel properties and broadening their application areas [85]. It involves the controlled freezing of aqueous sol, gel, or dispersion of BNNSs [86], followed by drying under reduced ambient pressure and post-treatments which strengthen the structure. During controlled freezing, phase separation occurs when water solidifies into ice, and the resulting solid phase acts as a template. For this reason, freeze casting is often also referred to as ice templating. The solidified solvent template is

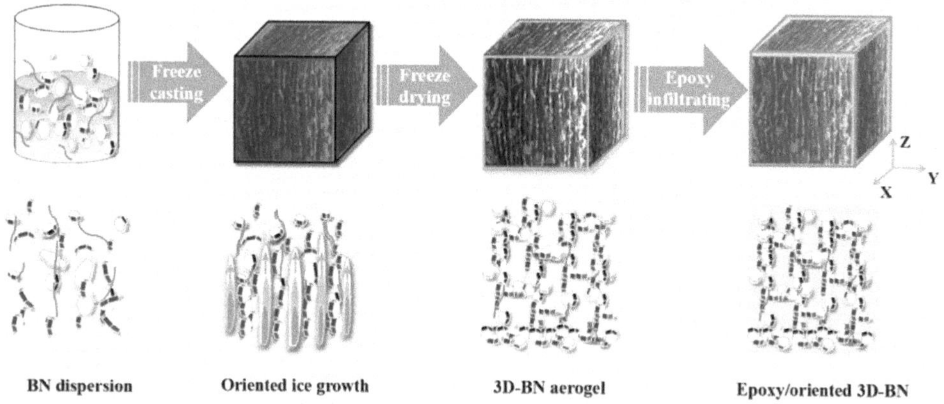

Figure 12.14 Typical process of the freeze drying method in synthesizing 3D BN nanostructures [84].

then removed by sublimation while the structural framework is retained, ultimately producing a well-shaped monolithic material. During this process, BNNSs together with the polymer component assemble on the formed ice template to form 3D structures. Changes in processing conditions may produce dramatic changes in the micro- and macro-structures of the obtained 3D structures. The microstructural properties of the stents, in terms of porosity and pore morphology (random [87], laminar [88], honeycomb [89], cellular [90], radial [91], etc.), can be customized by chemical and physical methods, whereas a wide range of geometries on the macroscopic scale can be achieved by combining freeze casting with other processing and techniques.

The freeze casting method produces a 3D scaffold with BNNSs highly oriented along the growth of ice while simultaneously building a 3D thermal conductive pathway that can be further processed to produce thermal management materials [92]. This scaffold is usually infiltrated with a polymer to strengthen the structure, thus forming a bulk composite with a continuous 3D thermal conductive pathway at a low-medium BNNS content. In this scaffold, BNNSs are oriented as the ice grows, and therefore, by controlling the assembly of such solidification, scaffolds of various structures can be formed, which can be further engineered to achieve different thermal conductive behaviors. Among the methods for controlling such processes, one of the commonly adopted strategies is adjusting the way the vectors of temperature gradients are applied because it controls the way ice grows. Zeng et. al. used single-direction freeze casting (Figure 12.15a) to fabricate a honeycomb-like BNNS structure [93] (Figure 12.15b,c). This highly aligned 3D scaffold was then infiltrated with epoxy to reach a high thermal conductivity of 2.85 $W{\cdot}m^{-1}{\cdot}K^{-1}$ at a low BNNS volume fraction of 9.29%. Hu et. al. used bidirectional freeze casting (Figure 12.15d) to assemble a laminar-structured BNNS aerogel where the BNNS aligned into parallelly arranged layers (Figure 12.15e,f) [94]. This scaffold can enhance phonon transport by restraining the out-of-plane motion of phonons, limiting phonon scattering in other directions. Therefore, after infiltrating polydimethylsiloxane (PDMS), the scaffold endows the fabricated composite with a high thermal conductivity of 3.28 $W{\cdot}m^{-1}{\cdot}K^{-1}$ at a filler content of 9.5 vol%. Huang et al. reported the fabrication of a radially aligned BNNS structure through introducing temperature gradients in both radial and vertical directions [91]. This structure enhances through-plane and in-plane thermal conductivities at a low filler content. Huang partially replaced reduced pressure drying with ambient drying and fabricated a ladder structure [95] with enhanced both in-plane and through-plane thermal conductivities. These studies reported the enhancement of thermal conductivity by embedding a BNNS scaffold inside the structure, which highlights the low filler content. This is because the low filler content indicates that the mechanical properties of the polymer composite are not greatly modified, which is favorable to applications demanding softness, flexibility, compressibility, and other typical polymer-like mechanical behaviors. Thermal interface materials, for example, require both enhanced thermal conductivity and the ability to fill space as mentioned above, making these 3D scaffold-embedded polymer composites more desirable. Besides infiltrating polymers, the scaffold can also be directly pressed into films, which also shows enhanced through-plane thermal conductivity [96].

Besides fabricating thermally conductive composites, these scaffolds, with proper modification of the components and structures, can directly produce BNNS-based aerogels [98] (Figure 12.16a). BNNS-based aerogels have an ultralight, highly porous structure where BNNSs and polymers are interconnected to create a 3D network with a very high air volume fraction (Figure 12.16b). For this structure, the strength of interconnection is very important to achieve free standing [99], which can be strengthened by covalently crosslinking polymers [99], introducing hydrogen bondings [100], and controlling the morphological characteristics and distribution of pores. The BNNS-based aerogels have excellent thermal insulating properties. The low

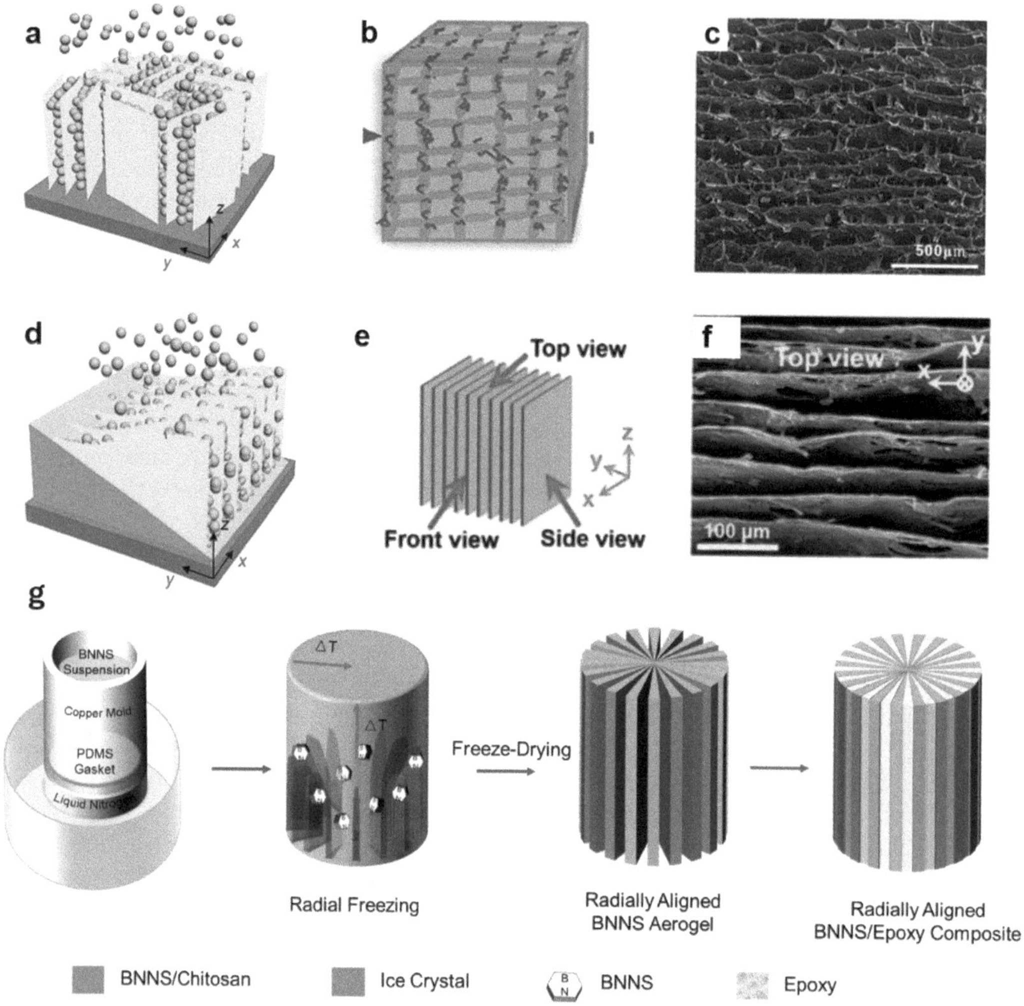

Figure 12.15 Schematics and the obtained structures of different freeze casting technologies. (a,d,g) Schematics of single-direction freeze (a), bidirectional freeze (d) [97], and radial freeze method [91] (g). (b,e) Schematics of the obtained structures using the single-direction freeze casting method [93] (b), which is a honeycomb structure, and the bidirectional freeze casting method [94] (e), which is a laminar structure. (c,f) SEM images of the obtained structures using the single-direction freeze casting method [93] (c) and the bidirectional freeze casting method [94] (f).

thermal conductivity of the large air volume fraction, together with low density and high porosity, significantly impedes the thermal conduction in the structure. Besides thermal conduction, thermal convection is also limited because isolated pores and the interconnected network in aerogels block the air flow [101]. The thermal radiation is limited by the small pore size [102]. In aerogels, BNNSs act as the structural support material for the network; besides, their incorporation improves the thermal stability of the aerogel due to their high thermal stability. Therefore, BNNS-based aerogels can be adopted as thermal insulators with a high operating temperature.

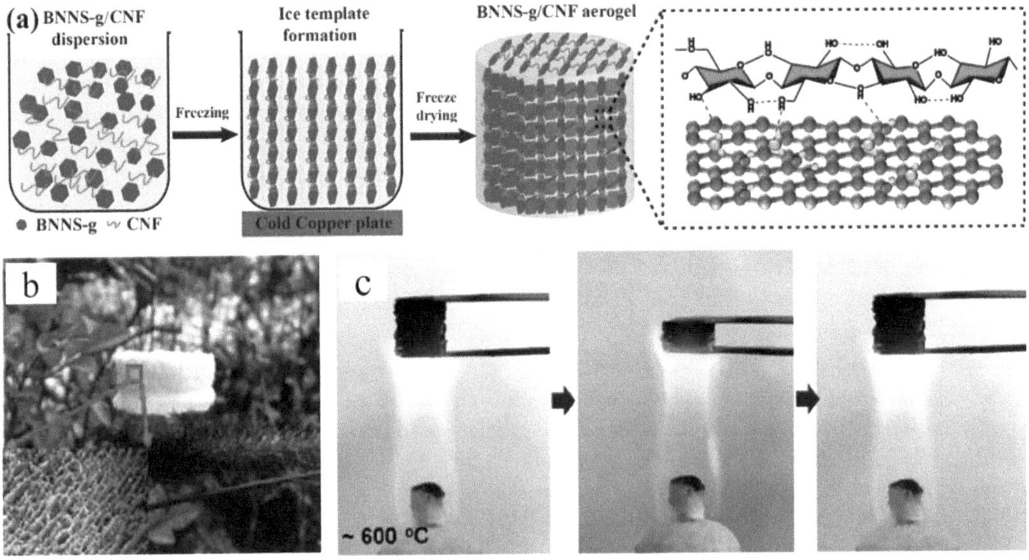

Figure 12.16 Aerogels fabricated using the freeze casting method. (a) Schematic of a typical synthesizing process of the BNNS-based aerogel [98]. (b) Ultralight [99] and (c) flame-retarded [104] properties of the aerogels.

Additionally, when they are utilized as a flame-retarded material, besides hindering the heat transfer, the BNNS network fixes the char formed from the carbonized polymer, thus preventing the further exposure of the polymer to flames. Therefore, they can also be used as flame-retarded materials (Figure 12.16c). Besides, the aerogel structure provides the substrate for phase change materials. When the phase change material melts, its flow is restrained by the capillary force of the tunnels in the network, preventing the overflow [103].

The freeze casting method provides great convenience in fabricating 3D scaffolds with different structures, which facilitates the investigation of structure designs in achieving different thermal conductive behaviors, while the fabrication of aerogels using this method produces high-performance thermal insulating and flame-retarded materials. However, the major limitation of this method lies in the low efficiency and high cost of the reduced pressure drying process. Therefore, it is usually adopted to provide insight into the structural design, but there is still a long way to go before producing thermal management materials.

12.5.1.2 Template Method

The template method involves fabricating BNNSs on a preformed template to form a 3D structure. It includes but is not limited to the abovementioned freeze casting methods. After BNNSs are fabricated on these templates, the freeze drying process is not necessary to achieve the removal of templates, leading to improved processing efficiency and decreased cost. Zuo et al. synthesized BNNSs on NaCl cubic, which was removed by dissolving in water, and produced PI films filled with hollow cubic-like particles [105], reaching a high out-of-plane thermal conductivity of 4.93 $W{\cdot}m^{-1}{\cdot}K^{-1}$ at 23.3 wt% loadings. Wang reported the fabrication of BNNSs on commercial melamine foam using layer-by-layer assembly to form a 3D thermally conductive network. The 3D BNNS network together with the foam template was then infiltrated by epoxy to fabricate a composite with a thermal conductivity of 0.6 $W \cdot m^{-1} \cdot K^{-1}$ at an ultralow BNNS loading of 1.1 vol% [106]. Yang et al. coated BNNSs on the commercial polyurethane foam, which was removed by pyrolysis in an argon atmosphere [87]. Hong reported the deposition of BNNSs on a structured PDMS mold, which is transferred by printing to a polymer substrate, leading to a stacked arrangement of BNNSs through the structure [107]. The templates used in these studies, such as NaCl and plastic foam, are easier to remove without damaging the structure, therefore allowing a higher processing efficiency. Besides, some polymer templates can be mechanically manipulated and deformed, during which the arrangement of 2D materials assembled on them is also modified, resulting in enhanced through-plane thermal conductivity [108]. With suitable templates applied, this method promises great potential in scalable production [109].

12.5.1.3 Field Aligning Method

The magnetic field method involves applying a field, i.e., fluidic field [110] and magnetic field [111], to align BNNSs, creating a highly ordered arrangement of hBN/BNNSs. The arranged hBN/BNNSs produce thermally conductive pathways and are further fabricated into polymer-based composites for thermal management materials. Magnetic field is mainly applied as the field to align hBN/BNNSs due to their advantage of precise and remote control of the orientation of anisotropic sheets. They also offer advances in aligning BNNSs over a long length, which is challenging for other aligning methods. In addition, different orientation angles or periodic arrangements can be easily programmed by simply changing the direction of the magnetic field [112]. In a typical process, hBN/BNNSs are modified by magnetic materials and dispersed in a dispersion with a viscosity. Then, a magnetic field is applied to align fillers in the dispersion, followed by post-treatments to solidify such alignment (Figure 12.17). Magnetic alignment requires the filler to respond to an external magnetic field [113]. Although hBN is not an intrinsically magnetic

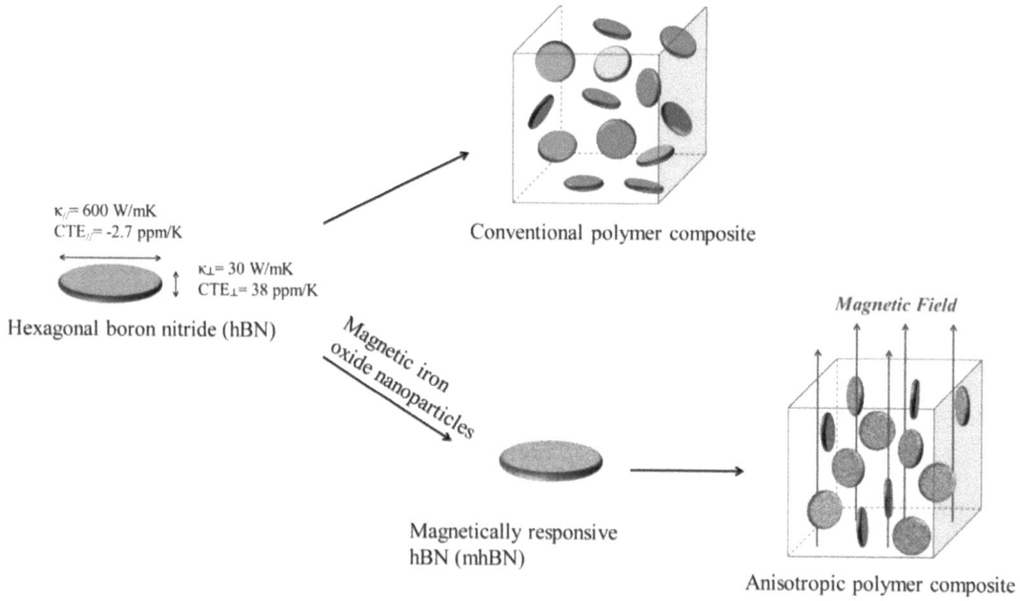

Figure 12.17 Typical process of the magnetic field aligning method in synthesizing 3D BN nanostructures [114].

material, it can be surface modified by using superparamagnetic Fe_3O_4 [114] or $ZnFeO_4$ [111]. When exposed to an external magnetic field, the modified hBN/BNNSs align themselves according to the field lines due to the dipole moments induced in the plane of nanosheets. During alignment, a high fluidity of the dispersion is required to prevent blockage of the filler movement; otherwise, a higher intensity of the applied magnetic field is required to align hBN/BNNSs, which, however, requires a very sophisticated and high-specification magnetic field generator. Besides, to guarantee the structural integrity of the alignment, solvents are usually not introduced to the dispersion. Consequently, polymers with a low degree of polymerization and monomers are usually used as the substrate in dispersion due to their low viscosities, which are cured to produce composites. Without solvents, the requirement of a high fluidity of dispersion also limits the maximum filler content of the final product [115]. The addition of magnetic responsive particles also compromises the insulating and dielectric properties of BNNSs. Fluidic field, generated by setting fluid in a certain state of flow, also offers the potential to align BNNSs. As mentioned above, BNNSs tend to arrange parallel to the velocity gradient vector due to their high aspect ratio. Therefore, through controlling the flowing conditions, the alignment and arrangement of BNNSs can be controlled and programmed using a fluidic field, just like using a magnetic field [116]. Bai reported the alignment of BNNSs in expansion flow [110] and produced a composite with a high BNNS content of 60 wt%.

12.5.1.4 Additive Manufacturing

Additive manufacturing, also known as 3D printing, refers to a process of making 3D objects by adding material layer by layer. This method offers the ability to create complex and intricate designs at little cost on process design and the advances in rapid prototyping [117], which has been widely used to build metal, alloys, ceramic, and polymer components. In a typical 3D printing process, hBN/BNNS is typically mixed with a polymer or other base material to form a printable filament or resin. This composite material is then used in a 3D printer, which lays down successive layers of the material according to a digital model, building the object from the bottom up [118]. When the mixture of hBN/BNNS and a polymer is extruded from the nozzle, the shear force generated in this process aligns hBN/BNNS along the extruding direction [90]. Through adjusting factors and parameters in this process, the quality of the alignment can thus be controlled. The as-extruded mixture has a rod-like structure, which is cured into a solid to form the final products. Currently, the 3D printing of hBN/BNNS is based on polymer substrates, and it produces composites in the specific shape or thermal conductive scaffold in the pre-designed structure, which holds great promise in designing and prototyping thermal management materials.

In conclusion, the investigation on 3D BN nanostructures produces high-performance thermally conductive materials by constructing designed thermally conductive pathways or aerogels with great thermal insulating and flame-retarded properties. It provides significant advances in the design of thermal management materials.

12.6 CONCLUSIONS AND OUTLOOK

This chapter presented a thorough review of BN nanostructures, ranging from 0D to 3D forms, and their pivotal role in enhancing heat dissipation in composite materials. The unique thermal properties of hBN and its derivatives, particularly their high thermal conductivity and electrical insulation, make them ideal candidates for advanced thermal management solutions.

The exploration of 0D BN nanostructures highlighted their potential as thermally conductive fillers, enabling isotropic thermal conductive pathways in composites. 1D BN nanostructures were shown to significantly enhance thermal and mechanical performance in various matrixes, while 2D BN nanostructures excel in spreading heat efficiently across planes, offering promising applications in electronics cooling. 3D BN nanostructures form intricate networks, providing enhanced pathways for thermal conductivity and laying the groundwork for innovative thermal management systems.

The insights gained not only advance the understanding of BN-based materials but also pave the way for their integration into a plethora of applications requiring efficient heat dissipation. From electronics to aerospace, these materials have immense potential to revolutionize thermal management.

Continued efforts are necessary to refine the synthesis processes of BN nanostructures, aiming for higher purity, better alignment, and enhanced thermal properties. There is a substantial opportunity to explore novel composite materials incorporating BN nanostructures, focusing on optimizing filler distribution, matrix interactions, and overall thermal performance. Future research should extend beyond traditional applications, exploring the use of hBN-based composites in emerging fields like renewable energy, where efficient heat management is crucial. Developing sustainable and scalable production methods for BN nanostructures will be vital to meet industry needs as the demand for advanced materials grows.

In conclusion, although significant strides have been made in understanding and utilizing BN nanostructures for heat dissipation, the journey is far from over. The continuous interplay between material science, engineering, and technology will be key to unlocking the full potential of these materials, fostering innovations that could redefine thermal management across industries.

ACKNOWLEDGMENTS

We are indebted to Dr. Minsu Liu for critical discussionsÚnd insights that greatly enriched this chapter. This work is sponsored by the National Natural Science Foundation of China (No. 5220020670), the Guangdong Basic and Applied Basic Research Foundation (No. 2020B0301030002), the Shenzhen Basic Research Project (No. JSGG20220831105402004,)We are thankful for their resources and support.

REFERENCES

1. Niu, H.; Guo, H.; Ren, Y.; Ren, L.; Lv, R.; Kang, L.; Bashir, A.; Bai, S. Spherical Aggregated BN/AlN Filled Silicone Composites with Enhanced through-Plane Thermal Conductivity Assisted by Vortex Flow. *Chem. Eng. J.* **2022**, *430*, 133155. https://doi.org/10.1016/j.cej.2021.133155.
2. Wang, R.; Cheng, H.; Gong, Y.; Wang, F.; Ding, X.; Hu, R.; Zhang, X.; He, J.; Tian, X. Highly Thermally Conductive Polymer Composite Originated from Assembly of Boron Nitride at an Oil–Water Interface. *ACS Appl. Mater. Interfaces* **2019**, *11* (45), 42818–42826. https://doi.org/10.1021/acsami.9b15259.
3. Pan, C.; Kou, K.; Zhang, Y.; Li, Z.; Ji, T.; Wu, G. Investigation of the Dielectric and Thermal Conductive Properties of Core–Shell Structured HGM@hBN/PTFE Composites. *Mater. Sci. Eng. B* **2018**, 238–239, 61–70. https://doi.org/10.1016/j.mseb.2018.12.015.
4. Jiang, Y.; Liu, Y.; Min, P.; Sui, G. BN@PPS Core-Shell Structure Particles and Their 3D Segregated Architecture Composites with High Thermal Conductivities. *Compos. Sci. Technol.* **2017**, *144*, 63–69. https://doi.org/10.1016/j.compscitech.2017.03.023.

5. Kong, N.; Tian, Y.; Huang, M.; Ye, C.; Yan, Y.; Peng, C.; Liu, J.; Han, F. Conformal BN Coating to Enhance the Electrical Insulation and Thermal Conductivity of Spherical Graphite Fillers for Electronic Encapsulation Field. *J. Alloys Compd.* **2023**, *968*, 172039. https://doi.org/10.1016/j.jallcom.2023.172039.
6. Cao, L.; Wang, J.; Dong, J.; Zhao, X.; Li, H.-B.; Zhang, Q. Preparation of Highly Thermally Conductive and Electrically Insulating PI/BNNSs Nanocomposites by Hot-Pressing Self-Assembled PI/BNNSs Microspheres. *Compos B Eng.* **2020**, *188*, 107882. https://doi.org/10.1016/j.compositesb.2020.107882.
7. Shi, X.; Zhang, R.; Ruan, K.; Ma, T.; Guo, Y.; Gu, J. Improvement of Thermal Conductivities and Simulation Model for Glass Fabrics Reinforced Epoxy Laminated Composites via Introducing Hetero-Structured BNN-30@BNNS Fillers. *J. Mater. Sci. Technol.* **2021**, *82*, 239–249. https://doi.org/10.1016/j.jmst.2021.01.018.
8. Ren, L.; Zeng, X.; Sun, R.; Xu, J.-B.; Wong, C.-P. Spray-Assisted Assembled Spherical Boron Nitride as Fillers for Polymers with Enhanced Thermally Conductivity. *Chem. Eng. J.* **2019**, *370*, 166–175. https://doi.org/10.1016/j.cej.2019.03.217.
9. Mun, H. J.; Park, J. Y.; Lim, M.; Cho, H.-B.; Choa, Y.-H. Preparation of H-BN Microspheres for Nanocomposites with High through-Plane Thermal Conductivity. *J. Am. Ceram. Soc.* **2023**, *106* (12), 7240–7250. https://doi.org/10.1111/jace.19219.
10. Wu, Z.; Dong, J.; Li, X.; Zhao, X.; Ji, C.; Zhang, Q. Enhancing Out-of-Plane Thermal Conductivity of Polyimide-Based Composites via the Construction of Inter-External Dual Heat Conduction Network by Binary Fillers. *Compos B Eng.* **2023**, *266*, 111001. https://doi.org/10.1016/j.compositesb.2023.111001.
11. Kittel, C. Interpretation of the Thermal Conductivity of Glasses. *Phys. Rev.* **1949**, *75* (6), 972–974. https://doi.org/10.1103/PhysRev.75.972.
12. Liu, H.; Gu, S.; Cao, H.; Li, X.; Li, Y. A Dense Packing Structure Constructed by Flake and Spherical Graphite: Simultaneously Enhanced in-Plane and through-Plane Thermal Conductivity of Polypropylene/Graphite Composites. *Compos. Commun.* **2020**, *19*, 25–29. https://doi.org/10.1016/j.coco.2020.02.007.
13. Slack, G. A. Nonmetallic Crystals with High Thermal Conductivity. *J. Phys. Chem. Solids* **1973**, *34* (2), 321–335. https://doi.org/10.1016/0022-3697(73)90092-9.
14. Cai, Q.; Scullion, D.; Gan, W.; Falin, A.; Zhang, S.; Watanabe, K.; Taniguchi, T.; Chen, Y.; Santos, E. J. G.; Li, L. H. High Thermal Conductivity of High-Quality Monolayer Boron Nitride and Its Thermal Expansion. *Sci. Adv.* **2019**, *5* (6), eaav0129. https://doi.org/10.1126/sciadv.aav0129.
15. Jo, I.; Pettes, M. T.; Kim, J.; Watanabe, K.; Taniguchi, T.; Yao, Z.; Shi, L. Thermal Conductivity and Phonon Transport in Suspended Few-Layer Hexagonal Boron Nitride. *Nano Lett.* **2013**, *13* (2), 550–554. https://doi.org/10.1021/nl304060g.
16. Li, C.; Bando, Y.; Zhi, C.; Huang, Y.; Golberg, D. Thickness-Dependent Bending Modulus of Hexagonal Boron Nitride Nanosheets. *Nanotechnology* **2009**, *20* (38), 385707. https://doi.org/10.1088/0957-4484/20/38/385707.
17. Liu, Y.; Pan, Y.; Yin, D.; Song, S.; Lin, L.; Zhang, M.; Qi, X.; Yao, J. Mechanical Properties and Thickness-Determined Fracture Mode of Hexagonal Boron Nitride Nanosheets under Nanoindentation Simulations. *Comput. Mater. Sci.* **2021**, *186*, 110047. https://doi.org/10.1016/j.commatsci.2020.110047.
18. Paul, D. R. Diffusion during the Coagulation Step of Wet-Spinning. *J. Appl. Polym. Sci.* **1968**, *12* (3), 383–402. https://doi.org/10.1002/app.1968.070120301.
19. Gao, T.; Yang, Z.; Chen, C.; Li, Y.; Fu, K.; Dai, J.; Hitz, E. M.; Xie, H.; Liu, B.; Song, J.; Yang, B.; Hu, L. Three-Dimensional Printed Thermal Regulation Textiles. *ACS Nano* **2017**, *11* (11), 11513–11520. https://doi.org/10.1021/acsnano.7b06295.
20. Wu, K.; Zhang, Y.; Gong, F.; Liu, D.; Lei, C.; Fu, Q. Highly Thermo-Conductive but Electrically Insulating Filament via a Volume-Confinement Self-Assembled Strategy for Thermoelectric Wearables. *Chem. Eng. J.* **2021**, *421*, 127764. https://doi.org/10.1016/j.cej.2020.127764.
21. Fang, Y.; Dong, J.; Zhao, X.; Chen, T.; Xiang, L.; Xie, Y.; Zhang, Q. Covalently Linked Polydopamine-Modified Boron Nitride Nanosheets/Polyimide Composite Fibers with Enhanced Heat Diffusion and Mechanical Behaviors. *Compos B Eng.* **2020**, *199*, 108281. https://doi.org/10.1016/j.compositesb.2020.108281.

22. Kong, D.; Zhang, J.; Hou, Z.; Zhao, X.; Ren, T.; Liu, C.; Tao, J.; Kong, N.; Gong, Y. Intensive and Thermally Conductive Boron Nitride/Aramid Nanofiber Composite Fibers Fabricated via a Wet Spinning Technique. *J. Mater. Chem. C* **2023.** https://doi.org/10.1039/D3TC02746E.
23. Lu, W.; Deng, Q.; Liu, M.; Ding, B.; Xiong, Z.; Qiu, L. Coaxial Wet Spinning of Boron Nitride Nanosheet-Based Composite Fibers with Enhanced Thermal Conductivity and Mechanical Strength. *Nano-Micro Lett.* **2023**, *16* (1), 25. https://doi.org/10.1007/s40820-023-01236-w.
24. Eom, W.; Lee, E.; Lee, S. H.; Sung, T. H.; Clancy, A. J.; Lee, W. J.; Han, T. H. Carbon Nanotube-Reduced Graphene Oxide Fiber with High Torsional Strength from Rheological Hierarchy Control. *Nat. Commun.* **2021**, *12* (1), 396. https://doi.org/10.1038/s41467-020-20518-0.
25. Zuo, X.; Zhang, X.; Qu, L.; Miao, J. Smart Fibers and Textiles for Personal Thermal Management in Emerging Wearable Applications. *Adv. Mater. Technol.* **2023**, *8* (6), 2201137. https://doi.org/10.1002/admt.202201137.
26. Yu, X.; Li, Y.; Wang, X.; Si, Y.; Yu, J.; Ding, B. Thermoconductive, Moisture-Permeable, and Superhy drophobic Nanofibrous Membranes with Interpenetrated Boron Nitride Network for Personal Cooling Fabrics. *ACS Appl. Mater. Interfaces* **2020**, *12* (28), 32078–32089. https://doi.org/10.1021/acsami.0c04486.
27. Miele, P.; Bernard, S.; Cornu, D.; Toury, B. Recent Developments in Polymer-Derived Ceramic Fibers (PDCFs): Preparation, Properties and Applications – A Review. *Soft Mater.* **2007**, *4* (2–4), 249–286. https://doi.org/10.1080/15394450701310228.
28. Cornu, D.; Miele, P.; Faure, R.; Bonnetot, B.; Mongeot, H.; Bouix, J. Conversion of B(NHCH 3) 3 into Boron Nitride and Polyborazine Fibres and Tubular BN Structures Derived Therefrom. *J. Mater. Chem.* **1999**, *9* (3), 757–761. https://doi.org/10.1039/A808213H.
29. Bernard, S.; Cornu, D.; Miele, P.; Vincent, H.; Bouix, J. Pyrolysis of Poly[2,4,6-Tri(Methylamino) Borazine] and Its Conversion into BN Fibers. *J. Organomet. Chem.* **2002**, *657* (1), 91–97. https://doi.org/10.1016/S0022-328X(02)01543-7.
30. Wideman, T.; Remsen, E. E.; Cortez, E.; Chlanda, V. L.; Sneddon, L. G. Amine-Modified Polyborazylenes: Second-Generation Precursors to Boron Nitride. *Chem. Mater.* **1998**, *10* (1), 412–421. https://doi.org/10.1021/cm970572r.
31. Duperrier, S.; Gervais, C.; Bernard, S.; Cornu, D.; Babonneau, F.; Miele, P. Controlling the Chemistry, Morphology and Structure of Boron Nitride-Based Ceramic Fibers through a Comprehensive Mechanistic Study of the Reactivity of Spinnable Polymers with Ammonia. *J. Mater. Chem.* **2006**, *16* (30), 3126–3138. https://doi.org/10.1039/B604482D.
32. Duperrier, S.; Gervais, C.; Bernard, S.; Cornu, D.; Babonneau, F.; Balan, C.; Miele, P. Design of a Series of Preceramic B-Tri(Methylamino)Borazine-Based Polymers as Fiber Precursors: Architecture, Thermal Behavior, and Melt-Spinnability. *Macromolecules* **2007**, *40* (4), 1018–1027. https://doi.org/10.1021/ma0623035.
33. Toury, B.; Bernard, S.; Cornu, D.; Chassagneux, F.; Létoffé, J.-M.; Miele, P. High-Performance Boron Nitride Fibers Obtained from Asymmetric Alkylaminoborazine. *J. Mater. Chem.* **2003**, *13* (2), 274–279. https://doi.org/10.1039/B207476A.
34. Miele, P.; Toury, B.; Cornu, D.; Bernard, S. Borylborazines as New Precursors for Boron Nitride Fibres. *J. Organomet. Chem.* **2005**, *690* (11), 2809–2814. https://doi.org/10.1016/j.jorganchem.2005.02.039.
35. Li, S.; Cao, Y.; Han, K.; Zhao, X.; Yu, M. Precursor Based on Polyborylborazine for BN Fiber: Preparation and Rheology Properties. *J. Macromol. Sci. Part B* **2009**, *48* (6), 1132–1142. https://doi.org/10.1080/00222340903275735.
36. Toury, B.; Miele, P.; Cornu, D.; Vincent, H.; Bouix, J. Boron Nitride Fibers Prepared from Symmetric and Asymmetric Alkylaminoborazines. *Adv. Funct. Mater.* **2002**, *12* (3), 228–234. https://advanced.onlinelibrary.wiley.com/doi/abs/10.1002/1616-3028(200203)12:3%3C228::AID-ADFM228%3E3.0.CO;2-U.
37. Zhang, X.; Zhang, H.; Li, D.; Xu, H.; Huang, Y.; Liu, Y.; Wu, D.; Sun, J. Highly Thermally Conductive and Electrically Insulating Polydimethylsiloxane Composites Prepared by Ultrasonic-Assisted Forced Infiltration for Thermal Management Applications. *Compos B Eng.* **2021**, *224*, 109207. https://doi.org/10.1016/j.compositesb.2021.109207.

38. Yao, Y.; Zeng, X.; Wang, F.; Sun, R.; Xu, J.; Wong, C.-P. Significant Enhancement of Thermal Conductivity in Bioinspired Freestanding Boron Nitride Papers Filled with Graphene Oxide. *Chem. Mater.* **2016**, *28* (4), 1049–1057. https://doi.org/10.1021/acs.chemmater.5b04187.
39. Song, N.; Wang, Q.; Jiao, D.; Pan, H.; Shi, L.; Ding, P. Highly Thermally Conductive SiO_2-Coated NFC/BNNS Hybrid Films with Water Resistance. *Compos. Part Appl. Sci. Manuf.* **2021**, *143*, 106261. https://doi.org/10.1016/j.compositesa.2020.106261.
40. Zhu, H.; Li, Y.; Fang, Z.; Xu, J.; Cao, F.; Wan, J.; Preston, C.; Yang, B.; Hu, L. Highly Thermally Conductive Papers with Percolative Layered Boron Nitride Nanosheets. *ACS Nano* **2014**, *8* (4), 3606–3613. https://doi.org/10.1021/nn500134m.
41. Chen, S.; Xu, R.; Liu, J.; Zou, X.; Qiu, L.; Kang, F.; Liu, B.; Cheng, H.-M. Simultaneous Production and Functionalization of Boron Nitride Nanosheets by Sugar-Assisted Mechanochemical Exfoliation. *Adv. Mater.* **2019**, *31* (10), 1804810. https://doi.org/10.1002/adma.201804810.
42. Lei, W.; Mochalin, V. N.; Liu, D.; Qin, S.; Gogotsi, Y.; Chen, Y. Boron Nitride Colloidal Solutions, Ultralight Aerogels and Freestanding Membranes through One-Step Exfoliation and Functionalization. *Nat. Commun.* **2015**, *6* (1), 8849. https://doi.org/10.1038/ncomms9849.
43. Smith McWilliams, A. D.; Martínez-Jiménez, C.; Matatyaho Ya'akobi, A.; Ginestra, C. J.; Talmon, Y.; Pasquali, M.; Martí, A. A. Understanding the Exfoliation and Dispersion of Hexagonal Boron Nitride Nanosheets by Surfactants: Implications for Antibacterial and Thermally Resistant Coatings. *ACS Appl. Nano Mater.* **2021**, *4* (1), 142–151. https://doi.org/10.1021/acsanm.0c02437.
44. Joni, I. M.; Balgis, R.; Ogi, T.; Iwaki, T.; Okuyama, K. Surface Functionalization for Dispersing and Stabilizing Hexagonal Boron Nitride Nanoparticle by Bead Milling. *Colloids Surf. Physicochem. Eng. Asp.* **2011**, *388* (1), 49–58. https://doi.org/10.1016/j.colsurfa.2011.08.007.
45. Coleman, J. N.; Lotya, M.; O'Neill, A.; Bergin, S. D.; King, P. J.; Khan, U.; Young, K.; Gaucher, A.; De, S.; Smith, R. J.; Shvets, I. V.; Arora, S. K.; Stanton, G.; Kim, H.-Y.; Lee, K.; Kim, G. T.; Duesberg, G. S.; Hallam, T.; Boland, J. J.; Wang, J. J.; Donegan, J. F.; Grunlan, J. C.; Moriarty, G.; Shmeliov, A.; Nicholls, R. J.; Perkins, J. M.; Grieveson, E. M.; Theuwissen, K.; McComb, D. W.; Nellist, P. D.; Nicolosi, V. Two-Dimensional Nanosheets Produced by Liquid Exfoliation of Layered Materials. *Science* **2011**, *331* (6017), 568–571. https://doi.org/10.1126/science.1194975.
46. Wu, Y.; Xue, Y.; Qin, S.; Liu, D.; Wang, X.; Hu, X.; Li, J.; Wang, X.; Bando, Y.; Golberg, D.; Chen, Y.; Gogotsi, Y.; Lei, W. BN Nanosheet/Polymer Films with Highly Anisotropic Thermal Conductivity for Thermal Management Applications. *ACS Appl. Mater. Interfaces* **2017**, *9* (49), 43163–43170. https://doi.org/10.1021/acsami.7b15264.
47. Wang, Z.-G.; Chen, M.-Z.; Liu, Y.-H.; Duan, H.-J.; Xu, L.; Zhou, L.; Xu, J.-Z.; Lei, J.; Li, Z.-M. Nacre-like Composite Films with High Thermal Conductivity, Flexibility, and Solvent Stability for Thermal Management Applications. *J. Mater. Chem. C* **2019**, *7* (29), 9018–9024. https://doi.org/10.1039/C9TC02845E.
48. Wu, K.; Fang, J.; Ma, J.; Huang, R.; Chai, S.; Chen, F.; Fu, Q. Achieving a Collapsible, Strong, and Highly Thermally Conductive Film Based on Oriented Functionalized Boron Nitride Nanosheets and Cellulose Nanofiber. *ACS Appl. Mater. Interfaces* **2017**, *9* (35), 30035–30045. https://doi.org/10.1021/acsami.7b08214.
49. Berni, A.; Mennig, M.; Schmidt, H. Doctor Blade. In *Sol-Gel Technologies for Glass Producers and Users*; Aegerter, M. A., Mennig, M., Eds.; Springer US: Boston, MA, **2004**; pp 89–92. https://doi.org/10.1007/978-0-387-88953-5_10.
50. Shen, H.; Guo, J.; Wang, H.; Zhao, N.; Xu, J. Bioinspired Modification of H-BN for High Thermal Conductive Composite Films with Aligned Structure. *ACS Appl. Mater. Interfaces* **2015**, *7* (10), 5701–5708. https://doi.org/10.1021/am507416y.
51. Teng, C.; Su, L.; Chen, J.; Wang, J. Flexible, Thermally Conductive Layered Composite Films from Massively Exfoliated Boron Nitride Nanosheets. *Compos. Part Appl. Sci. Manuf.* **2019**, *124*, 105498. https://doi.org/10.1016/j.compositesa.2019.105498.
52. Chen, L.; Xiao, C.; Tang, Y.; Zhang, X.; Zheng, K.; Tian, X. Preparation and Properties of Boron Nitride Nanosheets/Cellulose Nanofiber Shear-Oriented Films with High Thermal Conductivity. *Ceram. Int.* **2019**, *45* (10), 12965–12974. https://doi.org/10.1016/j.ceramint.2019.03.224.

53. LaChance, A. M.; Hou, Z.; Farooqui, M. M.; Carr, S. A.; Serrano, J. M.; Odendahl, C. E.; Hurley, M. E.; Morrison, T. E.; Kubachka, J. L.; Samuels, N. T.; Barrett, A. T.; Zhao, Y.; DeGennaro, A. M.; Sun, L.; Shaw, M. T. Doctor-Blade-Assisted Casting for Forming Thin Composite Coatings of Montmorillonite and Poly(Vinyl Alcohol). *Ind. Eng. Chem. Res.* **2022**, *61* (10), 3766–3774. https://doi.org/10.1021/acs.iecr.1c04381.
54. Patil, G. C. Doctor Blade: A Promising Technique for Thin Film Coating. In *Simple Chemical Methods for Thin Film Deposition: Synthesis and Applications*; Sankapal, B. R., Ennaoui, A., Gupta, R. B., Lokhande, C. D., Eds.; Springer Nature: Singapore, **2023**; pp 509–530. https://doi.org/10.1007/978-981-99-0961-2_12.
55. Zhou, Y.; Xu, L.; Liu, M.; Qi, Z.; Wang, W.; Zhu, J.; Chen, S.; Yu, K.; Su, Y.; Ding, B.; Qiu, L.; Cheng, H.-M. Viscous Solvent-Assisted Planetary Ball Milling for the Scalable Production of Large Ultrathin Two-Dimensional Materials. *ACS Nano* **2022**, *16* (7), 10179–10187. https://doi.org/10.1021/acsnano.1c11097.
56. Yu, C.; Gong, W.; Tian, W.; Zhang, Q.; Xu, Y.; Lin, Z.; Hu, M.; Fan, X.; Yao, Y. Hot-Pressing Induced Alignment of Boron Nitride in Polyurethane for Composite Films with Thermal Conductivity over 50 Wm^{-1} K^{-1}. *Compos. Sci. Technol.* **2018**, *160*, 199–207. https://doi.org/10.1016/j.compscitech.2018.03.028.
57. Hu, J.; Huang, Y.; Zeng, X.; Li, Q.; Ren, L.; Sun, R.; Xu, J.-B.; Wong, C.-P. Polymer Composite with Enhanced Thermal Conductivity and Mechanical Strength through Orientation Manipulating of BN. *Compos. Sci. Technol.* **2018**, *160*, 127–137. https://doi.org/10.1016/j.compscitech.2018.01.045.
58. Song, W.-L.; Wang, P.; Cao, L.; Anderson, A.; Meziani, M. J.; Farr, A. J.; Sun, Y.-P. Polymer/Boron Nitride Nanocomposite Materials for Superior Thermal Transport Performance. *Angew. Chem. Int. Ed.* **2012**, *51* (26), 6498–6501. https://doi.org/10.1002/anie.201201689.
59. Tu, H.; Xie, K.; Lin, X.; Zhang, R.; Chen, F.; Fu, Q.; Duan, B.; Zhang, L. Superior Strength and Highly Thermoconductive Cellulose/ Boron Nitride Film by Stretch-Induced Alignment. *J. Mater. Chem. A* **2021**, *9* (16), 10304–10315. https://doi.org/10.1039/D1TA00143D.
60. Kwon, O. H.; Ha, T.; Kim, D.-G.; Kim, B. G.; Kim, Y. S.; Shin, T. J.; Koh, W.-G.; Lim, H. S.; Yoo, Y. Anisotropy-Driven High Thermal Conductivity in Stretchable Poly(Vinyl Alcohol)/Hexagonal Boron Nitride Nanohybrid Films. *ACS Appl. Mater. Interfaces* **2018**, *10* (40), 34625–34633. https://doi.org/10.1021/acsami.8b12075.
61. Xu, L.; Zhan, K.; Ding, S.; Zhu, J.; Liu, M.; Fan, W.; Duan, P.; Luo, K.; Ding, B.; Liu, B.; Liu, Y.; Cheng, H.-M.; Qiu, L. A Malleable Composite Dough with Well-Dispersed and High-Content Boron Nitride Nanosheets. *ACS Nano* **2023**, *17* (5), 4886–4895. https://doi.org/10.1021/acsnano.2c11826.
62. Liu, Z.; Li, J.; Liu, X. Novel Functionalized BN Nanosheets/Epoxy Composites with Advanced Thermal Conductivity and Mechanical Properties. *ACS Appl. Mater. Interfaces* **2020**, *12* (5), 6503–6515. https://doi.org/10.1021/acsami.9b21467.
63. Yu, C.; Gong, W.; Zhang, J.; Lv, W.; Tian, W.; Fan, X.; Yao, Y. Hot Pressing-Induced Alignment of Hexagonal Boron Nitride in SEBS Elastomer for Superior Thermally Conductive Composites. *RSC Adv.* **2018**, *8* (45), 25835–25845. https://doi.org/10.1039/C8RA04700F.
64. Pan, X.; Shen, L.; Schenning, A. P. H. J.; Bastiaansen, C. W. M. Transparent, High-Thermal-Conductivity Ultradrawn Polyethylene/Graphene Nanocomposite Films. *Adv. Mater.* **2019**, *31* (40), 1904348. https://doi.org/10.1002/adma.201904348.
65. *3M Boron Nitride Cooling Fillers | 3M United Kingdom.* https://www.3m.co.uk/3M/en_GB/p/c/advanced-materials/ceramics/boron-nitride-cooling-fillers/ (accessed 2024-01-10).
66. Lv, L.; Dai, W.; Li, A.; Lin, C.-T. Graphene-Based Thermal Interface Materials: An Application-Oriented Perspective on Architecture Design. *Polymers* **2018**, *10* (11), 1201. https://doi.org/10.3390/polym10111201.
67. Song, Q.; Zhu, W.; Deng, Y.; Hai, F.; Wang, Y.; Guo, Z. Enhanced Through-Plane Thermal Conductivity and High Electrical Insulation of Flexible Composite Films with Aligned Boron Nitride for Thermal Interface Material. *Compos. Part Appl. Sci. Manuf.* **2019**, *127*, 105654. https://doi.org/10.1016/j.compositesa.2019.105654.
68. Yu, C.; Zhang, J.; Li, Z.; Tian, W.; Wang, L.; Luo, J.; Li, Q.; Fan, X.; Yao, Y. Enhanced Through-Plane Thermal Conductivity of Boron Nitride/Epoxy Composites. *Compos. Part Appl. Sci. Manuf.* **2017**, *98*, 25–31. https://doi.org/10.1016/j.compositesa.2017.03.012.

69. Kong, X.; Chen, Y.; Yang, R.; Wang, Y.; Zhang, Z.; Li, M.; Chen, H.; Li, L.; Gong, P.; Zhang, J.; Xu, K.; Cao, Y.; Cai, T.; Yan, Q.; Dai, W.; Wu, X.; Lin, C.-T.; Nishimura, K.; Pan, Z.; Jiang, N.; Yu, J. Large-Scale Production of Boron Nitride Nanosheets for Flexible Thermal Interface Materials with Highly Thermally Conductive and Low Dielectric Constant. *Compos B Eng.* **2024**, *271*, 111164. https://doi.org/10.1016/j.compositesb.2023.111164.
70. Hu, Q.; Bai, X.; Zhang, C.; Zeng, X.; Huang, Z.; Li, J.; Li, J.; Zhang, Y. Oriented BN/Silicone Rubber Composite Thermal Interface Materials with High Out-of-Plane Thermal Conductivity and Flexibility. *Compos. Part Appl. Sci. Manuf.* **2021**, 106681. https://doi.org/10.1016/j.compositesa.2021.106681.
71. Liu, J.; Li, W.; Guo, Y.; Zhang, H.; Zhang, Z. Improved Thermal Conductivity of Thermoplastic Polyurethane via Aligned Boron Nitride Platelets Assisted by 3D Printing. *Compos. Part Appl. Sci. Manuf.* **2019**, *120*, 140–146. https://doi.org/10.1016/j.compositesa.2019.02.026.
72. Lee, W.; Kim, J. Enhanced through-Plane Thermal Conductivity of Paper-like Cellulose Film with Treated Hybrid Fillers Comprising Boron Nitride and Aluminum Nitride. *Compos. Sci. Technol.* **2020**, *200*, 108424. https://doi.org/10.1016/j.compscitech.2020.108424.
73. Chi, H.; Liu, D.; Ma, C.; Song, M.; Zhang, P.; Dai, P. Simultaneously Enhanced In-Plane and Out-of-Plane Thermal Conductivity of a PI Composite Film by Tetraneedle-like ZnO Whiskers and BN Nanosheets. *ACS Appl. Polym. Mater.* **2023**, *5* (9), 6909–6919. https://doi.org/10.1021/acsapm.3c00958.
74. Dong, J.; Cao, L.; Li, Y.; Wu, Z.; Teng, C. Largely Improved Thermal Conductivity of PI/BNNS Nanocomposites Obtained by Constructing a 3D BNNS Network and Filling It with AgNW as the Thermally Conductive Bridges. *Compos. Sci. Technol.* **2020**, *196*, 108242. https://doi.org/10.1016/j.compscitech.2020.108242.
75. Wang, X.; Bian, H.; Ni, S.; Sun, S.; Jiao, L.; Dai, H. BNNS/PVA Bilayer Composite Film with Multiple-Improved Properties by the Synergistic Actions of Cellulose Nanofibrils and Lignin Nanoparticles. *Int. J. Biol. Macromol.* **2020**, *157*, 259–266. https://doi.org/10.1016/j.ijbiomac.2020.04.178.
76. Zhang, Y.; Zhao, Z.; Chen, M.; Wu, H.; Guo, S.; Qiu, J. Constructing Interconnected Network of MWCNT and BNNS in Electrospun TPU Films: Achieving Excellent Thermal Conduction yet Electrical Insulation Properties. *Carbon* **2024**, *218*, 118691. https://doi.org/10.1016/j.carbon.2023.118691.
77. Li, X.; Wu, B.; Li, Y.; Alam, M. M.; Chen, P.; Xia, R.; Lin, C.-T.; Qian, J. Construction of Oriented Interconnected BNNS Skeleton by Self-Growing CNTs Leading High Thermal Conductivity. *Adv. Mater. Interfaces* **2021**, *8* (6), 2001910. https://doi.org/10.1002/admi.202001910.
78. Zuo, S.; Lan, Y.; Luo, J.; Zhou, F.; Xu, L.; Xie, S.; Wei, X.; Zhou, L.; Ma, L.; Li, X.; Yin, C. Angular-Shaped Boron Nitride Nanosheets with a High Aspect Ratio to Improve the Out-of-Plane Thermal Conductivity of Polyimide Composite Films. *ACS Omega* **2022**, *7* (47), 43273–43282. https://doi.org/10.1021/acsomega.2c06013.
79. An, L.; Gu, R.; Zhong, B.; Wang, J.; Zhang, J.; Yu, Y. Quasi-Isotropically Thermal Conductive, Highly Transparent, Insulating and Super-Flexible Polymer Films Achieved by Cross Linked 2D Hexagonal Boron Nitride Nanosheets. *Small* **2021**, *17* (46), 2101409. https://doi.org/10.1002/smll.202101409.
80. Chen, J.; Huang, X.; Sun, B.; Jiang, P. Highly Thermally Conductive Yet Electrically Insulating Polymer/Boron Nitride Nanosheets Nanocomposite Films for Improved Thermal Management Capability. *ACS Nano* **2019**, *13* (1), 337–345. https://doi.org/10.1021/acsnano.8b06290.
81. Yang, X.; Guo, Y.; Han, Y.; Li, Y.; Ma, T.; Chen, M.; Kong, J.; Zhu, J.; Gu, J. Significant Improvement of Thermal Conductivities for BNNS/PVA Composite Films via Electrospinning Followed by Hot-Pressing Technology. *Compos B Eng.* **2019**, *175*, 107070. https://doi.org/10.1016/j.compositesb.2019.107070.
82. Teo, W. E.; Ramakrishna, S. A Review on Electrospinning Design and Nanofibre Assemblies. *Nanotechnology* **2006**, *17* (14), R89. https://doi.org/10.1088/0957-4484/17/14/R01.
83. Shao, G.; Hanaor, D. A. H.; Shen, X.; Gurlo, A. Freeze Casting: From Low-Dimensional Building Blocks to Aligned Porous Structures—A Review of Novel Materials, Methods, and Applications. *Adv. Mater.* **2020**, *32* (17), 1907176. https://doi.org/10.1002/adma.201907176.
84. Hu, J.; Huang, Y.; Yao, Y.; Pan, G.; Sun, J.; Zeng, X.; Sun, R.; Xu, J.-B.; Song, B.; Wong, C.-P. Polymer Composite with Improved Thermal Conductivity by Constructing a Hierarchically Ordered Three-Dimensional Interconnected Network of BN. *ACS Appl. Mater. Interfaces* **2017**, *9* (15), 13544–13553. https://doi.org/10.1021/acsami.7b02410.

85. Zhang, H.; Hussain, I.; Brust, M.; Butler, M. F.; Rannard, S. P.; Cooper, A. I. Aligned Two- and Three-Dimensional Structures by Directional Freezing of Polymers and Nanoparticles. *Nat. Mater.* **2005**, *4* (10), 787–793. https://doi.org/10.1038/nmat1487.
86. Chen, J.; Huang, X.; Zhu, Y.; Jiang, P. Cellulose Nanofiber Supported 3D Interconnected BN Nanosheets for Epoxy Nanocomposites with Ultrahigh Thermal Management Capability. *Adv. Funct. Mater.* **2017**, *27* (5), 1604754. https://doi.org/10.1002/adfm.201604754.
87. Yang, W.; Wang, Y.; Li, Y.; Gao, C.; Tian, X.; Wu, N.; Geng, Z.; Che, S.; Yang, F.; Li, Y. Three-Dimensional Skeleton Assembled by Carbon Nanotubes/Boron Nitride as Filler in Epoxy for Thermal Management Materials with High Thermal Conductivity and Electrical Insulation. *Compos B Eng.* **2021**, *224*, 109168. https://doi.org/10.1016/j.compositesb.2021.109168.
88. Mi, H.-Y.; Jing, X.; Politowicz, A. L.; Chen, E.; Huang, H.-X.; Turng, L.-S. Highly Compressible Ultra-Light Anisotropic Cellulose/Graphene Aerogel Fabricated by Bidirectional Freeze Drying for Selective Oil Absorption. Carbon **2018**, *132*, 199–209. https://doi.org/10.1016/j.carbon.2018.02.033.
89. Du, X.; Yang, W.; Zhu, J.; Fu, L.; Li, D.; Zhou, L. Aligning Diamond Particles inside BN Honeycomb for Significantly Improving Thermal Conductivity of Epoxy Composite. *Compos. Sci. Technol.* **2022**, *222*, 109370. https://doi.org/10.1016/j.compscitech.2022.109370.
90. Liang, Z.; Pei, Y.; Chen, C.; Jiang, B.; Yao, Y.; Xie, H.; Jiao, M.; Chen, G.; Li, T.; Yang, B.; Hu, L. General, Vertical, Three-Dimensional Printing of Two-Dimensional Materials with Multiscale Alignment. *ACS Nano* **2019**, *13* (11), 12653–12661. https://doi.org/10.1021/acsnano.9b04202.
91. Huang, T.; Li, Y.; Chen, M.; Wu, L. Bi-Directional High Thermal Conductive Epoxy Composites with Radially Aligned Boron Nitride Nanosheets Lamellae. *Compos. Sci. Technol.* **2020**, *198*, 108322. https://doi.org/10.1016/j.compscitech.2020.108322.
92. Wang, X.; Wu, P. 3D Vertically Aligned BNNS Network with Long-Range Continuous Channels for Achieving a Highly Thermally Conductive Composite. *ACS Appl. Mater. Interfaces* **2019**, *11* (32), 28943–28952. https://doi.org/10.1021/acsami.9b09398.
93. Zeng, X.; Yao, Y.; Gong, Z.; Wang, F.; Sun, R.; Xu, J.; Wong, C.-P. Ice-Templated Assembly Strategy to Construct 3D Boron Nitride Nanosheet Networks in Polymer Composites for Thermal Conductivity Improvement. *Small* **2015**, *11* (46), 6205–6213. https://doi.org/10.1002/smll.201502173.
94. Hu, B.; Zhang, W.; Guo, H.; Xu, S.; Li, Y.; Li, M.; Li, B. Nacre-Mimetic Elastomer Composites with Synergistic Alignments of Boron Nitride/Graphene Oxide towards High through-Plane Thermal Conductivity. *Compos. Part Appl. Sci. Manuf.* **2022**, 106891. https://doi.org/10.1016/j.compositesa.2022.106891.
95. Huang, T.; Yang, F.; Wang, T.; Wang, J.; Li, Y.; Huang, J.; Chen, M.; Wu, L. Ladder-Structured Boron Nitride Nanosheet Skeleton in Flexible Polymer Films for Superior Thermal Conductivity. *Appl. Mater. Today* **2022**, *26*, 101299. https://doi.org/10.1016/j.apmt.2021.101299.
96. Zhao, N.; Li, J.; Wang, W.; Gao, W.; Bai, H. Isotropically Ultrahigh Thermal Conductive Polymer Composites by Assembling Anisotropic Boron Nitride Nanosheets into a Biaxially Oriented Network. *ACS Nano* **2022**, *16* (11), 18959–18967. https://doi.org/10.1021/acsnano.2c07862.
97. Bai, H.; Chen, Y.; Delattre, B.; Tomsia, A. P.; Ritchie, R. O. Bioinspired Large-Scale Aligned Porous Materials Assembled with Dual Temperature Gradients. *Sci. Adv.* **2015**, *1* (11), e1500849. https://doi.org/10.1126/sciadv.1500849.
98. Liu, C.; Huang, C.; Li, Y.; Liu, Y.; Bian, H.; Xiang, Z.; Wang, H.; Wang, H.; Xiao, H. Freeze-Casting Production of Thermal Insulating and Fire-Retardant Lightweight Aerogels Based on Nanocellulose and Boron Nitride. *Int. J. Biol. Macromol.* **2023**, *252*, 126370. https://doi.org/10.1016/j.ijbiomac.2023.126370.
99. Zeng, X.; Ye, L.; Yu, S.; Sun, R.; Xu, J.; Wong, C.-P. Facile Preparation of Superelastic and Ultralow Dielectric Boron Nitride Nanosheet Aerogels via Freeze-Casting Process. *Chem. Mater.* **2015**, *27* (17), 5849–5855. https://doi.org/10.1021/acs.chemmater.5b00505.
100. Yang, W.; Yuen, A. C. Y.; Ping, P.; Wei, R.-C.; Hua, L.; Zhu, Z.; Li, A.; Zhu, S.-E.; Wang, L.-L.; Liang, J.; Chen, T. B. Y.; Yu, B.; Si, J.-Y.; Lu, H.-D.; Chan, Q. N.; Yeoh, G. H. Pectin-Assisted Dispersion of Exfoliated Boron Nitride Nanosheets for Assembled Bio-Composite Aerogels. *Compos. Part Appl. Sci. Manuf.* **2019**, *119*, 196–205. https://doi.org/10.1016/j.compositesa.2019.02.003.

101. Adegun, M. H.; Chan, K.-Y.; Yang, J.; Venkatesan, H.; Kim, E.; Zhang, H.; Shen, X.; Yang, J.; Kim, J.-K. Anisotropic Thermally Superinsulating Boron Nitride Composite Aerogel for Building Thermal Management. *Compos. Part Appl. Sci. Manuf.* **2023**, *169*, 107522. https://doi.org/10.1016/j.compositesa.2023.107522.
102. Zhan, H.-J.; Wu, K.-J.; Hu, Y.-L.; Liu, J.-W.; Li, H.; Guo, X.; Xu, J.; Yang, Y.; Yu, Z.-L.; Gao, H.-L.; Luo, X.-S.; Chen, J.-F.; Ni, Y.; Yu, S.-H. Biomimetic Carbon Tube Aerogel Enables Super-Elasticity and Thermal Insulation. *Chem* **2019**, *5* (7), 1871–1882. https://doi.org/10.1016/j.chempr.2019.04.025.
103. Wang, B.; Li, G.; Xu, L.; Liao, J.; Zhang, X. Nanoporous Boron Nitride Aerogel Film and Its Smart Composite with Phase Change Materials. *ACS Nano* **2020**, *14* (12), 16590–16599. https://doi.org/10.1021/acsnano.0c05931.
104. Feng, L.; Wei, P.; Song, Q.; Zhang, J.; Fu, Q.; Jia, X.; Yang, J.; Shao, D.; Li, Y.; Wang, S.; Qiang, X.; Song, H. Superelastic, Highly Conductive, Superhydrophobic, and Powerful Electromagnetic Shielding Hybrid Aerogels Built from Orthogonal Graphene and Boron Nitride Nanoribbons. *ACS Nano* **2022**, *16* (10), 17049–17061. https://doi.org/10.1021/acsnano.2c07187.
105. Zuo, S.; Li, M.; Xie, S.; Luo, J.; Xu, L.; Shi, Y.; Lan, N.; Zhou, L.; Yu, J.; Li, X.; Yin, C. Enhanced Out-of-Plane Thermal Conductivity of Polyimide Composite Films by Adding Hollow Cube-like Boron Nitride. *Ceram. Int.* **2023**, *49* (8), 12615–12624. https://doi.org/10.1016/j.ceramint.2022.12.124.
106. Wang, X.; Wu, P. Melamine Foam-Supported 3D Interconnected Boron Nitride Nanosheets Network Encapsulated in Epoxy to Achieve Significant Thermal Conductivity Enhancement at an Ultralow Filler Loading. *Chem. Eng. J.* **2018**, *348*, 723–731. https://doi.org/10.1016/j.cej.2018.04.196.
107. Hong, H.; Jung, Y. H.; Lee, J. S.; Jeong, C.; Kim, J. U.; Lee, S.; Ryu, H.; Kim, H.; Ma, Z.; Kim, T. Anisotropic Thermal Conductive Composite by the Guided Assembly of Boron Nitride Nanosheets for Flexible and Stretchable Electronics. *Adv. Funct. Mater.* **2019**, *29* (37), 1902575. https://doi.org/10.1002/adfm.201902575.
108. Dai, W.; Ma, T.; Yan, Q.; Gao, J.; Tan, X.; Lv, L.; Hou, H.; Wei, Q.; Yu, J.; Wu, J.; Yao, Y.; Du, S.; Sun, R.; Jiang, N.; Wang, Y.; Kong, J.; Wong, C.; Maruyama, S.; Lin, C.-T. Metal-Level Thermally Conductive yet Soft Graphene Thermal Interface Materials. *ACS Nano* **2019**, *13* (10), 11561–11571. https://doi.org/10.1021/acsnano.9b05163.
109. Yang, J.; Shen, X.; Yang, W.; Kim, J. Templating Strategies for 3D-Structured Thermally Conductive Composites: Recent Advances and Thermal Energy Applications. *Prog. Mater. Sci.* **2023**, *133*, 101054. https://doi.org/10.1016/j.pmatsci.2022.101054.
110. Niu, H.; Guo, H.; Kang, L.; Ren, L.; Lv, R.; Bai, S. Vertical Alignment of Anisotropic Fillers Assisted by Expansion Flow in Polymer Composites. *Nano-Micro Lett.* **2022**, *14* (1), 153. https://doi.org/10.1007/s40820-022-00909-2.
111. Shi, H.; Wang, H.; Lu, H.; Zhang, X.; Zhang, Z. Magnetic Field-Induced Orientation of Modified Boron Nitride Nanosheets in Epoxy Resin with Improved Flame and Wear Resistance. *Langmuir* **2021**, *37* (27), 8222–8231. https://doi.org/10.1021/acs.langmuir.1c00927.
112. He, H.; Peng, W.; Liu, J.; Chan, X. Y.; Liu, S.; Lu, L.; Le Ferrand, H. Microstructured BN Composites with Internally Designed High Thermal Conductivity Paths for 3D Electronic Packaging. *Adv. Mater.* **2022**, *34* (38), 2205120. https://doi.org/10.1002/adma.202205120.
113. Le Ferrand, H.; Bouville, F.; Niebel, T. P.; Studart, A. R. Magnetically Assisted Slip Casting of Bioinspired Heterogeneous Composites. *Nat. Mater.* **2015**, *14* (11), 1172–1179. https://doi.org/10.1038/nmat4419.
114. Lin, Z.; Liu, Y.; Raghavan, S.; Moon, K.; Sitaraman, S. K.; Wong, C. Magnetic Alignment of Hexagonal Boron Nitride Platelets in Polymer Matrix: Toward High Performance Anisotropic Polymer Composites for Electronic Encapsulation. *ACS Appl. Mater. Interfaces* **2013**, *5* (15), 7633–7640. https://doi.org/10.1021/am401939z.
115. Yuan, C.; Duan, B.; Li, L.; Xie, B.; Huang, M.; Luo, X. Thermal Conductivity of Polymer-Based Composites with Magnetic Aligned Hexagonal Boron Nitride Platelets. *ACS Appl. Mater. Interfaces* **2015**, *7* (23), 13000–13006. https://doi.org/10.1021/acsami.5b03007.
116. Kiriya, D.; Kawano, R.; Onoe, H.; Takeuchi, S. Microfluidic Control of the Internal Morphology in Nanofiber-Based Macroscopic Cables. *Angew. Chem. Int. Ed.* **2012**, *51* (32), 7942–7947. https://doi.org/10.1002/anie.201202078.

117. Guiney, L. M.; Mansukhani, N. D.; Jakus, A. E.; Wallace, S. G.; Shah, R. N.; Hersam, M. C. Three-Dimensional Printing of Cytocompatible, Thermally Conductive Hexagonal Boron Nitride Nanocomposites. *Nano Lett.* **2018**, *18* (6), 3488–3493. https://doi.org/10.1021/acs.nanolett.8b00555.
118. Gorur, M. C.; Doganay, D.; Durukan, M. B.; Cicek, M. O.; Kalay, Y. E.; Kincal, C.; Solak, N.; Unalan, H. E. 3D Printing of Hexagonal Boron Nitride Nanosheets/Polylactic Acid Nanocomposites for Thermal Management of Electronic Devices. *Compos B Eng.* **2023**, *265*, 110955. https://doi.org/10.1016/j.compositesb.2023.110955.

Chapter 13

Boron Nitride-Reinforced Thermally Conductive Polymeric Composites

Yuhang Meng, Dehong Yang, Dandan Sun, Xiangfen Jiang, and Xuebin Wang

13.1 INTRODUCTION

Over the past few decades, we have witnessed a significant and growing demand for electronic devices in various fields such as communication equipment, computing systems, automotive engineering, and aerospace. The trend toward integration and miniaturization of electronic components within these devices has become demanding. Unfortunately, this shift has resulted in a rapid increase in heat generation, negatively impacting the performance and service life of electronic components. Research showed that with every 2°C increase, the performance of electronic equipment will be reduced by 10% [1,2]. Therefore, improving the heat dissipation capability of electronic devices emerges as a critical imperative for their advancement. Lightweight, insulating, and processable polymers are commonly used substrates for electronic packaging and thermal interface materials (TIMs). However, the long-chain structure and complex chain connections of polymers result in significant phonon scattering, exhibiting lower TC, typically around 0.1–0.5 W m^{-1}·K^{-1} [3,4], which significantly inhibits thermal conduction within the polymers. Incorporating high TC fillers into these polymers to engineer thermally conductive composites represents an effective approach to designing thermal management materials. BN demonstrates outstanding thermal conductivity, with a theoretical TC of 1700–2000 W m^{-1}·K^{-1} [5]. Additionally, its stable chemical property, exceptional electrical insulation capability, high breakdown field strength, and low density make it an excellent thermally conductive filler in composite materials where electrical insulation and low dielectric performance are desired.

From the perspective of fillers, the TC of composites constructed with BN as the filler is mainly influenced by the following three factors: (1) the intrinsic TC of BN; (2) the interface between BN and the matrix as well as between BN particles themselves; (3) the three-dimensional (3D) filler network and alignment formed by BN within the polymer. Starting from these three influencing factors, this chapter explores the influence of BN's properties, including its inherent TC, atomic structure, crystallinity, defects, morphology, size, and scale, on the TC of composites. Additionally, it investigates the impact of processing strategies for BN within polymers, including dispersion, functionalization, construction of 3D filler networks, and alignment, on the TC of composites.

DOI: 10.1201/9781003647812-16

13.2 PROPERTIES OF BN FILLERS INFLUENCING TC

13.2.1 Intrinsic TC of BN Crystals

The band gap of BN is about 5–6 eV [6,7], which impedes the movement of electrons in BN; thus, BN primarily conducts heat through phonons. Given the presence of strong sp^2 B–N bonds, which are analogous to the C–C bonds responsible for efficient heat transfer through lattice vibrations in carbon nanotubes (CNTs) and graphene, BN is theoretically predicted to exhibit high TC [8]. At room temperature, the highest TC of bulk BN is about 400 W $m^{-1}{\cdot}K^{-1}$. However, theoretical prediction based on numerical solutions of the phonon Boltzmann transport equation suggests that single-layer BN could exhibit significantly higher TC, potentially exceeding 600 W $m^{-1}{\cdot}K^{-1}$ at room temperature [9]. This theoretical prediction was substantiated and even surpassed by experimental findings in 2019 when Li et al. demonstrated that high-quality single-layer BN can achieve an impressive TC of 751 W $m^{-1}{\cdot}K^{-1}$, while the TC of bilayer and trilayer BN decreases to 646 and 602 W $m^{-1}{\cdot}K^{-1}$, respectively [10]. One explanation for this phenomenon is that for multilayer materials, intrinsic phonon–phonon scattering and out-of-plane vibrations play an important role in determining the value of TC. The interlayer interactions and coupling in multilayer BN disrupt the two-dimensional (2D) selection rule, resulting in a significant reduction in TC compared to single-layer BN [9,11,12]. It has been confirmed that the TC of BN is significantly related to its number of layers, leading to a growing interest in the study of BN's across of different layers, which has driven the development of techniques for measuring BN's TC at the micro-nano scale. In 2013, Shi et al. measured the TC of 5-layer and 11-layer few-layer BN using a microbridge device with built-in resistance thermometers, which were found to be 250 and 360 W $m^{-1}{\cdot}K^{-1}$, respectively. It was observed that the presence of polymer residues, such as poly(methyl methacrylate) (PMMA), on the sample surface led to significant phonon scattering, and it is expected that as the layer thickness decreases, the impact of scattering on the TC will become increasingly pronounced [13]. In 2013, Yu et al. discovered that the TC of 9-layer BN is approximately 243 W $m^{-1}{\cdot}K^{-1}$ using a non-contact micro-Raman spectroscopy method [14]. In 2016, Li et al. measured the TC of 2-layer BN using pre-patterned microstructures with built-in platinum resistance thermometers and obtained the TC of 2-layer BN as approximately 484 W $m^{-1}{\cdot}K^{-1}$ [15]. The TC of few-layer BN shows obvious sensitivity to the change of the number of layers. However, when the number of layers exceeds a certain threshold, the change of TC becomes relatively small. In 2014, Alam et al. investigated the TC of BN with thicknesses of 10 nm and 20 nm (approximately 30 and 60 layers), and the TC values of both were about 100 W $m^{-1}{\cdot}K^{-1}$ [16].

In addition to thickness, the TC of BN is also influenced by its shape. The theoretical TC of BN nanoribbons can reach 2067 W $m^{-1}{\cdot}K^{-1}$ [17]. Moreover, it is theoretically predicted that the TC of single-walled BN nanotubes (BNNTs) may exceed 3000 W $m^{-1}{\cdot}K^{-1}$. Experimentally, the TC of 40 nm diameter BNNTs can reach 200–300 W $m^{-1}{\cdot}K^{-1}$ [18].

13.2.2 Atomic Structure of BN Crystals

BN has a hexagonal atomic arrangement similar to graphite, where B atoms and N atoms alternate to replace C atoms, forming the intralayer structure of BN. Therefore, BN is also referred to as 'white graphite'. Considering that both CNTs and graphene have sp^2 C–C bonds and exhibit excellent TC, it is predicted that BN with sp^2 B–N bonds may also have excellent heat transfer capabilities. However, apart from differences in constituent elements, there are also significant differences in isotope mixing between BN (19.9% ^{10}B, 80.1% ^{11}B) and graphite (98.9% ^{12}C, 1.1% ^{13}C) [9], which makes the impact of isotope scattering on the TC of BN more pronounced. Despite the identical electronic structure of substances composed of different isotopes, the variation in isotopic composition engenders alterations in lattice vibration mode. Eliminating isotopic disorder can increase phonon lifetime and enhance the TC of BN [19]. The simulation results demonstrate that, compared to BN composed of isotopes ^{10}B and ^{11}B, BN consisting of a single isotope exhibits higher TC [20]. The experimental results demonstrate that for monoisotopic ^{10}B bulk BN, the in-plane TC can reach 585 W $m^{-1}\cdot K^{-1}$, which is approximately 80% higher than that of isotopically disordered BN (52%:48% mixture of ^{10}B and ^{11}B) [21]. The TC of single-layer BN composed of pure ^{11}B can reach up to 1009 W $m^{-1}\cdot K^{-1}$, which is 140% larger than that of bulk BN [22], indicating that the isotope engineering mainly suppresses the out-of-plane optical phonon scatterings in BN, which subsequently reduces acoustic-optical scatterings between out-of-plane optical and transverse acoustic and longitudinal acoustic phonons.

13.2.3 Crystallinity of BN Crystals

Phonons represent the quantum expression of lattice vibrations. In BN, heat is transferred in the form of lattice vibrations, and its TC can be described by the Debye equation as follows [23]:

$$\lambda = \frac{C_p \upsilon l}{3} \tag{13.1}$$

where λ is the TC of BN, C_p is the specific heat capacity per unit volume of the material, υ is the group velocity of the phonon, and l is the mean free path of the phonon. High crystallinity makes a more ordered atomic arrangement in BN, resulting in smoother phonon transmission throughout the material (Figure 13.1). Consequently, the average mean free path of phonons increases, leading to an enhancement in the TC of BN. High-crystallinity and large-sized 2D BN has obvious advantages in improving TC, and developing a high-throughput synthesis method is of utmost importance [24]. The synthesis of highly crystalline BN can mainly be achieved through two approaches: top-down synthesis methods and bottom-up synthesis methods. Top-down synthesis methods mainly include mechanical or chemical exfoliation methods [24]. The exfoliation methods utilize mechanical force or chemical media to overcome the interlayer interaction forces of BN, achieving the high-quality exfoliation of BN from bulk to thin layers. Bottom-up approaches typically involve the self-assembly of smaller BN structural units into more complex and larger BN structural systems through strong chemical interactions, including chemical vapor deposition (CVD), physical vapor deposition, carbon substitution reactions, and others. Bottom-up methods can achieve atomic-level control, facilitating the formation of highly ordered structures and reducing the introduction of defects.

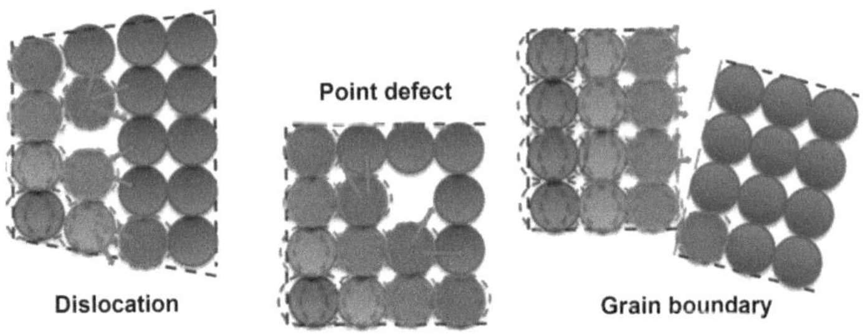

Figure 13.1 Phonon scattering in crystalline materials, due to various defects [23].

Regardless of the preparation method, the generation of defects (such as vacancies, dislocations, grain boundaries, etc.) in BN is an unavoidable issue [25]. Phonon motion generates phonon scattering at defects, which exerts a strong weakening effect on the TC of BN. As the defect concentration increases, the TC of BN will significantly decrease [26]. A random distribution of atomic vacancies at a concentration of 2% results in a 44% decrease in the TC of BN nanosheets (BNNSs) [25]. Wu et al. investigated the effects of different types of defects on the TC of monolayer BN. The results indicate that compared to Stone–Wales defects, monovacancy and divacancy defects have a more significant impact on the TC of monolayer BN [27]. BN is composed of two elements, B and N. The different electronegativities of B and N result in differences in the effects of B and N vacancies on the TC of BN. In BN, phonons travel faster along N atoms. Therefore, compared to BN containing B vacancies, BN with N vacancy defects exhibits lower TC [28].

13.2.4 Morphology of BN Crystals

When using BN as the filler to construct thermally conductive composites, it is necessary to consider its morphology. This consideration is not only because of the inherent variations in TC among BN structures of different morphologies, but also because of the disparate interfacial characteristics, dispersion patterns, and orientational tendencies exhibited by these diverse BN morphologies within the matrix. Tang et al. measured the TC of synthesized BN spherical nanoparticles (Figure 13.2a), BNNTs (Figure 13.2b), BN bamboo-structured nanotubes (Figure 13.2c), and collapsed nanotubes (Figure 13.2d). The corrected TC values for the four materials were found to be 14, 18, 17, and 46 W $m^{-1}{\cdot}K^{-1}$, respectively. These values are much lower than the TC of sintered bulk BN particles at room temperature (>200 W $m^{-1}{\cdot}K^{-1}$) [29]. This discrepancy arises from the significant contact thermal resistance between the nanoparticles during the assembly process of nanoscale BN particles. In addition, the number of interfaces formed by the four different morphologies of BN in creating heat transfer paths leads to the difference in TC between them (Figure 13.2e), whereas spherical particles create more interfaces, which results in higher interfacial thermal resistance and lower TC.

The morphology of BN can be classified into zero-dimensional (0D), one-dimensional (1D), 2D, and 3D according to dimensions [30,31]. In terms of 0D BN, it can mainly be divided into BN nanoparticles and BN fullerenes. The development of BN nanoparticles dates back to 1994, when nanoparticles with concentric BN layers were observed for the first time [32,33]. This discovery laid the groundwork for subsequent research, and high-throughput production of BN nanoparticles was realized in 2002 [34]. The 0D BN fullerene was first derived from electron-irradiated BN samples in 1998 [35]. 0D fillers have unique advantages in constructing isotropic thermally conductive composites. Simultaneously, their viscosity penetration threshold is considerably high, allowing for higher filler loading and better processing performance [36]. In terms of 1D BN, it mainly includes BNNTs and BN nanoribbons. Inspired by the synthesis of CNTs in 1991, BNNTs were achieved using arc discharge by Chopra et al. in 1995 [37]. This breakthrough stimulated further research interest, which eventually led to the large-scale production of BNNTs using a CVD process involving boron and MgO in 2002, marking an important milestone in BNNT research [38]. BN nanoribbons are also a type of 1D BN, consisting of single-layer or few-layered belt-like BN. In 2008, researchers achieved a significant advancement in the field by utilizing ZnS as a template for the synthesis of hollow BN nanoribbons, thereby expanding the family of 1D BN materials. The relatively high aspect ratio of 1D BN nanomaterials contributes to their enhanced axial TC. Theoretically, the TC of single-wall BNNTs could surpass 3000 W $m^{-1}{\cdot}K^{-1}$. Experimental results have demonstrated that BNNTs with a diameter of 40 nm exhibit an axial TC ranging from 200

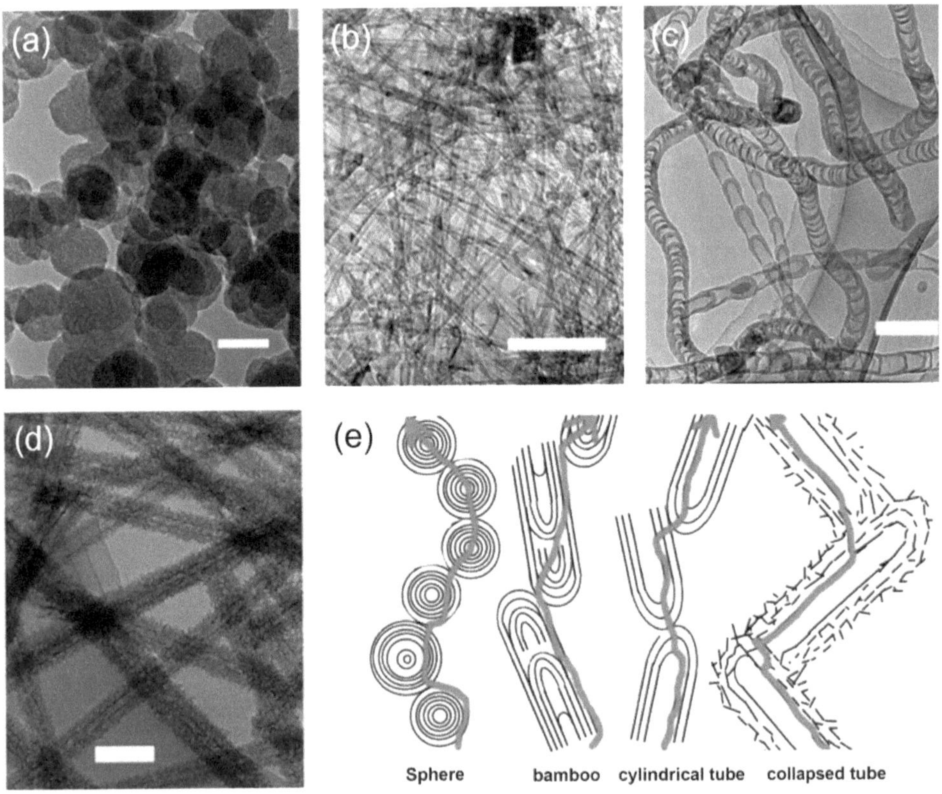

Figure 13.2 (a)–(d) Transmission electron microscopy images of the nanostructured BN materials used for TC measurements: (a) spherical nanoparticles; (b) structurally perfect nanotubes; (c) bamboo-structured nanotubes; (d) collapsed nanotubes. The scale bar is 200 nm. (e) Schematic illustration of the heat percolation passages for BN nanospheres, BN nanobamboo, multiwalled BNNTs, and collapsed multiwalled BNNTs. The coarse gray lines denote the heat percolation passages [29].

to 300 W $m^{-1} \cdot K^{-1}$ [18]. In addition to outstanding TC properties, 1D BN nanomaterials also possess the capability to form highly efficient 3D thermal conduction networks even at very low loading, endowing 1D BN nanomaterials with significant potential in thermal management applications. However, the tendency to aggregate and form bundles or clusters results in poor dispersion of 1D BN nanomaterials in the matrix, highlighting the significance of enhancing the dispersion as an important issue for researchers [39]. For 2D BN, it mainly includes BNNSs. The average TC of 2D BNNSs is higher than that of 0D nanoparticles and 1D nanotubes, since the two dimensionalities peculiar to the planes of nanosheets are free [40]. And 2D BNNSs disperse more easily in the filler compared to BNNTs. As early as 1995, monolayer BN films were epitaxially grown on the surfaces of Ni (111), Pd (111), and Pt (111) [41]. After the discovery of graphene through mechanical exfoliation [42], in 2005, Novoselov et al. prepared monolayer BNNSs using the same mechanical exfoliation method [43]. Subsequently, various synthesis methods continued to emerge. Shi et al. obtained large-area BN films with thicknesses ranging from 5 to 50 nm on polycrystalline nickel films using the CVD method in 2010 [44]. In 2011, Wang et al. developed the 'Chemical Blowing' method to achieve high-throughput production of large surface area, few-layer BNNSs [45]. Subsequently, high-quality BNNSs were mass-produced using biomass carbon thermal reduction, and their TC was increased by 14 times after being compounded with epoxy resin [40]. The 3D BN thermal conduction network typically comprises interconnected BN nanostructures, creating a continuous 3D framework that offers a pre-established, uninterrupted thermal conduction pathway. The distinctive 3D thermal conduction pathways can diminish interfacial contact between fillers, provide isotropic thermal conduction performance, and yield higher TC efficiency per unit mass [46].

13.2.5 Size of BN Crystals

According to the relationship between the phonon mean free path and the characteristic length of the crystal, the phonon transport modes can be divided into ballistic transport and diffusive transport [8]. Additionally, phonons also experience significant scattering at material grain boundaries [47], implying that the grain size has a certain influence on the TC of the material. From the perspective of the fillers themselves, certain materials, such as silicon and diamond, exhibit an increase in TC with an increase in grain size. When the size increases to a certain value, the TC approaches that of a single crystal, which is due to grain boundary and size effects [48]. However, the effect of grain size on the TC of BN is different. Chang et al. studied the change of the TC of BNNTs when the outer diameter changed from 30–40 nm by theoretical calculation, and the results showed that the TC decreased with the increase of the outer diameter [49]. Lu et al. also found through computational simulations that the TC of single-layer BNNS decreases with increasing size. This decline can be attributed to the heightened phonon scattering as the simulation size increases, resulting in a significant inhibitory effect on TC that surpasses the contribution of additional phonons to TC [50]. Considering the composite perspective, smaller filler sizes introduce more interfaces when establishing the complete thermal conduction paths, thereby diminishing the continuity of heat transfer. Consequently, a high Kapitza resistance is generated, which impedes the enhancement of the composites' TC [36,51]. Conversely, during the construction of thermal conduction networks, larger fillers with reduced specific surface areas and fewer interactions with the polymer can create a continuous heat transfer path, effectively reducing phonon–interface scattering and enhancing the TC of the material [52–54]. Kim et al. investigated the impact of varying particle sizes on the TC of BN/epoxy-terminated dimethylsiloxane (BN/ETDS) composites, demonstrating that at the same loading, TC increased with the increase of particle size [55]. Additionally, Moradi et al. observed that the TC of composites rose with increasing filler sizes (2, 30, and 180 μm), revealing the important effect of the size of BN crystals on TC [56]. Therefore, the size of the fillers needs to be considered comprehensively in the composite synthesis process.

The number of layers of BN also affects its TC. Section 13.2.1 has explored the relationship between the TC of BN and the number of layers. In instances with fewer layers, TC may be affected by residual polymers during the transfer process. However, overall, the TC of BN decreases with the increasing number of layers, primarily due to the increase in Umklapp scattering [8,10]. Notably, when used as the filler, thinner BNNSs will achieve higher performance compared to bulk BN [57–60].

13.3 THERMAL CONDUCTANCE OF POLYMER COMPOSITES REINFORCED BY BN

When BN fillers form composites with polymer matrixes, various additional properties must be taken into account beyond the inherent characteristics of BN and the matrix material, including the dispersion of BN, the interface between BN and the matrix, the 3D network formed by BN in the matrix, and the orientation of BN in the matrix. These properties can be exhibited by either BN or the polymer matrix but belong to the composite as a whole, and they have a significant impact on the overall thermal conductive performance.

13.3.1 Crystallinity of Polymer Matrixes

Before discussing composites, this section provides a brief introduction to polymers. In solid materials, heat transfer is predominantly facilitated by electrons and phonons. However, within polymeric structures, electron mobility is significantly constrained, thus rendering phonon-mediated heat transfer the primary mechanism [61,62]. Phonons are quantized expressions of lattice vibrations, and the ordered lattices are conducive to heat transfer. More degrees of freedom in amorphous polymer chains make heat easily dispersed to other chains, resulting in low TC (0.1–0.5 W $m^{-1}{\cdot}K^{-1}$) [3,4,23]. Increasing the crystallinity of polymers is a commonly used method to enhance their intrinsic TC [63]. When the crystallinity of polymers is enhanced, more direct heat conduction paths are formed within the crystalline regions of the polymer. Phonons can move rapidly along the direction of the polymer chains with fewer scattering events, leading to an increase in their mean free path and consequently enhancing their TC (Figure 13.3a and b) [47].

In recent years, extensive research has been conducted on the crystallinity and TC of polymers, with studies on polyethylene (PE) being particularly comprehensive. Molecular dynamics simulation results show that individual PE chains exhibit high TC, which has been validated experimentally: highly crystalline PE nanofibers prepared by local heating and stretching can achieve TC values up to 90 W $m^{-1}{\cdot}K^{-1}$ [64–66], while single-crystalline PE fibers can reach TC values as high as 104 W $m^{-1}{\cdot}K^{-1}$ [67]. In the study of epoxy resin, it has also been found that the TC is positively correlated with purity [68]. Additionally, the size of the crystalline regions also affects the TC of polymers. Studies on poly(butylene succinate) phosphorus-containing ionomers have shown that smaller crystalline regions compared to larger ones have more grain boundaries, leading to greater heat dissipation at the grain boundaries and lower TC [69]. Complex factors such as the side-chain structure of polymers and phonon scattering in crystalline and non-crystalline regions all influence the TC of polymer materials [36,61].

Designing highly crystalline polymers still presents certain challenges. Employing high TC fillers to construct composites is a simple method to enhance the TC of polymers. BN is widely adopted due to its high TC and insulating properties. Additionally, recent investigations suggest

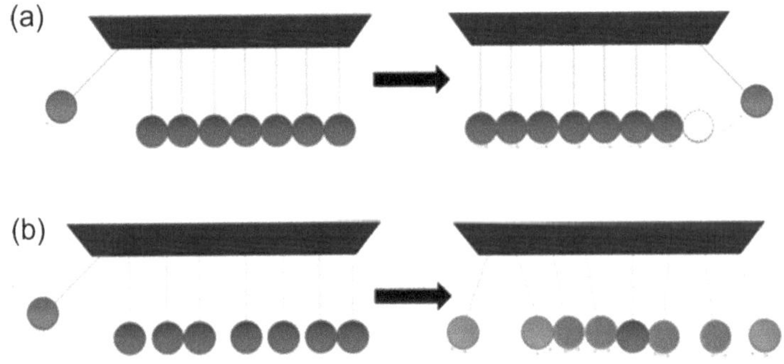

Figure 13.3 Schematic comparison by the Newton pendulum of TC in (a) crystalline materials and (b) polymers [23].

that BN can serve as a heterogeneous nucleating agent to enhance the crystallinity of polymer materials [70–72]. This versatility renders BN a particularly promising candidate for the development of high-performance polymer composites with enhanced thermal management capabilities.

13.3.2 Dispersion Techniques of BN Fillers in Polymers

Some studies suggest that uniform dispersion of fillers in the matrix may not necessarily lead to an increase in the TC of composites. In such composites, uniformly dispersed fillers are surrounded by the low TC matrix, preventing the formation of effective thermal conduction networks [23,47]. However, a majority of studies demonstrate that minimizing filler aggregation and enhancing dispersibility are beneficial for phonon transmission, thereby improving the TC of composites. Moreover, optimal material dispersion facilitates the transfer of applied stress from the matrix to the fillers, resulting in effective reinforcement and enhancing the mechanical properties of the composites [73]. Unfortunately, due to their high surface energy and hydrophobic nature, BN particles are challenging to uniformly disperse in polymer matrixes, which results in higher thermal resistance at the interfaces in composites [74]. So far, many techniques have been developed to enhance the dispersion of BN in matrixes, including plasma treatment, ball milling, liquid-phase exfoliation, heat treatment, and thermal substitution. These techniques typically focus on introducing hydrophilic functional groups to the highly active dangling bonds at the edges of BN, which facilitates better dispersion in hydrophilic polymers.

13.3.2.1 Plasma Processing

Plasma processing technology is characterized by its simplicity, rapid processing rates, and effective treatment outcomes, rendering it a highly promising dispersion technique for BN [75]. Its mechanism involves exciting the gas surrounding the sample and causing it to react with surrounding molecules, resulting in the formation of high-energy substances such as [OH], [O], and [H]. These reactive ions and radicals easily react with other chemicals present in the air and on the BN surface, leading to the breaking of existing chemical bonds and the formation of new ones. During plasma processing, Lewis acid–base interactions may occur, and B–N bonds may break and form new bonds at the edges or defects (Figure 13.4a–c) [76].

Currently, plasma processing technology has been used to amino-functionalize BN, enhancing its dispersion in polyvinyl alcohol (PVA). The TC of the composite formed by treated BN and PVA can reach a high through-plane TC of 2.4 W $m^{-1}{\cdot}K^{-1}$ (at 50 wt%) [77]. In addition to being used in a gaseous atmosphere, plasma–liquid interactions sometimes also occur in plasma processing. When using isopropanol (IPA) as a solvent for the plasma treatment of BN, new substances such as amorphous carbon layers may form due to the interaction between the plasma and IPA. These substances deposit on the surface of BN, aiding in the enhancement of the exfoliation and dispersion of BNNSs. This consequently improves the TC of BNNS/PVA composites which is twice that of nanocomposites containing non-plasma-treated BNNSs [60].

13.3.2.2 Ball Milling

Ball milling is a mechanochemical reaction strategy for introducing functional groups into BN. Under high-speed milling, significant shear forces can exfoliate bulk BN into BNNSs, exposing more active sites. Additionally, the higher energy during ball milling can cause the decomposition of auxiliary media, leading to the functionalization of BN (Figure 13.5). Considering the instability of B–N bonds at the boundary, these reactions often occur at the edges of BN, resulting

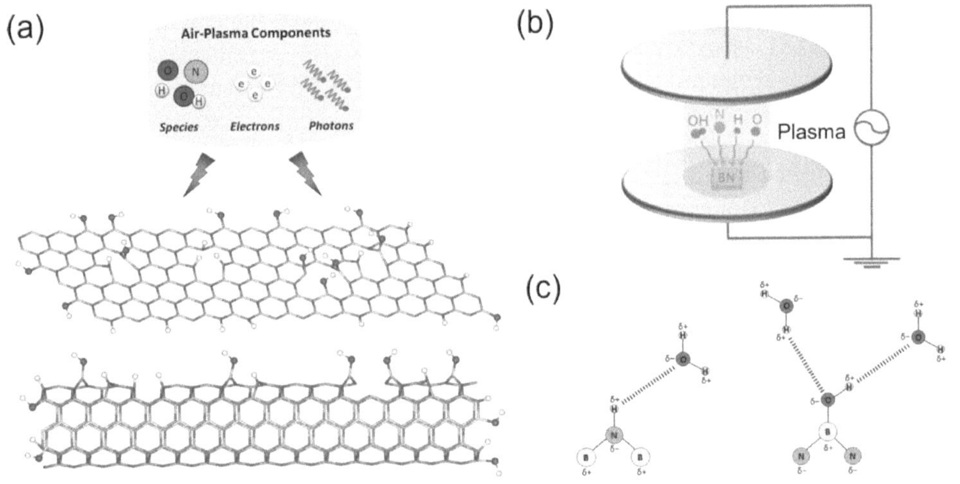

Figure 13.4 (a) Schematic representation of chemically engineered BN nanostructures by air-plasma treatment. (b) Schematics of the air-plasma device in which the samples were subject to direct bombardment of the plasma species. (c) Schematic diagram of hydroxyl-functionalized BN forming hydrogen bonds with water [76].

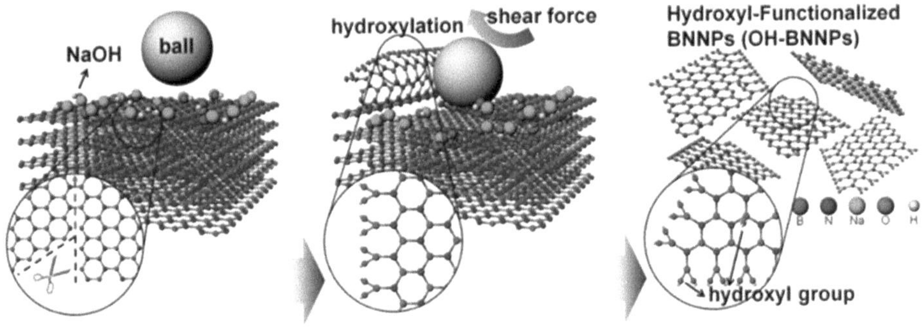

Figure 13.5 Schematic diagram of the ball milling mechanism of BN [84].

in localized functionalization that enhances dispersion in the matrix without compromising the intrinsic in-plane structure [78]. At present, various organic and inorganic substances have been selected to assist the functionalization of BN. In the process of high-speed ball milling, organic matter such as glucose can be used as a lubricant, which has natural lubricating properties, can buffer the strong shear force generated by the ball, and can prevent mechanical damage to the BNNS surface [86]. At the same time, high energy will lead to the decomposition of organic compounds such as urea, dopamine, and sugars, causing amino and hydroxyl groups adhere to the edge of BN [79,80,85,87]. The functionalized BN can form hydrogen bonds with the surrounding polymer and increase its interaction with the polymer [81]. In addition, some inorganic substances can introduce not only hydroxyl and amino groups in BN [83,84], but also other groups in BN to improve the properties of composites. For example, Yu et al. employed dry ice-assisted ball milling to functionalize graphene/BN edges with -COOH branches. The functionalized graphene/BN was observed to resist aggregation due to charge repulsion between nanosheets, exhibiting long-term stability in aqueous solutions. After 30 days, the maximum sedimentation was found to be only 1%–2% [82]. Liu et al. adopted $(NH_4)_3PO_4$-assisted ball milling to introduce amino groups and at the same time to functionalize the BN with phosphate groups, improving the flame retardant properties of the composite [85].

13.3.2.3 Liquid-Phase Exfoliation

In 2008, sonication exfoliation technology was introduced into the field of BN exfoliation [88,89]. Currently, sonication exfoliation is one of the most common liquid-phase exfoliation methods. Under ultrasonic conditions, the B–N bonds of BN at defect sites are prone to attack by oxygen atoms in water molecules, eventually causing large BN sheets to be torn into smaller fragments. -OH groups are introduced to the edges of BN, facilitating its good dispersion in water. To enhance the functionalization of BN, various solvents are chosen to assist in the sonication treatment. Nitric acid can covalently bond with defects in BN during ultrasonication, increasing active sites and branching more -OH groups [78]. Some scholars have adopted a similar approach, using concentrated sulfuric acid and nitric acid (1:3) to assist sonication [74]. In addition to acids, strong oxidizing agents such as hydrogen peroxide have also been used to assist in the functionalization of BN, enhancing its dispersion in the matrix [90]. Furthermore, in 2009, Zhi et al. achieved large-scale production of BNNSs using a strong polar solvent, N,N-dimethylformamide [91]. In 2011, Coleman et al. discovered that N-methyl-pyrrolidone and IPA, due to their surface energy similarity with BN, were effective solvents for exfoliating BN [92]. Mixed solvents usually exhibit better efficiency than single solvents; thus, solvent combinations are also widely used by researchers to exfoliate bulk BN [93]. In addition to introducing covalent functional groups on the surface of BN, some researchers have utilized cellulose nanofibers (CNFs) as solvents for the non-covalent functionalization of BN during sonication. During sonication, CNFs can be attached to the surface of BN through hydrophobic–hydrophobic interactions. CNFs have both hydrophilic and hydrophobic moieties, which can aid in the dispersion of BN [94,95]. Liquid-phase high-speed stirring is another liquid-phase exfoliation method. During the exfoliation process, BN is subjected to strong shear forces, causing it to be torn apart and exposing dangling bonds. Water can attack the dangling bonds at the edges of BNNSs, ultimately forming -OH groups [96].

13.3.2.4 Heat Treatment

Although BN exhibits strong antioxidative properties, high-temperature heat treatment in air can allow oxygen to penetrate the BN lattice. Subsequent agitation of the heat-treated BN in an aqueous solution can lead to its exfoliation and hydroxylation, thereby enhancing its dispersion in the solution [97]. When hot steam flows through BN, BN can react with high-temperature steam to generate NH_3. Then, exposed boron atoms with dangling bonds can react with hydroxyl groups, leading to the introduction of hydroxyl groups to the edge of BN (Figure 13.6) [98]. Hydroxyl-modified BN can form hydrogen bonds with the polymer matrix, uniformly dispersing within the matrix.

13.3.3 Functionalization of BN Fillers and Filler-and-Matrix Interfaces

In addition to the inherent thermal resistance of fillers and polymers, the contact thermal resistance and interfacial thermal resistance generated by poor contact between fillers and polymers are also important factors affecting the TC of composites. The surfaces of BN fillers and polymer matrixes are not perfectly smooth. Therefore, tiny gaps are generated upon contact. Additionally, the hydrophobic nature of BN results in difficulty in expelling air bubbles between it and the

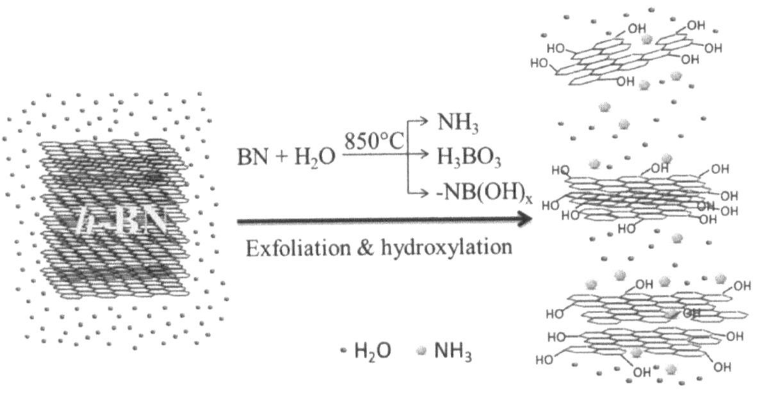

Figure 13.6 Schematic illustration of hydrolysis-assisted exfoliation and hydroxylation of BN powder in hot steam, where H_2O reacts with BN, forming NH_3 and -NB(OH)$_x$, and the highly energetic H_2O and NH_3 assist the exfoliation through diffusion between BN layers [98].

polymer during composite construction, leading to more gaps between the two phases. The TC of these gaps is often extremely low, for instance, air (~0.026 W m^{-1}·K^{-1}) [99], resulting in contact thermal resistance. Moreover, when BN contacts the polymer matrix, there is a mismatch in the vibrational modes of phonons at the interface, leading to significant phonon scattering and a reduction in the mean free path of phonons, thereby decreasing the TC of the composites. This is known as interface thermal resistance, also referred to as Kapitza resistance [100]. Functionalizing BN is an efficient strategy to reduce the thermal resistance between BN and the polymer matrix. Currently, BN functionalization methods can mainly be divided into covalent functionalization and non-covalent functionalization. Functionalized BN can enhance compatibility with the matrix through interactions such as hydrogen bonding, thereby avoiding pore formation and alleviating phonon mode mismatch.

13.3.3.1 Covalent Functionalization

Covalent functionalization is a method that utilizes the difference in electronegativity between nitrogen and boron in BN and the unsaturated chemical bonds, enabling active groups to be covalently bonded to these active sites (Figure 13.7). This functionalization method can increase the TC of composites by 5%–90% (Table 13.1). Covalent functionalization of BN can include in-plane functionalization and edge functionalization. While in-plane functionalization can improve its interface contact with the matrix when used as a filler, it also introduces defects, leading to redundant phonon scattering, which is detrimental to enhancing the TC of the composite. Wu et al. provided experimental validation of this phenomenon through the preparation of in-plane hydroxyl-functionalized BN and edge hydroxyl-functionalized BN as fillers. Their findings revealed that in-plane hydroxyl-functionalized BN/CNF exhibited only a modest increase in TC (10.2%) at low BNNS loading and even displayed a decreasing TC trend at higher loadings. In contrast, edge hydroxyl-functionalized BN/CNF achieved a substantial 97.8% increase in TC at a 60 wt% BN loading [78]. To date, numerous functional groups, including -OH and -NH_2, have been experimentally introduced into BN to improve its interfacial interaction with the matrix [73,85]. Wu et al. confirmed through molecular dynamics calculations that the TC of BNNTs functionalized with -NH_2 groups improved when forming composites with epoxy. At a 30 wt%

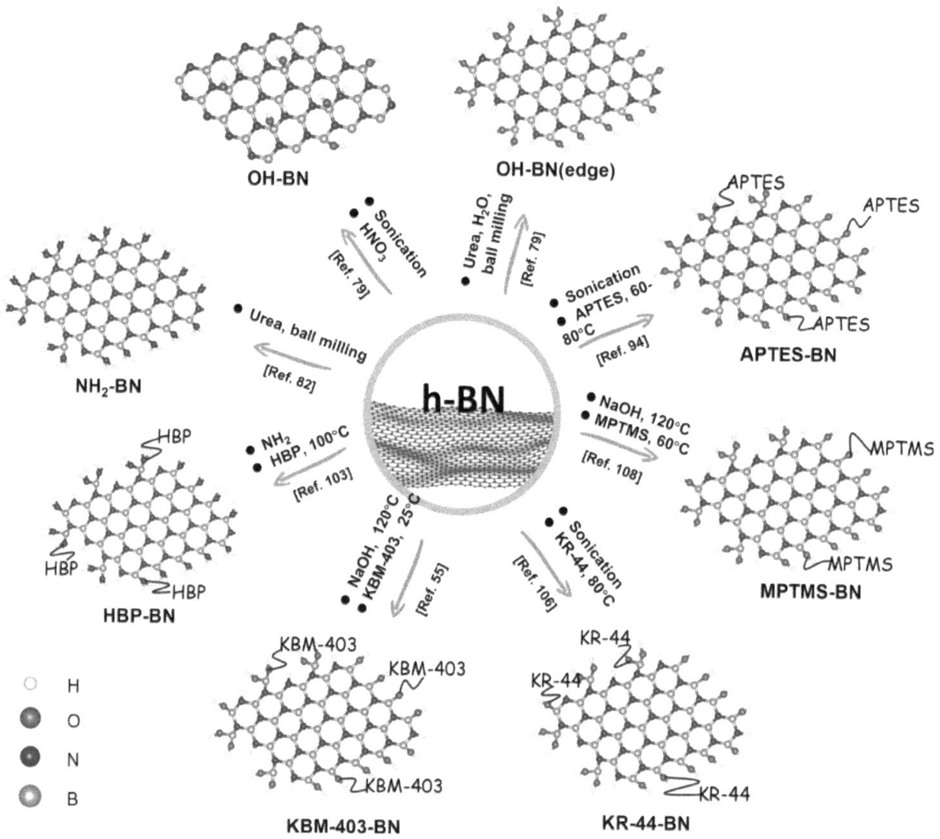

Figure 13.7 Summary of covalent functionalization strategies of BN.

TABLE 13.1 COMPARISON OF TC IMPROVEMENT VIA COVALENT FUNCTIONALIZATION

Matrix	Modification methods	Loading (wt%)	Unmodified TC (W m⁻¹·K⁻¹)	Modified TC (W m⁻¹·K⁻¹)	Enhancement (%)
CNF [78]	-OH	60	12.27	24.27	97.8
Epoxy [83]	-OH, $-NH_2$ and phosphate groups	30	0.55	1.00	81.8
Epoxy [93]	APTES	40	3.03	5.86	93.4
Epoxy [74]	APTES	30	1.04	1.18	13.5
Epoxy [79]	$-NH_2$	10	0.28	0.38	35.7
Epoxy [102]	HBP	5	0.27	0.33	22.2
Epoxy [103]	APS	10	0.9	1.62	80.0
Epoxy [104]	γ-Aminopropyl triethoxy silane	60	0.99	1.05	5.8

(Continued)

TABLE 13.1 (*CONTINUED*) COMPARISON OF TC IMPROVEMENT VIA COVALENT FUNCTIONALIZATION

Matrix	Modification methods	Loading (wt%)	Unmodified TC ($W\ m^{-1}{\cdot}K^{-1}$)	Modified TC ($W\ m^{-1}{\cdot}K^{-1}$)	Enhancement (%)
PVA [96]	-OH	50	9.03	14.6	61.7
PVA [77]	$-NH_2$	50	1.50	2.40	60.0
PI [105]	KR-44	50	0.74	0.86	16.2
PI [106]	γ-Glycidoxypropyl trimethoxysilane and aminopropylisobutyl polyhedral oligomeric silsesquioxane	30	0.69	0.71	2.9
ETDS [55]	KBM-403	70	2.92	4.11	40.8
ETDS [55]	KBM-703	70	2.92	3.88	32.9
Carboxylated styrene-butadiene rubber [107]	3-Mercaptopropyl trimethoxysilane	47	0.57	0.66	15.8

loading, the TC of the composite formed by 5% amino-functionalized BNNTs was 2.04 $W\ m^{-1}{\cdot}K^{-1}$, whereas the TC of the composite composed of unfunctionalized BNNTs was only 1.42 $W\ m^{-1}{\cdot}K^{-1}$. This enhancement can be attributed to the ability of functionalized BNNTs to form covalent bonds with the epoxy matrix, resulting in the alignment of phonon vibration modes between the two components, thus reducing phonon scattering at the interface. Additionally, $-NH_2$ could crosslink with epoxy resin, forming thermal bridges to facilitate heat transfer [101].

Following covalent modification with surface-OH groups, more complex functional groups can also be introduced to enhance the filler–matrix interaction. Various silane coupling agents have been introduced to the surface of BN through this method. Utilizing 3-mercaptopropyltrimethoxysilane for the functionalization of BN enables the introduction of sulfur groups onto the surface of BNNSs. The presence of sulfur-functionalized BNNSs optimizes the interface with carboxylated styrene-butadiene rubber, wherein the cross-linking reinforcement restricts the movement of polymer molecules. Carboxylated styrene-butadiene rubber exhibits a more uniform orientation, leading to improved mechanical properties [107]. For 3-amino-propyl-3-ethoxy-silane (APTES), it initially undergoes hydrogen bonding-mediated physical adsorption onto the surface of BNNSs and then dehydrates at 100°C to form covalent bonds. The amino groups in APTES are capable of reacting with epoxy groups, linking BNNSs to epoxy resin, thereby enhancing the interface compatibility between BNNSs and epoxy and reducing interfacial thermal resistance [93]. Simultaneously, modifying BN samples with APTES can also enhance the maximum filling capacity in epoxy [108]. In addition to 3-mercaptopropyltrimethoxysilane and APTES, other commonly used silane coupling agents include γ-aminopropyl triethoxy silane [104], γ-glycidoxypropyltrimethoxysilane [106], and aminopropylisobutyl polyhedral oligomeric silsesquioxane [106]. In addition to utilizing silane coupling agents, [102] other coupling agents, such as titanates [105], can also enhance the compatibility between BN and fillers, reducing the thermal resistance of composites.

It is noteworthy that covalent functionalization not only enhances the interfacial compatibility between fillers and the polymer matrix but also improves the filler-filler interactions compatibility through controlled modification. Kim et al. modified hydroxyl-rich BN with amino groups using APTES silane and functionalized multiwalled CNTs with carboxyl groups. By exploiting the reactivity between amino and carboxyl groups, they successfully fabricated BN-multiwalled CNTs hybrid fillers, forming a heat flow bridge between fillers to reduce interfacial thermal resistance [109].

13.3.3.2 Non-Covalent Functionalization

The covalent functionalization process involves the breaking and formation of chemical bonds, which often require significant energy input. As discussed in Section 13.3.2, high-energy ball milling, sonication, plasma treatment, and other methods have been applied to the covalent functionalization process, which may introduce defects to BN, increasing phonon scattering during the heat transfer process. In contrast, non-covalent functionalization of BN predominantly relies on physical interactions, including intermolecular forces, π–π stacking, and hydrogen bonding. This non-covalent functionalization process is relatively gentle to BN, requiring lower energy inputs and preserving the intrinsic structure of BN, thus maintaining its inherent properties. Furthermore, this functionalization method can occur on the surface of BN, enabling a wider range of interactions with polymers.

A common approach involves leveraging the spontaneous polymerization process of organic molecules to coat the BN surface with other organic molecules that exhibit excellent affinity, enhancing their affinity with the polymer matrix. Inspired by the composition of adhesive proteins in mussels, Messersmith et al. developed a surface modification method based on the self-polymerization of dopamine on material surfaces, also known as the dopamine coating method [110]. The polydopamine (PDA) coating can be formed on virtually all types of material surfaces, making this method widely applicable. Currently, many scholars have conducted research on the non-covalent functionalization of BN using this method. At a pH of 8.5, dopamine tends to spontaneously polymerize and wrap the BN surface through π–π interactions (Figure 13.8a–c). The PDA coating significantly improves the dispersion of BN in the matrix, optimizes the interfacial contact between the two phases, and contributes to enhancing the thermal and mechanical properties of composites [111–113]. It is important to note that due to the amorphous structure of PDA thin films, their intrinsic TC is relatively poor, which may potentially hinder effective heat transfer between BN and the polymer matrix (Table 13.2). Similar to PDA, tannic acids (TA) also oxidize and self-polymerize to oligomers at pH 7.8, coating BNNSs through physical interactions, which is faster and more cost-effective than PDA (Figure 13.8d–f). Moreover, the abundance of phenolic hydroxyl groups on TA can promote its connection with the polymer matrix and reduce the interface thermal resistance [114]. In addition, other substances such as cetyltrimethylammonium 4-vinylbenzoate, polyrhodanine (PRh), and similar compounds have also been employed in similar approaches to achieve the non-covalent functionalization of BN. For instance, cetyltrimethylammonium 4-vinylbenzoate can undergo polymerization of PRh monomers on the surface of BNNTs using a free radical initiator, ensuring a high degree of stability for BNNTs over prolonged periods (Figure 13.8g–i) [115]. For PRh, Fe^{3+} ions adsorbed onto the surface of BNNSs through charge interactions can serve as both initiators and oxidants, enabling in situ polymerization on the BNNS surface (Figure 13.8j–l). The wettability between PRh@BNNSs and styrene-butadience rubber is enhanced due to the functionalization of BNNSs. Additionally, the PRh coating reduces the polar component of the surface energy of BNNSs, resulting in PRh@BNNSs exhibiting interface properties close to those of the matrix [116].

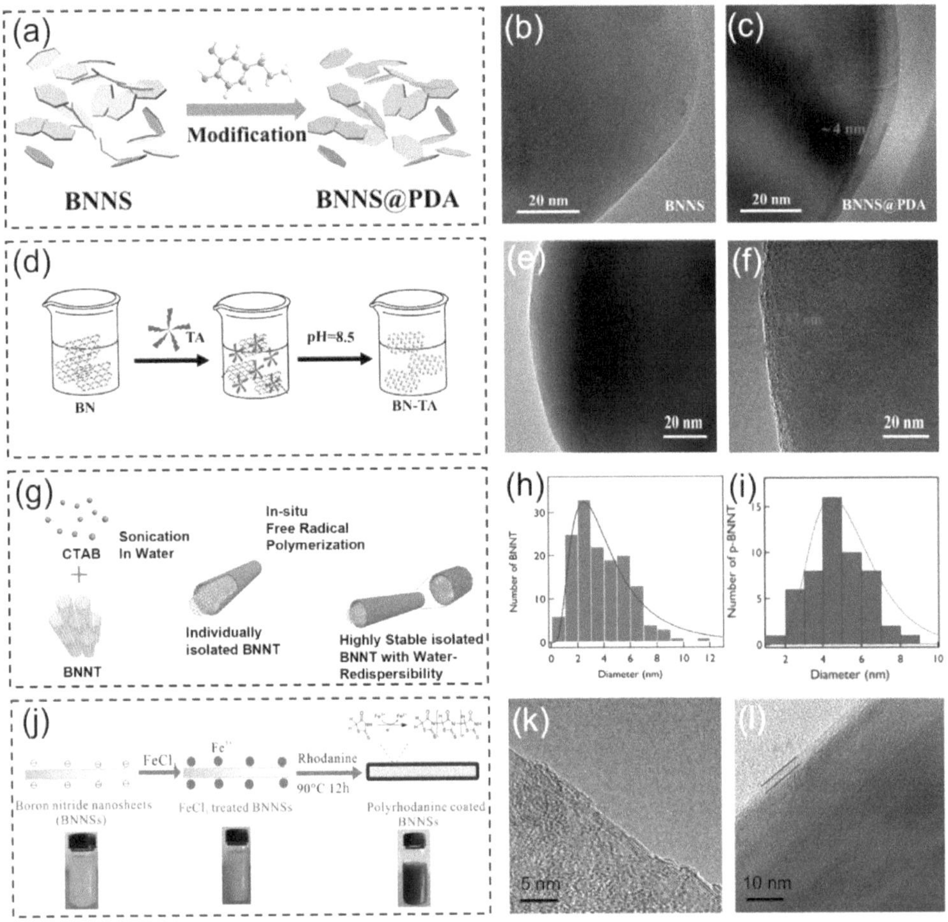

Figure 13.8 (a) Schematic illustration of PDA non-covalent functionalization [111]. TEM images of (b) BNNSs and (c) BNNSs@PDA [111]. (d) Schematic illustration of TA non-covalent functionalization [114]. HR-TEM images of (e) BN and (f) BN@TA [114]. (g) Schematic illustration of cetyltrimethylammonium 4-vinylbenzoate non-covalent functionalization [115]. Diameter distribution of (h) BNNTs and (i) functionalized BNNTs [115]. (j) Schematic illustration of PRh non-covalent functionalization [116]. TEM images of (k) BNNSs and (l) BNNSs@PRh [116].

In addition to achieving surface modification through polymerization on the surface of BN, methods utilizing physical forces such as hydrogen bonding, van der Waals forces, and Lewis acid–base interactions have also been widely adopted. Liu et al. revealed through density functional theory calculations that nitrogen-containing polymers can form strong interactions with BN via hydrogen bonding. By utilizing guanidine-like organic compounds to form bidirectional hydrogen bonds with PVA and BN, thermal bridges can be constructed [117]. Additionally, the N atoms in BN are electron-rich, while the B atoms are electron-deficient, allowing BN to act as either a Lewis base or acid, interacting with certain substances through a Lewis acid–base reaction. At present, among the studies on the Lewis acid–base functionalization of BN, there are many studies on using B atoms as the Lewis acid for functionalization [118]. The amino groups

TABLE 13.2 COMPARISON OF TC IMPROVEMENT VIA NON-COVALENT FUNCTIONALIZATION

Matrix	Modification methods	Loading	Unmodified TC ($W\,m^{-1}\cdot K^{-1}$)	Modified TC ($W\,m^{-1}\cdot K^{-1}$)	Enhancement (%)
BECy [112]	PDA	15 vol%	0.55	0.50	–9.1
PP [113]	PDA	25 wt%	0.48	0.43	–10.4
ANF [111]	PDA	50 wt%	3.33	3.94	18.3
PVA [124]	PDA	30 vol%	7.3	8.8	20.5
Styrene-butadience rubber [116]	PRh	27.5 vol%	0.55	0.75	36.3
XNBR [114]	TA	30 vol%	0.36	0.42	16.7
Polycarbonate [125]	CPAM	20 wt%	0.62	0.73	17.7
PMMA [123]	Chlorosulfonic acid	24.5 wt%	0.52	0.96	84.6

in octadecylamine or amine-terminated oligomeric polyethylene glycol can donate electrons and act as a Lewis base to interact with BN [119]. Moreover, some scholars have bifunctionalized BN to improve its compatibility with the matrix. Lee et al. used pyrene-tethered poly(4-vinylpyridine) to bifunctionalize BNNSs with Lewis acid–base and π–π interactions. Bifunctionalized BNNSs exhibited higher dispersion in alcoholic solvents than tert-butyl and amino-functionalized BNNSs [120].

As mentioned earlier, BNNSs exhibit a significant enhancement in TC compared to bulk BN, making the exfoliation of BN an indispensable step in the preparation of high TC composites. Similar to the covalent functionalization approach, one-step functionalization during the exfoliation process can lead to higher preparation efficiency. With the assistance of ultrasound, some ionic liquids can interact with BN through their charge characteristics and facilitate the non-covalent functionalization and exfoliation of BN, among which anion/cation–π interaction plays an important role in improving the functionalization ratio [121,122]. Morishita et al. used 1-butyl-3-methylimidazolium hexafluorophos-phate ([bmim][PF_6]) for the non-covalent functionalization exfoliation of BNNSs and then compounded them with PMMA. With a BNNS loading of 24.5 wt%, the through-plane TC of the composites increased by 86.5% [121]. In addition, Takuya Morishita et al. employed chlorosulfonic acid to achieve this goal. Chlorosulfonic acid, owing to its highly acidic protons, is anticipated to interact with the N atoms on the BN surface, leading to electrostatically stable exfoliation. Simultaneously, its vigorous reaction with water generates gaseous HCl and H_2SO_4, aiding in exfoliation. The SO_3Cl ions in chlorosulfonic acid also exhibit a high affinity for the PMMA matrix. Through this method, simultaneous exfoliation of BN and enhancement of its compatibility with the matrix interface can be achieved [123].

13.3.3.3 Thermal Resistance Evaluation Model

To quantitatively evaluate the thermal resistance between fillers and between fillers and the matrix, various models such as the Foygel model, the Effective Medium Theory, and the Hashin-Shtrikman model have been proposed. These models can provide information on the thermal resistance of composites by obtaining parameters from some samples in experiments, guiding experimental efforts.

13.3.3.3.1 Foygel Model

Foygel et al. proposed the Foygel model based on Monte Carlo simulations of CNTs with both ends capped, to calculate the critical percolation volume fraction at the onset of percolation. The TC of composites and the contact thermal resistance between fillers can be described using the following equation [126–128]:

$$\lambda - \lambda_m = K_0 \left(\frac{V_f - V_c}{1 - V_c} \right)^{\beta} \tag{13.2}$$

$$R_c = \frac{1}{K_0 L V_c^{\beta}} \tag{13.3}$$

where λ is the TC of the composites, λ_m is the TC of the matrix, K_0 is a pre-exponential factor ratio that is related to BN fillers, V_f is the volume fraction of BN, V_c is the critical volume fraction of BN, β is the percolation exponent, R_c is the contact resistance between BN, and L is the size of BN.

13.3.3.3.2 Effective Medium Theory

The Effective Medium Theory, proposed by Nan et al., incorporates the basic concept of Kapitza resistance and is tailored for various composite models. This model can be mainly categorized into four types, with one of the more commonly used formulations shown as follows, originally devised to describe the TC enhancement in CNT composites [100,103,127,129–132].

$$\frac{\lambda}{\lambda_m} = 1 + \frac{V_f p}{3} \frac{\frac{\lambda_f}{\lambda_m}}{p + \frac{2\lambda_f R_k}{d}} \tag{13.4}$$

In which p and d are respectively the aspect ratio and the diameter of BN, λ_f is the TC of BN, and R_k is the interfacial thermal resistance between BN fillers and the polymer matrix.

13.3.3.3.3 Hashin-Shtrikman Model

When computing composites consisting of multiple fillers, the Hashin-Shtrikman model is particularly applicable. This model can be utilized to calculate the overall thermal resistance of fillers and polymers. It can be described by the following equation [100,106,133,134]:

$$\frac{\lambda}{\lambda_m} = \frac{1 + 2K \sum_{i=1}^{n} E_i V_i}{1 - K \sum_{i=1}^{n} E_i V_i} \tag{13.5}$$

$$K = 13.3347 \exp\left(-13.2701 \frac{R_c \lambda_m}{L} \right) \tag{13.6}$$

$$E_i = \frac{\frac{\lambda_i}{\lambda_m} - 1}{\frac{\lambda_i}{\lambda_m} + 1} \quad (13.7)$$

where K is a constant, n is the number of types of BN fillers, V_i n is the number of types of BN fillers, R_c is the total thermal resistance of the polymer composites, and L represents the characteristic length of the unit model.

13.3.4 3D Network Construction of BN Fillers in Polymer Matrixes

In fact, in composites, fillers serve as the primary carriers for heat conduction. Fillers are interconnected to form a thermal conduction path with a network structure, namely the 3D filler network. According to the time of constructing the 3D network, it can be divided into two forms: (1) forming a 3D filler network during the construction of composites, using dispersed BN as the filler; (2) pre-constructing fillers with a 3D structure, then jointly constructing composites with the polymer matrix [135].

13.3.4.1 3D Thermal Conduction Networks with Dispersed Fillers

BN nanomaterials exist in various dimensions, including 0D, 1D, 2D, and 3D. For the construction of 3D filler networks in composites, 0D, 1D, and 2D BN are chosen as thermally conductive fillers. 0D BN fillers mainly consist of BN fullerenes and BN nanoparticles. In 1998, D. Golberg et al. first discovered BN fullerene nanomaterials through electron irradiation in transmission electron microscope (TEM) [136]. However, due to the lack of effective synthesis and purification routes, research on BN fullerenes in the field of thermal management is still relatively limited. BN nanoparticles represent another 0D structure of BN, distinct from BN fullerenes. As early as 1988, P. Bujard performed a composite of BN particles and epoxy, achieving the TC of 2.3 $W\ m^{-1}{\cdot}K^{-1}$ at 31 vol% BN loading [137]. Compared to the other dimensions of BN, 0D BN has unique advantages for reducing the viscosity of composites and achieving high filling levels. Moreover, their morphology engenders isotropic thermal conduction properties, which contribute to reducing the degree of anisotropy in the thermal conduction property of composites (Figure 13.9a and b). Despite these inherent advantages, achieving uniform dispersion within the filler remains a necessary condition for obtaining isotropic composites. Logakis et al. used a nanoparticle–microparticle mixture (with additional silane coupling and ball milling) to improve the dispersion of BN in epoxy. It showed an effective improvement: TC of 20 wt% BN/epoxy was 4 times higher than that of blank epoxy [138]. Muratov et al. used APTES to modify BN. The functionalized BN had better dispersion in polypropylene (PP). The TC of such composites was doubled compared to unfunctionalized BN/PP [108].

BNNTs are a typical 1D BN structure. Since the method of preparing CNTs by arc discharge evaporation was developed in 1991 [139], the preparation of BNNTs has also garnered considerable attention. In 1994, L. Corkill et al. theoretically predicted that structural analogs of CNTs and BNNTs could also exist stably [140]. In 1995, Zettl et al. reported the successful synthesis of BNNTs. They utilized a non-carbon plasma arc discharge technique to fabricate multiwalled tubes with inner diameters ranging from 1 to 3 nm and lengths of up to 200 nm between tungsten rods and copper electrodes [37]. Subsequently, BNNTs sparked significant research interest among many researchers. In terms of thermal conduction property, BNNTs offer the advantage of maximizing the exceptional TC of the (002) plane. Additionally, the high aspect ratios of 1D BNNTs contribute to higher efficiency in forming 3D thermal conduction networks compared to other shapes of fillers, enabling the formation of continuous networks at very low loading (Figure 13.9c and g) [36]. Dmitri Golberg et al. compared the performance differences between BNNTs and BN nanospheres as nanofillers. It was found that at a loading of 6 vol%, the TC of the nanofluid formed by BNNTs with water was 1.54 $W\ m^{-1}\cdot K^{-1}$, whereas for nanospheres, it was only 0.95 $W\ m^{-1}\cdot K^{-1}$ [141]. Huang et al. selected CNFs as the matrix, and at a loading of 25 wt%, the in-plane TC of BNNT/CNFs could reach 21.39 $W\ m^{-1}\cdot K^{-1}$, which is three times that of BNNS/CNF (Figure 13.9d–f) [94].

Inspired by the discovery of graphene [42], Novoselov et al. peeled single-layered BNNSs by mechanical exfoliation in 2005 [43]. Subsequently, various synthesis methods for BNNSs were developed. Compared with 0D and 1D BN nanomaterials, BNNSs can perform high TC in two dimensions. The key problem of constructing thermal conduction networks using 2D BNNSs as fillers lies in the exfoliation, dispersion, and orientation. The dispersion technique has been discussed in Section 13.3.2, and the orientation technique will be discussed in Section 13.3.5. This subsection focuses on the exfoliation of BN. During the exfoliation process of BN, significant energy can attack the active sites on the edges of BN, resulting in the formation of edge-functionalized BN, which has been demonstrated to be a highly efficient thermal conductive filler. Methods for BN exfoliation include ball milling, sonication, liquid-phase mechanical exfoliation, and gas-phase exfoliation. Ball milling is a commonly used mechanical exfoliation method for BNNSs, where the pressure and shear forces between steel balls can separate 2D materials connected by van der Waals forces (Figure 13.10c). In ball milling, shear forces can cut the layered materials from their outer surfaces, and compressive forces can peel thin 2D

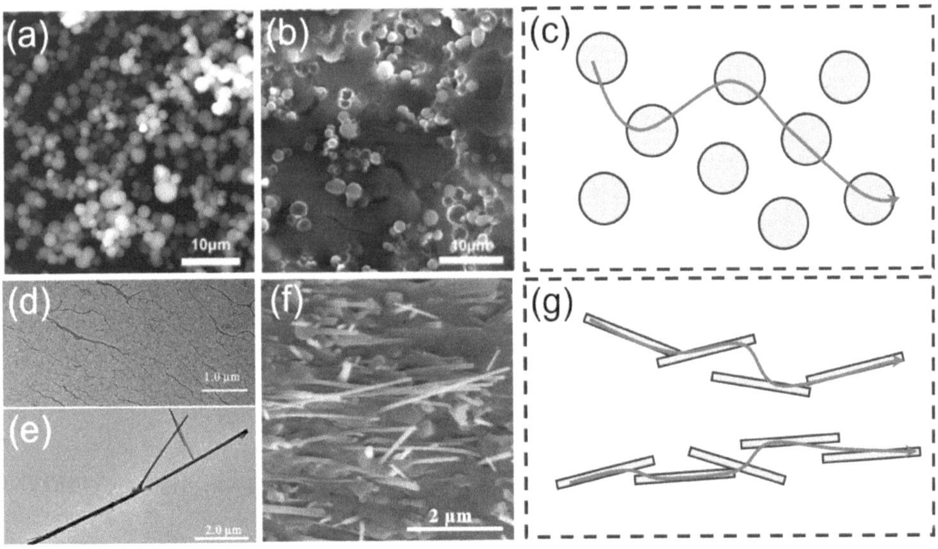

Figure 13.9 (a) SEM images of BN microspheres [142]. (b) SEM images of BN microsphere/Epoxy composites [142]. (c) The thermal conduction path formed by 0D BN spheres [141]. TEM images of (d) CNFs and (e) BNNTs [94]. (f) Typical cross-section morphology of BNNT/CNF composites [94]. (g) The thermal conduction path formed by 1D BNNTs [141].

nanosheets from the edges of layered materials. However, excessive ball milling energy can cause the thin layers of prepared 2D materials to fracture. Li et al. implemented a low-energy ball milling approach, combining BN powder with benzyl benzoate in a planetary mill under a high-pressure N_2 atmosphere, yielding thinner nanosheets that exhibited transparency under scanning electron microscopy (SEM) electron beam irradiation [143]. Recent advancements have seen the development of exfoliation methods assisted by substances such as urea [81], sodium hydroxide [84], and liquid metals [144]. Some scholars have also switched from high-energy ball milling to mechanical stirring to reduce the substantial energy input of ball milling and utilize shear forces to exfoliate BN [96]. Sonication is a liquid-phase exfoliation method. Compared to the high yield of about 60%–90% achieved by ball milling [145], its yield is relatively low. Nevertheless, its convenience and safety ensure its widespread application. In 2009, Zhi et al. achieved large-scale preparation of BNNSs using the strong polar solvent N,N-dimethylformamide [91]. In 2011, Nicolosi et al. exfoliated various 2D materials, including BN, using ultrasound, and found promising solvents such as N-methyl-pyrrolidone and IPA that could minimize exfoliation energy [92]. Subsequently, many efforts have been made to develop a series of promising solvents for exfoliation [146]. In addition to ball milling and liquid-phase exfoliation, gas-based BN exfoliation methods have also been developed. Different gas sources, including high-temperature steam [98], nitrogen gas produced by low-temperature liquid nitrogen gasification (Figure 13.10d) [147], supercritical gases (Figure 13.10e) [59], and compressible gases [148], have been introduced into the field of BN exfoliation. In addition to these top-down approaches for BNNS preparation, numerous researchers have developed bottom-up growth methods, such as CVD, physical vapor deposition, and biomass-directed on-site synthesis, for synthesizing high-quality, large-area BN (Figure 13.10a and b), which are not further elaborated here.

The fabrication of composites with a 3D thermal conduction structure within the matrix can be efficiently achieved through the simple process of mixing fillers with the polymer matrix. Upon reaching a certain loading, the fillers establish interconnections, thereby forming 3D thermal conduction pathways. During the preparation process, filler alignment can be achieved through methods such as hot pressing and rolling, thus optimizing the TC of the material in the desired orientation. In addition to utilizing single-dimensional fillers, the hybridization of multidimensional fillers presents an effective strategy for constructing 3D networks [149–151]. Compared with single-dimensional fillers, multidimensional filler hybrids offer superior filler dispersion, facilitating the bridging of adjacent fillers to establish a thermal conduction network while simultaneously reducing void content in composites, resulting in synergistic heat conduction enhancement [47]. One prevalent approach involves the utilization of hybrid fillers comprising 0D and 2D components. In this case, 2D fillers can fill the gaps between 0D fillers. Hong et al. observed that when aluminum nitride (AlN) and BN were combined in a 1:1 ratio, BN effectively filled the gaps between AlN particles, yielding composites with optimal TC [152]. In addition, 0D fillers functioned as connective bridges between 2D fillers [153,154]. Wang et al. found that due to the bridging connections of silver nanoparticles when the BNNS loading was 25.1 vol%, the TC of the composite increased from 1.63 to 3.06 W $m^{-1}\cdot K^{-1}$ [155]. Using hybrid fillers composed of 1D and 2D fillers is another common approach, where 1D BNNTs can be used to promote bridging between 2D BNNSs [156,157]. Su et al. found that compared with BNNT/epoxy and BNNS/epoxy composites, the TC values of BNNTs and BNNS/epoxy composites were higher at the same loading [158].

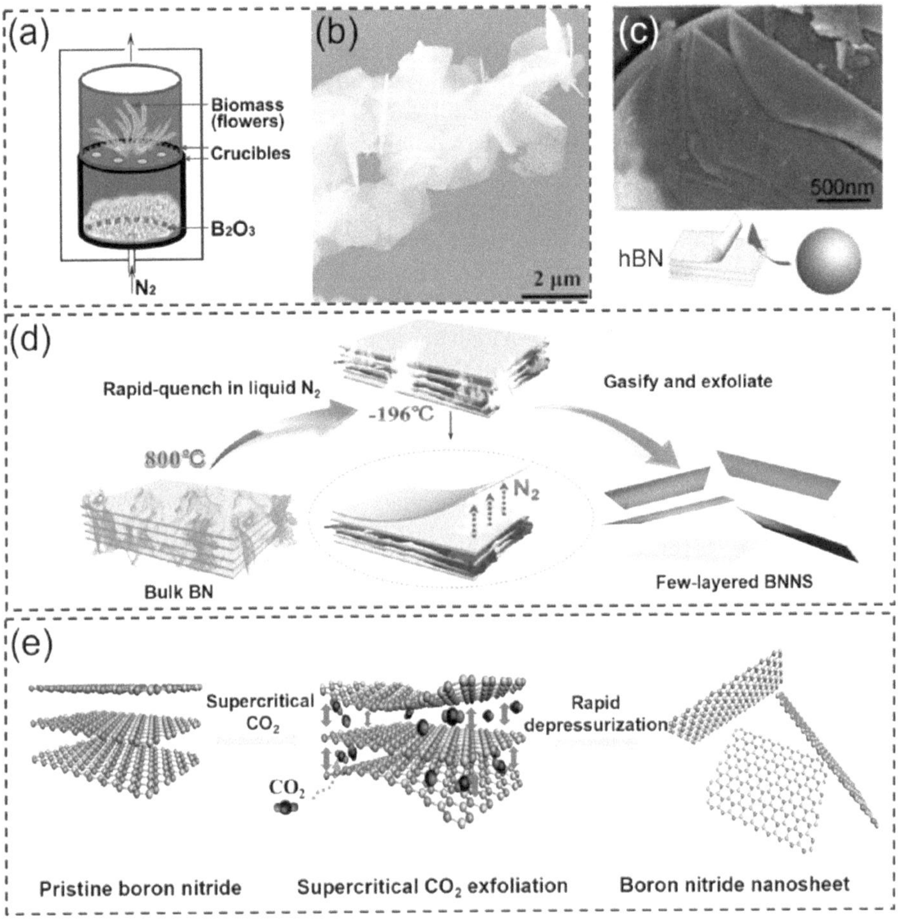

Figure 13.10 (a) Experimental scheme for biomass-directed on-site synthesis of BNNSs [40]. (b) SEM image of a BN-nanosheet-assembled barb [40]. (c) SEM images and corresponding diagrams under the shear force created by milling balls [143]. (d) Gas exfoliation of h-BN triggered by thermal expansion [147]. (e) Scheme of BNNS synthesis by shear-assisted supercritical CO_2 exfoliation [59].

13.3.4.2 3D Thermal Conduction Networks with Preconstructed 3D Frameworks

Achieving high TC at low filler loading while ensuring composite processability is a desirable objective [159]. Constructing 3D fillers in advance, can minimize interfacial contact among fillers and exhibit higher heat conduction efficiency per unit mass compared to randomly dispersed filler systems [46,160,161]. The approach of pre-constructing fillers with a 3D structure before composite fabrication often involves the pre-assembly of 0D, 1D, and 2D materials into 3D architectures. In recent years, methods for achieving higher TC at low loadings by pre-constructing 3D networks have begun to receive attention. Currently, techniques such as templating, electrospinning, and 3D printing have been developed for this purpose.

The template method is an important approach for constructing 3D networks of BN, which can mainly be divided into the foam template method, sacrificial template method, and ice template method. The foam template method involves using polymer foam as a template. Melamine foam (MF), characterized by its low density (4–12 mg cm^{-3}) and exceptionally high porosity (~99%), has emerged as an important candidate for polymer foam applications. The approach of growing 3D graphene on MF surfaces has been applied [162]. As a graphene analog, Wu et al. utilized a method of repeated layer-by-layer assembly to prepare 3D BN on MF. Through the electrostatic interaction between positively charged polyethyleneimine and negatively charged BNNSs, BN could be coated onto the MF, ultimately impregnated in epoxy to form composites with a 3D BN structure (Figure 13.11a). The 3D BN structure provides excellent thermal conduction paths for the composite. The epoxy resin composite obtained under an ultra-low BNNS loading of 1.1 vol% exhibited a 233% increase in TC [163]. Wei et al. utilized melamine-formaldehyde sponge (MS) as a template and developed composites with a hierarchical thermal conduction network featuring a leaf venation structure. The first-level thermal conduction network was formed by coating BN using a multilayer coating technique with the assistance of polyethyleneimine, while the second-level thermal conduction network consisted of BN dispersed in epoxy (Figure 13.11b). Together, these two-level thermal conduction networks formed a high-density phonon transport path, enabling efficient thermal conduction. At a loading of 35.9 wt%, the in-plane TC of the composites reached 10.2 W $m^{-1}\cdot K^{-1}$ [164]. Furthermore, utilizing surfactants to assist in the construction of gelatin foam also contributes to enriching the variety of templates, thereby achieving the construction of 3D BN thermal conduction networks [165,166].

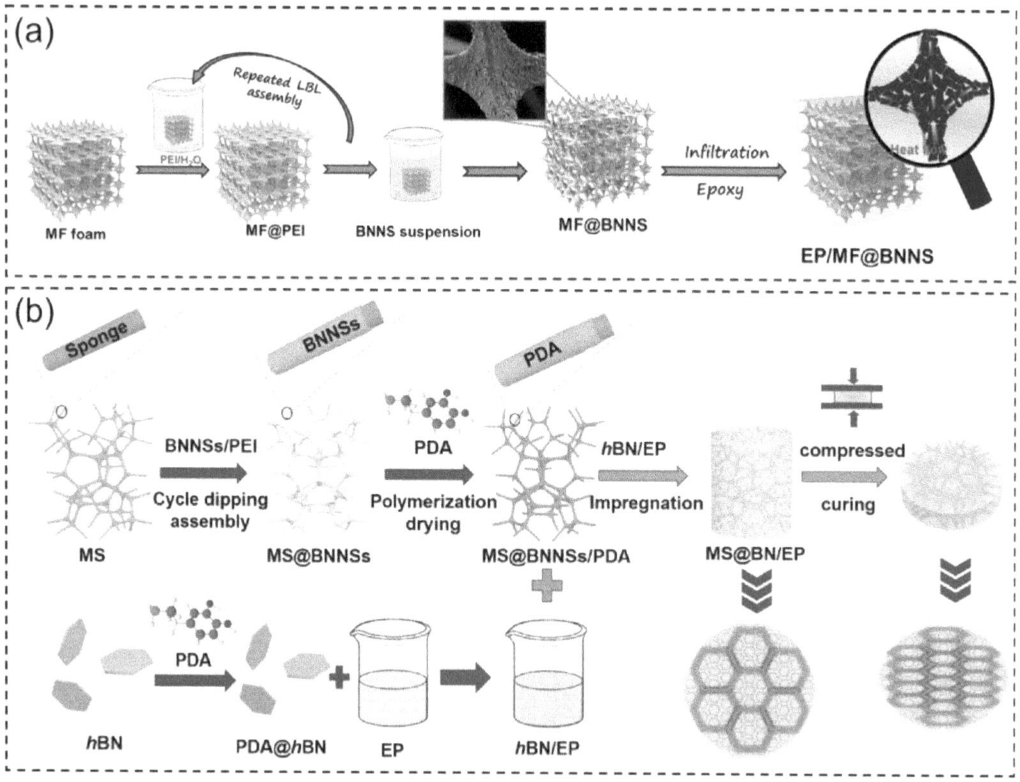

Figure 13.11 (a) Schematic illustration of the formation process of MF@BNNS foam and MF@BNNS/Epoxy composites [163]. (b) Schematic of the fabrication of MS@BN/Epoxy composites [164].

The sacrificial template method, which involves the removal of the template after the formation of 3D BN structures, offers advantages in reducing filler loading while maintaining high TC. Among the various sacrificial templates, salt templates have emerged as particularly advantageous due to their ubiquity, cost-effectiveness, facile removal process, and minimal environmental impact. NaCl and NH_4HCO_3 are frequently employed salt templates [160,167,168]. The fundamental principle underlying the salt template-based sacrificial method involves utilizing salt crystals to facilitate the construction of a BN-salt composite with the desired morphology. Subsequent removal of the salt template yields a 3D BN network (Figure 13.12a–d). Specifically, NaCl templates can be removed through aqueous solution, while NH_4HCO_3 templates can be removed through high temperature (80°C–90°C). 3D BN fillers with different structures exhibit

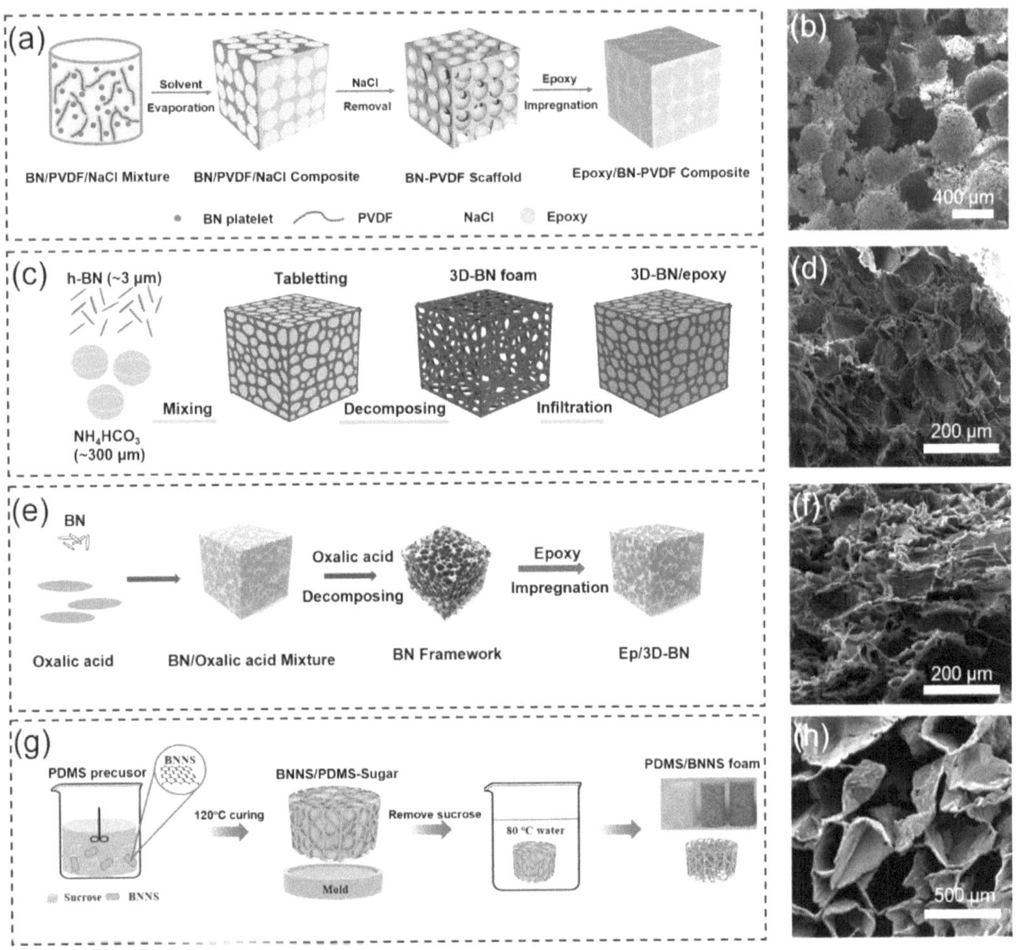

Figure 13.12 (a) Schematic diagram of 3D BN constructed with NaCl assistance [171]. (b) SEM image of 3D BN formed with NaCl as a template [171]. (c) Schematic diagram of 3D BN constructed with NH_4HCO_3 assistance [168]. (d) SEM image of 3D BN formed with NH_4HCO_3 as a template [168]. (e) Schematic diagram of 3D BN constructed with oxalic acid assistance [172]. (f) SEM image of 3D BN formed with oxalic acid as a template [172]. (g) Schematic diagram of 3D BN constructed with sugar assistance [173]. (h) SEM image of 3D BN formed with sugar as a template [173].

different characteristics. By using salt templates to construct BN with a hollow spherical shell structure, composites with excellent in-plane and through-plane TC can be obtained [167]. Meanwhile, when BN and 3D BN dispersed in the matrix form a 3D thermal conduction network together, composites with two-level heat transfer networks can be achieved, facilitating efficient thermal dissipation [169]. Additionally, to enhance the continuity between BNNS sheets on the template and their compatibility with the matrix, cross-linked poly(ethylene-co-vinyl acetate) and poly(vinylidene difluoride) can be utilized as an adhesive [170,171]. Furthermore, apart from salt templates, based on similar principles, acids and sugar substances can also serve as templates to construct 3D BN via the sacrificial template method (Figure 13.12e–h) [172,173].

Ice templating, also known as freeze casting or directional freezing, offers greater controllability compared to conventional sacrificial template methods. This technique has garnered significant attention across diverse fields, including insulation [174], biomedical [175,176], and energy [177] due to its superior manipulability. When there is a temperature gradient, the growth of the ice layer can induce the oriented alignment of BN along the direction of ice growth, promoting TC enhancement in a single direction (Figure 13.13a and b) [178]. When temperature gradients exist in two directions, the morphology of the material becomes even more distinctive: with both radial and vertical temperature gradients in the polymer, a unique circinate vane-type 3D structure can form. This structure provides continuous thermal conduction paths both in-plane and through-plane, achieving TC enhancement simultaneously in both in-plane and through-plane (Figure 13.13c and d) [179,180]. When temperature gradients exist in both horizontal and vertical directions, under the induction of temperature gradients, the material acquires a long-range layered structure (Figure 13.13e and f). BN/epoxy composites with such a structure can achieve a TC of 6.54 W $m^{-1}{\cdot}K^{-1}$ in-plane, along the direction of BN layer growth [181]. On the basis of ice templates, the method of nanowire bridging can enhance the 3D thermal conduction paths, further improving the TC of the composites [46,182,183].

Electrospinning technology involves the acceleration of charged polymer droplets under a high-voltage electric field to form fibers when jetted out in a stream, and the solvent evaporates, resulting in fibrous products. BNNSs can closely adhere to the fiber arrangement, forming a homogeneous orientation, thereby possessing a natural advantage in synthesizing thermally conductive paths with a 3D structure. Functionalizing BN can enhance its interface compatibility with the matrix [185]. PDA-modified BN can form strong hydrogen bonds with the polyvinylidene fluoride matrix, improving interfacial adhesion and reducing the interfacial thermal resistance of the fiber film [186]. Additionally, thermal pressing of the prepared composites can make the product film denser, increasing its orientation [187]. When preparing thermally conductive composites through electrospinning, it often requires pre-mixing of the BN filler and polymer matrix to prepare precursor solutions uniformly, which limits the content of BN in the composites. When the BN content is too high, successful spinning becomes challenging. Some scholars have addressed the issue of the upper limit of BN loading by combining electrospinning with electrostatic spraying, where BNNSs can be adsorbed onto PVA fibers through strong electrostatic interactions [187]. Furthermore, similar to electrospinning, 'extrusion' 3D printing technology can regulate the extruded structure to achieve the preparation of composites with a 3D thermal conduction network, and 3D printing technology offers stronger structural controllability and can achieve higher BN loading [53,188].

While the 3D network formed by self-assembly enhances the interconnection between BN fillers, the thermal conduction paths it establishes remain imperfect, resulting in significant phonon scattering at filler interfaces, thereby diminishing the material's TC [135,159]. Direct

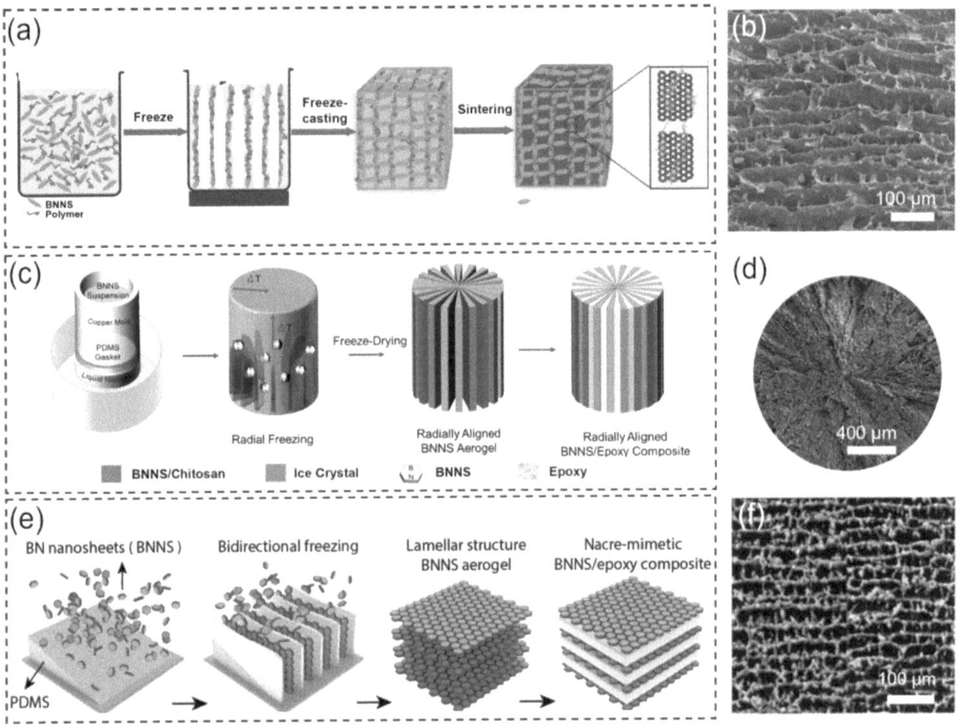

Figure 13.13 (a) Schematic diagram of the preparation of 3D BNNS aerogels [184]. (b) SEM image of the 3D BNNS aerogel in the perpendicular direction to ice growth [184]. (c) Schematic diagram for the fabrication route of the radially aligned BNNS/epoxy composite using a radial freeze casting method [179]. (d) SEM images of the BNNS aerogel [179]. (e) Schematic illustration of the fabrication route using a bidirectional freezing technique together with epoxy resin infiltration [181]. (f) The cross-sectional SEM images of the BNNS aerogel [181].

synthesis of 3D BN structures via CVD on templates further reduces the filler–filler interface thermal resistance generated by BNNS self-assembly [189]. Wong et al. synthesized 3D BN foam directly through CVD using nickel foam as a template, resulting in a 164.5% increase in TC compared to the polydimethylsiloxane matrix when incorporated into composites with BNNSs and polydimethylsiloxane [190].

13.3.5 Alignment of BN Fillers and Anisotropic Thermal Properties of Composites

The unique layered structure of BN imparts significant anisotropy to its thermal conduction property, with an in-plane TC reaching 751 W $m^{-1}\cdot K^{-1}$, while in the direction perpendicular to the plane, it is only 2 W $m^{-1}\cdot K^{-1}$ [10,13]. Consequently, the orientation plays a crucial role in the overall thermal performance of materials when used as fillers in composites. To maximize the TC of composites, according to the thermal conduction path theory, the thermal conductivity of composites significantly improves when the direction of heat flow is parallel to the thermal conduction network. Therefore, our goal is to achieve orientation as uniformly as possible, and this

TABLE 13.3 THE TC OF COMPOSITES CONSTRUCTED USING DIFFERENT ORIENTATION METHODS

				TC ($W\ m^{-1}\cdot K^{-1}$)		
Alignment	**Filler**	**Matrix**	**Loading**	**In**	**Through**	**References**
Magnetic field	BNNBs	Epoxy	40 wt%	/	3.3	[191]
Magnetic field	BNNSs	Epoxy	30 wt%	/	14.55	[192]
Magnetic field	BN	Epoxy	20 wt%	/	0.85	[193]
Electric field	BN	Polysiloxane	15 vol%	1.56		[194]
Electric field	BN	Epoxy	17.57 wt%	/	0.65	[195]
Hot pressing	BN	Polycarbonate	18.5 vol%	3.09	0.86	[196]
Hot pressing	BNNSs	WPU@EP	64%–72%	81.49	/	[197]
Stretching	BNNPs	PE	15 wt%	106.2	/	[198]
Shearing	BNNSs	Silicone rubber	30.8 vol%	5.47	/	[199]
Filtration	BN	Epoxy	15.4 vol%	/	2.01	[200]
Filtration	BNNTs	CNF	25 wt%	21.39	/	[94]
Filtration	BN	NFC	50 wt%	145.7	/	[156]

enhancement in TC is direction-specific. Indeed, the desired orientation varies depending on specific requirements. For instance, the TC of composites often exhibits anisotropic characteristics during the synthesis process, with in-plane TC exceeding through-plane TC due to different fabrication methods. To minimize anisotropy, some researchers orient BNNSs perpendicular to the interface direction, thereby enhancing TC through the plane. Others choose to strengthen the orientation along the in-plane direction to maximize the in-plane TC of composites, allowing for diffusion of heat across the entire plane when encountering point heat sources, thus avoiding heat accumulation. In summary, achieving a uniform arrangement of fillers in composites is crucial. Various methods have been developed to achieve a uniform arrangement of BN, including magnetic field alignment, electric field alignment, and mechanical force field, among others (Table 13.3).

13.3.5.1 Magnetic Field Alignment

The core of the magnetic field alignment method lies in imparting magnetic properties to BN fillers. Commonly used magnetic materials include Fe_3O_4, Fe_2O_3, etc., and functionalizing BN with magnetic materials is essential for achieving magnetic alignment. Functionalizing magnetic particles with cationic functional groups can result in positively charged particle surfaces, while BN tends to carry negative charges in solution, thus enabling the self-assembly of magnetic particles through electrostatic interactions [201,202]. In addition to adsorbing magnetic materials on the filler surface via surfactants, magnetic materials can also be directly grown on the filler surface. For Fe_3O_4, it can be formed by the coprecipitation of Fe^{2+} and Fe^{3+} ions and directly grown on the surface of BN [192]. For Fe_2O_3, high-temperature decomposition of organic iron compounds can lead to growth on the surface and edges of BN [203]. Due to the strong controllability of magnetic fields, there is significant flexibility in the alignment of BN after imparting magnetic properties. By controlling the applied magnetic field, it is possible to selectively enhance the in-plane or through-plane TC of composites (Figure 13.14) [191,203]. It is worth

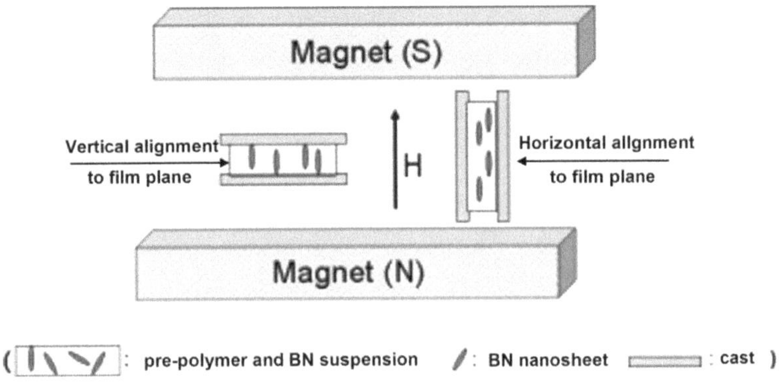

Figure 13.14 Schematic diagram of BN alignment achieved through a magnetic field [203].

mentioning that when the magnetic field intensity is strong, BN may exhibit a phenomenon of aggregated distribution, resulting in localized high TC on one side of the composites, which may lead to a decrease in overall TC [192]. Additionally, magnetic field-induced alignment devices often require the oriented material to be ferromagnetic, which affects the insulation properties of composites, resulting in a deterioration of insulation performance, making them unsuitable for applications requiring high insulation requirements [204].

13.3.5.2 Electric Field Alignment

Under the induction of an electric field, the BN in the composites becomes polarized, and the charge density at the edges of BNNSs increases [194]. Consequently, under adjacent BN dipole–dipole interactions, the alignment of the filler adjusts with the electric field (Figure 13.15). During this process, the alignment of BN is mainly influenced by electric field forces, torque, gravity, and polymer viscosity [205]. As the electric field intensity increases, the alignment of filler BN becomes more uniform, leading to a gradual improvement in the performance of composites. However, it is essential to control the electric field to avoid exceeding the breakdown voltage of the polymer; exceeding this voltage can destroy the organic matrix. Nevertheless, due to the insulating nature of BN, a significant voltage is required for induction, which can be addressed by using nanosecond-pulse electric fields to overcome these limitations [206]. The torque of BN is influenced by its size and the dielectric constant of the polymer. For a highly anisotropic alignment of BNNSs, BNNSs with a shorter longitudinal diameter and a polymer matrix with a higher dielectric constant would be preferred [194]. Regarding the viscosity of the polymer, lower viscosity is often desired, as it can facilitate the orientation of BN particles [205,206].

13.3.5.3 Mechanical Force Field

Several mechanical forces, including pressure, tensile force, and shear force, can achieve alignment of BN fillers in composites. Hot pressing is a common method for increasing material density and arranging fillers in a regular pattern, particularly prevalent in the preparation of thin film materials [185,187,207,208]. This method applies a force field perpendicular to the BN plane. When the sample is relatively thick, the fillers in the sample have greater freedom, leading to differences in alignment, which manifest as a certain downward trend in TC with increasing thickness of thin film samples. After hot pressing treatment, the fillers becomes denser under the action of compressive stress, and the freedom of fillers decreases, resulting in a more uniform

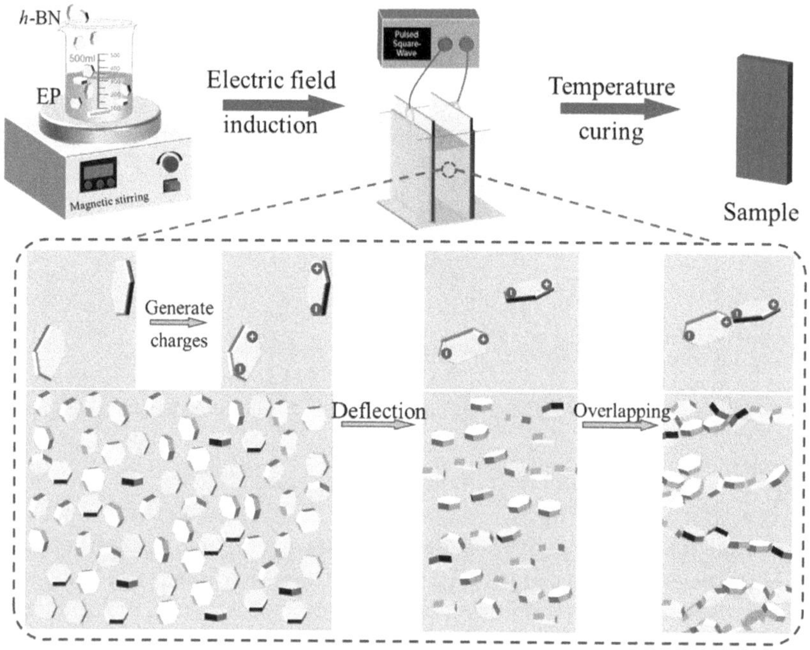

Figure 13.15 Schematic diagram of BN alignment achieved through an electric field [204].

alignment (Figure 13.16a) [196]. A further advancement in hot pressing technology is vacuum hot pressing, where vacuum conditions reduce pores in the material and increase material density. Apart from hot pressing, coating methods can also enhance filler orientation by providing pressure stress (Figure 13.16b) [197]. Unlike the hot pressing and coating processes, stretching aligns BN fillers in the direction of force. Additionally, stretching can induce an ordered structure in polymer chains, further optimizing heat transfer paths (Figure 13.16c). Compared to BNNP/PE composites without stretching treatment, stretching can result in more than a 40-fold improvement in its TC [198]. Shear force is another mechanical force for achieving uniform filler alignment. Oscillating shear applied to composites on a twin-roll mill can internally exert shear force, enhancing filler alignment (Figure 13.16e). Experimental evidence shows that mechanical shear can increase the degree of alignment of BN in a silicone rubber matrix by 67% (Figure 13.16d) [199].

13.4 OVERVIEW OF POLYMER COMPOSITES REINFORCED BY BN

Polymers are widely used in various electronic devices due to their characteristics, such as lightweight, ease of processing, and good insulation properties. However, the TC of polymers generally ranges between 0.1 and 0.5 W $m^{-1} \cdot K^{-1}$ [3,4]. The heat accumulation problem caused by low TC is detrimental to the performance of electronic devices. Using insulating and high TC BN as the filler with polymers to construct composites with high TC provides a reliable solution for the comprehensive performance utilization of devices. This section will discuss the effectiveness of BN as the filler in PVA, cellulose, epoxy resin, and polyimide (PI).

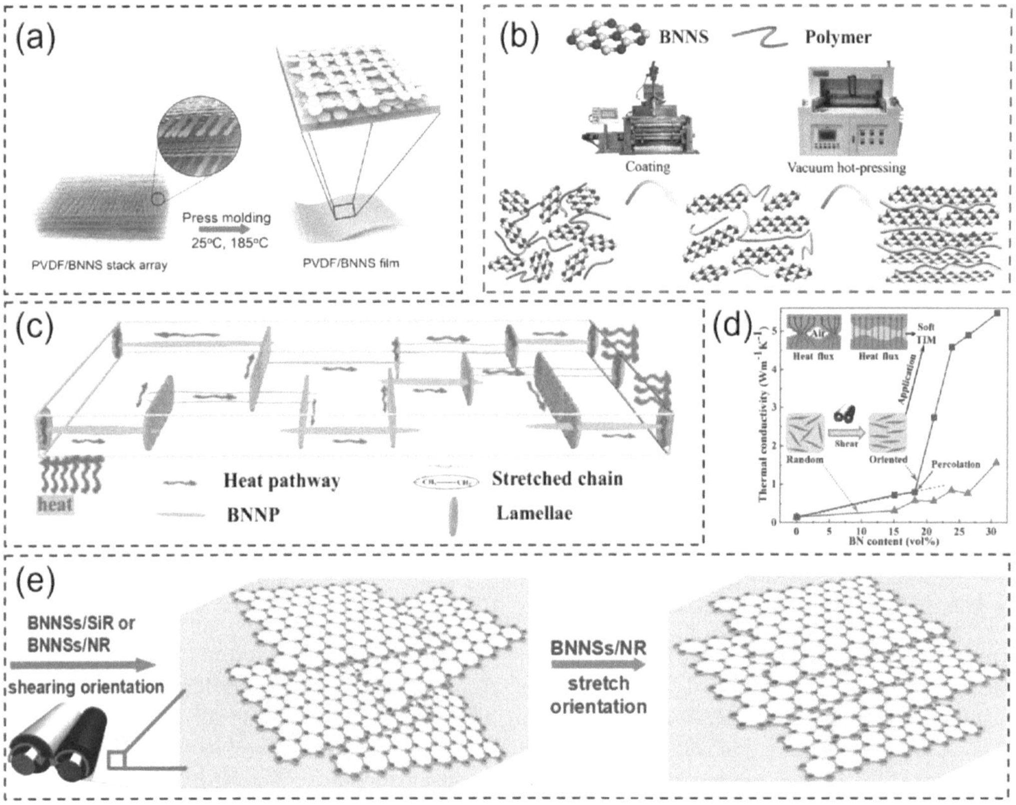

Figure 13.16 (a) Schematic diagram of hot pressing treatment [207]. (b) Schematic illustration of the formation process of the composites with ultrahigh TC using directional coating formation and vacuum pressure technology [197]. (c) Schematic showing potential thermal conduction mechanisms within the stretched nanocomposite film [198]. (d) Comparison diagram of TC of composites composed of oriented BNNS fillers and random BNNS fillers [199]. (e) Schematic diagram of the BNNS orientation process achieved through shear force [199].

13.4.1 BN/PVA Composites

PVA is a white, flocculent, or powdered solid organic compound characterized as a thermoplastic polymer with distinct processing attributes. Upon heating, it softens and can subsequently re-solidify into a new shape after cooling. This material can be molded into diverse shapes through thermal processing and can undergo multiple cycles of shaping. PVA serves as an ideal substrate for electronic and optoelectronic applications, demonstrating significant potential in the development of flexible devices [209]. However, PVA exhibits a low TC of only 0.15 W $m^{-1}{\cdot}K^{-1}$, resulting in inadequate heat dissipation [210]. To address this limitation, BN processed through techniques such as exfoliation, functionalization, and 3D structuring has been employed to enhance the overall performance of PVA. For example, after microplasma exfoliation, the BN and PVA composites exhibit an in-plane TC of approximately 2.1 W $m^{-1}{\cdot}K^{-1}$, which can increase to 13 W $m^{-1}{\cdot}K^{-1}$ at a loading of 50 wt% [60]. Various methods, including plasma treatment [77], ultrasonic exfoliation [96], and sugar-assisted ball milling [86], have been used to functionalize BN with amino and hydroxyl groups. Notably, hydroxyl covalent functionalization of BN can elevate the in-plane TC of BN/PVA composites to 14.6 W m^{-1}·K^{-1} at a BN loading of 50 wt%, as it enhances the interaction between BNNSs and the PVA matrix, effectively reducing phonon scattering. Moreover, that due to the enhanced interactions, the thermal stability of the composite has also improved, with the decomposition temperature increasing from 237°C (pure PVA) to 281°C [96]. In addition to covalent modifications, non-covalently modified substances, such as guanidine compounds [117], and PDA [124] were selected to functionalize BN to enhance the TC of PVA-based composites. Due to the strong hydrogen bonds formed simultaneously between guanidine substances, BN, and PVA, the BN/PVA composite paper treated with these substances can achieve an in-plane TC of 66.8 W m^{-1}·K^{-1} at a BN loading of 89.3 wt% [117]. Crucially, vacuum filtration and cold pressing processes were employed during synthesis, which facilitates the effective orientation of BN within the PVA matrix and contributes to obtaining high TC while ensuring that the composites maintain good flexibility even at high loadings [117]. In addition to these two processes, electrospinning technology is also commonly used to construct BN/PVA composite films with specific orientation [187], which can achieve an in-plane TC of 18.63 W m^{-1}·K^{-1} at a loading of 30 wt%, while the through-plane TC is lower, at around 1 W m^{-1}·K^{-1} [208].

13.4.2 BN/Cellulose Composites

Among the cellulose family, CNFs and cellulose nanocrystals stand out due to their strong functionality, good biodegradability, excellent mechanical strength, and flexibility, presenting promising prospects for application in next-generation green electronic products [85,211]. However, the TC of pure cellulose is relatively low, with a TC of only 0.035 W m^{-1}·K^{-1}. By manipulating the crystallinity or microcrystal size of cellulose, it can effectively modulate its TC [99]. The tuned high aspect ratio CNF films demonstrate a high in-plane TC of up to 1.24 W m^{-1}·K^{-1}, while the through-plane TC is significantly lower, approximately measuring 0.19 W m^{-1}·K^{-1} [211]. To enhance the overall TC of cellulose, both BNNSs and BNNTs have been used as fillers to construct composites. The inherent abundance of hydroxyl groups in cellulose facilitates strong interactions with BN, making cellulose an effective dispersant that promotes uniform distribution of BN within the composite matrix [95]. Experimental results have shown that the in-plane TC of CNF-matrix composites can reach 21.39 W m^{-1}·K^{-1} at a BNNT loading of 25 wt% [94]. The TC of composites composed of CNFs and 50 wt% BNNSs can even reach 145.7 W m^{-1}·K^{-1}, comparable to the TC of aluminum alloys [156]. Ball milled and functionalized BN is widely used to enhance the TC of CNFs or cellulose nanocrystals. Substances such as PDA, xylose, and urea are selected to assist in exfoliation and functionalization processes. For instance, BN functionalized with PDA can increase the TC of cellulose nanocrystals to 40 W m^{-1}·K^{-1} at a loading of 94 wt% [85]. Following urea-assisted ball milling, BN is edge-functionalized with hydroxyl groups, resulting in an increase in the TC of CNFs to 30.25 W m^{-1}·K^{-1} at a 70 wt% loading [80]. Additionally, BN obtained through xylose-assisted ball milling can enhance the TC of CNFs to 14.74 W m^{-1}·K^{-1} at a relatively low loading of 20 wt%, representing nearly an elevenfold increase [212]. In addition to ball milling, ultrasonic exfoliation and functionalization of BNNSs can also significantly enhance the TC of CNFs. With a 25 wt% loading, the in-plane TC of BNNSs/CNFs constructed using this method can reach 22.67 W m^{-1}·K^{-1} [213]. BN and CNFs are often used to construct composite films or papers, so efficient filtration techniques, such as vacuum filtration, greatly aid in their construction [78,80,85,94,156,212,213].

13.4.3 BN/Epoxy Composites

Epoxy resin is a type of thermosetting polymer that can cure into a robust, heat-resistant material through chemical and physical cross-linking during heating. This type of polymer is typically pliable at room temperature, but once heated to a certain temperature, cross-linking and other curing reactions occur, forming a 3D network structure that effectively hardens and stabilizes the material. Currently, electronic packaging can be categorized into three main types: plastic packaging, ceramic packaging, and metal packaging, with over 90% of integrated circuits utilizing plastic packaging. Notably, epoxy resin is favored as the primary packaging material due to its low cost and ease of processing, accounting for more than 90% in plastic packaging [214]. However, the TC of epoxy resin is only 0.2 W $m^{-1}{\cdot}K^{-1}$, which severely limits its application and development in electronic devices due to the resultant heat accumulation issues.

To address the issue of low TC inherent in epoxy resin, many design strategies based on BN fillers have been developed to optimize the characteristics of epoxy resin. For BN fillers of varying sizes, larger-sized fillers can ensure the continuity of heat conduction in composites, reduce interface phonon scattering, and provide better thermal performance. One study demonstrated that among BN sizes of 2, 30, and 180 μm, the 180 μm BN filler exhibited the most significant enhancement in TC for epoxy resin, achieving a value of 4.3 W $m^{-1}{\cdot}K^{-1}$ at a loading of 44.7 vol% [56]. Additionally, hybridization of multiple sizes can maximize the advantages of BN. Small-sized BN can fill the voids between large-sized BN, forming a denser filler network that enhances TC efficiency [79]. Functionalizing BN can further enhance the performance of BN/epoxy composites, with commonly used methods including hydroxyl, amino, and silane coupling agents (such as KH550, HBP, APTES) for covalent functionalization. BN covalently modified with APTES can achieve an in-plane TC of 5.86 W $m^{-1}{\cdot}K^{-1}$ and a through-plane TC of around 1.74 W $m^{-1}{\cdot}K^{-1}$ (40 wt%), which is 25 times higher than that of pure epoxy [93]. Furthermore, its thermal stability and storage modulus are also improved, with dielectric breakdown performance similar to epoxy resin, even under 1000 V [74]. To further enhance the TC of epoxy-based composites, researchers have controlled the alignment of BN fillers. Magnetic field control can align BN along the direction of heat flow, and at a 30 wt% loading, composites controlled by magnetic field can achieve a through-plane TC of 14.55 W $m^{-1}{\cdot}K^{-1}$ [192], which is nearly 72 times higher than pure epoxy resin. Additionally, various 3D BN fillers constructed as fillers using sacrificial templates, fiber templates, ice templates, and foam templates exhibit improved thermal conduction efficiency. Composites prepared using the ice templates method can achieve a through-plane TC of 4.02 W $m^{-1}{\cdot}K^{-1}$ and an in-plane TC of 3.87 W $m^{-1}{\cdot}K^{-1}$ at a 15 vol% of BN loading [179]. Bidirectional freezing technology can further increase the TC of composites to 6.07 W $m^{-1}{\cdot}K^{-1}$ (15 vol%) [181]. Using cellulose templates, composites can achieve a TC of 3.13 W $m^{-1}{\cdot}K^{-1}$ at a low loading of 9.6 vol% [161]. Furthermore, composites utilizing foam templates can reach a TC of 0.6 W $m^{-1}{\cdot}K^{-1}$ at an extremely low loading of 1.1 vol%, which is a 233% increase compared to pure epoxy [163].

13.4.4 BN/PI Composites

PI refers to polymers with imide rings in the main chain, which exhibit excellent thermal stability, mechanical properties, and insulation performance. As a specialty engineering material, PI has been widely used in fields such as automotive, aerospace, and electronic packaging [215,216]. However, its TC is low, typically ranging from 0.1 to 0.2 W $m^{-1}{\cdot}K^{-1}$, which limits its application. The development of PI/BN composites has emerged as a promising strategy to enhance the utility of PI [217].

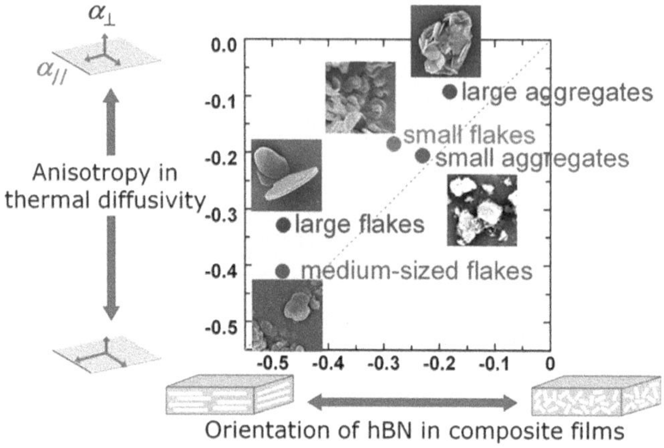

Figure 13.17 Correlations between orientation function and anisotropy in thermal diffusivity [216].

When constructing composites, the fillers with different particle sizes have different effects on the mechanism of action in PI. Larger aspect ratio samples, when used as fillers in PI, exhibit greater planar orientation, thus resulting in significant anisotropy in the TC of the composites (Figure 13.17). Ultimately, the in-plane TC of composites formed by medium-sized flakes and PI can reach 17.5 W $m^{-1}{\cdot}K^{-1}$ (60 vol%), while the through-plane TC of composites formed by large aggregates of BN and PI can reach 5.4 W $m^{-1}{\cdot}K^{-1}$ (60 vol%) [216]. In addition to using single-sized BN, a combination of different-sized BN fillers has also been employed to further enhance the TC of PI. When using BN fillers with sizes of 70 nm and 1 μm, the TC of the composites they form at a 30 wt% loading is 0.47 and 0.57 W $m^{-1}{\cdot}K^{-1}$, respectively. However, when employing a mixture of BN sizes as fillers at the same BN loading, the TC of the composites can be increased to 1.2 W $m^{-1}{\cdot}K^{-1}$ [218].

Similar to other composites, various modification techniques have been employed to enhance the affinity between BN and PI, thereby improving the TC of the resulting composites. After covalent modification with KH550, BN can attach to the PI molecular chains through participation in the PI polymerization, forming a regular arrangement. At a 30 wt% loading, the in-plane TC of the composite prepared using this method can reach 4.43 W $m^{-1}{\cdot}K^{-1}$ [219]. Additionally, non-covalent functionalization of BN can be achieved through PDA coating. The PDA coated on the BN surface can react with PI, allowing BN to covalently bond to the PI molecular chains. The TC of the composite fibers prepared using this method can reach 3.44 W $m^{-1}{\cdot}K^{-1}$ at a 10 wt% loading [220].

13.5 APPLICATIONS OF POLYMER COMPOSITES REINFORCED BY BN

The miniaturization, integration, and multi-functionality of equipment in industries such as electronics and automotive have led to a growing demand for thermal management technologies. This is because the accumulation of excessive undesirable heat can severely compromise the operational performance and service life of such devices. BN/polymer nanocomposites are particularly well-suited for these applications due to their high TC, low electrical conductivity, ease of manufacturing, lightweight, high corrosion resistance, and enhanced mechanical properties. This section will discuss several emerging applications of BN/polymer nanocomposites related to thermal management and heat dissipation.

In order to promptly dissipate the accumulated heat within electronic devices, TIMs are typically employed to facilitate surface contact between the electronics and the heat sink. The utilization of BN/polymer nanocomposites as TIMs presents a promising solution. These TIMs can fill the gaps within electronic components, reducing the thermal interface resistance between contacting surfaces, mitigating temperature increases, and ensuring the proper operation of electronics within their working temperature range. Various BN nanostructures, such as BN quantum dots, BNNSs, BNNTs, and 3D BN frameworks, have been developed as high TC fillers for polymer matrixes to enhance the thermal transport performance of TIMs. Chen et al. fabricated aligned BNNS/poly(vinylidene difluoride) nanocomposites via electrospinning and hot pressing. When employed as TIMs between metal oxide semiconductor field effect transistor and heat sinks, the equilibrium temperature is only around 65°C, resulting in a reduction of approximately 11°C compared to commercial silicone pads [207]. Zhou et al. reported a flexible BNNS/polyamide-imide composite film and further introduced boron nitride quantum dots (BNQDs) through an evaporation-induced self-assembly method. The BNQDs filled the gaps between the fillers and the matrix in the PAI/BNNS composite film and synergistically interacted with the BNNSs in the matrix, playing a crucial role in the formation of continuous heat conduction paths [221]. Zhao developed a BN/PU composite with a bidirectional stretched network through an improved bidirectional freezing and hot pressing process. This delicate thermal conduction network endows the composite with an exceptionally high in-plane TC (~39.0 W $m^{-1} \cdot K^{-1}$), exhibiting excellent cooling efficiency and maintaining outstanding thermal stability even after 1000 heating and cooling cycles [222]. Overall, BN-based polymer nanocomposites with high TC, high flexural strength, fatigue resistance, and electrical insulation properties have been considered as one of the most promising TIMs in the field of electronic devices.

Light-emitting diodes (LEDs) possess the advantages of low operating voltage, high luminous efficiency, pollution-free operation, and extended lifespan, making them a prominent focus of research as the fourth-generation lighting source. However, the ever-increasing demand for enhanced illumination has led to rising requirements for LED power supplies, resulting in increased power density and chip temperature, which in turn reduces efficiency. Therefore, the selection of materials with efficient thermal management systems, capable of rapidly dissipating heat along the heat transfer path, has become a key concern for researchers in recent years. BN/polymer nanocomposites can serve as ideal heatsink materials for LED devices, as they are environmentally friendly, lightweight, and electrically insulating and can mitigate thermal expansion mismatch during packaging. Zhang et al. first ball milled and amine-functionalized BN and then used carboxylated nanofibrillated cellulose (COOH-CNF) to disperse the BNNSs [223]. The BNNSs were uniformly dispersed in the CNF matrix through amide bonding, reducing BNNS agglomeration and the presence of voids, promoting the formation of effective thermal paths. The resultant filtration membrane, when utilized as a heat dissipation device for LED chips, is capable of rapidly reducing the temperature of the chip from 50°C to 35°C in just 50 seconds. Additionally, LED bulb sockets also play a crucial role in dissipating the heat generated within the structure. Typically, these components are made of low TC thermoplastic polymers, such as polycarbonate and polybutylene terephthalate. Utilizing BN-based nanocomposites as alternatives may further enhance the heat dissipation from LED bulbs and improve their lifespan [224]. Gorur et al. used BNNSs as a filler for a polylactic acid (PLA) matrix nanocomposite in 3D printing. They first prepared the nanocomposite filament and then used it to 3D print heatsinks and LED bulb holders. The study found that the addition of 40 vol% BNNS increased the TC of the

3D-printed PLA by 400%. The thermal performance of the BNNS/PLA nanocomposite was nearly comparable to that of commercial aluminum heatsinks. Furthermore, the 3D-printed BNNS/PLA nanocomposite LED bulb holders were more effective in dissipating the excess heat generated by the LEDs compared to commercial products [225].

The rapid development of new energy power has driven the growth of energy storage, particularly the large-scale expansion of energy storage devices. However, as the demand for energy density continues to increase, the safety of energy storage devices has become a major challenge. Specifically, due to uneven current distribution, internal short circuits, or inadequate heat dissipation, localized temperature increases may occur, significantly compromising safety and cycle life. The utilization of thermally conductive BN-filled polymer nanocomposites has been proven to be an effective solution to address these issues. Li et al. demonstrated that BNNS-modified lithium-ion battery separators possess rapid thermal response capabilities. This unique integration of thermal conduction and heat storage allows the functional separators to have the appealing ability to mitigate hot spots and suppress lithium dendrite growth during cycling, thereby reducing the risk of thermal runaway, which will aid in the development of high-safety and energy-dense energy storage devices [226]. To address the challenges of polysulfide shuttling, lithium dendrite growth, sluggish redox kinetics, and safety issues under high metallic lithium concentrations in lithium-sulfur batteries, Zhu et al. constructed a highly thermally conductive separator by weaving aligned BNNSs into a highly oriented CNT framework. The highly aligned and segregated structure can rapidly dissipate the heat accumulated due to internal resistance during continuous operation. The BNNSs provide abundant adsorption sites in the composite separator, while the highly graphitized CNT scaffold accelerates the catalytic conversion of high-order to low-order polysulfides. The array design can regulate the local current density and ion flux, significantly mitigating lithium dendrite growth and promoting the smooth deposition of lithium metal, opening a new path to overcome the safety issues of lithium-sulfur batteries [227]. Furthermore, in addition to thermally conductive separators, BN-based composite polymer electrolytes have also been proven to be fast lithium-ion transport paths, significantly improving ionic conductivity, optimizing thermal management, and weakening lithium dendrite growth, making them promising candidates for all-solid-state battery electrolytes [228].

The global energy consumption structure is transitioning from being primarily oil-based to a more diversified energy mix, necessitating greater demands on energy conversion devices such as fuel cells and solar cells. During the operation of these energy conversion systems, large amounts of heat are generated, which must be promptly transferred and dissipated through effective thermal management strategies, or else the equipment may overheat and become damaged. In the case of fuel cells, cooling plates and specific coolant media are typically used to absorb the heat from the fuel cells on both sides. BN/polymer nanocomposites have become suitable candidates for passive cooling plates due to their lightweight and high in-plane TC [229]. On the other hand, due to the high chemical stability, excellent mechanical strength, thermal stability, and proton permeability through the hexagonal rings of BN/polymer composites in the presence of solvents, they are often developed as proton exchange membranes in fuel cells. To date, materials such as BN/sulfonated polysulfone/polyvinyl phosphonic acid [230], BN/PAMPS-NMPA-based [231], and BN/PTFE-based [232], proton exchange membranes have demonstrated good proton conductivity, low hydrogen permeation current, high cell performance, and excellent thermal stability, making them potential solutions to address the safety, ion transport, and performance issues of fuel cells. Solar cells directly convert sunlight into electrical energy, but a significant portion of the solar energy is converted into heat during this process, leading to a reduction in the relative conversion efficiency and degradation of the solar cell performance. There is an urgent need to develop effective strategies to control the heat flow in photovoltaic solar cells to improve their performance. Yang et al. introduced highly thermally conductive BN materials at the grain boundaries of perovskite absorbers, which can form heat transfer channels to accelerate heat dissipation without compromising the solar cell efficiency. This approach has improved the overall lifetime of the solar cells, maintaining a high efficiency (over 85%) even after prolonged operation (2400 hours) [233].

13.6 CONCLUSIONS

In the future, devices will continue to evolve toward integration and miniaturization, resulting in increasingly higher demands for heat dissipation in materials. Materials with high TC will become more prevalent in people's fields of vision. The unique structure of BN not only provides excellent insulation performance but also possesses strong heat conduction capabilities, which will attract more attention. Utilizing BN as a filler in polymers is an effective strategy for enhancing their TC while ensuring the fundamental properties of the polymers, thereby addressing the heat dissipation issues in devices.

BN possesses excellent electrical insulation properties, high TC, low density, good mechanical properties, and mature preparation processes, making it a strong candidate for high TC polymer-based composites. However, achieving composites with superior comprehensive performance still requires improvement from multiple perspectives, including the preparation, dispersion, functionalization of BN, 3D filler network design, and alignment control.

i. Preparation of BN fillers. The selection of BN fillers primarily involves three aspects: defects, morphology, and size. When BN is used as a high TC filler, it is essential to ensure high crystallinity as much as possible and avoid the occurrence of defects. Defects can lead to severe phonon scattering, which is detrimental to heat transfer. Regarding the morphology of BN, various dimensions can be chosen as fillers, offering variable options for different application scenarios; however, overall, 3D fillers exhibit higher heat transfer efficiency. Concerning the size of BN, larger dimensions and thinner thicknesses offer greater advantages.
ii. Dispersion and functionalization of BN fillers. In the preparation of composites, it is necessary to avoid the agglomeration of BN and achieve uniform dispersion. Achieving uniform dispersion often relies on functionalizing BN to enhance its compatibility with the polymer matrix. Compared to in-plane covalent functionalization, edge covalent functionalization of BN can improve compatibility with the matrix while ensuring its inherent TC. Additionally, compared to covalent functionalization, non-covalent functionalization of BN can further avoid damage to the in-plane structure and reduce phonon scattering. However, non-covalent functionalization may introduce new limitations to heat conduction in composites due to the low TC of the organic molecules coating the surface of BN.
iii. 3D network design of BN fillers. Fillers in the polymer matrix need to form a 3D thermal conduction network. When using 0D, 1D, and 2D fillers, poor contact between fillers leads to interface thermal resistance, which negatively affects the TC of composites. In contrast, preconstructed 3D fillers have higher thermal transfer efficiency, and template methods are effective for pre-constructing 3D fillers. Additionally, directly synthesized 3D BN further reduces filler interface thermal resistance compared to assembled 3D BN.

iv. Alignment control of BN fillers. Alignment control of BN in composites can maximize the enhancement of TC in a specific direction, which is of great significance in practical applications. During synthesis, the intervention or control of magnetic field force, electric field force, mechanical force, gravity, and other force fields can effectively arrange BN.

With the rapid development of computational technologies, utilizing calculation and simulation techniques has become a crucial means of improving design efficiency, capability, and cost-effectiveness. In the construction of polymeric thermally conductive composites, simulations based on physical models play an important role in calculating the intrinsic TC of BN, the analysis of the functionalization mechanism of BN, and the structural design of BN in composites. In addition, data-driven machine learning is gradually being applied in the field of composite construction. With the support of a large volume of theoretical realization data, machine learning will provide new ideas for the rational design and preparation of high TC composites.

Although significant progress has been made in BN-filled polymeric composites, several challenges remain. The synthesis of high-quality BN continues to be difficult, including issues such as large-scale production, size and defect control, and cost. Moreover, the high-precision design and synthesis of 3D networks in composites remain challenging. Additionally, limitations in mechanical properties at high loadings of BN fillers pose further challenges in composite construction. In the future, a combination of experimental and simulation approaches will be needed to address these issues and further unlock the potential of BN-filled polymeric thermally conductive composites.

ACKNOWLEDGMENTS

It is supported by the National Natural Science Foundation of China (U23B2075, 52272039, 51972168), Key Research and Development Program in Jiangsu Province (BE2023085), Natural Science Foundation of Jiangsu Province of China (BK20231406), Fundamental Research Funds for the Central Universities – CEMaC GeoX Interdisciplinary Program (2024ZD07), and Innovative Talent (2024300328).

REFERENCES

1. Z. Wang et al., A roadmap review of thermally conductive polymer composites: Critical factors, progress, and prospects. *Advanced Functional Materials* **33**, 2301549 (2023).
2. Y. Guo et al., Constructing fully carbon-based fillers with a hierarchical structure to fabricate highly thermally conductive polyimide nanocomposites. *Journal of Materials Chemistry C* **7**, 7035–7044 (2019).
3. J. V. Bondareva et al., Thermal and electrical properties of additively manufactured polymer-boron nitride composite. *Polymers* **15**, 1214 (2023).
4. X. J. Zha et al., Enhanced thermal conductivity and balanced mechanical performance of PP/BN composites with 1 vol% finely dispersed MWCNTs assisted by OBC. *Advanced Materials Interfaces* **6**, 1900081 (2019).
5. S. Roy et al., Structure, properties and applications of two-dimensional hexagonal boron nitride. *Advanced Materials* **33**, 2101589 (2021).
6. D. Wickramaratne, L. Weston, C. G. Van de Walle, Monolayer to bulk properties of hexagonal boron nitride. *The Journal of Physical Chemistry C* **122**, 25524–25529 (2018).

7. C. Elias et al., Direct band-gap crossover in epitaxial monolayer boron nitride. *Nature Communications* **10**, 2639 (2019).
8. L. An et al., Hexagonal boron nitride nanosheets: Preparation, heat transport property and application as thermally conductive fillers. *Progress in Materials Science* **138**, 101154 (2023).
9. L. Lindsay, D. A. Broido, Enhanced thermal conductivity and isotope effect in single-layer hexagonal boron nitride. *Physical Review B* **84**, 155421 (2011).
10. Q. Cai et al., High thermal conductivity of high-quality monolayer boron nitride and its thermal expansion. *Science Advances* **5**, eaav0129 (2019).
11. L. Lindsay, D. A. Broido, Theory of thermal transport in multilayer hexagonal boron nitride and nanotubes. *Physical Review B* **85**, 035436 (2012).
12. K. Zhang, Y. Feng, F. Wang, Z. Yang, J. Wang, Two dimensional hexagonal boron nitride (2D-hBN): Synthesis, properties and applications. *Journal of Materials Chemistry C* **5**, 11992–12022 (2017).
13. I. Jo et al., Thermal conductivity and phonon transport in suspended few-layer hexagonal boron nitride. *Nano Letters* **13**, 550–554 (2013).
14. H. Zhou et al., High thermal conductivity of suspended few-layer hexagonal boron nitride sheets. *Nano Research* **7**, 1232–1240 (2014).
15. C. Wang et al., Superior thermal conductivity in suspended bilayer hexagonal boron nitride. *Scientific Reports* **6**, 25334 (2016).
16. M. T. Alam, M. S. Bresnehan, J. A. Robinson, M. A. Haque, Thermal conductivity of ultra-thin chemical vapor deposited hexagonal boron nitride films. *Applied Physics Letters* **104**, 013113 (2014).
17. T. Ouyang et al., Thermal transport in hexagonal boron nitride nanoribbons. *Nanotechnology* **21**, 245701 (2010).
18. C. Zhi et al., Towards thermoconductive, electrically insulating polymeric composites with boron nitride nanotubes as fillers. *Advanced Functional Materials* **19**, 1857–1862 (2009).
19. E. Janzen et al., Boron and nitrogen isotope effects on hexagonal boron nitride properties. *Advanced Materials* **36**, e2306033 (2023).
20. C. Sevik, A. Kinaci, J. B. Haskins, T. Çağın, Influence of disorder on thermal transport properties of boron nitride nanostructures. *Physical Review B* **86**, 075403 (2012).
21. C. Yuan et al., Modulating the thermal conductivity in hexagonal boron nitride via controlled boron isotope concentration. *Communications Physics* **2**, 43 (2019).
22. Q. Cai et al., Outstanding thermal conductivity of single atomic layer isotope-modified boron nitride. *Physical Review Letters* **125**, 085902 (2020).
23. N. Burger et al., Review of thermal conductivity in composites: Mechanisms, parameters and theory. *Progress in Polymer Science* **61**, 1–28 (2016).
24. Z. Tian et al., Crystalline boron nitride nanosheets by sonication-assisted hydrothermal exfoliation. *Journal of Advanced Ceramics* **8**, 72–78 (2019).
25. A. Zare, P. Sedigh, A. Montazeri, Enhancing multi-functional capabilities of boron nitride nanosheets through defect engineering. *Journal of Materials Science* **55**, 12995–13007 (2020).
26. A. I. Khan, I. A. Navid, M. Noshin, S. Subrina, Thermal transport characterization of hexagonal boron nitride nanoribbons using molecular dynamics simulation. *AIP Advances* **7**, 105110 (2017).
27. X. Wu, Q. Han, Thermal conductivity of monolayer hexagonal boron nitride: From defective to amorphous. *Computational Materials Science* **184**, 109938 (2020).
28. J. F. N. Dethan, Thermal conductivity of hydrogenated h-BN nanosheets: A reactive force field study. *Soft Materials* **21**, 271–279 (2023).
29. C. Tang et al., Thermal conductivity of nanostructured boron nitride materials. *Journal of Physical Chemistry B* **110**, 10354–10357 (2006).
30. X. F. Jiang et al., Recent progress on fabrications and applications of boron nitride nanomaterials: A review. *Journal of Materials Science & Technology* **31**, 589–598 (2015).
31. C. Xu, M. Miao, X. Jiang, X. Wang, Thermal conductive composites reinforced via advanced boron nitride nanomaterials. *Composites Communications* **10**, 103–109 (2018).
32. F. Banhart, M. Zwanger, H. J. Muhr, The formation of curled concentric-shell clusters in boron-nitride under electron-irradiation. *Chemical Physics Letters* **231**, 98–104 (1994).

33. K. Suenaga et al., Synthesis of nanoparticles and nanotubes with well-separated layers of boron nitride and carbon. *Science* **278**, 653–655 (1997).
34. C. Tang, Y. Bando, D. Golberg, Large-scale synthesis and structure of boron nitride sub-micron spherical particles. *Chemical Communications*, 2826–2827 (2002).
35. O. Stephan et al., Formation of small single-layer and nested BN cages under electron irradiation of nanotubes and bulk material. *Applied Physics a-Materials Science & Processing* **67**, 107–111 (1998).
36. H. Chen et al., Thermal conductivity of polymer-based composites: Fundamentals and applications. *Progress in Polymer Science* **59**, 41–85 (2016).
37. N. G. Chopra et al., Boron nitride nanotubes. *Science* **269**, 966–967 (1995).
38. C. Tang, Y. Bando, T. Sato, K. Kurashima, A novel precursor for synthesis of pure boron nitride nanotubes. *Chemical Communications*, **12**, 1290–1291 (2002).
39. A. Zandieh et al., Molecular engineering of the surface of boron nitride nanotubes for manufacture of thermally conductive dielectric polymer composites. *Applied Surface Science* **587**, 152779 (2022).
40. X. B. Wang et al., Biomass-directed synthesis of 20g high-quality boron nitride nanosheets for thermoconductive polymeric composites. *ACS Nano* **8**, 9081–9088 (2014).
41. A. Nagashima, N. Tejima, Y. Gamou, T. Kawai, C. Oshima, Electronic structure of monolayer hexagonal boron nitride physisorbed on metal surfaces. *Physical Review Letters* **75**, 3918–3921 (1995).
42. K. S. Novoselov et al., Electric field effect in atomically thin carbon films. *Science* **306**, 666–669 (2004).
43. K. S. Novoselov et al., Two-dimensional atomic crystals. *Proceedings of the National Academy of Sciences of the United States of America* **102**, 10451–10453 (2005).
44. Y. Shi et al., Synthesis of few-layer hexagonal boron nitride thin film by chemical vapor deposition. *Nano Letters* **10**, 4134–4139 (2010).
45. X. Wang et al., "Chemical blowing" of thin-walled bubbles: High-throughput fabrication of large-area, few-layered BN and C_x-BN nanosheets. *Advanced Materials* **23**, 4072–4076 (2011).
46. Y. Liu et al., Boron nitride-nanosheet enhanced cellulose nanofiber aerogel with excellent thermal management properties. *Carbohydrate Polymers* **241**, 116425 (2020).
47. Y. Guo, K. Ruan, X. Shi, X. Yang, J. Gu, Factors affecting thermal conductivities of the polymers and polymer composites: A review. *Composites Science and Technology* **193**, 108134 (2020).
48. H. Dong, B. Wen, R. Melnik, Relative importance of grain boundaries and size effects in thermal conductivity of nanocrystalline materials. *Scientific Reports* **4**, 7037 (2014).
49. C. W. Chang et al., Isotope effect on the thermal conductivity of boron nitride nanotubes. *Physical Review Letters* **97**, 085901 (2006).
50. J. Lu, N. Yang, r. Sun, C. Wong, Effect of domain size on thermal conductivity of boron nitride monolayer in equilibrium molecular dynamics with green-kubo method. *Journal of Integration Technology* **7**, 12–21 (2018).
51. M. Pokharel, H. Zhao, Z. Ren, C. Opeil, Grain boundary Kapitza resistance analysis of nanostructured $FeSb_2$. *International Journal of Thermal Sciences* **71**, 32–35 (2013).
52. J. Huang, Y. Guo, Y. Fan, Y. Liang, Molecular dynamics investigation of the thermal properties in single-walled boron nitride nanotube. *Materials Research Express* **7**, 025025 (2020).
53. J. Li, J. Leng, Y. Jiang, J. Zhang, Experimental characterization of 3D printed PP/h-BN thermally conductive composites with highly oriented h-BN and the effects of filler size. *Composites Part A: Applied Science and Manufacturing* **150**, 106586 (2021).
54. Y. Wang et al., The influence of boron nitride shape and size on thermal conductivity, rheological and passive cooling properties of polyethylene composites. *Composites Part A: Applied Science and Manufacturing* **161**, 107117 (2022).
55. K. Kim, M. Kim, Y. Hwang, J. Kim, Chemically modified boron nitride-epoxy terminated dimethylsiloxane composite for improving the thermal conductivity. *Ceramics International* **40**, 2047–2056 (2014).
56. S. Moradi, Y. Calventus, F. Roman, J. M. Hutchinson, Achieving high thermal conductivity in epoxy composites: Effect of boron nitride particle size and matrix-filler interface. *Polymers* **11**, 1156 (2019).
57. B. Zhong et al., The effects of the hexagonal boron nitride nanoflake properties on the thermal conductivity of hexagonal boron nitride nanoflake/silicone rubber composites. *Composites Part A: Applied Science and Manufacturing* **127**, 105629 (2019).

58. Z. Lin, A. McNamara, Y. Liu, K.-S. Moon, C.-P. Wong, Exfoliated hexagonal boron nitride-based polymer nanocomposite with enhanced thermal conductivity for electronic encapsulation. *Composites Science and Technology* **90**, 123–128 (2014).
59. X. Tian et al., Shear-assisted production of few-layer boron nitride nanosheets by supercritical CO_2 exfoliation and its use for thermally conductive epoxy composites. *Scientific Reports* 7, 17794 (2017).
60. R.-C. Zhang et al., Microplasma processed ultrathin boron nitride nanosheets for polymer nanocomposites with enhanced thermal transport performance. *ACS Applied Materials & Interfaces* **8**, 13567–13572 (2016).
61. Z. Han, A. Fina, Thermal conductivity of carbon nanotubes and their polymer nanocomposites: A review. *Progress in Polymer Science* **36**, 914–944 (2011).
62. X. Xu, J. Chen, B. Li, Phonon thermal conduction in novel 2D materials. *Journal of Physics: Condensed Matter* **28**, 483001 (2016).
63. M. K. Maurya, J. Wu, M. K. Singh, D. Mukherji, Thermal conductivity of semicrystalline polymer networks: Crystallinity or cross-linking? *ACS Macro Letters* **11**, 925–929 (2022).
64. A. Henry, G. Chen, High thermal conductivity of single polyethylene chains using molecular dynamics simulations. *Physical Review Letters* **101**, 235502 (2008).
65. J. Yu, B. Sundqvist, B. Tonpheng, O. Andersson, Thermal conductivity of highly crystallized polyethylene. *Polymer* **55**, 195–200 (2014).
66. R. Shrestha et al., Crystalline polymer nanofibers with ultra-high strength and thermal conductivity. *Nature Communications* **9**, 1664 (2018).
67. S. Shen, A. Henry, J. Tong, R. Zheng, G. Chen, Polyethylene nanofibres with very high thermal conductivities. *Nature Nanotechnology* **5**, 251–255 (2010).
68. G. Lv, E. Jensen, C. M. Evans, D. G. Cahill, High thermal conductivity semicrystalline epoxy resins with anthraquinone-based hardeners. *ACS Applied Polymer Materials* **3**, 4430–4435 (2021).
69. Y. Xu, S. Zhang, P. Wang, J. Wang, Synthesis of poly(butylene succinate) phosphorus-containing ionomers for versatile crystallization and improved thermal conductivity. *Polymer* **154**, 258–271 (2018).
70. Z. Li, J. Kong, L. Han, H. Zhang, L. Dong, Effect of crystallinity on the thermal conductivity of poly(3-hydroxybutyrate)/BN composites. *Polymer Bulletin* **75**, 1651–1666 (2017).
71. S. Wang, B. Wen, Effect of functional filler morphology on the crystallization behavior and thermal conductivity of PET resin: A comparative study of three different shapes of BN as heterogeneous nucleating agents. *Composites Science and Technology* **222**, 109346 (2022).
72. Y. Lei, M. Liang, Y. Chen, S. Zhou, H. Zou, Crystallization and thermal conductivity of poly (vinylidene fluoride)/boron nitride nanosheets composites. *Polymer-Plastics Technology and Materials* **59**, 1552–1561 (2020).
73. T. Sainsbury et al., Oxygen radical functionalization of boron nitride nanosheets. *Journal of the American Chemical Society* **134**, 18758–18771 (2012).
74. J. Hou et al., Preparation and characterization of surface modified boron nitride epoxy composites with enhanced thermal conductivity. *RSC Advances* **4**, 44282–44290 (2014).
75. V. Guerra, C. Wan, T. McNally, Thermal conductivity of 2D nano-structured boron nitride (BN) and its composites with polymers. *Progress in Materials Science* **100**, 170–186 (2019).
76. A. Pakdel, Y. Bando, D. Golberg, Plasma-assisted interface engineering of boron nitride nanostructure films. *ACS Nano* **8**, 10631–10639 (2014).
77. C. Xu et al., Interfacial thermal conductance enhancement of BN/PVA composites via plasma activations of fillers. *Composites Communications* **28**, 100963 (2021).
78. K. Wu et al., Preparation of a thermally conductive biodegradable cellulose nanofiber/hydroxylated boron nitride nanosheet film: The critical role of edge-hydroxylation. *Journal of Materials Chemistry A* **6**, 11863–11873 (2018).
79. W. Wang, M. Zhao, D. Jiang, X. Zhou, J. He, Amino functionalized boron nitride and enhanced thermal conductivity of epoxy composites via combining mixed sizes of fillers. *Ceramics International* **48**, 2763–2770 (2022).
80. K. Wu et al., Achieving a collapsible, strong, and highly thermally conductive film based on oriented functionalized boron nitride nanosheets and cellulose nanofiber. *ACS Applied Materials & Interfaces* **9**, 30035–30045 (2017).

81. W. Lei et al., Boron nitride colloidal solutions, ultralight aerogels and freestanding membranes through one-step exfoliation and functionalization. *Nature Communications* **6**, 8849 (2015).
82. H. Zhao et al., Green synthesis of graphene/boron nitride composites for ultrahigh thermally conductive fluids. *ACS Sustainable Chemistry & Engineering* **7**, 14266–14272 (2019).
83. G. Han et al., Highly flame-retardant epoxy-based thermal conductive composites with functionalized boron nitride nanosheets exfoliated by one-step ball milling. *Chemical Engineering Journal* **407**, 127099 (2021).
84. D. Lee et al., Scalable exfoliation process for highly soluble boron nitride nanoplatelets by hydroxide-assisted ball milling. *Nano Letters* **15**, 1238–1244 (2015).
85. C. Yu et al., One-step in situ ball milling synthesis of polymer-functionalized few-layered boron nitride and its application in high thermally conductive cellulose composites. *ACS Applied Nano Materials* **1**, 4875–4883 (2018).
86. Y. Li, T. Huang, M. Chen, L. Wu, Simultaneous exfoliation and functionalization of large-sized boron nitride nanosheets for enhanced thermal conductivity of polymer composite film. *Chemical Engineering Journal* **442**, 136237 (2022).
87. S. Chen et al., Simultaneous production and functionalization of boron nitride nanosheets by sugar-assisted mechanochemical exfoliation. *Advanced Materials* **31**, 1804810 (2019).
88. C. Gautam, S. Chelliah, Methods of hexagonal boron nitride exfoliation and its functionalization: Covalent and non-covalent approaches. *RSC Advances* **11**, 31284–31327 (2021).
89. W. Q. Han, L. J. Wu, Y. M. Zhu, K. Watanabe, T. Taniguchi, Structure of chemically derived mono- and few-atomic-layer boron nitride sheets. *Applied Physics Letters* **93**, 223103 (2008).
90. B.-H. Xie, X. Huang, G.-J. Zhang, High thermal conductive polyvinyl alcohol composites with hexagonal boron nitride microplatelets as fillers. *Composites Science and Technology* **85**, 98–103 (2013).
91. C. Zhi, Y. Bando, C. Tang, H. Kuwahara, D. Golberg, Large-scale fabrication of boron nitride nanosheets and their utilization in polymeric composites with improved thermal and mechanical properties. *Advanced Materials* **21**, 2889–2893 (2009).
92. J. N. Coleman et al., Two-dimensional nanosheets produced by liquid exfoliation of layered materials. *Science* **331**, 568–571 (2011).
93. Z. Liu, J. Li, X. Liu, Novel functionalized BN nanosheets/epoxy composites with advanced thermal conductivity and mechanical properties. *ACS Applied Materials & Interfaces* **12**, 6503–6515 (2020).
94. X. Zeng et al., A combination of boron nitride nanotubes and cellulose nanofibers for the preparation of a nanocomposite with high thermal conductivity. *ACS Nano* **11**, 5167–5178 (2017).
95. Y. Li et al., Nanocellulose as green dispersant for two-dimensional energy materials. *Nano Energy* **13**, 346–354 (2015).
96. X. Shi et al., Few-layer hydroxyl-functionalized boron nitride nanosheets for nanoscale thermal management. *ACS Applied Nano Materials* **3**, 2310–2321 (2020).
97. Z. Cui, A. J. Oyer, A. J. Glover, H. C. Schniepp, D. H. Adamson, Large scale thermal exfoliation and functionalization of boron nitride. *Small* **10**, 2352–2355 (2014).
98. F. Xiao et al., Edge-hydroxylated boron nitride nanosheets as an effective additive to improve the thermal response of hydrogels. *Advanced Materials* **27**, 7196–7203 (2015).
99. K. Uetani, T. Okada, H. T. Oyama, Crystallite size effect on thermal conductive properties of nonwoven nanocellulose sheets. *Biomacromolecules* **16**, 2220–2227 (2015).
100. K. Ruan, X. Shi, Y. Guo, J. Gu, Interfacial thermal resistance in thermally conductive polymer composites: A review. *Composites Communications* **22**, 100518 (2020).
101. S. Zhang et al., Surface NH_2-functionalized by C doping of boron nitride nanotube to improve the thermal conductivity of epoxy composites. *Composites Part B: Engineering* **223**, 109106 (2021).
102. J. Yu et al., Interfacial modification of boron nitride nanoplatelets for epoxy composites with improved thermal properties. *Polymer* **53**, 471–480 (2012).
103. C. Zhang et al., Aminopropyltrimethoxysilane-functionalized boron nitride nanotube based epoxy nanocomposites with simultaneous high thermal conductivity and excellent electrical insulation. *Journal of Materials Chemistry A* **6**, 20663–20668 (2018).
104. J. Gu, Q. Zhang, J. Dang, C. Xie, Thermal conductivity epoxy resin composites filled with boron nitride. *Polymers for Advanced Technologies* **23**, 1025–1028 (2012).

105. M. H. Tsai, I. H. Tseng, J. C. Chiang, J. J. Li, Flexible polyimide films hybrid with functionalized boron nitride and graphene oxide simultaneously to improve thermal conduction and dimensional stability. *ACS Applied Materials & Interfaces* **6**, 8639–8645 (2014).
106. Y. Guo et al., Enhanced thermal conductivities and decreased thermal resistances of functionalized boron nitride/polyimide composites. *Composites Part B: Engineering* **164**, 732–739 (2019).
107. Y. Zhang, Y. Fan, U. Kamran, S. J. Park, Improved thermal conductivity and mechanical property of mercapto group-activated boron nitride/elastomer composites for thermal management. *Composites Part A: Applied Science and Manufacturing* **156**, 106869 (2022).
108. D. S. Muratov, D. V. Kuznetsov, I. A. Il'inykh, I. N. Burmistrov, I. N. Mazov, Thermal conductivity of polypropylene composites filled with silane-modified hexagonal BN. *Composites Science and Technology* **111**, 40–43 (2015).
109. K. Kim, H. Oh, J. Kim, Fabrication of covalently linked exfoliated boron nitride nanosheet/multi-walled carbon nanotube hybrid particles for thermal conductive composite materials. *RSC Advances* **8**, 33506–33515 (2018).
110. H. Lee, S. M. Dellatore, W. M. Miller, P. B. Messersmith, Mussel-inspired surface chemistry for multifunctional coatings. *Science* **318**, 426–430 (2007).
111. T. Ma et al., Highly thermal conductivities, excellent mechanical robustness and flexibility, and outstanding thermal stabilities of aramid nanofiber composite papers with nacre-mimetic layered structures. *ACS Applied Materials & Interfaces* **12**, 1677–1686 (2020).
112. H. Wu, M. R. Kessler, Multifunctional cyanate ester nanocomposites reinforced by hexagonal boron nitride after noncovalent biomimetic functionalization. *ACS Applied Materials & Interfaces* **7**, 5915–5926 (2015).
113. L. Chen et al., Thermal conductivity performance of polypropylene composites filled with polydopamine-functionalized hexagonal boron nitride. *PLoS ONE* **12**, e0170523 (2017).
114. L. Yu, S. Gao, D. Yang, Q. Wei, L. Zhang, Improved thermal conductivity of polymer composites by noncovalent modification of boron nitride via tannic acid chemistry. *Industrial & Engineering Chemistry Research* **60**, 12570–12578 (2021).
115. S. H. Kang, S. W. Jeon, S. Y. Moon, Y. J. Yoon, T. H. Kim, Fabrication of noncovalently functionalized boron nitride nanotubes with high stability and water-redispersibility. *Journal of Physical Chemistry Letters* **11**, 4511–4516 (2020).
116. X. Wu, Z. Yang, W. Kuang, Z. Tang, B. Guo, Coating polyrhodanine onto boron nitride nanosheets for thermally conductive elastomer composites. *Composites Part A: Applied Science and Manufacturing* **94**, 77–85 (2017).
117. S. Yu et al., Hydrogen-bond-mediated surface functionalization of boron nitride micro-lamellae toward high thermal conductive papers. *Advanced Materials Interfaces* **10**, 2202196 (2023).
118. J. Yu, H. Mo, P. Jiang, Polymer/boron nitride nanosheet composite with high thermal conductivity and sufficient dielectric strength. *Polymers for Advanced Technologies* **26**, 514–520 (2015).
119. Y. Lin, T. V. Williams, J. W. Connell, Soluble, exfoliated hexagonal boron nitride nanosheets. *The Journal of Physical Chemistry Letters* **1**, 277–283 (2009).
120. H. Lee et al., Dual functionalization of hexagonal boron nitride nanosheets using pyrene-tethered poly(4-vinylpyridine) for stable dispersion and facile device incorporation. *ACS Applied Nano Materials* **3**, 7633–7642 (2020).
121. T. Morishita, N. Takahashi, Highly thermally conductive and electrically insulating polymer nanocomposites with boron nitride nanosheet/ionic liquid complexes. *RSC Advances* **7**, 36450–36459 (2017).
122. X. Wang, P. Wu, Aqueous phase exfoliation of two-dimensional materials assisted by thermoresponsive polymeric ionic liquid and their applications in stimuli-responsive hydrogels and highly thermally conductive films. *ACS Applied Materials & Interfaces* **10**, 2504–2514 (2018).
123. T. Morishita, H. Okamoto, Facile exfoliation and noncovalent superacid functionalization of boron nitride nanosheets and their use for highly thermally conductive and electrically insulating polymer nanocomposites. *ACS Applied Materials & Interfaces* **8**, 27064–27073 (2016).
124. H. Shen, J. Guo, H. Wang, N. Zhao, J. Xu, Bioinspired modification of h-BN for high thermal conductive composite films with aligned structure. *ACS Applied Materials & Interfaces*7, 5701–5708 (2015).

125. J. Wang et al., Noncovalent functionalization of boron nitride and its effect on the thermal conductivity of polycarbonate composites. *Journal of Applied Polymer Science* **134**, 44978 (2017).
126. M. Foygel, R. D. Morris, D. Anez, S. French, V. L. Sobolev, Theoretical and computational studies of carbon nanotube composites and suspensions: Electrical and thermal conductivity. *Physical Review B* **71**, 104201 (2005).
127. C. Xiao et al., Epoxy composite with significantly improved thermal conductivity by constructing a vertically aligned three-dimensional network of silicon carbide nanowires/boron nitride nanosheets. *Composites Part B: Engineering* **187**, 107855 (2020).
128. J. Hu et al., Polymer composite with improved thermal conductivity by constructing a hierarchically ordered three-dimensional interconnected network of BN. *ACS Applied Materials & Interfaces* **9**, 13544–13553 (2017).
129. C.-W. Nan, G. Liu, Y. Lin, M. Li, Interface effect on thermal conductivity of carbon nanotube composites. *Applied Physics Letters* **85**, 3549–3551 (2004).
130. C.-W. Nan, R. Birringer, D. R. Clarke, H. Gleiter, Effective thermal conductivity of particulate composites with interfacial thermal resistance. *Journal of Applied Physics* **81**, 6692–6699 (1997).
131. C. W. Nan, X. P. Li, R. Birringer, Inverse problem for composites with imperfect interface: Determination of interfacial thermal resistance, thermal conductivity of constituents, and microstructural parameters. *Journal of the American Ceramic Society* **83**, 848–854 (2000).
132. F. Deng, Q. S. Zheng, L. F. Wang, C. W. Nan, Effects of anisotropy, aspect ratio, and nonstraightness of carbon nanotubes on thermal conductivity of carbon nanotube composites. *Applied Physics Letters* **90**, 021914 (2007).
133. I.-L. Ngo, S. V. Prabhakar Vattikuti, C. Byon, A modified Hashin-Shtrikman model for predicting the thermal conductivity of polymer composites reinforced with randomly distributed hybrid fillers. *International Journal of Heat and Mass Transfer* **114**, 727–734 (2017).
134. Z. Hashin, S. Shtrikman, A variational approach to the theory of the effective magnetic permeability of multiphase materials. *Journal of Applied Physics* **33**, 3125–3131 (1962).
135. F. Zhang, Y. Feng, W. Feng, Three-dimensional interconnected networks for thermally conductive polymer composites: Design, preparation, properties, and mechanisms. *Materials Science and Engineering: R: Reports* **142**, 100580 (2020).
136. D. Golberg, Y. Bando, O. Stéphan, K. Kurashima, Octahedral boron nitride fullerenes formed by electron beam irradiation. *Applied Physics Letters* **73**, 2441–2443 (1998).
137. P. Bujard, InterSociety Conference on Thermal Phenomena in the Fabrication and Operation of Electronic Components (I-THERM '88), in *InterSociety Conference on Thermal Phenomena in the Fabrication and Operation of Electronic Components. I-THERM '88*, San Diego, California, USA (1988), pp. 41–49.
138. M. Donnay, S. Tzavalas, E. Logakis, Boron nitride filled epoxy with improved thermal conductivity and dielectric breakdown strength. *Composites Science and Technology* **110**, 152–158 (2015).
139. S. Iijima, Helical microtubules of graphitic carbon. *Nature* **354**, 56–58 (1991).
140. A. Rubio, J. L. Corkill, M. L. Cohen, Theory of graphitic boron nitride nanotubes. *Physical Review B: Condens Matter* **49**, 5081–5084 (1994).
141. C. Zhi, Y. Xu, Y. Bando, D. Golberg, Highly thermo-conductive fluid with boron nitride nanofillers. *ACS Nano* **5**, 6571–6577 (2011).
142. C. Liu et al., Controllable preparation of boron nitride microspheres and their epoxy-based thermoconductive composites. *Ceramics International* **49**, 27138–27146 (2023).
143. L. H. Li et al., Large-scale mechanical peeling of boron nitride nanosheets by low-energy ball milling. *Journal of Materials Chemistry* **21**, 11862–11866 (2011).
144. Z. Zhu et al., Liquid metal-assisted high-efficiency exfoliation of boron nitride for electrically insulating heat-spreader film. *ACS Applied Materials & Interfaces* **14**, 54256–54265 (2022).
145. Z. Wang, Y. Zhu, D. Ji, Z. Li, H. Yu, Scalable preparation of 2D boron nitride nanosheets and enhancement of UV absorption and thermal conductivity of cellulose nanofibers membrane. *Journal of Materials Science* **55**, 13815–13823 (2020).
146. Y. Wang, Z. Shi, J. Yin, Boron nitride nanosheets: Large-scale exfoliation in methanesulfonic acid and their composites with polybenzimidazole. *Journal of Materials Chemistry* **21**, 11371–11377 (2011).

147. W. Zhu et al., Controlled gas exfoliation of boron nitride into few-layered nanosheets. *Angewandte Chemie. International Edition in English* **55**, 10766–10770 (2016).
148. R. Rizvi et al., High-throughput continuous production of shear-exfoliated 2D layered materials using compressible flows. *Advanced Materials* **30**, e1800200 (2018).
149. X. Ge et al., Low melting-point alloy-boron nitride nanosheet composites for thermal management. *ACS Applied Nano Materials* **3**, 3494–3502 (2020).
150. X. Zheng, C. W. Park, Thermal and mechanical properties of carbon fiber-reinforced resin composites with copper/boron nitride coating. *Composite Structures* **220**, 494–501 (2019).
151. X. Zheng, S. Kim, C. W. Park, Enhancement of thermal conductivity of carbon fiber-reinforced polymer composite with copper and boron nitride particles. *Composites Part A: Applied Science and Manufacturing* **121**, 449–456 (2019).
152. J. P. Hong et al., High thermal conductivity epoxy composites with bimodal distribution of aluminum nitride and boron nitride fillers. *Thermochimica Acta* **537**, 70–75 (2012).
153. C. Chen et al., Construction of 3D boron nitride nanosheets/silver networks in epoxy-based composites with high thermal conductivity via in-situ sintering of silver nanoparticles. *Chemical Engineering Journal* **369**, 1150–1160 (2019).
154. C. Liu, Y. He, C. Chen, M. Chen, W. Yu, Synergistic enhancement of heat transfer between irregular shaped particles and boron nitride flakes at low loading. *Results in Physics* **15**, 102599 (2019).
155. F. Wang et al., Silver nanoparticle-deposited boron nitride nanosheets as fillers for polymeric composites with high thermal conductivity. *Scientific Reports* **6**, 19394 (2016).
156. H. Zhu et al., Highly thermally conductive papers with percolative layered boron nitride nanosheets. *ACS Nano* **8**, 3606–3613 (2014).
157. W. Yang et al., Three-dimensional skeleton assembled by carbon nanotubes/boron nitride as filler in epoxy for thermal management materials with high thermal conductivity and electrical insulation. *Composites Part B: Engineering* **224**, 109168 (2021).
158. J. Su, Y. Xiao, M. Ren, Enhanced thermal conductivity in epoxy nanocomposites with hybrid boron nitride nanotubes and nanosheets. *Physica Status Solidi (a)* **210**, 2699–2705 (2013).
159. J. Yang, X. Shen, W. Yang, J. K. Kim, Templating strategies for 3D-structured thermally conductive composites: Recent advances and thermal energy applications. *Progress in Materials Science* **133**, 101054 (2023).
160. J. Wang et al., Preparation and properties of three-dimensional boron nitride submicron tube/epoxy resin composites with high thermal conductivity. *Polymer Composites* **44**, 3477–3486 (2023).
161. J. Chen, X. Huang, Y. Zhu, P. Jiang, Cellulose nanofiber supported 3D interconnected BN nanosheets for epoxy nanocomposites with ultrahigh thermal management capability. *Advanced Functional Materials* **27**, 1604754 (2017).
162. L. Li, K. Wang, Z. Huang, C. Zhang, T. Liu, Highly ordered graphene architectures by duplicating melamine sponges as a three-dimensional deformation-tolerant electrode. *Nano Research* **9**, 2938–2949 (2016).
163. X. Wang, P. Wu, Melamine foam-supported 3D interconnected boron nitride nanosheets network encapsulated in epoxy to achieve significant thermal conductivity enhancement at an ultralow filler loading. *Chemical Engineering Journal* **348**, 723–731 (2018).
164. H. Yu et al., Highly thermally conductive polymer composite enhanced by two-level adjustable boron nitride network with leaf venation structure. *Composites Science and Technology* **222**, 109406 (2022).
165. Z. Tian et al., A thermal interface material based on foam-templated three-dimensional hierarchical porous boron nitride. *Journal of Materials Chemistry A* **6**, 17540–17547 (2018).
166. J. Li et al., Jelly-inspired construction of the three-dimensional interconnected BN network for lightweight, thermally conductive, and electrically insulating rubber composites. *ACS Applied Electronic Materials* **2**, 1661–1669 (2020).
167. C. Xiao et al., Preparation of highly thermally conductive epoxy resin composites via hollow boron nitride microbeads with segregated structure. *Composites Part A: Applied Science and Manufacturing* **121**, 330–340 (2019).
168. X. Xu et al., 3D boron nitride foam filled epoxy composites with significantly enhanced thermal conductivity by a facial and scalable approach. *Chemical Engineering Journal* **397**, 125447 (2020).

169. Z. Wang et al., Highly thermally conductive polymer composite enhanced by constructing a dual thermal conductivity network. *Polymer Composites* **44**, 5734–5745 (2023).
170. Y. Cui et al., Fabrication of EVA connected 3D BN network for enhancing the thermal conductivity of epoxy composites. *Composites Part B: Engineering* **224**, 109203 (2021).
171. X. Chen et al., Salt template assisted BN scaffold fabrication toward highly thermally conductive epoxy composites. *ACS Applied Materials & Interfaces* **12**, 16987–16996 (2020).
172. X. Wu, W. Liu, F. g. Shi, L. Yang, C. Zhang, Constructing three-dimensional boron nitride network for highly thermally conductive epoxy resin composites. *Polymer Composites* **43**, 1711–1717 (2022).
173. Q.-h. Lin et al., Construction of a 3D interconnected boron nitride nanosheets in a PDMS matrix for high thermal conductivity and high deformability. *Composites Science and Technology* **226**, 109528 (2022).
174. S. Deville, Wood-like polymeric materials by ice templating. *National Science Review* **6**, 184–185 (2019).
175. F. Diaz, N. Forsyth, A. R. Boccaccini, Aligned ice templated biomaterial strategies for the musculoskeletal system. *Advanced Healthcare Materials* **12**, e2203205 (2023).
176. J. Zhang et al., Green mass production of pure nanodrugs via an ice-template-assisted strategy. *Nano Letters* **19**, 658–665 (2019).
177. C. Huang, M. Dontigny, K. Zaghib, P. S. Grant, Low-tortuosity and graded lithium ion battery cathodes by ice templating. *Journal of Materials Chemistry A* **7**, 21421–21431 (2019).
178. H. Shen et al., Fabrication of oriented hBN scaffolds for thermal interface materials. *RSC Advances* **6**, 16489–16494 (2016).
179. T. Huang, Y. Li, M. Chen, L. Wu, Bi-directional high thermal conductive epoxy composites with radially aligned boron nitride nanosheets lamellae. *Composites Science and Technology* **198**, 108322 (2020).
180. S. He et al., Multi-directionally thermal conductive epoxy/boron nitride composites based on circinate vane type network. *Composites Communications* **25**, 100744 (2021).
181. J. Han, G. Du, W. Gao, H. Bai, An anisotropically high thermal conductive boron nitride/epoxy composite based on nacre-mimetic 3D network. *Advanced Functional Materials* **29**, 1900412 (2019).
182. H. Li et al., Ice-templated assembly strategy to construct three-dimensional thermally conductive networks of BN nanosheets and silver nanowires in polymer composites. *Composites Communications* **25**, 100601 (2021).
183. H. L. Gao et al., Macroscopic free-standing hierarchical 3D architectures assembled from silver nanowires by ice templating. *Angewandte Chemie. International Edition in English* **53**, 4561–4566 (2014).
184. X. Zeng et al., Ice-templated assembly strategy to construct 3D boron nitride nanosheet networks in polymer composites for thermal conductivity improvement. *Small* **11**, 6205–6213 (2015).
185. J. Gu et al., Dielectric thermally conductive boron nitride/polyimide composites with outstanding thermal stabilities via in -situ polymerization-electrospinning-hot press method. *Composites Part A: Applied Science and Manufacturing* **94**, 209–216 (2017).
186. D.-L. Zhang et al., Enhanced thermal conductivity and mechanical property through boron nitride hot string in polyvinylidene fluoride fibers by electrospinning. *Composites Science and Technology* **156**, 1–7 (2018).
187. G. Yang et al., Highly thermally conductive polyvinyl alcohol/boron nitride nanocomposites with interconnection oriented boron nitride nanoplatelets. *Composites Science and Technology* **201**, 108521 (2021).
188. L. M. Guiney et al., Three-dimensional printing of cytocompatible, thermally conductive hexagonal boron nitride nanocomposites. *Nano Letters* **18**, 3488–3493 (2018).
189. J. Yin, X. Li, J. Zhou, W. Guo, Ultralight three-dimensional boron nitride foam with ultralow permittivity and superelasticity. *Nano Letters* **13**, 3232–3236 (2013).
190. H. Fang, S.-L. Bai, C. P. Wong, Thermal, mechanical and dielectric properties of flexible BN foam and BN nanosheets reinforced polymer composites for electronic packaging application. *Composites Part A: Applied Science and Manufacturing* **100**, 71–80 (2017).

191. J. Wang et al., Magnetically oriented 3D-boron nitride nanobars enable efficient heat dissipation for 3D-integrated power packaging. *ACS Applied Nano Materials* **6**, 18508–18517 (2023).
192. Y. He et al., Achieving high out-of-plane thermal conductivity for boron nitride nano sheets/epoxy composite films by magnetic orientation. *Composites Part A: Applied Science and Manufacturing* **157**, 106933 (2022).
193. Z. Lin et al., Magnetic alignment of hexagonal boron nitride platelets in polymer matrix: Toward high performance anisotropic polymer composites for electronic encapsulation. *ACS Applied Materials & Interfaces* **5**, 7633–7640 (2013).
194. H.-B. Cho et al., Insulating polymer nanocomposites with high-thermal-conduction routes via linear densely packed boron nitride nanosheets. *Composites Science and Technology* **129**, 205–213 (2016).
195. X. Li, Q. Xu, Z. Lei, Z. Chen, Electrostatic flocking assisted aligned boron nitride platelets scaffold for enhancing the through-plane thermal conductivity of flexible thermal interface materials. *Ceramics International* **49**, 22623–22629 (2023).
196. N. Sun et al., Hot-pressing induced orientation of boron nitride in polycarbonate composites with enhanced thermal conductivity. *Composites Part A: Applied Science and Manufacturing* **110**, 45–52 (2018).
197. X. Yu et al., Flexible boron nitride composite membranes with high thermal conductivity, low dielectric constant and facile mass production. *Composites Science and Technology* **222**, 109400 (2022).
198. R. C. Zhang et al., Uniaxially stretched polyethylene/boron nitride nanocomposite films with metal-like thermal conductivity. *Composites Science and Technology* **196**, 108154 (2020).
199. Z. Kuang et al., Fabrication of highly oriented hexagonal boron nitride nanosheet/elastomer nanocomposites with high thermal conductivity. *Small* **11**, 1655–1659 (2014).
200. Z. Wang et al., The investigation of the effect of filler sizes in 3D-BN skeletons on thermal conductivity of epoxy-based composites. *Nanomaterials* **12**, 446 (2022).
201. C. Yuan et al., Thermal conductivity of polymer-based composites with magnetic aligned hexagonal boron nitride platelets. *ACS Applied Materials & Interfaces* **7**, 13000–13006 (2015).
202. Z. Su et al., Fabrication of thermal conductivity enhanced polymer composites by constructing an oriented three-dimensional staggered interconnected network of boron nitride platelets and carbon nanotubes. *ACS Applied Materials & Interfaces* **10**, 36342–36351 (2018).
203. H.-B. Cho et al., Modification of BN nanosheets and their thermal conducting properties in nanocomposite film with polysiloxane according to the orientation of BN. *Composites Science and Technology* **71**, 1046–1052 (2011).
204. L. Liang et al., High thermal conductivity composite h-BN/EP obtained by pulsed square-wave electric field induction. *Polymer* **290**, 126491 (2024).
205. Y. Han, S. Lv, C. Hao, F. Ding, Y. Zhang, Thermal conductivity enhancement of BN/silicone composites cured under electric field: Stacking of shape, thermal conductivity, and particle packing structure anisotropies. *Thermochimica Acta* **529**, 68–73 (2012).
206. H.-B. Cho et al., Epoxy resin-based nanocomposite films with highly oriented BN nanosheets prepared using a nanosecond-pulse electric field. *Materials Letters* **65**, 2426–2428 (2011).
207. J. Chen, X. Huang, B. Sun, P. Jiang, Highly thermally conductive yet electrically insulating polymer/boron nitride nanosheets nanocomposite films for improved thermal management capability. *ACS Nano* **13**, 337–345 (2019).
208. X. Yang et al., Significant improvement of thermal conductivities for BNNS/PVA composite films via electrospinning followed by hot-pressing technology. *Composites Part B: Engineering* **175**, 107070 (2019).
209. I. Saini et al., Grafted SiC nanocrystals: For enhanced optical, electrical and mechanical properties of polyvinyl alcohol. *Journal of Alloys and Compounds* **714**, 172–180 (2017).
210. M. Owais et al., Scalable fabrication of thermally conductive layered nacre-like self-assembled 3D BN-based PVA aerogel framework nanocomposites. *Polymers* **14**, 3316 (2022).
211. K. Zhang, P. Tao, Y. Zhang, X. Liao, S. Nie, Highly thermal conductivity of CNF/AlN hybrid films for thermal management of flexible energy storage devices. *Carbohydrate Polymers* **213**, 228–235 (2019).
212. Y. Xu et al., Enhancing thermal conductivity and toughness of cellulose nanofibril/boron nitride nanosheet composites. *Carbohydrate Polymers* **296**, 119938 (2022).

213. Z. Hu et al., An aqueous-only, green route to exfoliate boron nitride for preparation of high thermal conductive boron nitride nanosheet/cellulose nanofiber flexible film. *Composites Science and Technology* **168**, 287–295 (2018).
214. Y. Wen et al., Advances on thermally conductive epoxy-based composites as electronic packaging underfill materials—a review. *Advanced Materials* **34**, 2201023 (2022).
215. T. Huang et al., Modified graphene/polyimide nanocomposites: Reinforcing and tribological effects. *ACS Applied Materials & Interfaces* **5**, 4878–4891 (2013).
216. M. Tanimoto, T. Yamagata, K. Miyata, S. Ando, Anisotropic thermal diffusivity of hexagonal boron nitride-filled polyimide films: Effects of filler particle size, aggregation, orientation, and polymer chain rigidity. *ACS Applied Materials & Interfaces* **5**, 4374–4382 (2013).
217. S.-W. Jin et al., Eco-friendly preparation and characterization of highly thermally conductive polyimide/boron nitride composites. *Composites Part A: Applied Science and Manufacturing* **166**, 107396 (2023).
218. T. L. Li, S. L. Hsu, Enhanced thermal conductivity of polyimide films via a hybrid of micro- and nano-sized boron nitride. *The Journal of Physical Chemistry B* **114**, 6825–6829 (2010).
219. B. Wan, X. Li, X. Zeng, J.-W. Zha, Covalent-assisted construction of "scale-like" boron nitride/polyimide thermal interface materials with high thermal conductivity. *Composites Communications* **45**, 101803 (2024).
220. Y. Fang et al., Covalently linked polydopamine-modified boron nitride nanosheets/polyimide composite fibers with enhanced heat diffusion and mechanical behaviors. *Composites Part B: Engineering* **199**, 108281 (2020).
221. S. Zhou et al., High thermal conductivity property of polyamide-imide/boron nitride composite films by doping boron nitride quantum dots. *Journal of Materials Chemistry C* **7**, 13896–13903 (2019).
222. N. Zhao, J. Li, W. Wang, W. Gao, H. Bai, Isotropically ultrahigh thermal conductive polymer composites by assembling anisotropic boron nitride nanosheets into a biaxially oriented network. *ACS Nano* **16**, 18959–18967 (2022).
223. S. Zhang et al., Highly exfoliated boron nitride nanosheets via carboxyl nanocellulose for thermally conductive nanocomposite films. *ACS Applied Nano Materials* **7**, 2971–2982 (2024).
224. J. Jakovenko et al., in *2011 12th International Conference on Thermal, Mechanical & Multi-Physics Simulation and Experiments in Microelectronics and Microsystems*, Linz, Austria (2011), pp. 1/4–4/4.
225. M. C. Gorur et al., 3D printing of hexagonal boron nitride nanosheets/polylactic acid nanocomposites for thermal management of electronic devices. *Composites Part B: Engineering* **265**, 110955 (2023).
226. P. Li et al., Fast thermal responsive separators toward long-life and safe lithium metal batteries. *Nano Research* **17**, 2746–2754 (2024).
227. K. Zhu et al., "Three-in-one" strategy: Heat regulation and conversion enhancement of a multifunctional separator for safer lithium-sulfur batteries. *Carbon Energy* **5**, e352 (2023).
228. J. Ji et al., Highly dispersed and functionalized boron nitride nanosheets contribute to ultra-stable long-life all-solid-state batteries. *Journal of Materials Chemistry A* **11**, 11298–11309 (2023).
229. Q. Tao et al., Experimental investigation on heat transfer enhancement of a radiator in a simulated pemfc cooling system using hexagonal boron nitride nanofluids. *Heat Transfer Research* **52**, 31–45 (2021).
230. K. Z. Ekinci, S. Ünügür Çelik, A. Bozkurt, Enhancing the anhydrous proton conductivity of sulfonated polysulfone/polyvinyl phosphonic acid composite membranes with hexagonal boron nitride. *International Journal of Polymeric Materials and Polymeric Biomaterials* **64**, 683–689 (2015).
231. M. S. Tutgun, D. Sinirlioglu, S. U. Celik, A. Bozkurt, Preparation and characterization of hexagonal boron nitride and PAMPS-NMPA-based thin composite films and investigation of their membrane properties. *Ionics* **21**, 2871–2878 (2015).
232. Z. Huang et al., Ultra-thin h-BN doped high sulfonation sulfonated poly (ether-ether-ketone) of PTFE-reinforced proton exchange membrane. *Journal of Membrane Science* **644**, 120099 (2022).
233. N. Yang et al., Improving heat transfer enables durable perovskite solar cells. *Advanced Energy Materials* **12**, 2200869 (2022).

Section IV

The Energy Applications

Boron Nitride Nanosheets in Storage and Harvesting

Chapter 14

Boron Nitride Nanomaterials Solving the Critical Issues of Next-Generation Batteries

Baozhi Yu and Donggun Kim

14.1 INTRODUCTION

Developing safe and high-performance energy storage systems has become an urgent issue because of the rapid growth of new industry applications, such as electric vehicles, large-scale energy storage stations, high-end communication terminals, aerospace, *etc.* [1,2]. Lithium-ion batteries (LIBs) are mature battery systems widely applied in various electronic products [3–5]. A significant improvement in the performance and price of LIBs has been achieved since 1991 [6,7]. However, LIBs have nearly approached their limits and cannot fulfill the current safer and higher energy density demands [8,9]. Therefore, developing so-called 'beyond Li-ion' technologies has been emphasized over the past few years [10–13]. For example, lithium metal is being used as an anode to enable high-energy batteries because of its extremely high theoretical specific capacity of 3860 mAh g^{-1}, low mass density of 0.534 g cm^{-3}, and the lowest negative electrochemical potential (−3.040 V *vs.* standard hydrogen electrode) [14–16]. Emerging new battery systems employing lithium metal anodes include lithium-sulfur batteries [9,17,18], lithium metal batteries [19–22], and lithium–oxygen ($Li\text{-}O_2$) batteries [23–26], which exhibit much higher energy densities than LIBs. However, several critical issues, such as dendrite growth, shuttle effects, and poor cycling stability, make these systems difficult for practical exploitation. In addition, because the leakage of the flammable organic electrolyte in batteries causes serious safety problems, solid-state batteries are under development to solve the safety issue. Nevertheless, compared to the batteries with liquid electrolytes, the performance of the solid-state batteries still needs to be improved substantially. To overcome these challenging issues, employing advanced nanomaterials is a promising approach. This section introduces the role of BN in solving these challenges.

BN has been demonstrated in an extensive range of applications, including water remediation [27–29], gas purification [30,31], and drug delivery [32–34] due to unique properties such as excellent oxidation and corrosion resistance, exceptional thermal stability and conductivity, wide-bandgap semiconducting property, and low mass density. In 1842, the BN structure was synthesized for the first time with boric acid (H_3BO_3) and potassium cyanide [35]. Generally, the electric insulating BN ceramic is not considered for developing electrode technology in battery systems. Actually, this unique property, plus the excellent physical and chemical properties (*i.e.*, thermal stability, thermal conductivity, mechanical chemical stability), gives BN materials a significant advantage over other materials to be able to survive in a battery's tough environment of

DOI: 10.1201/9781003647812-18

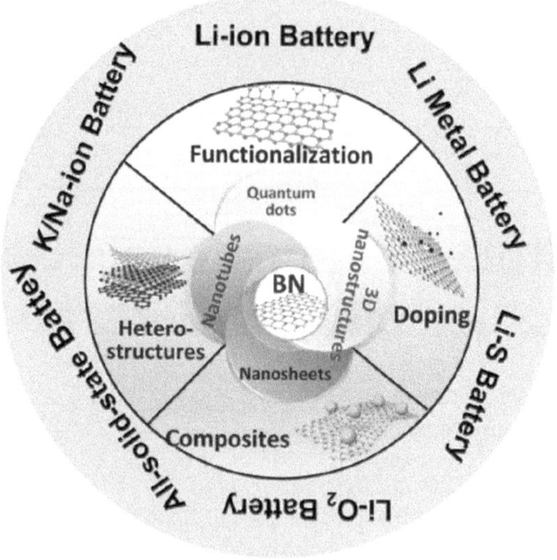

Figure 14.1 BN nanomaterials and their applications in LIBs and beyond.

high temperature, acidic corrosive, and robust electric field and helps to overcome some critical issues. BN nanomaterials, especially nanosheets and nanotubes, are attracting a lot of attention due to their unique design and new properties [36–38]. Also, more novel properties are also expected to emerge from BN through physical or chemical functionalization. In addition, doping is another excellent strategy to bring brand-new functionalities, features, and thus new applications. Literature found in recent years on the Web of Science by searching for the keywords of 'Boron Nitride' and 'Battery' has shown that the number of papers published is ~300. One-third of articles are published in the past two years, the highest number among the numbers for 20 years, indicating increasing usage of BN nanomaterials for high-performance batteries. Many review papers were published, which focus on the synthesis and structural design strategies of BN nanomaterials [27,36,39–49]. However, there are very few overviews on how BN nanomaterials help the energy storage systems, especially LIBs and beyond, to overcome some performance and safety challenges.

Here is a comprehensive summary of recent advances in applying BN nanomaterials on LIBs and 'beyond lithium-ion' batteries (Figure 14.1). At first, the main challenges of the battery's performance and safety are introduced. Second, the novel properties of BN nanomaterials, including BN nanostructures, doped BN, and BN heterostructures, are reviewed. Finally, the most recent advances in real battery applications are summarized.

14.2 BN FOR CAPACITY AND ENERGY DENSITY IMPROVEMENTS

The first objective that must be realized for next-generation batteries is enhanced energy output. Numerous types of research have demonstrated impressive battery performances, such as high specific capacity or cycling stability. However, when considering energy density, most reports still cannot overcome the commercialized Li-ion batteries of about ~250 Wh kg^{-1}. BN nanomaterials could be applied to the battery system, and BN could improve these energy output shortages.

14.2.1 Additives in the Electrodes

The performance of the battery system mainly depends on the quality of the electrodes. The electric insulation of BN implies that BN alone cannot be used as a host material. However, BN nanomaterials sometimes show unique properties as additives for developing electrode technology in battery systems. Examples include that B_4C and BN are used as additives in MnO_2 cathodes for alkaline batteries [50]. BN is found to be beneficial to increasing the initial capacity and voltage of the alkaline batteries with the MnO_2 cathode because BN can act as the electron donor to change the intercalation cell potential during the discharge period (Figure 14.2a). Bhauriyal et al. reported a similar phenomenon observed in Al batteries [51]. Graphene/BN (G/BN) heterostructure was employed as the cathode material for Al batteries that exhibited a higher working voltage and capacity than conventional Al batteries. The results indicate that because of the interaction between the *h*-BN and graphene, $AlCl_4$ has higher binding energy with the G/BN heterostructure (−1.19 eV) than pristine BN (−0.92 eV). Therefore, the G/*h*-BN heterostructure can improve the battery's overall stability during charge/discharge cycles. Moreover, the G/BN heterostructure provides a voltage of 2.14 V, which is larger than that of monolayer graphene (1.99 V) (Figure 14.2b, c), because opening up the partially closed interlayer spaces requires a larger energy change.

Therefore, the G/BN heterostructure can provide a comparatively higher voltage value. The two examples above clearly show that BN nanomaterials can be used as additives in the electrodes to increase the working voltage of the batteries. Furthermore, BN nanomaterials can also benefit the capacity of the electrodes when used as additives in the electrodes. The reduced graphene oxide/BN (rGO/BN) composites are reported to be the anodes of LIBs [52,53] (Figure 14.2d). The rGO/BN composite with 2 wt% BN content exhibited good performance, including high rate capability, a high capacity of 278 mAh g^{-1} at a high current density of 100 mA g^{-1}, and high capacity retention over the first 200 cycles. The improved performance can be attributed to the inhibition of the restacking of rGO layers by BN nanosheets (BNNSs). The electrostatic interactions and weak van der Waals between BN and rGO nanosheets contribute to the excessive expansion of the lithiated rGO/BN interlayer distance. However, it is also suggested that more BN is not beneficial for capacity improvement because of the increased resistance and less electron exchange at the electrodes.

14.2.2 High-Performance Electrocatalysts

Rechargeable lithium–oxygen batteries have attracted significant investigations recently since they were first reported in 1996 by Abraham and Jiang because of their extremely high theoretical energy density of 3500 Wh kg^{-1} [54]. BN nanomaterials have shown great potential in Li-O_2 battery applications. Early research reported by Uosaki et al. [55] shows the possibility of BN-supported Au to achieve oxygen production in the battery cell. This theoretical calculation predicted a four-electron process for water and a two-electron process for hydrogen peroxide with *h*-BN/Au surface modification. Lee et al. [56] employed *h*-BN/Ni heterostructure as the cathode for Li-O_2 batteries and analyzed the reaction mechanism of this cathode. On the other hand, Zhao et al. [57] described carbon-doped BNNSs as metal-free electrocatalysts. It is found that C-singly doping into *h*-BN nanosheets can contribute to the high spin density and charge density and reduce the energy gap, resulting in the enhancement of O_2 adsorption. Later, further research by Chowdhury et al. [58] showed the effect of different doping elements (C, Si, P, O) of BN for Li-O_2 batteries. The different doping elements and doping positions produce different charge

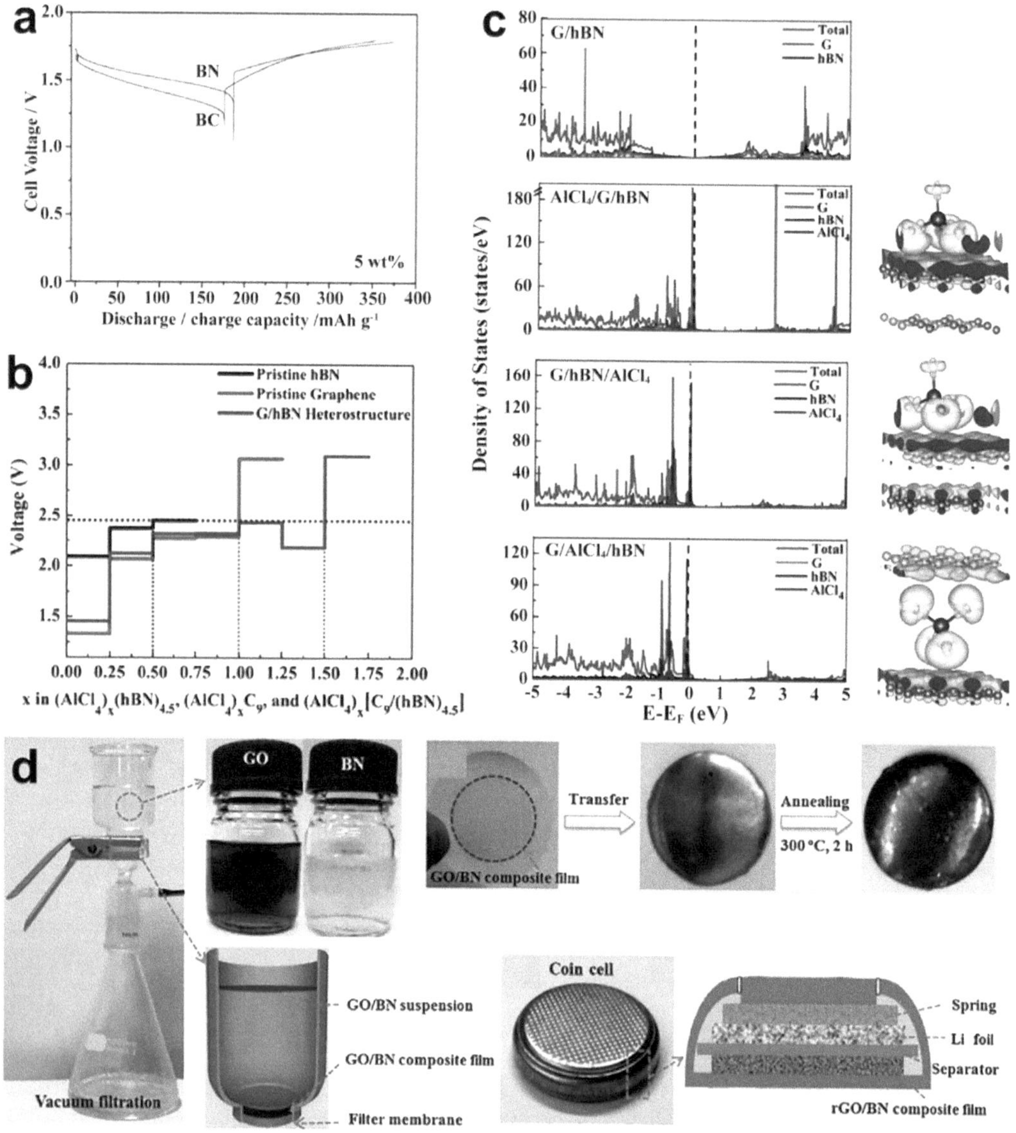

Figure 14.2 (a) The discharge/charge behavior of MnO_2 cathode alkaline batteries with BC and BN as the additives. (Reprinted with permission [50]. Copyright 2010, Elsevier B.V.) (b) The voltage behavior of the G/*h*-BN heterostructure with $AlCl_4$. (c) TDOS and PDOS of pristine G/*h*-BN, $AlCl_4$/G/hBN, G/hBN/$AlCl_4$, and G/$AlCl_4$/hBN. (Reprinted with permission [51]. Copyright 2019, American Chemical Society.) (d) The synthesis, microstructure, and performance in batteries of rGO/BN composites. (Reprinted with permission [52]. Copyright 2015, Elsevier Ltd.)

and discharge potential levels. The Si-doped BN at the boron position showed the highest charge potential of 6.48 V, and C-doped BN at the N position showed the highest discharge potential of 2.20 V. These clearly indicate that doped BN nanomaterials could benefit the development of ultrahigh energy density Li-O_2 batteries.

14.3 BN FOR RATE PERFORMANCE IMPROVEMENTS

The rate performance of the battery systems is mainly related to the transport properties of electrons and ions in the batteries. As is known, the high wettability of a battery separator can significantly accelerate ion transport between the electrodes, thus reducing the internal resistance of the cells. De Moraes et al. [59] developed BNNS inks that can be coated on separators for rechargeable Li-ion batteries. Ethyl cellulose was used as a polymer dispersant and stabilizer to prepare the ink via a liquid-phase exfoliation method. Then, the polymer volatilization occurs when the printed structures are thermally annealed. The porous microstructure and nano-scale carbon coating layer are formed on the BNNS, which contributes to the high wettability of the electrolyte. The results show that the Li-ion battery cells with the blade-coated *h*-BN separator showed capacities of 109 and 41 mAh g^{-1} at 0.2°C and 15°C, respectively, while the cells with the Celgard separator only possess capacities of 99 and 4 mAh g^{-1} at 0.2°C and 15°C, respectively. The LIBs based on printed BN separators demonstrated an excellent rate performance due to the high wettability, porosity, and ionic conductivity of BN separators, which promoted the rapid diffusion kinetics of lithium ions and enhanced the transport of lithium ions. The *h*-BN separators have a porosity of more than 60%, which is much higher than that of the commercial separators (38%). The high porosity and organic electrolyte wettability of *h*-BN separators result in the high ionic conductivity (over 1 mS cm^{-1}) of the *h*-BN separators, which is more than 2 times better than commercial separators at room temperature. Li et al. [60] also reported that BNNSs are suitable components for quasi-liquid solid electrolytes because BNNSs can absorb tenfold the weight of Li-ion liquid (IL) solution compared with its weight. The BNNSs possess a high specific surface area because of the nanoporous and few-layered structure of BNNSs. When a small amount of Li-IL is loaded on BNNSs, the outside surface of the nanolayers could be covered by Li-IL. As the amount of the Li-IL increases, the Li-IL will enter the interlayer of the BNNS gradually, which is confined in the nanopore space. The ordered lithium-ion transport pathways between and within BN nanolayers and the enormous absorption are the main reasons for the high ionic conductivity. This is developed as the Li-IL solution confining effect. Hyun et al. [61] produced a gel electrolyte with high-modulus *h*-BN nanoplatelets for solid-state LIBs. The obtained gel electrolyte with the exfoliated *h*-BN nanoplatelets retains high ionic conductivity at room temperature (>1 mS cm^{-1}) but possesses a much-improved mechanical property by 2 orders of magnitude (shear storage modulus ~5 MPa). More importantly, similar to Li's report [60], the large surface area of *h*-BN nanoplatelets significantly contributes to the adsorption of Li-IL, which stabilizes the Li-IL at high voltages. In addition, it is found that the exfoliated *h*-BN gel electrolyte suppressed the decomposition reaction of 1-ethyl-3-methylimidazolium cations at 4.3 V (vs. Li/Li^+), which is consistent with the degradation of the NMC|Li-IL|Li cell, and extended electrochemical stability up to 5.3 V (vs. Li/Li^+). Shim et al. [62] developed a multifunctional additive, perfluoropolyether-functionalized 2D BN nanoflakes, for gel polymer electrolytes (GPEs). A sonication-assisted exfoliation method was employed to prepare the perfluoropolyether-functionalized BN nanoflakes (FBN), and the nano-sized BN powder was functionalized with perfluoropolyether-functionalized pyrene molecules. The properties of GPE, including the ionic conductivity, Li^+ transference number, the mechanical modulus, and dendrite-suppressing capability, are substantially improved via adding a small amount (0.5 wt%) of BN nanoflakes into the GPE. As the FBN content in the GPE increases from 0 to 0.5 wt%, the ionic conductivity of the GPE increases due to the increase of electrolyte uptake, which is the same mechanism as previous examples. However, the electrolyte uptake in the GPE exhibited a decreasing trend when the FBN content was larger than 0.5 wt%, which further led to a decrease in the ionic conductivity. The effective electrolyte uptake is reduced with the formation of a crystalline structure in the matrix due to the inhomogeneous dispersion and the aggregation of excess FBN. Based on these reports, it can be concluded that the BN-based materials can benefit the absorption of the electrolyte or liquid components of the quasi-liquid electrolytes, leading to an improvement in the ion transport properties and stability of the battery systems.

The development of functional additives for SSEs has shown great potential for improving the mobility of lithium ions in SSEs. Zhang et al. [63] employed BN as an additive in garnet-type $Li_{6.25}Al_{0.25}La_3Zr_2O_{12}$ (LALZO) ceramic SSE and investigated the content effect of BN additives in the garnet-type ceramic SSE (Figure 14.3a). The LALZO electrolyte with a small percentage of 1 wt% BN exhibits 30 times higher ionic conductivity (6.21×10^{-5} S cm^{-1}) than the pure LALZO electrolyte,

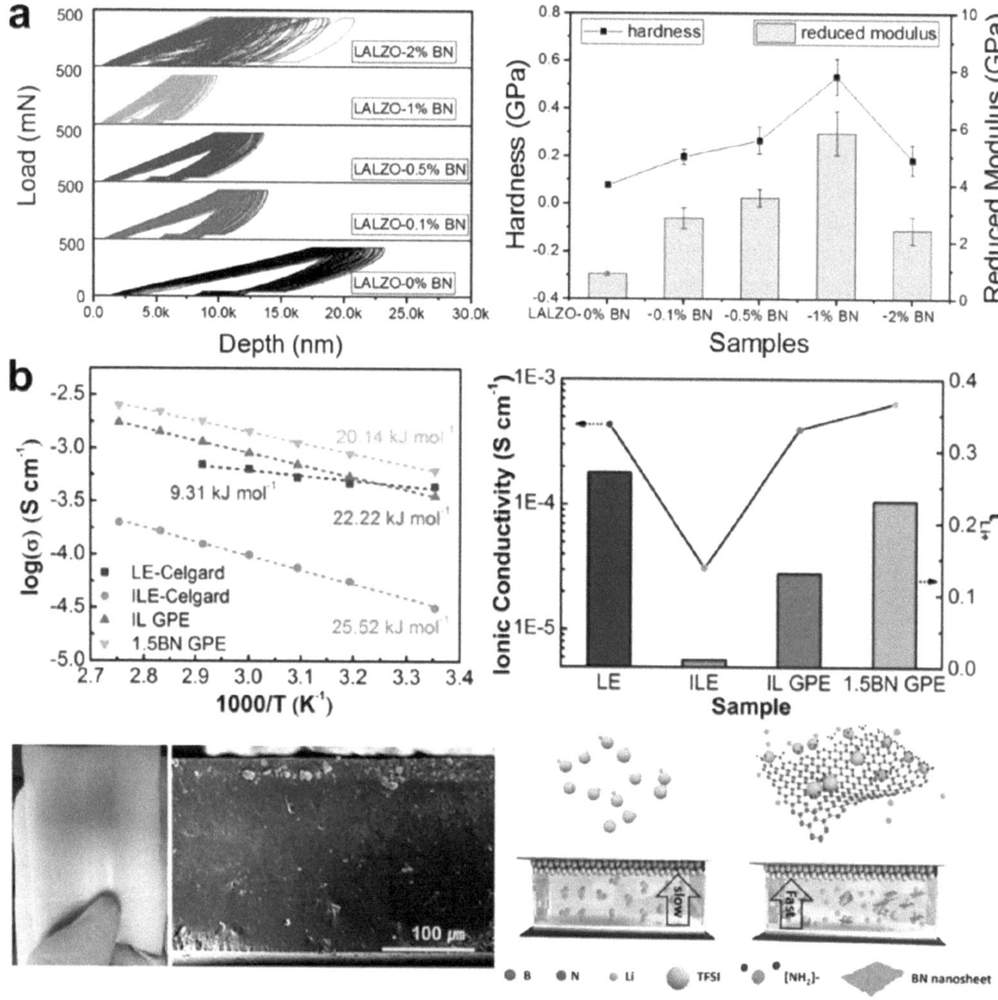

Figure 14.3 (a) The structure, physical, and conductivity behavior of the *h*-BN-added LALZO SSE. (Reprinted with permission [63]. Copyright 2019, American Chemical Society.) (b) The structure, conductivity, and mechanism of amine group-functionalized BNNSs added robust ion gel SSEs. (Reprinted with permission [64]. Copyright 2020, Wiley-VCH GmbH.)

but a higher weight ratio of BN (2 wt%) leads to a significant decrease in the ion conductivity. More importantly, Shim et al. [62] found another unique property of FBN, the anion-trapping ability. The Li^+ transference number of the GPE, which is another critical Li-ion transport property of the electrolytes, can be increased by adding FBN in the GPE because of the anion-trapping ability of FBN. In particular, an IL-based GPE shows very low lithium-ion transference numbers because of the strong ionic nature of ILs as a solvent, which strongly limits the application of IL electrolytes. Kim et al. [64] employed amine group-functionalized (AFBNNSs) as an additive for robust GPEs (Figure 14.3b) and improved the Li-ion transport. In this research, 0.5–2.5 wt% amine group-functionalized BN nanosheets (AFBNNSs) were added into an IL GPE and compared with a standard commercial liquid electrolyte, IL electrolyte, and pure IL GPE. The lithium-ion transference number of the GPE is increased from 0.12 to 0.23 due to the addition of amine-functionalized BNNSs. More importantly, it is found that there is a strong interaction between the amine groups on the BNNSs and the bis(trifluoromethanesulfonyl)imide anions, which significantly limits the anion mobility and consequently increases lithium-ion mobility. This mechanism was revealed with nuclear magnetic resonance analysis for the first time. The AFBNNS addition increased the ionic conductivity to 6.47×10^{-4} S cm^{-1} at 25°C, compared with 3.98×10^{-4} S cm^{-1} of pure GPEs. However, the ionic conductivity decreased after adding AFBNNSs more than 1.5 wt%. This is because more AFBNNSs blocked the conduction pathway and inhibited the conduction of Li ions.

14.4 BN FOR CYCLING STABILITY IMPROVEMENTS

From refurbishing the Li metal anode, interfacial instabilities by high reactivity of Li metal and dendritic Li deposition by uneven Li-ion flux at the interface significantly reduce battery cycling stability. Mechanically robust and electronically insulating BN assists in alleviating non-uniform Li plating and improves interfacial properties between Li metal and electrolytes.

14.4.1 Preventing Dendrite Growth

Dendrite growth, which causes the short circuit and capacity decay of batteries, always happens in metal battery systems during the charge/discharge process. Several strategies have been developed and applied to prevent the metal dendrite growth. Among them, reducing the concentration gradient of Li ions is considered as a promising strategy to prevent metal dendrite growth. Wu et al. employed BNNSs as the additive in a liquid electrolyte to inhibit the Li protrusions [65] (Figure 14.4a). B atoms can interact with Lewis basic anions in the electrolyte, leading to a reduced Li concentration gradient and unified Li deposition to avoid dendrite formation. The reduced Li concentration gradient can be attributed to the increased lithium-ion conductivity and transference number from 0.47 to 0.55 (blank electrolyte and 7 mg mL^{-1} BNNS concentration). The increasing transference number of Li ions caused the elimination of the space charge-induced electric field near the electrolyte/electrode interface. However, the Li-ion transference number drops to 0.50 when the concentration of BNNSs is increased to 10 mg mL^{-1} because fewer edge B atoms are exposed to interact with bis(trifluoromethanesulfonyl) imide anions in the high-concentration BNNS dispersion liquid. As a result, both hysteresis and cycle stability of the symmetric lithium cells with the higher content of BNNS (10 mg mL^{-1}) are worse compared to the cells with lower contents of BNNS (2 and 7 mg mL^{-1}). The reason is that the unstable and non-uniform Li deposition is caused by the BNNS aggregation in the high-concentration dispersion. Besides controlling the ion concentration gradient, the construction

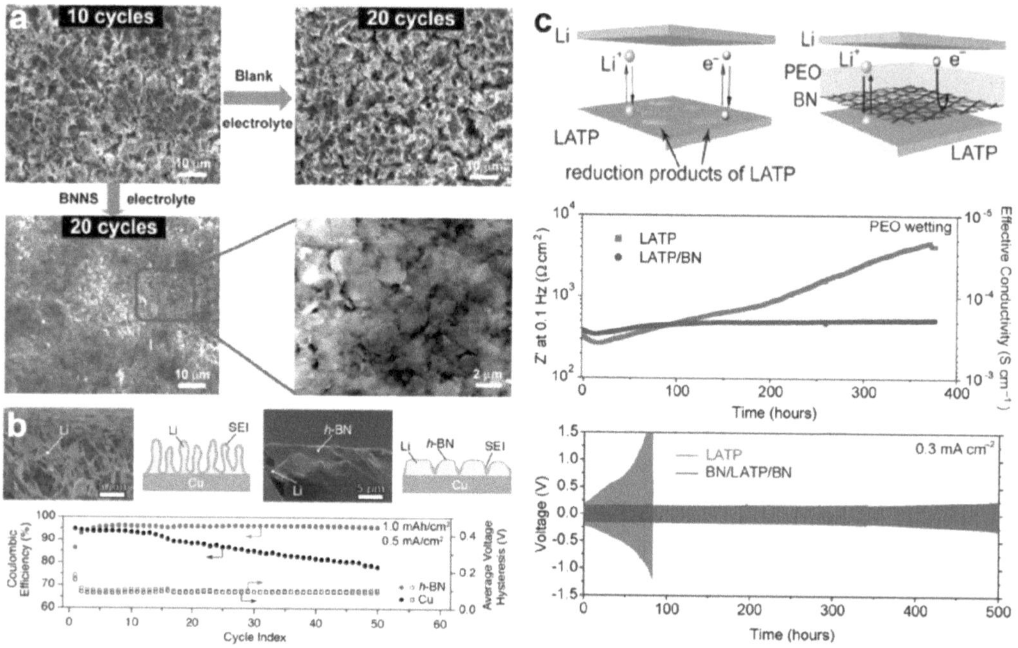

Figure 14.4 (a) The mechanism of lithium dendrite inhibition by BNNSs and the electrochemical performance. (Reprinted with permission [63]. Copyright 2020, Elsevier Ltd.) (b) The lithium dendrite inhibition of *h*-BN SEI and the coulombic efficiency behavior. (Reprinted with permission [66]. Copyright 2019, American Chemical Society.) (c) BN-based nanocomposite layer coating anode and the inhibition mechanism of electronic-conduction-induced dendrite growth. (Reprinted with permission [53]. Copyright 2019, Elsevier Inc.)

of an artificial solid electrolyte interphase (SEI) layer on metal anodes can also benefit the metal dendrite growth inhibition. Yan et al. [66] reported the application of *h*-BN nanofilms as an artificial SEI to prevent lithium dendrite formation (Figure 14.4b). The BN-Li-Cu interlayer structure can effectively inhibit the formation of dendrites and mossy lithium, reduce the reaction with electrolytes, and improve circulation efficiency. This research indicated that the *h*-BN, with chemically inert, stable, and strong mechanical strength, covered the surface of Li metal and prevented the reaction between Li metal and the electrolyte, which contributes to lower electrolyte loss and less dendrite growth [67]. In the study, because of the lithium-ion conductive ability of *h*-BN SEI, the lithium metal can be deposited under *h*-BN after cycling. The high mechanical strength of BN also supported the stability of the anode surface and relieved the damage of traditional SEI at the bare lithium surface caused by the volume change.

In solid-state metal battery systems, the growth of metal dendrites also occurs during cycling and causes the short circuit of batteries. Therefore, increasing interfacial ionic conductivity and applying an artificial SEI layer are considered to be useful strategies to support SSE systems in avoiding dendrite growth. For example, Wen et al. [68] reported that BNNSs were designed with lithium as a composite anode, preventing dendrite growth. In this study, the Li_3N-BN SEI, which *in situ* forms at the interface between SSE and Li metal, provides high ionic conductivity and strong thermodynamic stability, which facilitates interfacial ion transport. This significantly contributes to the suppression of the dendrite formation. Furthermore, the existence of insulator BNNSs with a high Young's modulus will minimize the volume change of lithium anodes, which is critical to obtain dendrite-free lithium anodes. The low electronic conductivity of BNNSs can provide dendrite growth inhibition, according to the electronic-conduction-induced dendrite formation mechanism. Cheng et al. also applied the lower electronic conductivity of BN materials to produce a BN-based nanocomposite layer for the anode coating. It is proven that the BN layer helped the SSE system avoid dendrite growth from the electronic-conduction-induced mechanism [53] (Figure 14.4c). The electronic-conduction-induced dendrite formation mechanism relates to the electronic conduction behavior of SSE and shows the importance of insulation ability at the interface between SSE and lithium.

Furthermore, the unstable surface temperature of the electrodes leads to unstable metal crystallized speed and electrolyte concentration, resulting in dendrite growth. Therefore, uniform thermal distribution on the metal anode is critical to obtain stable metal anodes. Hu et al. [69] reported that the BN flake layer coated on the Zn metal anode in a zinc-based flow battery inhibited the dendrite growth because of a more uniform plating/stripping process. Because of the high thermal conductivity of BN flakes, the homogenizing thermal distribution can be achieved, leading to the uniform plating/stripping process. In addition, the high thermal stability of BN flake also supports the high-temperature application of zinc-based flow batteries.

14.4.2 Stable Interfaces between Lithium Anodes and Solid-State Electrolytes

The interfaces between the SSE and electrodes strongly affect the cycling life of SSE battery systems. Some of the electrodes reacted with the SSE and produced an SEI film with lower ionic conductivity than the SSE and at the same time consumed the active materials on the electrode, resulting in a capacity decay after several cycles. Xu et al. [70] reported that adding BN nanoflakes improved the interfacial behavior between the lithium metal and $Li_7P_3S_{11}$ SSE. In this research, the layered structure of BN flakes partially isolated the direct contact of LPS and Li metal. This isolation inhibited the reaction between the Li metal and SSE but allowed the transfer of Li ions, which improved the interface stability. This interface stability improvement led to a higher initial Coulomb efficiency of 80.4% (LPS-2wt% BN) in Li-S SSE batteries. Furthermore, the cells with LPS-2wt% BN demonstrated a specific capacity of 185.4 mAh g^{-1} after 50 cycles, while the cells with pure LPS showed a specific capacity of 91.5 mAh g^{-1}. Theoretically, more BN in the LPS leads to more suppression of the interface reaction. This interface reaction suppression is beneficial for the electrochemical performance of solid-state Li-S batteries. However, the lower ionic conductivity of BN negatively affects the electrochemical performance when the content of BN is higher than 2 wt%. Therefore, a trade-off BN content exists to achieve a better rate performance of the BN-doped $Li_7P_3S_{11}$ SSE battery.

In ceramic SSE systems, the poor contact between the SSE and the metal anode is more serious because of the poor mechanical flexibility of ceramic SSEs. A possible strategy to solve the contact problem before cycling is to increase the adhesion ability between SSE and melted anode materials. Wen et al. indicated that the viscosity increase in the BNNS-added molten Li increased the adhesion ability between the SSE and melted Li anode [68] (Figure 14.5). This good interfacial connection contributes to lower contact resistance and benefits the performance of batteries. A higher specific capacity of 150 mAh g^{-1} is achieved by batteries with BNNS–lithium anodes, compared with 120 mAh g^{-1} of pure lithium anode batteries at 0.2°C. Furthermore, the dendrite inhibition ability of the BNNS-added anode systems we discussed before also enhanced the interfacial stability and achieved a 100-cycle stable charge/discharge process at 0.5°C. However, similar to the report of Xu et al. [70], the concentration of BNNS cannot increase to higher than 5% because of the lower ionic conductivity of BN.

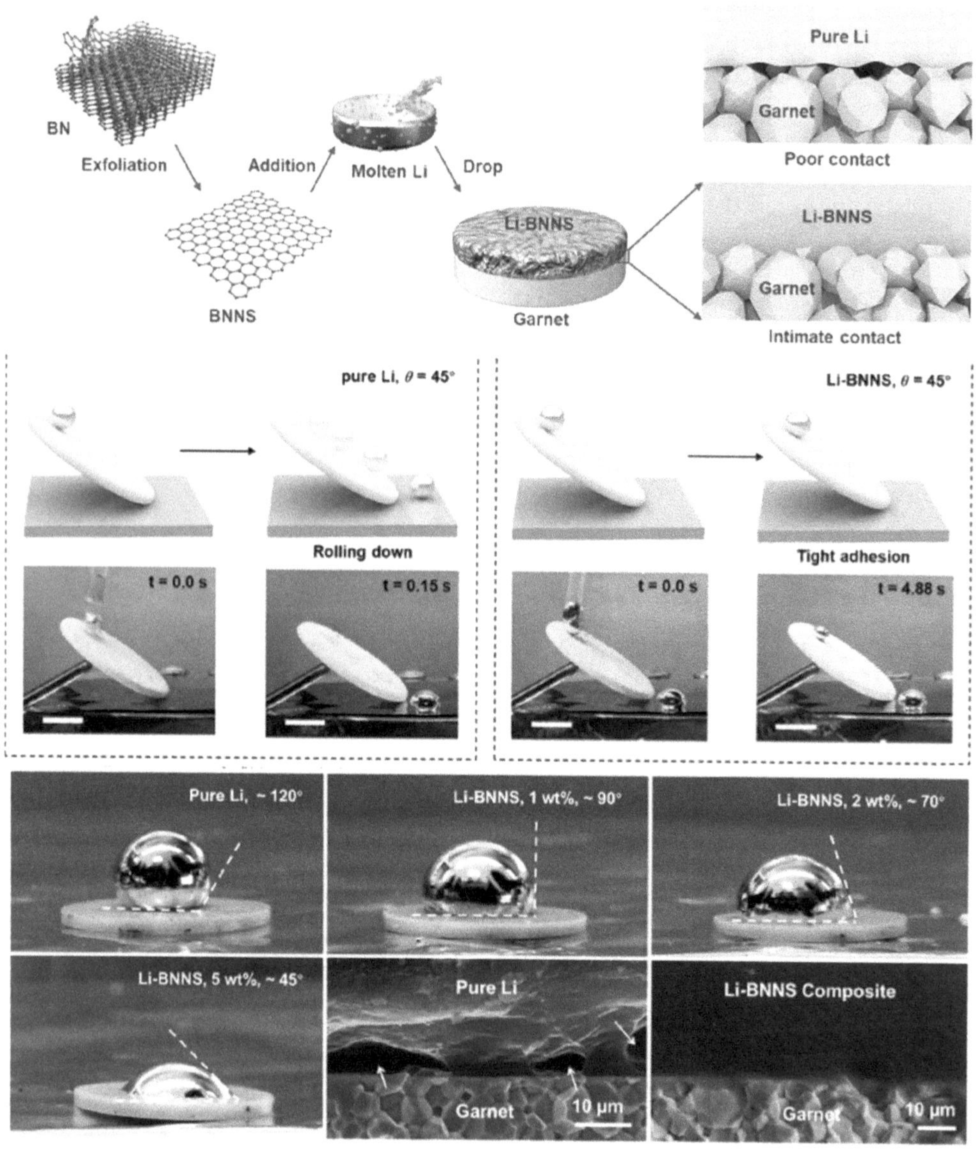

Figure 14.5 The interface behavior, infiltration, and electrochemical behavior of melted Li with the BNNS additive. (Reprinted with permission [68]. Copyright 2019, American Chemical Society.)

14.5 BN FOR BATTERY SAFETY IMPROVEMENTS

Apart from the electrochemical performance issues, future battery demands also require improved safety and reliability. Adopting flammable liquid electrolytes and relevant thermal runaways is still an unresolved issue in batteries. Moreover, recent tragic reports on burned-down electric vehicles emphasize the safety issues. BN with exceptional thermal properties could solve the aforementioned safety crisis.

14.5.1 Separator Protection

Coating a protective layer on the separator is an efficient way to improve thermal stability. Ceramic-enhanced separators show high mechanical strength, very high melting points, and relatively high thermal conductivity [71–74]. Several strategies are generally applied to prepare these separators, including coating or synthesizing the ceramic layers, such as Al_2O_3 and SiO_2, on the separator surfaces or by embedding ceramic powders in the polymeric materials. However, the ceramic layers are heavy and will significantly increase the weight of the separators. Compared with ceramics, BN nanomaterials are lightweight and possess the same thermal stability, which makes them more suitable as the protection layer for separators. Rahman et al. [75] reported a BNNT-protected separator prepared with properly designed BNNTs. The porous channels of separators for lithium-ion diffusion were not blocked (Figure 14.6a). Better thermal stability up to 150°C was achieved, which ensures the safer operation of the cells at high temperatures and also maintains the good performance of the battery. Because the extra heat was absorbed and spread through BNNTs, the rate capability of the cell is improved dramatically by the BNNT-protected separator. Alternately, directly producing the separator with BN is another way to obtain highly thermal stable separators. Waqas et al. [76] reported a highly thermal stable poly (hexagonal BN/poly (vinylidene fluoride-hexafluoropropylene) (PE-BN/PVDF-HFP) microporous separator with a bilayer structure. A wet chemical method was developed to prepare the PE-BN/PVDF-HFP separators. Because of superior mechanical properties and large interaction surface area of the BN, the addition of BN particles promotes the interfacial interaction between PE and PVDF-HFP layers, prevents the separation between layers, and improves the resistance to the damage caused by the dendrite growth. When the separator is annealed at 140°C for 1 hour, the thermal shrinkage rate is 6.6% and the electrolyte absorption rate is 348%.

De Moraes et al. [59] developed the *h*-BN ink formulation with adjustable viscosity, which is compatible with various printing methods, from high viscosity blade coating to low viscosity inkjet printing, as described in the previous section. They also found that the composite

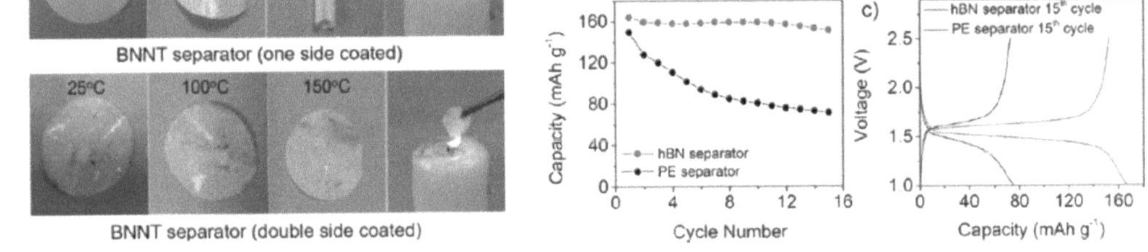

Figure 14.6 (a) The temperature behavior of the BNNT-protected polyolefin separator. (Reprinted with permission [75]. Copyright 2019, Elsevier B.V.) (b) Carbon-coated *h*-BN nanosheets and PVDF separators. (Reprinted with permission [77]. Copyright 2019, American Chemical Society.)

separators with carbon-coated BNNSs and PVDF polymers had high thermal stability, electrolyte wettability, and porosity [77]. BNNSs also provide high thermal stability and ensure a safe operation at 120°C (Figure 14.6b).

14.5.2 Thermal Stability

In solid-state batteries, the SSE is one of the most important components. Because polymer electrolytes are poor thermal conductors, thermal changes in polymer electrolytes can generate localized heat accumulated sites. Uneven heat distribution induces non-uniform lithium-ion flux in the SSE and interfaces. Moreover, localized high-temperature sites accelerate the local lithium plating rate [78]. The local current density is considerably higher in the localized high-temperature sites; consequently, the cell experiences unstable lithium deposition and produces lithium dendrites. In this case, BN nanomaterials could be used as an additive in the solid-state electrolytes to overcome these issues due to the high thermal stability and high thermal conductivity of BN. Yin et al. [79] developed an electrolyte based on polyethylene oxide and 2D BN nanoflakes for improving the thermal response of solid-state Li-S batteries (Figure 14.7). When BN nanoflakes were added into the polyethylene oxide-based SSE, thermal diffusivity (α) and corresponding thermal conductivity (κ) were significantly increased due to the BN's highly thermal conductive nature. This BN nanoflake-added SSE quickly reached the temperature

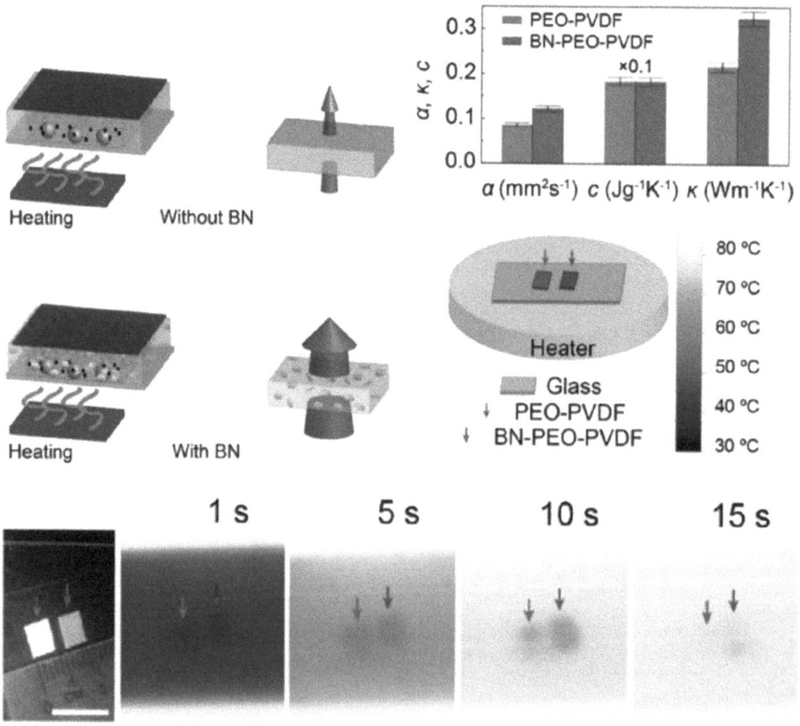

Figure 14.7 Design of the SSE with the BN nanoflake additive and heat transfer test using infrared images. (Reprinted with permission [79]. Copyright 2019, Wiley-VCH GmbH.)

equilibrium under sudden injection of the heating source. Faster thermal equilibrium attained by SSE with BN nanoflakes validated accelerated thermal conduction by BN additives. Moreover, it is shown that the ionic transport properties and mechanical properties of the electrolyte were also improved with the addition of the BNNS (12 wt%). More reversible cathode reactions and uniform lithium deposition were achieved, resulting in better battery performance. Therefore, BN improves the thermal stability of the system and maintains high electrochemical performance due to the excellent thermal stability and chemical stability of BN nanomaterials. Rodrigues et al. [80] reported a composite electrolyte, which is the key to extending the performance of LIBs to high temperatures. The composite was prepared by mixing BN powder and Li bis(trifluoromethanesulfonyl)imide in the ionic liquids. Combined with conventional electrodes, the composite electrolyte can be stably operated over 600 times at 120°C, only with a capacity decay of less than 3%. These results clearly indicate that BN nanomaterials can protect SSE at high temperatures. When the surface of BN nanomaterials is modified, such as SiO_2, the as-prepared surface-modified BN could enhance its fast thermal conduction ability in the SSE [81]. This is because surface modification could reduce interfacial thermal resistance between BN nanomaterials. SSE with surface-modified BN addition displayed the fastest thermal relaxation and retained its shape under high-temperature exposure, demonstrating both thermal stability and heat dissipation.

14.5.3 Heat Management

Based on the consideration of safety, battery thermal management (BTM) systems need to keep batteries in the optimal temperature range to provide an excellent balance between aging and performance. Currently, the BTM systems can quickly transfer and absorb the heat generated by the battery to keep battery packs within the recommended operating temperature range. However, extensive research has been conducted to satisfy the requirements through developing advanced BTM systems. Advanced BTM systems can be classified according to heat transfer media into various types, such as phase change materials, liquid cooling, air cooling, and their combinations. In addition, heat pipes or fans are added to some BTM systems to improve heat transfer efficiency further, ensuring that the battery's heat can be quickly transmitted to the surrounding environment. Traditional heat management with water cooling systems tends to be more complex and cumbersome, which requires more auxiliary equipment such as tanks, pumps, pipes, *etc.* In addition, liquid cooling system maintenance is relatively more complex and expensive.

Various strategies have been investigated to develop advanced BTM systems. Coating the material with excellent thermal conductivity on the surface of the cell is an efficient strategy. Bo et al. [82] developed a sandwich structure of vertically oriented graphene (VG) and BNNSs. BNNSs inhibit electron transfer and benefit electrical insulation (Figure 14.8a). VGs utilize graphene's ultrahigh plane thermal conductivity to adjust the heat dissipation performance by controlling the morphology, which can be further explained by Maxwell-Garnett's effective medium approximation and finite element simulation. In particular, the sandwich structure has good thermal conductivity and good electrical insulation. In addition, it has good thermal stability (up to 100°C) and glass transition temperature up to 119.7°C. Infrared thermal imaging showed that the heating rate of composites was about four times higher than that of polymers. In the actual thermal management process, the sandwich structure can effectively reduce the operating temperature of the light-emitting diode and LIB by ~16.6°C and ~10°C, respectively. As a result, during the charging/discharging process, the battery's capacity retention rate is

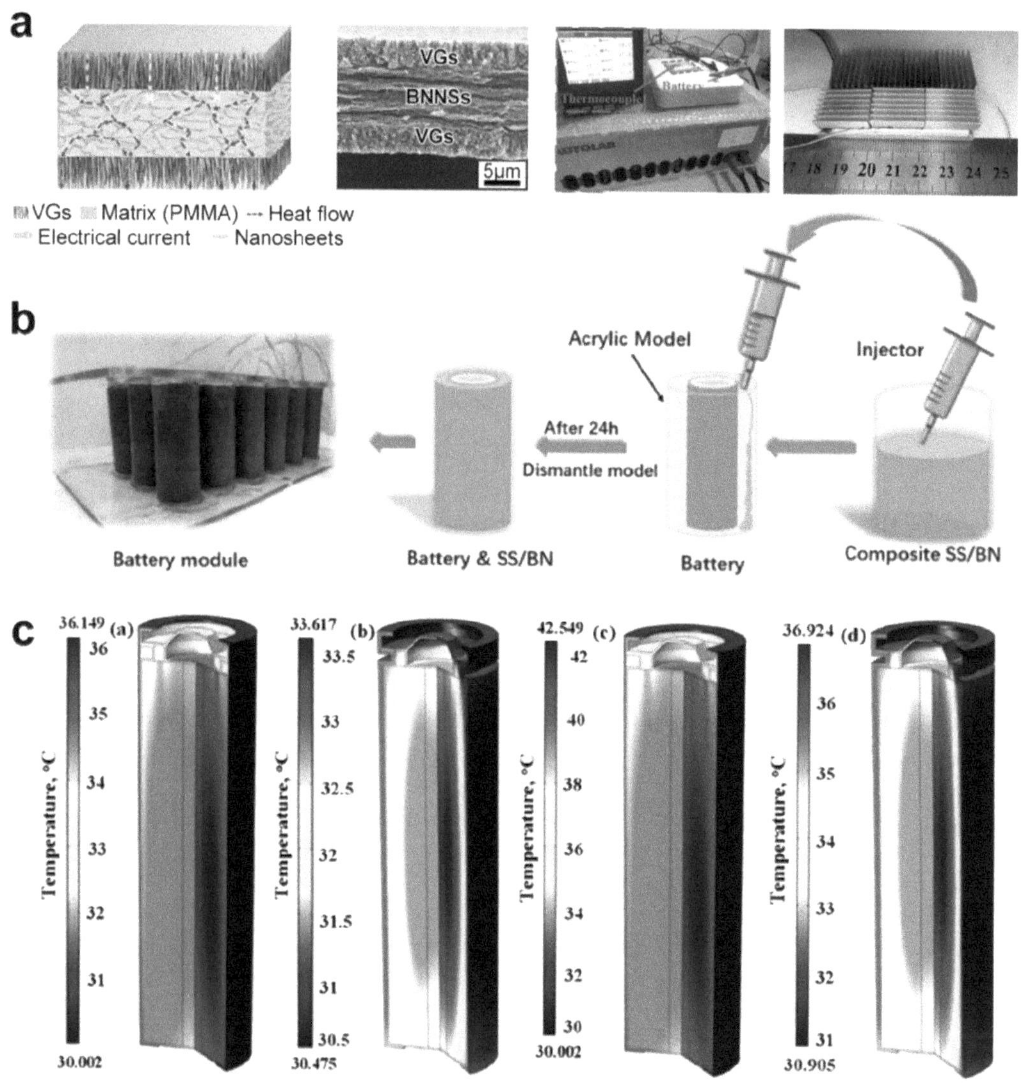

Figure 14.8 (a) VGs-BNNSs-VGs/PMMA. (Reprinted with permission [82]. Copyright 2019, Elsevier Ltd.) (b) Schematic diagram of the preparation process of the battery coating process. (Reprinted with permission [83]. Copyright 2020, Elsevier B.V.) (c) Internal temperature distribution of the 18,650 cells. (Reprinted with permission [84]. Copyright 2014, Elsevier Ltd.)

significantly improved by about 22.2%. Li et al. [83] reported a silicone sealant (SS)/BN composite material, which is cast on the cell surface and then soaked in water (Figure 14.8b). SSs with different BN weight ratios were prepared in the laboratory, specifically 0, 5, 10, and 15 wt%, and their performance was studied through various tests. Compared with the water cooling BTM system, the immersion system has the advantages of long-term anti-corrosion and high-efficiency heat dissipation and is also economical and practical. The performance of forced ventilation cooling, natural air cooling, and immersion cooling methods is compared during the charge/discharge measurement. The results indicate that the maximum temperature of the battery can be significantly reduced by the SS/BN immersion system, and the temperature in the module is balanced at different discharge rates. In the process of 3°C discharge, the maximum temperatures of the three immersed SS modules are all less than 35°C. In addition, a tiny temperature difference within 0.5°C can be kept in the immersed SS/BN module. This study provides a theoretical basis for designing and optimizing the BTM system, especially in underwater or wet conditions. Saw et al. [84] replaced the polymer insulation material with BNNS coating on the battery case via the Taguchi method (Figure 14.8c). The effects of coating thickness, casing surface roughness, and their interactions were investigated with the L9 orthogonal array. The results showed that the batteries are more prone to thermal runaway and thermal aging than the batteries with BN coating. In addition, coating the battery surface with BNNSs also allows the cold plate to attach directly to the battery case for effectively cooling the battery.

14.6 SUMMARY AND OUTLOOK

This chapter summarizes the development of BN nanomaterials for next-generation batteries. BN nanomaterials, including BN nanostructures, doped, functionalized BN, and BN heterostructures, demonstrated new properties and functions, which can help batteries overcome performance and safety challenges. These include but are not limited to (1) electrodes with high capacity and working voltage; (2) high-temperature electrodes; (3) inhibiting dendrite growth; (4) stable interfaces between lithium anodes and SSEs; (5) preventing the shuttle effect; (6) the electrocatalysts for Li-O_2 batteries; (7) improving thermal stability in separators, electrodes, and electrolytes; and (8) more efficient BTM systems. Previous research shows that BN can be used in all components of a battery device and improve their performance and functionalities, especially helping to solve major issues in next-generation batteries such as energy density, cycling life, and safety. Although BN nanomaterials have broad and promising prospects, the progress in related fields is still rather modest, especially compared to carbon analogs. The inertness and high chemical stability of BN prevent immediate post-reaction or modification. So far, a variety of post-synthetic methods have been developed for BN nanomaterials. However, the efficiency and scope are still rarely satisfactory. To tackle this challenge, more advanced strategies are needed. Similar to the rapid development in new battery technologies, new BN structures and properties are created continuously. For example, chemical catalysts from defective BN have been found recently, which could accelerate electrochemical reactions and make the battery more efficient [85]. VdW structures containing BNNSs exhibit superior combination properties than individual nanosheets and have been predicted theoretically to help batteries' capacity and cycling performance. Experimental research needs to be conducted to test these predictions.

Based on the consideration of the above limitations and challenges regarding BN nanomaterials, next-generation batteries, and their future applications, more effort should be put into developing more efficient, economical, and scalable strategies for BN nanomaterials. This will provide an excellent opportunity for the precise design of functional BN nanomaterials based on the property orientation to fulfill the growing need for future high-performance electronic, biological, photoelectronic, and electrochemical devices in the environment, energy, and composite fields.

REFERENCES

1. Y. Guo, H. Li, T. Zhai, Reviving lithium-metal anodes for next-generation high-energy batteries. *Advanced Materials* **29**, 1700007 (2017).
2. X.-G. Yang, T. Liu, C.-Y. Wang, Thermally modulated lithium iron phosphate batteries for mass-market electric vehicles. *Nature Energy* **6**, 176–185 (2021).
3. X. Zeng et al., Commercialization of lithium battery technologies for electric vehicles. *Advanced Energy Materials* **9**, 1900161 (2019).
4. C. Costa et al., Recycling and environmental issues of lithium-ion batteries: Advances, challenges and opportunities. *Energy Storage Materials* **37**, 433–465 (2021).
5. J. Li et al., Polymers in lithium-ion and lithium metal batteries. *Advanced Energy Materials* **11**, 2003239 (2021).
6. B. Scrosati, History of lithium batteries. *Journal of Solid State Electrochemistry* **15**, 1623–1630 (2011).
7. M. Li, J. Lu, Z. Chen, K. Amine, 30 years of lithium-ion batteries. *Advanced Materials*, **30**(33), e1800561 (2018).
8. M. M. Thackeray, C. Wolverton, E. D. Isaacs, Electrical energy storage for transportation—approaching the limits of, and going beyond, lithium-ion batteries. *Energy & Environmental Science* **5**, 7854–7863 (2012).
9. R. Fang et al., More reliable lithium-sulfur batteries: Status, solutions and prospects. *Advanced Materials* **29**, 1606823 (2017).
10. S. Sripad, V. Viswanathan, Evaluation of current, future, and beyond Li-ion batteries for the electrification of light commercial vehicles: Challenges and opportunities. *Journal of the Electrochemical Society* **164**, E3635 (2017).
11. Y. S. Meng, Introduction: Beyond Li-ion battery chemistry. *Chemical Reviews* **120**, 6327 (2020).
12. Y. Tian et al., Promises and challenges of next-generation "beyond Li-ion" batteries for electric vehicles and grid decarbonization. *Chemical Reviews*, **121**, 1623–1669 (2020).
13. K. Kubota, M. Dahbi, T. Hosaka, S. Kumakura, S. Komaba, Towards K-ion and Na-ion batteries as "beyond Li-ion". *The Chemical Record* **18**, 459–479 (2018).
14. J. Qian et al., High rate and stable cycling of lithium metal anode. *Nature Communications* **6**, 1–9 (2015).
15. H. Yang et al., Recent progress and perspective on lithium metal anode protection. *Energy Storage Materials* **14**, 199–221 (2018).
16. D. Lin, Y. Liu, Y. Cui, Reviving the lithium metal anode for high-energy batteries. *Nature Nanotechnology* **12**, 194 (2017).
17. A. Manthiram, Y. Fu, S.-H. Chung, C. Zu, Y.-S. Su, Rechargeable lithium–sulfur batteries. *Chemical Reviews* **114**, 11751–11787 (2014).
18. Z. W. Seh, Y. Sun, Q. Zhang, Y. Cui, Designing high-energy lithium–sulfur batteries. *Chemical Society Reviews* **45**, 5605–5634 (2016).
19. D. Aurbach, E. Zinigrad, H. Teller, P. Dan, Factors which limit the cycle life of rechargeable lithium (metal) batteries. *Journal of the Electrochemical Society* **147**, 1274 (2000).
20. B. Liu, J.-G. Zhang, W. Xu, Advancing lithium metal batteries. *Joule* **2**, 833–845 (2018).
21. J. Liu et al., Pathways for practical high-energy long-cycling lithium metal batteries. *Nature Energy* **4**, 180–186 (2019).
22. S. Xia, X. Wu, Z. Zhang, Y. Cui, W. Liu, Practical challenges and future perspectives of all-solid-state lithium-metal batteries. *Chem* **5**, 753–785 (2019).
23. Y.-C. Lu et al., Lithium–oxygen batteries: Bridging mechanistic understanding and battery performance. *Energy & Environmental Science* **6**, 750–768 (2013).
24. R. Cao et al., The mechanisms of oxygen reduction and evolution reactions in nonaqueous lithium–oxygen batteries. *ChemSusChem* **7**, 2436–2440 (2014).
25. W.-J. Kwak et al., Lithium–oxygen batteries and related systems: Potential, status, and future. *Chemical Reviews* **120**, 6626–6683 (2020).
26. P. Wang et al., Atomically dispersed cobalt catalyst anchored on nitrogen-doped carbon nanosheets for lithium-oxygen batteries. *Nature Communications* **11**, 1–11 (2020).

27. I. Ihsanullah, Boron nitride-based materials for water purification: Progress and outlook. *Chemosphere*, **263**, 127970 (2020).
28. D. Liu, M. Zhang, L. He, Y. Chen, W. Lei, Layer-by-layer assembly fabrication of porous boron nitride coated multifunctional materials for water cleaning. *Advanced Materials Interfaces* **4**, 1700392 (2017).
29. W. Lei, D. Portehault, D. Liu, S. Qin, Y. Chen, Porous boron nitride nanosheets for effective water cleaning. *Nature Communications* **4**, 1–7 (2013).
30. H. Cho et al., Scalable gas-phase purification of boron nitride nanotubes by selective chlorine etching. *Chemistry of Materials* **32**, 3911–3921 (2020).
31. S. Lin et al., Hidden porous boron nitride as a high-efficiency membrane for hydrogen purification. *Physical Chemistry Chemical Physics* **22**, 22778–22784 (2020).
32. I. V. Sukhorukova et al., Boron nitride nanoparticles with a petal-like surface as anticancer drug-delivery systems. *ACS Applied Materials & Interfaces* **7**, 17217–17225 (2015).
33. Z. Khatti, S. M. Hashemianzadeh, Boron nitride nanotube as a delivery system for platinum drugs: Drug encapsulation and diffusion coefficient prediction. *European Journal of Pharmaceutical Sciences* **88**, 291–297 (2016).
34. G. Ciofani, Potential applications of boron nitride nanotubes as drug delivery systems. *Expert Opinion on Drug Delivery* **7**, 889–893 (2010).
35. W. Balmain, XLVI. Observations on the formation of compounds of boron and silicon with nitrogen and certain metals. *The London, Edinburgh, and Dublin Philosophical Magazine and Journal of Science* **21**, 270–277 (1842).
36. D. Golberg et al., Boron nitride nanotubes and nanosheets. *ACS Nano* **4**, 2979–2993 (2010).
37. J. Guan et al., Epoxy resin nanocomposites with hydroxyl (OH) and amino (NH_2) functionalized boron nitride nanotubes. *Nanocomposites* **4**, 10–17 (2018).
38. L. H. Li, Y. Chen, A. M. Glushenkov, Boron nitride nanotube films grown from boron ink painting. *Journal of Materials Chemistry* **20**, 9679–9683 (2010).
39. R. Arenal, A. Lopez-Bezanilla, Boron nitride materials: An overview from 0D to 3D (nano) structures. *Wiley Interdisciplinary Reviews: Computational Molecular Science* **5**, 299–309 (2015).
40. M. Foroutan, S. J. Fatemi, S. M. Fatemi, A mini-review on dispersion and functionalization of boron nitride nanotubes. *Journal of Nanostructure in Chemistry*, **10**, 1–10 (2019).
41. L. H. Li, Y. Chen, Atomically thin boron nitride: Unique properties and applications. *Advanced Functional Materials* **26**, 2594–2608 (2016).
42. N. Izyumskaya et al., Recent development of boron nitride towards electronic applications. *Advanced Electronic Materials* **3**, 1600485 (2017).
43. S. Yu et al., Boron nitride-based materials for the removal of pollutants from aqueous solutions: A review. *Chemical Engineering Journal* **333**, 343–360 (2018).
44. A. Merlo, V. Mokkapati, S. Pandit, I. Mijakovic, Boron nitride nanomaterials: Biocompatibility and bio-applications. *Biomaterials Science* **6**, 2298–2311 (2018).
45. D. Gonzalez-Ortiz, C. Salameh, M. Bechelany, P. Miele, Nanostructured boron nitride–based materials: Synthesis and applications. *Materials Today Advances* **8**, 100107 (2020).
46. J. D. Caldwell et al., Photonics with hexagonal boron nitride. *Nature Reviews Materials* **4**, 552–567 (2019).
47. Q. Weng, X. Wang, X. Wang, Y. Bando, D. Golberg, Functionalized hexagonal boron nitride nanomaterials: Emerging properties and applications. *Chemical Society Reviews* **45**, 3989–4012 (2016).
48. R. Han et al., Functionalised hexagonal boron nitride for energy conversion and storage. *Journal of Materials Chemistry A* **8**, 14384–14399 (2020).
49. V. Sharma, H. L. Kagdada, P. K. Jha, P. Śpiewak, K. J. Kurzydłowski, Thermal transport properties of boron nitride based materials: A review. *Renewable and Sustainable Energy Reviews* **120**, 109622 (2020).
50. M. Minakshi, M. G. Blackford, Electrochemical characteristics of B_4C or BN added MnO_2 cathode material for alkaline batteries. *Materials Chemistry and Physics* **123**, 700–705 (2010).
51. P. Bhauriyal, G. Bhattacharyya, K. S. Rawat, B. Pathak, Graphene/hBN heterostructures as high-capacity cathodes with high voltage for next-generation aluminum batteries. *The Journal of Physical Chemistry C* **123**, 3959–3967 (2019).

52. H. Li, R. Y. Tay, S. H. Tsang, W. Liu, E. H. T. Teo, Reduced graphene oxide/boron nitride composite film as a novel binder-free anode for lithium ion batteries with enhanced performances. *Electrochimica Acta* **166**, 197–205 (2015).
53. Q. Cheng et al., Stabilizing solid electrolyte-anode interface in Li-metal batteries by boron nitride-based nanocomposite coating. *Joule* **3**, 1510–1522 (2019).
54. K. Abraham, Z. Jiang, A polymer electrolyte-based rechargeable lithium/oxygen battery. *Journal of the Electrochemical Society* **143**, 1 (1996).
55. K. Uosaki et al., Boron nitride nanosheet on gold as an electrocatalyst for oxygen reduction reaction: Theoretical suggestion and experimental proof. *Journal of the American Chemical Society* **136**, 6542–6545 (2014).
56. M. Lee, Y. Hwang, K.-H. Yun, Y.-C. Chung, Cathode reaction mechanism on the h-BN/Ni (111) heterostructure for the lithium-oxygen battery. *Journal of Power Sources* **307**, 379–384 (2016).
57. J. Zhao, Z. Chen, Carbon-doped boron nitride nanosheet: An efficient metal-free electrocatalyst for the oxygen reduction reaction. *The Journal of Physical Chemistry C* **119**, 26348–26354 (2015).
58. C. Chowdhury, A. Datta, Doped boron nitride surfaces: Potential metal free bifunctional catalysts for non-aqueous Li–O_2 batteries. *Physical Chemistry Chemical Physics* **20**, 16485–16492 (2018).
59. A. C. M. de Moraes et al., Ion-conductive, viscosity-tunable hexagonal boron nitride nanosheet inks. *Advanced Functional Materials* **29**, 1902245 (2019).
60. M. Li et al., Graphene-analogues boron nitride nanosheets confining ionic liquids: A high-performance quasi-liquid solid electrolyte. *Small* **12**, 3535–3542 (2016).
61. W. J. Hyun et al., High-modulus hexagonal boron nitride nanoplatelet gel electrolytes for solid-state rechargeable lithium-ion batteries. *ACS Nano* **13**, 9664–9672 (2019).
62. J. Shim et al., 2D boron nitride nanoflakes as a multifunctional additive in gel polymer electrolytes for safe, long cycle life and high rate lithium metal batteries. *Energy & Environmental Science* **10**, 1911–1916 (2017).
63. Z. Zhang, A. R. Gonzalez, K. L. Choy, Boron nitride enhanced garnet-type ($Li_{6.25}A_{l0.25}La_3Zr_2O_{12}$) ceramic electrolyte for an all-solid-state lithium-ion battery. *ACS Applied Energy Materials* **2**, 7438–7448 (2019).
64. D. Kim et al., Amine-functionalized boron nitride nanosheets: A new functional additive for robust, flexible ion gel electrolyte with high lithium-ion transference number. *Advanced Functional Materials* **30**, 1910813 (2020).
65. J. Wu et al., Electrolyte with boron nitride nanosheets as leveling agent towards dendrite-free lithium metal anodes. *Nano Energy* **72**, 104725 (2020).
66. K. Yan et al., Ultrathin two-dimensional atomic crystals as stable interfacial layer for improvement of lithium metal anode. *Nano Letters* **14**, 6016–6022 (2014).
67. K. J. Harry, D. T. Hallinan, D. Y. Parkinson, A. A. MacDowell, N. P. Balsara, Detection of subsurface structures underneath dendrites formed on cycled lithium metal electrodes. *Nature Materials* **13**, 69–73 (2014).
68. J. Wen et al., Highly adhesive Li-BN nanosheet composite anode with excellent interfacial compatibility for solid-state Li metal batteries. *ACS Nano* **13**, 14549–14556 (2019).
69. J. Hu, M. Yue, H. Zhang, Z. Yuan, X. Li, A boron nitride nanosheets composite membrane for a long-life zinc-based flow battery. *Angewandte Chemie International Edition* **59**, 6715–6719 (2020).
70. X. Xu, L. Wang, H. Fei, L. Ci, Boron nitride doped $Li_7P_3S_{11}$ solid electrolyte with improved interfacial compatibility and application in all-solid-state Li/S battery. *Journal of Materials Science: Materials in Electronics* **30**, 19119–19125 (2019).
71. X. Qi et al., Covalent grafting interface engineering to prepare highly efficient and stable polypropylene/mesoporous SiO_2 separator for Li-ion batteries. *Applied Surface Science* **541**, 148405 (2021).
72. J. Liang et al., A nano-shield design for separators to resist dendrite formation in lithium-metal batteries. *Angewandte Chemie International Edition* **59**, 6561–6566 (2020).
73. E. Ceylan Cengiz et al., Investigation of the effect of using Al_2O_3–nafion barrier on room-temperature Na–S batteries. *The Journal of Physical Chemistry C* **121**, 15120–15126 (2017).
74. X. Jiang, X. Zhu, X. Ai, H. Yang, Y. Cao, Novel ceramic-grafted separator with highly thermal stability for safe lithium-ion batteries. *ACS Applied Materials & Interfaces* **9**, 25970–25975 (2017).

75. M. M. Rahman et al., High temperature and high rate lithium-ion batteries with boron nitride nanotubes coated polypropylene separators. *Energy Storage Materials* **19**, 352–359 (2019).
76. M. Waqas et al., High-performance PE-BN/PVDF-HFP bilayer separator for lithium-ion batteries. *Advanced Materials Interfaces* **6**, 1801330 (2019).
77. A. C. M. De Moraes et al., Phase-inversion polymer composite separators based on hexagonal boron nitride nanosheets for high-temperature lithium-ion batteries. *ACS Applied Materials & Interfaces* **12**, 8107–8114 (2020).
78. Y. Zhu et al., Fast lithium growth and short circuit induced by localized-temperature hotspots in lithium batteries. *Nature Communications* **10**, 2067 (2019).
79. X. Yin et al., Thermal conductive 2D boron nitride for high-performance all-solid-state lithium–sulfur batteries. *Advanced Science* **7**, 2001303 (2020).
80. M. T. F. Rodrigues et al., Hexagonal boron nitride-based electrolyte composite for Li-ion battery operation from room temperature to 150°C. *Advanced Energy Materials* **6**, 1600218 (2016).
81. X. Zhang, W. Guo, L. Zhou, Q. Xu, Y. Min, Surface-modified boron nitride as a filler to achieve high thermal stability of polymer solid-state lithium-metal batteries. *Journal of Materials Chemistry A* **9**, 20530–20543 (2021).
82. Z. Bo et al., Bifunctional sandwich structure of vertically-oriented graphenes and boron nitride nanosheets for thermal management of LEDs and Li-ion battery. *Applied Thermal Engineering* **150**, 1016–1027 (2019).
83. X. Li et al., Evaluation of lithium battery thermal management using sealant made of boron nitride and silicone. *Journal of Power Sources* **451**, 227820 (2020).
84. L. Saw, Y. Ye, A. Tay, Feasibility study of boron nitride coating on lithium-ion battery casing. *Applied Thermal Engineering* **73**, 154–161 (2014).
85. Z. Liu et al., Boron radicals identified as the source of the unexpected catalysis by boron nitride nanosheets. *ACS Nano* **13**, 1394–1402 (2019).

Chapter 15

Functionalized Boron Nitride Nanosheets for Developing Lithium–Sulfur Batteries

Ye Fan, Tao Tao, Md Mokhlesur Rahman, and Ying Ian Chen

15.1 INTRODUCTION

In case the current Li-ion battery technology falls short in meeting the demands of high-energy and high-power applications, such as future transportation and renewable energy storage, the development of rechargeable energy storage systems with higher energy density is imperative [1]. Li-S batteries with 2600 Wh kg^{-1} theoretical energy density [2,3] represent a promising frontier in electrochemical energy storage systems, offering attributes such as high-power density, long cycle life, efficiency, and cost-effectiveness. Sulfur, known for its multi-electron transfer redox chemistry, presents a compelling option for higher power density compared to single-electron reaction systems. The redox pathway between the lithium metal anode and the sulfur cathode can be described below:

$$S_8 + Li^+ + e^- \rightarrow Li_2S_x \left(2.4 - 2.1\,V\right) \tag{15.1}$$

$$Li_2S_x + Li^+ + e^- \rightarrow Li_2S_2 \text{ and/or } Li_2S \left(2.1 - 1.5\,V\right) \tag{15.2}$$

In organic electrolytes, lithium PSs form in the solid sulfur cathode through Equation (15.1) and are subsequently converted into insoluble lithium PSs (Li_2S_2 and/or Li_2S) at the end of the discharge process [4] via Equation (15.2). Consequently, Li-S batteries boast a significantly higher theoretical energy density among various energy storage devices. However, challenges persist, including the 'shuttle effect', which leads to capacity decay and lower Coulombic efficiency [5–8]. Moreover, issues like poor conductivity of sulfur and volume variation during cycling further hinder Li-S battery performance. The key issues of Li-S batteries are as follows:

i. The low conductivity of sulfur and its sulfides restricts electron transport in the cathode, leading to a reduced initial specific capacity [3,9].
ii. The PS shuttle effect, where sulfur species move between electrodes, causes poor Coulombic efficiency and short cycle life [3,9–11].
iii. The substantial volume changes in the sulfur electrode during cycling disrupt the composite electrode's structure, resulting in significant capacity degradation [12–14].

DOI: 10.1201/9781003647812-19

To address these challenges, considerable research has focused on modifying electrode architecture [15–28]. Despite efforts to enhance Li-S batteries, such as developing interlayers and separators, success remains limited. Thus, significant challenges and opportunities persist in Li-S battery development [29–40]. In this context [41–43], FBN emerges as a promising solution. FBN, with its stable structure and various functional groups like amino, offers the potential to mitigate the shuttle effect effectively. By introducing FBN-based interlayers and separators, researchers aim to improve Li-S battery performance. Similarly, in recent years, attention has turned to developing separators capable of inhibiting the shuttle effect [34,44,45]. The introduction of an FBN (carbonate) separator, coated onto a Celgard 2400 separator, represents a significant advancement of this multifunctional separator, which shows promise in mitigating the shuttle effect and enhancing lithium-ion diffusion in Li-S batteries.

Furthermore, to address the insulating nature of sulfur and sulfides [46–48], researchers have explored liquid cathode systems, such as Li-PS liquid cathodes [49–53]. In traditional Li-S batteries, the detailed reactions are described below.

$$S_8 + 2Li \rightarrow Li_2S_8 \quad (15.3)$$

$$Li_2S_8 + 2Li \rightarrow Li_2S_{8-x} + Li_2S_x \quad (15.4)$$

$$2Li_2S_x + (2x-4)Li \rightarrow nLi_2S_2 \quad (15.5)$$

$$Li_2S_x + (2x-2)Li \rightarrow nLi_2S \quad (15.6)$$

$$Li_2S_2 + 2Li \rightarrow 2Li_2 \quad (15.7)$$

To avoid forming sulfur with extremely low conductivity (5.0×10^{-30} S cm^{-1} at 25°C), the liquid electrode mechanism is assumed to proceed only via Equations (15.4)–(15.7) during discharge and Equations (15.7) to (15.4) during charge. Based on these reactions, the theoretical capacity of the Li-PS system is calculated to be 1470 mA h g^{-1}, equivalent to 87.5% of sulfur capacity (1680 mA h g^{-1}), with the dischargeable S_8^{2-} ion as the functional material. However, the PS 'shuttle effect' makes it challenging to confine PS ions to the cathode side. Consequently, sulfur loading on the electrode remains essential, even when traditional methods such as electrode modification are applied to suppress the shuttle effect. The Li-PS batteries employ the FBN (carbonate) separator to elucidate their mechanism, possibility, and performance compared to traditional Li-S solid cathode systems.

In summary, the development of Li-S batteries stands at a critical juncture, with advancements in BN materials offering promising avenues to overcome existing challenges and unlock the full potential of this transformative energy storage technology.

15.2 AMINO-FUNCTIONALIZED BN (A-FBN) NANOSHEETS FOR LI-S BATTERIES

Li-S batteries face a well-known challenge known as the 'shuttle effect', which leads to rapid capacity decay and hinders their future application. In this chapter, a-FBN is shown to have PS absorption ability, which helps inhibit the shuttle effect and enhances the electrochemical performance of Li-S batteries.

15.2.1 Synthesis of a-FBN Nanosheets

a-FBN was synthesized following a previously reported procedure [54]. In summary, h-BN (Momentive Performance Materials Inc.) and urea (Sigma-Aldrich) were mixed at a weight ratio of 1:60 in a steel milling container and processed using a planetary ball mill (Pulverisette 7, Fritsch) under a nitrogen atmosphere. The milling was carried out at 700 rpm for 20 hours at room temperature, enabling efficient large-scale exfoliation of h-BN due to the high rotation speed. Variations in the weight ratio (1:20 and 1:100) and milling time (10 and 30 hours) were also investigated. The final product was designated as a-FBN.

15.2.2 Interaction between a-FBN Nanosheets and PS

The adsorption behavior of PS onto various BN materials, including commercial BN powder (CBN), porous BN (PBN), and a-FBN nanosheets, was examined. Scanning electron microscopy (SEM) images (Figure 15.1) revealed distinctive morphologies: CBN comprised stacked particles sized between 2 and 10 μm (Figure 15.1a), while PBNs exhibited a porous nanosheet structure, and a-FBN displayed a fluffy and lamellar morphology (Figure 15.1b and c). Among them, PBNs demonstrated the highest surface area, reaching up to 1427 $m^2 g^{-1}$.

In theory, the materials with higher surface area exhibit higher adsorption ability in physical. However, the PS adsorption onto different BN materials, which was evaluated using UV spectroscopy (Figure 15.2a), produced a different result from the theory. Spectral analysis between 200 and 800 nm exhibited peaks solely within the 200–300 nm range, associated with a high degree of PS (S_8^{2-}) ions. Comparing peak intensities for various BN materials, CBN showed no shift, indicating minimal PS adsorption due to its large particle size, small surface area, and unfunctionalized surfaces. PBN exhibited weaker PS peaks, suggesting partial PS absorption owing to its nanosheet structure and high surface area. Notably, PS peaks almost disappeared for a-FBN, indicating substantial PS ion absorption. Compared with CBN and PBN, a-FBN had a moderate surface area but numerous positive functional groups (NH_2) attached to its surface and edges, which seemed pivotal for PS ion adsorption. Additionally, color contrast photos further demonstrated a-FBN's superior adsorption capability for PS (S_8^{2-}) ions compared to other materials (Figure 15.2b).

The maximum adsorption capacity of PS (S_8^{2-}) ions by a-FBN was determined using the Langmuir adsorption isotherm model (Figure 15.2c), yielding a value of 487 mg g^{-1}. Raman spectroscopy further confirmed the adsorption of PS ions by a-FBN (Figure 15.2d). All samples exhibited a characteristic peak at 1364 cm^{-1}, corresponding to the BN vibration mode (E_{2g}) [55]. Compared to pure BN, notable changes were observed in the Raman spectrum of a-FBN with adsorbed PS, indicating interactions between a-FBN and sulfur. Peaks below 600 cm^{-1} were primarily attributed to sulfur signals from sulfur-containing compounds [56]. Additionally, the peaks at 822 and 874 cm^{-1} corresponded to B–S bonds [57], confirming the presence of sulfur-based compounds in a-FBN. The X-ray diffraction patterns of a-FBN with adsorbed PS exhibited peaks corresponding to $S_7(NH)$, $S_9(NH)$, and $S_3(NH)$ around 16°, 21°, 24°, and 25°, further confirming the PS ion absorption mechanism with a-FBN (Figure 15.3).

To comprehend the mechanism of negatively charged PS ion adsorption by a-FBN, various ion adsorption studies were conducted (Figure 15.4). Typical examples of positive and negative ions, Cu^{2+} and MnO_4^{2-}, were chosen. A new peak corresponding to Cu^{2+} emerges between 400 and 500 nm after the addition of a-FBN, attributed to the increased local concentration of Cu^{2+} resulting from the positively charged groups introduced by a-FBN (Figure 15.4a). In contrast, the peak

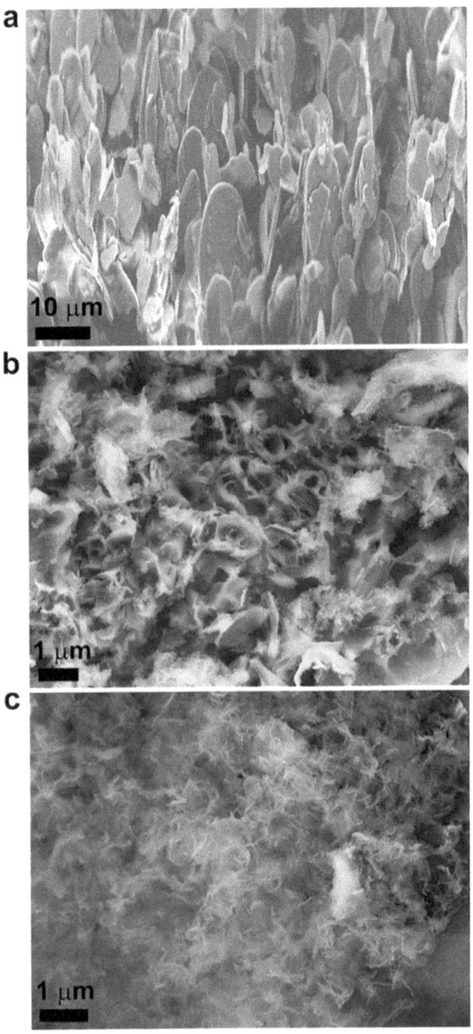

Figure 15.1 SEM images of (a) CBN, (b) PBN, and (c) a-FBN. (Reprinted with permission from Ref [41]. Copyright 2018, John Wiley & Sons, Inc.)

of MnO_4^{2-} decreases after adding a-FBN, similar to the behavior observed with PS (Figure 15.4b). This indicates that negatively charged ions can be effectively adsorbed by the functionalized a-FBN due to its positively charged groups.

15.2.3 Performance of Li-S Batteries with an a-FBN Nanosheet Interlayer

The configuration of the Li–S cell with an a-FBN/G interlayer is shown in Figure 15.5a. This thin, lightweight interlayer is applied to the cathode surface by coating with an a-FBN/G film, where a-FBN comprises 40 wt% of the hybrid layer, with a CNT/S cathode composite serving as the host material [58]. The SEM image in Figure 15.5b displays a cross-section of the uniform interlayer structure, approximately 1 μm thick. As shown in the SEM image from the interlayer surface

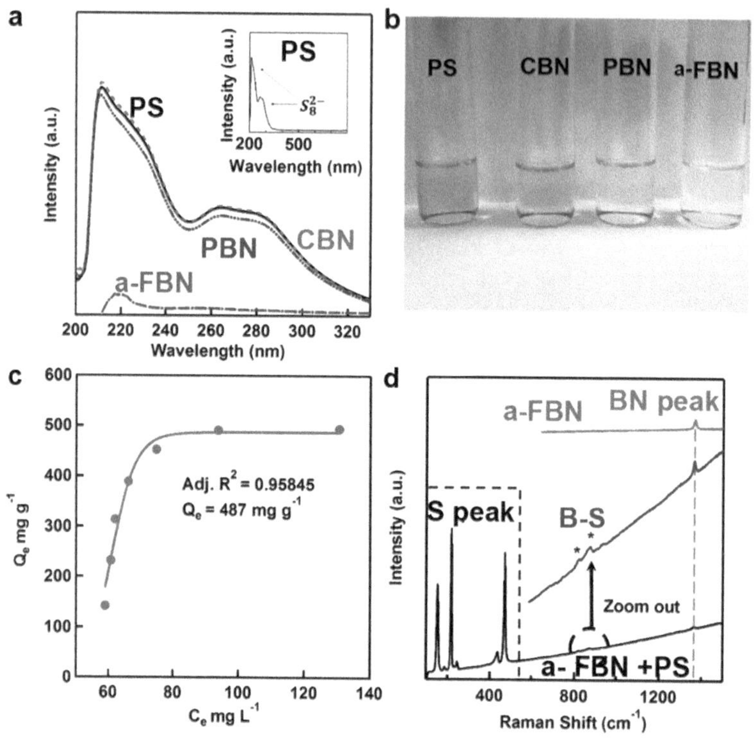

Figure 15.2 PS adsorption performance of different BN materials. (a) UV spectra of a PS solution before and after adsorption by CBN powder, PBN nanosheets, and a-FBN. (b) Visual comparison of PS in DME/DOL (1:1) solution before and after adsorption by CBN, PBN, and a-FBN. (c) Maximum adsorption capacity of a-FBN for S_8^{2-} ions. (d) Raman spectra of a-FBN before and after PS adsorption. (Reprinted with permission from Ref [41]. Copyright 2018, John Wiley & Sons, Inc.)

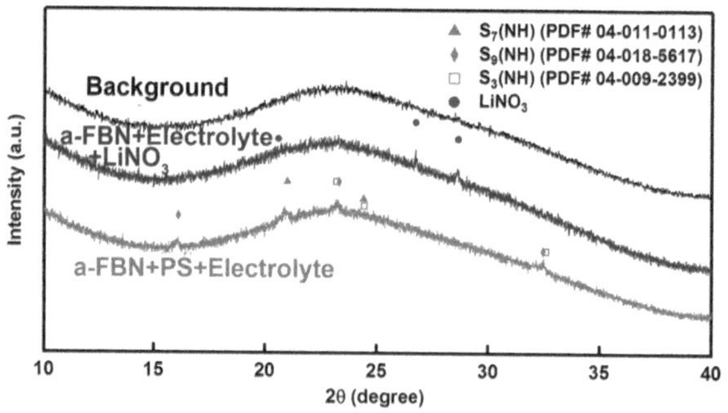

Figure 15.3 X-ray diffraction patterns for PS adsorbed by a-FBN and a-FBN with electrolyte. (Reprinted with permission from Ref [41]. Copyright 2018, John Wiley & Sons, Inc.)

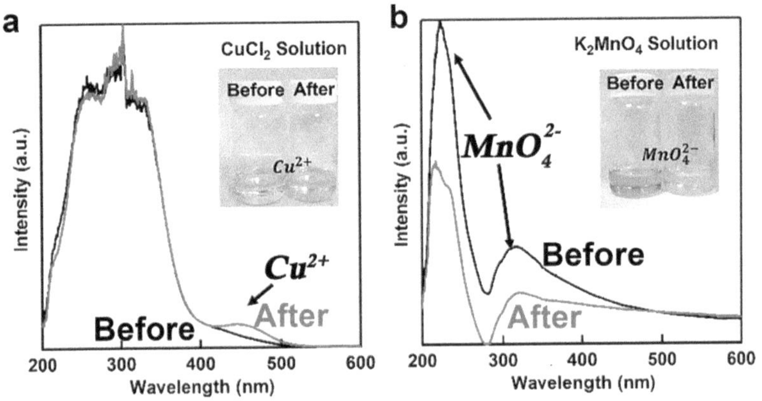

Figure 15.4 Characterization of ion-selective adsorption by FBN nanosheets. (a) UV spectra of $CuCl_2$ solution in DME/DOL (1:1) before and after adsorption by a-FBN. (b) UV spectra of K_2MnO_4 solution in DME/DOL (1:1) before and after adsorption by a-FBN. (Reprinted with permission from Ref [41]. Copyright 2018, John Wiley & Sons, Inc.)

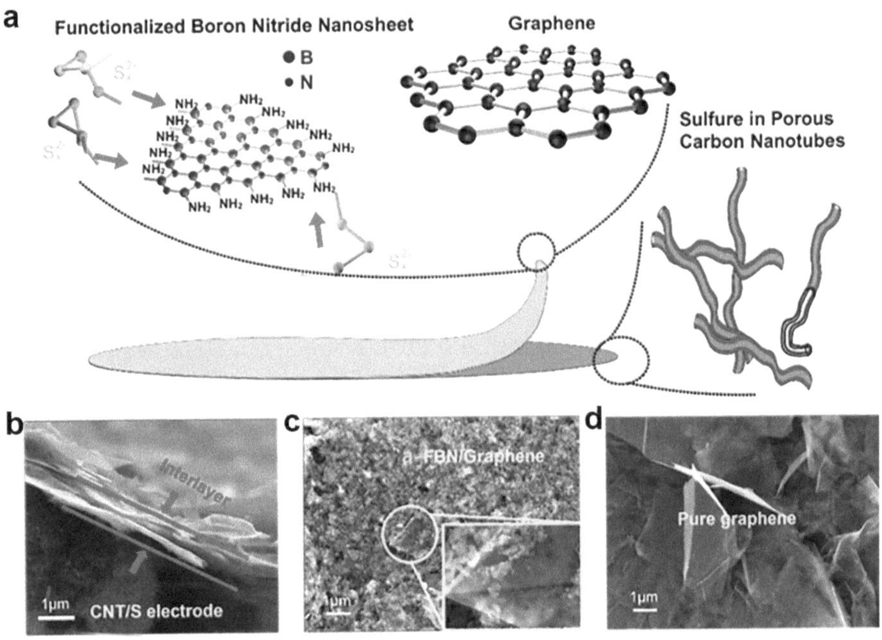

Figure 15.5 Electrode Structure. (a) Schematic illustration of a Li–S cell featuring an a-FBN/G interlayer. (b) SEM image showing the cross-section of the a-FBN/G interlayer and cathode. (c) SEM image of the surface of the a-FBN/G interlayer. (d) SEM image of the surface of a pure graphene interlayer. (Reprinted with permission from Ref [41]. Copyright 2018, John Wiley & Sons, Inc.)

(Figure 15.5c), the a-FBNs are well mixed with graphene sheets. The small a-FBNs possess abundant surfaces and edges, providing numerous adsorption sites to trap PS anions during cycling. Additionally, the SEM images in Figure 15.5c and d reveal that the graphene flakes are significantly larger, about 50 times the size of the a-FBNs.

To evaluate electrochemical performance, Li–S cells underwent galvanostatic charge–discharge within the voltage range of 1.6–2.8 V. Figure 15.6a illustrates the rate performance of cells with varying a-FBN/G interlayer contents and without an interlayer. At a low current density of 0.2C, the initial discharge capacity of the cells increased with higher a-FBN content. The capacities reached 800 mA h g^{-1} for 10 wt% a-FBN, 950 mA h g^{-1} for 20 wt% a-FBN, 1000 mA h g^{-1} for 30 wt% a-FBN, 1100 mA h g^{-1} for 40 wt% a-FBN, and 1100 mA h g^{-1} for 50 wt% a-FBN, respectively. Among these, the interlayer with 40 wt% a-FBN exhibited the best rate capability.

Further investigation in Figure 15.6b compares the rate capabilities of cells with a 40 wt% a-FBN/G interlayer and without an interlayer. While the cell without an interlayer suffered from significant capacity decay, the cell with 40 wt% a-FBN/interlayer demonstrated superior high-rate capability, retaining capacities of 1125, 881, 777, and 668 mA h g^{-1} at current rates of 0.2, 0.5, 1, and 2C, respectively. Even at a high current rate of 6C, the cell maintained a capacity of 556 mA h g^{-1} and a high Coulombic efficiency of 98.4%. Upon returning to 0.2C, a capacity of 912 mA h g^{-1} was recovered, indicating high-capacity retention.

Discharge/charge voltage profiles obtained at different C rates for selected cycles (1st, 5th, 10th, 15th, and 20th) are depicted in Figure 15.6c and d. All cells displayed two distinct plateaus at lower rates of 0.2C and 0.5C, corresponding to the conversion of sulfur to long-chain lithium PSs and the formation of solid lithium sulfides, typical in Li–S batteries with carbon/sulfur cathodes. At higher rates, such as 4C, the charge plateau was not observed in the cell without an interlayer, whereas the a-FBN/G interlayer cell showed clear plateaus, indicating better high-rate performance.

To assess long cycling performance, cells with and without an a-FBN/G interlayer were tested at 1C and 3C for 1000 cycles (Figure 15.7a). The C rates in this study are based on the mass and theoretical capacity of sulfur (1C = 1675 mA g^{-1}, 3C = 5025 mA g^{-1}). The cell without an interlayer showed significant capacity loss, with the capacity dropping to 568, 400, 325, and 278 mA h g^{-1} at the 100th, 300th, 600th, and 1000th cycles, respectively, yielding a capacity retention of just ≈48%, typical due to the shuttle effect. In contrast, the cell with an a-FBN/G interlayer maintained an initial discharge capacity of about 1100 mA h g^{-1} (specific energy density 2500 W h kg^{-1}), with a moderate decline in capacity during the first 100 cycles, attributed to the activation process inherent in Li-S batteries [33]. After this, the cycling performance remained stable with a high Coulombic efficiency. The capacity stayed above 700 mA h g^{-1} (specific energy density 1440 W h kg^{-1}) at 1C, with an average Coulombic efficiency of 99.2% over 1000 cycles, resulting in a minimal average capacity decay rate of 0.0067% per cycle. When tested at a higher current of 3C (5025 mA g^{-1}), the cell with the a-FBN/G interlayer exhibited stable cycling, maintaining 558 mA h g^{-1} after 1000 cycles.

The enhanced cycling stability of the cells with the a-FBN/G interlayer is also evident in their charge–discharge curves (Figure 15.7b). The galvanotactic charge/discharge behaviors of the cells with and without the a-FBN/G interlayer at 3C are consistent with the typical charge/discharge profiles of the Li–S batteries (Figure 15.7b and c), corresponding to a two-stage reduction of elemental sulfur to lithium PSs (Li_2S_{4-8}) and then to Li_2S_2/Li_2S, respectively. Discharge curves of the cell with the a-FBN/G interlayer obtained at a higher current (3C) still clearly contain two plateaus at the 250th and 500th cycles, indicating that the electrochemical reactions at higher charge/discharge rates follow the processes similar to those occurring at lower rates (*e.g.*, 1C) (Figure 15.7b). In addition, the overlap of the discharge plateaus during the cycling tests also suggests the good

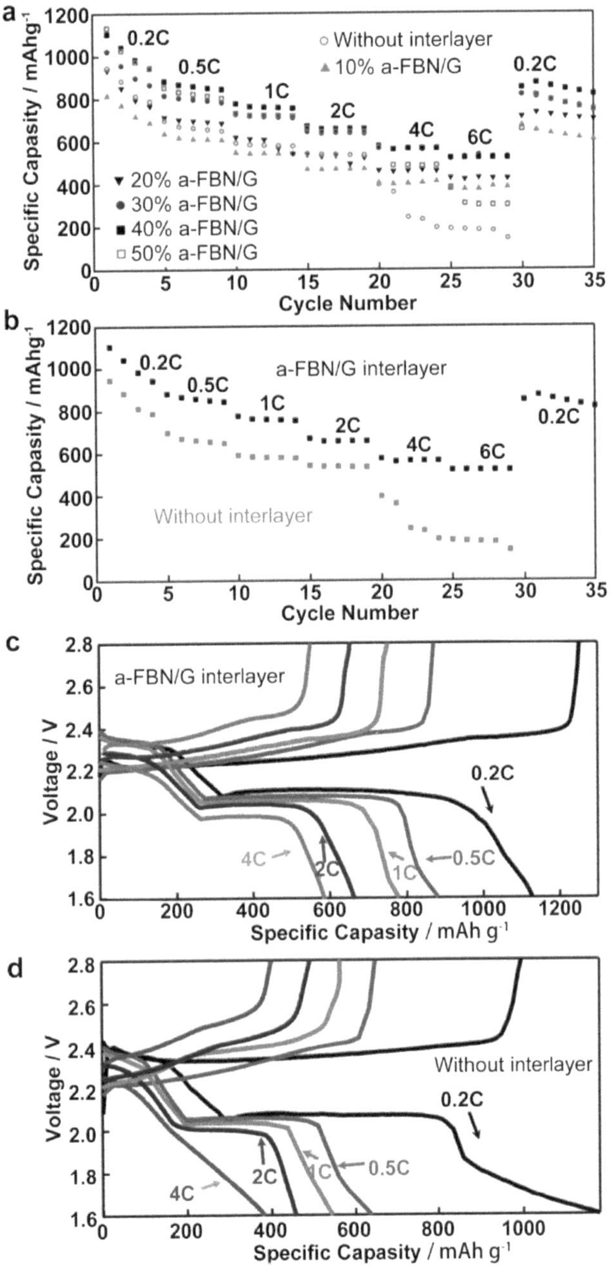

Figure 15.6 Rate performance of the cells. (a) Rate performance of cells with a 10%–50% a-FBN/G interlayer compared to those without an interlayer. (b) Comparison of rate performance between cells with a 40 wt% a-FBN/G interlayer and those without. (c) Charge and discharge profiles of Li–S cells with an a-FBN/G interlayer at 0.2, 0.5, 1, 2, and 4C. (d) Charge and discharge profiles of Li–S cells without an interlayer at 0.2, 0.5, 1, 2, and 4C. (Reprinted with permission from Ref [41]. Copyright 2018, John Wiley & Sons, Inc.)

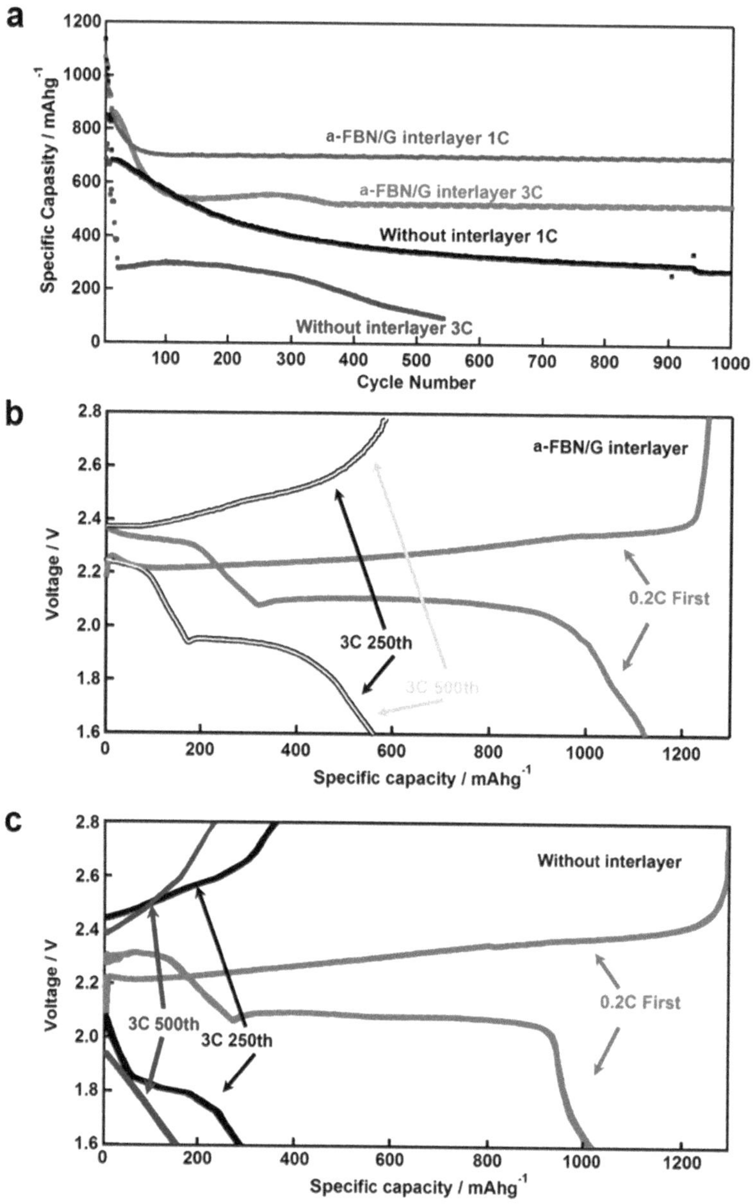

Figure 15.7 Long cycle performance of the cells. (a) Cycling stability of a cell with an a-FBN/G interlayer and without an interlayer at 1C and 3C. (b) The charge and discharge curves of the Li–S cell with an a-FBN/G interlayer in the 1st, 250th, and 500th cycles. (c) The charge and discharge curve of the Li-S cell without an interlayer in the 1st, 250th, and 500th cycles. (Reprinted with permission from Ref [41]. Copyright 2018, John Wiley & Sons, Inc.)

stability and reversibility of the electrode. In contrast, the discharge curves of the cell without an interlayer are different and unable to maintain their typical features (Figure 15.7c).

Cyclic voltammograms performed on cells after cycling, shown in Figure 15.8, further highlighted the electrochemical performance difference between cells with and without an interlayer. While cells without an interlayer exhibited decreasing redox peak intensities with increasing cycle numbers, cells with a 40% a-FBN/G interlayer maintained consistent redox peak intensities, consistent with charge–discharge curve observations.

These findings further underscore that a-FBN with positively charged groups can adsorb PS ions due to strong electrostatic interactions, while graphene in the composite coating assists in forming a conductive and flexible film. Therefore, the enhanced high-rate capability and excellent lithium storage reversibility of cells with an a-FBN/G interlayer can be attributed to three factors: (1) Reduced charge transfer resistance, demonstrated by impedance analysis (Figure 15.9); (2) stabilization of migrating PSs by a-FBN with positive NH_2 groups, which prevents their dissolution into the electrolyte and preserves electrode integrity and stability during

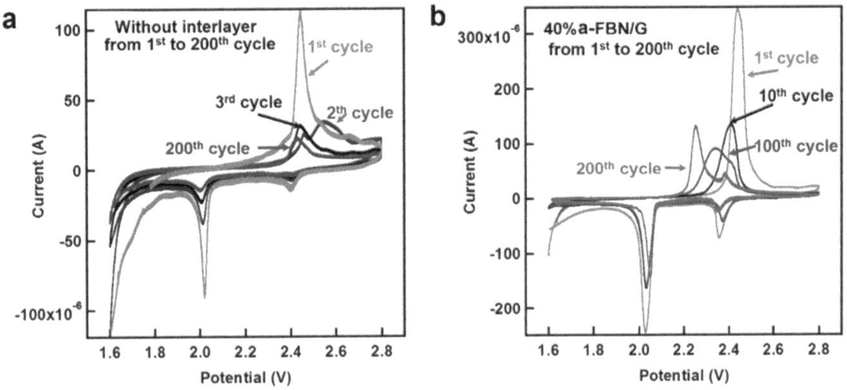

Figure 15.8 Cyclic voltammogram of the cycled cells without and with an a-FBN/G interlayer. (Reprinted with permission from Ref [41]. Copyright 2018, John Wiley & Sons, Inc.)

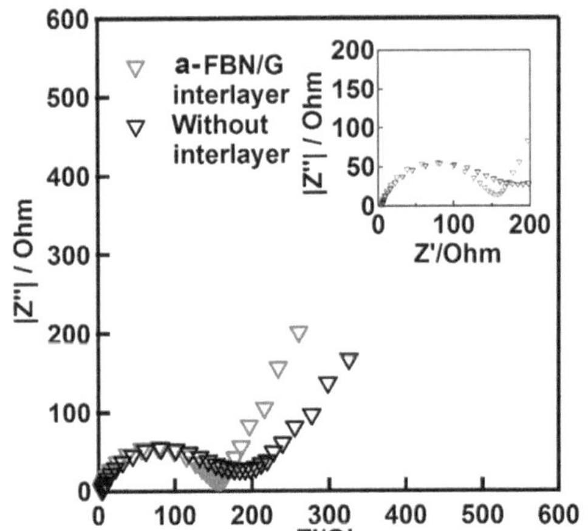

Figure 15.9 EIS of the cells with an a-FBN/G interlayer and without an interlayer before the cycle. (Reprinted with permission from Ref [41]. Copyright 2018, John Wiley & Sons, Inc.)

redox reactions; (3) confinement of PS in the cathode region, facilitating the reuse of precipitated Li_2S_2/Li_2S throughout cycling. This combination of lower charge transfer resistance, minimized PS shuttling, and enhanced conductivity makes a-FBN/G films highly effective as interlayers for Li–S batteries, improving cycling stability and Coulombic efficiency.

In conclusion, a novel lightweight interlayer for Li–S batteries has been developed, comprising a thin film of a-FBN nanosheets and graphene coated onto the cathode for the first time. This interlayer not only reduces charge transfer resistance but also mitigates the shuttling problem. The positive functional groups of a-FBN serve as ion-attracting sites for negatively charged PS products via electrostatic interaction. As a result, the addition of an a-FBN/G film improves the Li–S battery capacity from 382 mA h g^{-1} (with only graphene as the interlayer) to over 560 mA h g^{-1} at 3C. It also reduces the cyclic capacity decay rate from 0.032% to 0.0037% per cycle. Remarkably, a minimal capacity degradation rate of 0.0067% per cycle is achieved at 1C over 1000 cycles. This novel a-FBN/G film could offer a new design strategy for other energy storage systems, such as lithium-ion batteries, lithium–air batteries, and supercapacitors, where controlling ion transport is essential.

15.3 CARBONATE-FUNCTIONALIZED BN (C-FBN) NANOSHEETS FOR LI-S BATTERIES

The development of a-FBN demonstrated the potential of FBN in Li-S batteries. However, due to the insulating nature of BN, a-FBN (positively charged groups), with the PS absorption ability, still requires graphene as a conductor. This reduced the concentration of functional groups in the interlayer. To address this, c-FBN nanosheets, possessing a negatively charged group with PS repelling ability, are applied to the separator in Li-S batteries in this chapter. The c-FBN layer on the separator creates an electrical barrier that inhibits the shuttle effect and enhances the electrochemical performance of Li-S batteries.

15.3.1 Synthesis of c-FBN Nanosheets

To obtain c-FBNs, BN nanosheets and guanidine carbonate (Sigma-Aldrich) were ball-milled in a ball mill under a nitrogen atmosphere for 24 hours [59], with a mass ratio of 1:1. The milled product was collected, washed with deionized water, and lyophilized to obtain the final c-FBN product.

For the preparation of c-FBN-coated separators, a solution was made by dissolving 1 wt% polyvinylidene fluoride in N-methyl-2-pyrrolidone. Subsequently, 10 mg of c-FBN was dispersed in 100 mL of the stock solution. Controlled amounts of this dispersion were then poured onto Celgard 2400 separators to achieve different coating densities. The coating layer was formed using a vacuum filtration system. Excess polyvinylidene fluoride and methyl-2-pyrrolidone were removed by thorough washing with deionized water. The c-FBN-coated separators were dried in a vacuum oven at 60°C for 24 hours. To achieve varying coating thicknesses, five different coating suspensions were prepared by adjusting the weight of c-FBN.

15.3.2 Interaction between c-FBN Nanosheets and PS

To illustrate the c-FBN separator's ability to reject PSs, an H-type cell with interchangeable separators was employed (Figure 15.10a and b). These cells contained a yellow-colored PS solution in one chamber and a colorless electrolyte in the other. After 24 hours, the c-FBN separator showed

Figure 15.10 (a) Color contrast of the PS solution in the left chamber and the colorless electrolyte in the right chamber of an H-type cell with a c-FBN separator after 24 hours; (b) color contrast of the PS solution and colorless electrolyte in an H-type cell with a Celgard separator after 24 hours; (c) comparison of separators (soaked in colorless electrolyte) with and without c-FBN, taken from cycled cells. (Reprinted with permission from Ref [42]. Copyright 2020, American Chemical Society.)

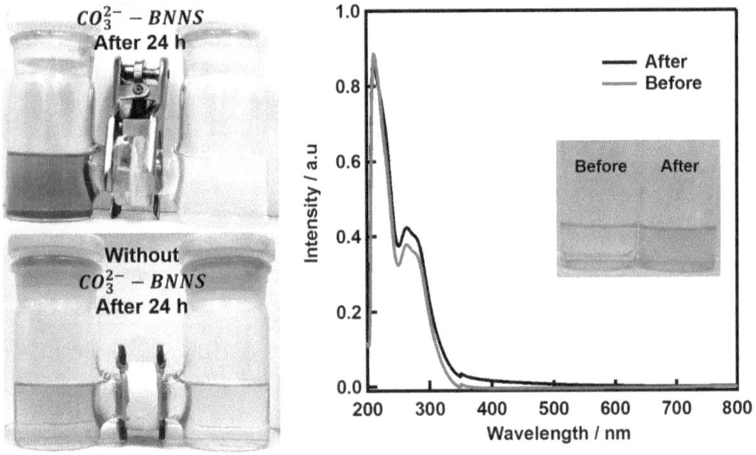

Figure 15.11 Ultraviolet spectrum of the PS solution (without c-FBN powder) and the PS solution after the addition of c-FBN powder. (Reprinted with permission from Ref [42]. Copyright 2020, American Chemical Society.)

no color change in the electrolyte chamber (Figure 15.10a), while the Celgard separator resulted in the yellow electrolyte (Figure 15.10b), indicating significant PS rejection by the c-FBN separator. This was confirmed by soaking both separators (c-FBN and Celgard) in a colorless electrolyte solution after cycling. Remarkably, the c-FBN separator exhibited no color change in the electrolyte solution, confirming its rejection of negatively charged PSs during cycling (Figure 15.10c).

To understand the mechanism of PS rejection by the c-FBN separator, c-FBN nanosheets were added to an extremely dilute PS solution. UV absorption spectra of the original extremely dilute PS solution and c-FBN nanosheet-added solutions were compared (Figure 15.11). The peak

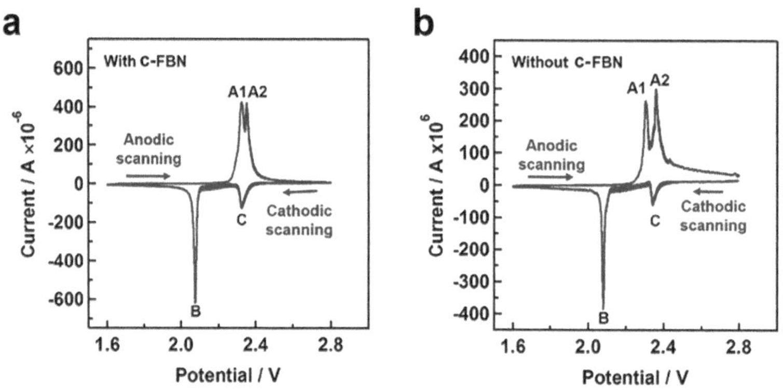

Figure 15.12 (a) Cell with a c-FBN separator and (b) cell with a Celgard separator (without c-FBN). (Reprinted with permission from Ref [42]. Copyright 2020, American Chemical Society.)

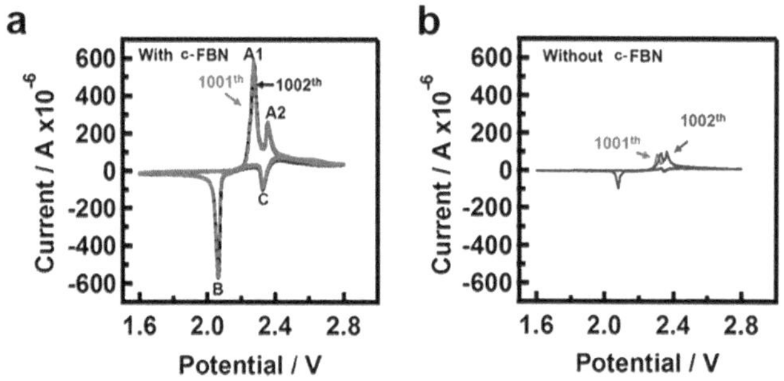

Figure 15.13 (a) CVs of the cells with and (b) without c-FBN separators after long-term cycling at 3C. (Reprinted with permission from Ref [42]. Copyright 2020, American Chemical Society.)

between 250 and 300 nm, linked to ions, showed a rise in UV intensity after adding c-FBN powder, indicating an increased local ion concentration caused by repulsion from the negatively charged CO_3 groups. This implies that c-FBN nanosheets with CO_3 groups efficiently repel PS anions through strong electrostatic interactions.

The electrochemical performance of cells with c-FBN separators is excellent due to PS migration rejection. The Li^+ ion permeability of cells with c-FBN separators is also crucial, especially for rate capability. Hence, Li^+ ion diffusion coefficients were calculated using the Randles–Sevcik equation based on cyclic voltammetry (CV) measurements (Figures 15.12a, b and 15.13a, b).

The calculated diffusion coefficients of Li ions with c-FBN separators were significantly higher than those without c-FBN separators, indicating faster Li^+ diffusion with c-FBN. The stability of Li-ion transfer throughout cycling was demonstrated by consistent diffusion coefficients even after 1000 cycles. This suggests stable Li^+ transfer and electrode integrity in cells with c-FBN separators.

The diffusion coefficient of lithium ions, D_{Li^+} ($cm^2 s^{-1}$), was evaluated by CV and calculated according to the Randles–Sevcik equation [44,60,61]:

$$I_p = 2.69 \times 10^5 n^{1.5} A D_{Li^+}^{0.5} C_{Li} \nu^{0.5} \tag{15.8}$$

In this equation, I_p (A) refers to the peak current, n is the number of electrons involved in the reaction (for Li-S cells, $n=2$), A (cm^2) is the electrode area (1.538 cm^2), C_{Li} (mol mL^{-1}) represents the lithium-ion concentration in the electrolyte, and v denotes the scanning rate (V s^{-1}).

The lithium-ion diffusion coefficient, D_{Li^+} ($cm^2 s^{-1}$), is used to compare the transference number of lithium ions according to the following equation:

$t_{Li^+} = \frac{D_{Li^+}}{D_{Li^+} + D_-}$ [62], in which t_{Li^+} is the transference number of lithium ions, and D_- is the average diffusion coefficient of the anion (including $TFSI^-$ and S_x^{2-}) in the electrolyte. The transference number of the whole cell is accorded to a standard calculation equation: $t_\pm = \frac{z_\pm \Delta n_a}{Q/F}$. In this equation, $t_\pm$ denotes the transference number of lithium ions (+) and other anions (–), $z_\pm$ is the charge number (with $z_+ = 1$ and $z_- = 1$ or 2), Δn_a represents the amount of substance transported (mol), which includes lithium ions and other anions, Q is the total charge passing through the entire cell (C), and F is the Faraday constant.

In this study, the equations $t_{Li^+} = \frac{D_{Li^+}}{D_{Li^+} + D_-}$, $t_{Li^+} = \frac{z_+ \Delta n_{Li^+}}{Q/F}$, and $t_- = \frac{z_- \Delta n_-}{Q/F}$ were employed to explain the anion diffusion process. Here, $D_- = D_{S_x^{2-}} + D_{TFSI^-}$, $\Delta n_- = \Delta n_{S_x^{2-}} + \Delta n_{TFSI^-}$. Q remains consistent during charge–discharge cycles at the same current density in both cells with and without the c-FBN separator. As D_{Li^+} increases, both Δn_{Li^+} and t_{Li^+} increase as well. Conversely, a reduction in the anion transference number ($t_{Li^+} + t_- = 1$) leads to a slower transport rate per unit time, in accordance with the diffusion coefficient definition. This behavior is visible in the cathodic and anodic peaks in the CV curves, marked as B, C, and A1, A2 (Figures 15.14a, b and 15.15a, b). Using the Randles–Sevcik Equation (15.8), the diffusion coefficients of Li ions with the c-FBN separator at various CV voltage regions are as follows: $D(A1)_{c\text{-FBN}} = 1.22 \times 10^{-8}$, $D(A2)_{c\text{-FBN}} = 1.22 \times 10^{-8}$, $D(B)_{c\text{-FBN}} = 2.71 \times 10^{-8}$, and $D(C)_{c\text{-FBN}} = 8.82 \times 10^{-10}$ cm s^{-1} for the first cycle. In comparison, for the Celgard separator (without c-FBN), the diffusion coefficients are as follows: $D(A1)_C = 6.56 \times 10^{-9}$, $D(A2)_C = 6.13 \times 10^{-9}$, $D(B)_C = 9.45 \times 10^{-9}$, and $D(C)_C = 3.57 \times 10^{-10}$ cm s^{-1} during the first cycle. Diffusion coefficients in the cell with the c-FBN separator are approximately three times greater than in the cell with the Celgard separator during the first discharge. After 1000 cycles, the diffusion coefficients of Li ions in the c-FBN cell are as follows: $D(A1) = 2.62 \times 10^{-8}$, $D(A2) = 3.21 \times 10^{-9}$, $D(B) = 2.54 \times 10^{-8}$, and $D(C) = 1.05 \times 10^{-9}$ cm s^{-1}. This demonstrates stable Li^+ diffusion in the c-FBN separator cell from the initial discharge cycle to the 1001st cycle.

The diffusion coefficients from peaks B and C indicate the ion transport dynamics and PS redox process during discharge [4]. These results suggest stable Li-ion transport and electrode material integrity in the c-FBN separator cell throughout the entire cycling process. The Li^+ diffusion rate in the cell with a c-FBN separator is significantly faster than in the one without it. Additionally, the transference number, t_{Li^+}, is higher in the cell with the c-FBN separator because the c-FBN material, a Lewis base, enhances the transference number [63–66]. This suggests a decrease in the diffusion of anions like $D_{S_x^{2-}}$ and D_{TFSI^-} due to the relationship $t_+ + t_- = 1$. The overall electrochemical performance (capacity, cycling stability, and rate capability) of the Li-S cell with a c-FBN separator is much better than the cell without a c-FBN separator.

The higher transference number in cells with c-FBN separators suggests enhanced Li^+ diffusion due to c-FBN's Lewis base properties. This leads to decreased diffusion of anions such as S and Li_2S_x, indicating improved electrochemical performance in terms of capacity, cycling stability, and rate capability compared to cells without c-FBN separators.

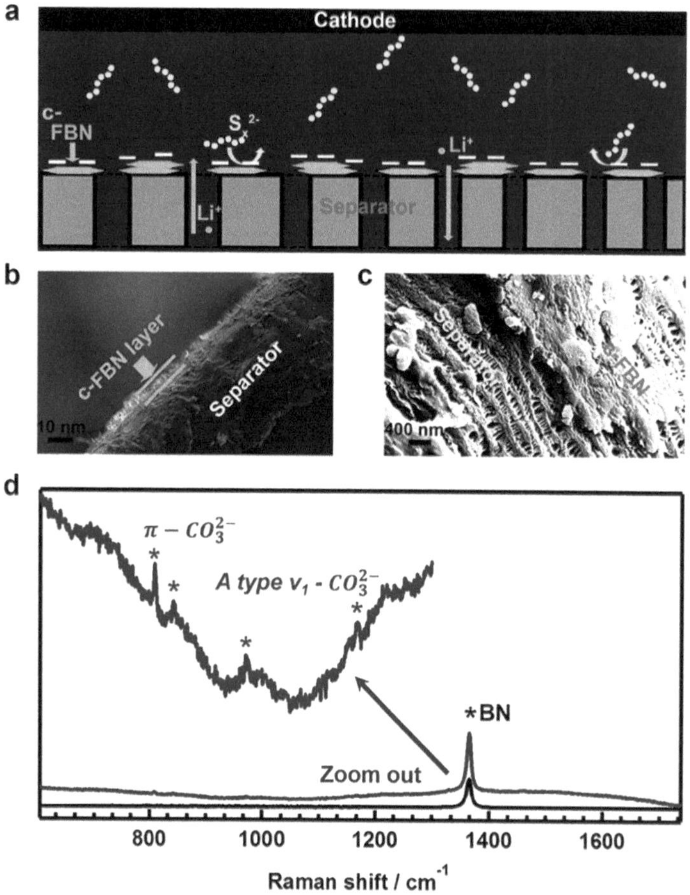

Figure 15.14 (a) Diagram of a sandwich-structured Li-S cell with a c-FBN separator; (b) SEM image of the cross-section of the c-FBN nanosheet coating on the separator; (c) high-magnification SEM image of the c-FBN separator surface; (d) Raman spectrum of the c-FBN nanosheets. (Reprinted with permission from Ref [42]. Copyright 2020, American Chemical Society.)

15.3.3 Performance of Li-S batteries with the c-FBN Nanosheet Separator

A schematic depiction of a Li-S cell featuring a c-FBN separator is presented in Figure 15.14a. The SEM image in Figure 15.14b showcases a cross-sectional view of the coating layer with a c-FBN content of $0.21 \pm 0.01\,mg\,cm^{-2}$ on a separator. A higher magnification SEM image (Figure 15.14c) reveals that c-FBN nanosheets were coated onto the surface of the Celgard 2400 separator without entirely obstructing the lithium-ion transport channels. Furthermore, Raman spectroscopy analysis (Figure 15.14d) confirmed the presence of functional groups in c-FBN nanosheets. Zeta potential analysis showed that c-FBN nanosheets had a significant negative potential of $-44 \pm 3\,mV$, promoting the repulsion of negative ions [67–70].

To assess the electrochemical performance of the c-FBN separator, cycling performance was initially examined. Figure 15.15 compares the cycling performance between cells with and without c-FBN separators at a current density of 0.2C ($335\,mA\,g^{-1}$). The Li-S cell with the c-FBN

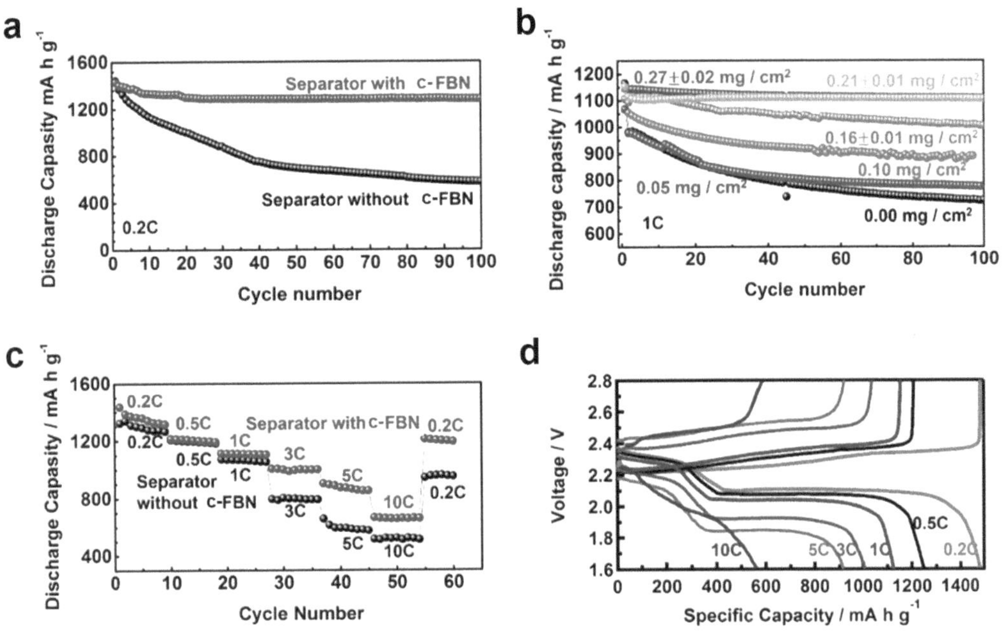

Figure 15.15 (a) Comparison of cycling stability between cells with and without the c-FBN separator at 0.2C for 100 cycles; (b) performance of cells with c-FBN separators using varying c-FBN nanosheet concentrations; (c) rate capability of cells with and without the c-FBN separator; and (d) charge–discharge profiles of cells with c-FBN at different current rates. (Reprinted with permission from Ref [42]. Copyright 2020, American Chemical Society.)

separator showed an initial discharge capacity of 1485 mA h g^{-1}, which experienced a slight decline during the first 10 cycles due to the activation process. Following this, the cell stabilized at a high discharge capacity of 1250 mA h g^{-1} after 100 cycles. In comparison, the cell lacking the c-FBN separator exhibited considerable capacity loss, with a retention of just 498 mA h g^{-1} after 100 cycles. This demonstrates the capability of c-FBN-coated Celgard separators in mitigating the PS shuttle effect. To further refine the c-FBN coating, five separators with varying coating densities were prepared, and their cycling performance was assessed at a current density of 1C (Figure 15.15b). Increasing the c-FBN content in the coating suspension enhanced cycling stability, but exceeding 0.21 mg cm^{-2} did not yield further improvement. Therefore, a c-FBN coating with a nanosheet content of 0.21 mg cm^{-2} was selected as the model system for further experimental work.

Figure 15.15c demonstrates the rate capability performance between cells with and without c-FBN separators. The cell with a c-FBN separator exhibited superior rate capability, retaining capacities of 1486, 1232, 1194, 1008, and 904 mA h g^{-1} at current densities of 0.2, 0.5, 1, 3, and 5C, respectively. Even at an extremely high current of 10C, the cell delivered a capacity of 582 mA h g^{-1} with a high Coulombic efficiency of >98%. Furthermore, when the current density reverted to 0.2C, the capacity recovered to 1225 mA h g^{-1} after 60 cycles (~83% retention of the initial capacity measured at 0.2C), indicating stability under high current charging and discharging conditions. In contrast, the cell with a Celgard separator exhibited significant capacity decay at high current densities, likely due to the severe PS shuttle effect. Additionally, discharge–charge potential profiles obtained at different C rates of the cell with a c-FBN separator for selected cycles are shown in Figure 15.15d, demonstrating stable discharge–charge potential profiles.

Figure 15.16 presents the long-term cycling stability of cells at high current densities. A comparison between cells with and without c-FBN separators at 1C and 3C is shown. The cells incorporating c-FBN separators exhibited consistent performance, retaining capacities of 1027 mA h g^{-1} at 1C and 897 mA h g^{-1} at 3C after 1000 cycles. The enhanced cycling performance of cells with c-FBN separators was further evaluated at a very high current rate of 10C (Figure 15.16b), demonstrating the excellent stability of c-FBN-incorporated separators.

CV was conducted to examine the electrochemical reaction kinetics in cells with and without c-FBN separators after prolonged cycling. The CV curves presented in Figures 15.13a and b, corresponding to cells after 1001 and 1002 discharge–charge cycles at 3C, demonstrate that cells with c-FBN separators maintain stable and reversible redox reactions.

To examine the interfacial behavior of electrodes with and without c-FBN separators, electrochemical impedance spectroscopy (EIS) was conducted. The resulting Nyquist plots for cycled cells (first discharge state) are shown in Figure 15.17a and b. These plots feature a depressed semicircle in the high-frequency region, another in the medium-frequency region, and a straight line in the low-frequency region. To analyze the impedance, the plots were evaluated using *Z*-view software, and a simplified equivalent circuit was proposed (shown in the inset of Figure 15.17a). The circuit includes R_e, representing electrolyte resistance, $R_{int}//CPE_{int}$ for interphase contact resistance and capacitance, $R_{ct}//CPE_{ct}$ for charge transfer resistance and capacitance, and CPE_{dif}, indicative of diffusion impedance associated with lithium-ion movement.

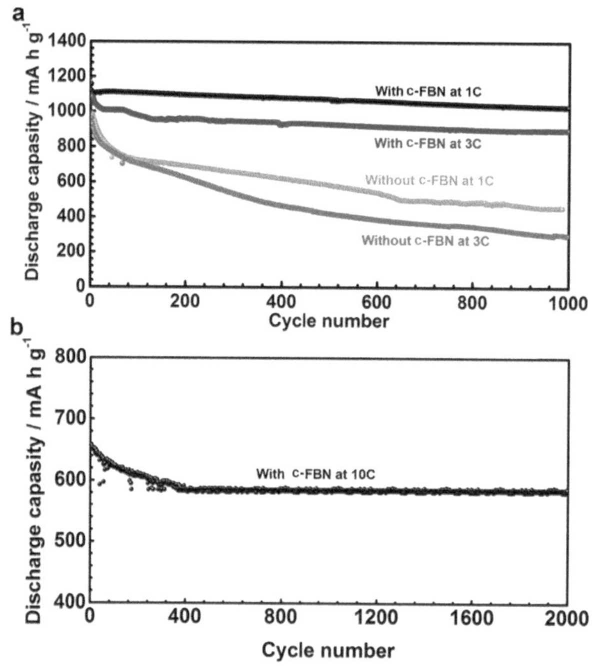

Figure 15.16 (a) Long-term cycling performance of cells with and without c-FBN separators at 1C and 3C for 1000 cycles; (b) cycling stability of cells with c-FBN separators at 10C for up to 2000 cycles. (Reprinted with permission from Ref [42]. Copyright 2020, American Chemical Society.)

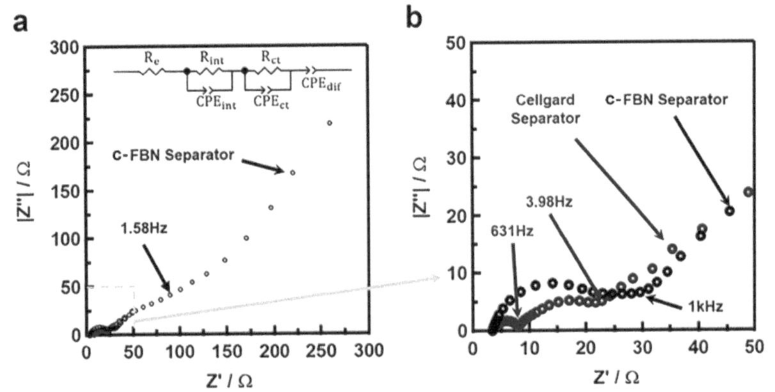

Figure 15.17 (a, b) EIS (equivalent circuit model is depicted in the inset of Figure 15.17a) of the cells with and without c-FBN separators at the end of the first discharge. (Reprinted with permission from Ref [42]. Copyright 2020, American Chemical Society.)

The initial semicircle in the high-frequency region corresponds to the interphase contact resistance and capacitance ($R_{int}//CPE_{int}$), representing electron conduction from the current collector to the reaction sites. The formation of insulating reduction products on the sulfur electrode surface during discharge can increase the interphase contact resistance, slightly enlarging the high-frequency semicircle. The second semicircle in the medium-frequency region reflects charge transfer resistance and capacitance ($R_{ct}//CPE_{ct}$) at the interface between the conductive agent and electrolyte.

In the Nyquist plots, the semicircle diameter for the cell with a Celgard separator is smaller compared to the one with c-FBN, indicating higher resistance in the c-FBN-containing cell. Specifically, interphase contact resistance (R_{int}) is 4.16 Ω for the Celgard cell and 24.4 Ω for the c-FBN cell. Similarly, the charge transfer resistance (R_{ct}) is higher for the c-FBN separator (88.7 Ω) than for the Celgard separator (12 Ω). The increased contact resistance in the c-FBN cell is attributed to the incorporation of insulating c-FBN nanosheets into the Celgard separator, directly contacting the sulfur cathode surface [35,37]. The higher R_{ct} value in the c-FBN cell further suggests that the separator inhibits PS transfer and helps accumulate insulating discharge products such as Li_2S and Li_2S_2 [71]. This indicates that c-FBN nanosheets enhance the suppression of PS shuttle and internal redox processes.

In conclusion, a novel separator has been engineered for Li-S batteries by applying a thin layer of c-FBN nanosheets onto a commercial Celgard separator. This innovative separator addresses the issue of internal redox shuttle, a significant limitation of traditional separators. Incorporating c-FBN nanosheets into Celgard separators resulted in a significant capacity improvement, increasing from 456 to 1027 mA h g^{-1} at 1C, and from 295 to 897 mA h g^{-1} at 3C after 1000 cycles, with minimal capacity loss.

Remarkably, cells utilizing c-FBN separators demonstrate exceptional stability, maintaining a capacity of 585 mA h g^{-1} with retention of approximately 90% at a high rate of 10C over 2000 cycles. This enhanced performance may be attributed to the increased Li-ion diffusion coefficient of the cell and the rejection of PSs. The innovative separator restricts the migration of dissolved PSs by leveraging strong electrostatic interactions between the negatively charged PS and c-FBN nanosheets. This interaction enhances ion movement, mitigates internal shuttle effects, and results in cells with superior capacity, excellent rate performance, and enhanced long-term stability. Consequently, this separator provides a valuable strategy for improving the performance of Li-S batteries.

15.4 c-FBN NANOSHEETS FOR Li-PS CATHOLYTE BATTERIES

The inherent insulating nature of sulfur inhibits improvements in sulfur loading and rate performance in Li-S batteries. High-conductivity PS ions are difficult to retain on the cathode side. However, with the implementation of the c-FBN separator, Li-PS catholyte batteries exhibit high sulfur loading and enhanced rate performance.

15.4.1 Mechanism of Li-PS Catholyte Batteries

The Li-PS cells were tested through galvanostatic charge–discharge at varying current densities and potential ranges. The c-FBN separator, which contains CO_3^{2-} groups, effectively prevents the passage of S_8^{2-} ions due to the electrostatic repulsion between the negatively charged ions (Figure 15.18a). Based on the discharge characteristics of Li-S batteries [4,72,73], the theoretical specific capacity of S_8^{2-} is 1470 mA h g^{-1}, approximately 87.7% of sulfur's theoretical capacity (1675 mA h g^{-1}). For comparison, the Li-PS cell used the standard calculation of 1C = 1675 mA g^{-1}, typical for Li-S cells. The sulfur content was found to be around 25.6 mg (equivalent to about 92 wt% sulfur content in Li-S cells) for the 0.5M Li-PS cell, and 2.56 mg (approximately 53 wt% sulfur content in Li-S cells) for the 0.05M Li-PS cell. Importantly, the sulfur content in the 0.5M Li-PS cell exceeds that found in most conventional Li-S cells [4,43,47–53,73–79].

Figure 15.18b shows the charge–discharge curves of the 0.5M Li-PS cell at a current density of 0.2C (with the second discharge and first charge plotted for clarity). In this experiment, one cell was charged to 2.35 V, while the other was charged to 2.38 V, with the discharge cut-off voltage fixed at 1.6 V for both cells. For the 2.35 V charged cell, a single discharge plateau is observed at around 2.1 V, along with a slope between 2.1 and 2.3 V. The plateau at 2.1 V corresponds to the conversion from S_x^{2-} to Li_2S_2/Li_2S, while the slope between 2.1 and 2.3 V reflects the transformation between S_8^{2-} and S_x^{2-} [53,72,73]. In contrast, when charged to 2.38 V, two discharge plateaus at 2.1 and 2.3 V are observed, along with the same slope between 2.1 and 2.3 V. A new discharge plateau forms at 2.3 V during the second discharge, attributed to the conversion of solid sulfur to S_8^{2-} ions [53,72,73]. The results indicate that sulfur can only form in Li-PS cells when the charging voltage exceeds 2.35 V, suggesting that Li-PS cells should operate within a restricted potential window of 1.6–2.35 V to maintain the liquid cathode nature.

Supporting this finding, CV measurements (Figure 15.18c) were performed on the same cells charged to 2.35 V and collected at the end of the second discharge (as shown in Figure 15.18b). The CV measurements were conducted within the potential range of 1.6–2.35 V at a scan rate of 0.05 mV s^{-1}. In this range, a clear single pair of redox peaks is observed, with anodic and cathodic peaks around 2.3 and 2.05 V, respectively. The anodic peak corresponds to the conversion from Li_2S_2/Li_2S to S_8^{2-}, while the cathodic peak at 2.05 V reflects the transformation from S_8^{2-} to other PSs. In comparison, the CV curve of the Li-S cell exhibits two cathodic peaks around 2.3 and 2.05 V, with the 2.3 V peak attributed to the transformation from sulfur to S_8^{2-} ions [53,72,73]. This confirms that by limiting the charge potential to 2.35 V, the Li-PS cell can avoid the transformation of sulfur to S_8^{2-} ions. Based on these observations, it is concluded that the Li-PS system follows the reaction mechanisms in Equations (15.10) and (15.11), while the Li-S system follows

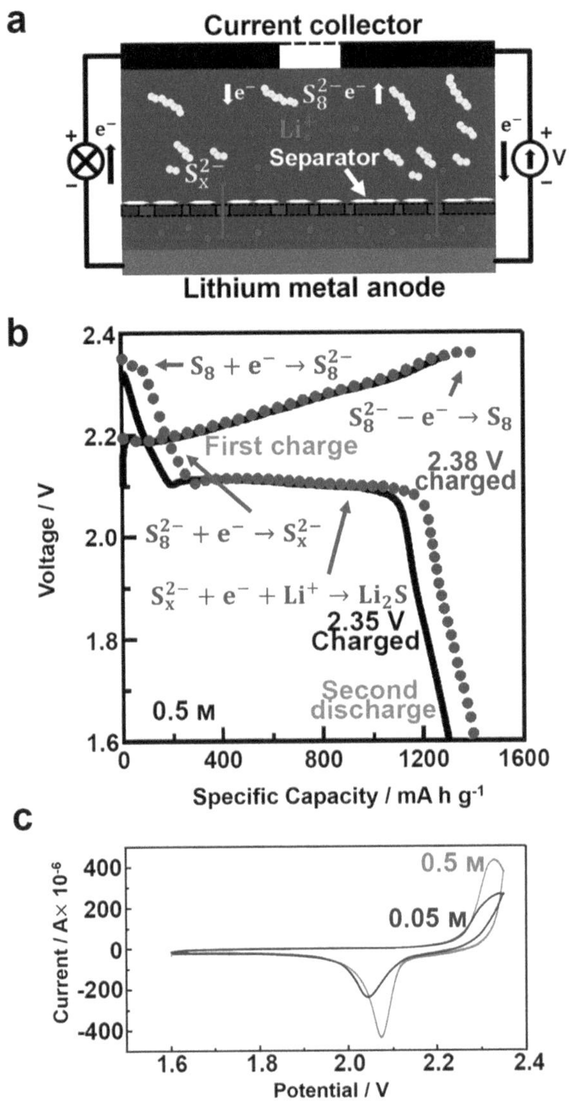

Figure 15.18 (a) Diagram illustrating the configuration and working principle of the Li-PS cell. (b) First charge and second discharge profiles of the 0.5M Li-PS cells at 0.2C, within cut-off potentials of 1.6–2.35 V and 1.6–2.38 V. (c) Cyclic voltammograms of the cells, measured after the second discharge, with the test data shown in Figure 15.18b. (Reprinted with permission from Ref [43]. Copyright 2020, Elsevier.)

Equations (15.9)–(15.11). As a result, no sulfur production occurs at the end of the charge process in the Li-PS system. The reaction mechanism during the cycling process in Li-S cells can be outlined as follows:

$$S_8(\text{Solid}) + 2e^- \leftrightarrow S_8^{2-}\ (\text{Liquid})(\text{discharge}\ \ 2.3\ \text{V}) \tag{15.9}$$

$$xS_8^{2-}(\text{Liquid}) + (16-2x)e^- \leftrightarrow 8S_x^{2-}\ (\text{Liquid})(\text{discharge}\ \ 2.3-2.1\ \text{V}) \tag{15.10}$$

$$2S_x^{2-}\left(\text{Liquid}\right)+2x\text{Li}^+ + (2x-4)e^- \leftrightarrow x\text{Li}_2\text{S}_2\ \left(\text{Liquid}\right)\ \text{and/or}$$
$$S_x^{2-}\left(\text{Liquid}\right)+2x\text{Li}^+ + (2x-2)e^- \leftrightarrow x\text{Li}_2\text{S}\ (\text{Solid})\left(\text{dicharge}\ 2.1-1.6\ \text{V}\right) \quad (15.11)$$

15.4.2 Performance of Li-PS Catholyte Batteries with a c-FBN Nanosheet Separator

The 0.5M Li-PS cells exhibit remarkable rate performance across various current densities and charge cut-off voltages. At rates ranging from 0.2C to 20C, reversible capacities of 1302, 1232, 1194, 1089, 1048, 956, and 763 mA h g^{-1} were achieved, respectively (Figure 15.19a). Upon returning to 0.2C, a notable capacity retention of approximately 99.7% was observed, with a recovered capacity of 1298 mA h g^{-1} after 48 cycles, highlighting the advantageous characteristics of the liquid cells. Adjusting the charging potential window for each current rate proved crucial for optimizing the true rate performance of the 0.5M Li-PS cells, as suggested by previous findings.

The charge–discharge potential profiles of the 0.5M Li-PS cells obtained at various current densities and charge cut-off voltages are depicted in Figure 15.19b. Notably, distinct slopes and plateaus are evident across different current densities, with the 0.5M Li-PS cell retaining a typical long discharge plateau even at a high rate of 20C, a feature absent in Li-S cells operated at such high rates. This observation underscores the ability of the 0.5M Li-PS cell to operate effectively above 20C within a wide potential window, facilitating ultrafast charge–discharge processes.

Figure 15.20 compares the cycling performance of 0.5M and 0.05M Li-PS cells at varying current rates. Both cells demonstrate stable capacity retention at 0.2C, with the 0.5M Li-PS cell achieving an initial discharge capacity of 1302 mA h g^{-1} and retaining 1300 mA h g^{-1} after 100 cycles, approaching nearly 99.9% capacity recovery. Despite its lower initial discharge capacity, the 0.05M Li-PS cell exhibits a remarkable capacity retention of 97% after 100 cycles. The superior cycling stability of Li-PS cells compared to most reported Li-S cells is attributed to their ability to prevent activation decay, a common issue in Li-S systems.

Furthermore, the 0.5M Li-PS cell demonstrates high discharge capacities at 5C and 10C after 500 cycles, with capacity retentions of ~96% and ~80%, respectively (Figure 15.20b). In contrast, the 0.05M Li-PS cell exhibits even better cycling stability at 10C, with a capacity retention of ~90% after 500 cycles. The observed capacity decay at high current rates is associated with the production of insulating Li_2S, particularly pronounced in the 0.5M Li-PS cell due to its higher sulfur loading.

To evaluate the interfacial behavior and conductivity of 0.5M and 0.05M Li-PS cells, EIS was performed. Figure 15.21a and b compares the Nyquist plots of two fresh cells (before cycling) of 0.5M Li-PS and 0.05M Li-PS. The plot for the 0.05M Li-PS cell reveals two semi-circles in the high-to-medium-frequency region. The first semicircle is attributed to the resistance of the aluminum–carbon surface, while the second corresponds to the resistance between carbon and S_8^{2-} ions [33,80,81]. The first semicircle diameter shows little variation between the cells, but the second semicircle for the 0.5M Li-PS cell is notably smaller or potentially absent, suggesting high PS ion conductivity in the 0.5M Li-PS cell. This accounts for the cell's high initial capacity due to efficient activation.

Figure 15.21c and d displays the conductivity of the 0.5M Li-PS cell in both charge and discharge states. The cell was first discharged to 1.6 V, followed by an EIS measurement. After charging to 2.8 V, EIS was again performed. The charge state shows a larger diameter in the first

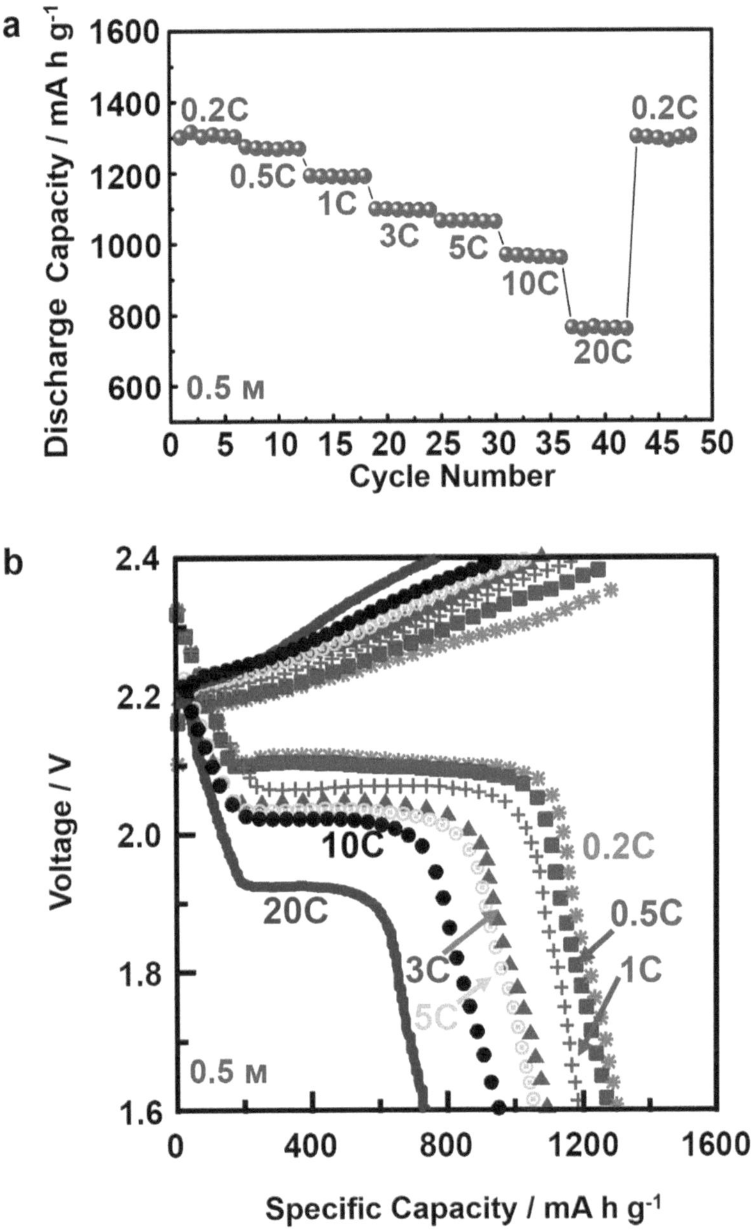

Figure 15.19 (a) Rate capability of the 0.5M Li-PS cell at current densities from 0.2C to 20C and across different charge cut-off voltages (2.35, 2.38, 2.39, and 2.4 V). (b) Charge–discharge curves for the 0.5M Li-PS cell at various current rates and charge cut-off voltages. (Reprinted with permission from Ref [43]. Copyright 2020, Elsevier.)

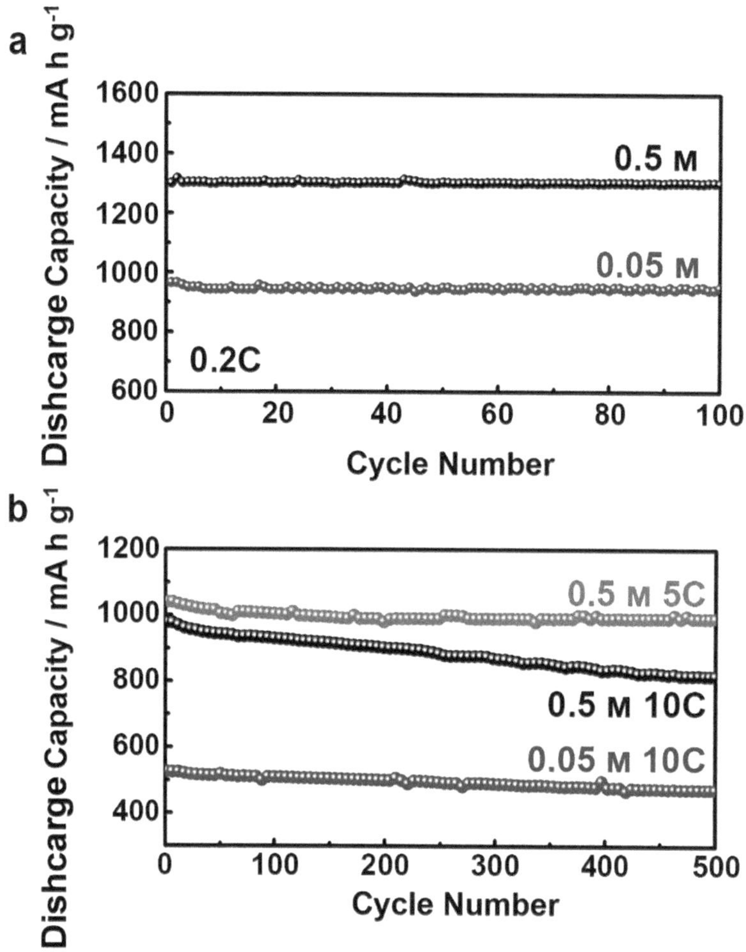

Figure 15.20 (a) Long-term cycling performance of the 0.05M and 0.5M Li-PS cells at 0.2C for up to 100 cycles. (b) Performance of the 0.05M and 0.5M Li-PS cells at 5C and 10C over 500 cycles. (Reprinted with permission from Ref [43]. Copyright 2020, Elsevier.)

semicircle compared to the discharge state, indicating increased resistance due to the formation of solid sulfur at 2.8V. In contrast, the discharge state (1.6V) forms solid Li_2S, which has better conductivity than sulfur, contributing to the superior rate performance of the Li-PS cells over conventional Li-S cells [82–84].

Li-PS cells present significant advantages over Li-S systems. In the latter, maintaining a high sulfur-carbon ratio on the cathode is difficult, especially when sulfur content exceeds 75wt%, [43,47,49,50,52,53,76,77,85] due to increased resistance from thick sulfur layers and the detachment of large sulfur particles. Li-PS cells, however, can easily maintain a higher sulfur content (92wt%), ensuring greater energy density. Li-PS cells also operate efficiently at high current rates. For instance, a flat discharge plateau at 2.0V is observed at 10C, indicating excellent performance above 10C, unlike Li-S systems, which show a sloped discharge curve at high currents [86,87]. At 10C, Li-PS cells achieve 780mA h g^{-1}, while most Li-S cells fall below 650mA h

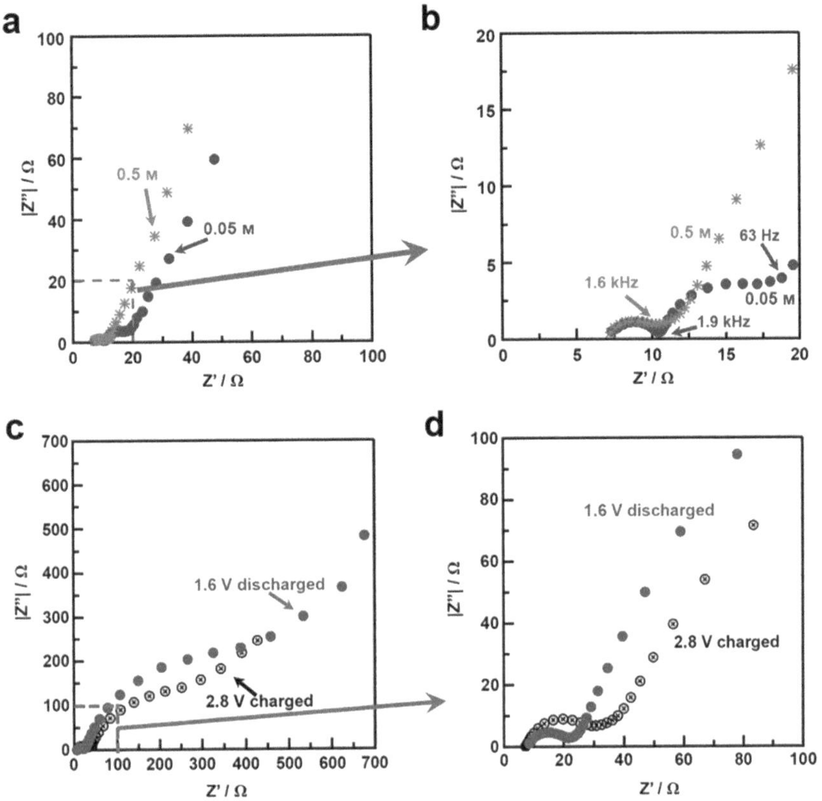

Figure 15.21 (a) and (b) Nyquist plots of the freshly prepared 0.5M and 0.05M Li-PS cells. (c) and (d) Nyquist plots of the 0.5M Li-PS cells after charging to 2.8 V and discharging to 1.6 V. (Reprinted with permission from Ref [43]. Copyright 2020, Elsevier.)

g^{-1} [88,89]. Additionally, Li-PS cells exhibit a 75% recovery in capacity when the current rate is reduced from 10C to 0.2C, outperforming most Li-S cells, which show a recovery of less than 65% [87,90–92]. This superior performance is attributed to the high sulfur loading and the enhanced conductivity from the soluble PS ions [48,73,78].

Furthermore, the cycling performance of Li-PS cells exceeds that of Li-S cells, owing to the suppression of the shuttle effect in the Li-PS system [33,91]. While Li-S cells typically exhibit a capacity retention of 77% at 0.2C after 150 cycles and 94% at 1C after 200 cycles [86–89,91], the Li-PS cells in this study demonstrated exceptional retention of 99.9% at 0.2C after 100 cycles and 96% at 5C after 500 cycles. The exceptional capacity retention observed in Li-PS cells is primarily due to the soluble PS ions, which help reduce volume expansion during the phase transformation between sulfur and S_8^{2-} ions, as well as during sulfur rearrangement in the charging process [33,91].

Although the 0.5M Li-PS cell exhibits strong electrochemical performance, its capacity retention at high current densities is not as favorable when compared to some other liquid systems, particularly redox flow batteries [93,94]. The challenge arises from the discharge products, Li_2S_2 and Li_2S, formed in the Li-PS system. While soluble Li_2S_2 [95] is beneficial, the formation of insoluble Li_2S can cause certain discharge products to detach from the conductive network, hindering their ability to react back into PS ions at high current densities. Redox flow batteries systems, on the other hand, avoid this issue since their final products are typically soluble ions.

In summary, the Li-PS liquid cells, with a sulfur content of ~92 wt%, represent a novel advancement in overcoming the 'shuttle effect' in traditional Li-S cells. The cells exhibit remarkable electrochemical performance, including high capacity, outstanding cycling stability, extended cycle life, and excellent rate capability. Specifically, the Li-PS cells achieved 1300 mA h g^{-1} at 0.2C (99.9% retention) after 100 cycles, 998 mA h g^{-1} at 5C (96% retention) after 500 cycles, and 780 mA h g^{-1} at 10C (80% retention) after 500 cycles outperforming most reported Li-S cells. Although some problems still exist in Li-PS cells, this novel attempt opens up new possibilities to develop a more robust sulfur-based system with ultrafast charging–discharging and high energy density.

15.5 CONCLUSIONS AND CHALLENGES

In conclusion, the development of Li-S batteries holds significant promise for addressing the energy storage demands of high-power and high-energy applications. The unique redox chemistry of sulfur offers a higher theoretical energy density compared to conventional lithium-ion batteries. However, challenges such as the 'shuttle effect', poor conductivity of sulfur, and volume variation during cycling hinder the performance of Li-S batteries. FBN emerges as a promising solution to address these challenges. By incorporating FBN-based interlayers and separators, researchers aim to mitigate the shuttle effect and improve the performance of Li-S batteries. FBN demonstrates exceptional adsorption capability for PS ions, enhancing the battery's cycling stability and Coulombic efficiency.

Moreover, the development of liquid cathode systems, such as Li-PS batteries, further expands the possibilities for improving Li-S battery performance. By utilizing FBN-coated separators, Li-PS batteries show remarkable rate capability, cycling stability, and high energy density, outperforming traditional Li-S cells. Overall, the integration of FBN materials represents a significant advancement in the development of Li-S batteries, offering a pathway to overcome existing challenges and unlock their full potential in energy storage technology.

However, despite these advancements, several challenges remain to be addressed, such as optimizing the FBN coating density and further enhancing the stability of Li-PS cells at high current densities. Additionally, scalability and cost-effectiveness are critical factors that need to be considered for the widespread adoption of Li-S battery technology. The next step in research could involve conducting larger-scale experiments to validate the performance of FBN-coated separators in practical Li-S battery systems. Furthermore, exploring novel materials and fabrication techniques could lead to further improvements in energy density, cycling stability, and cost-effectiveness, ultimately advancing the commercial viability of Li-S batteries for real-world applications.

REFERENCES

1. P. G. Bruce, S. A. Freunberger, L. J. Hardwick, J.-M. Tarascon, Li–O_2 and Li–S batteries with high energy storage. *Nature Materials* **11**, 19–29 (2012).
2. J. R. Akridge, Y. V. Mikhaylik, N. White, Li/S fundamental chemistry and application to high-performance rechargeable batteries. *Solid State Ionics* **175**, 243–245 (2004).
3. X. Ji, L. F. Nazar, Advances in Li–S batteries. *Journal of Materials Chemistry* **20**, 9821–9826 (2010).

4. S. S. Zhang, Liquid electrolyte lithium/sulfur battery: Fundamental chemistry, problems, and solutions. *Journal of Power Sources* **231**, 153–162 (2013).
5. V. Kolosnitsyn, E. Karaseva, Lithium-sulfur batteries: Problems and solutions. *Russian Journal of Electrochemistry* **44**, 506–509 (2008).
6. Y. V. Mikhaylik, J. R. Akridge, Polysulfide shuttle study in the Li/S battery system. *Journal of the Electrochemical Society* **151**, A1969–A1976 (2004).
7. Y. V. Mikhaylik, I. Kovalev, R. Schock, K. Kumaresan, J. Xu, J. Affinito, High energy rechargeable Li-S cells for EV application: Status, remaining problems and solutions. *Ecs Transactions* **25**, 23–34 (2010).
8. Y. Diao, K. Xie, S. Xiong, X. Hong, Analysis of polysulfide dissolved in electrolyte in discharge-charge process of Li-S battery. *Journal of the Electrochemical Society* **159**, A421–A425 (2012).
9. S. Evers, L. F. Nazar, New approaches for high energy density lithium–sulfur battery cathodes. *Accounts of Chemical Research* **46**, 1135–1143 (2012).
10. A. Manthiram, Y. Fu, Y.-S. Su, Challenges and prospects of lithium–sulfur batteries. *Accounts of Chemical Research* **46**, 1125–1134 (2012).
11. R. Demir-Cakan, M. Morcrette, A. Guéguen, R. Dedryvère, J.-M. Tarascon, Li–S batteries: Simple approaches for superior performance. *Energy & Environmental Science* **6**, 176–182 (2013).
12. Y. Yang, G. Zheng, Y. Cui, Nanostructured sulfur cathodes. *Chemical Society Reviews* **42**, 3018–3032 (2013).
13. J. Nelson, S. Misra, Y. Yang, A. Jackson, Y. Liu, H. Wang, H. Dai, J. C. Andrews, Y. Cui, M. F. Toney, In operando X-ray diffraction and transmission X-ray microscopy of lithium sulfur batteries. *Journal of the American Chemical Society* **134**, 6337–6343 (2012).
14. H.-S. Ryu, H.-J. Ahn, K.-W. Kim, J.-H. Ahn, J.-Y. Lee, Discharge process of Li/PVdF/S cells at room temperature. *Journal of Power Sources* **153**, 360–364 (2006).
15. X. Ji, K. T. Lee, L. F. Nazar, A highly ordered nanostructured carbon–sulphur cathode for lithium–sulphur batteries. *Nature Materials* **8**, 500–506 (2009).
16. B. Zhang, X. Qin, G. Li, X. Gao, Enhancement of long stability of sulfur cathode by encapsulating sulfur into micropores of carbon spheres. *Energy & Environmental Science* **3**, 1531–1537 (2010).
17. N. Li, M. Zheng, H. Lu, Z. Hu, C. Shen, X. Chang, G. Ji, J. Cao, Y. Shi, High-rate lithium–sulfur batteries promoted by reduced graphene oxide coating. *Chemical Communications* **48**, 4106–4108 (2012).
18. J.-J. Chen, Q. Zhang, Y.-N. Shi, L.-L. Qin, Y. Cao, M.-S. Zheng, Q.-F. Dong, A hierarchical architecture S/MWCNT nanomicrosphere with large pores for lithium sulfur batteries. *Physical Chemistry Chemical Physics* **14**, 5376–5382 (2012).
19. K. T. Lee, R. Black, T. Yim, X. Ji, L. F. Nazar, Surface-initiated growth of thin oxide coatings for Li-sulfur battery cathodes. *Advanced Energy Materials* **2**, 1490–1496 (2012).
20. L. Yin, J. Wang, F. Lin, J. Yang, Y. Nuli, Polyacrylonitrile/graphene composite as a precursor to a sulfur-based cathode material for high-rate rechargeable Li–S batteries. *Energy & Environmental Science* **5**, 6966–6972 (2012).
21. G. C. Li, G. R. Li, S. H. Ye,X. P. Gao, A polyaniline-coated sulfur/carbon composite with an enhanced high-rate capability as a cathode material for lithium/sulfur batteries. *Advanced Energy Materials* **2**, 1238–1245 (2012).
22. F. Wu, J. Chen, R. Chen, S. Wu, L. Li, S. Chen,T. Zhao, Sulfur/polythiophene with a core/shell structure: Synthesis and electrochemical properties of the cathode for rechargeable lithium batteries. *The Journal of Physical Chemistry* C **115**, 6057–6063 (2011).
23. Z. W. Seh, W. Li, J. J. Cha, G. Zheng, Y. Yang, M. T. McDowell, P.-C. Hsu, Y. Cui, Sulphur–TiO_2 yolk–shell nanoarchitecture with internal void space for long-cycle lithium–sulphur batteries. *Nature Communications* **4**, 1331 (2013).
24. B. Ding, L. Shen, G. Xu, P. Nie, X. Zhang, Encapsulating sulfur into mesoporous TiO_2 host as a high performance cathode for lithium–sulfur battery. *Electrochimica Acta* **107**, 78–84 (2013).
25. F.-F. Zhang, X.-B. Zhang, Y.-H. Dong, L.-M. Wang, Facile and effective synthesis of reduced graphene oxide encapsulated sulfur via oil/water system for high performance lithium sulfur cells. *Journal of Materials Chemistry* **22**, 11452–11454 (2012).

26. H. Wang, Y. Yang, Y. Liang, J. T. Robinson, Y. Li, A. Jackson, Y. Cui, H. Dai, Graphene-wrapped sulfur particles as a rechargeable lithium–sulfur battery cathode material with high capacity and cycling stability. *Nano Letters* **11**, 2644–2647 (2011).
27. S. Evers, L. F. Nazar, Graphene-enveloped sulfur in a one pot reaction: A cathode with good coulombic efficiency and high practical sulfur content. *Chemical Communications* **48**, 1233–1235 (2012).
28. M.-S. Park, J.-S. Yu, K. J. Kim, G. Jeong, J.-H. Kim, Y.-N. Jo, U. Hwang, S. Kang, T. Woo, Y.-J. Kim, One-step synthesis of a sulfur-impregnated graphene cathode for lithium–sulfur batteries. *Physical Chemistry Chemical Physics* **14**, 6796–6804 (2012).
29. S.-H. Chung, A. Manthiram, A hierarchical carbonized paper with controllable thickness as a modulable interlayer system for high performance Li–S batteries. *Chemical Communications* **50**, 4184–4187 (2014).
30. Y.-S. Su, A. Manthiram, A new approach to improve cycle performance of rechargeable lithium–sulfur batteries by inserting a free-standing MWCNT interlayer. *Chemical Communications* **48**, 8817–8819 (2012).
31. S. H. Chung, A. Manthiram, Carbonized eggshell membrane as a natural polysulfide reservoir for highly reversible Li-S batteries. *Advanced Materials* **26**, 1360–1365 (2014).
32. T.-G. Jeong, Y. H. Moon, H.-H. Chun, H. S. Kim, B. W. Cho, Y.-T. Kim, Free standing acetylene black mesh to capture dissolved polysulfide in lithium sulfur batteries. *Chemical Communications* **49**, 11107–11109 (2013).
33. S. Bai, X. Liu, K. Zhu, S. Wu, H. Zhou, Metal–organic framework-based separator for lithium–sulfur batteries. *Nature Energy* **1**, 16094 (2016).
34. J. Sun, Y. Sun, M. Pasta, G. Zhou, Y. Li, W. Liu, F. Xiong, Y. Cui, Entrapment of polysulfides by a black-phosphorus-modified separator for lithium-sulfur batteries. *Advanced Materials* **28**, 9797–9803 (2016).
35. C.-H. Chang, S.-H. Chung, S. Nanda, A. Manthiram, A rationally designed polysulfide-trapping interface on the polymeric separator for high-energy Li–S batteries. *Materials Today Energy* **6**, 72–78 (2017).
36. Y. Jiang, F. Chen, Y. Gao, Y. Wang, S. Wang, Q. Gao, Z. Jiao, B. Zhao, Z. Chen, Inhibiting the shuttle effect of Li–S battery with a graphene oxide coating separator: Performance improvement and mechanism study. *Journal of Power Sources* **342**, 929–938 (2017).
37. C. Oh, N. Yoon, J. Choi, Y. Choi, S. Ahn, J. K. Lee, Enhanced Li–S battery performance based on solution-impregnation-assisted sulfur/mesoporous carbon cathodes and a carbon-coated separator. *Journal of Materials Chemistry A* **5**, 5750–5760 (2017).
38. G. Yang, S. Shi, J. Yang, Y. Ma, Insight into the role of $Li_2 S_2$ in Li–S batteries: A first-principles study. *Journal of Materials Chemistry A* **3**, 8865–8869 (2015).
39. H. Wang, C. Zhang, Z. Chen, H. K. Liu, Z. Guo, Large-scale synthesis of ordered mesoporous carbon fiber and its application as cathode material for lithium–sulfur batteries. *Carbon* **81**, 782–787 (2015).
40. S. Chen, B. Sun, X. Xie, A. K. Mondal, X. Huang, G. Wang, Multi-chambered micro/mesoporous carbon nanocubes as new polysulfides reserviors for lithium–sulfur batteries with long cycle life. *Nano Energy* **16**, 268–280 (2015).
41. Y. Fan, Z. Yang, W. Hua, D. Liu, T. Tao, M. M. Rahman, W. Lei, S. Huang, Y. I. Chen, Functionalized boron nitride nanosheets/graphene interlayer for fast and long-life lithium–sulfur batteries. *Advanced Energy Materials* **7**, 1602380 (2017).
42. Y. Fan, D. Liu, M. M. Rahman, T. Tao, W. Lei, S. Mateti, B. Yu, J. Wang, C. Yang, Y. I. Chen, Repelling polysulfide ions by boron nitride nanosheet coated separators in lithium–sulfur batteries. *ACS Applied Energy Materials* **2**, 2620–2628 (2019).
43. Y. Fan, M. M. Rahman, T. Tao, Y. I. Chen, Ultra-fast and high-energy density polysulfide-eight ion batteries. *Journal of Power Sources* **477**, 229018 (2020).
44. J.-Q. Huang, T.-Z. Zhuang, Q. Zhang, H.-J. Peng, C.-M. Chen, F. Wei, Permselective graphene oxide membrane for highly stable and anti-self-discharge lithium–sulfur batteries. *ACS Nano* **9**, 3002–3011 (2015).

45. Y. Zhang, L. Miao, J. Ning, Z. Xiao, L. Hao, B. Wang, L. Zhi, A graphene-oxide-based thin coating on the separator: An efficient barrier towards high-stable lithium–sulfur batteries. *2D Materials* **2**, 024013 (2015).
46. M. Barghamadi, A. S. Best, A. I. Bhatt, A. F. Hollenkamp, M. Musameh, R. J. Rees, T. Rüther, Lithium–sulfur batteries—the solution is in the electrolyte, but is the electrolyte a solution? *Energy & Environmental Science* **7**, 3902–3920 (2014).
47. S. S. Zhang, Improved cyclability of liquid electrolyte lithium/sulfur batteries by optimizing electrolyte/sulfur ratio. *Energies* **5**, 5190–5197 (2012).
48. S. S. Zhang, Does the sulfur cathode require good mixing for a liquid electrolyte lithium/sulfur cell? *Electrochemistry Communications* **31**, 10–12 (2013).
49. S. Han, X. Pu, X. Li, M. Liu, M. Li, N. Feng, S. Dou, W. Hu, High areal capacity of Li-S batteries enabled by freestanding CNF/rGO electrode with high loading of lithium polysulfide. *Electrochimica Acta* **241**, 406–413 (2017).
50. G. Zhou, E. Paek, G. S. Hwang, A. Manthiram, Long-life Li/polysulphide batteries with high sulphur loading enabled by lightweight three-dimensional nitrogen/sulphur-codoped graphene sponge. *Nature Communications* **6**, 7760 (2015).
51. X. Liang, C. Hart, Q. Pang, A. Garsuch, T. Weiss, L. F. Nazar, A highly efficient polysulfide mediator for lithium–sulfur batteries. *Nature Communications* **6**, 5682 (2015).
52. Y. Qiu, W. Li, W. Zhao, G. Li, Y. Hou, M. Liu, L. Zhou, F. Ye, H. Li, Z. Wei, High-rate, ultralong cycle-life lithium/sulfur batteries enabled by nitrogen-doped graphene. *Nano Letters* **14**, 4821–4827 (2014).
53. S. Urbonaite, P. Novák, Importance of 'unimportant' experimental parameters in Li–S battery development. *Journal of Power Sources* **249**, 497–502 (2014).
54. W. Lei, V. N. Mochalin, D. Liu, S. Qin, Y. Gogotsi, Y. I. Chen, Boron nitride colloidal solutions, ultralight aerogels and freestanding membranes through one-step exfoliation and functionalization. *Nature Communications* **6**, 8849 (2015).
55. M. Shanmugam, R. J. Gedrim, C. Durcan, B. Yu, 2D layered insulator hexagonal boron nitride enabled surface passivation in dye sensitized solar cells. *Nanoscale* **5**, 11275–11282 (2013).
56. L. Zhang, L. Ji, P. Glans, Y. Zhang, J. Zhu, J. Guo, Electronic structure and chemical bonding of a graphene oxide–sulfur nanocomposite for use in superior performance lithium–sulfur cells. *Physical Chemistry Chemical Physics* **14**, 13670–13675 (2012).
57. J. D. Carpenter, B. S. Ault, Matrix isolated study of the reaction of diborane with methanethiol: Spectroscopic characterization of methylmercaptoborane, H_2BSCH_3. *The Journal of Physical Chemistry* **97**, 3697–3700 (1993).
58. Z. Xiao, Z. Yang, H. Nie, Y. Lu, K. Yang, S. Huang, Porous carbon nanotubes etched by water steam for high-rate large-capacity lithium–sulfur batteries. *Journal of Materials Chemistry A* **2**, 8683 (2014).
59. D. Liu, W. Lei, S. Qin, K. D. Klika, Y. I. Chen, Superior adsorption of pharmaceutical molecules by highly porous BN nanosheets. *Physical Chemistry Chemical Physics* **18**, 84–88 (2016).
60. J. Wang, F. Lin, H. Jia, J. Yang, C. W. Monroe, Y. NuLi, Towards a safe lithium–sulfur battery with a flame-inhibiting electrolyte and a sulfur-based composite cathode. *Angewandte Chemie International Edition* **53**, 10099 (2014).
61. S. Das, S. Majumder, R. Katiyar, Kinetic analysis of the Li^+ ion intercalation behavior of solution derived nano-crystalline lithium manganate thin films. *Journal of Power Sources* **139**, 261–268 (2005).
62. A. J. D'Angelo, M. J. Panzer, Enhanced lithium ion transport in poly (ethylene glycol) diacrylate-supported solvato ionogel electrolytes via chemically cross-linked ethylene oxide pathways. *The Journal of Physical Chemistry B* **121**, 890–895 (2017).
63. K.-H. Oh, D. Lee, M.-J. Choo, K. H. Park, S. Jeon, S. H. Hong, J.-K. Park, J. W. Choi, Enhanced durability of polymer electrolyte membrane fuel cells by functionalized 2D boron nitride nanoflakes. *ACS Applied Materials & Interfaces* **6**, 7751–7758 (2014).
64. S.-Y. Xie, W. Wang, K. S. Fernando, X. Wang, Y. Lin, Y.-P. Sun, Solubilization of boron nitride nanotubes. *Chemical Communications* **29**, 3670–3672 (2005).
65. Y. Lin, T. V. Williams, J. W. Connell, Soluble, exfoliated hexagonal boron nitride nanosheets. *The Journal of Physical Chemistry Letters* **1**, 277–283 (2009).

66. J. Shim, H. J. Kim, B. G. Kim, Y. S. Kim, D.-G. Kim, J.-C. Lee, 2D boron nitride nanoflakes as a multifunctional additive in gel polymer electrolyte for safe, long cycle life and high rate lithium metal batteries. *Energy & Environmental Science* **10**, 1911–1916 (2017).
67. D. Aurbach, O. Chusid, In situ FTIR spectroelectrochemical studies of surface films formed on Li and nonactive electrodes at low potentials in Li salt solutions containing CO_2. *Journal of the Electrochemical Society* **140**, L155–L157 (1993).
68. C. Chadefaux, A.-S. Le Hô, L. Bellot-Gurlet, I. Reiche, Curve-fitting micro-ATR-FTIR studies of the amide I and II bands of type I collagen in archaeological bone materials. *e-PRESERVATION Science* **6**, 129–137 (2009).
69. G. Penel, G. Leroy, C. Rey, E. Bres, MicroRaman spectral study of the PO_4 and CO_3 vibrational modes in synthetic and biological apatites. *Calcified Tissue International* **63**, 475–481 (1998).
70. I. Li, C. H. Su, H. S. Sheu, H. C. Chiu, Y. W. Lo, W. T. Lin, J. H. Chen, C. S. Yeh, $Gd_2O(CO_3)_2{\cdot}H_2O$ particles and the corresponding Gd_2O_3: Synthesis and applications of magnetic resonance contrast agents and template particles for hollow spheres and hybrid composites. *Advanced Functional Materials* **18**, 766–776 (2008).
71. Z. A. Ghazi, X. He, A. M. Khattak, N. A. Khan, B. Liang, A. Iqbal, J. Wang, H. Sin, L. Li, Z. Tang, MoS_2/celgard separator as efficient polysulfide barrier for long-life lithium–sulfur batteries. *Advanced Materials* **29**, 1606817 (2017).
72. Y. Li, H. Zhan, S. Liu, K. Huang, Y. Zhou, Electrochemical properties of the soluble reduction products in rechargeable Li/S battery. *Journal of Power Sources* **195**, 2945–2949 (2010).
73. S. S. Zhang, J. A. Read, A new direction for the performance improvement of rechargeable lithium/sulfur batteries. *Journal of Power Sources* **200**, 77–82 (2012).
74. S.-Y. Chung, J. T. Bloking, Y.-M. Chiang, Electronically conductive phospho-olivines as lithium storage electrodes. *Nature Materials* **1**, 123–128 (2002).
75. A. Manthiram, Materials challenges and opportunities of lithium ion batteries. *The Journal of Physical Chemistry Letters* **2**, 176–184 (2011).
76. L. Qie, A. Manthiram, A facile layer-by-layer approach for high-areal-capacity sulfur cathodes. *Advanced Materials* **27**, 1694–1700 (2015).
77. X. Gao, J. Li, D. Guan, C. Yuan, A scalable graphene sulfur composite synthesis for rechargeable lithium batteries with good capacity and excellent columbic efficiency. *ACS Applied Materials & Interfaces* **6**, 4154–4159 (2014).
78. Y. V. Mikhaylik, J. R. Akridge, Polysulfide shuttle study in the Li/S battery system. *Journal of the Electrochemical Society* **151**, A1969 (2004).
79. H. Yamin, E. Peled, Electrochemistry of a nonaqueous lithium/sulfur cell. *Journal of Power Sources* **9**, 281–287 (1983).
80. G. Zhou, D.-W. Wang, L.-C. Yin, N. Li, F. Li, H.-M. Cheng, Oxygen bridges between NiO nanosheets and graphene for improvement of lithium storage. *ACS Nano* **6**, 3214–3223 (2012).
81. M. Reddy, T. Yu, C.-H. Sow, Z. X. Shen, C. T. Lim, G. Subba Rao, B. Chowdari, α-Fe_2O_3 nanoflakes as an anode material for Li-ion batteries. *Advanced Functional Materials* **17**, 2792–2799 (2007).
82. S. Zeng, L. Li, L. Xie, D. Zhao, N. Zhou, N. Wang, S. Chen, Graphene-supported highly crosslinked organosulfur nanoparticles as cathode materials for high-rate, long-life lithium-sulfur battery. *Carbon* **122**, 106–113 (2017).
83. J. S. Lee, W. Kim, J. Jang, A. Manthiram, Sulfur-embedded activated multichannel carbon nanofiber composites for long-life, high-rate lithium–sulfur batteries. *Advanced Energy Materials* **7**, 1601943 (2017).
84. C. Zheng, S. Niu, W. Lv, G. Zhou, J. Li, S. Fan, Y. Deng, Z. Pan, B. Li, F. Kang, Propelling polysulfides transformation for high-rate and long-life lithium–sulfur batteries. *Nano Energy* **33**, 306–312 (2017).
85. Y. Li, J. Fan, J. Zhang, J. Yang, R. M. Yuan, J.-K. Chang, M. S. Zheng, Q. F. Dong, A honeycomb-like Co@N–C composite for ultrahigh sulfur loading Li–S batteries. *ACS Nano* **11**, 11417–11424 (2017).
86. D. Su, M. Cortie, G. Wang, Fabrication of N-doped graphene–carbon nanotube hybrids from Prussian blue for lithium–sulfur batteries. *Advanced Energy Materials* **7**, 1602014 (2017).
87. H. Pan, Z. Cheng, Z. Xiao, X. Li, R. Wang, The fusion of imidazolium-based ionic polymer and carbon nanotubes: One type of new heteroatom-doped carbon precursors for high-performance lithium–sulfur batteries. *Advanced Functional Materials* **27**, 1703936 (2017).

88. X. Li, K. Ding, B. Gao, Q. Li, Y. Li, J. Fu, X. Zhang, P. K. Chu, K. Huo, Freestanding carbon encapsulated mesoporous vanadium nitride nanowires enable highly stable sulfur cathodes for lithium-sulfur batteries. *Nano Energy* **40**, 655–662 (2017).
89. Z. Xiao, Z. Yang, L. Zhou, L. Zhang, R. Wang, Highly conductive porous transition metal dichalcogenides via water steam etching for high-performance lithium–sulfur batteries. *ACS Applied Materials & Interfaces* **9**, 18845–18855 (2017).
90. H. Tang, S. Yao, X. Shen, X. Xi, K. Xiao, Lithium–sulfur batteries with high rate and cycle performance by using multilayered separators coated with Ketjen Black. *Energy Technology* **5**, 623–628 (2017).
91. D. Guo, X. A. Chen, H. Wei, M. Liu, F. Ding, Z. Yang, K. Yang, S. Wang, X. Xu, S. Huang, Controllable synthesis of highly uniform flower-like hierarchical carbon nanospheres and their application in high performance lithium–sulfur batteries. *Journal of Materials Chemistry A* **5**, 6245–6256 (2017).
92. Y. Liu, G. Li, J. Fu, Z. Chen, X. Peng, Strings of porous carbon polyhedrons as self-standing cathode host for high-energy-density lithium–sulfur batteries. *Angewandte Chemie* **129**, 6272–6276 (2017).
93. L. Wei, T. Zhao, G. Zhao, L. An, L. Zeng, A high-performance carbon nanoparticle-decorated graphite felt electrode for vanadium redox flow batteries. *Applied Energy* **176**, 74–79 (2016).
94. I. Mustafa, I. Lopez, H. Younes, R. A. Susantyoko, R. A. Al-Rub, S. Almheiri, Fabrication of freestanding sheets of multiwalled carbon nanotubes (Buckypapers) for vanadium redox flow batteries and effects of fabrication variables on electrochemical performance. *Electrochimica Acta* **230**, 222–235 (2017).
95. N. Ding, X. Li, S. W. Chien, Z. Liu, Y. Zong, In situ monitoring the viscosity change of an electrolyte in a Li–S battery. *Chemical Communications* **53**, 10152–10155 (2017).

Chapter 16

Boosting Lithium Storage Capacity by Introducing Defects in Graphene and Boron Nitride

Khagesh Tanwar

16.1 INTRODUCTION

The most common anode, graphitic carbon, used in lithium batteries delivers a specific capacity of 372 mA h g^{-1} [1,2] when used with cathodes such as $LiCoO_2$ (274 mA h g^{-1}) [3] and $LiFePO_4$ (170 mA h g^{-1}) [4]. However, due to its limited storage capacity, graphitic carbon as an anode is incompatible with high-capacity batteries such as lithium–sulfur (Li-S) batteries [2]. An improvement in the performance of lithium batteries requires exploration of promising alternative anode materials. Lithium metal as an anode material with the highest specific capacity, 3860 mA h g^{-1} [5,6], is an ideal candidate. However, its practical application is limited by dendritic growth during charge–discharge cycles, which poses safety risks and degrades performance. To overcome this limitation, alternative high-capacity anode materials such as silicon [7,8], black phosphorus [9], metal sulfides [8], and metal oxides [10–12] were explored both theoretically and experimentally [13,14]. Among these, silicon-based anodes have demonstrated significant promise, despite undergoing substantial volume changes during lithiation and delithiation processes [9,15]. Recent advancements in material development have successfully enhanced battery performance, mitigating some of these challenges [16–18]. In case of metal oxides or sulfides, a wide range of materials were studied, and some promising candidates were proposed [10]. However, due to their higher intercalation potential and low practical reversible capacity, little practical utilization was observed [10].

Recent advancements in nanoscale device fabrication using Van der Waals homostructures and heterostructures as anodes have enabled ultradense lithium storage in Li-ion batteries [19–21]. Absorption of lithium ions on 2D monolayer structures such as graphene [22], hexagonal BN (*h*-BN) [23], MoS_2 [24,25], and ZrS_2 [26] was widely explored for their use as anode materials [27]. Bilayer graphene was reported to exhibit multilayer stacked lithium storage with a hexagonal close packing which exceeded the capacity of graphitic carbon (LiC_6) [20]. Besides, in bilayer homostructures, MoS_2-MoS_2 was studied, and lithiation and phase transition upon lithium intercalation were revealed by in-situ transmission electron microscopy [28]. The promising results observed in the homostructures further encouraged the exploration of heterostructures, where

DOI: 10.1201/9781003647812-20

the suitable properties can be tailor-made by selecting various monolayers. For instance, metal sulfide monolayers possess weak mechanical strength and low electronic conductivity, which can be significantly improved by forming heterostructures of metal sulfides with graphene. With the idea of tunability, Van der Waals heterostructures such as graphene–MoS_2 [29,30], graphene–(hexagonal BN) [23], graphene–MoO_2 [31], (hexagonal BN)–MoS_2 [29], and WS_2–$NbSe_2$ [32] were explored as anode materials by first-principles methods. Interestingly, in most of the explored homo/heterostructures, the lithium storage capacity exceeded the limit of bulk graphite structure (LiC_6).

Recently, with defect engineering, 2D bilayer structures were modified to fine-tune their electronic properties [33,34]. In fact, the heterostructures are also a type of 2D defect in Van der Waals structures, which happen to improve the lithium storage capacity. Therefore, it is plausible to ask if lithium storage can be further enhanced in Van der Waals homostructures and heterostructures by introducing controlled defects. It is especially interesting considering the possibility of creating controlled point defects for practical devices by employing modern techniques [33]. In this chapter, using density functional theory (DFT)-based calculations, BN-BN, G-G, and G-BN 2D bilayer structures with point defects were studied for their energetic stability and lithium storage capacity. Point defects in these structures induce ultradense lithium ordering, significantly surpassing the storage capacity of previously studied similar materials. For example, the storage capacity of a G-BN heterostructure with a point defect was found to be 11 times greater than that of the same heterostructure without a defect. The immense increase of lithium storage in G-BN heterostructures with point defects nominates them as potential anode materials for high-capacity lithium batteries.

16.2 METHODS OF DFT CALCULATIONS

First-principles DFT calculations were carried out using the VASP 5.4.4 package [35,36]. The Generalized Gradient Approximation, as formulated by Perdew, Burke, and Ernzerhof, was employed [37]. Ion-electron interactions were modeled using projector-augmented wave potentials with a plane-wave cutoff energy set to 500 eV [38,39]. Geometry optimization was performed with a force convergence threshold of 0.01 eV $Å^{-1}$ per atom and an energy convergence threshold of 10^{-5} eV for energy. The Brillouin zone was sampled using a $5 \times 5 \times 1$ Γ-centered κ-points grid. Van der Waals interactions were accounted for using the DFT-D3 method developed by Grimme [40]. Structures including graphene (G), BN, BN-BN, G-G, and G-BN were modeled with a $2\sqrt{3} \times 2\sqrt{3}$ surface unit cell. These designs were created using the VESTA visualization software [41]. The G-BN heterostructure exhibits a 1.88% strain due to lattice mismatch. A 20 Å vacuum was applied in the z-direction to prevent interactions between periodic images. While AB-type stacking is generally more stable in pristine 2D bilayer structures, AA-type stacking was chosen for the base configurations due to its higher stability following lithium intercalation. To introduce point defects, one boron atom was removed from the top layer of the BN-BN and G-BN structures, while one carbon atom was removed from the top layer of the G-G structure. The removal of a single atom from a 48-atom 2D bilayer structure corresponds to a defect density of 2.08 at %. Lithium atoms were incrementally added at the most energetically stable sites between the layers until maximum intercalation was achieved. Charge transfer between lithium and the 2D bilayer structures was evaluated using Bader charge population analysis [42], with DDEC6-type net atomic charges determined through the Chargemol program [43].

16.3 LITHIUM INTERCALATION SCHEME IN BILAYER 2D MATERIALS

The distribution of lithium atoms in bilayer structures was optimized using a G-BN heterostructure with a point defect (boron defect) as shown in Figure 16.1. The optimized structure was realized by finding the most stable lithium intercalation site in the structure using energetic stability based on the intercalation energy calculated by the formula:

$$E_I = \frac{1}{n}\left[E\left(\text{Li}_n X\right) - E\left(X\right) - nE\left(\text{Li}\right)\right] \tag{16.1}$$

where E_I represents the intercalation energy, $E\,(\text{Li}_n X)$ is the energy of the structure containing 'n' lithium atoms, $E\,(X)$ is the energy of the base structure without any lithium (X corresponds to BN-BN, G-BN, or G-G with or without a point defect), and $E\,(\text{Li})$ denotes the energy per atom of lithium in bulk body-centered cubic lithium.

To understand the intercalation in bilayer structures, two different types of sites, namely, favorable sites and stable sites, were considered. The favorable sites were selected based on symmetry and distance from the other lithium atoms in the structure. Then, the site with the highest intercalation energy among the favorable sites was considered as a stable site. For instance, as presented in Figure 16.1 (1-Li), for the first lithium, two different favorable sites were considered; first, just below the defect, and the second farthest from the defect, as highlighted by dashed black circles. The lithium at the site below the defect is more stable (−3.24 eV) as compared to at the site far from the defect (−1.61 eV), as plotted in Figure 16.2 (1-Li). Therefore, the site below

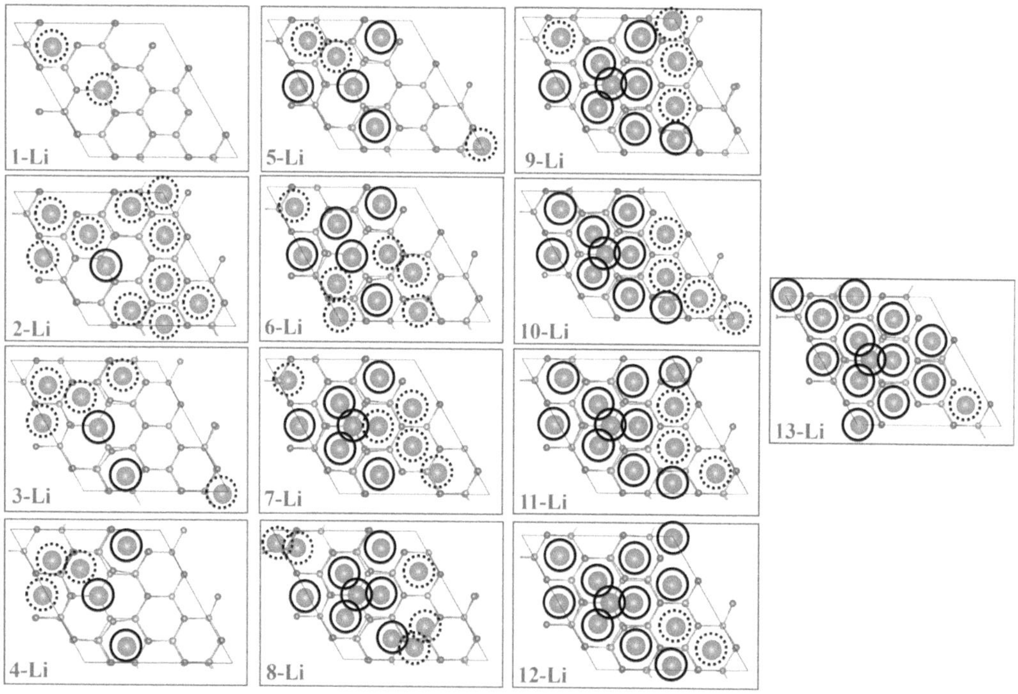

Figure 16.1 Lithium intercalation scheme in G-BN structure with a point defect; the favorable sites are highlighted by dashed black circles; the most stable ($n-1$) sites are highlighted by solid black circles. (Reprinted with permission [44]. Copyright 2021 American Chemical Society.)

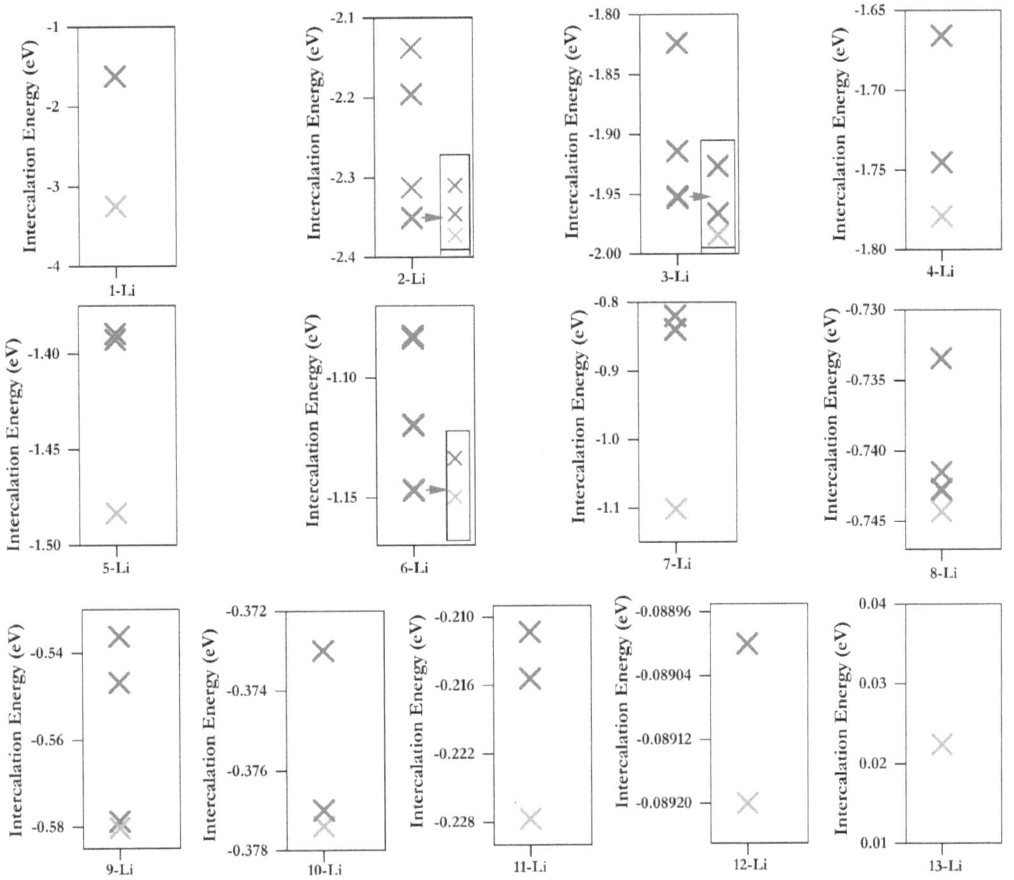

Figure 16.2 Intercalation energy for favorable sites in G-BN structure with one defect; the most stable sites are highlighted by green points; the total intercalation energy is positive for 13-Li; therefore, it was not considered for comparison. (Reprinted with permission [44]. Copyright 2021 American Chemical Society.)

the defect was considered as the stable site for the first lithium. For the second lithium, the first lithium was kept fixed (highlighted by a solid back circle in Figure 16.1 (2-Li)), and the intercalation energies for all the other favorable sites were calculated. The most stable site for 2-Li was found based on the lowest intercalation energy in Figure 16.2 (2-Li). For the third lithium, two lithium atoms at stable sites were kept fixed (Figure 16.1 (3-Li)), and other favorable sites were explored. This way, a stable site for the nth lithium was found by keeping $(n-1)$ lithium fixed and exploring the favorable sites. The intercalation energy of the stable site for the nth lithium among the explored favorable sites is highlighted by green color in Figure 16.2.

16.4 LITHIUM STORAGE IN BILAYER 2D MATERIALS

Figure 16.3a–c illustrates the schematic of the maximum lithium intercalation in BN-BN, G-G, and G-BN structures without point defects. The BN-BN, G-G, and G-BN structures accommodate zero, five, and one lithium atom(s), respectively, in their defect-free configurations.

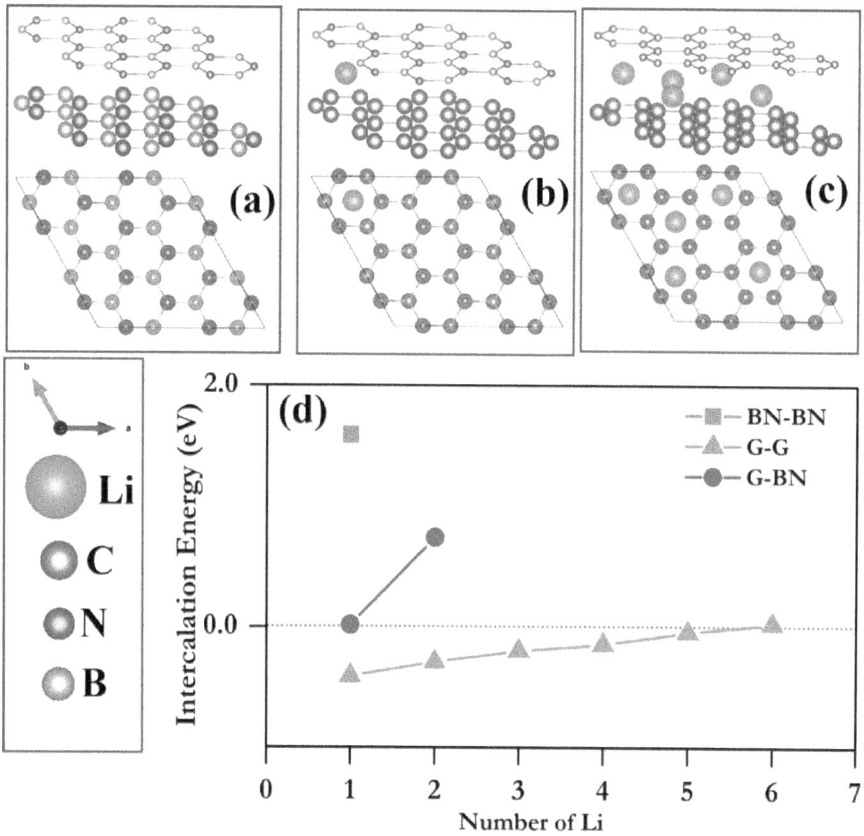

Figure 16.3 (a–c) Side and top views of the maximum lithium intercalation in (a) BN-BN, (b) G-G, and (c) G-BN structures without a point defect. (d) Intercalation energy as a function of lithium concentration in BN-BN, G-G, and G-BN structures without point defects. (Reprinted with permission [44]. Copyright 2021 American Chemical Society.)

16.5 LITHIUM STORAGE IN BILAYER 2D MATERIALS WITH POINT DEFECTS

Figure 16.4a–f shows the relaxed BN-BN, G-G, and G-BN structures without and with a point defect, respectively. As shown in Figure 16.4g (left axis), the binding energy of structures with a point defect is higher compared to those without a point defect. This indicates that the structures become more stable upon introducing a point defect. The total charge on the top layer transferred from the bottom layer was calculated by Bader charge analysis. It is apparent from Figure 16.4g (right axis) that the introduction of a point defect increases the charge transfer between the layers. The stability of the structures based upon binding energy is in correlation with the net charge transfer upon defect creation, as observed from Figure 16.4g.

To analyze the impact of point defects at the molecular scale, the band structures of all the studied configurations were computed. The incorporation of point defects into these structures results in the emergence of defect states within the band structure. Figure 16.5a–f compares the band structures of the structures without or with a point defect. The BN-BN structure exhibits

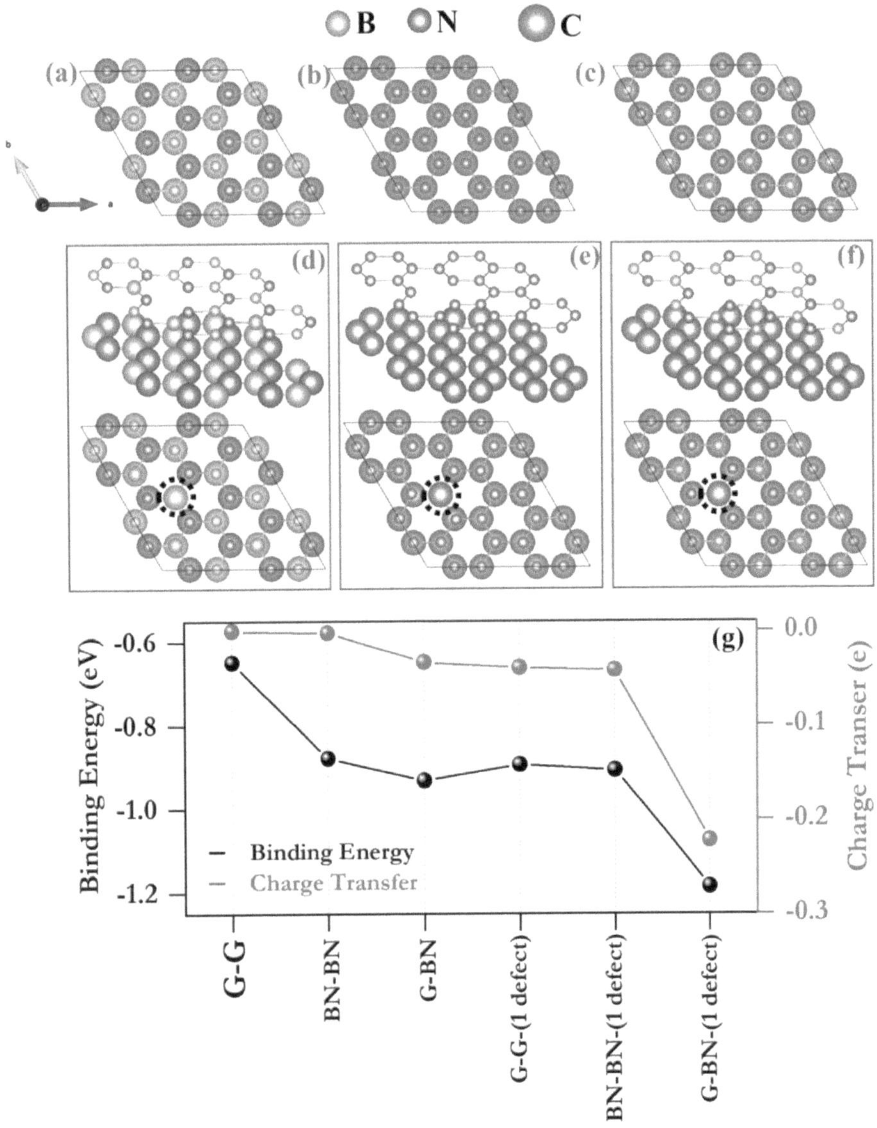

Figure 16.4 (a–c) Top views of relaxed BN-BN, G-G, and G-BN structures without a point defect. (d–f) Side and top views of relaxed BN-BN, G-G, and G-BN structures with a point defect, with the defect site highlighted in a dashed black circle. (g) Binding energy and charge transfer (to the top layer) of various homostructures and heterostructures with a single point defect. (Reprinted with permission [44]. Copyright 2021 American Chemical Society.)

a wide bandgap in the absence of point defects. However, the introduction of a point defect significantly narrows the bandgap, attributed to the appearance of defect states near the Fermi level (Figure 16.5a and b). Analogously, in the band structures of G-G and G-BN structures, the emergence of several extra bands close to the Fermi level was observed.

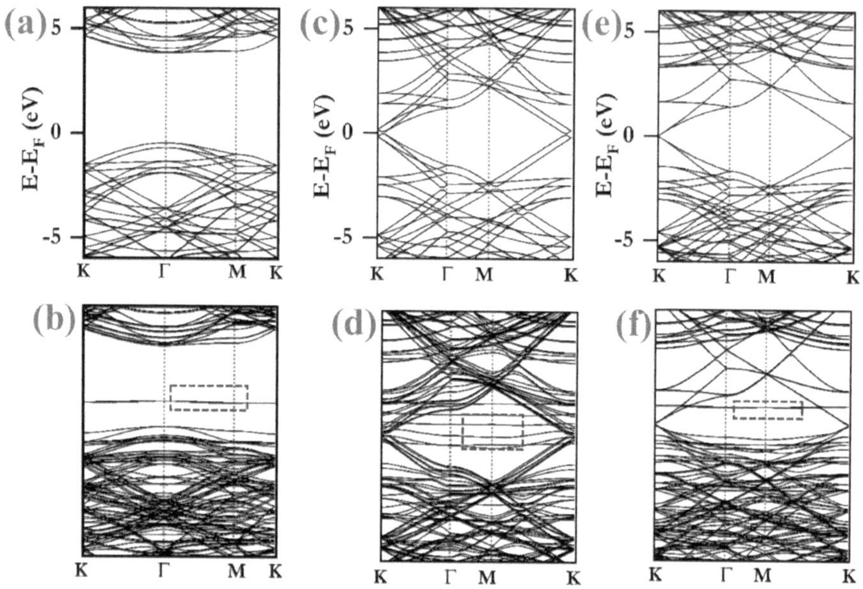

Figure 16.5 (a, b) Band structures of BN-BN without and with point defects. (c, d) Band structures of G-G without and with point defects. (e, f) Band structures of G-BN without and with point defects. (Reprinted with permission [44]. Copyright 2021 American Chemical Society.)

To examine the impact of point defects on lithium storage in bilayer structures, the energetic stability of BN-BN, G-G, and G-BN configurations with a point defect was analyzed after lithium intercalation and compared to their defect-free counterparts. Figure 16.6a–c illustrates the schematics of maximum lithium intercalation in BN-BN, G-G, and G-BN structures with a point defect, respectively. The defect-free BN-BN, G-G, and G-BN structures can accommodate zero, five, and one lithium atoms, as shown in Figure 16.3a–c, which increases to 4, 10, and 12 after introducing a point defect in the structure, as observed in Figure 16.6a–c.

Figure 16.6d illustrates the change in intercalation energy as lithium is incrementally added, one atom at a time, to the most energetically stable site. Interestingly, the G-BN structure without a point defect can accommodate only one lithium atom, whereas introducing a point defect increases its capacity to hold up to 12 lithium atoms, an 11-fold enhancement in lithium storage. As shown in Figure 16.6d, further lithium insertion renders the structure energetically unfavorable, resulting in positive intercalation energy. Bilayer graphene (G-G structure) accommodates a maximum of five lithium atoms, consistent with previously reported studies [23]. Introducing a carbon defect in the top layer of the G-G structure doubles its lithium storage capacity, allowing it to accommodate up to 10 lithium atoms, as shown in Figure 16.6d. In contrast, the BN-BN structure is unstable even with a single lithium atom, as seen in Figure 16.3d. However, creating a boron defect enables the structure to hold up to four lithium atoms. Although, as evident in Figure 16.6g, the intercalation energy of the BN-BN structure with a point defect remains negative even after eight lithium insertions, according to the intercalation energy criteria, the structure can store more than eight lithium atoms. However, as discussed in Section 16.6, the interlayer distance in the BN-BN structure with a point defect changes significantly, indicating the instability of the structure after the intercalation of four lithium atoms.

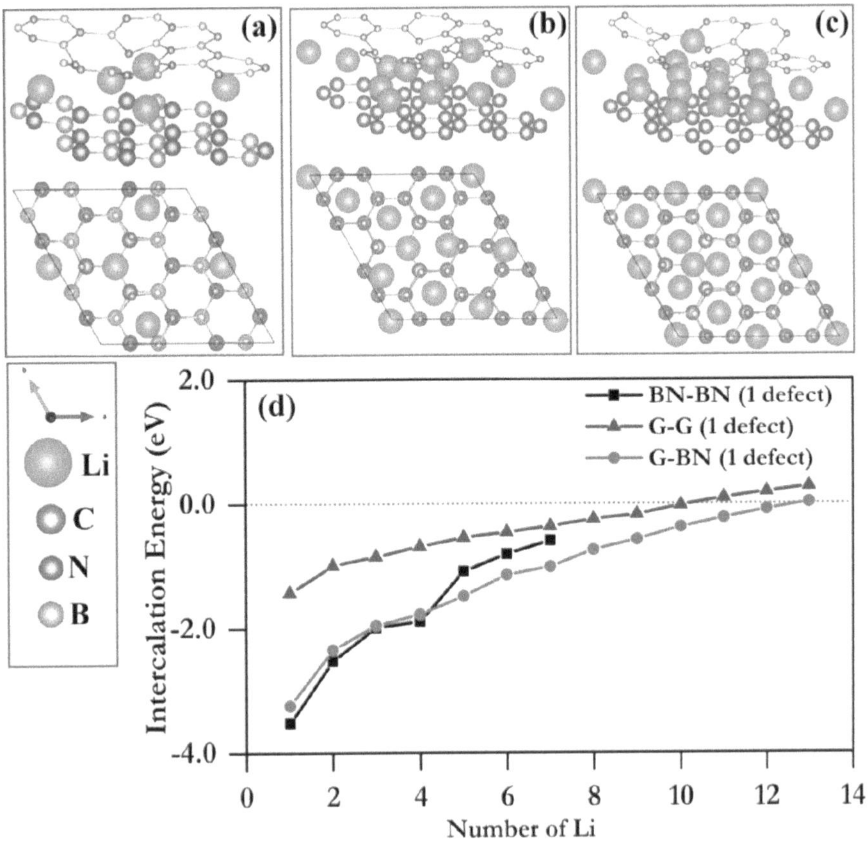

Figure 16.6 (a–c) (side and top views): Maximum lithium intercalation in BN-BN (a), G-G (b), and G-BN (c) structures with a point defect. (d) Intercalation energy as a function of lithium concentration in BN-BN, G-G, and G-BN structures with a point defect. (Reprinted with permission [44]. Copyright 2021 American Chemical Society.)

16.6 EFFECTS OF HIGH DEFECT DENSITY ON LITHIUM INTERCALATION

To examine the impact of increasing defects on lithium storage, two defects, one in the top layer and one in the bottom layer, were introduced in all the structures considered. As shown in Figure 16.7a, the structures with two defects are slightly more stable than their counterparts with a single point defect. The G-BN structure with two point defects is the most stable configuration (Figure 16.7a); thus, it was selected for further investigation into its lithium storage capacity.

Figure 16.7b presents the schematics of the G-BN structure with two point defects, one located in the top layer and the other in the bottom layer. As shown in Figure 16.7c, the intercalation energy in the G-BN structure with two point defects decreases in a manner similar to the structure with one point defect. After the intercalation of 12 lithium atoms, the intercalation energy in the G-BN structure with one point defect turns positive. Interestingly, in the G-BN structure

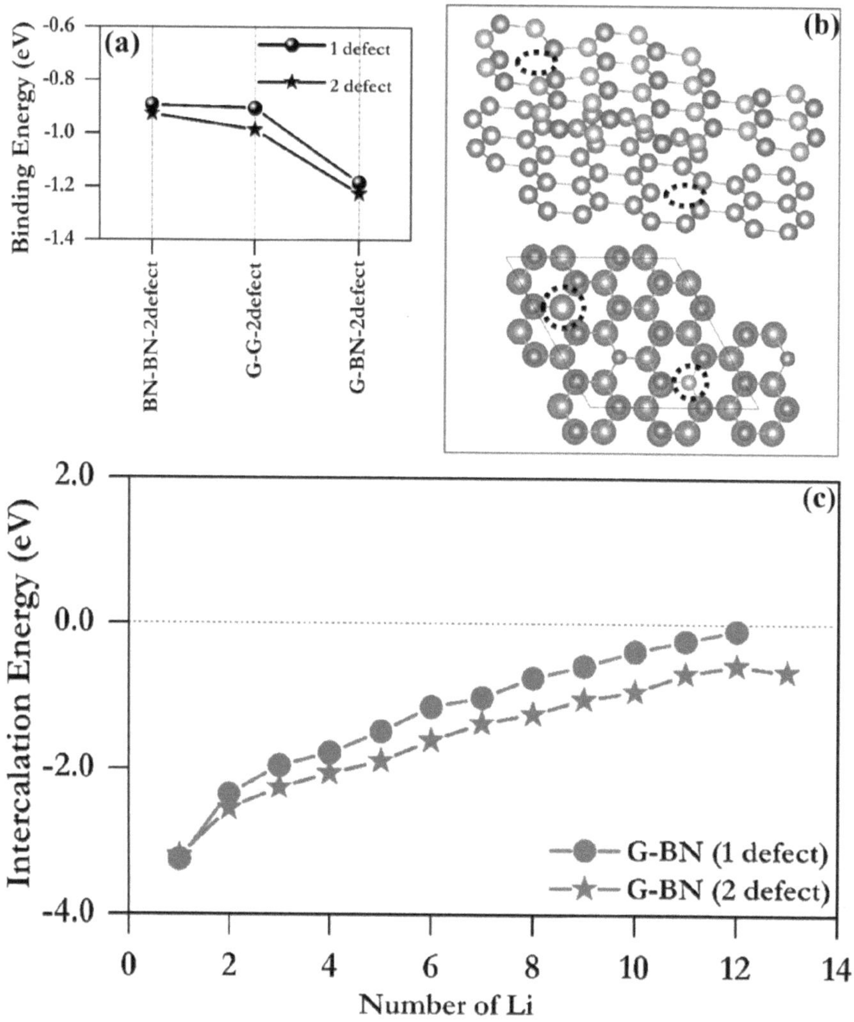

Figure 16.7 (a) The binding energy of different structures with one and two point defects. (b) (Side and top views) The relaxed G-BN structure with two point defects (one in the top layer and one in the bottom layer), with defect sites highlighted by dashed black circles. (c) The intercalation energy of lithium atoms in the G-BN structure with one and two point defects. (Reprinted with permission [44]. Copyright 2021 American Chemical Society.)

with two point defects, the intercalation energy becomes even more negative after the addition of the 12th lithium atom, as illustrated in Figure 16.7c, therefore suggesting that more than 13 Li atoms are stable in the structure. However, the interlayer distance in the G-BN structure with two point defects increases abruptly by 68% after the intercalation of 13 lithium atoms, indicating the breakdown of layer binding, as discussed in detail in Section 16.6. Therefore, it was concluded that the G-BN structure with two point defects can accommodate 12 lithium atoms, the same as the structure with one point defect. This suggests that increasing the defect density in these structures does not lead to a further increase in lithium storage capacity.

16.7 LITHIUM INTERCALATION MECHANISM IN BILAYER 2D MATERIALS WITH POINT DEFECTS

To investigate the mechanism behind lithium storage, Bader charge analysis was used to calculate the total charge transfer from lithium atoms to the structures. Figure 16.8 presents the variation in charge transfer for different configurations as the lithium concentration increases. Notably, in both the G-G and G-BN structures with one point defect, the charge transfer increases steadily with lithium insertion. In contrast, for the BN-BN structure with one point defect, the charge transfer decreases after the intercalation of four lithium atoms.

Figure 16.9a–c shows the side and top views of differential charge density corresponding to varying lithium concentration for BN-BN, G-G, and G-BN structures with a point defect, respectively. In the BN-BN structure with a point defect (Figure 16.9a), the charge primarily transfers to the top layer after eight lithium intercalations. In contrast, for the G-G and G-BN structures with a point defect (Figure 16.9b, c), the charge is distributed more evenly across both layers.

In the BN-BN bilayer structure with a point defect, the charge is mainly transferred to the top layer (the layer with the defect) up to four lithium intercalations. This occurs because the pristine BN layer (without a point defect) does not accept any charge from lithium, as shown in Figure 16.3d (BN-BN structure without a point defect). After four lithium atoms are intercalated, the top layer becomes saturated and can no longer accept additional charge from the lithium atoms. As a result, the dominant internal repulsion between the lithium atoms leads to layer separation after the addition of the 4th lithium in the BN-BN structure with a point defect. Therefore, it is concluded that the BN-BN structure with a point defect can accommodate only four lithium atoms, which represents a notable increase compared to the BN-BN structure without a point defect.

The higher lithium storage capacity in the G-BN structure with a point defect, compared to the G-G structure with a point defect, can be attributed to the formation of a potential well in the structure with a point defect. A larger potential well results in a higher binding energy for the first intercalated lithium atom. As shown in Figure 16.6d, the intercalation energy for the first lithium atom is –1.42 eV in the G-G structure with a point defect and –3.52 eV in the G-BN structure with a point defect. This confirms the presence of a larger potential well in the G-BN structure, leading to its higher lithium storage capacity compared to the G-G structure. To gain further insights into the lithium intercalation mechanism in the considered configurations, the variation of distance between the layers with increasing lithium concentration was further scrutinized, as

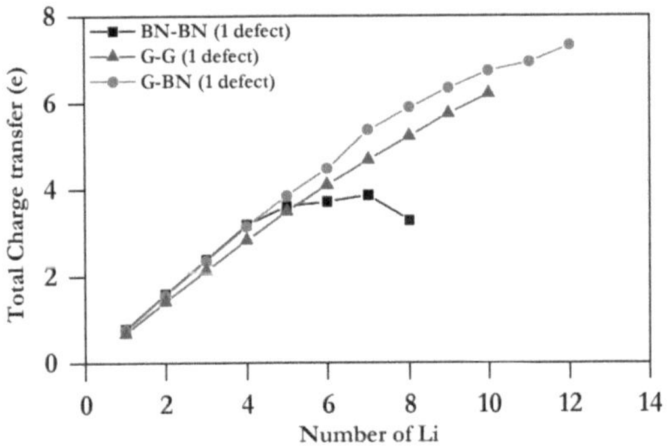

Figure 16.8 Total charge transfer by the lithium atoms to the BN-BN, G-G, and G-BN structures with a point defect. (Reprinted with permission [44]. Copyright 2021 American Chemical Society.)

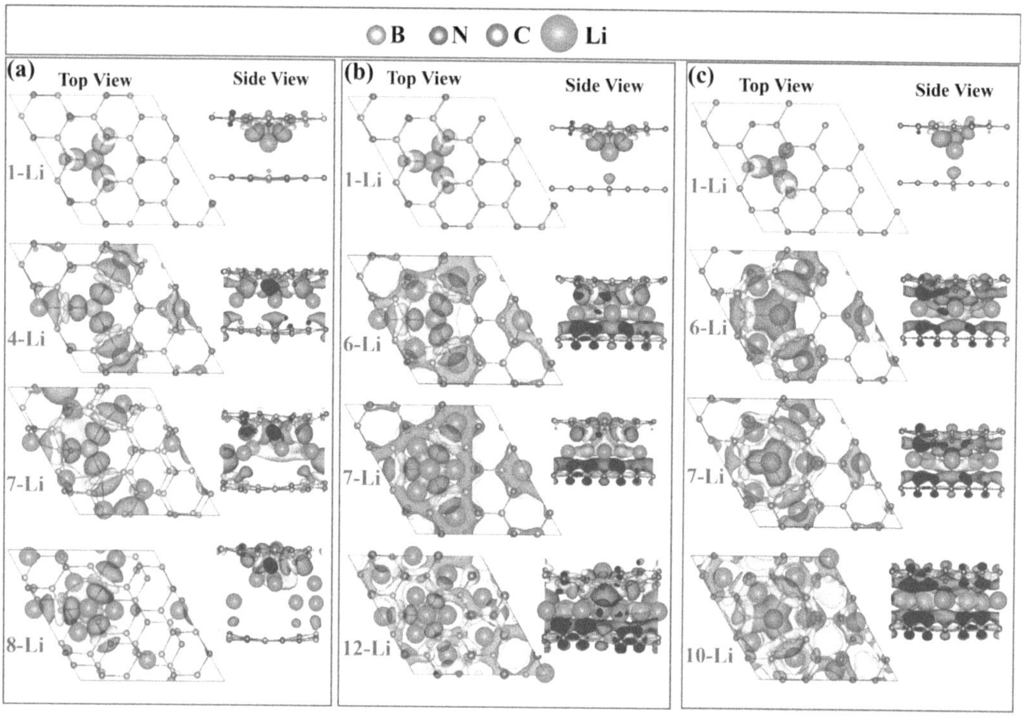

Figure 16.9 (a–c) Top views and side views of differential charge density with varying lithium concentration for BN-BN, G-G, and G-BN structures with a point defect, respectively. (Reprinted with permission [44]. Copyright 2021 American Chemical Society.)

presented in Figure 16.10. The intercalation of the first lithium atom causes a slight increase in the interlayer distance across all the structures with a point defect. Interestingly, in the G-G and G-BN structures with a point defect, the interlayer distance remains unchanged after subsequent lithium intercalation, up to the maximum capacity. However, in the BN-BN structure with a point defect, the distance increases significantly after the intercalation of four lithium atoms. From the perspective of lithium storage capacity and intercalation reversibility, the minimal change in the interlayer distance in the G-G and G-BN structures with a point defect indicates limited volume expansion upon lithium intercalation, suggesting easier lithium diffusion during charge and discharge cycles.

For better visualization, the schematics of BN-BN, G-G, G-BN structures with a point defect, and G-BN structure with two defects with varying lithium concentration are presented in Figure 16.11a–d, respectively.

To further clarify the instability of the BN-BN structure with a point defect after four lithium intercalations, the change in interlayer distance was calculated. After the insertion of four lithium atoms, the distance between the layers in the BN-BN structure with a point defect increases by 43%, as shown in Figures 16.10 and 16.11a. This suggests a breakdown of the BN-BN structure after four lithium intercalations, which is in correlation with the charge transfer mechanism.

Similarly, in the G-BN structure with two point defects, the interlayer distance increases significantly after the intercalation of 12 lithium atoms, as depicted in Figures 16.10 and 16.11d. However, in the G-BN structure with one point defect, the charge is transferred homogeneously

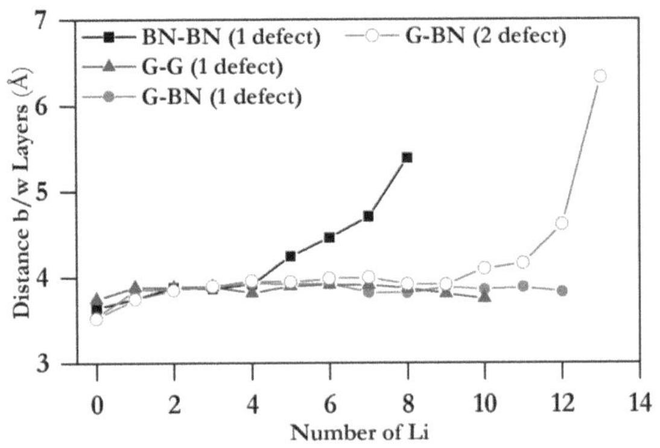

Figure 16.10 Variation of the distance with lithium intercalation between the layers of BN-BN, G-G, and G-BN structures with a point defect, and G-BN structure with two point defects. (Reprinted with permission [44]. Copyright 2021 American Chemical Society.)

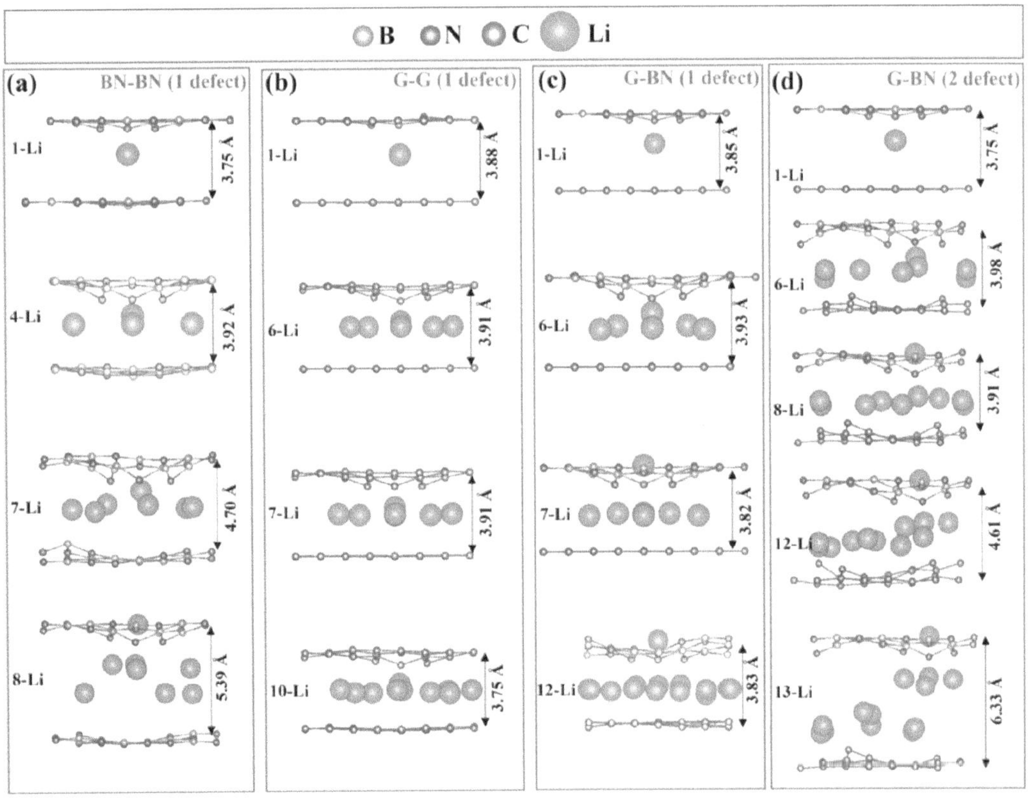

Figure 16.11 Schematics of the distance between the layers at various lithium concentrations in the (a) BN-BN structure with a point defect. (b) G-G structure with a point defect. (c) G-BN structure with a point defect, and (d) G-BN structure with two point defects. (Reprinted with permission [44]. Copyright 2021 American Chemical Society.)

to both layers until maximum intercalation (Figures 16.10 and 16.11c). As a result, it is anticipated that in the G-BN structure with two point defects, the charge will be evenly distributed across both layers up to the maximum intercalation. This suggests that the separation between the layers in the G-BN structure with two point defects cannot be caused by a preferential charge transfer to one layer, as observed in Figures 16.10 and 16.11d. However, it is likely that after the intercalation of 12 lithium atoms, the density of lithium atoms exceeds the highest packing density achievable in the G-BN structure with two point defects. As a result, the internal repulsion between the lithium atoms leads to layer separation and the breakdown of layer binding in the G-BN structure with two point defects.

16.8 CONCLUSIONS

In this chapter, the BN-BN, G-G, and G-BN bilayer structures were modified by defect engineering that improves the lithium storage capacity by an order of magnitude. Binding energy calculations indicate that structures with a point defect are energetically more stable than their defect-free counterparts. Intercalation energy calculations demonstrate that lithium storage can be significantly enhanced by introducing controlled point defects in 2D bilayer structures. The creation of a point defect in the G-BN structure resulted in an 11-fold increase in lithium storage capacity. Through band-structure calculations and the analysis of the charge transfer mechanism, the lithium storage limitations of BN-BN structures with a point defect were thoroughly examined. Bader charge population analysis and the changes in interlayer distance highlight the lithium storage mechanism in 2D bilayer structures with a point defect.

REFERENCES

1. R. Yazami, P. Touzain, A reversible graphite-lithium negative electrode for electrochemical generators. *Journal of Power Sources* **9**, 365–371 (1983).
2. Z. Liu, Q. Yu, Y. Zhao, R. He, M. Xu, S. Feng, S. Li, L. Zhou, L. Mai, Silicon oxides: A promising family of anode materials for lithium-ion batteries. *Chemical Society Reviews* **48**, 285–309 (2019).
3. J. B. Goodenough, Y. Kim, Challenges for rechargeable Li batteries. *Chemistry of Materials* **22**, 587–603 (2010).
4. A. K. Padhi, K. S. Nanjundaswamy, J. B. Goodenough, Phospho-olivines as positive-electrode materials for rechargeable lithium batteries. *Journal of the Electrochemical Society* **144**, 1188–1194 (1997).
5. N. Nitta, F. Wu, J. T. Lee, G. Yushin, Li-ion battery materials: Present and future. *Materials Today* **18**, 252–264 (2015).
6. W. Xu, J. Wang, F. Ding, X. Chen, E. Nasybulin, Y. Zhang, J. -G. Zhang, Lithium metal anodes for rechargeable batteries. *Energy & Environmental Science* **7**, 513–537 (2014).
7. M. W. Verbrugge, D. R. Baker, X. Xiao, Q. Zhang, Y. T. Cheng, Experimental and theoretical characterization of electrode materials that undergo large volume changes and application to the lithium-silicon system. *Journal of Physical Chemistry* C, **119**, 5341–5349 (2015).
8. J. Lu, Z. Chen, Z. Ma, F. Pan, L.A. Curtiss, K. Amine, The role of nanotechnology in the development of battery materials for electric vehicles. *Nature Nanotechnology* **11**, 1031–1038 (2016).
9. C.-M. Park, H.-J. Sohn, Black phosphorus and its composite for lithium rechargeable batteries. *Advanced Materials* **19**, 2465–2468 (2007).
10. M. V. Reddy, G. V. Subba Rao, B. V. R. Chowdari, Metal oxides and oxysalts as anode materials for Li ion batteries. *Chemical Reviews* **113**, 5364–5457 (2013).

11. M. M. Rahman, M. Zhou, I. Sultana, S. Mateti, A. Falin, L. H. Li, Z.-S. Wu, Y. Chen, Additive-free Nb_2O_5-TiO_2 hybrid anode towards low-cost and safe lithium-ion batteries: A green electrode material produced in an environmentally friendly process. *Batteries & Supercaps* **2**, 160–167 (2019).
12. B. Yu, T. Tao, S. Mateti, S. Lu, Y. Chen, Nanoflake arrays of lithiophilic metal oxides for the ultra-stable anodes of lithium-metal batteries. *Advanced Functional Materials* **28**, 1803023 (2018).
13. J. A. Yuwono, P. Burr, C. Galvin, A. Lennon, Atomistic insights into lithium storage mechanisms in anatase, rutile, and amorphous TiO_2 electrodes. *ACS Applied Materials & Interfaces* **13**, 1791–1806 (2021).
14. B. J. Morgan, G. W. Watson, GGA+U description of lithium intercalation into anatase TiO_2. *Physical Review B* **82**, 144119 (2010).
15. M. N. Obrovac, L. Christensen, Structural changes in silicon anodes during lithium insertion/extraction. *Electrochemical and Solid State Letters* **7**, A93 (2004).
16. X. Zhang, D. Wang, X. Qiu, Y. Ma, D. Kong, K. Müllen, X. Li, L. Zhi, Stable high-capacity and high-rate silicon-based lithium battery anodes upon two-dimensional covalent encapsulation. *Nature Communications* **11**, 1–9 (2020).
17. W. W. Tipton, C. R. Bealing, K. Mathew, R. G. Hennig, Structures, phase stabilities, and electrical potentials of Li-Si battery anode materials. *Physical Review B* **87**, 1–7 (2013).
18. C. E. L. Foss, S. Müssig, A. M. Svensson, P. J. S. Vie, A. Ulvestad, J. P. Mæhlen, A. Y. Koposov, Anodes for Li-ion batteries prepared from microcrystalline silicon and enabled by binder's chemistry and pseudo-self-healing. *Scientific Reports* **10**, 1–8 (2020).
19. K. Ji, J. Han, A. Hirata, T. Fujita, Y. Shen, S. Ning, P. Liu, H. Kashani, Y. Tian, Y. Ito, J. I. Fujita, Y. Oyama, Lithium intercalation into bilayer graphene. *Nature Communications* **10**, 1–10 (2019).
20. M. Kühne, F. Börrnert, S. Fecher, M. Ghorbani-Asl, J. Biskupek, D. Samuelis, A. V. Krasheninnikov, U. Kaiser, J. H. Smet, Reversible superdense ordering of lithium between two graphene sheets. *Nature* **564**, 234–239 (2018).
21. E. Pomerantseva, Y. Gogotsi, Two-dimensional heterostructures for energy storage. *Nature Energy* **2**, 17089 (2017).
22. J. Zheng, Z. Ren, P. Guo, L. Fang, J. Fan, Diffusion of Li + ion on graphene: A DFT study. *Applied Surface Science* **258**, 1651–1655 (2011).
23. S. N. Shirodkar, E. Kaxiras, Li intercalation at graphene/hexagonal boron nitride interfaces. *Physical Review B* **93**, 245438 (2016).
24. H. J. Chen, J. Huang, X. L. Lei, M. S. Wu, G. Liu, C. Y. Ouyang, B. Xu, Adsorption and diffusion of lithium on MoS_2 monolayer: The role of strain and concentration. *International Journal of Electrochemical Science* **8**, 2196–2203 (2013).
25. B. Xu, L. Wang, H. J. Chen, J. Zhao, G. Liu, M. S. Wu, Adsorption and diffusion of lithium on 1T-MoS_2 monolayer. *Computational Materials Science* **93**, 86–90 (2014).
26. S. U. Rehman, A. Samad, M. Saeed, B. Amin, M. Hafeez, I. A. Mir, L. Zhu, Computational insight of ZrS_2/graphene heterobilayer as an efficient anode material. *Applied Surface Science* **551**, 149304 (2021).
27. A. K. Geim, I. V. Grigorieva, Van der Waals heterostructures. *Nature* **499**, 419–425 (2013).
28. L. Wang, Z. Xu, W. Wang, X. Bai, Atomic mechanism of dynamic electrochemical lithiation processes of MoS_2 nanosheets. *Journal of the American Chemical Society* **136**, 6693–6697 (2014).
29. D. T. Larson, I. Fampiou, G. Kim, E. Kaxiras, Lithium intercalation in graphene-MoS_2 heterostructures. *Journal of Physical Chemistry* C **122**, 24535–24541 (2018).
30. R. H. Miwa, W. L. Scopel, Lithium incorporation at the MoS_2/graphene interface: An ab initio investigation. *Journal of Physics Condensed Matter* **25**, 445301 (2013).
31. J. Ma, J. Fu, M. Niu, R. Quhe, MoO_2 and graphene heterostructure as promising flexible anodes for lithium-ion batteries. *Carbon* **147**, 357–363 (2019).
32. H. Liu, Z. Huang, G. Wu, Y. Wu, G. Yuan, C. He, X. Qi, J. Zhong, A novel WS_2/$NbSe_2$ vdW heterostructure as an ultrafast charging and discharging anode material for lithium-ion batteries. *Journal of Materials Chemistry A* **6**, 17040–17048 (2018).
33. J. Jiang, T. Xu, J. Lu, L. Sun, Z. Ni, Defect engineering in 2D materials: Precise manipulation and improved functionalities. *Research* 2019, 1–14 (2019).

34. J.S. Park, S. Kim, Z. Xie, A. Walsh, Point defect engineering in thin-film solar cells. *Nature Reviews Materials* **3**, 194–210 (2018).
35. G. Kresse, J. Furthmüller, Efficiency of ab-initio total energy calculations for metals and semiconductors using a plane-wave basis set. *Computational Materials Science* **6**, 15–50 (1996).
36. G. Kresse, J. Furthmüller, Efficient iterative schemes for ab initio total-energy calculations using a plane-wave basis set. *Physical Review B* **54**, 11169–11186 (1996).
37. J. P. Perdew, K. Burke, M. Ernzerhof, Generalized gradient approximation made simple. *Physical Review Letters* **77**, 3865–3868 (1996).
38. P. E. Blöchl, Projector augmented-wave method. *Physical Review B* **50**, 17953–17979 (1994).
39. D. Joubert, From ultrasoft pseudopotentials to the projector augmented-wave method. *Physical Review B* **59**, 1758–1775 (1999).
40. S. Grimme, J. Antony, S. Ehrlich, H. Krieg, A consistent and accurate ab initio parametrization of density functional dispersion correction (DFT-D) for the 94 elements H-Pu. *The Journal of Chemical Physics* **132**, 154104 (2010).
41. K. Momma, F. Izumi, VESTA 3 for three-dimensional visualization of crystal, volumetric and morphology data. *Journal of Applied Crystallography* **44**, 1272–1276 (2011).
42. G. Henkelman, A. Arnaldsson, H. Jónsson, A fast and robust algorithm for Bader decomposition of charge density. *Computational Materials Science* **36**, 354–360 (2006).
43. T. A. Manz, N. G. Limas, Introducing DDEC6 atomic population analysis: Part 1. Charge partitioning theory and methodology. *RSC Advances* **6**, 47771–47801 (2016).
44 K. Tanwar et al., Huge lithium storage in 2D bilayer structures with point defects, *The Journal of Physical Chemistry C* **125**, 23957 (2021).

Chapter 17

Selective Ion Transport and Energy Harvesting from Restacked Boron Nitride Nanosheets

Si (Alex) Qin

17.1 INTRODUCTION

In the last two decades, the rise of 2D nanomaterials has revolutionized various scientific and technological fields. These materials, characterized by their atomically thin structures, such as graphene, BN, and transition metal dichalcogenides, have demonstrated remarkable mechanical, electrical, and thermal properties. Their exceptional surface area and tunable electronic characteristics have paved the way for significant advancements in nanotechnology, including enhanced energy storage, efficient catalysis, and high-performance electronic devices. The unique properties of 2D nanomaterials have also led to breakthroughs in medical diagnostics, drug delivery, and environmental remediation, establishing them as pivotal components in next-generation technologies.

Initially, these atomically thin materials such as graphene were considered ideal for interfacial barriers because they are impermeable to gases and liquids [1]. However, such large 2D nanomaterials are challenging to synthesis; hence, larger structures of such materials are always constructed from smaller sheets for applications. Researchers soon realized that the structures such as films and membranes made from stacked 2D nanomaterials are not as tight as they thought. Nair et al. reported that water molecules can permeate restacked graphene oxide films several orders of magnitude faster than small gas molecules such as helium due to the existence of nanofluidic channels when the film is hydrated [2]. This phenomenon was reported on many 2D nanomaterials since then, including on films constructed from graphene oxide [3–7], BN [8–11], $Ti_3C_2T_x$ MXene [9,12–15], molybdenum disulfide (MoS_2) [16,17], and clay nanosheets [18,19].

The nanofluidic channels formed by the spacing between stacked sheets have highly uniform sizes, comparable to those of hydrated ions, typically ranging from sub-nanometer to a few nanometers. They form an interconnected network of nanochannels, enabling gas molecules, liquid molecules, and ions to pass through. This discovery prompts researchers to investigate how these nanofluidic channels can be utilized. Membrane-based osmotic energy harvesting is a promising technology to harvest the energy from ion concentration gradients (blue energy),

DOI: 10.1201/9781003647812-21

such as that between seawater and freshwater. It is a clean energy with zero carbon footprint and complements existing renewable energy resources such as solar, wind, and hydroelectric power, which are pivotal means in the renewable energy mix to ensure energy diversification and energy security. Furthermore, the simple energy harvesting process presents scalability potential to satisfy both domestic and industrial energy requirements. Meanwhile, portable and micro energy harvesting devices can be used to power implantable devices by utilizing different streams inside the body or integrated as wearable power sources for portable analytical devices [20,21].

BN, as a representative member of the 2D nanomaterials family, possesses many unique properties, such as chemical stability, thermal stability, and high thermal conductivity, combined with a highly charged surface. These properties make it an interesting candidate for constructing nanofluidic membranes. This chapter mainly focuses on providing a high-level overview of the research progress in this field using BN-based membranes and discusses the challenges and opportunities faced by the researchers.

17.2 NANOFLUIDIC ION TRANSPORT

Ions are ubiquitously present in aqueous environments. The transport of ions and the charges they carry is essential to many electrochemical reactions and biological processes, such as those occurring in batteries, supercapacitors, fuel cells, and cellular membranes [22,23]. Understanding transport of ions and their interaction at materials' interfaces not only deepens our understanding on electrochemical and biological processes, but also helps us design and develop better material structures for various applications in which ion transport plays a crucial role, including but not limit to energy conversion such as batteries and supercapacitors, enabling new applications such as DNA sequencing [24], salinity energy harvesting [25], water desalination and purifications [26], nanofiltration, and membrane distillation [27].

17.2.1 Nanofluidic Channels Formed by Stacking of Nanosheets

In classic fluidic systems, solutions are typically considered to be a homogeneous medium where equal molecules and ions are uniformly dispersed and can move freely. These assumptions generally work well for macro- and microfluidic systems. Most of the phenomena can be well described using the Navier–Stokes equations. However, in nanofluidic systems, the size of the channels is comparable to the hydrated/solvation ions and solvent molecules. Under this condition, in addition to the classic Navier–Stokes model, transport of ions and associated solvent molecules is largely dictated by the interaction between ions and the channel walls, which is influenced by factors such as pore size, ion size and charge, surface charge, and surface functional groups of the channel materials.

17.2.1.1 Pore Geometry

Nanochannels within restacked 2D nanomaterials reside within the spacing between neighboring nanosheets. This means that one of their dimensions is limited by the *d*-spacing of the stacked structure. The spacing can range from sub-nanometer to a few nanometers, large enough to accommodate water and some molecules to enter and form a continuous hydration layer. Because of the presence of this solution layer between the nanosheets, ions can move easily along the plane direction and are hindered along the direction perpendicular to the plane.

17.2.1.2 Channel Size

The size and shape of the nanopores influence ion transport. Pores that are too small can restrict the transport of larger ions due to size exclusion, while the right pore size can enhance the transport for certain ion types (cations or anions). In addition, as intuitively illustrated in Figure 17.1, nanofluidic ion transport behavior is strongly affected by the electric double layer (EDL) structure close to the channel surface. When the channel is close to or smaller than the size of the EDL, it can often be conceptually described by the Debye screening length:

$$\kappa^{-1} = \sqrt{\frac{\varepsilon_r \varepsilon_0 RT}{2e^2 I}} \tag{17.1}$$

where ε_0 is the permittivity of free space, ε_r is the dielectric constant of the electrolyte, R is the gas constant, T is the absolute temperature in Kelvin, e is the elementary charge, and I is the ionic strength of the electrolyte. In aqueous electrolyte at room temperature, Equation 17.1 can be further simplified as:

$$\kappa^{-1} = \frac{0.304}{\sqrt{I(M)}} \tag{17.2}$$

where $I(M)$ is the ionic strength of the electrolyte in molar.

We can clearly see that the Debye length is inversely proportional to the square root of ionic strength (ionic concentration). This indicates that the nanofluidic ion transport behavior was only observed when the ion concentration is lower [28]. It also provided researchers with an opportunity to tune ion transport behavior by engineering suitable channel geometry and size [3,29] (Figure 17.2).

17.2.1.3 Surface Charge and Surface Functional Groups

The surface charge and the charge density of the channel walls play a significant role in ion transport behavior. Higher surface charge densities enhance the EDL, which increases the selectivity for ions of opposite charge (counterions) [7,28,30]. The surface charge on the channels is determined by the surface chemistry of the materials. The introduction of functional

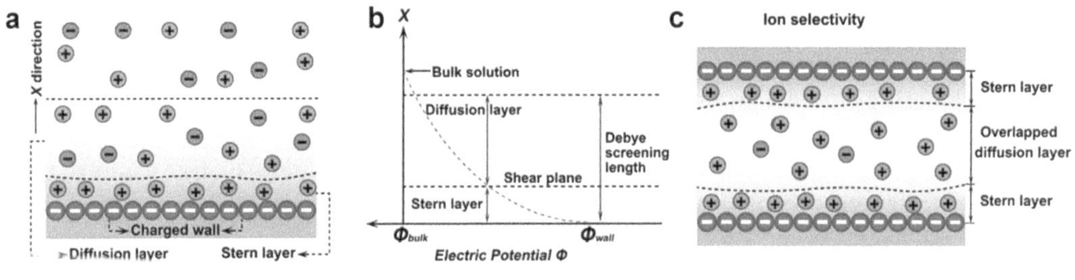

Figure 17.1 Schematic illustration of ion transport in nanofluidic channels. Schematics showing (a) the structure and (b) electrical distribution of the EDL next to a negatively charged wall. (c) Schematic showing the origin of ion selectivity in a surface-charged nanochannel. The electrical double layers (Stern layer and diffusion layer) near the surface overlapped when the diameter of the nanochannel is comparable with the Debye length, resulting in the exclusion of co-ions from the channel, while counterions can pass through the channel easily. Adapted with permission [23]. Copyright Elsevier 2019.

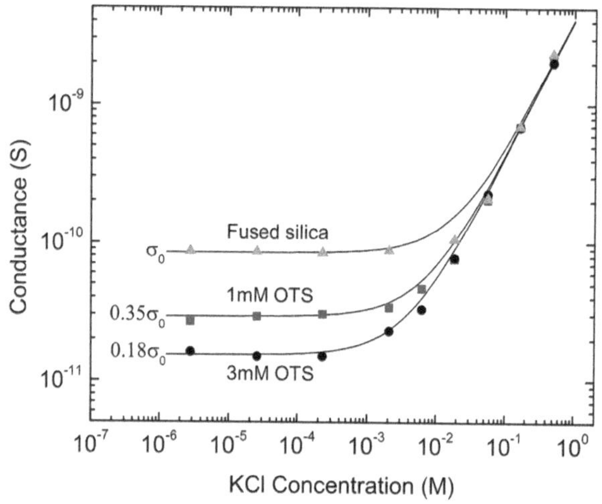

Figure 17.2 Effect of surface charge density and ionic concentration on ion transport in fused silica channels. Reused with permission [28]. Copyright 2004 American Physical Society.

groups on the channel walls can enhance ion selectivity by providing specific binding sites for target ions. Functionalization can be tailored to improve selectivity for applications. Some functional groups will also undergo protonation/deprotonation under different pH, which offers another avenue for controlling surface charge and subsequently controls the ion transport behavior.

17.2.1.4 Ion Size and Solvation

Last but not least, ion species also play an important role in the ion transport through nanofluidic channels. First, different ion species with different valence states have different electric charges. Second, ions are always present in solution in their solvated form (*i.e.*, they move in the electrolyte with a shell consisting of solvent molecules); interaction of ions and solvent molecules also will affect the ion transport dynamics. Some studies also suggest that solvated ions may go through partial desolvation when they pass through very small channels and have to shed some of the solvent molecules along the way [31].

In summary, nanofluidic ion transport behavior can be controlled mainly by two approaches: first, by changing physical properties (affected by channel size, ion concentration, pore geometry), and the other is surface charge (influenced by surface functionalization). These characteristics highlight the versatility and potential of nanofluidic channels formed by stacking nanosheets, making them suitable for a wide range of applications in nanotechnology and materials science.

17.2.2 Selective Ion Transport in Nanofluidic Channels

Selective ion transport in nanofluidic channels refers to the ability of these channels to preferentially allow certain ions to pass through while blocking others. This selectivity is due to the selective ion–channel interactions. We previously discussed the surface charge-governed ion transport behavior. In addition to size exclusion, the surface charge of the channel walls plays a crucial role here in enabling selective ion transport. Briefly, when the channel size is comparable to or smaller than the Debye length, EDLs from each channel wall will start to overlap (As shown in Figure 17.1c). Ions passing through the channel are constantly moving under a strong electric field. This means that ions with opposite charge (counterions) to the channel will be attracted, while ions with the same charge (co-ions) will be repelled from the channel. As a result, these charged nanofluidic channels should have a preferential selectivity for counterions.

The ability to control ion movement at the nanoscale opens up possibilities for more efficient and selective processes [32]. This selectivity is crucial for enabling many applications in energy conversion, water purification, and biosensing. For example, in nanofiltration membranes, selective ion transport is critical for removing specific ions and contaminants from water, providing high efficiency and low energy consumption in water treatment processes [33]. It is also employed in some biosensors to detect specific ions or molecules, enabling rapid and accurate detection in medical diagnostics and environmental monitoring [34]. Last but not least, this selective ion transport is also utilized in nanofluidic osmotic energy harvesting, where electricity energy can be harvesting using external circuitry from this selective ion transport, leading to efficient energy conversion from salinity gradients. The following sections provide an overview on the development and progress of energy harvesting applications using membrane constructed from BN nanosheets.

17.3 ENERGY HARVESTING FROM RESTACKED BN NANOSHEETS

Scientists have long been interested in electric eels—a species that uses electric charge for defense, hunting, communication, and navigation. They can generate powerful electric shocks up to several hundred volts through their electric organs to stun the prey. Nanofluidic energy harvesting uses a similar mechanism, 'pumping' cations (or anions) through ion selective membranes under external driving forces. A corresponding electric current can be harvested using an external circuit connected to two sides of the membrane.

Lei et al. first reported the preparation of surface-functionalized BN nanosheets, which are water dispersible, enabling long-time storage and solution processing of these BN nanosheets [35]. Chen et al. demonstrated that functionalized BN nanosheets can be fabricated into multifunctional membranes showing semi-permeability with various organic solvents, outstanding performance in molecular separation while retaining ultrahigh flux [36,37]. Dou et al. reported BN membranes with abundant percolating nanochannels for highly efficient ethylene/ethane separation [38].

These results demonstrated that BN nanosheets with high surface charge and exceptional thermal and chemical stability can be prepared. Later, Qin et al. reported BN membranes fabricated *via* vacuum-assisted filtration and featuring integrated 2D nanofluidic channels (Figure 17.3). The prepared BNN membrane demonstrated stable ionic conductivity. In low-concentration electrolytes (<0.1 mM), the conductivity of the BN membranes is several orders of magnitude higher than that of the bulk solution, indicating a surface charge-governed ion transport behavior. Due to the outstanding thermal and chemical stability, the BN membrane ionic conductivity remains stable even in highly acidic and alkaline environments and at high temperatures (up to 90°C) [8]. This section explores two primary mechanisms of energy harvesting in these systems: pressure-induced energy harvesting and osmotic energy harvesting.

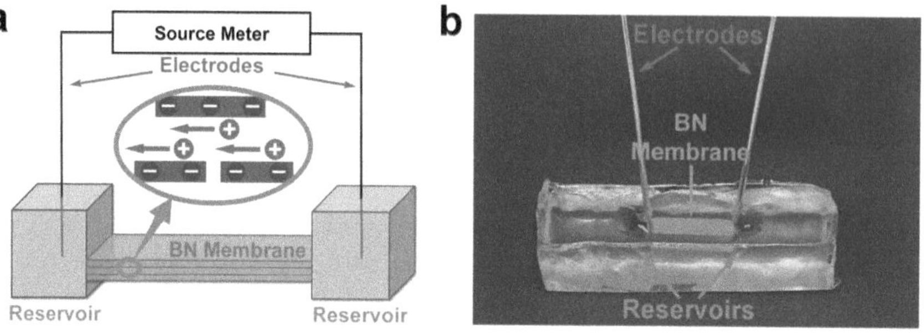

Figure 17.3 (a) Schematic showing the BN nanofluidic channels and the ionic conductivity testing setup. (b) Photo of the BN membrane fixed in PDMS, and one electrode was connected to the reservoir on each side for measuring ion conductivity through the BN membrane. Reproduced with permission [8]. Copyright 2017 American Chemical Society.

17.3.1 Energy Harvesting from Pressure Gradient

Energy harvesting through nanofluidic channels is fundamentally a selective ion transport process. When an ion selective membrane is placed in a homogeneous electrolyte, external forces are required to drive the ions through these selective channels. One approach to achieve a net ion flux is by the application of external pressure (*e.g.*, gas pressure). This requires the membrane to be able to withstand the pressure and stress without being damaged, which often limits the pressure one can apply in the real world.

Qin et al. reported the preparation of BN membranes from functionalized BN nanosheets (Figure 17.4a) [11]. The membrane was mounted between two cells filled with 0.1 M sodium chloride (NaCl) electrolyte. When a gas pressure gradient (5 kPa) was applied to one side of the membrane, the ions were pushed through the membranes under this pressure, and a current response 12.1 nA is generated in the 0.1 M NaCl solution (Figure 17.4b–e), which translates to a maximum output power of 0.64 m W m^{-2} [11]. Due to the small size of BN nanosheets, pure BN nanosheet membranes usually exhibit poor mechanical properties for a freestanding membrane. Chen et al. introduced ultra-strong aramid nanofibers into functional BN nanosheets and successfully fabricated aramid/BN composite membranes. These membranes can generate up to 47 nA on a membrane area of 3.14 mm^2 in a 0.1 M NaCl solution. In general, the application of external pressure is cumbersome. Challenges generally include the design and fabrication of a mechanically robust membrane that can tolerate extra applied pressure cycles, and a special membrane system that can be operated when pressurized. This added cost and complexity limit its design flexibility and feasibility for applications.

17.3.2 Energy Harvesting from Osmotic Gradient

Another method to achieve a net ion flux through the membrane is by employing an osmotic gradient across the membrane. By placing the membrane between two solutions with different concentrations, ions from the high concentration side will have a tendency to flow to the low-concentration side. In a nanofluidic channel, the difference in ion concentration generates an osmotic pressure gradient within the diffuse layer, which leads to a diffusio-osmotic flow

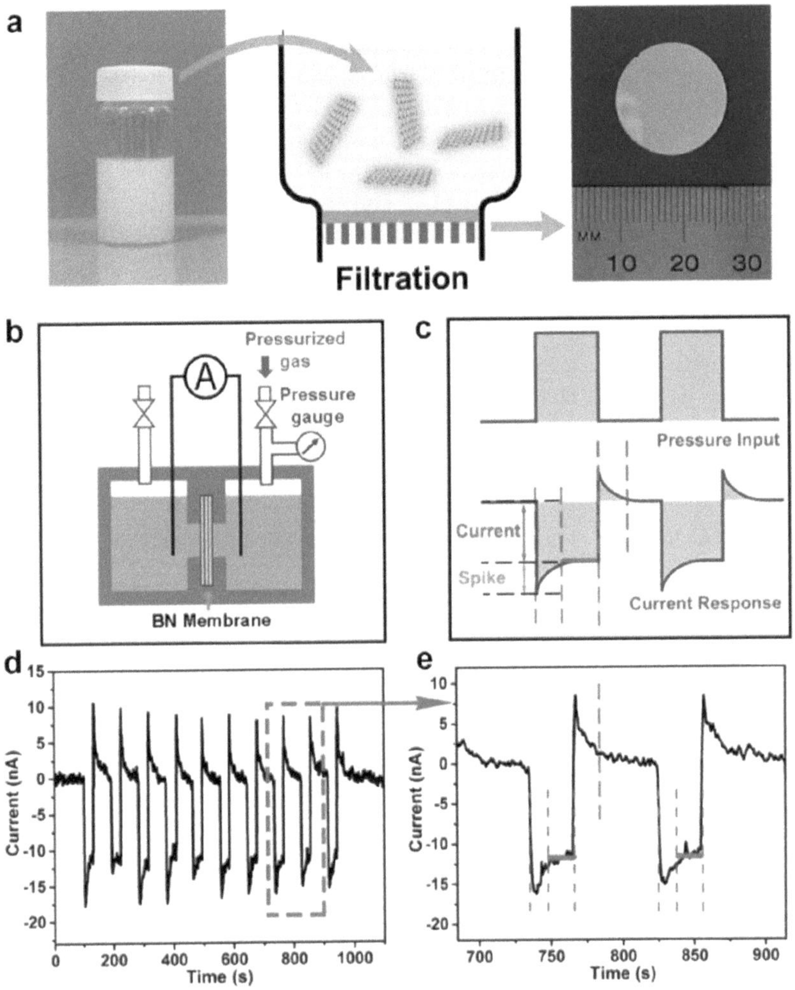

Figure 17.4 (a) A schematic showing the preparation process of the BN membrane, including a photograph of the BN colloidal dispersion vacuum filtration step, and the resulting BN membrane with a glossy surface. (b) Schematic showing the pressure-driven energy harvesting experiment setup. (c) Schematic drawing showing the shape of the current response. (d and e) Current response showing the energy harvesting when the BN membrane is driven using a 5 kPa pressure gradient. Adapted with permission [11]. Copyright 2018 Elsevier.

inside the nanofluidic channels. The velocity of the diffusion-osmotic flow comes from a combination of viscous flow and osmotic pressure gradient within the Debye layer. The osmotic pressure can be calculated from the equation:

$$\Pi = i\frac{n}{V}RT = Icrt \tag{17.3}$$

where R is the ideal gas constant, T is the temperature in Kelvin, i is the van't Hoff factor, n is the number of moles of solute present, V is the volume of the solution, and C is the molar concentration of solute. Using this estimation, the osmotic pressure difference between a typical 0.01/0.5 M NaCl solution at room temperature can be calculated to be as high as >2300 kPa. This is orders of magnitude higher than that achievable using a gas pressure system.

A better equation and model used to describe the ion dynamics through nanochannels under the influence of EDL was given by Prieve et al. [39]. The diffusio-osmotic velocity of the ions can be described as follows:

$$V_{DO} = -\frac{\varepsilon}{4\pi\eta}\frac{kT}{Ze}\left[\beta\zeta - 2\frac{kT}{Ze}\ln\left(1-\gamma^2\right)\right]\frac{d\ln C_\infty}{dx} \tag{17.4}$$

where ε is the permittivity, η is the viscosity, ζ is the surface zeta potential, T is the temperature, e is the elemental charge, Z is the number of element charge of the ions, k is the Boltzmann constant, $\beta = (D_+ - D_-)/(D_+ + D_-)$ is the diffusivity difference between cations and anions, $\gamma = \tanh\frac{Ze\zeta}{4kT}$, and $\frac{d\ln C_\infty}{dx}$ describes the concentration gradient along the nanochannel, where C_∞ is the bulk ion concentration. This equation clearly illustrates the factors that influence nanofluidic ion transport.

Yazda et al. reported that coating hexagonal BN nanosheets on a perforated silicon nitride membrane can achieve high osmotic power density [40,41]. They used a method called tip-controlled local breakdown on an atomic force microscope to fabricate arrays of nanopores on a silicon nitride membrane before coating BN nanosheets on the membrane. The high osmotic power density was attributed to the high surface charge density on the h-BN surface, which enhanced the selective ionic transport through nanopores with a size of ~10 nm (diameter). They obtained a maximum power density of ~15 W m^{-2} on a 3×3 array with ~6 nm nanopores of ~6 nm in diameter and 500 nm pore-to-pore spacing.

Synthesizing large amounts of large BN nanosheets remains a great technical challenge until now. Because of this, neat BN membranes are difficult to obtain and often suffer from weak mechanical properties. Other methods that utilize BN nanosheets for energy harvesting have been explored, such as by loading BN into a nanoporous membrane or by blending BN into a composite membrane. Pendse et al. grew BN nanoparticles using chemical vapor deposition on anodized aluminum oxide membranes [32]. These membranes contain a high nanopore density; ~10^8 pores cm^{-2} provide abundant pathways for ion transport. A maximum power of 105 W m^{-2} was observed, corresponding to a membrane areal power density of 165 mW m^{-2}.

Blending BN with other 1D or 2D nanomaterials has also been explored [30]. Yang et al. reported a 2D/2D composite membrane by assembling BN and MXene nanosheets for energy harvesting (Figure 17.5) [9]. The BN/MXene composite membranes presented good stability and enhanced output power density when the BN content reached 44 wt%, reaching 2.3 W m^{-2}, almost twice that of a pristine MXene membrane. At higher temperatures, the output power can be further increased to 6.2 W m^{-2} at 336 K and stabilizes for 10 h at 321 K, revealing excellent structure and thermal stability of the membrane. Zhang et al. also reported a 3D BN/cellulose composite membrane capable of achieving a high power density (10.6 W m^{-2} on a small membrane area of 0.0314 mm^2) and an average power density of ~1 W m^{-2} on a larger area [33].

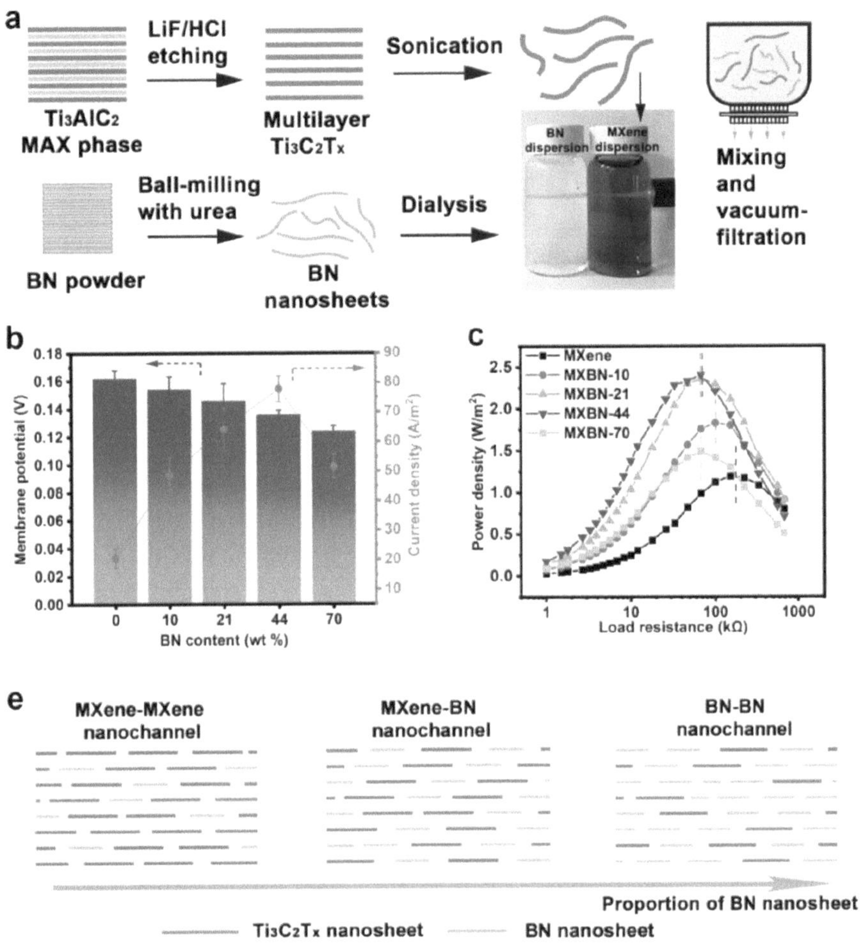

Figure 17.5 (a) Schematic diagram of the fabrication of BN/MXene membranes. Inset: Photos of BN dispersion and MXene dispersion showing a Tyndall effect. (b) BN/MXene membrane potential and current density with different BN contents. (c) Output power density of the BN/MXene membranes with different BN contents as a function of load resistance. (e) Schematic of three hypothetical models of nanochannels: MXene–MXene nanochannel, MXene–BN nanochannel, and BN–BN nanochannel. Adapted with permission [9]. Copyright 2021 American Chemical Society.

17.4 FUTURE DEVELOPMENT AND PERSPECTIVES

Energy harvesting from nanofluidic channels in restacked nanosheets represents a promising avenue for sustainable energy technologies. Both pressure-induced and osmotic energy harvesting mechanisms offer unique advantages and challenges, with ongoing research focused on optimizing materials and designs for enhanced performance. As advancements continue, these systems are expected to play a significant role in the development of efficient and sustainable energy solutions for applications in water treatment, energy harvesting, and chemical sensing.

17.4.1 Challenges

BN as a material itself has many unique properties and advantages when compared with other 2D nanomaterials. Its excellent thermal and chemical stability makes it suitable for various demanding applications. As a building block for nanofluidic systems, it faces some of the common challenges in the field.

17.4.1.1 Large-Scale Synthesis and Functionalization

The synthesis of BN nanosheets, particularly in their hexagonal form (*h*-BN), often requires complex processes such as chemical vapor deposition and high-pressure high-temperature processes. These processes are not only energy-intensive but also require precise control over reaction parameters. Further exfoliation of these as-synthesized bulk BN particles into nanosheets is also challenging and difficult to scale up. The high costs associated with the raw materials, energy consumption, and complex equipment needed for BN synthesis contribute to the overall production costs. Reducing these costs is crucial for making BN economically viable for large-scale applications. Due to their chemical inertness and lack of surface functional groups, BN nanosheets are generally difficult to disperse in solvents. Functionalization is often needed to make BN nanosheets dispersible in a solvent and suitable for solution processing. Functionalization of BN nanosheets with specific chemical groups that can selectively interact with target ions will be highly desirable for many applications. However, most of the functionalization methods involve additional processing steps, which can further complicate large-scale production. Ensuring that BN can be effectively integrated into composite materials or devices while maintaining its desirable properties adds another layer of complexity to the production process.

17.4.1.2 Balancing Selectivity and Permeability

Balancing selectivity and permeability in BN membranes and, in general, other nanofiltration membranes remains a main challenge that is critical for their effective application in nanofluidic systems. BN membranes are renowned for their high thermal and chemical stability, but optimizing their ion selectivity while ensuring adequate permeability requires innovative approaches in material design and fabrication. The intrinsic properties of BN, such as its limited flexibility and rigid crystalline structure, can hinder the efficient fabrication of robust and strong BN-based membranes. Development of composite membranes that combine BN with other materials aims to reduce the cost and enhance the robustness of such membranes. This can sometimes lead to affecting the channel characteristics, reduced porosity, and permeability. Careful material selection and further structure engineering and optimization are often required to enhance both selectivity and permeability. Designing BN-based composite membranes that incorporate materials with high permeability, such as other 1D and 2D nanomaterials, can potentially offer the best of both worlds: the robustness and stability of BN, combined with the exceptional transport properties of other nanomaterials.

17.4.2 Future Opportunities

Despite the challenges faced by the field, research on nanofluidic channels in restacked nanosheets for energy harvesting is rapidly evolving, with numerous opportunities for future development. Future research should focus on solving some of the key challenges of nanofluidic energy harvesting systems to improve the scalability, stability, and cost-effectiveness of these systems and broaden their practical applications.

17.4.2.1 Enhancing Ion Selectivity and Transport

Ion selectivity is the key property of any ion selective membrane. Enhancing ion selectivity in BN-based nanofluidic membranes presents a multitude of opportunities. One significant avenue lies in the functionalization of BN nanosheets. By attaching specific chemical groups to the surface of BN nanosheets, researchers can create membranes with tailored ion selectivity, enabling the selective toward targeted ions, thereby enhancing the specificity and efficiency of ion transport through the membrane. In addition, advancements in nanofabrication techniques offer the potential to precisely engineer the pore structure and surface characteristics of BN membranes. Techniques such as atomic layer deposition and molecular layer deposition can be used to create membranes with finely tuned pore sizes and surface chemistries, optimizing them for specific ion separation applications.

17.4.2.2 Integration with Other Technologies

With the integration of BN-based nanofluidic membranes with other cutting-edge technologies, we can potentially create multifunctional systems that leverage the strengths of each component for enhanced performance and novel applications. For instance, the combination of BN nanofluidic membranes with thermoelectric systems can pave the way for innovative devices that utilize selective ion transport for efficient thermal-to-electrical energy conversion. This integration could lead to the development of advanced energy harvesting systems that capitalize on the stability and selectivity of BN membranes while exploiting thermoelectric effects to generate sustainable energy from temperature gradients.

Similarly, integrating BN membranes with photothermal technologies holds significant promise for environmental applications. Photothermal systems, which convert light into heat, can be combined with BN membranes to create devices capable of selective ion transport under various environmental conditions. Such systems could be employed in water purification, desalination, or pollutant removal, leveraging the stability and functionality of BN to enhance the efficiency and specificity of these processes.

Moreover, the integration of BN nanofluidic membranes with electronic and sensor technologies could lead to the development of highly sensitive and selective detection systems. By utilizing BN membranes' ion selectivity and chemical stability, these integrated devices could achieve superior performance in detecting and measuring specific ions or molecules, with applications ranging from environmental monitoring to medical diagnostics.

Future research on nanofluidic channels in restacked nanosheets for energy harvesting holds significant promise, focusing on improving scalability, stability, and cost-effectiveness, thus broadening their practical applications. Enhancing ion selectivity in BN-based nanofluidic membranes through functionalization and advanced nanofabrication techniques presents numerous opportunities. Integration of BN membranes with technologies such as thermoelectric systems, photothermal technologies, and electronic and sensor technologies could lead to multifunctional systems with innovative applications in energy conversion, water purification, pollutant removal, and medical diagnostics. By exploiting the stability, selectivity, and unique properties of BN, these advancements could pave the way for highly efficient and sustainable energy solutions, contributing to advancements in both environmental and technological fields.

17.5 CONCLUSIONS

Future research on nanofluidic channels in restacked nanosheets for energy harvesting holds significant promise, focusing on improving scalability, stability, and cost-effectiveness, thus broadening their practical applications. Enhancing ion selectivity in BN-based nanofluidic membranes through functionalization and advanced nanofabrication techniques presents numerous opportunities. Integration of BN membranes with technologies such as thermoelectric systems, photothermal technologies, and electronic and sensor technologies could lead to multifunctional systems with innovative applications in energy conversion, water purification, pollutant removal, and medical diagnostics. By exploiting the stability, selectivity, and unique properties of BN, these advancements could pave the way for highly efficient and sustainable energy solutions, contributing to advancements in both environmental and technological fields.

REFERENCES

1. O. Leenaerts, B. Partoens, F. M. Peeters, Graphene: A perfect nanoballoon. *Applied Physics Letters* **93**, 193107 (2008).
2. R. R. Nair, H. A. Wu, P. N. Jayaram, I. V. Grigorieva, A. K. Geim, Unimpeded permeation of water through Helium-leak–tight graphene-based membranes. *Science* **335**, 442–444 (2012).
3. Y. Qian et al., Boosting osmotic energy conversion of graphene oxide membranes via self-exfoliation behavior in nano-confinement spaces. *Journal of the American Chemical Society* **144**, 13764–13772 (2022).
4. Z. Zhang et al., Vertically transported graphene oxide for high-performance osmotic energy conversion. *Advanced Science* **7**, 2000286 (2020).
5. P.-P. Yan et al., Two-dimensional nanofluidic membranes with intercalated in-plane shortcuts for high-performance blue energy harvesting. *Small* **19**, 2205003 (2023).
6. S.-H. Liu et al., Topologically programmed graphene oxide membranes with bioinspired superstructures toward boosting osmotic energy harvesting. *Advanced Functional Materials* **33**, 2211532 (2023).
7. K. Raidongia, J. Huang, Nanofluidic ion transport through reconstructed layered materials. *Journal of the American Chemical Society* **134**, 16528–16531 (2012).
8. S. Qin et al., High and stable ionic conductivity in 2D nanofluidic ion channels between boron nitride layers. *Journal of the American Chemical Society* **139**, 6314–6320 (2017).
9. G. Yang et al., Stable $Ti_3C_2T_x$ MXene–boron nitride membranes with low internal resistance for enhanced salinity gradient energy harvesting. *ACS Nano* **15**, 6594–6603 (2021).
10. C. Chen et al., Bio-inspired nanocomposite membranes for osmotic energy harvesting. *Joule* **4**, 247–261 (2020).
11. S. Qin et al., Nanofluidic electric generators constructed from boron nitride nanosheet membranes. *Nano Energy* **47**, 368–373 (2018).
12. G. Yang et al., Ultrathin $Ti_3C_2T_x$ (MXene) membrane for pressure-driven electrokinetic power generation. *Nano Energy* **75**, 104954 (2020).
13. J. Lao et al., Aqueous stable Ti_3C_2 MXene membrane with fast and photoswitchable nanofluidic transport. *ACS Nano* **12**, 12464–12471 (2018).
14. G. Yang et al., Simultaneous electrokinetic energy conversion and organic molecular sieving by two-dimensional confined nanochannels. *Chemical Engineering Journal* **446**, 136870 (2022).
15. G. Yang et al., Advanced Janus membrane (MXene/CoAl-LDH) for efficient asymmetric ion transport and nanofluidic energy harvesting. *Nano Energy* **118**, 108972 (2023).
16. M. Deng, K. Kwac, M. Li, Y. Jung, H. G. Park, Stability, molecular sieving, and ion diffusion selectivity of a lamellar membrane from two-dimensional molybdenum disulfide. *Nano Letters* **17**, 2342–2348 (2017).

17. C. Zhu et al., Metallic two-dimensional MoS_2 composites as high-performance osmotic energy conversion membranes. *Journal of the American Chemical Society* **143**, 1932–1940 (2021).
18. L. Cao et al., Lamellar porous vermiculite membranes for boosting nanofluidic osmotic energy conversion. *Journal of Materials Chemistry A* **9**, 14576–14581 (2021).
19. T. Huang et al., Two-dimensional sodium channels with high selectivity and conductivity for osmotic power generation from wastewater. *ACS Nano* **17**, 17245–17253 (2023).
20. W.-S. Hsu, A. Preet, T.-Y. Lin, T.-E. Lin, Miniaturized salinity gradient energy harvesting devices. *Molecules* **26**, 5469 (2021).
21. X.-C. Chen, H. Zhang, S.-H. Liu, Y. Zhou, L. Jiang, Engineering polymeric nanofluidic membranes for efficient ionic transport: Biomimetic design, material construction, and advanced functionalities. *ACS Nano* **16**, 17613–17640 (2022).
22. H. Daiguji, Ion transport in nanofluidic channels. *Chemical Society Reviews* **39**, 901–911 (2010).
23. K. Xiao, L. Jiang, M. Antonietti, Ion transport in nanofluidic devices for energy harvesting. *Joule* **3**, 2364–2380 (2019).
24. Z. Cao, P. Yadav, A. Barati Farimani, Which 2D material is better for DNA detection: Graphene, MoS_2, or MXene? *Nano Letters* **22**, 7874–7881 (2022).
25. Z. Zhang et al., Cation-selective two-dimensional polyimine membranes for high-performance osmotic energy conversion. *Nature Communications* **13**, 3935 (2022).
26. A. Ali, R. A. Tufa, F. Macedonio, E. Curcio, E. Drioli, Membrane technology in renewable-energy-driven desalination. *Renewable & Sustainable Energy Reviews* **81**, 1–21 (2018).
27. S. Kim et al., Extreme ion-transport inorganic 2D membranes for nanofluidic applications. *Advanced Materials* **35**, 2206354 (2023).
28. D. Stein, M. Kruithof, C. Dekker, Surface-charge-governed ion transport in nanofluidic channels. *Physical Review Letters* **93**, 035901 (2004).
29. C. Cheng et al., Ion transport in complex layered graphene-based membranes with tuneable interlayer spacing. *Science Advances* **2**, e1501272 (2016).
30. X. Jia et al., Enhanced selective ion transport in highly charged bacterial cellulose/boron nitride composite membranes for thermo-osmotic energy harvesting. *Nano Letters* **24**, 2218–2225 (2024).
31. Z. Wang et al., Pore-assisted lithium deposition in hierarchically porous and hollow carbon textile for highly stable lithium anode. *Journal of Power Sources* **489**, 229464 (2021).
32. A. Pendse et al., Highly efficient osmotic energy harvesting in charged boron-nitride-nanopore membranes. *Advanced Functional Materials* **31**, 2009586 (2021).
33. R. Zhang et al., Construction of cellulose 3D network composite membrane supported by hydroxylated boron nitride with high surface charge density to achieve high-efficiency osmotic energy harvesting. *Chemical Engineering Journal* **500**, 157000 (2024).
34. L. Yu et al., 3D-printed mechanically strong and extreme environment adaptable boron nitride/cellulose nanofluidic macrofibers. *Nano Research* **16**, 7609–7617 (2023).
35. W. Lei et al., Boron nitride colloidal solutions, ultralight aerogels and freestanding membranes through one-step exfoliation and functionalization. *Nature Communications* **6**, 8849 (2015).
36. C. Chen et al., Functionalized boron nitride membranes with multipurpose and super-stable semi-permeability in solvents. *Journal of Materials Chemistry A* **6**, 21104–21109 (2018).
37. C. Chen et al., Functionalized boron nitride membranes with ultrafast solvent transport performance for molecular separation. *Nature Communications* **9**, 1902 (2018).
38. H. Dou et al., Boron nitride membranes with a distinct nanoconfinement effect for efficient ethylene/ethane separation. *Angewandte Chemie International Edition* **58**, 13969–13975 (2019).
39. D. C. Prieve, J. L. Anderson, J. P. Ebel, M. E. Lowell, Motion of a particle generated by chemical gradients. Part 2. Electrolytes. *Journal of Fluid Mechanics* **148**, 247–269 (1984).
40. K. Yazda et al., High osmotic power generation via nanopore arrays in hybrid hexagonal boron nitride/silicon nitride membranes. *Nano Letters* **21**, 4152–4159 (2021).
41. A. Siria et al., Giant osmotic energy conversion measured in a single transmembrane boron nitride nanotube. *Nature* **494**, 455–458 (2013).

Section V

Greener Future

Boron Nitride Nanosheets for Environmental Protection

Chapter 18

Understanding and Controlling Surface Chemistry of Boron Nitride Nanosheets

Alireza Dibaji, Dulini Fernando, Zhen Liu, Motilal Mathesh, and Wenrong Yang

18.1 INTRODUCTION

BNNSs stand out as a highly promising and extensively investigated family of two-dimensional (2D) materials. Their synthesis dates back to 2004, following the synthesis of graphene, the first 2D nanomaterial [1]. BNNSs consist of a few atomic thick layers of bulk boron nitride (BN). Despite their thin composition, these BNNSs are commonly acknowledged for their chemical inertness. The remarkable chemical inertness and resistance to oxidation pose challenges in the realm of functionalization, distinguishing BNNSs as a unique and resilient material in the 2D material landscape.

18.1.1 Unique Properties of 'White Graphene'

BNNSs share a structural resemblance with graphene, where carbon in graphene is substituted by boron and nitrogen in BNNSs (see Figure 18.1). The bond length in B–N measures 1.44 Å, slightly larger than the C–C bond length in graphene at 1.42 Å [2,3]. This subtle distinction

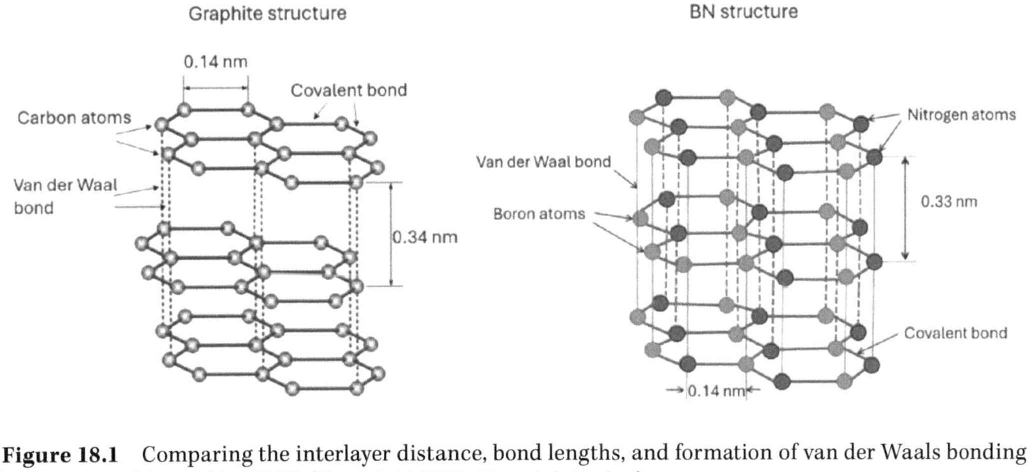

Figure 18.1 Comparing the interlayer distance, bond lengths, and formation of van der Waals bonding between graphite and h-BN [6]. (Copyright 2021, Materials Today.)

DOI: 10.1201/9781003647812-23

imparts BNNSs with a structure often likened to 'white graphene', showcasing properties akin to graphene.

The calculated Young's modulus for BNNSs falls within the range of 0.71–0.97 TPa, closely aligning with experimental values for graphene's Young's modulus, which is 1.0 TPa. Moreover, BNNSs exhibit a breaking strength ranging from 120 to 165 GPa, [4] a parameter in close proximity to graphene's breaking strength of 130 GPa [5]. These similarities underscore the remarkable parallels between BNNSs and graphene in terms of structural features and mechanical properties.

Despite sharing some similarities with carbon-based nanomaterials like graphene and MXenes, BNNSs possess unique properties and applications not found in these counterparts. Notably, while graphene and MXenes are electrical conductors, BNNSs exhibit electrical insulation. Unlike graphene, BNNSs are partially ionic due to the stronger electronegativity of nitrogen atoms. The resulting limited π bond strength and broader bandgap make BNNSs electrically insulating [7]. Although they cannot transfer electrons across or between sheet-like structures, BNNSs find applications as dielectric materials and heat dissipater fillers in advanced electronic circuits, preventing electric leakage and maintaining safe temperatures.

Furthermore, BNNSs exhibit strong resistance to extreme chemical conditions, elevated temperatures, and humidity, leading to superior oxidation and corrosion stability compared to graphene. Their higher thermal conductivity, coupled with electrical insulation, also makes BNNSs more effective than graphene in preventing galvanic corrosion. These properties make BNNSs reliable protective coatings for metals and surfaces susceptible to oxidation and corrosion.

Regarding their morphology, BNNSs typically form multilayer structures due to their distinctive B–N stacking, which helps stabilize the overall configuration. The interlayer spacing between the layers of multilayer BNNSs or between tube walls/sheet layers generally measures 0.33–0.34 nm. While some studies suggest that BNNSs have larger interlayer distances than bulk hexagonal boron nitride (h-BN), this discrepancy may be due to internal stresses within the bent layers [8].

Density functional theory (DFT) calculations by Alem et al. [9] reveal the natural formation of B–N covalent bonds between adjacent BN bilayers, known as 'closed' edge termination. The energy gain per interlayer bond is calculated to be −4.9 eV. Unlike the sp^2 hybrid bonds in the bulk h-BN, the curvatures at the edges of BNNSs lead to hybridization in both σ and π symmetry states. The stability of AB stacking over AA stacking is attributed to electrostatic interlayer forces, guiding adjacent BN layers to adopt specific stacking types. DFT calculations emphasize the role of van der Waals forces in maintaining a specific distance between BN layers [10].

18.1.2 BN Nanomaterials and Surface Modification

Like graphene, the first exfoliated 2D nanomaterial, BNNSs offer potential for functionalization with reactive intermediates such as carbenes, radicals, arynes, and nitrenes. The covalent functionalization of graphene involves various reactions, including nucleophilic substitution, electrophilic addition, condensation, and addition [11]. For example, the interaction of reactive epoxy resin with graphene oxide surfaces can result in edge-functionalized graphene [11]. These sites, which are reactive, provide advantages beyond the limitations of BN-based materials. While attempts have been made to pre-functionalize BNNSs, they exhibit less reactivity than graphene. The substitution of C–C in graphene with C–O or oxidation introduces open gates for surface modification, unlike BNNSs, which require heteroatom doping due to their wide bandgap. Non-covalent functionalization of graphene, utilizing π–π bond interaction, hydrogen bonding, ionic bonding, and electrostatic interaction modifications, has been well-established [12]. The superior dispersion of graphene oxide in aqueous solutions makes non-covalent functionalization more achievable and uniform than with BNNSs.

Despite being chemically inert and impermeable to gases and ions, BNNSs require functionalization for practical applications. BNNSs resist oxidation and remain stable up to ~850°C in air [13]. However, covalent modification through oxidation reactions remains challenging. The surface chemistry of BN nanostructures is intricately tied to their semi-iconicity. Functionalization of BNNSs becomes feasible by capitalizing on the imperfect electronegativity of the N position or the electronegativity of the fragmented positive B position. A diverse array of functional groups, including hydroxyl (OH), ether (OR), amino (NH_2 and NHR), acyl (COR), alkyl (R), halogen (X) groups, and heteroatoms (C, O, and S), have been employed for the purpose of functionalizing BNNS surfaces [14]. Hydroxylation is a crucial functionalization that enhances water processability and constructs a platform for further functional group attachments. NH_2 and NH_R groups are also commonly employed for nucleophilic functionalization, rendering resulting BNNSs highly soluble in water [15]. The OR group, containing an alkyl group bonded to an oxygen atom, can be directly attached to BN basal planes and BN nanoribbons (BNNRs). Alkyl groups can be attached through reactive intermediates or directly reaching the B position. The Billups–Birch reaction has been reported for efficient BN alkylation [16].

Other groups, such as hydrogen, can also be attached to BNNSs, influencing their band gap [17]. Hydrogenation results in a band gap of ~3.0 eV [18], while semi-hydrogenation yields an even lower band gap of ~2.24 eV [19]. An O edge-functionalized BNNR displays a peroxide-like structure with a band gap of 2.8 eV, while OH-BNNR exhibits a polyol-like structure with a band gap of 4.2 eV [20]. Carbon-doped BNNSs have a narrowed band gap of 2.1 eV, making them effective under visible-light illumination [18]. Doping with O increases the electrical conductivity of BN films [19]. The bandgap of BNNSs can be effectively tuned through different functionalization approaches.

This review concisely highlights the challenges and issues related to the surface functionalization and characterization of BNNSs. It presents potential strategies, explores underlying mechanisms, and critically examines the relationships between processing, structure, and properties. The primary goal is to spark new ideas and functionalization approaches by gaining a fundamental understanding of surface properties and engineering BNNSs with designed structures and predictable behaviors.

18.2 DEFECT ENGINEERING IN SURFACE CHEMISTRY OF BNNSs

The precise identification and quantification of defects in BNNSs represent a critical challenge in 2D nanomaterial research. As highlighted earlier, defect formation is an inherent outcome during the synthesis of 2D nanomaterials, significantly influencing their surface properties. Effectively recognizing and managing these defects is crucial for the tailored and efficient engineering of these materials. One common structural imperfection found in sp^2 carbon allotropes is the Stone–Wales defect, resulting from a 90° rotation of the C–C bond. This rotation transforms the hexagonal graphene lattice into pentagonal or heptagonal structures, as illustrated in Figure 18.2. The high activation barrier associated with this bond rotation results in a defect with a relatively low density [21].

In BNNRs, the emergence of Stone–Wales defects and the rotation of B–N bonds alter the local curvature at defect sites. DFT calculations delineate the zigzag and armchair BNNRs, revealing

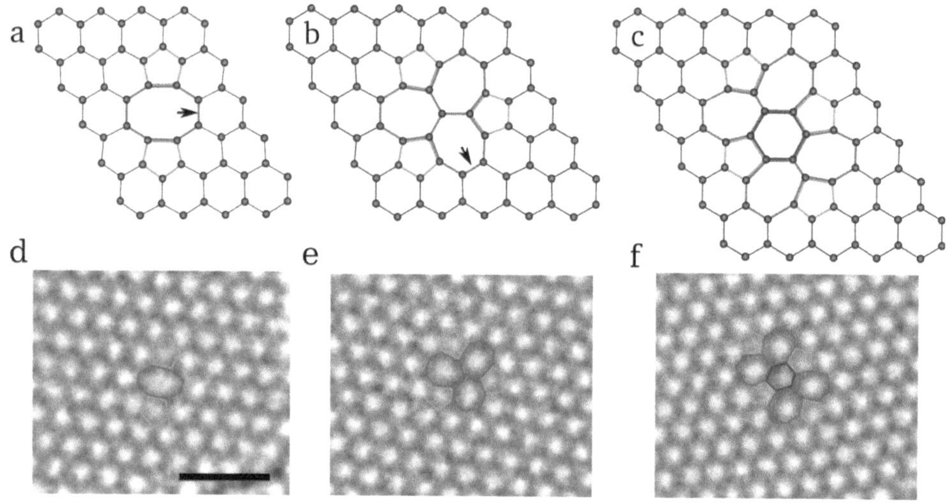

Figure 18.2 (a–c) Atomic structures of reconstructed double vacancy defects in graphene; (d–f) TEM images of the same structures. (Reproduced with permission [21]. Copyright 2011, The American Chemical Society.)

the induced pyramidalization angle at defect sites, indicative of their sp^3 hybridization degree. The formation of this angle is a result of structural deformation and the extension of π bonds out of the planar structure. Consequently, defect-free BNNRs without curvature and Stone–Wales defects exhibit no taper angle, maintaining a 0° angle between the σ and π bonds. The presence of heptagon–heptagon rings resulting from B–N rotation introduces a pyramidalization angle greater than 0°, signifying the formation of Stone–Wales defects. Furthermore, DFT computations demonstrate enhanced exothermic reactions at Stone–Wales defect sites compared to reactions at the center of BNNRs. The newly formed B–B and N–N bonds in the fusion of pentagon–heptagon rings at the edges are identified as the most reactive sites [22]. While theoretical calculations have predicted the formation and stabilization of defects in the order of square–octagon (4–8) pairs in BN nanostructures, it is commonly accepted that Stone–Wales defects are thermodynamically unfavorable in h-BN layers due to the necessity for the formation of B–B or N–N bonds [23].

Point defect is another type found in BN single and double layers. High-resolution transmission electron microscopy images can reveal triangular vacancies, often caused by the knock-on effect from an electron beam interacting with specimen atoms. Zobelli et al. noted that beam energies below 74 keV are insufficient to create defect-free BNNTs since they do not remove atoms or create vacancies. However, as the energy increases from 74 to 84 keV, B atoms are removed, and at 84 keV, even atoms above N can be dislodged by electron beam bombardment [24]. Since B atoms have a lower threshold for electron beam energy compared to N atoms, the resulting vacancies are typically B vacancies surrounded by N-terminated edges. These defects can also display optical activity by interacting with excitons [25]. Research has shown that defects in 2D materials and diamonds can act as single-photon emitters, making them promising candidates for nanoscale quantum information applications [26]. At cryogenic temperatures, defects in transition metal dichalcogenides serve as efficient photon emitters, while at room

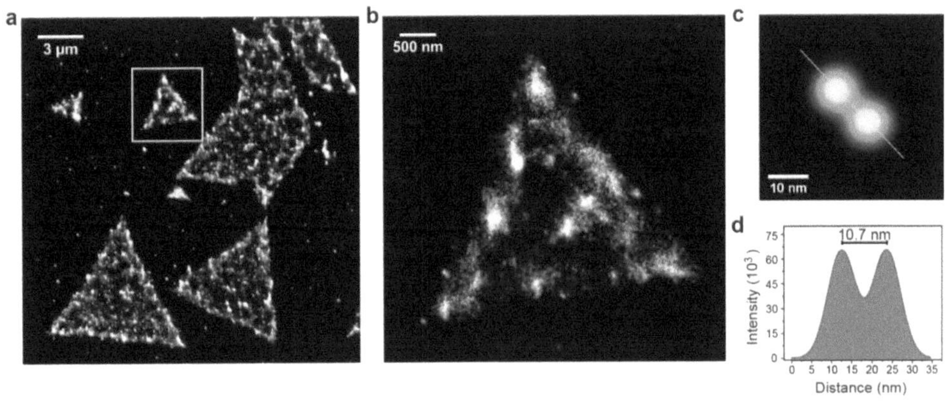

Figure 18.3 (a) Reconstructed single-molecule localization microscopy image of defects in h-BN. (b) Selected region in (a). (c) Two distinguishing close emitters. (d) Distance profile shown in (c). (Reproduced with permission [68]. Copyright 2018, The American Chemical Society.)

temperature, h-BN defects show superior photon emission with remarkable sensitivity [27]. Pan et al. [28] introduced a novel method for quantifying defects by leveraging their optical activity on monolayer BNNS surfaces. They used single-molecule localization microscopy to precisely identify and quantify individual optically active defects within a range of 10.7 nm (Figure 18.3). This method was further applied to activate defects on h-BN surfaces through protonation in water, where proton mobility and defect movement across the h-BN monolayer made them fluorescently distinguishable, enabling real-time monitoring for potential applications [29].

18.2.1 Synthesis Techniques of BNNSs

The synthesis of BNNSs is crucial for their practical use, with scalable and cost-efficient methods being a key requirement. The synthesis process directly impacts the surface functionalization and properties of BNNSs, with top-down and bottom-up methods being the primary approaches.

Top-down methods include mechanical and liquid exfoliation, where BNNSs are separated from bulk h-BN. Mechanical exfoliation, though straightforward, often results in multilayered BNNSs and requires high-energy treatments to achieve monolayers [30,31]. Ball-milling is another top-down method that, by adjusting parameters, can produce BNNSs with various functional groups and low vertical dimensions [32]. Liquid exfoliation, using solvents and ultrasonic vibrations, offers scalability but faces challenges with nanosheet stability [33,34].

Bottom-up methods, such as chemical vapor deposition (CVD) and chemical synthesis, create BNNSs from precursors [35–37]. CVD offers high-quality BNNSs with controllable dimensions and crystallinity, while chemical synthesis is simpler but can result in variable product dimensions and chemical structures [8,38].

Other methods, like plasma-assisted processes, provide high efficiency and well-controlled functional groups but require high energy [39]. Each synthesis method has distinct advantages and limitations, making the choice dependent on specific research goals. The synthesis process influences BNNS characteristics like layer number, dimensions, defect density, and functionalization potential, making it critical for optimizing BNNS performance.

18.2.2 Solvation and Covalent Functionalization of BNNSs

Covalent functionalization of BNNSs involves breaking polarized π bonds and sp^3 hybridized σ bonds, which is challenging due to their high bond energies. When a functional group is attached to one B or N atom, the opposing N or B atom must also be functionalized to maintain charge balance. This process can be carried out in both liquid and solid phases.

18.2.2.1 Solid State Functionalization

Solvent-free surface functionalization of CVD-grown BNNSs has been accomplished through controlled plasma-assisted hydroxylation [40]. While the surface topography stays mostly unchanged, micro-Raman spectra reveal chemical composition changes in samples treated with Ar/N2 and Ar/O2 plasmas. Further plasma treatment with Ar/NH3 adjusts the incorporation of oxygen into BNNSs, as confirmed by X-ray Photoelectron Spectroscopy (XPS) data. The wettability of these nanomaterials is notably affected by the –OH groups attached to plasma-treated BNNS surfaces, rather than by other oxygen-containing groups. Although controlling surface oxygen functionalization is beneficial, the reaction mechanism remains unclear, and the impact of surface chemistry is somewhat limited [41].

18.2.2.2 Liquid Phase Functionalization

Most BNNS functionalization occurs in the liquid phase because it allows for uniform surface modification, is easier to execute, and is more energy-efficient. BNNSs, with their high specific surface area from individually separated nanosheets, tend to restack due to thermodynamic favorability. Solvation helps prevent this restacking in bulk assemblies, addressing a significant challenge in BNNS applications. Pure BNNSs generally have poor dispersion in aqueous solutions and minimal dispersion in toxic, corrosive, or volatile organic solvents. Pre-functionalization with –OH groups improves water dispersibility and adds reaction sites. Hydroxylation of BNNSs involves covalently attaching –OH groups to electrophilic B sites through methods such as ball-milling with NaOH [42], plasma-assisted interface engineering [43], hydrolysis by ultrasonication [44], and reactive radical attack methods [45,46]. For example, nitric acid can attach –OH groups to BNNS surfaces, bonding to marginal and planar B atoms [47]. OH-BNNSs enhance both the breakdown strength and dielectric constant of poly(vinylidene fluoride) due to hydrogen bonding with poly(vinylidene fluoride). Likewise, OH-BNNSs integrated with thermoplastic polyamide-imide form a 3D interconnected network that improves thermal conductivity through effective solvation and interfacial interactions [48]. Qin et al. [49] fabricated a stable nanofluidic ion channel using OH-BN layers, demonstrating ion passage through interlayer spaces in hydrated BN membranes. OH groups on BNNSs reduce proton transport barriers, enabling operation in harsh conditions and high temperatures. Thus, solvation is often the initial step in BNNS functionalization, providing water processability and multiple bonding sites for further chemical reactions.

18.2.2.3 Functionalization with $-NH_2$ Groups

$-NH_2$ groups are attached to BNNS surfaces through various techniques to produce NH_2-BNNSs [50]. For instance, urea-assisted ball-milling introduces a significant amount of $-NH_2$ groups into the edges of BNNSs [51]. Aqueous colloidal solutions of NH_2-BNNSs, with concentrations up to 30 mg/mL, remain stable for up to two weeks. These solutions can be used to create ultralight aerogels through freeze-drying and highly transparent, freestanding membranes via cryo-drying and filtration. The high concentration of NH_2-BNNSs in solution is due to hydrogen bonding between adjacent nanosheets, which aids in fabricating BN membranes and aerogels. These membranes are stable in corrosive liquids and organic solvents, [52] making them suitable for molecular separation and permeation applications. NH_2-BNNS combined with cross-linked thermally rearranged polyimide (XTR) reduces H_2 permeability but significantly improves H_2 gas selectivity, enabling practical H_2 separation [53]. The introduction of NH_2-BNNSs forms FBN-XTR clusters with rough surfaces, which affects the packing of XTR chains. NH_2-BNNS membranes also facilitate ionic transport through interlayer spaces [49]. Due to their chemical and thermal stability, these membranes can operate in corrosive solutions and at high temperatures. Kim et al. [54] recently developed robust, flexible ion gel electrolytes by incorporating NH_2-BNNSs to improve lithium ion (Li^+) transport properties. The ion gel electrolyte, created through in situ thermal polymerization, shows enhanced mechanical properties and increased Li^+ transference number. Strong interactions between NH_2-BNNSs and bis(trifluoromethanesulfonyl)imide anions (TFSI–) reduce anion rotational movements, thereby boosting Li^+ mobility.

18.2.2.4 Heteroatomic Doping

Heteroatomic doping involves substituting B or N atoms across the BNNS basal plane, significantly altering surface properties. The primary objective is to narrow the wide bandgap of BNNSs, opposite to graphene's bandgap opening. Efforts to dope C atoms into BNNSs during chemical synthesis or post-processing have yielded two forms: C2 dopants or aromatic rings attached to BN lattices. Theoretical studies predict energy-favorable structures of B–C–N at various doping levels [55]. Experimentally, C-doped BNNSs have been synthesized via methods including CVD [55], hydrothermal reactions [56], thermal annealing [57], thermal degradation of C precursors [58], and pyrolysis of graphitic carbon nitride (C3N4) [59]. Recently, defective C-doped BNNSs were synthesized using a template-based method with [Bmin]BF4 as the carbon source, forming porous templates where defective C-doped BNNSs grow via hydrothermal reaction [60]. Doping carbon atoms narrows the bandgap, making BNNSs suitable for electrochemical applications. C-doped BNNSs, used as electrocatalysts for nitrogen reduction reactions, achieve an NH3 production of 36.7 $\mu g\ h^{-1} mg^{-1}$ and a faradic efficiency of 6.51%, significantly outperforming undoped BNNSs. The catalytic activity is attributed to unsaturated boron radicals responsible for N_2 adsorption and reduction. O atom doping has been studied theoretically and experimentally [61,62]. O doping occurs at N positions, increasing π electron density and reducing BNNSs' bandgap, akin to graphene N doping. O-doped BNNSs, with adjustable bandgaps, are promising for applications such as magnetic manipulation and electrocatalysis.

18.2.2.5 BNNS/Graphene Heterostructures

Combining BNNS and graphene monolayers through covalent bonding leverages the advantages of both materials. This is particularly attractive in microelectronics, where BNNSs act as an insulator and graphene as a superconductor. BNNS/graphene heterostructures are produced in three steps: CVD growth on a substrate, etching to obtain the desired pattern, and subsequent growth of graphene/BNNS on the etched area [63]. Graphene nucleates at BNNS grain corners, growing on Cu–Ni alloys with good control to form epitaxial high-quality graphene/BNNS in-plane heterostructures, [64] maintaining consistent electrical properties across the composite material's interface.

18.2.2.6 Other Covalent Functionalization Methods

Various methods have been used to functionalize BNNSs for a wide range of applications. Sulfate-modified BNNSs demonstrate strong chemical adsorption capabilities for NO_2 gas, making them ideal for high-performance gas sensors [65]. Cyclopentadienyl-functionalized BNNSs are effective in adsorbing water pollutants [66], while polyethyleneimine functionalized BNNSs, with their high amine density, enhance CO_2 capture efficiency [67]. Covalent functionalization primarily aims to (1) improve BNNS dispersibility and processability in water, (2) leverage BNNS's large surface area for composite materials, and (3) modify the electronic structure of semiconductors for use in nanoelectronics.

18.2.2.7 Non-Covalent Functionalization Methods

Non-covalent functionalization of BNNSs is a promising technique that enhances their potential for various applications without altering their intrinsic properties [69]. Unlike covalent modifications, which involve forming strong chemical bonds that can disrupt the structure of BNNSs, non-covalent functionalization relies on weaker interactions such as van der Waals forces, π–π stacking, and electrostatic interactions [70]. This approach preserves the mechanical strength, thermal stability, and chemical inertness of BNNSs, which are crucial for applications in harsh environments. Non-covalent functionalization allows for the attachment of functional molecules, such as enzymes or polymers, onto the surface of BNNSs, thereby imparting new properties while maintaining the original characteristics of the nanosheets [71,72].

The process of non-covalent functionalization can significantly enhance the solubility and dispersion of BNNSs in various solvents, which is critical for their use in biological and medical applications. For instance, the use of peptides or proteins to wrap around BNNSs can improve their solubility and facilitate their self-assembly into hierarchical structures [73]. This technique has been shown to improve the catalytic performance of enzymes when immobilized on BNNSs, as demonstrated with microperoxidase-11 (MP-11) [74]. The non-covalent attachment of MP-11 to BNNSs results in strong electron coupling between the enzyme and the nanosheets, enhancing the enzyme's catalytic activity. Such functionalization strategies open up new possibilities for using BNNSs in biocatalysis, drug delivery, and as carriers for biomolecules, highlighting their potential in advancing nanotechnology and biomedicine.

18.3 DELVING FURTHER INTO THE REALM OF SURFACE CHEMISTRY OF BNNSS

BNNSs share a fundamentally similar 2D structure to graphene, consisting of single or few layers arranged in a hexagonal network of sp_3-bonded B and N atoms. Recent years have witnessed a substantial focus on advancing the surface functionalization of BNNSs, aiming to enhance their scientific and commercial applicability.

18.3.1 Unanticipated Catalytic Properties of BNNSs

Contrary to their reputation for chemical inertness, BNNSs exhibit unexpected catalytic activity in both pristine and functionalized forms [75]. This catalytic activity spans a diverse range of chemical reactions, including oxidative dehydrogenation, nitrogen reduction reactions, photocatalytic degradation, hydrogen evolution reactions, and hydrochlorination [32,75]. The inherent chemical stability and mechanical strength of BNNSs contribute to their efficiency and sustainability as catalysts.

The presence of unsaturated boron (B) radicals, especially at defects and edges of BNNSs, is hypothesized to serve as active sites in various catalytic processes [76]. Boron-centered radicals, identified as key players in these processes, are formed at defects and edges of BNNSs. Exfoliation through sonication generates further defects and unsaturated boron atoms by cleaving boron–nitrogen bonds [76]. Nanosheets with different termination and functional groups, such as amino and hydroxyl groups, exhibit similar reactivity, supporting the idea that the radicals are centered on boron.

In catalytic applications, BNNSs facilitate the oxidation of compounds such as 3,5,3′,5′-tetramethylbenzidine in the presence of hydrogen peroxide (H_2O_2). BNNSs promote the decomposition of H_2O_2 to generate hydroxyl radicals, which then react with 3,5,3′,5′-tetramethylbenzidine, resulting in a blue-colored charge transfer complex [76]. The generation of these radicals is a key mechanism underlying the catalytic performance of BNNSs [77].

Further expanding on this, porous BN nanorods have shown excellent catalytic performance for the degradation of organic compounds via peroxymonosulfate activation [78]. This non-radical oxidation is driven by defects on the porous BN, leading to the generation of singlet oxygen ($^1O_2$). The adsorption energy between peroxymonosulfate and BN, as well as the peroxymonosulfate activation kinetics, is a critical factor in this process. The evolution of $^1O_2$ is confirmed by electron spin resonance and chemical scavenging experiments, highlighting the robustness of this catalytic pathway across different water matrixes and its ability to produce less toxic intermediates [78].

Moreover, porous h-BN nanosheets have demonstrated high catalytic activity for the epoxidation of styrene using tert-butyl hydroperoxide as an oxidant [79]. The activity is attributed to oxygen-containing functional groups at the edges of porous h-BN. These groups, including B–OH and B–O–O–B edges, act as active sites for the reaction. The transformation of these groups with temperature changes, as monitored by in situ Fourier transform infrared spectroscopy, reveals the dynamics of the catalytic sites [79].

Additionally, boron carbon oxynitride quantum dots leverage boron radical active sites for catalyzing peroxydisulfate ($S_2O_8^{2-}$), significantly enhancing electrochemiluminescence efficiency [80]. This system demonstrates environmental friendliness and biocompatibility, making it suitable for sensitive bioanalysis applications.

These insights suggest that the catalytic properties of BNNSs and their derivatives are closely linked to the presence and manipulation of boron radicals. Enhancing the density and stability of these reactive sites could further improve the catalytic capabilities of BN-based materials. Additionally, exploring the potential of BNNSs in various catalytic processes and environmental applications could lead to significant advancements in industrial catalysis (Figure 18.4).

18.3.2 BNNSs as Superior Lubricant Additives

BNNSs are expected to play a significant role in the water-based lubricant industry, offering benefits over graphene and molybdenum disulfide additives. However, their limited dispersibility in water or lubricant oil has been a barrier to their widespread use. This issue is addressed through surface functionalization, where BNNSs are modified with Pebax. Pebax provides both rigidity from its polyamide blocks and elasticity from its polyether blocks [81] and is recognized as a biocompatible material [82]. Pebax is grafted onto BNNSs by bonding with the unsaturated atoms created during exfoliation. This combination shows great potential as an environmentally friendly water-based lubricant, effectively reducing friction and wear rates [83].

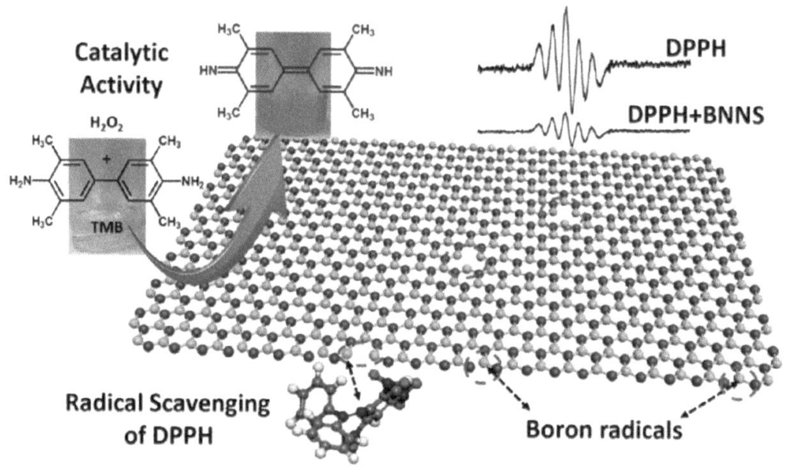

Figure 18.4 Radical scavenging method for the detection of boron radicals on BNNSs and the catalytic activity of BNNSs for the oxidation of 3,5,3′,5′-tetramethylbenzidine [76]. (Copyright 2019, The American Chemical Society.)

18.3.3 Biomaterials

BNNSs have undergone extensive biocompatibility assessments in various cells, demonstrating non-toxicity in vivo. Cytotoxicity evaluations, including apoptosis assays [84] and dosage tests across different cell lines, indicate the absolute safety of BNNSs up to 0.1 mg/mL [85]. Beyond biocompatibility, BNNSs exhibit exceptional flexibility and stability in biological environments, positioning them as an efficient drug carrier nanomaterial [86,87]. Furthermore, applications in tissue engineering have been explored [88,89]. The cosmetic industry has embraced BN-based nanomaterials, leveraging their transparency and biocompatibility as excellent skincare fillers, particularly when combined with amino acids.

18.3.4 Construction of 3D Superstructures

Integrating nanomaterials into bulk materials while preserving their exceptional 2D properties is a significant challenge in nanoscience. The delicate nature of ultrathin 2D nanosheets complicates their handling, making the processability of 3D superstructures more favorable. These structures, which maintain nearly the same performance as 2D nanosheets, can be created through self-assembly in liquids or by substituting elements on templates. The self-assembly of solvated BNNSs leads to the formation of 3D hydrogels, driven by hydrophilic interactions among functionalized nanosheets. While water molecules increase flexibility, they may also weaken mechanical strength. Alternatively, Qiao et al. achieved a notable breakthrough by creating a spherical superstructure composed of BNNSs (SS-BNNSs) using crystalline metal–organic framework particles as a template [90]. The metal–organic framework undergoes conversion into SS-BNNSs through hydrolysis and nitrogenization reactions, while preserving its initial spherical superstructure morphology. This 3D structure exhibits enhanced oxidative dehydrogenation activity compared to other h-BN materials. However, the synthesis of BNNSs using templates brings about unavoidable impurities, including oxygen and carbon. Subsequent research efforts should delve into understanding the effects of these impurities on the properties of BNNSs. Exploring alternative templates, such as C_3N_4 nanoparticles and melamine resins, could offer avenues for synthesizing 3D BNNS superstructures through element substitution.

18.3.5 Addressing the Challenge of Single-Molecule Transport: Tracing Proton Movement at the h-BN Crystals/Water Interface with Single-Molecule Localization Microscopy

Understanding single-molecule transport across interfaces is a complex task due to the lightweight, small size, and high reactivity of these entities. However, recent breakthroughs have enabled the successful tracking of protons, the smallest units, at the interface of h-BN crystals and aqueous solutions at room temperature [29]. Employing single-molecule localization microscopy, researchers observed individual proton trajectories, relying on the protonation and activation of optically active defects on the surface of h-BN crystals. The interface between h-BN and water holds promise as a pathway for single-molecule transport, demonstrating broad potential for molecular charge transfer. The research suggests the possibility of observing or quantifying other ions, functional groups (such as OH and NH_2), or physically adsorbed molecules (such as N_2 and O_2) at this solid/water interface. Additionally, introducing different types or quantities of structural defects holds promise for improving the performance of the h-BN/water interface. In a connected advancement, the elongation of single-stranded DNA emerges as a critical aspect in genome analysis. Notably, single-stranded DNA molecules demonstrate spontaneous migration from the graphene surface to h-BN nanochannels, and substantial self-stretching (approximately 6.25 Å) has been documented in in-plane graphene/h-BN/graphene heterostructures by Luan and Zhou [91].

18.3.6 Advanced Thermal Management and Insulation for Electronic Devices: Harnessing the Potential of 2D BNNSs

Integrated circuits represent the backbone of modern electronic devices, requiring a meticulous assembly of various components. This assembly involves the integration of both passive elements, such as capacitors, and active components like transistors. The reliable functioning of these components crucially depends on a well-defined insulating layer. Ensuring the thermal efficiency of electronic devices is equally vital, necessitating advanced materials that can efficiently conduct heat while providing electrical insulation.

In the realm of materials that exhibit high thermal conductivity and electrical insulating properties, only a select few, such as diamond and BN, have proven effective. While diamond is a known contender, BN emerges as a more practical option due to factors like its abundance, cost-effectiveness, and ease of processing. However, utilizing bulk BN crystals directly in devices or as packaging materials is impractical.

This is where 2D BNNSs come into play, offering a versatile solution. Thanks to their high aspect ratio, BNNSs can be seamlessly incorporated into composites alongside polymer-based packaging materials and device components.

This integration facilitates effective heat management, a critical aspect for preventing device overheating and maintaining optimal operating temperatures. The distinctive characteristics of BNNSs, such as flexibility, inherent insulating properties, exceptional thermal conductivity, and chemical stability, establish them as superb dielectric materials. An exemplary application of BNNSs in electronic components is demonstrated by Worsley et al. [92], who successfully fabricated h-BN/graphene inkjet-printed capacitors for integrated circuits using biocompatible graphene and h-BN inks. This innovative approach showcases the adaptability and potential of BNNSs in advancing electronic component fabrication.

Beyond their role in thermal management, BNNS films with triangular pores exhibit remarkable efficacy in seawater desalination. These films demonstrate robust resistance to oxidation and corrosion, coupled with exceptional reusability [93]. This multifaceted utility of BNNSs highlights their versatility and potential in diverse technological applications.

While the field of surface functionalization for BNNSs is rapidly progressing, challenges persist. Addressing these challenges necessitates not just technical advancements in materials handling but also a profound understanding of the complex interplay among processing, structure, and properties. Future research endeavors should prioritize the development of machinable materials assembled from single layers of BNNSs without compromising their intrinsic performance. This holistic approach will undoubtedly drive the continued evolution of BNNSs and their integration into advanced electronic devices.

18.3.7 A Facile Approach for Promoting Non-covalent Interactions between BNNSs and Gold Nanoparticles

The interaction between nanomaterials and their surrounding medium is often governed by the formation of an electrical double layer (EDL) at the surface of colloidal particles. In conductive colloids, the EDL is typically described by Maxwell–Wagner interfacial polarization, while in non-conductive colloids, such as BNNSs, the EDL formation differs due to their semi-ionic nature [94]. The modulation of EDL thickness is crucial for controlling colloidal interactions, especially when non-covalent forces such as van der Waals interactions, hydrogen bonding, and hydrophobic/hydrophilic interactions are involved.

Figure 18.5 presents a novel approach to modulate the EDL of BNNSs and gold nanoparticles (AuNP) using controlled pH conditions. By reducing the EDL, BNNSs can interact with AuNPs and prevent their aggregation in the presence of L-cysteine (L-Cys), a known zwitterion [95]. The results highlight the potential of BNNSs as a scaffold for stabilizing functionalized AuNPs in aqueous dispersions. The non-covalent interactions explored in this work offer a straightforward and effective method for material synthesis, with promising implications for the development of advanced sensing devices and multifunctional materials in various scientific and technological fields.

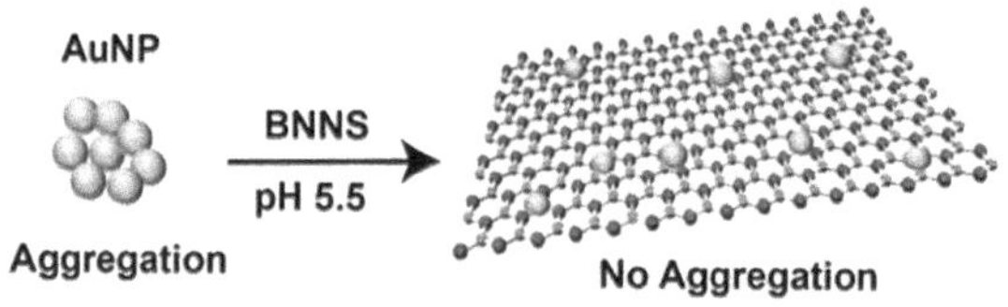

Figure 18.5 Schematic illustration of modulating the EDL of BNNSs and gold nanoparticles (AuNP) using controlled pH conditions [95]. (Copyright 2020, The American Chemical Society.)

18.4 CONCLUSIONS

BNNSs possess remarkable properties that make them highly desirable for advanced technological applications, yet their chemical inertness poses significant challenges to surface functionalization. Through extensive research efforts, progress has been made in overcoming these challenges by employing innovative approaches in surface engineering and defect manipulation. This chapter has provided a detailed exploration of BNNSs, addressing their intrinsic properties and the complexities of their functionalization. By delving into surface chemistry and defect engineering, key strategies for tailoring BNNS surface functionalities have been elucidated. Furthermore, the potential applications of BNNSs in diverse domains—ranging from catalysis and lubrication to biomaterials, sensor technology, and electronic device insulation—underscore their versatility and transformative potential. Future advancements in BNNS functionalization will undoubtedly expand their applicability, paving the way for groundbreaking innovations in science and industry.

REFERENCES

1. K. S. Novoselov et al., Electric field effect in atomically thin carbon films. *Science* **306**, 666–669 (2004).
2. L. H. Li et al., Large-scale mechanical peeling of boron nitride nanosheets by low-energy ball milling. *Journal of Materials Chemistry* **21**, 11862–11866 (2011).
3. M. Corso et al., Boron nitride nanomesh. *Science* **303**, 217–220 (2004).
4. A. Bosak et al., Elasticity of hexagonal boron nitride: Inelastic X-ray scattering measurements. *Physical Review B—Condensed Matter and Materials Physics* **73**, 041402 (2006).
5. C. Lee, X. Wei, J. W. Kysar, J. Hone, Measurement of the elastic properties and intrinsic strength of monolayer graphene. *Science* **321**, 385–388 (2008).
6. Z. Liu et al., Challenges and solutions in surface engineering and assembly of boron nitride nanosheets. *Materials Today* **44**, 194–210 (2021).
7. Q. Weng, X. Wang, X. Wang, Y. Bando, D. Golberg, Functionalized hexagonal boron nitride nanomaterials: Emerging properties and applications. *Chemical Society Reviews* **45**, 3989–4012 (2016).
8. D. Golberg et al., Boron nitride nanotubes and nanosheets. ACS *Nano* **4**, 2979–2993 (2010).
9. N. Alem et al., Subangstrom edge relaxations probed by electron microscopy in hexagonal boron nitride. Physical *Review Letters* **109**, 205502 (2012).
10. N. Marom et al., Stacking and registry effects in layered materials: The case of hexagonal boron nitride. Physical *Review Letters* **105**, 046801 (2010).
11. T. Kuila et al., Chemical functionalization of graphene and its applications. *Progress in Materials Science* **57**, 1061–1105 (2012).
12. D. C. Elias et al., Control of graphene's properties by reversible hydrogenation: Evidence for graphane. *Science* **323**, 610–613 (2009).
13. L. H. Li, J. Cervenka, K. Watanabe, T. Taniguchi, Y. Chen, Strong oxidation resistance of atomically thin boron nitride nanosheets. ACS *Nano* **8**, 1457–1462 (2014).
14. T. Ikuno, T. Sainsbury, D. Okawa, J. M. J. Fréchet, A. Zettl, Amine-functionalized boron nitride nanotubes. *Solid State Communications* **142**, 643–646 (2007).
15. W. Lei et al., Boron nitride colloidal solutions, ultralight aerogels and freestanding membranes through one-step exfoliation and functionalization. *Nature Communications* **6**, 1–8 (2015).
16. C. A. de los Reyes et al., Tunable alkylation of white graphene (hexagonal boron nitride) using reductive conditions. *The Journal of Physical Chemistry* C **123**, 19725–19733 (2019).
17. X. Huang et al., Polyhedral oligosilsesquioxane-modified boron nitride nanotube based epoxy nanocomposites: An ideal dielectric material with high thermal conductivity. *Advanced Functional Materials* **23**, 1824–1831 (2013).
18. C. Huang et al., Carbon-doped BN nanosheets for metal-free photoredox catalysis. *Nature Communications* **6**, 7698 (2015).

19. R. Sevak Singh et al., Band gap effects of hexagonal boron nitride using oxygen plasma. *Applied Physics Letters* **104**, (2014).
20. A. Lopez-Bezanilla, J. Huang, H. Terrones, B. G. Sumpter, Structure and electronic properties of edge-functionalized armchair boron nitride nanoribbons. *The Journal of Physical Chemistry* C **116**, 15675–15681 (2012).
21. F. Banhart, J. Kotakoski, A. V. Krasheninnikov, Structural defects in graphene. ACS *Nano* **5**, 26–41 (2011).
22. R. Gao et al., High-yield synthesis of boron nitride nanosheets with strong ultraviolet cathodoluminescence emission. *The Journal of Physical Chemistry* C **113**, 15160–15165 (2009).
23. Y. Lin, J. W. Connell, Advances in 2D boron nitride nanostructures: Nanosheets, nanoribbons, nanomeshes, and hybrids with graphene. *Nanoscale* **4**, 6908–6939 (2012).
24. A. Zobelli, A. Gloter, C. Ewels, G. Seifert, C. Colliex, Electron knock-on cross section of carbon and boron nitride nanotubes. *Physical Review B—Condensed Matter and Materials Physics* **75**, 245402 (2007).
25. S. Gupta, W. Wu, S. Huang, B. I. Yakobson, Single-photon emission from two-dimensional materials, to a brighter future. *The Journal of Physical Chemistry Letters* **14**, 3274–3284 (2023).
26. J.-W. Pan et al., Multiphoton entanglement and interferometry. *Reviews of Modern Physics* **84**, 777–838 (2012).
27. R.-C. Zhang et al., Microplasma processed ultrathin boron nitride nanosheets for polymer nanocomposites with enhanced thermal transport performance. *ACS Applied Materials & Interfaces* **8**, 13567–13572 (2016).
28. F. Pan et al., Fabrication of gold nanoparticles/l-cysteine functionalized graphene oxide nanocomposites and application for nitrite detection. *Journal of Alloys and Compounds* **744**, 51–56 (2018).
29. J. Comtet et al., Direct observation of water-mediated single-proton transport between hBN surface defects. *Nature Nanotechnology* **15**, 598–604 (2020).
30. K. S. Novoselov et al., Two-dimensional atomic crystals. Proceedings of the National Academy of Sciences *of the United States of America* **102**, 10451–10453 (2005).
31. J. C. Meyer, A. Chuvilin, G. Algara-Siller, J. Biskupek, U. Kaiser, Selective sputtering and atomic resolution imaging of atomically thin boron nitride membranes. *Nano Letters* **9**, 2683–2689 (2009).
32. B. Liu et al., Insight into catalytic mechanisms for the reduction of nitrophenol via heterojunctions of gold nanoclusters on 2D boron nitride nanosheets. *ChemNanoMat* **5**, 784–791 (2019).
33. C. Zhi, Y. Bando, C. Tang, H. Kuwahara, D. Golberg, Large-scale fabrication of boron nitride nanosheets and their utilization in polymeric composites with improved thermal and mechanical properties. Advanced *Materials* **21**, 2889–2893 (2009).
34. J. N. Coleman et al., Two-dimensional nanosheets produced by liquid exfoliation of layered materials. *Science* **331**, 568–571 (2011).
35. M. H. Khan et al., Few-atomic-layered hexagonal boron nitride: CVD growth, characterization, and applications. *Materials Today* **20**, 611–628 (2017).
36. A. L. Gibb et al., Atomic resolution imaging of grain boundary defects in monolayer chemical vapor deposition-grown hexagonal boron nitride. *Journal of the American Chemical Society* **135**, 6758–6761 (2013).
37. Y. Shi et al., Synthesis of few-layer hexagonal boron nitride thin film by chemical vapor deposition. Nano *Letters* **10**, 4134–4139 (2010).
38. P. Wu et al., A template-free solvent-mediated synthesis of high surface area boron nitride nanosheets for aerobic oxidative desulfurization. *Chemical Communications* **52**, 144–147 (2016).
39. H. Chen et al., Construction of a nanoporous highly crystalline hexagonal boron nitride from an amorphous precursor for catalytic dehydrogenation. *Angewandte Chemie* **131**, 10736–10740 (2019).
40. H. Achour et al., Plasma surface functionalization of boron nitride nano-sheets. *Diamond and Related Materials* **77**, 110–115 (2017).
41. Z. Liu et al., Probing the tunable surface chemistry of graphene oxide. *Chemical Communications* **51**, 10969–10972 (2015).
42. D. Lee et al., Scalable exfoliation process for highly soluble boron nitride nanoplatelets by hydroxide-assisted ball milling. *Nano Letters* **15**, 1238–1244 (2015).

43. A. Pakdel, Y. Bando, D. Golberg, Plasma-assisted interface engineering of boron nitride nanostructure films. ACS *Nano* **8**, 10631–10639 (2014).
44. Y. Lin et al., Aqueous dispersions of few-layered and monolayered hexagonal boron nitride nanosheets from sonication-assisted hydrolysis: Critical role of water. *The Journal of Physical Chemistry* C **115**, 2679–2685 (2011).
45. T. Sainsbury et al., Oxygen radical functionalization of boron nitride nanosheets. *Journal of the American Chemical Society* **134**, 18758–18771 (2012).
46. T. Sainsbury et al., Covalently functionalized hexagonal boron nitride nanosheets by nitrene addition. *Chemistry–A European Journal* **18**, 10808–10812 (2012).
47. L. Wu et al., Largely enhanced energy storage density of poly (vinylidene fluoride) nanocomposites based on surface hydroxylation of boron nitride nanosheets. *Journal of Materials Chemistry A* **6**, 7573–7584 (2018).
48. F. Jiang, S. Cui, N. Song, L. Shi, P. Ding, Hydrogen bond-regulated boron nitride network structures for improved thermal conductive property of polyamide-imide composites. *ACS Applied Materials & Interfaces* **10**, 16812–16821 (2018).
49. S. Qin et al., High and stable ionic conductivity in 2D nanofluidic ion channels between boron nitride layers. *Journal of the American Chemical Society* **139**, 6314–6320 (2017).
50. R. Kumar et al., Functionality preservation with enhanced mechanical integrity in the nanocomposites of the metal–organic framework, ZIF-8, with BN nanosheets. *Materials Horizons* **1**, 513–517 (2014).
51. W. Lei et al., Boron nitride colloidal solutions, ultralight aerogels and freestanding membranes through one-step exfoliation and functionalization. *Nature Communications* **6**, 8849 (2015).
52. C. Chen et al., Functionalized boron nitride membranes with ultrafast solvent transport performance for molecular separation. *Nature Communications* **9**, 1902 (2018).
53. Y. Wang et al., Functionalized boron nitride nanosheets: A thermally rearranged polymer nanocomposite membrane for hydrogen separation. *Angewandte Chemie* **130**, 16288–16293 (2018).
54. D. Kim et al., Amine-functionalized boron nitride nanosheets: A new functional additive for robust, flexible ion gel electrolyte with high lithium-ion transference number. *Advanced Functional Materials* **30**, 1910813 (2020).
55. J. Lu et al., Order–disorder transition in a two-dimensional boron–carbon–nitride alloy. *Nature Communications* **4**, 2681 (2013).
56. L. Chen et al., Template-free synthesis of carbon-doped boron nitride nanosheets for enhanced photocatalytic hydrogen evolution. *Applied Catalysis B: Environmental* **241**, 246–255 (2019).
57. M. Fan et al., Doping nanoscale graphene domains improves magnetism in hexagonal boron nitride. *Advanced Materials* **31**, 1805778 (2019).
58. C. Zhao et al., Carbon-doped boron nitride nanosheets with ferromagnetism above room temperature. *Advanced Functional Materials* **24**, 5985–5992 (2014).
59. D. Portehault et al., High-surface-area nanoporous boron carbon nitrides for hydrogen storage. *Advanced Functional Materials* **20**, 1827–1833 (2010).
60. Z. Liu et al., Defective carbon-doped boron nitride nanosheets for highly efficient electrocatalytic conversion of N_2 to NH_3. *ACS Sustainable Chemistry & Engineering* **8**, 5278–5286 (2020).
61. W. Lei et al., Oxygen-doped boron nitride nanosheets with excellent performance in hydrogen storage. *Nano Energy* **6**, 219–224 (2014).
62. J. Wu, W. Zhang, Tuning the magnetic and transport properties of boron-nitride nanotubes via oxygen-doping. *Solid State Communications* **149**, 486–490 (2009).
63. L. Britnell et al., Field-effect tunneling transistor based on vertical graphene heterostructures. *Science* **335**, 947–950 (2012).
64. L. Ci et al., Atomic layers of hybridized boron nitride and graphene domains. *Nature Materials* **9**, 430–435 (2010).
65. Y. He et al., High-performance NO_2 sensors based on spontaneously functionalized hexagonal boron nitride nanosheets via chemical exfoliation. *Nanoscale* **11**, 21909–21916 (2019).
66. S. Daneshnia, A. Shams, D. Daraei, M. Abdouss, M. Daneshmayeh, Novel thin film nanocomposite membrane modified with boron nitride nanosheets for water treatment. *Colloids and Surfaces A: Physicochemical and Engineering Aspects* **651**, 129768 (2022).

67. K. Huang et al., Aminopolymer functionalization of boron nitride nanosheets for highly efficient capture of carbon dioxide. *Journal of Materials Chemistry A* **5**, 16241–16248 (2017).
68. J. Feng et al., Imaging of optically active defects with nanometer resolution. Nano *Letters* **18**, 1739–1744 (2018).
69. J. Niskanen et al., Dually-functionalized boron nitride nanotubes to target glioblastoma multiforme. *Materials Today Chemistry* **16**, 100270 (2020).
70. N. S. Mishra, P. Saravanan, A review on the synergistic features of hexagonal boron nitride (white graphene) as adsorbent-photo active nanomaterial. *ChemistrySelect* **3**, 8023–8034 (2018).
71. J. Yang et al., Preparation and characterization of boron nitride nanosheet ferric oxide composite (BNNS@ Fe_3O_4) through the double stabilization of PVP and its adsorption to Congo red. *Journal of Polymer Research* **28**, 1–11 (2021).
72. H. E. Çamurlu, S. Mathur, O. Arslan, E. Akarsu, Modification of hexagonal boron nitride nanoparticles with fluorosilane. *Ceramics International* **42**, 6312–6318 (2016).
73. R. Wang et al., Microperoxidase-11 functionalized nanozyme with enhanced peroxidase-mimicking activities for visual detection of cysteine. *Analytica Chimica Acta* **1267**, 341386 (2023).
74. R. Li et al., Non-covalent surface modification of boron nitride nanotubes for enhanced catalysis. *Chemical Communications* **50**, 225–227 (2014).
75. J. T. Grant et al., Selective oxidative dehydrogenation of propane to propene using boron nitride catalysts. *Science* **354**, 1570–1573 (2016).
76. Z. Liu et al., Boron radicals identified as the source of the unexpected catalysis by boron nitride nanosheets. ACS *Nano* **13**, 1394–1402 (2019).
77. D. Esposito, Radically active. *Nature Catalysis* **2**, 105–105 (2019).
78. Y. Bao et al., Unexpected intrinsic catalytic function of porous boron nitride nanorods for highly efficient peroxymonosulfate activation in water treatment. *ACS Applied Materials & Interfaces* **14**, 18409–18419 (2022).
79. H. Fu et al., Understanding the catalytic sites in porous hexagonal boron nitride for the epoxidation of styrene. *ACS Catalysis* **11**, 8872–8880 (2021).
80. M.-F. Sun et al., High-efficient electrochemiluminescence of BCNO quantum dot-equipped boron active sites with unexpected catalysis for ultrasensitive detection of MicroRNA. *Analytical Chemistry* **92**, 14723–14729 (2020).
81. A. Car, C. Stropnik, W. Yave, K.-V. Peinemann, Pebax®/polyethylene glycol blend thin film composite membranes for CO_2 separation: Performance with mixed gases. *Separation and Purification Technology* **62**, 110–117 (2008).
82. C. Hoffendahl, G. Fontaine, S. Bourbigot, Flame retardancy of bio-based polyether-block-amide polymer (PEBAX). *Polymer Degradation and Stability* **98**, 1247–1255 (2013).
83. L. An et al., Simultaneous production and functionalization of hexagonal boron nitride nanosheets by solvent-free mechanical exfoliation for superlubricant water-based lubricant additives. *npj 2D Materials and Applications* **3**, 1–9 (2019).
84. L. H. Li, Y. Chen, Atomically thin boron nitride: Unique properties and applications. *Advanced Functional Materials* **26**, 2594–2608 (2016).
85. M. Kıvanç et al., Effects of hexagonal boron nitride nanoparticles on antimicrobial and antibiofilm activities, cell viability. *Materials Science and Engineering*: C **91**, 115–124 (2018).
86. S. Feng, H. Zhang, C. Zhi, X.-D. Gao, H. Nakanishi, pH-responsive charge-reversal polymer-functionalized boron nitride nanospheres for intracellular doxorubicin delivery. *International Journal of Nanomedicine* **13**, 641–652 (2018).
87. M. Emanet, Ö. Şen, M. Çulha, Evaluation of boron nitride nanotubes and hexagonal boron nitrides as nanocarriers for cancer drugs. *Nanomedicine* **12**, 797–810 (2017).
88. B. Farshid, G. Lalwani, M. Shir Mohammadi, J. Simonsen, B. Sitharaman, Boron nitride nanotubes and nanoplatelets as reinforcing agents of polymeric matrices for bone tissue engineering. *Journal of Biomedical Materials Research Part B: Applied Biomaterials* **105**, 406–419 (2017).
89. C. Shuai et al., Akermanite scaffolds reinforced with boron nitride nanosheets in bone tissue engineering. *Journal of Materials Science: Materials in Medicine* **26**, 1–9 (2015).

90. C. Qiao et al., Tuning surface electronic structure of two-dimensional cobalt-based hydroxide nanosheets for highly efficient water oxidation. *ChemCatChem* **12**, 2823–2832 (2020).
91. B. Luan, R. Zhou, Spontaneous ssDNA stretching on graphene and hexagonal boron nitride in plane heterostructures. *Nature Communications* **10**, 4610 (2019).
92. R. Worsley et al., All-2D material inkjet-printed capacitors: Toward fully printed integrated circuits. *Acs Nano* **13**, 54–60 (2018).
93. H. Gao et al., Rational design and strain engineering of nanoporous boron nitride nanosheet membranes for water desalination. *The Journal of Physical Chemistry* C **121**, 22105–22113 (2017).
94. Y. Xiao, M. R. Wiesner, "Nanoparticle aggregation and deposition in porous media" in Baoshan Xing, Chad D. Vecitis, Nicola Senesi (eds), *Engineered Nanoparticles and the Environment: Biophysicochemical Processes and Toxicity* (2016), pp. 224–244.
95. A. Dibaji et al., Modulating electrical double layers: Facile approach for promoting noncovalent interactions between boron nitride nanosheets and gold nanoparticles. Langmuir **40**, 19799–19803 (2024).

Chapter 19

Boron Nitride-Based Nanomaterials for Absorption and Separation of Oil, Ions, and Organic Matters

Guoliang Yang, Lifeng Wang, Dan Liu, and Weiwei Lei

19.1 INTRODUCTION

For the absorption of various substances, including oil, organic matter, and ions, materials with large specific surface area, low density, and special surface chemistry are highly desired. Due to the outstanding chemical stability and surface physicochemical properties, BN materials have attracted much attention in environmental remediation, oil cleanup, and heavy metal ion removal as well as ion-related energy applications. Over the past decade, the synthesis of BN materials has developed rapidly, yielding various nanostructures, including nanotubes, nanosheets, nanoribbons, and 3D mesostructures. Additionally, BN-based composites have been developed to improve the stability and performance of adsorbent and separation materials for oil cleanup, water purification, and clean energy harvesting. This chapter begins with a summary of BN material synthesis for oil absorption and separation. In the second part, ion absorption and dye removal on BN-based composites are discussed. In the final section, theoretical studies about ion and water transport through BN materials are reviewed.

19.2 BN NANOMATERIALS FOR OIL ABSORPTION AND SEPARATION

19.2.1 Nanostructures of BN for Oil Absorption

A large specific surface area and a hydrophilic surface are two critical features for developing high-performance oil adsorbents. Lei et al. reported that porous BN nanosheets (BNNSs) exhibit excellent absorption performance for a wide range of oils and solvents (Figure 19.1a and b) [1]. The porous BNNSs absorb up to 33 times their own weight in ethylene glycol and 29 times their own weight in used engine oil (Figure 19.1c). These uptakes are much higher than those of commercial activated carbon, BN particles, and nonporous BNNSs, indicating significant performance enhancement due to porosity (Figure 19.1d). Importantly, the absorbed oil can be removed from the nanosheets by directly burning them in air, allowing the nanosheets to be reused several times. Lin et al. reported a template-free method to prepare 3D BN foam directly. The brief procedures include ultrasonic-assisted self-assembly, freeze-drying, and pyrolysis processes [2].

DOI: 10.1201/9781003647812-24

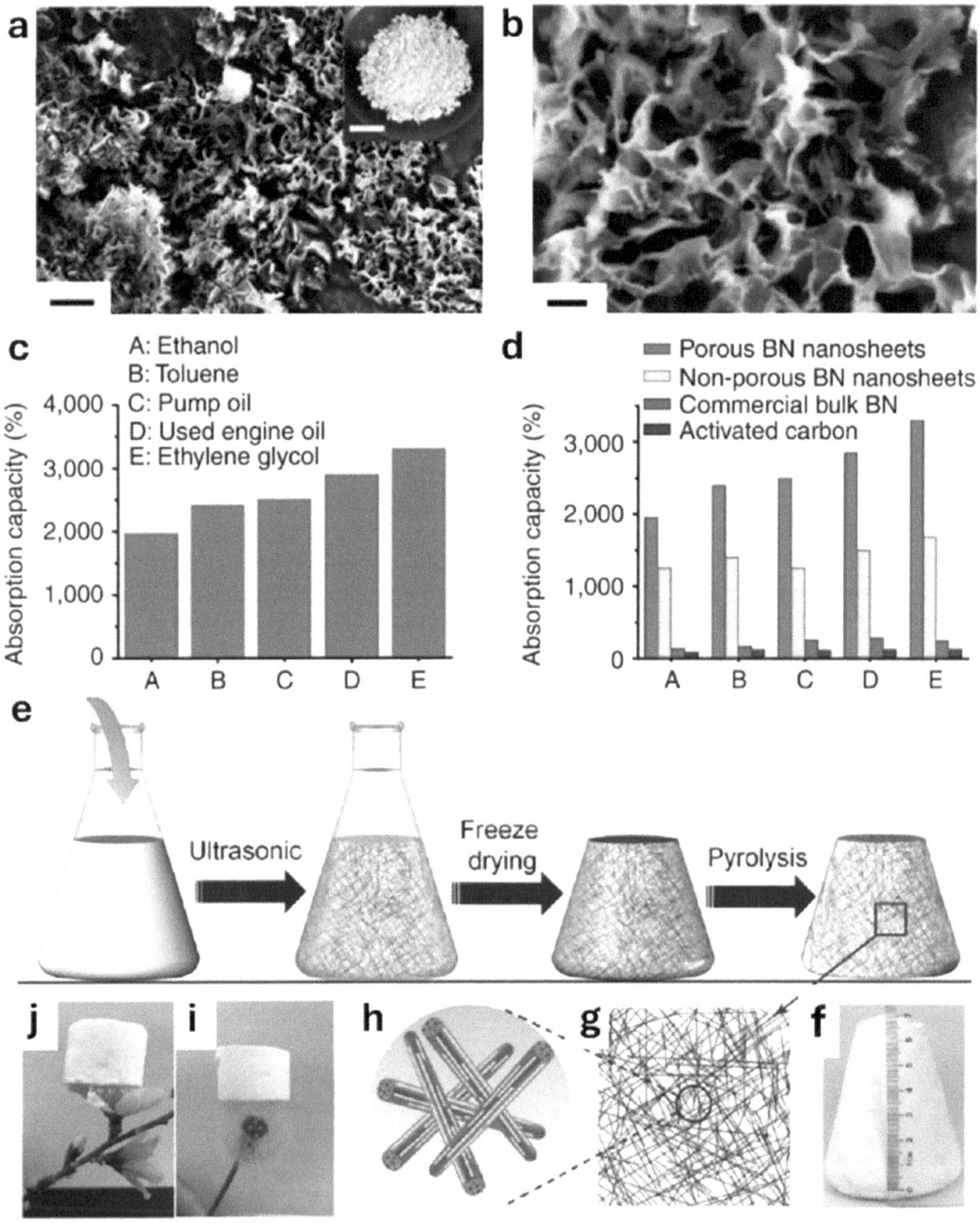

Figure 19.1 (a, b) SEM image of the porous BNNSs. The inset image shows synthetic white BN powder. Scale bar: (a) 2 μm, inset image of (a) 1 cm and (b) 200 nm. (c) Absorption capacities of porous nanosheets for five solvents and oils. (d) Comparison of absorption capacities for porous BNNSs with other materials. (Reproduced with permission from [1].) (e) The process of preparation of BN foam. (f) The photograph of the BN foam. (g, h) The structure of BN foam. (i, j) The photographs of low-density BN foams. (Reproduced with permission from [2].)

The as-prepared BN foam with low density, high porosity, and excellent resilience can be formed into various shapes with large sizes (Figure 19.1e–j). Remarkably, their absorption capacities for numerous organic solvents and oils can reach up to around 8000%, coupled with a high recovery efficiency of ~94%. Gautam et al. proposed a low-temperature (~300°C) solid-state synthesis approach to fabricate low-density porous 3D h-BN foam [3]. These samples not only demonstrated structural and mechanical stability but also exhibited excellent oil absorption properties and high resistance to acid corrosion. Pham et al. introduced a single-step, cost-effective, and scalable method to fabricate crinkled h-BN nanofoam through carbothermic reduction and nitrogenization of boron oxide and graphite [4]. The resulting nanofoams featured sheet-like structures with many wrinkles and edges. Importantly, a large number of nanochannels are created during the transformation process, resulting in a high surface area ($41.2 \pm 0.8\,m^2g^{-1}$) conducive to the absorption of oil.

Xia et al. reported a thermal annealing method with excess nitrogen as the porogen agent for the fabrication of BN aerogels with controlled microstructures and properties (Figure 19.2) [5]. By adjusting the mole ratio of boron to nitrogen, the physicochemical properties of the as-prepared BN foam can be enhanced. Notably, the BN foam with a B/N mole ratio of 2:6 displayed significantly higher specific surface area ($938\,m^2g^{-1}$), compared to the foams with the B/N mole ratios of 2:3 and 2:12. As a result, the BN aerogels with micropores exhibited high capacities for liquid phase absorption, adsorptive desulfurization, and CO_2 capture. Additionally, Pham et al. presented a carbothermic reduction approach for producing 3D aerogels containing highly crystalline sp^2-bonded BN layers [6]. The specific aerogel surface showed superhydrophobic characteristics and demonstrated a remarkable oil absorption capacity of approximately 1500 wt%. Furthermore, Hao et al. devised a template-assisted atomic layer deposition method to prepare hollow BN nanotube (BNNT)-based

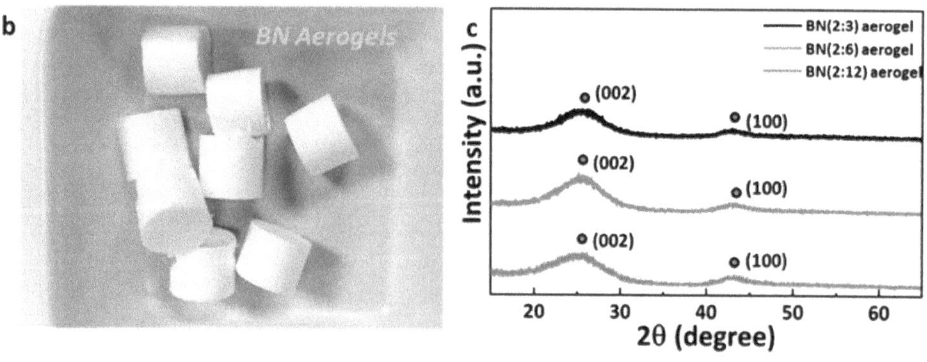

Figure 19.2 (a) Schematic diagram of the fabrication of the BN aerogels. (b) Photographs of BN aerogels. (c) X-ray diffraction data of BN aerogels with different B/N mole ratios. (Reproduced with permission from [5].)

aerogels for water purification. The BNNT aerogels, featuring superhydrophobicity, displayed exceptional stability in air, acidic, and alkaline conditions and high absorption capability for various organic solvents and oils, such as pump oil, mineral oil, and toluene [7].

Song et al. also developed a template-assisted and catalyst-free chemical vapor deposition (CVD) route for the synthesis of BN aerogels. In their innovative approach, the carbon templates were used for synthesizing the BNNTs, subsequently removed through thermal treatment in an oxygen atmosphere at 600°C. Compared with conventional porous BN materials, the prepared BN foam with the continuous 3D network structure exhibited an ultralight feature with a high specific surface area of around 1051 $m^2 g^{-1}$ and a mass density of 0.6 mg cm^{-3}. These highly hydrophobic aerogels displayed an extraordinary oil absorption capability, absorbing up to 160 times their own weight [8]. Xue et al. developed a two-step CVD conversion method to prepare hollow BNNTs, which could be assembled to ultralight foams with interconnected tubular structures. The quasi-nanoscale tubular foam exhibited remarkable shape recovery properties even after undergoing compressive deformations of up to 90%. Moreover, their structural integrity and superhydrophobicity were maintained up to 30 days, showcasing excellent corrosion resistance in harsh acidic and alkaline environments [9].

19.3 BN-BASED NANOCOMPOSITES FOR OIL ABSORPTION AND SEPARATION

19.3.1 Absorption of Oil and Dyes

BN materials are typically prepared through an annealing process, which endows the as-prepared BN materials with hydrophobic and excellent flame-retardant properties. Based on these specific features, Liu et al. demonstrated a two-step strategy to fabricate BNNS-based membranes and sponges with superhydrophobic surface and outstanding antioxidized stability (Figure 19.3). Initially, hydrophobic BNNSs were synthesized via the thermal treatment of a boron trioxide and guanidine hydrochloride mixture. Subsequently, the prepared BNNSs and polydimethyl diallyl ammonium chloride were coated on to the melamine sponge using a simple layer-by-layer assembly process. The BN-based foam with low water adhesion and density, high porosity, and robustness displayed ultrafast oil/water separation and high oil absorption capability (112 times its own weight) [10].

Graphene oxide (GO) is a critical and widely studied two-dimensional material known for its negative surface charge and hydrophilic properties. Li et al. reported a facile direct freeze-drying method to prepare porous 3D reduced GO (rGO)/BN foam for oil absorption (Figure 19.4). The resulting foam can be fabricated on a large scale, exhibiting strong hydrophobicity and

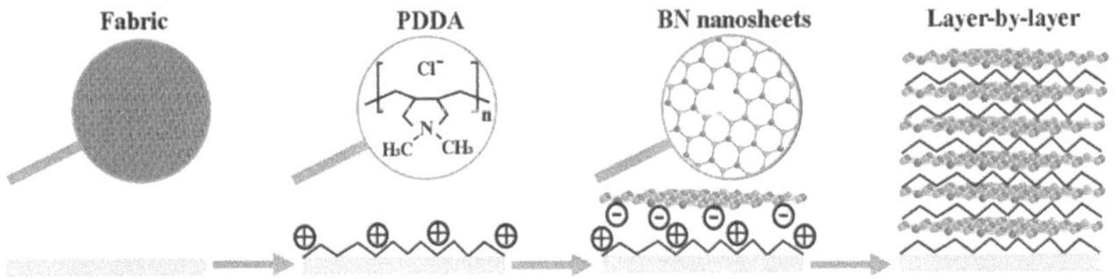

Figure 19.3 Fabrication of the BN/polydimethyl diallyl ammonium chloride coating by a layer-by-layer approach. (Reproduced with permission from [10].)

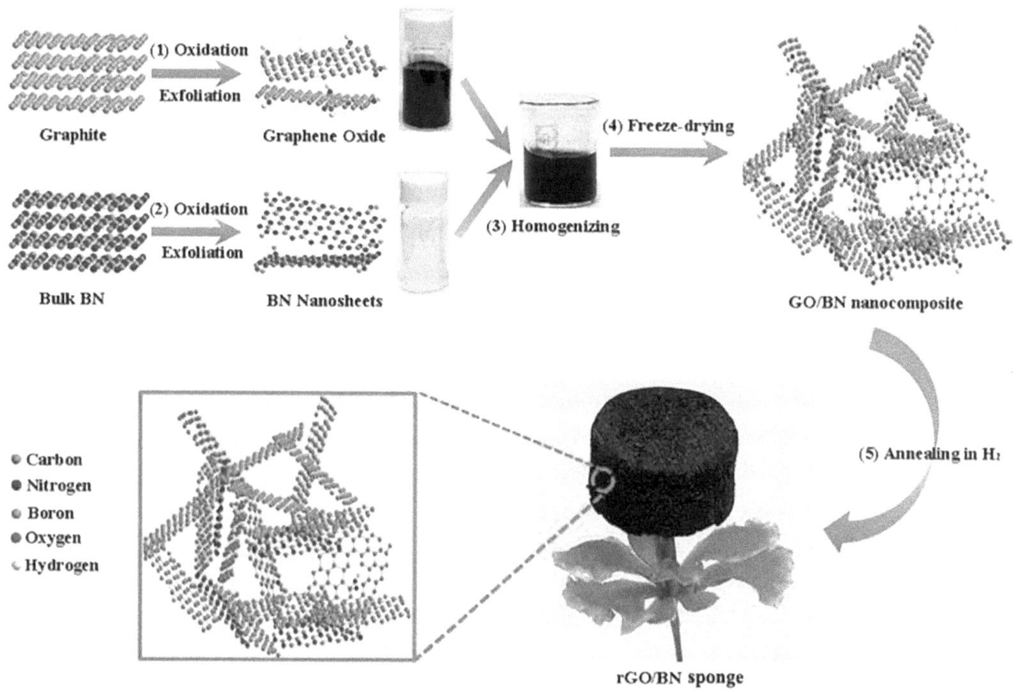

Figure 19.4 Schematic diagram of preparation steps of rGO/BN sponge. (Reproduced with permission from [11].)

superoleophilicity, making it promising for oil adsorption under practical conditions. The unique 3D interconnected porous network of the rGO/BN foam imparts it with light weight, high compressibility and excellent compressive recoverability in both air and organic solvents [11].

As a natural, low-cost, and environmentally friendly material, carboxymethylcellulose with a great number of surface groups is an ideal material for assembling nanocomposites. Chen et al. developed a sinter-free and low-energy-consuming method to fabricate porous BN aerogels using carboxymethylcellulose as a crosslinking agent [12]. Bulk BN powders were first treated by a ball milling process to functionalize h-BN with hydrophilic groups, facilitating uniform mixing of BNNSs and carboxymethylcellulose. The resulting aerogels with a low density of ~0.035 g mL^{-1} and large pores of 38–86 μm exhibited a high oil absorption capacity of 31.55 g g^{-1} with good cyclicity. Gupta et al. developed a microstructured aerogel composed of GO and h-BN nanoplatelets and polyvinyl alcohol (PVA) as a crosslinking agent via a lyophilization (deep freeze-drying) route [13]. Due to the abundant surface oxygen groups of h-BN and GO nanosheets, PVA polymer chains crosslinked with the nanosheets uniformly in the hydrogels and the resulting aerogels. The interaction between PVA chains, GO, and h-BN was found to strengthen the mechanical stability. By adjusting the weight ratio of GO and h-BN, these aerogels exhibited good shape stability and high absorption capacity for various oils and solvents.

Hao et al. developed a coprecipitation mixing and supercritical CO_2 drying method to prepare porous polycaprolactone/BN nanocomposites [14]. The integration of BN nanoplatelets (BNNPs) into the polycaprolactone matrix increased the porosity of the resulting foams and enhanced their compatibility with oils and nonpolar organic solvents. By exploiting a synergistic interplay between surface morphology and interfacial tension effects, polycaprolactone/BNNP

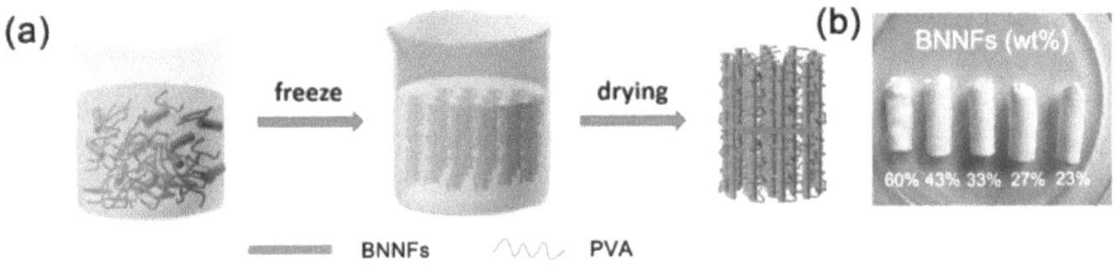

Figure 19.5 (a) The schematic diagram of fabrication steps of BNNFs/PVA composites. (b) The photograph of the as-prepared BNNFs/PVA composite foams with different contents of BNNFs. (Reproduced with permission from [17].)

composites exhibited significant hydrophobicity with a contact angle of 135°, as well as strong oleophilicity, showing high absorption capacities for oil and nonpolar organic solvents.

Xue et al. reported an in-situ polymerization approach to prepare thermally responsive hydrogels by incorporating few-layered amino functionalized-BNNS (BNNS-NH_2) in poly(N-isopropylacrylamide) hydrogels. The resulting composite hydrogels display outstanding sensitivity to temperature changes with a good absorption capacity of 3.7 g g^{-1} at the lower critical solution temperature [15]. Sam et al. developed a BN/SiO_2/polyurethane (PU) foam with hydrophobicity, low density, low cost, and unique wettability properties. The as-prepared foam displayed high absorption capacity for oil and organic solvents, with absorption capacity ranging from 1300% to 4000%, along with good recyclability [16]. Gao et al. fabricated porous hydrophobic BN nanofibers (BNNFs)/PVA composite foams by freeze-drying the mixed BNNFs and PVA solution (Figure 19.5) [17]. The BNNFs/PVA composite aerogels, featuring a microstructure, hydrophobic, and negatively charged surface, exhibited excellent performance in oil–water separation and adsorption of cationic dyes. Additionally, it was found that the inserted BNNSs improved the flexibility and hydrophobic properties of the composite aerogels. In another study, Xie et al. prepared TiO_2/BNNTs composites for dye decolorization in a simulated cotton dyeing effluent, achieving better performance for reactive Yellow 161 (RY161) compared to both synthesized TiO_2 and commercial TiO_2 (P25) [18].

Ponnamma et al. developed a robust nanocomposite comprising h-BN nanosheets (h-BNNSs), grape bunch-like cobalt oxide cubic crystalline nanoparticles, and polystyrene (PS) using an electrospinning approach [19]. The hybrid h-BN/cobalt oxide nanoparticles fillers were first synthesized by the coprecipitation method. During the electrospinning process, gamma irradiation treatment was utilized to enhance crosslinking efficiency and improve surface roughness with higher mechanical strength and Young's modulus. The resulting nanocomposites showed hydrophobic properties with a contact angle of 152±2°. Consequently, these nanocomposites exhibited a good oil absorption ability and efficiency in oil–water solution with long-term stability. Furthermore, 3D-cross-linked PS-polymethyl methacrylate/divinyl benzene-2D-BN nanosheet (PS-PMMA/DVB-2D-nanosheet) polymer composites were also fabricated by a bulk copolymerization route [20]. Thanks to the hydrophobic and oleophilic properties, the resulting nanocomposites exhibited a high absorption capacity and faster absorption rate for oil and organic solvents.

19.3.1.1 Separation of Oil

Membrane separation plays a key role in oil removal due to its continuous operation, contrasting with oil absorption methods. Liu et al. designed a high-flux polyvinylidene fluoride (PVDF)/BNNS nanocomposite membrane by a phase-inversion approach [21]. With the insertion of BNNSs, the composite membranes showed a higher permeation rate of water than the pristine

PVDF membrane. The composite membranes with a high specific surface area (1427 $m^2 g^{-1}$) exhibited outstanding separation performance in an oil/water mixture (>99.99%) and absorption of organic molecules with high water permeance (~8×104 L μm $m^{-2} h^{-1} bar^{-1}$). Another study investigated the separation mechanisms of PVDF/BN membranes, concluding that the B and N atoms facilitated the formation of a water layer on the membrane's inner surface, promoting water flow in the oil/water separation process [22].

Due to its superior properties, including high mechanical strength, chemical and thermal stability, and excellent compatibility, polyacrylonitrile (PAN) is widely employed in the membrane separation industry. He et al. developed BNNS/PAN nanofibrous membranes using a simple electrospinning approach [23]. The separation performance and water flux of the membranes can be adjusted by varying the amount of BNNS added. The membranes exhibited a maximum permeation flux of 12,885±131 L μm $m^{-2} h^{-1} bar^{-1}$ with a separation efficiency of 99%, significantly higher than that of the original PAN membrane (64.6±2.2%) for pump oil/water solutions. Zhang et al. designed h-BN/poly(arylene ether nitrile) nanofibrous composite membranes with high fluxes and excellent stability [24]. Due to the hydrophilic properties of h-BNNSs, dopamine was modified on their surface to improve their hydrophilicity. The modified nanosheets were coated on the poly(arylene ether nitrile) nanofibrous supporting layers by a vacuum-assisted filtration method. The composite membrane with hydrophilic channels showed a high water flux of up to 663.5 L $m^{-2} h^{-1}$ with a rejection ratio over 99.0%.

In addition to the traditional sponges, some BN-based magnetic nanocomposites are designed for oil absorption with recycling capability. Therefore, magnetic BN was prepared by modifying iron oxide particles on BN materials for crude oil recovery (Figure 19.6) [25]. The lyophilic spongelike materials with high surface area and porosity absorbed crude oil up to 53 times of their weight. Importantly, the composite foam can be easily separated under a magnetic field. Habibi et al. developed a low-cost dip-coating approach to prepare compressible, magnetic, and superhydrophobic PU/polyaniline/hexagonal BN@Fe_3O_4/polyacrylicoleic acid resin sponge (PU/PANI/h-BN@Fe_3O_4/AR) foam and PU/carbon black (CB)/h-BN (h-BN)@Fe_3O_4 (PU/CB/h-BN@Fe_3O_4) foam. These magnetic foams not only exhibited excellent oil adsorption performance, but also could be removed by a magnet easily, showing great potential in practical applications [26,27].

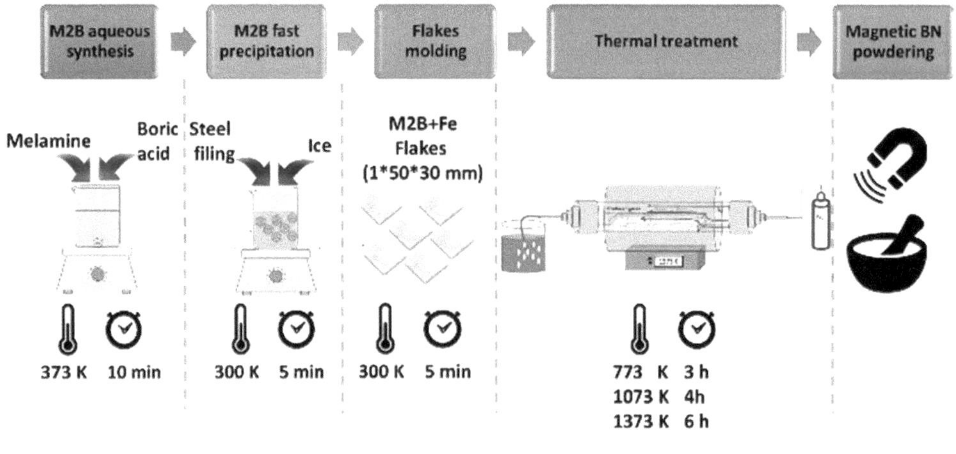

Figure 19.6 Schematic diagram of the porous magnetic BN preparation process. (Reproduced with permission from [25].)

19.4 BN-BASED NANOMATERIALS FOR ABSORPTION AND SEPARATION OF DYES AND IONS

19.4.1 BN-Based Nanomaterials for Dye and Ion Absorption

Over the past few decades, the rapid industrial development has led to a significant discharge of wastewater into water bodies. This wastewater contains harmful substances like organic dyes, toxic metal ions, and cleaning agents, posing severe threats to all forms of aquatic life. In particular, certain dyes used in industries such as plating, textiles, and printing are notoriously difficult to break down biologically due to their complex aromatic structures. These dyes are also resistant to degradation by light, heat, and oxidizing agents. Moreover, even in minimal amounts, synthetic organic dyes can visibly pollute water, reduce its gas solubility, and pose toxic, carcinogenic, and mutagenic risks to humans. Consequently, removing these dye pollutants from effluents is crucial for ensuring water safety and protecting human health.

Mahdizadeh et al. explored a rapid synthesis method to prepare BN nanomaterials using a microwave-assisted method [28]. The synthesis process is both quick and cost-effective, involving readily available materials like borax and ammonium chloride, which are heated in a microwave for just 10 minutes. The resulting h-BNNSs exhibited a surface area of 177.07 $m^2 g^{-1}$ and displayed a band gap of 5.92 eV, making them suitable for photonic applications. They can rapidly adsorb high concentrations of dyes, such as methylene blue (MB) and Rhodamine B from solutions, achieving an adsorption capacity of 472.4 g g^{-1} for MB. This high capacity is attributed to the large surface area and the unique electronic properties of h-BN. Additionally, the h-BNNSs showed remarkable selectivity in adsorption, effectively removing MB from mixed dye solutions. Chen et al. explored a cationic-surfactant-assisted solvothermal exfoliation method that facilitates the preparation of BNNSs (Figure 19.7) [29]. These BNNSs are functionalized with diamide phenanthroline groups, enhancing their capability for the selective and efficient adsorption of uranium (VI) from lanthanides, which is crucial for reducing radioactive threats and enhancing the sustainability of uranium resources. The prepared BNNS achieved a high maximum adsorption capacity of 313 mg g^{-1} and ultrafast adsorption kinetics, reaching equilibrium in less than a minute. The current method offers significant selectivity for uranium over various lanthanides, with selectivity coefficients exceeding 120 at high acid concentrations.

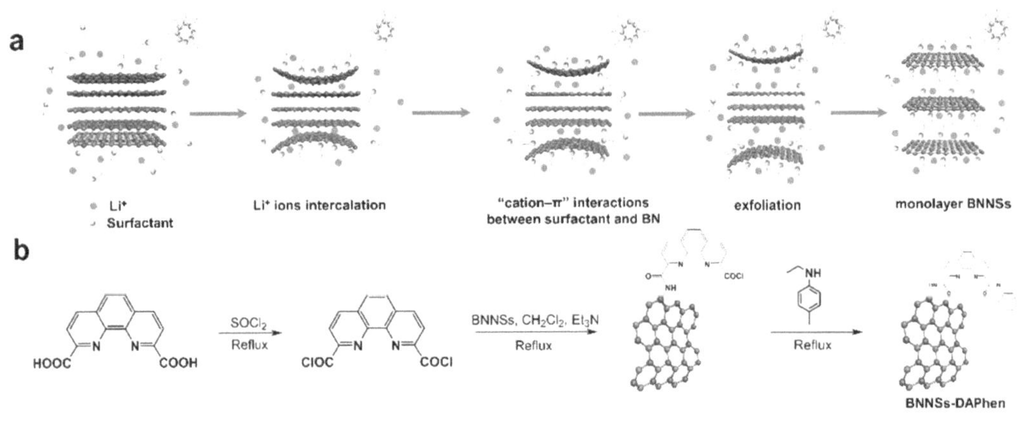

Figure 19.7 (a) Schematic images showing the preparation process. (b) Modification process. (Reproduced with permission from [29].)

Global industrial activities, including mining, processing, and a waste disposal, have led to widespread contamination of water and soil with hazardous substances like toxic heavy metals and organic compounds. For instance, arsenic (As) is recognized as particularly dangerous. Research indicates that around 140 million people globally are exposed to drinking water containing dangerously high levels of arsenic ions. Prolonged exposure can harm the central nervous system, cause skin darkening, and increase the risk of various cancers, including those of the skin, lungs, liver, kidneys, and prostate. Consequently, the World Health Organization has set a safety limit for arsenic ions in drinking water at 10 parts per billion (ppb). However, studies show that in several regions of India and neighboring Bangladesh, the levels of arsenic in drinking water significantly exceed this threshold. Therefore, there is an urgent need to control metal ion levels in drinking water.

Bangari et al. prepared BNNS-Fe_3O_4 nanocomposite adsorbents using a solvothermal process for enhanced arsenic removal from water [30]. The nanocomposites showed a high arsenic adsorption capacity (26.3 mg g^{-1}), significantly outperforming traditional BNNSs (5.3 mg g^{-1}). This was attributed to the increased surface area and enhanced interaction sites provided by Fe_3O_4 nanoparticles. These nanocomposites can reduce arsenic levels from 856 ppb to below 10 ppb, meeting the World Health Organization's drinking water standards. BNNS-Fe_3O_4 nanocomposites are not only effective but also regenerable, maintaining their adsorption capacity after several cycles, which is crucial for sustainable application.

Konopatsky et al. utilized the polyol process to synthesize FePt nanoparticles and distributed them on the surface of BN nanomaterials (Figure 19.8) [31]. These nanoparticles underwent a temperature-activated phase transformation, enhancing their magnetic properties. The FePt nanoparticles act as catalytic centers for hydrogen dissociation, with the adsorbed hydrogen transferred to the h-BN, facilitating the fcc to fct phase transformation and allowing control over the magnetic properties. Density functional theory (DFT) calculations supported these observations, indicating effective hydrogen absorption by the h-BN substrate from the FePt nanoparticles. Additionally, the FePt/h-BN nanomaterials demonstrated a high adsorption capacity for various organic dyes, suggesting potential environmental applications in water purification. This study presented a significant advancement in the development of nanomaterials combining unique magnetic and environmental functionalities, with promising applications in catalysis and pollutant removal.

Han et al. synthesized composite aerogels based on graphene/BN nanomaterials at a low temperature for removing ciprofloxacin (CIP) from water [32]. The composite aerogels showed

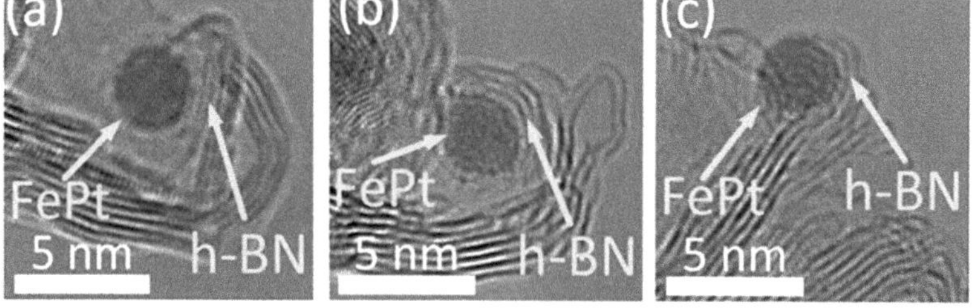

Figure 19.8 High-resolution transmission electron microscopy (HRTEM) images of individual FePt NPs before (a and c) and after prolonged electron beam irradiation for several minutes (b). (Reproduced with permission from [31].)

a density of 28–34 mg cm^{-3}, approximately 99% porosity, and enables compressive strength of 40–52 kPa. Their unique composition enables rapid adsorption rates and effective CIP removal efficiency up to 99% under optimal conditions. The graphene/BN aerogels demonstrated a maximum adsorption capacity of up to 195 mg g^{-1} and a normalized capacity of 2.03 mg m^{-2}, which are competitive with traditional adsorbents like activated carbon. The removal of CIP is mainly through the interaction of the BN and graphene with the functional groups of CIP, confirmed by DFT calculations. More importantly, the aerogels exhibited excellent recyclability with effective performance maintained over multiple cycles, highlighting their potential for sustainable application in water treatment.

19.5 BN-BASED MEMBRANES FOR WATER PURIFICATION

Nanofiltration (NF) technology is extensively employed in the treatment of saline wastewater because of its distinct separation capabilities. Nonetheless, the development of NF membranes, which are central to NF technology, is hindered by a fundamental compromise between permeability and selectivity. Due to their excellent chemical and thermal stability, BN-based membranes have attracted much attention. BN-based membranes can be fabricated by direct deposition methods on a support [33–36] and indirect assembly approaches [37,38].

19.5.1 BNNS-Based Membranes

Over the past ten years, advancements in exfoliation techniques have enabled the fabrication of bulk BN powder into few-layer, even single-layer nanosheets, grafted with various surface groups, thereby bestowing good water-dispersion ability upon the BNNSs (Figure 19.9a and b) [39,40]. Due to their ultrathin structure and hydrophilic properties, BNNSs are excellent candidates for preparing separation membranes with outstanding separation performance and high solvent permeation rates [41,42].

Chen et al. reported a facile vacuum-assisted filtration method to assemble a thin BN layer on the nylon support. (Figure 19.9c–e) The thin BN membrane (400 nm) exhibited a high solvent permeance of 5950 L $m^{-2}h^{-1}$ bar^{-1} for acetone and approximately 2700 L $m^{-2}h^{-1}$ bar^{-1} for water. In terms of separation performance, the BN-based membranes also displayed a high rejection rate (>99%) for Au particles and Evans blue in organic solvents with high solvent permeability (over 560 L $m^{-2}h^{-1}$ bar^{-1}). Additionally, due to their outstanding antioxidative properties, the BN membranes demonstrated chemical stability and maintained performance in strongly acidic, alkaline, or oxidative environments for at least one month [38]. Furthermore, to enhance performance, several modification methods were designed to adjust the surface charge and chemical properties of BN materials [18,43–45].

19.5.2 BNNSs as Nanofillers in NF Membranes

Additionally, the BNNSs with high surface charge and various surface groups can also serve as fillers in the polymer-based NF membranes, enhancing their permeability and selectivity. Zheng et al. developed high-performance NF membranes using 2D BNNSs incorporated into reticulated polyamide thin-film nanocomposite (TFN) membranes (Figure 19.10) [46]. This research investigated the application of tannic acid (TA) for exfoliating and dispersing BNNSs. This led to a new method for integrating BNNSs into a polyamide (PA) layer, which was used to create high-performance TFN

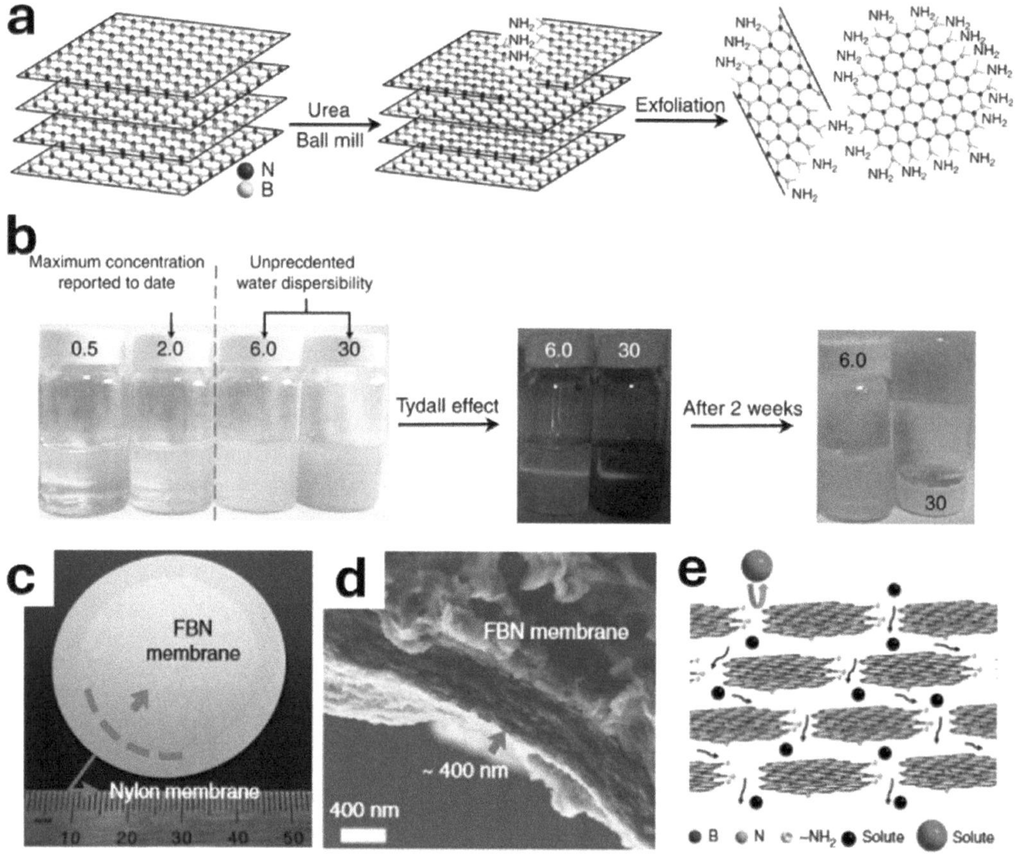

Figure 19.9 (a) Diagram of the exfoliation process via ball milling. (b) Photos of colloidal BN solutions (0.5–30 mg/mL), showing the Tyndall effect and stability. Reproduced with permission from [39] (c) A photograph of the FBN membrane on a commercial nylon membrane. (d) The cross-sectional SEM image of an FBN-0.4 membrane. (e) Schematic diagram of the separation mechanism. (Reproduced with permission from [38].)

membranes on a poly (m-phenylene isophthalamide) (PMIA) base. The introduction of exfoliated BNNSs effectively regulated the interfacial polymerization process, facilitating the formation of a highly permselective layer, resulting in improved water flux and stable solute rejection for inorganic salts. More importantly, the water flux of the PA-TA@BNNSs-10%/PMIA membrane reached up to $109.1 \pm 2.49\,L\,m^{-2}$, while maintaining a high rejection rate of $97.5 \pm 0.38\%$ for Na_2SO_4. Moreover, the long-term stability of these membranes has been assessed, revealing outstanding performance in prolonged filtration processes.

Abdikheibari et al. designed nanocomposite NF membranes with enhanced water flux and organic fouling resistance by the integration of few-layered $BN(NH_2)$ nanosheets (Figure 19.11) [47]. The modified membranes demonstrated a 59% enhancement in flux, a 50% improvement in total fouling resistance, and maintained natural organic matter separation above 92%. The study also displayed the long-term operational stability and enhanced recoverability of the modified membranes. In addition, the membranes based on $BN(NH_2)$ nanosheets demonstrated long stability.

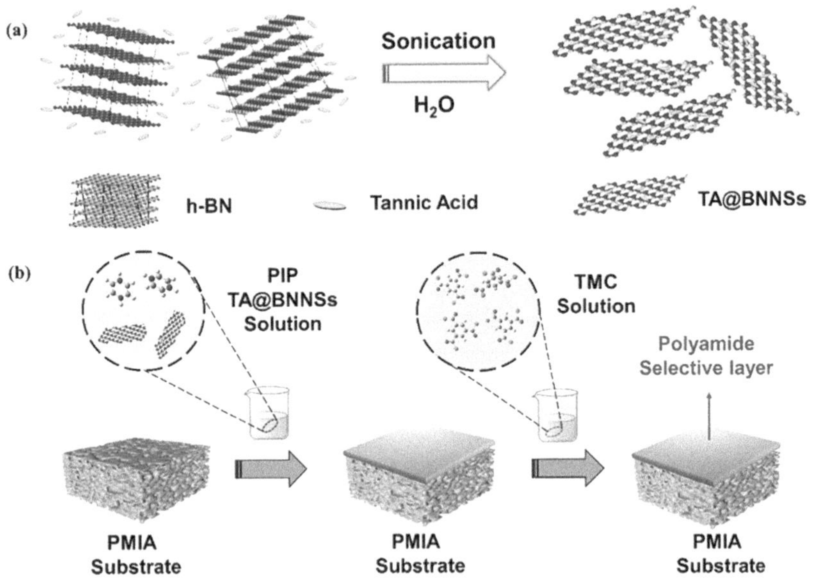

Figure 19.10 (a) Preparation of h-BNNSs. (b) Preparation of composite NF membranes. (Reproduced with permission from [46].)

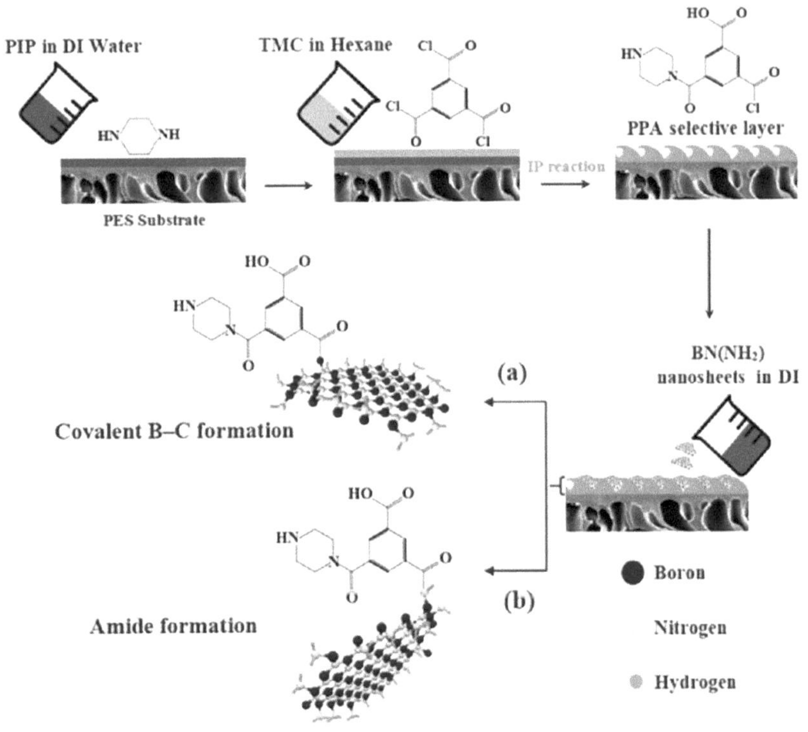

Figure 19.11 Schematic illustration of the fabrication procedure for nanocomposite membranes featuring $BN(NH_2)$: (a) nanosheets react with accessible groups in the PIP-TMC network to form a covalent B–C bond, or (b) an amide structure. (Reproduced with permission from [47].)

In order to improve the membranes' water permeability and antifouling properties, which are essential for efficient water process engineering applications, Kallem et al. utilized the non-solvent-induced phase separation process to integrate the (hydroxyapatite) HAp/BN nanocomposites into the polyethersulfone (PES) membranes (Figure 19.12) [48]. The resulting composite membranes exhibited significantly improved water flux and fouling resistance. For instance, the membrane with 4% HAp/BN demonstrated a 97% increase in water flux and 90% effective rejection of bovine serum albumin, showcasing high antifouling properties and improved water permeation. The enhancements in the membranes' performance are attributed to the increased surface hydrophilicity and the introduction of additional pathways for water transport, provided by the high surface area of the BNNSs and the hydrophilic nature of the HAp.

Kulshrestha et al. used the chemically modified BN to improve its integration into sulfonated poly ether ether ketone (SPEEK) to create the cation exchange membranes, aimed at improving salt removal from brackish water (Figure 19.13) [49]. These modifications involved a sulfonation process that enhances the hydrophilicity and structural stability of the composite membranes. Enhanced properties like higher ionic conductivity and ion exchange capacity are

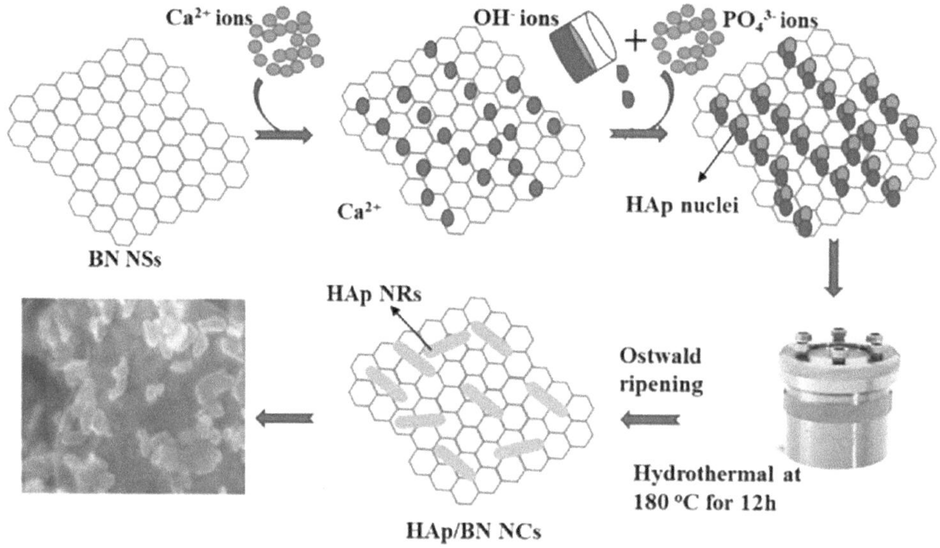

Figure 19.12 HAp/BN nanocomposite synthesis scheme. (Reproduced with permission from [48].)

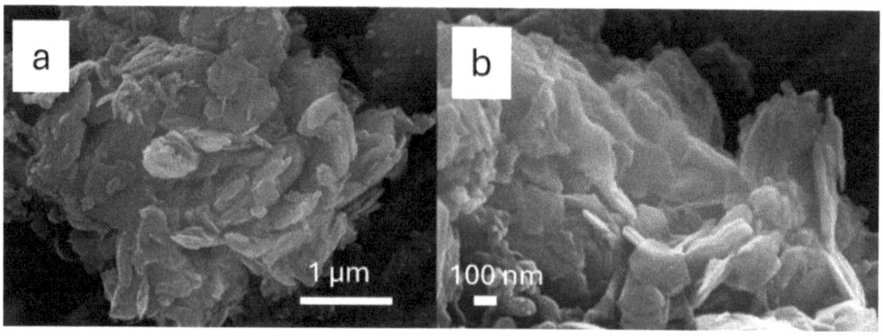

Figure 19.13 SEM images for hBN (a) and SBN (b). (Reproduced with permission from [49].)

demonstrated, with improvements quantified through various analyses such as transmission electron microscopy (TEM) and scanning electron microscopy (SEM) for morphological studies and ionic measurements. Significantly, the SBN-10/SPEEK membrane showed superior performance in terms of current efficiency and power consumption, making it an excellent candidate for industrial-scale water desalination.

Sert et al. conducted research to enhance the antifouling properties of PES ultrafiltration membranes [50]. They achieved this by integrating BN quantum dots (BNQD) into the membranes, resulting in composites that exhibit improved biological activity and superior filtration performance. Different concentrations of BNQD were blended into PES using a non-solvent induced phase separation method, and various properties of the composite membranes, such as porosity, hydrophilicity, and pore size, were examined. They found that BNQD improved the water flux and biofouling resistance of the membranes. For instance, a membrane containing 2 wt% BNQD exhibited a significant increase in water flux compared to the pristine membrane. The composite membranes also showed enhanced antimicrobial and antibiofilm activities, with higher DNA cleavage and cell viability inhibition ability. Overall, the integration of BNQD into PES membranes not only improved their antifouling capabilities but also introduced functional properties that could be beneficial in various applications, including medical and environmental technologies.

Vatanpour et al. designed the polyamide thin-film composite NF membranes with the incorporation of BNQDs to improve their permeability, antifouling properties, and separation efficiency [51]. They noted that the inclusion of BNQDs at varying concentrations (0.001–0.02 wt%) during the interfacial polymerization of membranes resulted in a smoother, thinner, and more hydrophilic surface, which contributed to increased water flux and improved antifouling capabilities. Specifically, the membrane with 0.01 wt% BNQDs exhibited the best separation efficiency, demonstrating high rejection rates for various salts and increased water flux by up to 170% compared to the unmodified version, which can be attributed to the enhanced negative surface charge.

Abdikheibari et al. fabricated nanocomposite NF membranes modified with functionalized BNNSs to enhance fouling resistance and water permeability for effective drinking water treatment [52]. The nanosheets were incorporated into the polypiperazine amide layer of the membranes during the interfacial polymerization process. With 0.004 wt% $BN(NH_2)$, the modified membrane (PPA-BN-4) demonstrated a 13.4% increase in permeation rate without significantly compromising natural organic matter rejection. It retained a normalized flux of 0.97 $L\ m^{-2}\ h^{-1}$ during filtration, indicating strong resistance to fouling. This membrane outperformed a bare polypiperazine amide (PPA) membrane and showed superior water treatment capabilities, maintaining high natural organic matter rejection rates and enhanced permeate flow.

Ahmadian-Alam et al. used the deposition approach to fabricate the hybrid multilayer thin-film membranes by incorporating nanofillers like BNNSs and metal–organic framework nanoparticles for water treatment (Figure 19.14) [53]. The research results demonstrated that h-BNNSs significantly enhance the membranes' effectiveness compared to metal–organic framework nanoparticles, with improved rejection rates and permeation for certain dyes and sulfate ions. Additionally, the inclusion of h-BN in the membranes' negatively charged layers resulted in better retention of dyes like Acid Blue and Bromophenol Blue, with water flux improvements of 60% and 45%, respectively, for each dye solution. Despite these enhancements, the rejection of sulfate ions remained stable at around 78%, indicating effective separation capabilities.

Vatanpour et al. utilized the PES membranes enhanced with a nanocomposite consisting of BN, silver, and GO to improve their separation and antibacterial capabilities for water treatment [54]. The nanocomposite was synthesized through functionalization and blending processes and then incorporated into the PES membrane at varying concentrations. This modification led to

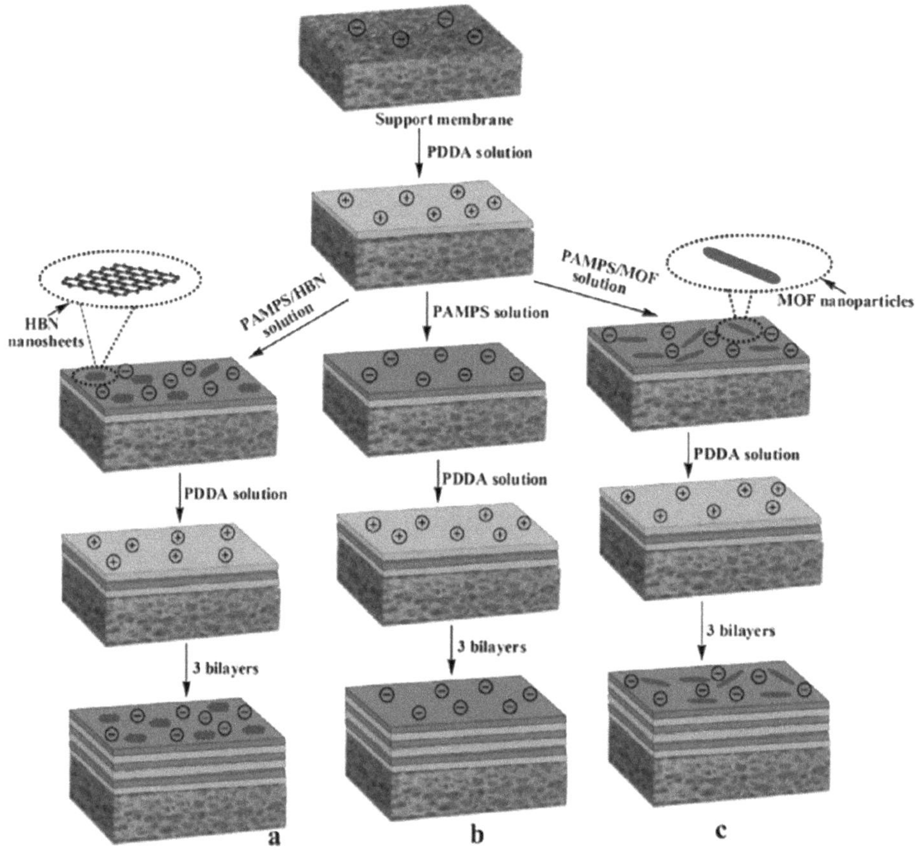

Figure 19.14 (a) Preparing the hybrid multilayer thin film. (b) The multilayer thin film without nanofillers. (c) The hybrid multilayer thin film with metal–organic framework nanoparticles. (Reproduced with permission from [53].)

significant improvements in hydrophilicity and porosity of the membranes, increasing water permeability by 40% and enhancing dye removal and fouling resistance. For example, the modified membrane demonstrated an 88.9% rejection rate for reactive black dye and a 77.7% rejection rate for reactive red dye, attributable to an increased negative surface charge. These enhancements make the modified PES membranes highly suitable for practical applications in water treatment, contributing to the sustainability and efficacy of environmental engineering practices.

Liu et al. designed PVDF-g-PNIPAAm/BN@PDA-Ag nanocomposite membranes, which are responsive to temperature and light, offering innovative solutions for wastewater separation and membrane cleaning [55]. The integration of BN@PDA-Ag into the membrane enhanced its photothermal response, allowing rapid heating under light exposure, which leads to a contraction of the PNIPAAm chains and an increase in pore size, thereby facilitating more effective cleaning and fouling control. The membranes demonstrated retention rates of 82%, 99%, and 95% for bovine serum protein (BSA), oil–water emulsions, and various dyes, respectively, with flux recovery rates exceeding 90% following light cleaning. The membranes maintain excellent separation performance with easy switchable functionality between separation and cleaning modes, driven by external thermal or light stimuli.

Nafti Mateur et al. developed porous gelatin membranes using a novel Pickering emulsion technique, stabilized with h-BNNSs for polyelectrolyte-enhanced ultrafiltration, targeting the removal of iron ions from aqueous solutions [56]. Various parameters, such as crosslinking times with glutaraldehyde and drying times, significantly influenced the membranes' properties. By manipulating these conditions, membranes with desired pore sizes and filtration capabilities were developed. Initial tests without polyelectrolytes showed low iron ion removal efficiency. However, incorporating poly(acrylic acid) as a polyelectrolyte significantly enhanced iron ion removal, achieving high removal rates of up to 97% at certain pH levels and pressures. The research shows a sustainable approach to membrane fabrication that could revolutionize water purification techniques, particularly in removing contaminants through a biodegradable and efficient method.

Ning et al. designed the pH-responsive BNNS/polyacrylic acid NF membranes to enhance water purification technologies through improved dye separation and tunable water permeance [57]. The introduction of polyacrylic acid into BNNS membranes led to increased hydrophilicity and electronegativity, resulting in a substantial boost in water flux (up to 228 L $m^{-2}h^{-1}$) and dye rejection rates (up to 98% for MB), significantly outperforming traditional BNNS membranes. These membranes exhibited size-selective filtration, effectively separating various dyes and dye/salt mixtures from water. The unique pH-responsive characteristics of these membranes allow for adjustable filtration performance in different pH environments, showcasing their versatility in water treatment applications. Moreover, the membranes demonstrated excellent environmental and cyclic stability, maintaining high performance over extended periods and under various conditions.

Wang et al. designed temperature-responsive textiles using poly[N,N-dimethyl(methacryloy lethyl)ammonium propane sulfonate], specifically for efficient oil–water separation. This technology utilizes the unique properties of BN and a vinylsilane coupling agent to enhance the hydrophobic characteristics of cotton fibers. The preparation process involved a two-step modification: initially coating the cotton with BN and vinylsilane to increase hydrophobicity, followed by enzymatic graft polymerization of poly[N,N-dimethyl(methacryloylethyl)ammonium propane sulfonate] to introduce temperature-responsive hydrophilic–hydrophobic switching capabilities [44]. These fabrics demonstrated the ability to selectively separate oil and water based on temperature changes. Below 28°C, the hydrophobic layer dominates, facilitating oil passage while blocking water. Above this temperature, the hydrophilic properties are enhanced, allowing water to pass and blocking oil. This switchable wettability under different temperature conditions enables effective and customizable oil–water separation. Specially, the textiles showed high separation efficiency and flux, with values like 542.57 L $m^{-2}h^{-1}$ for oil and 477.26 L $m^{-2}h^{-1}$ for water. The fabric also shows excellent reusability and stability under mechanical stress and varying pH levels, indicating its potential for practical applications in environmental cleanup and industrial processes.

BNNSs are increasingly valued as multifunctional nanofillers for polymer composites, notable for their wide band gap, electrical insulation, high thermal conductivity, and exceptional stability. Integrating BNNSs into polymer matrixes is emerging as a key method to boost the thermal conductivity of polymers. Undoubtedly, large-scale production of BNNSs is vital for the advancement of BNNS/polymer composites.

Yin et al. prepared h-BNNSs using sulfanilic acid (SA), which facilitates the integration of BNNSs into the PVA matrix to form the composite membrane (Figure 19.15) [58]. The membranes demonstrated remarkable mechanical strength and thermal properties. With a 5 wt% loading of SA-functionalized BNNS (f-BNNS), the membranes achieve tensile strengths of approximately 11.37 MPa—an 84.2% improvement over neat PVA membranes. Additionally, the membranes exhibited a significant enhancement in thermal conductivity, nearly 6.47 times greater than that

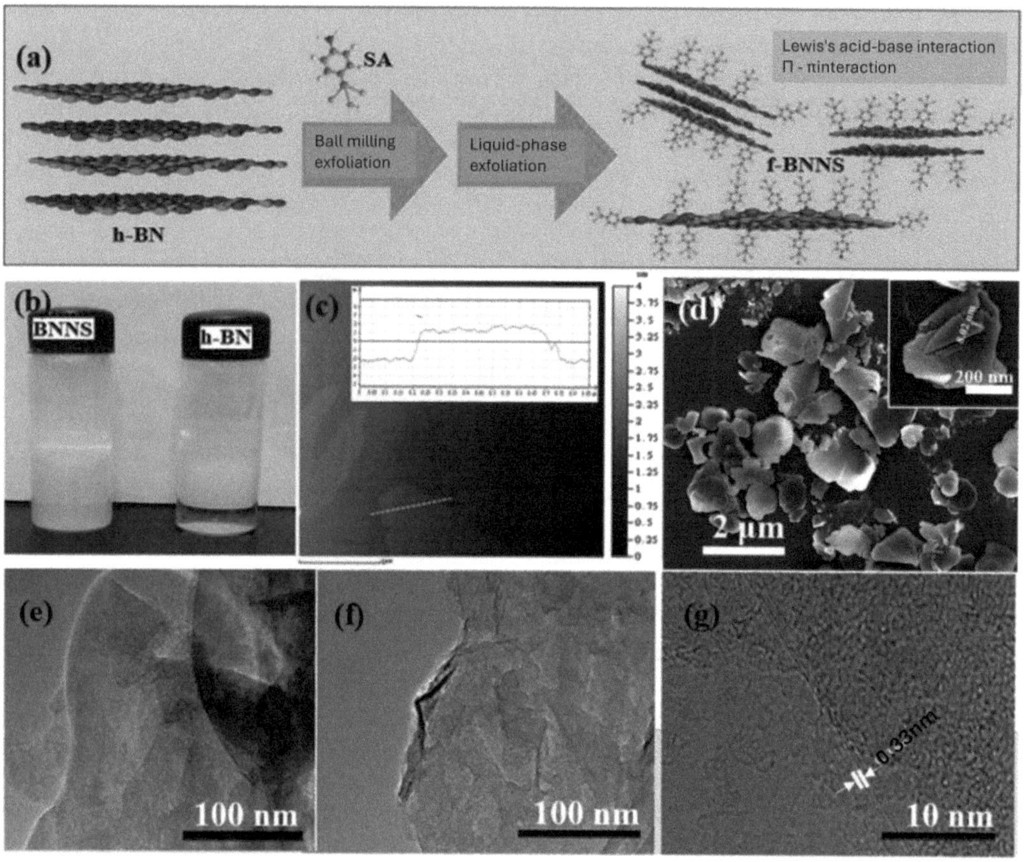

Figure 19.15 (a) Schematic showing the preparation process of h-BN. (b) Photograph showing the Tyndall effect of the samples. (c) AFM. (d) SEM. (e, f) TEM images, and (g) HRTEM image. (Reproduced with permission from [58].)

of pure PVA, making them ideal for applications requiring heat management. The prepared composite nanofiber membranes displayed a significant advancement. These materials hold great promise for use in electronic packaging and water treatment, thanks to their improved mechanical, thermal, and filtration characteristics.

Jiang et al. studied the synthesis and application of BN by loading it into membranes for removing organic pollutants from water [59]. Using PAN as the base polymer, they explored the creation of BN-PAN electrospun nanofibrous membranes with a focus on enhancing their absorption capacity for dyes and pollutants in water. They utilized the electrospinning and spray techniques to integrate 3D BN into the PAN fibers. The membranes demonstrated enhanced removal capabilities due to the incorporation of 3D BN, which could be thermally regenerated and reused multiple times without a decrease in performance. The absorption rates for water pollutants including Congo red, basic yellow 1, and Rhodamine B were tested and found to be 46%, 53%, and 45%, respectively. The research showed that these membranes could be effective for water purification applications, offering a reusable and economically advantageous solution for the treatment of polluted water.

19.6 THEORETICAL SIMULATION OF BN NANOMATERIALS FOR ABSORPTION OF OIL AND IONS

Apart from the experimental studies, theoretical simulations of BN-based materials for ion transport and absorption were reported. Some traditional methods, including molecular dynamics simulations and DFT calculations, and new approaches, such as machine learning techniques, are used for exploring ion rejection, water transport, and absorption performance of BN-based materials [60–62].

Cyanide, known for its high toxicity, is widely used in metallurgical processes and chemical synthesis, resulting in significant environmental challenges due to its presence in industrial effluents. Azamat and Khataee utilized the molecular dynamics simulation method to investigate the BNNTs as nanostructure membranes for the removal of cyanide from aqueous solutions [60]. They found that the smaller (5,5) BNNT was effective at blocking cyanide ions due to its limited diameter but allowed water molecules to pass. In contrast, larger nanotubes (6,6), (7,7), and (8,8) allowed both water and smaller quantities of cyanide ions to pass, depending on the pressure and electric field applied. The simulations revealed how water molecules organize differently inside each type of nanotube, impacting the flow and filtration process. The arrangement of water molecules was detailed through radial distribution function analyses. This research suggests that BNNTs, with their unique structural properties, could be tailored for effective removal of specific contaminants like cyanide, making them promising candidates for environmental remediation technologies.

Lu et al. presented a study on the use of machine learning techniques to simulate the adsorption of organic pollutants onto porous BN nanoclusters [61]. They studied the adsorption kinetics of organic dyes at different temperatures using a computational approach instead of traditional kinetic models. The ADA-LR model yielded the lowest MSE and highest R2 scores, indicating it as the most accurate model for predicting the adsorption kinetics at both low and high temperatures. The study also demonstrated the potential of machine learning to enhance understanding and prediction of adsorption processes, providing a robust alternative to conventional methods. Finally, the research underscores the effectiveness of machine learning in handling complex datasets and improving the prediction accuracy of adsorption processes, thus contributing to more efficient and cost-effective solutions for environmental technologies.

Vafa et al. provided a comprehensive study on BN nanomaterial membranes (BNSMs) for wastewater treatment, specifically for removing heavy metal ions from water (Figure 19.16) [63]. Using molecular dynamics simulations, various structures and conditions of BNSMs were analyzed to determine their effectiveness in ion rejection and water purification performance. They found that smaller slit widths in rigid monolayer BNSMs showed near 100% ion rejection due to geometric constraints which effectively block ion transport. Conversely, larger slits allowed more water flow but reduced ion rejection rates. Second, adding layers to the membrane generally increased ion rejection rates due to increased barriers to ion transport. For instance, rigid bilayer BNSMs exhibited higher rejection rates due to additional physical barriers formed by overlapping slits. Third, flexible monolayer BNSMs demonstrated higher water flow rates but lower ion rejection percentages compared to their rigid counterparts. This behavior was attributed to the flexibility of the membrane allowing for a larger slit opening under pressure, facilitating more water and ion transport. Finally, they demonstrated the potential of using BNSMs in advanced water treatment technologies, where manipulating membrane characteristics such as slit width, layering, and flexibility can optimize performance for specific purification needs.

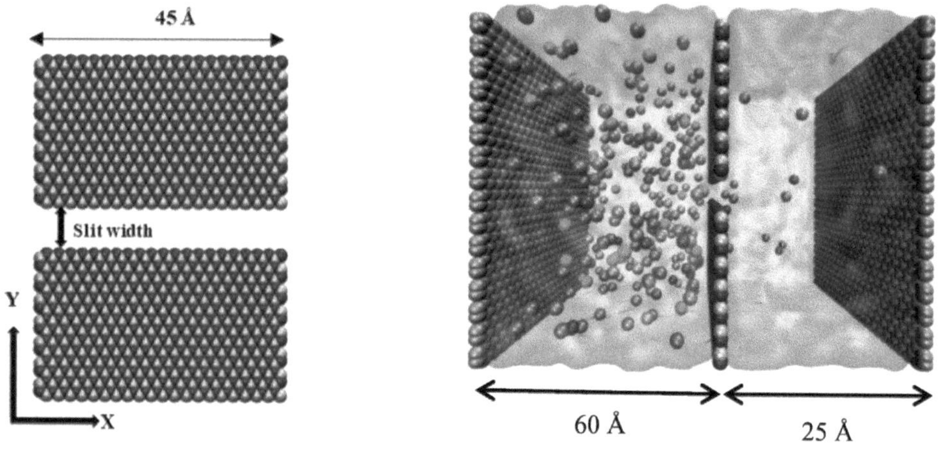

Figure 19.16 (a) Top view of the BNSM surface containing a nanoslit in the mid-section. (b) A schematic representation of the simulation cell. (Reproduced with permission from [63].)

Du et al. explored the efficacy of h-BN nanoporous films for removing mercury (Hg^{2+}) from wastewater using molecular dynamics simulations [64]. It delved into the relationships between water flux, pore chemistry, pore size, and applied pressure, as well as the mechanisms of Hg^{2+} filtration. They found that high Hg^{2+} rejection rates were observed with smaller pore sizes that block Hg^{2+} while allowing water to pass. Membranes like B5 achieved a 100% rejection rate for Hg^{2+}, making them ideal for mercury removal. The study also highlighted the importance of the hydration shell of Hg^{2+} ions, noting that dehydration is required for ions to pass through smaller pores, a process facilitated by the membrane's charge properties. The simulation results indicated that water flux increased with pore size and applied pressure. Membranes with larger pores demonstrated greater water permeability, and nitrogen-edged pores showed higher water flux compared to boron-edged pores due to stronger interactions with water molecules. Overall, this research underscored the potential of using h-BN nanoporous membranes in advanced wastewater treatment applications, particularly for removing toxic heavy metal ions like mercury, by exploiting the unique properties of nanoporous materials.

Azamat et al. investigated the removal of cadmium ions (Cd (II)) from water using carbon nanotubes, BNNTs, and silicon carbide nanotubes through molecular dynamics simulations [65]. The simulations demonstrated that the armchair nanotubes used could effectively remove Cd (II) from wastewater. The performance varied among the nanotube types due to differences in their structural and electrical properties. The removal of Cd (II) ions was facilitated by the applied electric field, which helped drive the ions through the nanotubes. This process was analyzed using the potential of mean force calculations, providing insights into the energy barriers involved. They also found that differences in ion removal efficiency were observed among carbon nanotubes, BNNTs, and silicon carbide nanotubes. This variability is attributed to the unique properties of each nanotube type, such as charge distribution and interaction with water molecules. This research is significant due to the environmental and health hazards posed by cadmium in water. Nanotubes were evaluated for their effectiveness in ion separation under an applied electric field, an innovative approach in enhancing water treatment technologies.

Nguyen and Beskok examined water desalination using h-BN/graphene membranes through molecular dynamics simulations [66]. Comparing the desalination performance of these materials, it was found that graphene membranes, particularly when charged, perform more efficiently in terms of ion rejection and pressure requirements than h-BN membranes. A pore diameter of 15.9 Å with a total fixed charge of 12e on graphene provides optimal results, achieving nearly 100% rejection for Na^+ and 98% for Cl^- ions, with a 51.8% lower pressure drop. The study emphasizes the dominant role of electrostatic interactions facilitated by charging the membrane over steric and hydrodynamic effects, allowing for efficient salt rejection even with larger pore sizes. Overall, the findings offer a promising approach for increasing efficiency while managing operational pressures.

Kommu and Singh conducted research on extracting ethanol from water using graphene and h-BN slit pores [67]. They used molecular dynamics simulations to explore this approach, emphasizing the materials' distinct properties and extensive surface area as key advantages. The study found that 9 Å pores are most effective for ethanol selectivity, while 7 Å pores showed the least effectiveness, likely due to molecular sieving effects. Compared to graphene, h-BN pores generally demonstrated higher ethanol selectivity, attributed to better interaction dynamics with ethanol molecules. Ethanol shows a higher adsorption rate in hBN pores across all considered pore widths, suggesting a more favorable interaction compared to graphene. The diffusion coefficients of ethanol and water molecules decrease with reduced pore width, which affects their mobility within the pores. Both ethanol and water molecules exhibit longer residence times in narrower pores, with hBN surfaces enhancing ethanol's residence time over graphene. This detailed insight into the structural and dynamical behavior of confined ethanol–water mixtures indicates the potential of h-BN and graphene in enhancing the efficiency of ethanol–water separation, leveraging the distinctive properties of these nanomaterials to improve separation processes.

Sharma and Govind Rajan investigated how grain boundaries and electrostatic interactions affect the desalination capabilities of nanoporous hBN membranes [62]. They employed molecular dynamics simulations and DFT calculations in their study, as illustrated in Figure 19.17. Three configurations of nanoporous bi-crystalline hBN with varying misorientation angles 13.2°, 21.8°, and 32.2° are examined. The findings revealed that these grain boundaries modify the nanopore areas and shapes, which subsequently affect both the water flow and ion rejection

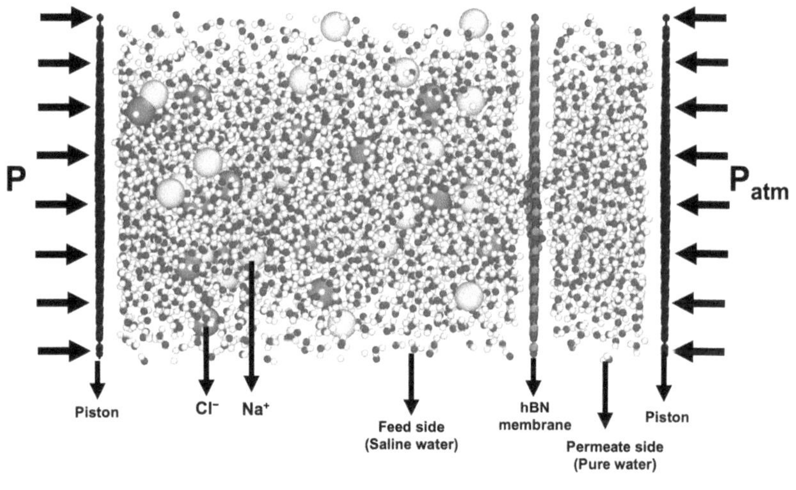

Figure 19.17 Schematic showing the water desalination simulation setup confined by graphene pistons. (Reproduced with permission from [62].)

rates. They found a significant increase in water flow by approximately 30% for a 13.2° misorientation, albeit with a decrease in Na^+ ion rejection by about 6%, compared to monocrystalline hBN. Elongated nanopores in the 21.8° and 32.2° configurations manage to achieve similar water permeation rates to those of larger, less elongated pores but with a minor reduction in ion rejection efficiency. Moreover, changes in the partial charges of boron (B) and nitrogen (N) atoms, as determined through DFT calculations, markedly influence the water flow rates, showing the highest rates when considering modified charges compared to uncharged or bulk partial charges. The current research highlights the importance of nanopore geometry and electrostatics in optimizing the desalination capabilities of hBN membranes, suggesting potential improvements in water purification technology, although balancing water permeability with ion rejection remains a challenge.

Omidirad and Azizi utilized DFT with dispersion corrections to analyze the adsorption capabilities of h-BNNSs, both perfect and defected, toward various solvents like ethanol, acetone, and water [68]. This research aims to enhance the understanding of how defects can improve adsorption effectiveness, crucial for applications in solvent recovery and environmental cleanup. Initially, they found that perfect h-BNNSs exhibit relatively weak adsorption for these solvents, with energy of adsorption (E_{ads}) ranging from −4 to −8 kcal/mol, indicating only modest interaction strengths. However, when defects were introduced into the h-BN structure, such as Stone–Wales and vacancy defects, there was a significant increase in adsorption strength. For example, E_{ads} for ethanol on defected h-BN jumped to about −20 kcal/mol, and for water, it increased to about −12 kcal/mol. This enhancement is attributed to stronger interactions between the polar solvent molecules and the defected sites on the h-BN surface. Further analysis through charge density profiles and Hirshfeld charge analysis indicated minimal charge transfer in these interactions, emphasizing the predominance of non-covalent interactions in solvent adsorption. The study's computational approach effectively mirrors experimental and other theoretical findings, confirming the accuracy and reliability of the methods used.

19.7 CONCLUSIONS

In this chapter, we have comprehensively examined the recent advancements in the synthesis of BN materials and BN-based composites, specifically focusing on their applications in oil cleanup, ion absorption, dye removal, and ion-related separation applications. The remarkable chemical stability and mechanical properties of BN material-based membranes and foams demonstrate significant potential for prolonged use in practical scenarios, even under severe conditions such as strongly acidic, alkaline, and high-temperature environments. Furthermore, theoretical simulations have emerged as invaluable tools for investigating BN materials in distinct states, thereby providing indispensable insights for the future design of high-performance membranes and absorbents tailored for efficient oil, ion, and dye removal. These findings not only underscore the promising prospects of BN material-based materials in addressing pressing environmental challenges but also pave the way for the development of innovative solutions with enhanced efficacy and sustainability.

REFERENCES

1. W. Lei, D. Portehault, D. Liu, S. Qin, Y. Chen, Porous boron nitride nanosheets for effective water cleaning. *Nature Communications* **4**, 1777 (2013).
2. J. Lin et al., Self-assembly of porous boron nitride microfibers into ultralight multifunctional foams of large sizes. *ACS Applied Materials & Interfaces* **9**, 44732–44739 (2017).

3. C. Gautam et al., Synthesis of low-density, carbon-doped, porous hexagonal boron nitride solids. ACS *Nano* **9**, 12088–12095 (2015).
4. T. Pham et al., One-step conversion of graphite to crinkled boron nitride nanofoams for hydrophobic liquid absorption. *ACS Applied Nano Materials* **4**, 3500–3507 (2021).
5. D. Xia et al., Compressive and thermally stable boron nitride aerogels as multifunctional sorbents. *Dalton Transactions* **51**, 836–841 (2022).
6. T. Pham et al., Nanoscale structure and superhydrophobicity of sp^2-bonded boron nitride aerogels. *Nanoscale* **7**, 10449–10458 (2015).
7. W. Hao, C. Marichy, A. Brioude, Promising properties of ALD boron nitride nanotube mats for water purification. *Environmental Science: Nano* **4**, 2311–2320 (2017).
8. Y. Song et al., Ultralight boron nitride aerogels via template-assisted chemical vapor deposition. Scientific *Reports* **5**, 10337 (2015).
9. Y. Xue et al., Multifunctional superelastic foam-like boron nitride nanotubular cellular-network architectures. ACS *Nano* **11**, 558–568 (2017).
10. D. Liu, M. Zhang, L. He, Y. Chen, W. Lei, Layer-by-layer assembly fabrication of porous boron nitride coated multifunctional materials for water cleaning. *Advanced Materials Interfaces* **4**, 1700392 (2017).
11. H. Li et al., Multifunctional and highly compressive cross-linker-free sponge based on reduced graphene oxide and boron nitride nanosheets. *Chemical Engineering Journal* **328**, 825–833 (2017).
12. M. Chen, J. Xie, C. Xiong, H. Wang, Synthesis and characterization of hexagonal BN-based aerogels for absorbing oils. *Ceramics International* **47**, 19970–19977 (2021).
13. K. Gupta, P. Joshi, O. P. Khatri, h-BN and graphene-based ultralight hybrid aerogels: Highly efficient sorbent for recovery of hydrocarbon oils and organic solvents. *Journal of Environmental Chemical Engineering* **9**, 106788 (2021).
14. L. Hao et al., Nanocomposite foam involving boron nitride nanoplatelets and polycaprolactone: Porous structures with multiple length scales for oil spill cleanup. *Industrial & Engineering Chemistry Research* **56**, 14670–14677 (2017).
15. S. Xue et al., Boron nitride nanosheets/PNIPAM hydrogels with improved thermo-responsive performance. *Materials* **11**, 1069 (2018).
16. E. K. Sam, D. K. Sam, J. Chen, X. Lv, J. Liu, Hydrophobic porous BN/SiO_2@ PU as ternary adsorbents for efficient oil/water separation. *Journal of Porous Materials* **27**, 1149–1158 (2020).
17. X. Gao et al., Preparation of boron nitride nanofibers/PVA composite foam for environmental remediation. *Colloids and Surfaces A: Physicochemical and Engineering Aspects* **604**, 125287 (2020).
18. W. Xie et al., Reactive yellow 161 decolorization by TiO_2/porous boron nitride nanosheet composites in cotton dyeing effluent. *ACS Sustainable Chemistry & Engineering* **5**, 1392–1399 (2017).
19. D. Ponnamma et al., White graphene-cobalt oxide hybrid filler reinforced polystyrene nanofibers for selective oil absorption. *Polymers* **12**, 4 (2019).
20. M. R. Krishnan, Y. F. Aldawsari, E. H. Alsharaeh, 3D-poly (styrene-methyl methacrylate)/divinyl benzene-2D-nanosheet composite networks for organic solvents and crude oil spill cleanup. *Polymer Bulletin* **79**, 3779–3802 (2021).
21. D. Liu et al., Multifunctional polymer/porous boron nitride nanosheet membranes for superior trapping emulsified oils and organic molecules. *Advanced Materials Interfaces* **2**, 1500228 (2015).
22. K. C. Mashiloane, O. T. Mahlangu, M. J. Madito, M. M. Motsa, E. N. Nxumalo, Effect of hexagonal-boron nitride nanosheets (h-BNNSs) on the structural morphology and performance of polyvinylidene fluoride (PVDF) membranes for water-oil separation. *Materials Today Communications* **35**, 106107 (2023).
23. L. He, W. Lei, D. Liu, One-step facile fabrication of mechanical strong porous boron nitride nanosheets–polymer electrospun nanofibrous membranes for repeatable emulsified oil/water separation. *Separation and Purification Technology* **264**, 118446 (2021).
24. G. Zhang et al., Construction of superhydrophilic/underwater superoleophobic polydopamine-modified h-BN/poly (arylene ether nitrile) composite membrane for stable oil-water emulsions separation. *Polymers for Advanced Technologies* **31**, 1007–1018 (2020).
25. J. H. Ramirez Leyva, A. Hethnawi, G. Vitale, N. N. Nassar, Magnetic nanostructured white graphene for oil spill and water cleaning. *Industrial & Engineering Chemistry Research* **57**, 13065–13076 (2018).

26. N. Habibi, A. Pourjavadi, Magnetic, thermally stable, and superhydrophobic polyurethane sponge: A high efficient adsorbent for separation of the marine oil spill pollution. *Chemosphere* **287**, 132254 (2022).
27. N. Habibi, A. Pourjavadi, Thermally conductive and superhydrophobic polyurethane sponge for solar-assisted separation of high-viscosity crude oil from water. *ACS Applied Materials & Interfaces* **14**, 7329–7339 (2022).
28. A. Mahdizadeh, S. Farhadi, A. Zabardasti, Microwave-assisted rapid synthesis of graphene-analogue hexagonal boron nitride (h-BN) nanosheets and their application for the ultrafast and selective adsorption of cationic dyes from aqueous solutions. RSC *Advances* **7**, 53984–53995 (2017).
29. G. Chen et al., High-yield production of monolayer boron nitride nanosheets by cationic-surfactant-assisted solvothermal exfoliation for the ultrafast and selective separation of U (VI) from lanthanides. *Separation and Purification Technology* **278**, 119645 (2021).
30. R. S. Bangari, A. K. Singh, S. Namsani, J. K. Singh, N. Sinha, Magnetite-coated boron nitride nanosheets for the removal of arsenic (V) from water. *ACS Applied Materials & Interfaces* **11**, 19017–19028 (2019).
31. A. S. Konopatsky et al., Structure, magnetic and adsorption properties of novel FePt/h-BN heteromaterials. *Nano Research* **16**, 1473–1481 (2023).
32. L. Han et al., Graphene-boron nitride composite aerogel: A high efficiency adsorbent for ciprofloxacin removal from water. *Separation and Purification Technology* **278**, 119605 (2021).
33. J. Gao et al., Membranes based on porous hexagonal boron nitride nanorods for ultrafast and effective molecular separation. *Journal of Membrane Science* **647**, 120307 (2022).
34. R. Qian et al., Robust all-inorganic hydrophobic BN nanosheets coated β-sialon membrane for membrane distillation. *Journal of the European Ceramic Society* **42**, 2672–2677 (2022).
35. M. Weber et al., Boron nitride nanoporous membranes with high surface charge by atomic layer deposition. *ACS Applied Materials & Interfaces* **9**, 16669–16678 (2017).
36. C. Xu et al., Porous monolith of few-layered boron nitride for effective water cleanup. *Journal of Materials Chemistry A* **10**, 846–854 (2022).
37. C. Cao et al., Scalable exfoliation and gradable separation of boric-acid-functionalized boron nitride nanosheets. *2D Materials* **6**, 035014 (2019).
38. C. Chen et al., Functionalized boron nitride membranes with ultrafast solvent transport performance for molecular separation. *Nature Communications* **9**, 1902 (2018).
39. W. Lei et al., Boron nitride colloidal solutions, ultralight aerogels and freestanding membranes through one-step exfoliation and functionalization. *Nature Communications* **6**, 8849 (2015).
40. D. L. Keshebo et al., Simultaneous exfoliation and functionalization of hexagonal boron nitride in the aqueous phase for the synthesis of high-performance wastewater treatment membrane. *Journal of Cleaner Production* **314**, 128083 (2021).
41. C. Chen et al., Functionalized boron nitride membranes with multipurpose and super-stable semi-permeability in solvents. *Journal of Materials Chemistry A* **6**, 21104–21109 (2018).
42. C. Chen et al., Ultrafast, stable ionic and molecular sieving through functionalized boron nitride membranes. *ACS Applied Materials & Interfaces* **11**, 30430–30436 (2019).
43. A. Hafeez et al., Tuneable molecular selective boron nitride nanosheet ultrafiltration lamellar membrane for dye exclusion to remediate the environment. *Chemosphere* **303**, 135066 (2022).
44. B. Qi et al., Enzymatic construction of temperature-responsive PDMAPS-decorated textiles for oil-water separation. *Colloids and Surfaces A: Physicochemical and Engineering Aspects* **656**, 130340 (2023).
45. M. Xu et al., Charged boron nitride nanosheet membranes for improved organic solvent nanofiltration. *ACS Applied Materials & Interfaces* **15**, 12524–12533 (2023).
46. X. Zheng et al., Reticulated polyamide thin-film nanocomposite membranes incorporated with 2D boron nitride nanosheets for high-performance nanofiltration. *ACS Applied Materials & Interfaces* **15**, 28606–28617 (2023).
47. S. Abdikheibari, W. Lei, L. F. Dumée, A. J. Barlow, K. Baskaran, Novel thin film nanocomposite membranes decorated with few-layered boron nitride nanosheets for simultaneously enhanced water flux and organic fouling resistance. *Applied Surface Science* **488**, 565–577 (2019).

48. P. Kallem, G. Bharath, K. Rambabu, C. Srinivasakannan, F. Banat, Improved permeability and antifouling performance of polyethersulfone ultrafiltration membranes tailored by hydroxyapatite/boron nitride nanocomposites. *Chemosphere* **268**, 129306 (2021).
49. A. Rajput et al., Sulfonated poly ether ether ketone (SPEEK) based composite cation exchange membranes for salt removal from brackish water. *Colloids and Surfaces A: Physicochemical and Engineering Aspects* **614**, 126157 (2021).
50. B. Sert et al., Investigation of the antifouling properties of polyethersulfone ultrafiltration membranes by blending of boron nitride quantum dots. *Colloids and Surfaces B: Biointerfaces* **205**, 111867 (2021).
51. V. Vatanpour, M. Mehrabi, M. Masteri-Farahani, S. Amiri, Improvement of permeability and antifouling properties of polyamide thin-film nanocomposite nanofiltration membranes using boron nitride quantum dots. *Industrial & Engineering Chemistry Research* **63**, 5237–5248 (2024).
52. S. Abdikheibari, W. Lei, L. F. Dumée, N. Milne, K. Baskaran, Thin film nanocomposite nanofiltration membranes from amine functionalized-boron nitride/polypiperazine amide with enhanced flux and fouling resistance. *Journal of Materials Chemistry A* **6**, 12066–12081 (2018).
53. L. Ahmadian-Alam, H. Mahdavi, S. M. M. Davijani, Influence of structurally and morphologically different nanofillers on the performance of polysulfone membranes modified by the assembled PDDA/PAMPS-based hybrid multilayer thin film. *Journal of Environmental Management* **300**, 113809 (2021).
54. V. Vatanpour et al., Investigation of boron nitride/silver/graphene oxide nanocomposite on separation and antibacterial improvement of polyethersulfone membranes in wastewater treatment. *Journal of Environmental Chemical Engineering* **10**, 107035 (2022).
55. H. Liu et al., Temperature and photo sensitive PVDF-g-PNIPAAm/BN@ PDA-Ag nanocomposite membranes with superior wasterwater separation and light-cleaning capabilities. *Separation and Purification Technology* **331**, 125692 (2024).
56. M. Nafti Mateur et al., Porous gelatin membranes obtained from pickering emulsions stabilized with h-BNNS: Application for polyelectrolyte-enhanced ultrafiltration. *Membranes* **10**, 144 (2020).
57. D. Ning et al., pH-responsive BNNS/PAA nanofiltration membrane for tunable water permeance and efficient dye separation. *Separation and Purification Technology* **328**, 125023 (2024).
58. C.-G. Yin et al., Multifunctional boron nitride nanosheet/polymer composite nanofiber membranes. *Polymer* **162**, 100–107 (2019).
59. D.-H. Jiang et al., Facile 3D boron nitride integrated electrospun nanofibrous membranes for purging organic pollutants. *Nanomaterials* **9**, 1383 (2019).
60. J. Azamat, A. Khataee, Molecular dynamics simulations of removal of cyanide from aqueous solution using boron nitride nanotubes. *Computational Materials Science* **128**, 8–14 (2017).
61. Lu, Yin, et al. Computational simulation of organic pollutant removal from wastewater by different porous boron nitride (PBN) nanoclusters. *Journal of Molecular Liquids* **382** 121850 (2023).
62. B. B. Sharma, A. Govind Rajan, How grain boundaries and interfacial electrostatic interactions modulate water desalination via nanoporous hexagonal boron nitride. *The Journal of Physical Chemistry B* **126**, 1284–1300 (2022).
63. N. Vafa et al., Molecular dynamics simulation of hexagonal boron nitride slit membranes for wastewater treatment. *Journal of Molecular Liquids* **382**, 121842 (2023).
64. J. Du et al., Insights into water permeability and Hg^{2+} removal using two-dimensional nanoporous boron nitride. *New Journal of Chemistry* **44**, 18084–18091 (2020).
65. J. Azamat, B. Hazizadeh, Removal of Cd (II) from water using carbon, boron nitride and silicon carbide nanotubes. *Membrane and Water Treatment* **9**, 63–68 (2018).
66. C. T. Nguyen, A. Beskok, Water desalination performance of h-BN and optimized charged graphene membranes. *Microfluidics and Nanofluidics* **24**, 1–12 (2020).
67. A. Kommu, J. K. Singh, Separation of ethanol and water using graphene and hexagonal boron nitride slit pores: A molecular dynamics study. *The Journal of Physical Chemistry C* **121**, 7867–7880 (2017).
68. R. Omidirad, K. Azizi, Theoretical study of adsorption of ethanol and acetone molecules by perfect and defected h-BN nanosheet. *Superlattices and Microstructures* **139**, 106403 (2020).

Chapter 20

Exceptional Gas Storage in Boron Nitride Nanosheets via Mechanochemistry

Srikanth Mateti, Ying Ian Chen, Qiran Cai, Chunmei Zhang, Aijun Du, Selvakannan Periasamy, and Qi Han

20.1 INTRODUCTION

Specific adsorption may be a promising elective low-energy preparation for cryogenic refining for the separation of gases. Traditionally, the petrochemical industry employs an energy-intensive cryogenic distillation method to separate these gas mixtures, constituting 15% of global energy consumption [1]. With respect to vitality utilization, adsorption-based separation utilizing metal–organic systems and films consumes much less than 1/3 of the energy and is 10 times more energy efficient [2]. An assortment of permeable strong materials and atomic strainers and molecular sieves, such as zeolites [3,4] and metal–organic frameworks (MOFs) [5,6], have been created to separate gases through particular adsorption. MOFs known for their high surface area and tunable pore structure show tremendous adsorption capacities and partition efficiencies of hydrocarbon gases [7,8]. Two-dimensional materials (i.e., BN nanosheets) have been found to have amazing gas/molecular adsorption capabilities due to their expansive surface zones and conformational change adsorption qualities [9,10].

Hypothetical investigations on the conceivable gas partition, i.e., H_2/C_2H_4 [11] and CO/H_2, have been detailed on imperfect BN [12]. Simulation revealed that the defects in BN initiate a chemical process to trap external particles [13]. Liu et al. tentatively proposed that free boron radicals at the edges and defect sites of BN nanosheets exhibit catalytic behavior [14]. Nash et al. utilized a ball processing preparation to introduce defects in BN and found that the nitrogen vacancies are most likely the catalytically active sites in hydrogenation [15]. Permeable BN nanosheets have been endeavored for gas separation in a conventional gas adsorption process [16], but the result isn't comparable to that of MOFs. In this manner, novel, more proficient gas separation processes ought to be created. Mechanochemistry or responsive ball processing has been utilized to extend gas adsorption on metals and indeed advance chemical responses between metals and gases through the arrangement of hydrides, nitrides, and carbides [17–20].

Tackling climate change stands as an urgent priority, demanding immediate action to mitigate its severe consequences. Novel material development and more effective methods of storage are significant for the enhancement of carbon dioxide capture and separation from natural gas and nitrogen. Herein, we will introduce a new green method of storage and separation. We were able to capture a significant amount of carbon dioxide, 103.6 wt%, in BN using a mechanochemical

DOI: 10.1201/9781003647812-25

process with careful control. The stored carbon dioxide can be released upon heating, and this cycle can be repeated multiple times (minimum four), marking a transformative advancement in CO_2 gas capture and separation technology. This approach offers the advantages of being environmentally friendly, cost-effective, and efficient

This chapter illustrates that green mechanochemistry synthesis of BN nanosheets enables an energy-saving gas storage method. The gases include olefins, paraffins, and carbon dioxide. Moreover, it provides continuity in processing due to the recovery of adsorbed gas by the heating of only a BN-processing reactor. Without any use of high temperature and high pressure, there is considerable energy effect; it contrasts positively compared to sorption of gases and vapors on MOF materials and porous materials applied within different separation processes. Further, in BN mechanochemical processing, several different adsorption mechanisms for hydrocarbon and other gases are explored.

20.2 GAS STORAGE PHENOMENA

20.2.1 Hydrocarbon Storage

Hexagonal BN (h-BN) powders (PT110 grade, 99% purity) were procured from Momentive Performance Materials Inc and utilized as received in the experiment. Hydrocarbon gases, including acetylene (C_2H_2), ethylene (C_2H_4), ethane (C_2H_6), and methane (CH_4), and CO_2 with a purity of 99.99% were obtained from Coregas Pty Ltd, Australia. Acetylene filling tests were conducted at an industrial acetylene cylinder filling station, emphasizing the necessity for expert handling and caution, particularly when handling pressures exceeding 206 kPa [21].

$$n=(P.V)/(R.T) \tag{20.1}$$

20.2.1.1 Specific Milling Intensity Calculation

The specific milling intensity is calculated by Equation 20.2:

$$I=\frac{M_b V_{\max} f}{M_p} \tag{20.2}$$

where M_b is the mass of the ball, V_{max} is the maximum velocity of the ball, f is the impact frequency, and M_p is the mass of the powder [22].

Under the same milling condition (rotation speed), the intensity can be increased by changing the M_b/M_p ratio.

A vertical planetary ball mill was employed for this investigation, featuring a milling jar equipped with a pressure valve for atmosphere control. Gas pressure was monitored using a gas gauge attached to the jar. Further details regarding the milling system can be found in reference [23]. Gravimetric capacity calculations were performed as follows: The gravimetric capacity of gas adsorption on BN was determined using Equation 20.1, where n represents the number of moles, P denotes the pressure, V signifies the volume of the ball mill jar, R stands for the gas constant (8.314), and T represents the temperature. The number of moles was converted to liters and then divided by the mass of BN utilized in the ball mill.

Gas adsorption onto BN was achieved using a rotating ball milling apparatus (refer to Chapter 2). In this setup, an individual gas was introduced into a sealed milling reactor containing four grams of h-BN powder and several steel milling balls, maintaining a pressure of 400 kPa. Without milling action, the gas showed no adsorption onto the BN powder, and there was no observable pressure change. However, upon initiating milling, a notable reduction in pressure was observed due to gas adsorption, as depicted in Figure 20.1a. Within 20 hours of milling, the pressure of acetylene gas (C_2H_2) decreased from 400 to 20 kPa, indicating near-complete adsorption onto the BN. Similarly, for ethylene gas (C_2H_4), the pressure dropped to 95 kPa under the same milling conditions. Conversely, slower pressure reductions were observed for the two paraffin gases (CH_4 and C_2H_6). Figure 20.1b illustrates the gravimetric capacities for gas adsorption of the four hydrocarbon gases over time. After 20 hours of continuous milling, 282 $cm^3 g^{-1}$ of C_2H_2 and 228 $cm^3 g^{-1}$ of C_2H_4 were adsorbed onto the BN powder. The uptake capacities for the paraffin gases (ethane and methane) were comparatively lower, with 188 $cm^3 g^{-1}$ for C_2H_6 and 152 $cm^3 g^{-1}$ for CH_4. Notably, the capacity curves depicted in Figure 20.1b did not plateau after 20 hours, indicating that gas adsorption did not reach saturation within this timeframe. With prolonged milling, gas adsorption persisted at a reduced rate due to the declining gas pressure within the sealed milling reactor (as shown in Figure 20.1a).

In order to sustain a high adsorption rate, the milling reactor underwent refilling with fresh gas (up to 400 kPa) every 5 hours during the milling process, ensuring that the gas pressure within

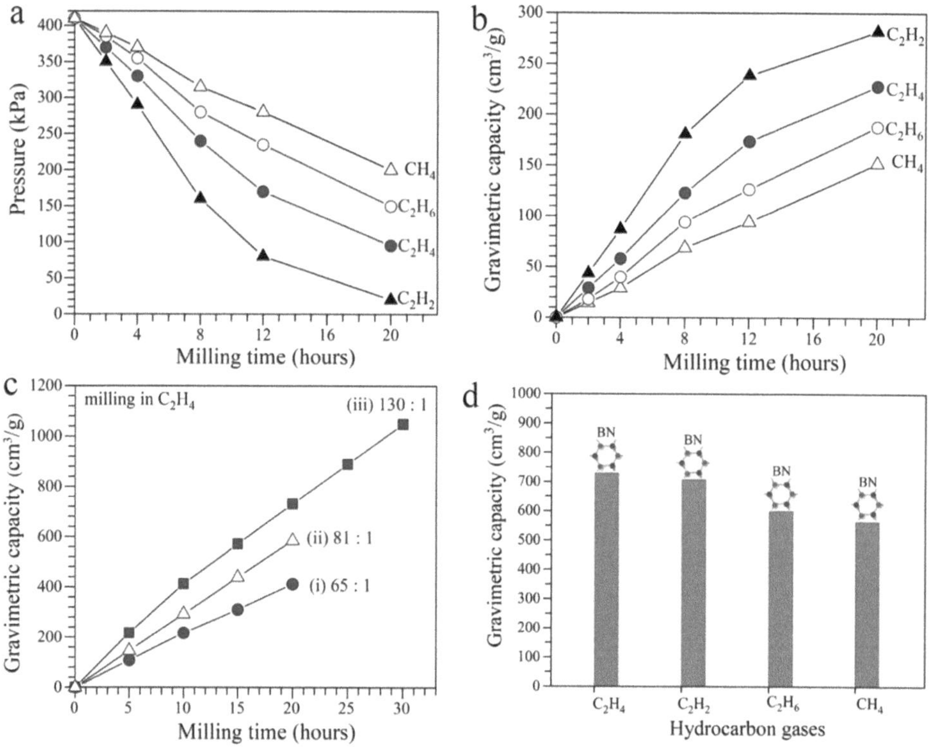

Figure 20.1 Gas adsorption capacity: (a) Hydrocarbon gas pressure reductions during ball milling and (b) Corresponding gravimetric uptake of the gases as a function of milling time. (c) The gravimetric capacity of BN for C_2H_4 as a function of milling time and intensity (different weight ratios of the milling balls to the BN as indicated). (d) Maximum gravimetric uptake of hydrocarbon gases after ball milling of BN for 20 hours. (Reprinted with permission from Ref. [21] Copyright 2018 Wiley-VCH.)

the reactor remained above 300 kPa. A significantly higher adsorption capacity of C_2H_4, reaching 412 cm^3g^{-1}, was achieved after 20 hours of milling when C_2H_4 was refilled at 5, 10, and 15 hours (see Figure 20.1c (i)). Furthermore, increasing the milling intensity by varying the weight ratio of milling balls to BN led to a higher adsorption rate. Elevated milling intensity imparted more energy into the BN, thus accelerating reaction kinetics during milling. Specifically, as the weight ratio increased from 65:1 to 81:1 and 130:1, the adsorption capacities after 20 hours increased to 585 and 730 cm^3g^{-1}, respectively (refer to Figure 20.1c (ii) and (iii)). Prolonging the milling treatment also augmented adsorption; for instance, extended milling in C_2H_4 gas to 30 hours resulted in an adsorption capacity of 1048 cm_3g^{-1} (see Figure 20.1c (iii)).

Even under mild milling conditions—such as 20 hours of milling and a weight ratio of 130:1—the milling treatment with three gas refills yielded an uptake capacity of 708 cm^3g^{-1} for C_2H_2 (as shown in Figure 20.1d). The total adsorption of CH_4 amounted to 564 cm^3g^{-1} after 20 hours of milling, while the uptake capacity for C_2H_6 reached 600 cm^3g^{-1}. The curves depicted in Figure 20.1c demonstrate that C_2H_4 gas adsorption maintained a high adsorption rate and did not saturate by 30 hours. Considering a reduced starting amount of BN to 1 g, a saturated gas adsorption of 534.9 cm^3g^{-1} was observed after 28 hours of milling. Thus, the adsorption capabilities are contingent upon various testing conditions, including gas pressure, BN quantity, milling intensity, and duration. To our knowledge, even under typical ball milling conditions—e.g., 4 g of BN milled for 20 hours with adequate gas—BN nanosheets have achieved higher uptake capacities for olefin and alkyne gases compared to all other reported materials.

The adsorption behavior of C_2H_4 molecules onto BN samples was further examined using Fourier Transform Infrared Spectroscopy (FTIR). The FTIR spectra, displayed in Figure 20.2a, encompass the fingerprint region (2700–3150 cm^{-1}) characteristic of hydrocarbon molecules. The initial BN powder exhibited a featureless FTIR spectrum (Figure 20.2a (i)), indicating the absence of adsorbed C_2H_4. Following 20 hours of continuous milling in the presence of C_2H_4, the FTIR spectrum of the BN sample revealed two peaks at 2850 and 2920 cm^{-1}, corresponding to sp^3 C–H stretching of adsorbed C_2H_4 molecules. The adsorption of the other hydrocarbon gases by BN during ball milling was also confirmed by FTIR analysis (Figure 20.2b).

To further validate C_2H_4 adsorption, the milled BN sample underwent heating from room temperature to 550°C at a rate of 25°C min^{-1} using a simultaneous thermal analysis-FTIR coupled system. The resulting spectrum of the gas collected during heating exhibited the same characteristic peaks attributed to C_2H_4 molecules, confirming the adsorption of C_2H_4 gas. The gas collected from the milling device after heating treatment exhibited a recovery rate of 82% for C_2H_4. Subsequent FTIR analysis of the heated BN indicated the removal of all adsorbed C_2H_4 from the BN surface. Re-milling the cleaned BN in C_2H_4 led to a similar gas pressure reduction, with subsequent FTIR confirming the re-adsorption of C_2H_4 molecules.

Similarly, FTIR analysis confirmed the adsorption of other hydrocarbon gases by BN during ball milling. Therefore, gas adsorption onto BN was achieved at room temperature through mechanical energy via ball milling, with the adsorbed gases released upon heating. The cleaned BN samples were found to be reusable for further adsorption. Under identical milling conditions, the adsorption rates varied among the different gases investigated (as shown in Figure 20.1b). The highest observed adsorption rate was 21.6 $cm^3g^{-1}h^{-1}$ for C_2H_2, nearly three times faster than that for CH_4 (7.2 $cm^3g^{-1}h^{-1}$). The trend in adsorption capacities indicated a weaker affinity of BN for paraffin gases (CH_4 and C_2H_6) compared to olefin gases (C_2H_2 and C_2H_4). These significant differences in adsorption capacities and rates suggest the potential for selective adsorption of these hydrocarbon gases by BN under controlled milling conditions. Moreover, the ball milling process appears suitable for separating other gas groups with varying adsorption rates.

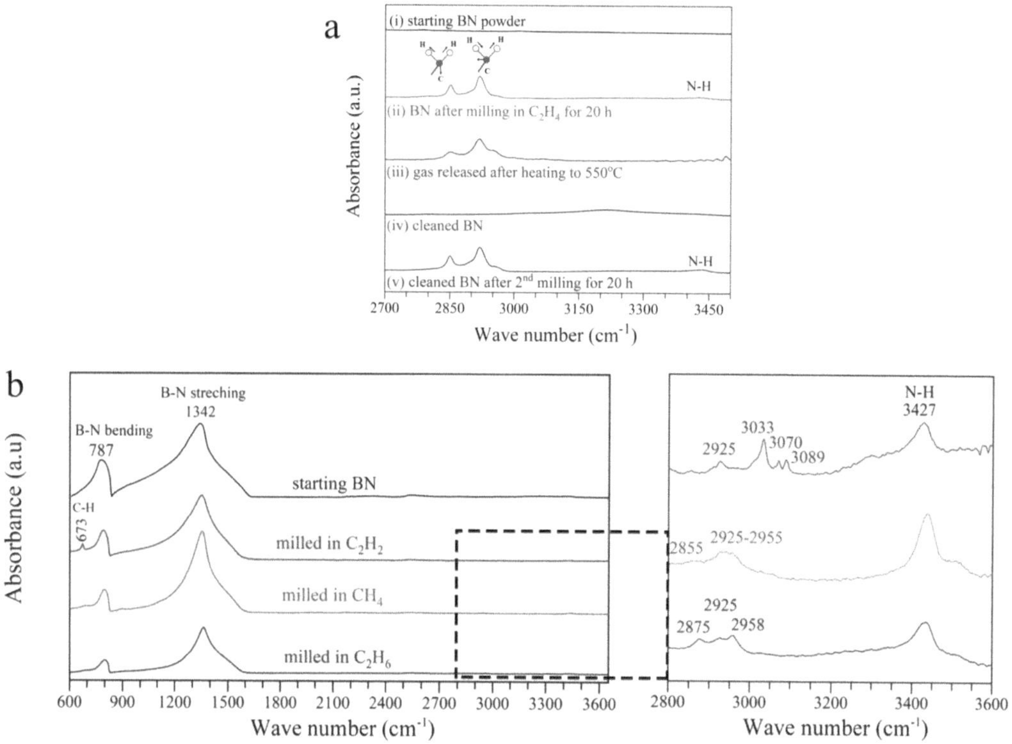

Figure 20.2 (a) FTIR spectra of the initial BN powder and the BN samples after various milling and heating treatments. (b) FTIR spectra of BN milled in hydrocarbon gases (and enlarged spectra). (Reprinted with permission from Ref. [21] Copyright 2018 Wiley-VCH.)

The mechanisms underlying gas adsorption on BN during ball milling were thoroughly investigated. The mechanical grinding of powders leads to a significant reduction in particle size and an increase in surface area, facilitating the physical adsorption of gas molecules by creating new surfaces. The Brunauer–Emmett–Teller surface area of the BN powder was analyzed, revealing a substantial increase in surface area after 20 hours of milling in CH_4, whereas a smaller change was observed for BN milled in C_2H_4 under identical conditions (Figure 20.3a). The surface area of BN increased by 623.6 $m^2 g^{-1}$ after 20 hours of milling in CH_4, while a smaller change in surface area (99.8 $m^2 g^{-1}$) was observed for the BN sample milled in C_2H_4 under the same milling conditions. While BN with the smaller surface area adsorbed much more C_2H_4, in spite of a greater increase in the surface area for the former, BN with a greater surface area adsorbed less CH_4. All these results indicate that the surface area of BN alone is not decisive in determining its adsorption capability and further support that physical adsorption is not a primary mechanism by which BN adsorbs olefin gases.

Distinct crystalline structures were observed in milled BN samples through X-ray diffraction (XRD) patterns. BN samples milled in C_2H_4 maintained the hexagonal crystalline structure, whereas those milled in CH_4 displayed a disordered structure (Figure 20.3b). Scanning electron microscopy (SEM) analysis further confirmed these differences, with BN powders milled in the presence of C_2H_4 exhibiting nanosheet-like morphologies, while those milled in CH_4 displayed spherical shapes. This difference in morphology suggests that different mechanisms are

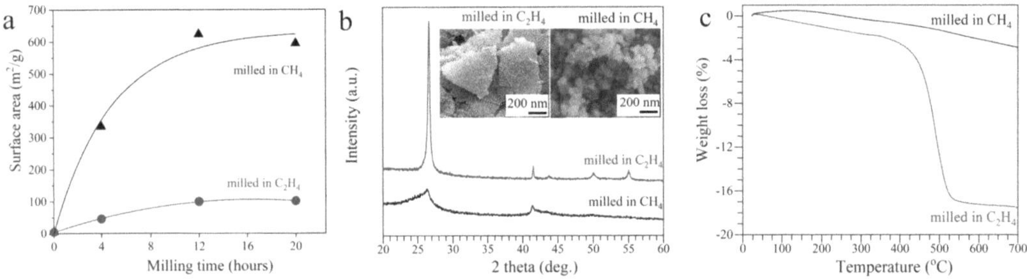

Figure 20.3 (a) Brunauer–Emmett–Teller surface area of BN milled in CH_4 and C_2H_4. (b) SEM images and XRD patterns of BN samples milled in C_2H_4 and CH_4. (c) TGA curves of the BN samples after milling in C_2H_4 and CH_4. (Reprinted with permission from Ref. [21] Copyright 2018 Wiley-VCH.)

responsible for the adsorption of olefin and paraffin gases. To elucidate these mechanisms, BN samples continuously milled for 20 hours in the presence of either CH_4 or C_2H_4 were subjected to thermal gravimetric analysis (TGA). For BN milled in C_2H_4, a gradual weight loss (Figure 20.3c) occurred up to 400°C, followed by a sharp curve change indicative of large and rapid weight loss from 400°C to 550°C, suggesting chemical adsorption of C_2H_4 on the BN nanosheets.

X-ray photoelectron spectroscopy (XPS) confirmed the formation of chemical bonds between C_2H_4 molecules and BN nanosheets, revealing N–C bonding between them (Figure 20.4).

The bond formation between gas molecules and BN was analyzed using XPS. For BN samples processed with C_2H_4, the B 1s spectrum in Figure 20.4a indicates two characteristic B–N peaks at binding energies of 190.3 eV and 191.1 eV [24,25]. The N 1s spectrum in Figure 20.4b shows a predominant N–B bond at 397.9 eV and an N–C bond at 398.6 eV [24], with a smaller peak at 400.1 eV likely due to N–H bonding [26]. In contrast, the B 1s and N 1s spectra of the original BN powder show no evidence of gas adsorption. The C 1s spectrum in Figure 20.4c displays a major peak at 284.5 eV, associated with C–C and C–H bonds from adsorbed C_2H_4 molecules, and a C–N bond peak at 285.3 eV [26,27]. The N 1s and C 1s spectra show N–C bonding peaks, while the B 1s spectrum lacks B–C bonding peaks, suggesting N–C bonding between C_2H_4 molecules and BN nanosheets. Additionally, a small amount of C_2H_4 is physically absorbed on BN nanosheets through N–H bonding, as indicated by the N–H peak in the N 1s spectrum and the corresponding FTIR spectra, which aligns with the slight weight loss observed below 400°C.

For the BN samples milled in CH4, the B 1s spectrum (Figure 20.4d) shows two B–N bonding peaks and a significant B–O bonding peak at 193.1 eV [28]. The milled BN samples have a large surface area, making them reactive with environmental gases and prone to oxygen contamination from the air during XPS analysis. The N 1s spectrum in Figure 20.4e displays the same N–B bonding peak and a broad, weak shoulder band centered at 399.5 eV, which may be attributed to N–H bonding [29]. The C 1s spectrum in Figure 20.3h features a major peak at 284.5 eV associated with C–C and C–H bonds, along with a C–O peak at 286.3 eV [27,30]. The prominent N–H bonding peak at 399.5 eV in Figure 20.4f suggests that CH_4 molecules are physically absorbed onto the BN surface through N–H bonding, which aligns with FTIR results.

In contrast, BN milled in CH_4 exhibited physical adsorption of CH_4 molecules onto BN surfaces via weak N–H bonding, as indicated by XPS results. The XPS analysis also suggests that paraffin molecules are adsorbed onto BN surfaces via weak N–H bonding, while olefin molecules form bonds with unsaturated carbon atoms in the olefin gas. However, the formation of C–N bonds on BN nanosheets requires the energy provided by ball milling actions, indicating a mechanochemical reaction between BN and gases.

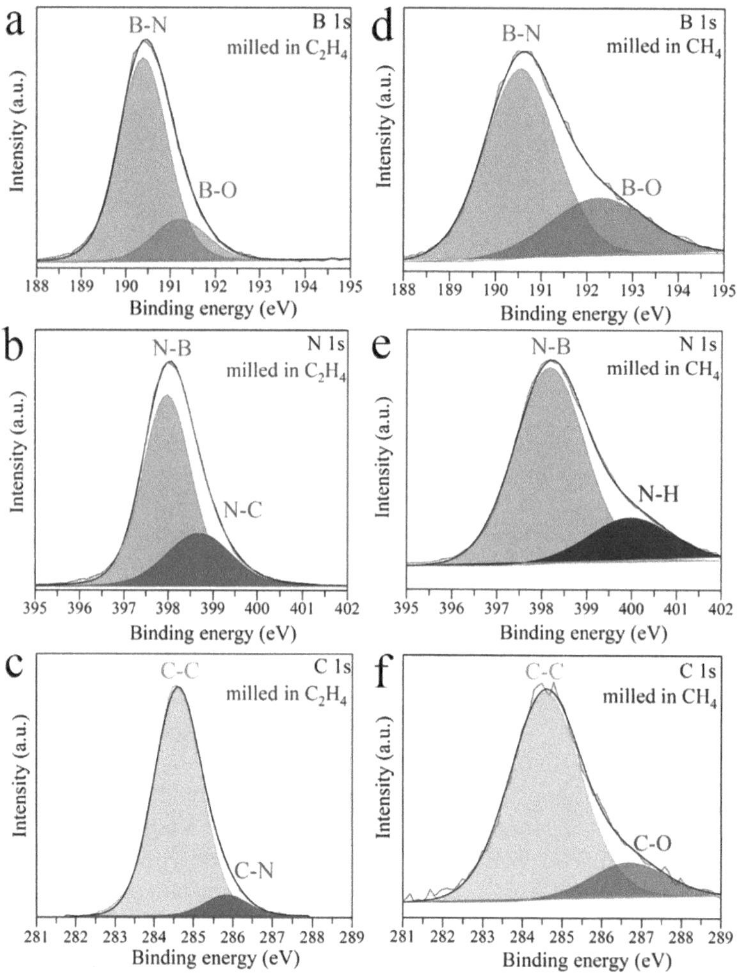

Figure 20.4 Analysis of gas adsorption mechanisms: B 1s, N 1s, and C 1s XPS spectra of the BN samples milled in C_2H_4 (a, b, c) and CH_4 (d, e, f). (Reprinted with permission from Ref. [21] Copyright 2018 Wiley-VCH.)

To confirm the adsorption mechanisms, theoretical calculations were employed to describe gas adsorption and potential separation on BN nanosheets using density functional theory. A model consisting of a 7×7 h-BN supercell with one nitrogen and three boron atom defects (V_{N+3B}) was utilized to ensure sufficient vacuum space, preventing interactions between adjacent unit cells (refer to Figure 20.5). The adsorption energy (binding energy), denoted as E_{ad}, was calculated using the equation:

$$E_{ad} = E_{tot} - E_{substrate} - E_{C_xH_y} \tag{20.3}$$

Here, E_{tot} represents the total energy of the *CxHy*/substrate system after geometry optimization, $E_{substrate}$ denotes the energy of the relaxed two-dimensional BN nanosheet with one nitrogen and three boron atom defects, and $E_{C_xH_y}$ represents the energy of the C_xH_y molecule.

The adsorption energy for C_2H_2 on the V_{N+3B} system is −5 eV (see Figure 20.5a), significantly lower than that of CH_4 (Figure 20.5b). Similarly, for group II gases, the adsorption energy for C_2H_4 (Figure 20.5c) is notably lower than that of C_2H_6 (Figure 20.5d). Chemical bonds are formed between the carbon atoms in C_2H_2 and C_2H_4 molecules and the nitrogen atoms located at the defect edge of the V_{N+3B} substrate, indicating chemical adsorption. Conversely, much longer adsorption distances (>3 Å) are observed for CH_4 and C_2H_6 adsorbed on the V_{N+3B} substrate, indicating physical adsorption rather than chemical bonding. This suggests that the V_{N+3B} system exhibits a strong affinity for unsaturated hydrocarbons like acetylene and ethylene, facilitating successful separation of these gas groups. Further research is necessary to fully elucidate the adsorption mechanisms during the dynamic milling process.

Other defects in BN nanosheets were also investigated, such as one boron and three nitrogen atom defects (V_{B+3N}). However, V_{B+3N}, terminated with electron-deficient boron atoms, cannot form bonds with C_2H_4, rendering it unable to effectively separate C_2H_4 and C_2H_6 gases.

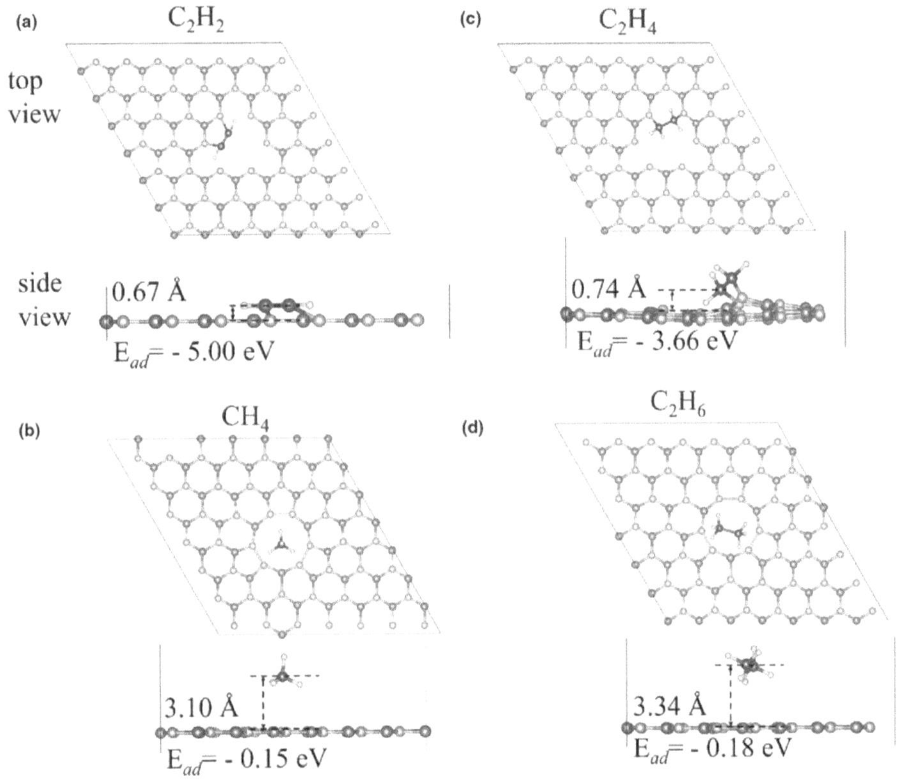

Figure 20.5 Theoretical density functional theory calculations of gas adsorption: Optimized geometry for C_xH_y (x=1–2; y=2–6) adsorbed on a 7×7 supercell h-BN sheet composed of 1 N and 3 B defects (V_{N+3B}). (a, b) The top side views of the group of C_2H_2 and CH_4 adsorbed on V_{N+3B}; (c, d) the top views and side views of the group of C_2H_4 and C_2H_6 adsorbed on V_{N+3B}; the brown, gray, white, and green atoms represent C, N, H, and B atoms, respectively. The adsorption distances are presented and defined as the vertical distance between the C atom and the V_{N+3B} substrate. (Reprinted with permission from Ref. [21] Copyright 2018 Wiley-VCH.)

20.2.2 Carbon Dioxide Storage

Storage and separation of carbon dioxide (CO_2) have emerged as pivotal areas of research, offering potential solutions to industrial expansion and improving human living standards. CO_2, primarily generated from fossil fuel combustion, stands as the main contributor to global warming and is also recognized as an acidic gas present in natural gas, biogas, and landfill gas, diminishing their calorific value [31]. According to the International Energy Agency in 2021, global CO_2 emissions surged to a record high of 36.3 Gt, signifying an alarming trend likely to persist due to escalating energy demands fueled by industrialization and human activities [32]. Undoubtedly, effective CO_2 separation and storage will be crucial in combating the greenhouse effect, facilitating the production of clean energy gases, and safeguarding ecosystems. Thus, the development of technologies capable of efficiently separating or capturing CO_2 is urgently needed. Over the past two decades, considerable research efforts have focused on selectively separating CO_2 from various gas mixtures, employing diverse materials, such as porous carbon [33] and metal–organic frameworks [34]. However, these materials encounter significant challenges and remain subject to ongoing investigation. For instance, zeolites' adsorption capacity is contingent upon charge density and size, while MOF-based adsorbents face issues related to chemical and thermal stability [35,36]. Large-scale implementation and high costs pose key hurdles in conventional adsorption-based gas storage and separation processes, aligning with the target set by the US Department of Energy to achieve $100 per ton of CO_2 capture or storage by 2035 [37].

Two grams of BN powder underwent ball milling in a sealed milling cell filled with either nitrogen or carbon dioxide gas. Figure 20.6a illustrates the pressure drop with milling time due to gas adsorption during ball milling. A very slight drop of 30 kPa was obtained for nitrogen gas after 20 hours of milling, whereas in the case of CO_2, a significant drop of 320 kPa was obtained for the same conditions. The corresponding gas uptake capacity for 2 g of BN powder was 10.6 mmol g^{-1} for CO_2 and 0.9 mmol g^{-1} for nitrogen after 20 hours of milling, Figure 20.6b. These findings indicate distinct adsorption rates of nitrogen and carbon dioxide on BN during milling, suggesting BN as an ideal candidate for CO_2 separation from nitrogen gas mixtures via mechanochemical processes. Notably, CO_2 uptake had not saturated by the end of the 20-hour milling period (Figure 20.6a), as evidenced by the ongoing pressure reduction. CO_2 uptake capacity in BN could be further enhanced by adjusting ball milling conditions, such as ball-to-powder ratio, milling speed, or regular refilling of CO_2 gas. Gas adsorption by 1 g of BN saturated at the end of 27 hours of milling with 20.1 mmol g^{-1}. Additionally, Figure 20.6c depicts the effect of various ball mill parameters on the CO_2 adsorption rate and capacity. Previous studies indicated that the particle size of initial BN is not a critical parameter for gas uptake capacity. At a lower ball-to-powder ratio of 130:1 and refilling fresh CO_2 gas to 600 kPa every 5 hours of milling, more CO_2 was stored compared to continuous milling with a lower ball-to-powder ratio of 65:1 and refilling with a lower pressure of 400 kPa. The maximum CO_2 storage capacity in BN reached 21.3 mmol g^{-1}, equivalent to 0.5 L of CO_2 stored in 1 g of BN, or 103.6 wt% of CO_2 stored in BN when milling duration was extended to 40 hours, although saturation was still not achieved. This is reportedly the highest CO_2 uptake by any material according to the literature. Figure 20.1d presents gas uptake by other reported materials.

The adsorption of CO_2 molecules was examined using FTIR. The FTIR spectra displayed in Figure 20.7a, with enlarged spectra in Figure 20.7b, demonstrate that the initial BN powder did not contain any adsorbed CO_2, as confirmed by the featureless FTIR spectrum. Peaks at 793 and 1363 cm^{-1} correspond to B–N stretching and B–N–B bending, respectively. However, in milled

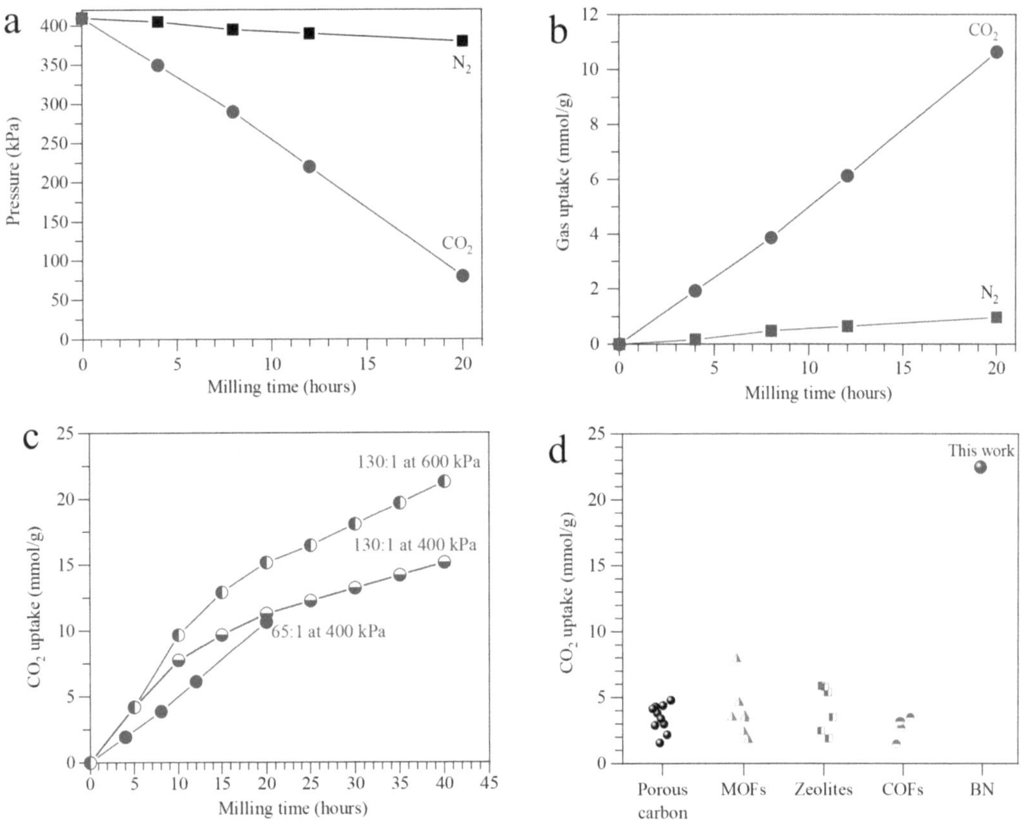

Figure 20.6 Gas storage capacity: (a) gas (CO_2 and N_2) pressure reductions during ball milling. (b) Corresponding uptake of the gases as a function of milling time. (c) The gravimetric capacity of BN for CO_2 as a function of milling time and intensity (different weight ratios of the milling balls to the BN as indicated). (d) Comparison of the CO_2 uptake capacity of other materials with BN. (Reprinted with permission from Ref. [38] Copyright 2018 Wiley-VCH.)

BN powder, the presence of a peak at 2279 cm^{-1} (for BN milled in CO_2) corresponding to N=C=O confirms CO_2 molecule attachment, absent in the starting BN, with no such adsorption observed for BN milled in N_2. CO_2 re-adsorption on cleaned BN was also studied (see Figure 20.7c). Accordingly, after 20 hours of milling (Figure 20.7c (ii)), new peaks at 2279 cm^{-1} were detected due to the adsorption of CO_2 [39]. The BN sample milled was then heated from room temperature to 800°C to remove the adsorbed CO_2, as shown by the FTIR spectra of the heated BN sample, revealing clean BN and the release of the adsorbed carbon species. The cleaned BN was used for further milling, and CO_2 re-adsorption was confirmed on FTIR spectra iv. This, therefore, illustrates that at room temperature, CO_2 molecules are being adsorbed onto BN and, through heating, are being released; cleaned BN can be reused for gas adsorption.

The XRD analysis (Figure 20.8a) was conducted to investigate the crystal structure of BN before and after milling in CO_2. The peak at 26° corresponding to (002) exhibited broadening, indicating the exfoliation of bulk BN layers into nanosheets. Accordingly, BN milled for 14 hours

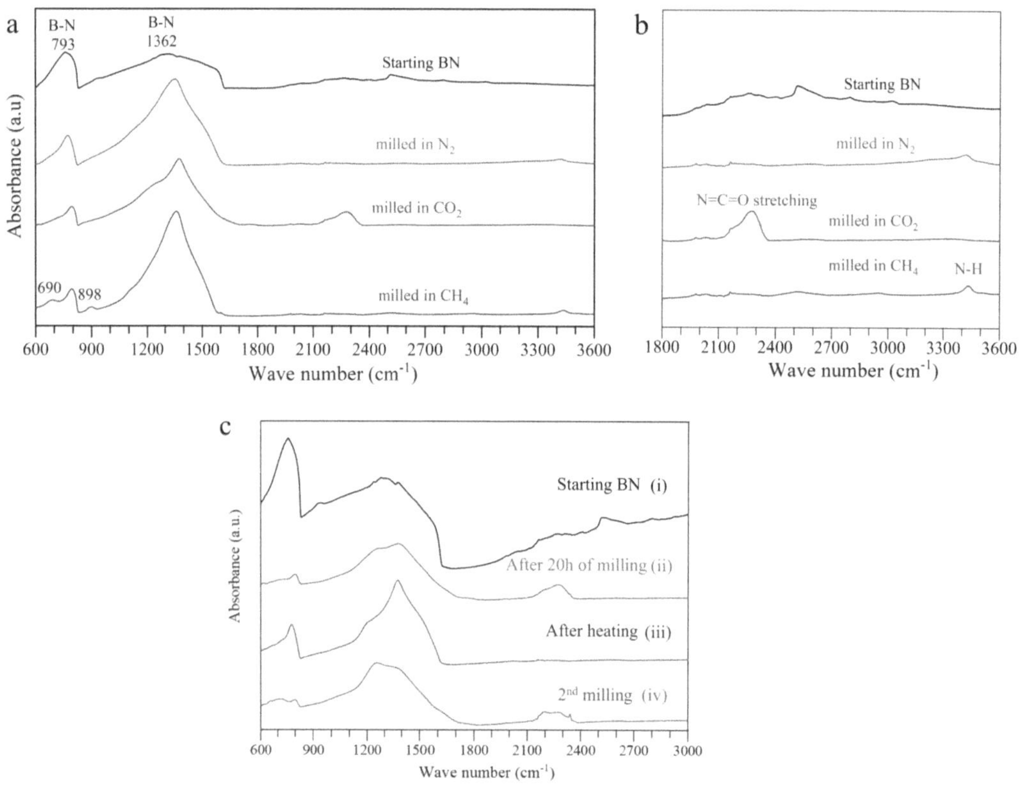

Figure 20.7 (a) FTIR spectra of the initial BN powder and the BN samples after milling in different gases and (b) enlarged version of FTIR spectra. (c) FTIR spectra of the initial BN powder and the BN samples after milling and heating treatments. (Reprinted with permission from Ref. [38] Copyright 2018 Wiley-VCH.)

showed the broadening of the (002) peak, which reflected the reduction of crystalline size, Figure 20.8b. The SEM images further supported the generation of BN nanosheets displaying a layered structure after milling even for 20 hours, Figure 20.8c and e. Finally, high-resolution transmission electron microscopy (TEM) images, Figure 20.8d, of 4 hours milled BN indicated that this sample indeed displays a layered structure and very high crystallinity. BN milled for 16 hours showed lattice fringes and selected area electron diffraction patterns confirming the nanocrystalline nature. Similarly, the surface area increased from 5.5 m^2 g^{-1} for the starting BN to 307.7 m^2 g^{-1} for BN milled for 12 hours. This in turn increases the CO_2 adsorption with increased milling time (Figure 20.8g). TGA curves (Figure 20.8h) confirmed that BN exhibited both physical and chemical adsorption. CO_2 began to desorb at 89°C, with an initial weight loss of 2.7 wt% observed until 211°C, indicating the release of physisorbed CO_2. Further weight loss (20.7 wt%) suggested chemical bonding between BN and CO_2.

The bond formation between CO_2 gas molecules and BN was investigated using XPS. The B1s spectrum confirmed characteristic B–N bonds at a binding energy of 190.3 eV, as well as B–O bonds at 191.3 eV [25] (Figure 20.9a). The dominant peak in the N1s spectra at a binding energy of 397.8 eV denoted N–B bonding, while the additional peak at 398.8 eV indicated bonding with

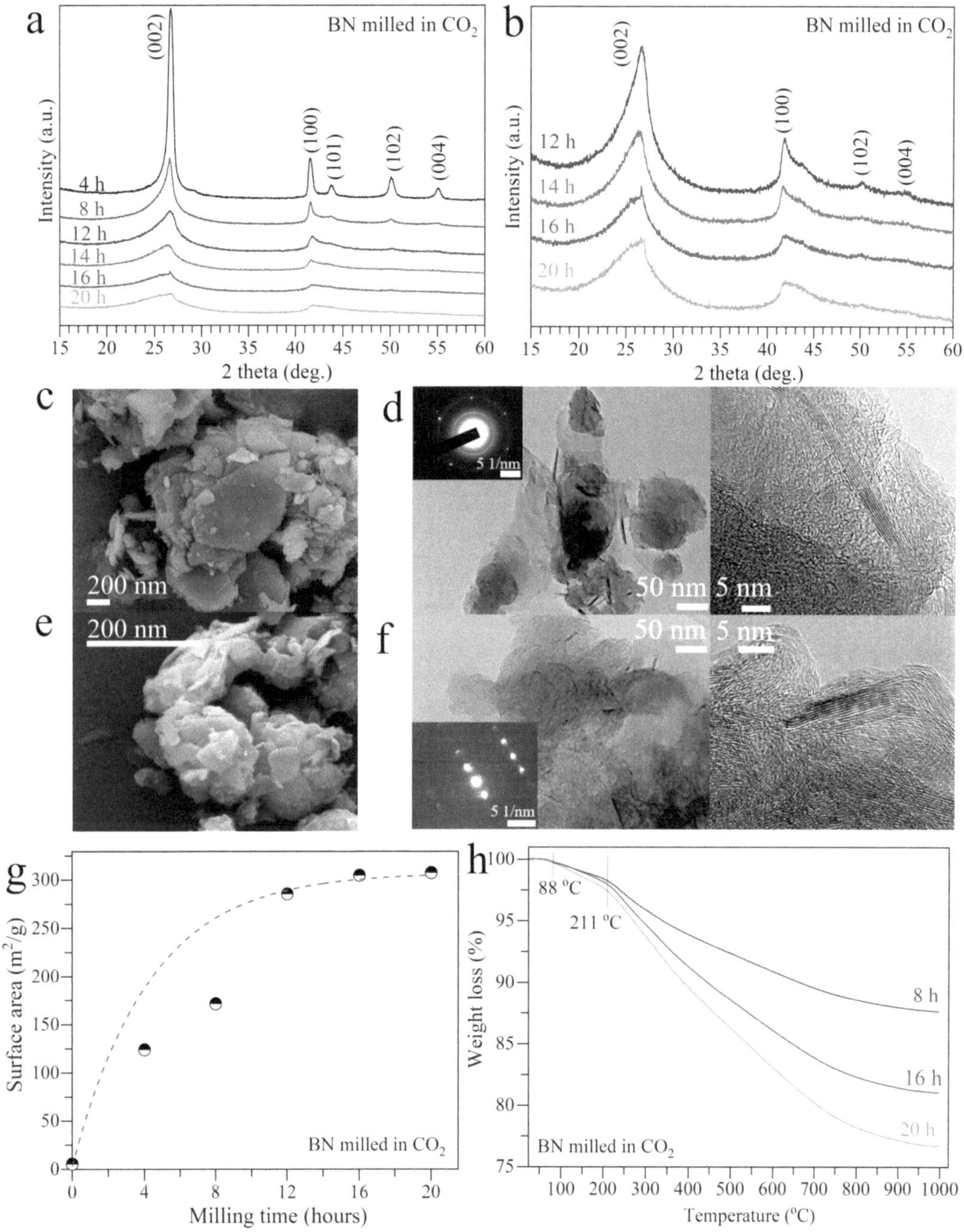

Figure 20.8 (a, b) XRD patterns of BN milled in CO_2 for different time intervals. (c, e) SEM images of milled BN for 4 hours and 16 hours. (d, f) HR TEM images and selected area electron diffraction (inset) of BN milled in CO_2 for 4 hours and 16 hours. (g) Brunauer–Emmett–Teller surface area and (h) TGA curves of BN milled in CO_2. (Reprinted with permission from Ref. [38] Copyright 2018 Wiley-VCH.)

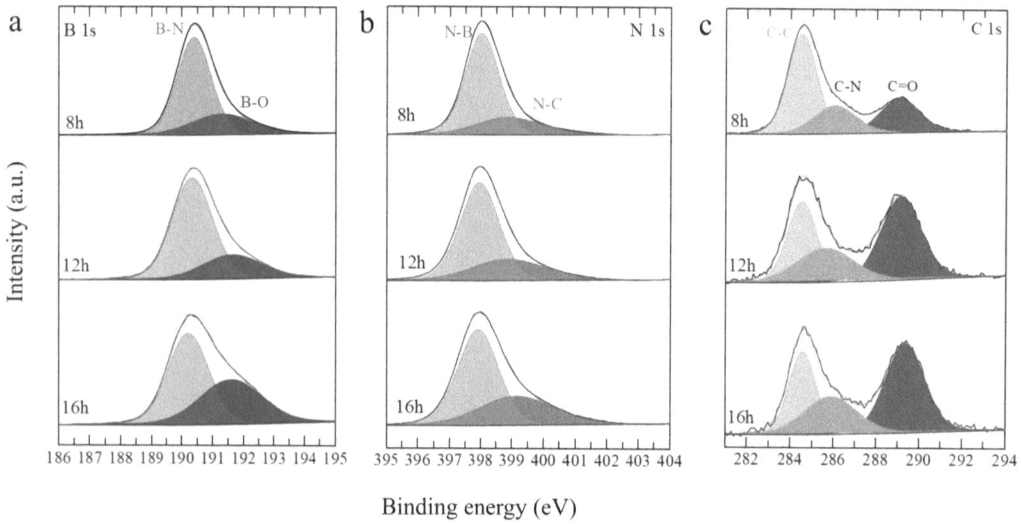

Figure 20.9 B 1s, N 1s, and C 1s XPS spectra of the BN samples milled in CO_2 (a–c). (Reprinted with permission from Ref. [38] Copyright 2018 Wiley-VCH.)

carbon, specifically N–C (Figure 20.9b). The C 1s spectrum (Figure 20.9c) exhibited a major peak at 284.5 eV [27] associated with two C–C bonds, along with a C–N bonding peak at 285.3 eV [26] and a C=O peak at 288.9 eV [40]. The presence of N–C bonding peaks in both N 1s and C 1s spectra suggested N–C bonding between CO_2 molecules and BN nanosheets.

20.3 CONCLUSIONS

In essence, ball milling initiates mechanochemical reactions, facilitating the formation of C–N bonds between alkyne and olefin molecules and BN nanosheets. This results in significantly enhanced adsorption capacity of BN for olefin gases. The varying adsorption rates of different gases highlight BN's high selectivity for gases in mechanochemical processes. The adsorbed gases can be recovered through heating, and this mechanochemical process is environmentally friendly, generating no by-products. Both milling and heating processes are scalable and energy-efficient compared to cryogenic distillation. BN can be reused after cleaning the adsorbed gases, further enhancing its cost-effectiveness and sustainability. These advantages, including substantial energy savings and low costs, make large-scale gas storage achievable with efficiency. Furthermore, this mechanochemical process has the potential to revolutionize CO_2 gas capture technology, offering the advantages of environmental friendliness, cost-effectiveness, and high efficiency.

REFERENCES

1. D. S. Sholl, R. P. Lively, Seven chemical separations to change the world. *Nature* **532**, 435–437 (2016).
2. L. Yang et al., Energy-efficient separation alternatives: Metal–organic frameworks and membranes for hydrocarbon separation. *Chemical Society Reviews* **49**, 5359–5406 (2020).

3. Y. Chai et al., Control of zeolite pore interior for chemoselective alkyne/olefin separations. *Science* **368**, 1002–1006 (2020).
4. P. J. Bereciartua et al., Control of zeolite framework flexibility and pore topology for separation of ethane and ethylene. *Science* **358**, 1068–1071 (2017).
5. E. D. Bloch et al., Hydrocarbon separations in a metal-organic framework with open iron (II) coordination sites. *Science* **335**, 1606–1610 (2012).
6. J. Y. Lin, Molecular sieves for gas separation. *Science* **353**, 121–122 (2016).
7. W. G. Cui, T. L. Hu, X. H. Bu, Metal–organic framework materials for the separation and purification of light hydrocarbons. *Advanced Materials* **32**, 1806445 (2020).
8. Y. Wang, S. B. Peh, D. Zhao, Alternatives to cryogenic distillation: Advanced porous materials in adsorptive light olefin/paraffin separations. *Small* **15**, 1900058 (2019).
9. H. W. Kim et al., Selective gas transport through few-layered graphene and graphene oxide membranes. *Science* **342**, 91–95 (2013).
10. Q. Cai et al., Boron nitride nanosheets improve sensitivity and reusability of surface-enhanced Raman spectroscopy. *Angewandte Chemie* **128**, 8545–8549 (2016).
11. Y. Zhang et al., Hexagonal boron nitride with designed nanopores as a high-efficiency membrane for separating gaseous hydrogen from methane. *The Journal of Physical Chemistry* C **119**, 19826–19831 (2015).
12. L. Li et al., Porous BN with vacancy defects for selective removal of CO from H_2 feed gas in hydrogen fuel cells: A DFT study. *Journal of Materials Chemistry A* **4**, 15631–15637 (2016).
13. N. Lvova, O. Y. Ananina, Theoretical study of the adsorption properties of porous boron nitride nanosheets. *Computational Materials Science* **115**, 11–17 (2016).
14. Z. Liu et al., Boron radicals identified as the source of the unexpected catalysis by boron nitride nanosheets. ACS *Nano* **13**, 1394–1402 (2019).
15. D. J. Nash et al., Heterogeneous metal-free hydrogenation over defect-laden hexagonal boron nitride. ACS *Omega* **1**, 1343–1354 (2016).
16. D. Saha et al., Nanoporous boron nitride as exceptionally thermally stable adsorbent: Role in efficient separation of light hydrocarbons. *ACS Applied Materials & Interfaces* **9**, 14506–14517 (2017).
17. Y. Chen, J. Williams, Formation of metal hydrides by mechanical alloying. *Journal of Alloys and Compounds* **217**, 181–184 (1995).
18. Y. Chen, J. S. Williams, Competitive gas-solid reactions realized by ball milling of Zr in ammonia gas. *Journal of Materials Research* **11**, 1500–1506 (1996).
19. A. Glushenkov, H.-Z. Zhang, Y. Chen, Reactive ball milling to produce nanocrystalline ZnO. Materials *Letters* **62**, 4047–4049 (2008).
20. A. Khorshidi, J. Violet, J. Hashemi, A. A. Peterson, How strain can break the scaling relations of catalysis. *Nature Catalysis* **1**, 263–268 (2018).
21. S. Mateti, C. Zhang, A. Du, S. Periasamy, Y. I. Chen, Superb storage and energy saving separation of hydrocarbon gases in boron nitride nanosheets via a mechanochemical process. *Materials Today* **57**, 26–34 (2022).
22. Y.-L. Chen, J.-Y. Chou, Selection of anti-caking agents through crystallization. *Powder Technology* **77**, 1–6 (1993).
23. Y. Chen, T. Halstead, J. Williams, Influence of milling temperature and atmosphere on the synthesis of iron nitrides by ball milling. *Materials Science and Engineering*: A **206**, 24–29 (1996).
24. F. Guo et al., Carbon-doped BN nanosheets for the oxidative dehydrogenation of ethylbenzene. *Angewandte Chemie* **129**, 8343–8347 (2017).
25. P. R. Kidambi et al., In situ observations during chemical vapor deposition of hexagonal boron nitride on polycrystalline copper. *Chemistry of Materials* **26**, 6380–6392 (2014).
26. J. C. Vickerman, "Molecular surface mass spectrometry by SIMS" in John C. Vickerman, Ian S. Gilmore (eds), *Surface Analysis: The Principal Techniques* (2009), John Wiley & Sons, Ltd, pp. 113–205.
27. L. Lin et al., One-pot synthesis of highly greenish-yellow fluorescent nitrogen-doped graphene quantum dots for pyrophosphate sensing via competitive coordination with Eu^{3+} ions. *Nanoscale* **7**, 15427–15433 (2015).

28. J. Qu, Q. Li, C. Luo, J. Cheng, X. Hou, Characterization of flake boron nitride prepared from the low temperature combustion synthesized precursor and its application for dye adsorption. *Coatings* **8**, 214 (2018).
29. A. Mohtasebi, T. Chowdhury, L. H. Hsu, M. C. Biesinger, P. Kruse, Interfacial charge transfer between phenyl-capped aniline tetramer films and iron oxide surfaces. *The Journal of Physical Chemistry* C **120**, 29248–29263 (2016).
30. Y.-L. Huang et al., Self-assembly of graphene onto electrospun polyamide 66 nanofibers as transparent conductive thin films. *Nanotechnology* **22**, 475603 (2011).
31. Z. Dai, L. Ansaloni, L. Deng, Recent advances in multi-layer composite polymeric membranes for CO2 separation: a review. *Green Energy & Environment* **1**, 102–128 (2016).
32. IEA, *Global Energy Review: CO2 Emissions in 2020* (IEA, 2021).
33. L. Rao et al., Nitrogen enriched porous carbons from d-glucose with excellent CO_2 capture performance. *Chemical Engineering Journal* **362**, 794–801 (2019).
34. Z. H. Rada et al., Functionalized UiO-66 by Single and Binary $(OH)_2$ and NO_2 Groups for Uptake of CO_2 and CH_4. *Industrial & Engineering Chemistry Research* **55**, 7924–7932 (2016).
35. S. Kumar, R. Srivastava, J. Koh, Utilization of zeolites as CO_2 capturing agents: advances and future perspectives. *Journal of CO_2 Utilization* **41**, 101251 (2020).
36. M. Minelli et al., Characterization of novel geopolymer–zeolite composites as solid adsorbents for CO_2 capture. *Chemical Engineering Journal* **341**, 505–515 (2018).
37. Z. Zhang et al., Emerging trends in sustainable CO_2-management materials. *Advanced Materials* **34**, 2201547 (2022).
38. S. Mateti et al., A mechanochemical process to capture and separate carbon dioxide from natural gas using boron nitride nanosheets. Materials Horizons **11**, 2950–2956 (2024).
39. E.-M. Köck, M. Kogler, T. Bielz, B. Klötzer, S. Penner, In situ FT-IR spectroscopic study of CO_2 and CO adsorption on Y_2O_3, ZrO_2, and yttria-stabilized ZrO_2. *The Journal of Physical Chemistry* C **117**, 17666–17673 (2013).
40. J. Y. Loh, N. P. Kherani, X-ray photospectroscopy and electronic studies of reactor parameters on photocatalytic hydrogenation of carbon dioxide by defect-laden indium oxide hydroxide nanorods. *Molecules* **24**, 3818 (2019).

Chapter 21

Mechanochemical Innovations in Gas Separation with Boron Nitride Nanosheets

Srikanth Mateti, Ying Ian Chen, Qiran Cai, Aijun Du, and Qi Han

21.1 INTRODUCTION

Gas separation is a critical process across numerous industries, driven by the need to purify products, protect the environment, ensure safety, and enhance economic efficiency. By separating gases from mixtures, industries can produce high-purity materials essential for applications in natural gas processing, hydrogen production, air pollution control, and cryogenic air separation. This process not only helps in meeting stringent environmental regulations and improving energy efficiency but also maximizes the value of resources and optimizes various industrial operations. As a result, gas separation technologies play an indispensable role in modern industrial practices, contributing significantly to sustainability and economic growth [1]. Mechanochemistry or responsive ball processing has been utilized to extend gas adsorption on metals and indeed advance chemical responses between metals and gases through the arrangement of hydrides, nitrides, and carbides [2–6]. This chapter discusses the green mechanochemistry synthesis of BN nanosheets, enabling an energy-saving separation process for the separation of hydrocarbon and carbon dioxide gas mixtures.

21.2 GAS SEPARATION

21.2.1 Olefin and Paraffin Separation

The efficient separation of gases, such as the removal of trace amounts of C_2H_2 from high-purity CH_4, is crucial for various industrial applications where CH_4 serves as a clean fuel or a feedstock in chemical production [7]. To evaluate the efficacy of our method in such separations, a mixture comprising 20 vol% C_2H_2 and 80 vol% CH_4 was introduced into a milling reactor containing 4 g of BN and 4 milling balls at 410 kPa. After milling for 2 hours with a weight ratio of 65:1, the gas mixture's pressure decreased to 350 kPa, indicating successful gas adsorption. Subsequent gas analysis via gas-phase Fourier transform infrared spectroscopy (g-FTIR) revealed the presence of only CH_4 in the remaining gas after 2 hours of milling (Figure 21.1a (ii)), confirming the effective removal of C_2H_2. Longer milling durations were required for higher C_2H_2 concentrations, with at least 4 hours (Figure 21.1a (iv)) of milling necessary to fully remove C_2H_2 from a mixture containing 30 vol% C_2H_2 and 70 vol% CH_4.

DOI: 10.1201/9781003647812-26

Similar gas separation techniques can be applied to separate C_2H_4 and C_2H_6, two other industrial products requiring purification (Figure 21.1b). BN powders were milled in the presence of a mixture comprising 20 vol% C_2H_4 and 80 vol% C_2H_6, resulting in a pressure decrease of 60 kPa after 2 hours of milling. Gas analysis via g-FTIR confirmed the successful removal of C_2H_4, with only C_2H_6 present in the remaining gas (Figure 21.1b (iv)). The gas separation process involves introducing a mixture of alkyne or olefin gas (C_2H_2 or C_2H_4) and paraffin gas (CH_4 or C_2H_6) into the milling reactor (Figure 21.2a). Mechanical milling activates the selective adsorption of alkyne and olefin gases onto BN powder, while paraffin gas stays in the reactor chamber and can be evacuated through the vacuum system. The adsorbed gases can be recovered by heating the milling reactor/BN, allowing for continuous or repeated processing as long as fresh gas is supplied.

This method effectively separates mixed hydrocarbon gases by removing olefin gases after a fine-tuned milling treatment, considering parameters such as milling intensity, time, and gas pressure. For mixtures containing more than two gases with different adsorption rates and capacities, a separation process involving multiple milling reactors is proposed (Figure 21.2b).

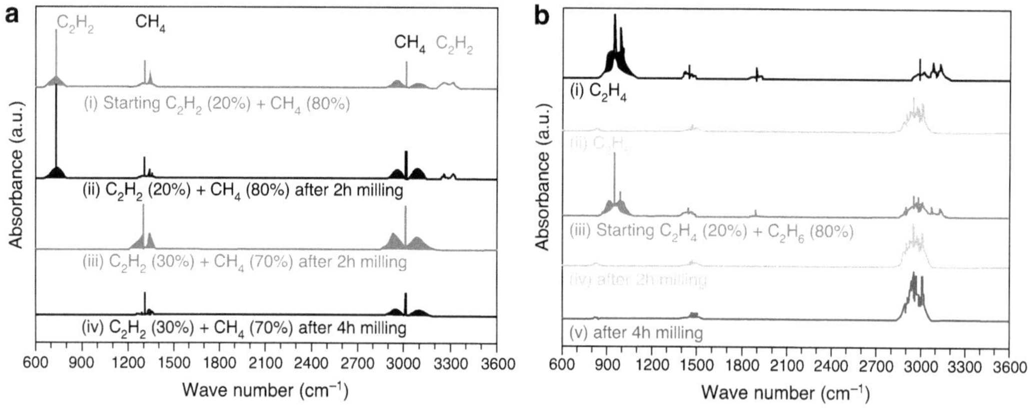

Figure 21.1 (a) g-FTIR spectra of the initial C_2H_2/CH_4 mixture and the remaining gas after milling for 2 hours. (b) g-FTIR spectra of pure C_2H_4 and C_2H_6, the initial C_2H_4/C_2H_6 gas mixture (20:80 by volume), and the remaining gas after milling for 2 and 4 hours. ((a, b) Reprinted with permission from Ref. [8], Copyright 2016 Springer Nature.)

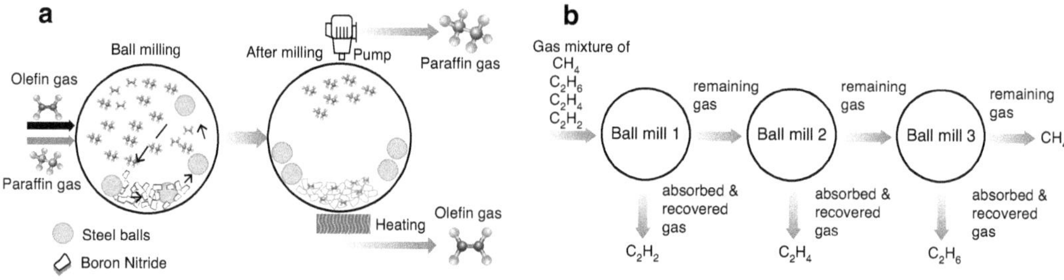

Figure 21.2 (a) Schematic of the ball milling and heating separation process for mixtures containing a single olefin gas and a single paraffin gas. (b) Proposed separation of olefin and paraffin gases by combined ball milling and heating. ((a, b) Reprinted with permission from Ref. [8], Copyright 2016 Springer Nature.)

The efficiency of gas separation depends on the selective adsorption capability of the BN. The ball milling gas adsorption process is similar to the gas adsorption process of porous materials, but operates in a specialized ball mill reactor. The additional cost incurred is the energy consumed by the milling process. For instance, the electricity cost for a 20-hour milling process is estimated at $0.32 USD. The estimated energy consumption of the ball milling gas adsorption process in separating 1000 L of an olefin/paraffin mixture is 76.8 KJ/s, much lower compared to traditional cryogenic distillation, which requires much more energy input. The ball milling method offers an energy-efficient alternative for gas separation, reducing the need for high-energy inputs typically associated with cryogenic techniques. This makes it a promising option for large-scale industrial applications, especially in terms of energy savings. Besides, the ball milling process can be continuously operated, further enhancing its efficiency in gas separation processes [9].

21.2.2 CO_2 Separation from CH_4 and N_2 Separation

Flue gas and air typically contain 8%–15% CO_2 along with nitrogen gas, necessitating the removal of CO_2 from nitrogen to facilitate separation [10]. Since nitrogen gas does not absorb infrared radiation, we utilized gas chromatography (GC) (Perkin Elmer Clarus 590 with a thermal conductivity detector, TCD) to analyze the gas mixtures. Initially, several grams of BN powder were loaded into the milling cell along with multiple milling balls. A gas mixture comprising 20% CO_2 (v/v) in N_2 was introduced into the milling jar at 400 kPa pressure. Following milling for 2.5 hours, the pressure decreased to 360 kPa. The gas mixture's composition was measured at successive time intervals using GC analysis, revealing that after milling for 2.5, 5, and 5.5 hours, 9.9%, 0.4%, and 0% of CO_2 gas, respectively, remained in the milling jar. The chromatograms obtained at these time points are depicted in Figure 21.3a, and the corresponding gas compositions after ball milling for a certain number of hours are shown in Figure 21.3b and c.

Initially, GC was calibrated with known concentrations of each gas. All data points exhibited a strong correlation with gas compositions, with R-squared values of the linear fits close to unity. After 5.5 hours of milling, CO_2 was completely adsorbed into the BN, effectively disappearing from the mixture. Additionally, the milled BN was analyzed using powder-FTIR (see Figure 21.4a and b), where the peak at 2279 cm^{-1} confirmed the exclusive adsorption of CO_2, absent in BN milled in nitrogen for the same duration [12].

Gas-phase analysis was performed using both qualitative and quantitative techniques. For qualitative analysis, gases were examined before and after milling with a gas cell attached to a Bruker Vertex 70 spectrometer, which scanned the range from 4000 to 600 cm^{-1} with a resolution of 1 cm^{-1}. For compositional analysis of CO_2/N_2 and CO_2/CH_4 mixtures, GC with a TCD was employed. The sample was first loaded into a 1 cc sample loop before being injected into the GC. The temperatures of the GC oven and TCD detector were set to 60°C and 200°C, respectively. To maintain a consistent flow rate of 30 mL/min, the carrier gas flow was controlled using programmable pneumatic control flow controllers. Before analyzing the gas mixture samples, the GC was calibrated, considering that the TCD detector measures the thermal conductivity difference between the sample and reference flow rather than the actual gas compositions. Calibration lines were established under fixed operating conditions by correlating the chromatographic peak areas with the known compositions of standard gases. Standard gas mixtures were prepared in-house, and for calibration, the known composition of the binary mixture was circulated into the sample loop to remove any residual traces from previous samples. After this, the standard gas was injected into the GC, and the resulting chromatographic peak areas were recorded and plotted against their respective compositions to establish a calibration curve.

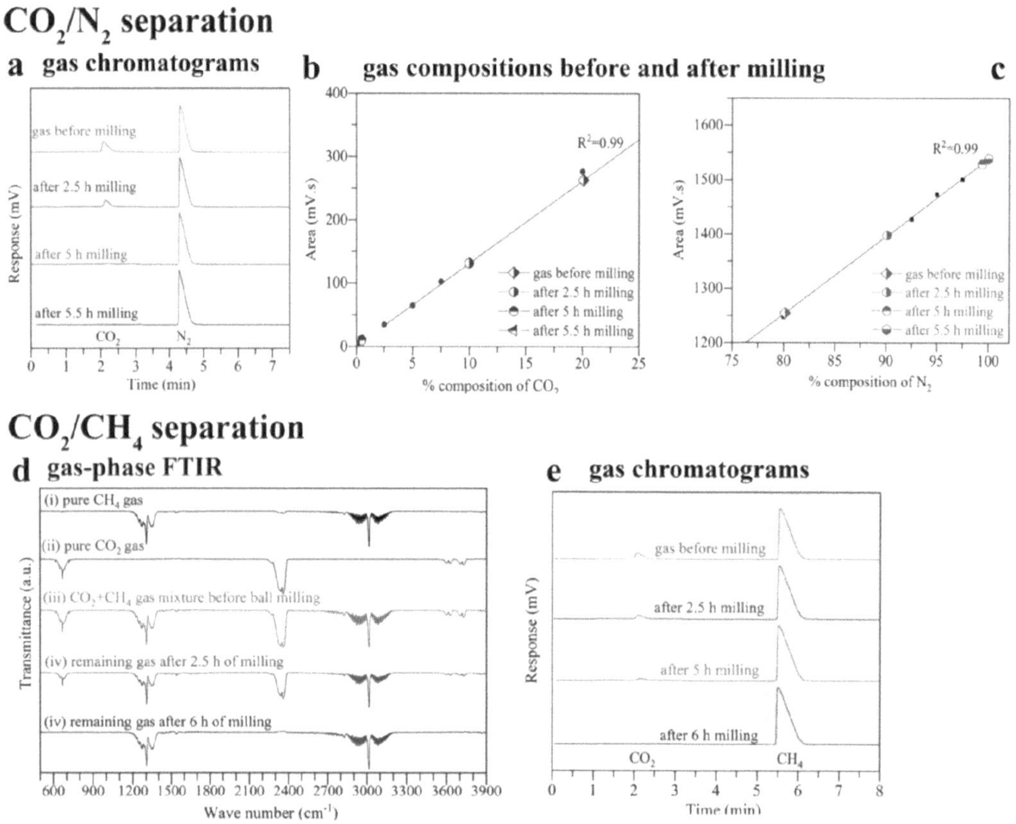

Figure 21.3 Separation of gas mixtures: (a) gas chromatograms of CO_2/N_2 gas mixtures before and after milling, gas composition (b–d) g-FTIR spectra of the initial CO_2/CH_4 mixture and the remaining gas after milling for 2.5 hours and 6 hours, and the corresponding gas chromatograms. ((a, b) Reprinted with permission from Ref. [11], Copyright 2016 Springer Nature.)

These operational conditions were maintained at a constant value for all the runs during the analysis of the unknown gas samples for gas composition measurement [11].

This mechanochemical process also facilitates the removal of CO_2 from natural gas mixtures. Gas-phase FTIR spectra of individual gases (CH_4 and CO_2), as well as the gas mixture (80%/20% by volume), are depicted in Figure 21.3d (i), (ii), and (iii), respectively. Following ball milling with BN for 2.5 hours (Figure 21.3), some residual CO_2 is still present in the ball mill chamber. However, after 6 hours of milling (Figure 21.3), all the CO_2 has been adsorbed onto the BN, leaving only CH_4 gas in the ball mill chamber. Gas chromatograms further confirm successful separation after 6 hours of milling (Figure 21.2e). Quantitative analysis reveals that after milling for different time intervals, the remaining CO_2 gas is reduced from 12.5% after 2.5 hours to 0% after 6 hours. Additionally, besides gas-phase analysis, we investigated the ball-milled BN under individual and gas mixtures for 6 hours. The FTIR spectra (see Figure 21.5a and b) indicate that BN milled in CH_4 for 6 hours exhibits peaks at 2950 and 3435 cm^{-1} corresponding to CH_4 adsorption

[8,13], whereas for BN milled in CO_2 and a mixture of CO_2/CH_4 for the same milling duration (see Figure 21.5 (ii), (iii)), the FTIR spectra display peak positions at 2279, 2544, and 3354 cm^{-1}, corresponding to CO_2 bonds respectively [14]. This demonstrates BN's selective absorption of CO_2 from CH_4 mixtures. Similarly, only CO_2 adsorption was observed (see Figure 21.4) for the CO_2/N_2 gas mixture.

It's worth noting that BN milled in any of the gas mixtures retains a layered structure. XRD patterns (Figures 21.4c and 21.5c) and SEM images (inset) confirm the hexagonal crystalline structure of the initial BN, which is further supported by TEM images and corresponding SAED (see Figures 21.4d and 21.5d). The actual selectivity is determined under mixed gas separation and calculated using Equation (21.1).

The gas separation capability of the mechanochemical process was evaluated by conducting mixed gas milling. CO_2/N_2 (20/80, v/v) and CO_2/CH_4 (20/80, v/v) were utilized to calculate the real selectivity [15].

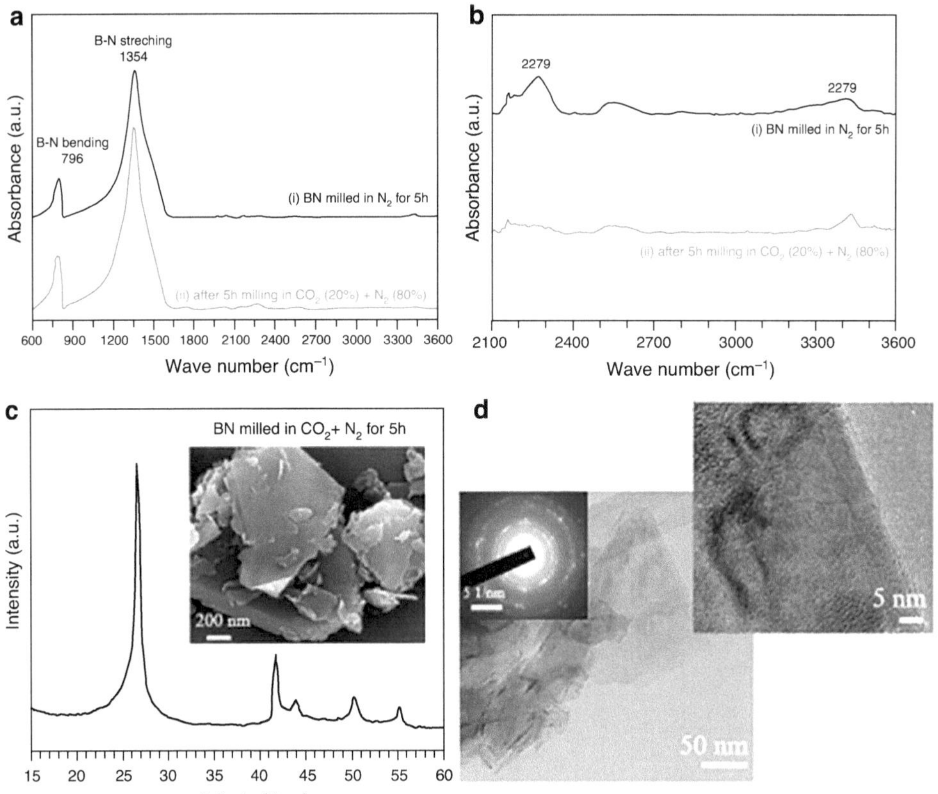

Figure 21.4 FTIR spectra of the BN milled for 5.5 hours in a gas mixture of $CO_2 + N_2$ (i) and in N_2 (ii); (b) Enlarged spectra of the same. (c) XRD patterns and SEM images (inset) of BN samples milled for 6 hours in CO_2/N_2 and CO_2/CH_4 mixtures, respectively. (d) TEM images of the BN milled (for 6 hours) in a gas mixture of $CO_{2+}N_2$; inset: SAED patterns. ((a, b) Reprinted with permission from Ref. [11], Copyright 2016 Springer Nature.)

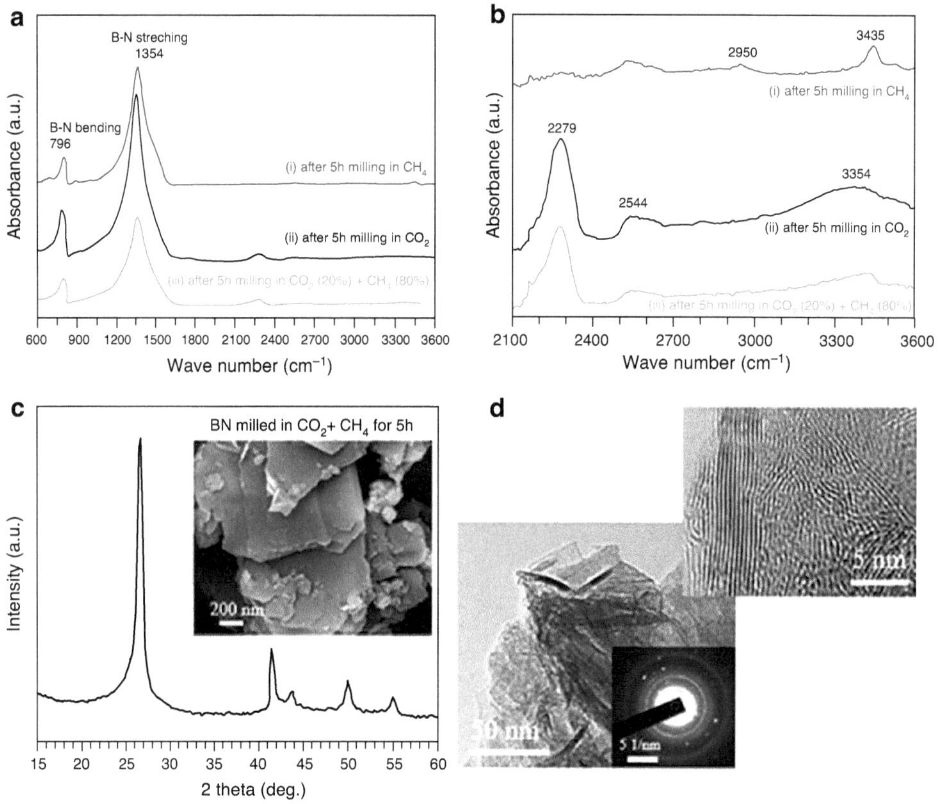

Figure 21.5 FTIR spectra of the BN milled for 5 hours in a gas mixture of CH_4 (i) and in N_2 (ii) and CH_4+CO_2 (iii). (b) Enlarged spectra of the same. (c) XRD patterns and SEM images (inset) of BN samples milled for 5 hours in CO_2/CH_4 mixtures, respectively. (d) TEM images of the BN milled (for 5 hours) in a gas mixture of CO_2+CH_4; inset: SAED patterns. ((a, b) Reprinted with permission from Ref. [11], Copyright 2016 Springer Nature.)

$$\alpha_{\text{real}} = \frac{x_a / y_a}{x_b / y_b} \tag{21.1}$$

where x_a and y_a are the mole fractions of gas '*a*' in the gas and adsorbent phases, respectively. The actual selectivities for CO_2/CH_4 and CO_2/N_2 gas mixtures are determined to be 202 and 616, respectively. The efficiency of any adsorption or membrane separation processes heavily relies on the partial pressures of the gas species involved. As the concentration of species decreases in the gas phase, it notably impacts the kinetics of adsorption and gradually reduces the selectivity. However, in the mechanochemical process, the dependency on partial pressure is minimal, and it predominantly hinges on the material's affinity for the gas. Consequently, achieving purities exceeding 99.99% in a single stage is feasible, whereas conventional separation methods often necessitate multiple stages to achieve similar purities. Therefore, this technique represents a clean and solvent-free approach to absorb CO_2 and effectively separate it from methane and nitrogen gas mixtures. Furthermore, the adsorbed gas can be released upon heating.

21.3 CONCLUSIONS

In essence, ball milling initiates mechanochemical reactions to selectively adsorb the gases, thereby separating gases from a gas mixture. The differing adsorption rates of olefin/alkyne gases compared to paraffin gases underscore the remarkable selectivity of BN for alkyne and olefin gases during mechanochemical processes. This selectivity allows for the efficient separation of these gases from paraffin gases, improving the overall gas separation process. Once adsorbed, the alkyne and olefin gases can be easily recovered through heating, making the process both practical and cost-effective. Furthermore, the mechanochemical process is environmentally friendly, as it does not produce any harmful by-products, ensuring a cleaner and more sustainable method for gas separation compared to traditional techniques. Further, this process also demonstrates efficient separation of CO_2 from a gas mixture containing natural gas (CH_4) or nitrogen due to the superior adsorption selectivity of CO_2 over the other gases. Upon heating, the stored CO_2 can be released, and this cycle can be repeated multiple times. This mechanochemical process has the potential to revolutionize CO_2 gas capture and separation technology. Both milling and heating processes are scalable and energy-efficient compared to cryogenic distillation. BN can be reused after cleaning the adsorbed gases, further enhancing its cost-effectiveness and sustainability. These advantages, including substantial energy savings and low costs, make large-scale gas separation achievable with efficiency.

REFERENCES

1. D. S. Sholl, R. P. Lively, Seven chemical separations to change the world. *Nature* **532**, 435–437 (2016).
2. Y. Chen, J. S. Williams, Competitive gas-solid reactions realized by ball milling of Zr in ammonia gas. *Journal of Materials Research* **11**, 1500–1506 (1996).
3. A. Glushenkov, H.-Z. Zhang, Y. Chen, Reactive ball milling to produce nanocrystalline ZnO. *Materials Letters* **62**, 4047–4049 (2008).
4. Y. Chen, J. Williams, Formation of metal hydrides by mechanical alloying. *Journal of Alloys and Compounds* **217**, 181–184 (1995).
5. G.-F. Han et al., Dissociating stable nitrogen molecules under mild conditions by cyclic strain engineering. *Science Advances* **5**, eaax8275 (2019).
6. A. Khorshidi, J. Violet, J. Hashemi, A. A. Peterson, How strain can break the scaling relations of catalysis. *Nature Catalysis* **1**, 263–268 (2018).
7. A. Luna-Triguero, J. Vicent-Luna, R. Madero-Castro, P. Gómez-Álvarez, S. Calero, Acetylene storage and separation using metal–organic frameworks with open metal sites. *ACS Applied Materials & Interfaces* **11**, 31499–31507 (2019).
8. S. Mateti, C. Zhang, A. Du, S. Periasamy, Y. I. Chen, Superb storage and energy saving separation of hydrocarbon gases in boron nitride nanosheets via a mechanochemical process. *Materials Today* **57**, 26–34 (2022).
9. A. Orga, M. Anusi, J. Obibuenyi, C. Njoku, Saving energy and tray numbers in ethylene-ethane distillation towers by vapor permeation membrane load sharing. *Chemical and Process Engineering Research* **42**, 32–43 (2016).
10. M. Songolzadeh, M. Soleimani, M. Takht Ravanchi, R. Songolzadeh, Carbon dioxide separation from flue gases: A technological review emphasizing reduction in greenhouse gas emissions. *The Scientific World Journal* **2014**, 828131 (2014).
11. S. Mateti et al., A mechanochemical process to capture and separate carbon dioxide from natural gas using boron nitride nanosheets. *Materials Horizons* **11**, 2950–2956 (2024).
12. J. C. Vickerman, "Molecular surface mass spectrometry by SIMS" in John C. Vickerman, Ian S. Gilmore (eds), *Surface Analysis: The Principal Techniques* (2009), John Wiley & Sons, Ltd, pp. 113–205.

13. D. Scarano, S. Bertarione, G. Spoto, A. Zecchina, C. O. Areán, FTIR spectroscopy of hydrogen, carbon monoxide, and methane adsorbed and co-adsorbed on zinc oxide. *Thin Solid Films* **400**, 50–55 (2001).
14. E.-M. Köck, M. Kogler, T. Bielz, B. Klötzer, S. Penner, In situ FT-IR spectroscopic study of CO_2 and CO adsorption on Y_2O_3, ZrO_2, and yttria-stabilized ZrO_2. *The Journal of Physical Chemistry C* **117**, 17666–17673 (2013).
15. S. Mane et al., Fabrication of microporous polymers for selective CO_2 capture: The significant role of crosslinking and crosslinker length. *Journal of Materials Chemistry A* **5**, 23310–23318 (2017).

Index